OXFORD GRADUATE TEXTS IN MATHEMATICS

OXFORD GRADUATE TEXTS IN MATHEMATICS

Stochastic Integration Theory

Péter Medvegyev

This book has been printed digitally and produced in a standard specification
in order to ensure its continuing availability

OXFORD
UNIVERSITY PRESS

Great Clarendon Street, Oxford OX2 6DP

Oxford University Press is a department of the University of Oxford.
It furthers the University's objective of excellence in research, scholarship,
and education by publishing worldwide in

Oxford New York

Auckland   Cape Town Dar es Salaam   Hong Kong   Karachi
Kuala Lumpur Madrid Melbourne Mexico City   Nairobi
New Delhi Shanghai Taipei Toronto
With offices in
Argentina Austria Brazil Chile Czech Republic France Greece
Guatemala Hungary Italy Japan South Korea Poland Portugal
Singapore Switzerland Thailand Turkey Ukraine Vietnam

Oxford is a registered trade mark of Oxford University Press
in the UK and in certain other countries

Published in the United States
by Oxford University Press Inc., New York

© Péter Medvegyev, 2007

ISBN 978-0-19-921525-6

To the memory of my father

Contents

Preface

I started to write this book a few years ago mainly because I wanted to understand the theory of stochastic integration. Stochastic integration theory is a very popular topic. The main reason for this is that the theory provides the necessary mathematical background for derivative pricing theory. Of course, many books purport to explain the theory of stochastic integration. Most of them concentrate on the case of Brownian motion, and a few of them discuss the general case. Though the first type of book is quite readable, somehow they disguise the main ideas of the general theory. On the other hand, the books concentrating on the general theory were, for me, of a bit sketchy. I very often had quite serious problems trying to decode what the ideas of the authors were, and it took me a long time, sometimes days and weeks, to understand some basic ideas of the theory. I was nearly always able to understand the main arguments but, looking back, I think some simple notes and hints could have made my suffering shorter.

The theory of stochastic integration is full of non-trivial technical details. Perhaps from a student's point of view the best way to study and to understand measure theory and the basic principles of modern mathematical analysis is to study probability theory. Unfortunately, this is not true for the general theory of stochastic integration. The reason for this is very simple: the general theory of stochastic integration contains too much measure theory. Perhaps the best way to understand the limits of measure theory is to study the general theory of stochastic integration. I think this beautiful theory pushes modern mathematics to its very limits. On the other hand, despite many technical details there are just a very few simple issues which make up the backbone of stochastic analysis.

1. The first one is, of course, *martingales* and *local martingales*. The basic concept of stochastic analysis is random noise. But what is the right mathematical model for the random noise? Perhaps the most natural idea would be the random walk, that is processes with stationary and independent increments: the so called Lévy processes, with mean value zero. But, unfortunately, this class of processes has some very unpleasant properties. Perhaps the biggest problem is that the sum of two Lévy process is not a Lévy process again. Modern mathematics is very much built on the idea of linearity. If there is not some very fundamental and very clear reason for it, then every reasonable class of mathematical objects should be closed under linear combinations. The concept of random noise comes

very much from applications. One of the main goals of mathematics is to build safe theoretical tools and, like other scientific instruments, mathematical tools should be both simple and safe, similar to computer tools. Most computer users never read the footnotes in computer manuals, they just have a general feeling about the limits of the software. It is the responsibility of the writer of the software to make the software work in a plausible way. If the behaviour of the software is not reasonable, then its use becomes dangerous, e.g. you could easily lose your files, or delete or modify something and make the computer behave unpredictably, etc. Likewise, if an applied mathematical theory cannot guarantee that the basic objects of the theory behave reasonably, then the theory is badly written, and as one can easily make hidden errors in it, its usage is dangerous. In our case, if the theory cannot guarantee that the sum of two random noises is again a random noise, then the theory is very dangerous from the point of view of sound applications. The main reason for introducing martingales is that from the intuitive point of view they are very close to the idea of a random walk, but if we fix the amount of observable information they form a linear space. The issue of local martingales is a bit more tricky. Of course local martingales and not just real martingales form the class of random noise. Without doubt, local martingales make life for a stochastic analyst very difficult. From an intuitive, applied point of view, local martingales and martingales are very close and that is why it is easy to make mistakes. Therefore, in most cases the mathematical proofs have to be very detailed and cautious. On the other hand the local martingales form a large and stable class, so the resulting theory is very stable and simple to use. As in elementary algebra, most of the problems come from the fact that one cannot divide by zero. In stochastic analysis most of the problems come from the fact that not every local martingale is a martingale and therefore one can take expected values only with care. Is there some intuitive idea why one should introduce local martingales? Perhaps, yes. First of all one should realize that not really martingales, but uniformly integrable martingales, are the objects of the theory. If we observe a martingale up to a fixed, finite moment of time we get a uniformly integrable martingale, but most of the natural moments of time are special random variables. The measurement of the time-line is, in some sense, very arbitrary. Traditionally we measure it with respect to some physical, astronomical movements. For some processes this coordinate system is rather arbitrary. It is more natural, for example, to say 'after lunch I called my friend' than to say 'I called my friend at twenty-three past and sometimes at twenty-two past one depending on the amount of food my wife gave me'. Of course the moment of time after lunch is a random variable with respect to the coordinate system generated by the relative position of the earth and the sun, but as a basis for observing my general habits this random time, 'after lunch', is the natural point of orientation. So, in some ways, it is very natural to say that a process is a random noise if one can define a sequence of random moments, so-called stopping times, $\tau_0 < \tau_1 < \ldots$ such that if we observe the random noise up to τ_k the truncated processes are uniformly integrable martingales, which is exactly the definition of local martingales. The idea that local martingales are the good mathematical models for random noise comes from the fact that sometimes we want to perturb the measurement of the time-line in an order-preserving way, and we want the class of 'random noise processes' to be invariant under these transformations.

2. The second-most important concept is *quadratic variation*. One can think of stochastic analysis as the mathematical theory of quadratic variation. In classical analysis one can define an integral only when the integrator has bounded variation. Even in this case, one can define two different concepts of integration. One is the Lebesgue–Stieltjes type of integration and the other is the Riemann–Stieltjes concept of integration. If the integrand is continuous, then the two concepts are equal. It is easy to see, that if the integrand is left-continuous and in the Riemann–Stieltjes type integrals one may choose only the starting point of the sub-intervals of the partitions as test-point, then for these type of approximating sums the integrals of Riemann–Stieltjes type will converge and they are equal to the Lebesgue–Stieltjes integrals. One may ask whether one can extend this trick to some more general class of integrators. The answer is yes. It turns out that the same concept works if the integrators are local martingales. There is just one new element: the convergence of the integrating sums holds only in probability. If the integrators are local martingales or if they have finite variation then for this integral, the so-called *integration by parts formula* is valid. In this formula, the most notable factor is the quadratic co-variation $[X, Y](t)$. If, for example, X is continuous and Y has finite variation then $[X, Y](t) = 0$ but generally $[X, Y](t) \neq 0$. As the stochastic integrals are defined only by convergence in probability the random variable $[X, Y](t)$ is defined only up to a measure-zero set. This implies that the trajectories of the process $t \mapsto [X, Y](t)$ are undefined. One can exert quite a lot of effort to show that there is a right-continuous process with limits from the left, denoted by $[X, Y]$ such that for every t the value of $[X, Y]$ at time t is a version of the random variable $[X, Y](t)$. The key observation in the proof of this famous theorem is that $XY - [X, Y]$ is a local martingale and it is the only process for which this property holds and the jump-process of the process $[X, Y]$ is the process $\Delta X \Delta Y$. The integration by parts formula is the prototype of *Itô's formula*, which is the main analytical tool of stochastic analysis. Perhaps it is not without interest to emphasize that the main difficulty in proving this famous formula, in the general case of discontinuous processes, is to establish the existence of the quadratic variation. It is worth mentioning that it is relatively easy to show the existence of the quadratic variation for the so-called locally square-integrable martingales. It is nearly trivial to show the existence of the quadratic variation when the trajectories of the process have finite variation. Hence, it is not so difficult to prove the existence of $[X] \overset{\circ}{=} [X, X]$ if process X has a decomposition $X = V + H$ where the trajectories of V have finite variation and H is a so-called locally square-integrable martingale. The main problem is that we do not know that every local martingale has this decomposition! To

prove that this decomposition exists one should show the *Fundamental The-orem of Local Martingales*, which is perhaps the most demanding result of the theory.

3. The third most important concept of the theory is *predictability*. There are many interrelated objects in the theory modified by the adjective predictable. Perhaps the simplest and most intuitive one is the concept of predictable stopping time. Stopping times describe the occurrence of random events. The occurrence of a random event is *predictable*, if there is a sequence of other events which announces the predictable event. That is, a stopping time τ is predictable if there is a sequence of stopping times (τ_n) with $\tau_n \nearrow \tau$ and $\tau_n < \tau$ whenever $\tau > 0$. This definition is very intuitive and appealing. If τ is a predictable stopping time, then one can say that the event

$$[\tau, \infty) \stackrel{\circ}{=} \{(t, \omega) : \tau(\omega) \le t\} \subseteq \mathbb{R}_+ \times \Omega$$

is also predictable. The σ-algebra generated by these type of predictable random intervals is called the σ-algebra of predictable events. One should agree that this definition of predictability is in some sense very close to the intuitive idea of predictability. Quite naturally, a stochastic process is called predictable if it is measurable with respect to the σ-algebra of the predictable events. It is an important and often useful observation that the set of predictable events is the same as the σ-algebra generated by the left-continuous adapted processes. Recall that a process is called *adapted* when its value for every moment of time is measurable with respect to the σ-algebra representing the amount of information available at that time. The values of left-continuous processes are at least infinitesimally predictable. One of the most surprising facts of stochastic integration theory is that in the general case the integrands of stochastic integrals should be predictable. Although it looks like a very deep mathematical observation, one should also admit that this is a very natural result. The best interpretation of stochastic integrals is that they are the net results of continuous-time trading or gaming processes. Everybody knows that in a casino one should play a trading strategy only if one decides about the stakes before the random events generating the gains occur. This means that the playing strategy should be predictable. An important concept related to predictability is the concept of the predictable compensator. If one has a risky stochastic process X, one can ask whether there is a compensator P for the risk of process X. The compensator should be 'simpler' than the process itself. Generally it is assumed that P is monotone or at least it has finite variation. The compensator P should be predictable and one should assume that $X - P$ is a totally random process, that is $X - P$ is a local martingale. This is of course a very general setup, but it appears in most of the applications of stochastic analysis. For a process X there are many compensators, that is there are many processes Y such that $X - Y$ is a local martingale. Perhaps the simplest one is X itself. But it is very important that the predictable compensator of X, if it exists and if it has finite variation, is in fact unique. The reason for this is that every predictable local martingale is continuous, and if the trajectories of a continuous local martingale have finite variation then the local martingale is constant.

4. Stochastic integration theory is built on probability theory. Therefore every object of the theory is well-defined only almost surely and this means that stochastic integrals are also defined almost surely. In classical integration theory, one first defines the integral over some fixed set and then defines the integral function. In stochastic integration theory this approach does not work as it is entirely non-trivial how one can construct the *integral process* from the almost surely defined separate integrals. Therefore, in stochastic integration theory one immediately defines the integral processes, so stochastic integrals are processes and not random variables.

5. There are basically two types of local martingales: *continuous* and *purely discontinuous* ones. The canonical examples of continuous local martingales are the Wiener processes, and the simplest purely discontinuous local martingales are the compensated Poisson processes. Every local martingale which has trajectories with finite variation is purely discontinuous, but there are purely discontinuous local martingales with infinite variation. Every local martingale has a unique decomposition $L = L(0) + L^c + L^d$, where L^c is a continuous local martingale and L^d is a purely discontinuous local martingale. A very important property of purely discontinuous local martingales is that they are sums of their *continuously compensated single jumps*. S_i, by definition, is a single jump if there is a stopping time τ such that every trajectory of S_i is constant before and after the random jump-time τ. The single jumps obviously have trajectories with finite variation, and as the compensators P_i, by definition, also have finite variation, the compensated single jumps $L_i \triangleq S_i - P_i$ also have trajectories with finite variation. Of course this does not imply that the trajectories of L, as infinite sums, should also have finite variation. If L is a purely discontinuous local martingale and $L = \sum_i L_i$ where L_i are continuously compensated single jumps, then one can think about the stochastic integral with respect to L as the sum of the stochastic integrals with respect to L_i. Every L_i has finite variation so, in this case, the stochastic integral, as a pathwise integral, is well-defined and if the integrand is predictable then the integral is a local martingale. Of course one should restrict the class of integrands as one has to guarantee the convergence of the sum of the already defined integrals. If the integrand is predictable then the stochastic integral with respect to a purely discontinuous local martingale is a sum of local martingales. Therefore it is also a local martingale.

6. The stochastic integral with respect to continuous local martingales is a bit more tricky. The fundamental property of stochastic integrals with respect to local martingales is that the resulting process is also a local martingale. The intuition behind this observation is that the basic interpretation of stochastic integration is that it is the cumulative gain of an investment process into a randomly changing price process. Every moment of time we decide about the size

of our investment, this is the integrand, and our short term gains are the product of our investment and the change of the random price-integrator. Our total gain is the sum of the short term gains. If we can choose our strategy only in a predictable way it is quite natural to assume that our cumulative gain process will be also totally random. That is, if the investment strategy is predictable and the random integrator price process is a local martingale, then the net, cumulative gain process is also a local martingale. How much is the quadratic variation of the resulting gain process? If $H \bullet L$ denotes the integral of H with respect to the local martingale L then one should guarantee the very natural identity $[H \bullet L] = H^2 \bullet [L]$, where the right-hand side expression $H^2 \bullet [L]$ denotes the classical pathwise integral of H^2 with respect to the increasing process $[L]$. The identity is really very natural as $[L]$ describes the 'volatility' of L along the time-line, and if in every moment of time we have H pieces of L then our short term change in 'volatility' will be $(H\Delta L)^2 \approx H^2 \cdot \Delta [L]$. So our aggregated 'volatility' is $\sum H^2 \Delta [L] \stackrel{\circ}{=} H^2 \bullet L$. It is a very nice observation that there is just one continuous local martingale, denoted by $H \bullet L$, for which $[H \bullet L, N] = H \bullet [L, N]$ holds for every continuous local martingale N. The stochastic integral with respect to a local martingale L is the sum of two integrals: the integral $H \bullet L^c$ with respect to the continuous and the integral $H \bullet L^d$ with respect to the purely discontinuous part of L.

7. As there are local martingales which have finite variation, one can ask whether the new and the classical definitions are the same or not? The answer is that if the integrand is predictable the two concepts of integration are not different. This allows us to further generalize the concept of stochastic integration. We say that process S is a *semimartingale* if $S = L + V$ where L is a local martingale and V is adapted and has finite variation. One can define the integral with respect to S as the sum of the integrals with respect L and with respect to V. A fundamental problem is that in the discontinuous case, as we have local martingales with finite variation, the decomposition is not unique. But as for processes with finite variation the two concepts of integration coincide, this definition of stochastic integral with respect to semimartingales is well-defined.

In the first chapter of the book we introduce the basic definitions and some of the elementary theorems of martingale theory. In the second chapter we give an elementary introduction to stochastic integration theory. Our introduction is built on the concept of Itô–Stieltjes integration. In the third chapter we shall discuss the structure of local martingales and in Chapter Four we shall discuss the general theory of stochastic integration. In Chapter Six we prove Itô's formula. In Chapter Seven we apply the general theory to the classical theory of processes with independent increments.

Finally it is a pleasure to thank to those who have helped me to write this book. In particular I would like to thank the efforts of Tamás Badics from University of Pannonia and Petrus Potgieter from University of South Africa. They read most of the book and without their help perhaps I would not have been able to finish the book. I wish to thank István Dancs and János Száz from Corvinus University for support and help. I would like to express my gratitude to the Magyar Külkereskedelmi Bank for their support.

Budapest, 2006

Medvegyev@math.bke.hu
medvegyev.uni-corvinus.hu

1

STOCHASTIC PROCESSES

In this chapter we first discuss the basic definitions of the theory of stochastic processes. Then we discuss the simplest properties of martingales, the Martingale Convergence Theorem and the Optional Sampling Theorem. In the last section of the chapter we introduce the concept of localization.

1.1 Random functions

Let us fix a probability space $(\Omega, \mathcal{A}, \mathbf{P})$. As in probability theory we refer to the set of real-valued $(\Omega, \mathcal{A})$-measurable functions as *random variables*. We assume that the space $(\Omega, \mathcal{A}, \mathbf{P})$ is *complete*, that is all subsets of measure zero sets are also measurable. This assumption is not a serious restriction but it is a bit surprising that we need it. We shall need this assumption many times, for example when we prove that the hitting times[1] of Borel measurable sets are stopping times[2]. When we prove this we shall use the so-called Projection Theorem[3] which is valid only when the space $(\Omega, \mathcal{A}, \mathbf{P})$ is complete. We shall also use the Measurable Selection Theorem[4] several times, which is again valid only when the measure space is complete. Let us remark that all applications of the completeness assumption are connected to the Predictable Projection Theorem, which is the main tool in the discussion of discontinuous semimartingales.

In the theory of stochastic processes, random variables very often have infinite value. Hence the image space of the measurable functions is not $\mathbb{R}$ but the set of extended real numbers $\overline{\mathbb{R}} \stackrel{\circ}{=} [-\infty, \infty]$. The most important examples of random variables with infinite value are stopping times. Stopping times give the random time of the occurrence of observable events. If for a certain outcome ω the event never occurs, it is reasonable to say that the value of the stopping time for this ω is $+\infty$.

---

[1]See: Definition 1.26, page 15.
[2]See: Definition 1.21, page 13.
[3]See: Theorem A.12, page 550.
[4]See: Theorem A.13, page 551.

1

1.1.1 Trajectories of stochastic processes

In the most general sense *stochastic processes* are such functions $X(t, \omega)$ that for any fixed parameter t the mappings $\omega \mapsto X(t, \omega)$ are random variables on $(\Omega, \mathcal{A}, \mathbf{P})$. The set of possible time parameters Θ is some subset of the extended real numbers. In the theory of continuous-time stochastic processes Θ is an interval, generally $\Theta = \mathbb{R}_+ \overset{\circ}{=} [0, \infty)$, but sometimes $\Theta = [0, \infty]$ and $\Theta = (0, \infty)$ is also possible. If we do not say explicitly what the domain of the definition of the stochastic process is, then Θ is $\mathbb{R}_+$. It is very important to append some remarks to this definition. In probability theory the random variables are equivalence classes, which means that the random variables $X(t)$ are defined up to measure zero sets. This means that in general $X(t, \omega)$ is meaningless for a fixed ω. If the possible values of the time parameter t are countable then we can select from the equivalence classes $X(t)$ one element, and fix a measure zero set, and outside of this set the expressions $X(t, \omega)$ are meaningful. But this is impossible if Θ is not countable[5]. Therefore, we shall always assume that $X(t)$ is a function already carefully selected from its equivalence class. To put it in another way: when one defines a stochastic process, one should fix the space of possible trajectories and the stochastic processes are function-valued random variables which are defined on the space $(\Omega, \mathcal{A}, \mathbf{P})$.

Definition 1.1 *Let us fix the probability space $(\Omega, \mathcal{A}, \mathbf{P})$ and the set of possible time parameters[6] Θ. The function X defined on $\Theta \times \Omega$ is a stochastic process over $\Theta \times \Omega$ if for every $t \in \Theta$ it is measurable on $(\Omega, \mathcal{A}, \mathbf{P})$ in its second variable.*

Definition 1.2 *If we fix an outcome $\omega \in \Omega$ then the function $t \mapsto X(t, \omega)$ defined over Θ is the trajectory or realization of X corresponding to the outcome ω. If all[7] the trajectories of the process X have a certain property then we say that the process itself has this property. For example, if all the trajectories of X are continuous then we say that X is continuous, if all the trajectories of X have finite variation then we say that X has finite variation, etc.*

Recall that in probability theory the role of the space $(\Omega, \mathcal{A}, \mathbf{P})$ is a bit problematic. All the relevant questions of probability theory are related to the joint distributions of random variables and the whole theory is independent of the specific space carrying the random variables having these joint distributions.

---

[5]This is what the author prefers to call the *revenge of the zero sets*. This is very serious and it will make our life quite difficult. The routine solution to this challenge is that all the processes which we are going to discuss have some sort of continuity property. In fact, we shall nearly always assume that the trajectories of the stochastic processes are *regular*, that is at every point all the trajectories have limits from both sides and they are either right- or left-continuous. As we want to guarantee that the martingales have proper trajectories we shall need the so-called *usual assumptions*.

[6]In most of the applications Θ is the time parameter. Sometimes the natural interpretation of Θ is not the time but some spatial parameter. See: Example 1.126, page 90. In continuous 'time' theory of stochastic processes Θ is an interval in the half-line $\mathbb{R}_+$.

[7]Not almost all trajectories. See: Definition 1.8, page 6, Example 1.11, page 8.

Of course it is not sufficient to define the distributions alone. For instance, it is very important to clarify the relation between the lognormal and the normal distribution, and we can do it only when we refer directly to random variables. Hence, somehow, we should assume that there is a measure space carrying the random variables with the given distributions: if ξ has normal distribution then $\exp(\xi)$ has lognormal distribution. This is a very simple and very important relation which is not directly evident from the density functions. The existence of a space $(\Omega, \mathcal{A}, \mathbf{P})$ enables us to use the power of measure theory in probability theory, but the specific structure of $(\Omega, \mathcal{A}, \mathbf{P})$ is highly *irrelevant*. The space $(\Omega, \mathcal{A}, \mathbf{P})$ contains the 'causes', but we see only the $\xi(\omega)$ 'consequences'. We never observe the outcome ω. We can see only its consequence $\xi(\omega)$. As the space $(\Omega, \mathcal{A}, \mathbf{P})$ is irrelevant one can define it in a 'canonical way'. In probability theory, generally, $\Omega \overset{\circ}{=} \mathbb{R}$, $\mathcal{A} \overset{\circ}{=} \mathcal{B}(\mathbb{R})$ and $\mathbf{P}$ is the measure generated by the distribution function of ξ or in the multidimensional case $\Omega \overset{\circ}{=} \mathbb{R}^n$ and $\mathcal{A} \overset{\circ}{=} \mathcal{B}(\mathbb{R}^n)$. In both cases Ω is the space of all possible realizations. Similarly in the theory of stochastic processes the only entities which one can observe are the trajectories. Sometimes it is convenient if Ω is the space of possible trajectories. In this case we say that Ω is given in its *canonical form*. It is worth emphasizing that in probability theory there is no advantage at all in using any specific representation. In the theory of stochastic processes the relevant questions are related to time and all the information about the time should be somehow coded in Ω. Hence, it is very plausible if we assume that the elements of Ω are not just abstract objects which somehow describe the information about the timing of certain events, but are also functions over the set of possible time values. That is, in the theory of stochastic processes, the canonical model is not just *one* of the possible representation: it is very often the *right* model to discuss certain problems.

1.1.2 Jumps of stochastic processes

Of course, the theory of stochastic processes is an application of mathematical analysis. Hence the basic mathematical tool of the theory of stochastic processes is measure theory. To put it another way, perhaps one of the most powerful applications of measure theory is the theory of stochastic processes. But measure theory is deeply sequential, related on a fundamental level to countable objects. We can apply measure theory to continuous-time stochastic processes only if we restrict the trajectories of the stochastic processes to 'countably determined functions'.

Definition 1.3 *Let $I \subseteq \mathbb{R}$ be an interval and let Y be an arbitrary topological space. We say that the function $f : I \to Y$ is regular if at any point $t \in I$, where it is meaningful, f has left-limits*

$$f(t-) \overset{\circ}{=} f_-(t) \overset{\circ}{=} \lim_{s \nearrow t} f(s) \in Y$$

and right-limits

$$f(t+) \overset{\circ}{=} f_+(t) \overset{\circ}{=} \lim_{s \searrow t} f(s) \in Y.$$

We say that f is right-regular *if it is regular and it is right-continuous. We say that f is* left-regular *if it is regular and it is left-continuous.*

If f is a real-valued function, that is if $Y \overset{\circ}{=} \mathbb{R}$ in the above definition, then the existence of limits means that the function has *finite* limits. As, in this book, stochastic processes are mainly real-valued stochastic processes, to make the terminology as simple as possible we shall always assume that regular processes have *finite limits.*

If the process X is regular and if t is an interior point of Θ then as the limits are finite it is meaningful to define the *jump*

$$\Delta X(t) \overset{\circ}{=} X(t+) - X(t-)$$

of X at t. It is not too important, but a bit confusing, that somehow one should fix the definition of jumps of the regular processes at the endpoints of the time interval Θ. If $\Theta = \mathbb{R}_+$ then what is the jump of the function χ_Θ at $t = 0$? Is it zero or one?

Definition 1.4 *We do not know anything about X before $t = 0$ so by definition we shall assume that $X(0-) \overset{\circ}{=} X(0)$. Therefore for any right-regular process on $\mathbb{R}_+$*

$$\Delta X(0) \overset{\circ}{=} X(0+) - X(0-) = 0. \qquad (1.1)$$

In a similar way, if, for example, $\Theta \overset{\circ}{=} [0, 1)$ and $X \overset{\circ}{=} \chi_\Theta$, then X is right-regular and does not have a jump at $t = 1$. Observe that in both examples the trajectories were continuous functions on Θ so it is a bit strange to say that the jump process of a continuous process is not zero[8]. It is not entirely irrelevant how we define the jump process at $t = 0$. If we consider process $F \overset{\circ}{=} \chi_{\mathbb{R}_+}$ as a distribution function of a measure then how much is the integral $\int_{[0,1]} 1 dF$? We shall assume that the distribution functions are right-regular and not left-regular. By definition[9] $\int_0^1 1 dF$ is the integral over $(0, 1]$ and as F is right-regular

---

[8]One can take another approach. In general: what is the value of an undefined variable? If X is the value process of a game and τ is some exit strategy, then what is the value of the game if we never exit from the game, that is if $\tau = \infty$? It is quite reasonable to say that in this case the value of the game is zero. Starting from this example one can say that once a variable is undefined then we shall assume that its value is zero. If one uses this approach then $X(0-) \overset{\circ}{=} 0$ and $\Delta X(0) = X(0+)$.

[9]In measure theory one can very often find the convention $\int_a^b f d\mu \overset{\circ}{=} \int_{[a,b)} f d\mu$. We shall assume that the integrator processes are right- and not left-continuous, so we shall use the convention $\int_a^b f d\mu \overset{\circ}{=} \int_{(a,b]} f d\mu$.

the measure of $(0, 1]$ is $F(1) - F(0) = 0$ so $\int_0^1 1 dF = 0$. According to our convention one can think that

$$\int_{[0,1]} 1 dF = F(1) - F(0-) = F(1) - F(0) = 1 - 1 \overset{\circ}{=} 0.$$

On the other hand one can correctly argue that

$$\int_{[0,1]} 1 dF \overset{\circ}{=} \int_{\mathbb{R}} \chi([0,1]) dF = 1.$$

To avoid these type of problems we shall never include the set $\{t = 0\}$ in the domain of integration.

The regular functions have many interesting properties. We shall very often use the next propositions:

Proposition 1.5 *Let f be a real-valued regular function defined on a finite and closed interval $[a, b]$. For any $c > 0$ the number of the jumps in $[a, b]$ bigger in absolute value then c is finite. The number of the jumps of f are at most countable.*

Proof. The second part of the proposition is an easy consequence of the first part. Assume that there is an infinite number of points (t_n) in $[a, b]$ for which $|\Delta f(t_n)| \geq c$. As $[a, b]$ is compact, one can assume that $t_n \to t^*$. Obviously we can assume that for an infinite number of points $t_n \leq t^*$ or $t^* \leq t_n$. Hence we can assume that $t_n \nearrow t^*$. But f has a left-limit at t^* so if $x, y < t^*$ are close enough to t^* then $|f(x) - f(y)| \leq c/4$. If t_n is close enough to t^* and $x < t_n < y$ are close enough to t_n and to t^* then

$$c \leq |f(t_n+) - f(t_n-)| \leq$$

$$\leq |f(t_n+) - f(y)| + |f(y) - f(x)| + |f(x) - f(t_n-)| \leq \frac{3}{4} c,$$

which is impossible. $\qquad\qquad\square$

Proposition 1.6 *If a function f is real valued and regular then it is bounded on any compact interval.*

Proof. Fix a finite closed interval $[a, b]$. If f were not bounded on $[a, b]$ then there would be a sequence (t_n) for which $|f(t_n)| \geq n$. As $[a, b]$ is compact one could assume, that $t_n \to t^*$. We could also assume that e.g. $t_n \nearrow t^*$ and therefore $f(t_n) \to f(t^*-) \in \mathbb{R}$ which is impossible. $\qquad\qquad\square$

Proposition 1.7 *Let f be a real valued regular function defined on a finite and closed interval $[a, b]$. If the jumps of f are smaller than c then for any $\varepsilon > 0$ there is a δ such that*

$$|f(t') - f(t'')| < c + \varepsilon \quad whenever \quad |t' - t''| \leq \delta.$$

Proof. If such a δ were not available then for some $\delta_n \searrow 0$ for all n there would be t'_n, t''_n such that $|t'_n - t''_n| \leq \delta_n$ and

$$|f(t'_n) - f(t''_n)| \geq c + \varepsilon. \tag{1.2}$$

As $[a, b]$ is compact, one could assume that $t'_n \to t^*$ and $t''_n \to t^*$ for some t^*. Notice that except for a finite number of indexes (t'_n) and (t''_n) are on different sides of t^*, since if, for instance, for an infinite number of indexes $t'_n, t''_n \geq t^*$ then for some subsequences $t'_{n_k} \searrow t^*$ and $t''_{n_k} \searrow t^*$ and as the trajectories of f are regular $\lim_{k\to\infty} f(t'_{n_k}) = \lim_{k\to\infty} f(t''_{n_k})$ which contradicts (1.2). So we can assume that $t'_n \nearrow t^*$ and $t''_n \searrow t^*$. Using again the regularity of f, one has $|\Delta f(t^*)| \geq c + \varepsilon$ which contradicts the assumption $|\Delta f| \leq c$. $\qquad\square$

1.1.3   When are stochastic processes equal?

A stochastic process X has three natural 'facets'. The first one is the process itself, which is the two-dimensional 'view'. We shall refer to this as $X(t, \omega)$ or just as X. With the first notation we want to emphasize that X is a function of two variables. For instance, the different concepts of measurability, like predictability or progressive measurability, characterize X as a function of two variables. We shall often use the notations $X(t)$ or sometimes X_t, which denote the random variable $\omega \mapsto X(t, \omega)$, that is the random variable belonging to moment t. Similarly we shall use the symbols $X(\omega)$, or X_ω as well, which refer to the trajectory belonging to ω, that is $X(\omega)$ is the 'facet' $t \mapsto X(t, \omega)$ of X.

Definition 1.8 *Let X and Y be two stochastic processes on the probability space $(\Omega, \mathcal{A}, \mathbf{P})$.*

1. *The process X is a modification of the process Y if for all $t \in \Theta$ the variables $X(t)$ and $Y(t)$ are almost surely equal, that is for all $t \in \Theta$*

$$\mathbf{P}(X(t) = Y(t)) \triangleq \mathbf{P}(\{\omega : X(t, \omega) = Y(t, \omega)\}) = 1.$$

   By this definition, the set of outcomes ω where $X(t, \omega) \neq Y(t, \omega)$, can depend on $t \in \Theta$.
2. *The processes X and Y are indistinguishable if there is a set $N \subseteq \Omega$ which has probability zero, and whenever $\omega \notin N$ then $X(\omega) = Y(\omega)$, that is $X(t, \omega) = Y(t, \omega)$ for all $t \in \Theta$ and $\omega \notin N$.*

Proposition 1.9 *Assume that the realizations of X and Y are almost surely continuous from the left or they are almost surely continuous from the right. If X is a modification of Y then X and Y are indistinguishable.*

Proof. Let N_0 be the set of outcomes where X and Y are not left-continuous or right-continuous. Let (r_k) be the set of rational points[10] in Θ and let

$$N_k \stackrel{\circ}{=} \{X(r_k) \neq Y(r_k)\} \stackrel{\circ}{=} \{\omega : X(r_k, \omega) \neq Y(r_k, \omega)\}.$$

X is a modification of Y hence $\mathbf{P}(N_k) = 0$ for all k. Therefore if $N \stackrel{\circ}{=} \cup_{k=0}^{\infty} N_k$ then $\mathbf{P}(N) = 0$. If $\omega \notin N$ then $X(r_k, \omega) = Y(r_k, \omega)$ for all k, hence as the trajectories $X(\omega)$ and $Y(\omega)$ are continuous from the same side $X(t, \omega) = Y(t, \omega)$ for all $t \in \Theta$. Therefore outside N obviously $X(\omega) = Y(\omega)$, that is X and Y are indistinguishable. $\qquad\square$

Example 1.10 With modification one can change the topological properties of trajectories.

In the definition of stochastic processes one should always fix the analytic properties like continuity, regularity, differentiability etc. of the trajectories. It is not a great surprise that with modification one can dramatically change these properties. For example, let $(\Omega, \mathcal{A}, \mathbf{P}) \stackrel{\circ}{=} ([0, 1], \mathcal{B}, \lambda)$ and $Y(t, \omega) \equiv 0$. The trajectories of Y are continuous. If $\chi_{\mathbb{Q}}$ is the characteristic function of the rational numbers, and $X(t, \omega) \stackrel{\circ}{=} \chi_{\mathbb{Q}}(t + \omega)$ then for all ω the trajectories of X are never continuous but X is a modification of Y. From the example it is also obvious that it is possible for X to be a modification of Y but for X and Y not to be indistinguishable. $\qquad\square$

If X and Y are stochastic processes then, unless we explicitly say otherwise, $X = Y$ means that X and Y are indistinguishable.

1.2 Measurability of Stochastic Processes

As we have already mentioned, the theory of stochastic processes is an application of measure theory. On the one hand this remark is almost unnecessary as measure theory is the cornerstone of every serious application of mathematical analysis. On the other hand it is absolutely critical how one defines the class of

---

[10]Recall that Θ is an interval in $\mathbb{R}$. If X and Y are left-continuous then left-continuity is meaningless in the left endpoint of Θ, so if Θ has a left endpoint then we assume that this left endpoint is part of (r_k). Similarly when X and Y are right-continuous and Θ has right endpoint then we assume that this endpoint is in (r_k).

measurable functions which one can use in stochastic analysis. Every stochastic process is a function of two variables, so it is obvious to assume that every process is product measurable.

Example 1.11 An almost surely continuous process is not necessarily product measurable.

Let $(\Omega, \mathcal{A}, \mathbf{P}) \doteq ([0,1], \mathcal{B}, \lambda)$ and let E be a subset of $[0,1]$ which is not Lebesgue measurable. The process

$$
X(t, \omega) \doteq \left\{ \begin{array}{lll} 0 & \text{if} & \omega \neq 0 \\ \chi_E(t) & \text{if} & \omega = 0 \end{array} \right.
$$

is almost surely continuous. X is not product measurable as by Fubini's theorem the product measurability implies partial measurability but if $\omega = 0$ then $t \mapsto X(t, \omega)$ is not measurable. Although the example is trivial it is not without any interest. Processes X and Y are considered to be equal if they are indistinguishable. So in theory it can happen that X is product measurable and $X = Y$ but Y is not product measurable. To avoid these type of measurability problems we should for example, assume that the different objects of stochastic analysis, like martingales, local martingales, or semimartingales etc. are right-regular and not just almost surely right-regular. Every trajectory of a Wiener processes should be continuous, but it can happen that it starts only almost surely from zero. □

1.2.1 Filtration, adapted, and progressively measurable processes

A fundamental property of time is its 'irreversibility'. This property of time is expressed with the introduction of the filtration.

Definition 1.12 *Let us fix a probability space* $(\Omega, \mathcal{A}, \mathbf{P})$. *For every* $t \in \Theta$ *let us select a* σ-*algebra* $\mathcal{F}_t \subseteq \mathcal{A}$ *in such a way that whenever* $s < t$ *then* $\mathcal{F}_s \subseteq \mathcal{F}_t$. *The correspondence* $t \mapsto \mathcal{F}_t$ *is called a* filtration *and we shall denote this correspondence by* $\mathcal{F}$. *The quadruplet* $(\Omega, \mathcal{A}, \mathbf{P}, \mathcal{F})$ *is called a* stochastic basis. *With the filtration* $\mathcal{F}$ *one can define the* σ-*algebras*

$$
\mathcal{F}_{t+} \doteq \cap_{s>t} \mathcal{F}_t, \quad \mathcal{F}_{t-} \doteq \sigma \left(\cup_{s<t} \mathcal{F}_t \right), \quad \mathcal{F}_\infty \doteq \sigma \left(\mathcal{F}_t : t \in \Theta \right).
$$

1. *The filtration* $\mathcal{F}$ *is* right-continuous, *if* $\mathcal{F}_t = \mathcal{F}_{t+}$ *for all* t.
2. *The filtration* $\mathcal{F}$ *is* left-continuous, *if* $\mathcal{F}_t = \mathcal{F}_{t-}$ *for all* t.
3. *We say that the filtration* $\mathcal{F}$ *satisfies the* usual conditions *if* $\mathcal{F}$ *is right-continuous and* $\mathcal{F}_t$ *contains for all* t *all the measure zero sets of* $(\Omega, \mathcal{A}, \mathbf{P})$.
4. *We say that the stochastic basis* $(\Omega, \mathcal{A}, \mathbf{P}, \mathcal{F})$ *satisfies the* usual conditions, *if* $(\Omega, \mathcal{A}, \mathbf{P})$ *is complete and the filtration* $\mathcal{F}$ *satisfies the usual conditions.*

It is obvious from the introduced terminology, that generally we shall assume that the filtration and the stochastic basis satisfy the usual conditions. The usual interpretation of the σ-algebra $\mathcal{F}_t$ is that it contains the events which occurred up to time t, that is $\mathcal{F}_t$ contains the information which is available at moment t. As $\mathcal{F}_t$ is the information at moment t one can interpret $\mathcal{F}_{t-}$ as the information available before t and $\mathcal{F}_{t+}$ is the information available just after[11] t.

A quite natural question is how one can define a filtration $\mathcal{F}$. Let X be a stochastic process, that is let X be a function of two variables. Assume that X is product measurable. In this case $X(t)$ is $\mathcal{A}$-measurable for all t. Let us define the σ-algebras $\mathcal{F}_t^X \subseteq \mathcal{A}$ generated by the sets

$$\{X(t_1) \in I_1, \ldots, X(t_n) \in I_n\} \tag{1.3}$$

where $t_1, \ldots, t_n \leq t$ are arbitrary elements in Θ and $I_1, \ldots, I_n$ are arbitrary intervals. Obviously if $s < t$ then $\mathcal{F}_s^X \subseteq \mathcal{F}_t^X$, hence $\mathcal{F}^X$ is really a filtration. $\mathcal{F}^X$ is called *the filtration generated by* X.

Example 1.13 If w is the canonical Wiener process then $\mathcal{F}^w$, the filtration generated by w, is not right-continuous.

Let w be the canonical Wiener process. By definition this means that the set of trajectories of w is the space

$$\Omega \stackrel{\circ}{=} \{f : f \in C(\mathbb{R}_+) \text{ and } f(0) = 0\}.$$

Let F be the set of outcomes $\omega \in \Omega$ for which there is an $\varepsilon > 0$ that on the interval $[0, \varepsilon]$ the trajectory $w(\omega)$ is zero. Obviously $F = \cup_n F_n$, where F_n is the set of outcomes ω, for which $w(\omega)$ is zero on the interval $[0, 1/n]$. F_n is measurable as it is equal to the set

$$\left\{w(r_n) = 0, r_n \in \left[0, \frac{1}{n}\right] \cap \mathbb{Q}\right\}.$$

Obviously $\mathbf{P}(F_n) = 0$, therefore $\mathbf{P}(F) = 0$. By definition $w(0) \equiv 0$, therefore $\mathcal{F}_0^w = \{\Omega, \emptyset\}$. Hence $F \notin \mathcal{F}_0^w$. If $t > 0$ and $1/n \leq t$, then obviously $F_n \in \mathcal{F}_t^w$, therefore $\cup_{1/n \leq t} F_n \in \mathcal{F}_t^w$. On the other hand for every $t > 0$ evidently $\cup_{1/n \leq t} F_n = F$, since obviously $\cup_{1/n \leq t} F_n \subseteq F$ and if $\omega \in F$ then $\omega \in F_n \subseteq \cup_{1/n \leq t} F_n$ for some index n. Hence $F \in \cap_{t>0} \mathcal{F}_t^w = \mathcal{F}_{0+}^w$, that is $\mathcal{F}_0^w \neq \mathcal{F}_{0+}^w$. Let us remark that, as we shall see later, if $\mathcal{N}$ is the collection of sets with

---

[11] One can observe that the interpretation of $\mathcal{F}_{t-}$ is intuitively quite appealing, but the interpretation of $\mathcal{F}_{t+}$ looks a bit unclear. It is intuitively not obvious that what type of information one can get in an infinitesimally short time interval after t or to put it in another way it is not too clear why one can get $\mathcal{F}_t \neq \mathcal{F}_{t+}$. Therefore from an intuitive point of view it is not a great surprise that we shall generally assume that $\mathcal{F}_t = \mathcal{F}_{t+}$.

measure-zero in $\mathcal{A}$ then the filtration $\mathcal{F}_t \overset{\circ}{=} \sigma\left(\mathcal{F}_t^w \cup \mathcal{N}\right)$ is right-continuous, so this extended $\mathcal{F}$ satisfies the usual conditions[12]. The σ-algebra $\mathcal{F}_0^w = \{\Omega, \emptyset\}$ is complete, which implies that to make $\mathcal{F}$ right-continuous one should add to the σ-algebra $\mathcal{F}_t^w$ all the null sets from $\mathcal{A}$, or at least the null sets of $\mathcal{F}_t^w$ for all t and it is not sufficient to complete the σ-algebras $\mathcal{F}_t^w$ separately. $\qquad\square$

Definition 1.14 *We say that process X is* adapted *to the filtration $\mathcal{F}$ if $X(t)$ is measurable with respect to $\mathcal{F}_t$ for every t. A set $A \subseteq \Theta \times \Omega$ is* adapted *if the process χ_A is adapted.*

In the following we shall fix a stochastic basis $(\Omega, \mathcal{A}, \mathbf{P}, \mathcal{F})$ and if we do not say otherwise we shall always assume that all stochastic processes are adapted with respect to the filtration $\mathcal{F}$ of the stochastic basis. It is easy to see that the set of adapted sets form a σ-algebra.

Example 1.15 If $\mathcal{F}_t \equiv \{\emptyset, \Omega\}$ for all t then only the deterministic processes are adapted. If $\mathcal{F}_t \equiv \mathcal{A}$ for all t then every product measurable stochastic process is adapted.

The concept of adapted processes is a dynamic generalization of partial measurability. The dynamic generalization of product measurability is progressive measurability:

Definition 1.16 *A set $A \subseteq \Theta \times \Omega$ is* **progressively measurable** *if for all $t \in \Theta$*

$$A \cap ([0, t] \times \Omega) \in \mathcal{R}_t \overset{\circ}{=} \mathcal{B}\left([0, t]\right) \times \mathcal{F}_t,$$

that is for all t the restriction of A to $[0, t] \times \Omega$ is measurable with respect to the product σ-algebra

$$\mathcal{R}_t \overset{\circ}{=} \mathcal{B}\left([0, t]\right) \times \mathcal{F}_t.$$

The progressively measurable sets form a σ-algebra $\mathcal{R}$. We say that a process X is progressively measurable *if it is measurable with respect to $\mathcal{R}$.*

It is clear from the definition that every progressively measurable process is adapted.

Example 1.17 Adapted process which is not progressively measurable.

---

[12]See: Proposition 1.103, page 67.

Let $\Omega \overset{\circ}{=} \Theta \overset{\circ}{=} [0,1]$ and let $\mathcal{F}_t \overset{\circ}{=} \mathcal{A}$ be the σ-algebra generated by the finite subsets of Ω. If $D \overset{\circ}{=} \{t = \omega\}$ then the function $X \overset{\circ}{=} \chi_D$ is obviously adapted. We prove that it is not product measurable. Assume that $\{X = 1\} = D \in \mathcal{B}(\Theta) \times \mathcal{A}$. By the definition of product measurability $Y \overset{\circ}{=} [0,1/2] \times \Omega \in \mathcal{B}(\Theta) \times \mathcal{A}$. So if $D \in \mathcal{B}(\Theta) \times \mathcal{A}$ then $Y \cap D \in \mathcal{B}(\Theta) \times \mathcal{A}$. Therefore by the projection theorem[13] $[0,1/2] \in \mathcal{A}$ which is impossible. Therefore $D \notin \mathcal{B}(\Theta) \times \mathcal{A}$. If $\mathcal{F}_t \overset{\circ}{=} \mathcal{A}$ for all t then X is adapted but not progressively measurable. $\square$

Example 1.18 Every adapted, continuous from the left and every adapted, continuous from the right process is progressively measurable[14].

Assume, for example, that X is adapted and continuous from the right. Fix a t and let $0 = t_0^{(n)} < t_1^{(n)} < \ldots < t_k^{(n)} = t$ be a partition of $[0,t]$. Let us define the processes

$$X_n(s) \overset{\circ}{=} \begin{cases} X(0) & \text{if} & s = 0 \\ X\left(t_k^{(n)}\right) & \text{if} & s \in \left(t_{k-1}^{(n)}, t_k^{(n)}\right] \end{cases} .$$

As X is adapted X_n is measurable with respect to the σ-algebra $\mathcal{R}_t \overset{\circ}{=} \mathcal{B}([0,t]) \times \mathcal{F}_t$. If the sequence of partitions $(t_k^{(n)})$ is infinitesimal, that is if

$$\lim_{n\to\infty} \max_k \left| t_k^{(n)} - t_{k-1}^{(n)} \right| = 0$$

then as X is right-continuous $X_n \to X$. Therefore the restriction of X to $[0,t]$ is $\mathcal{R}_t$-measurable. Hence X is progressively measurable. $\square$

Example 1.19 If X is regular then ΔX is progressively measurable.

Like the product measurability, the progressive measurability is also a very mild assumption. It is perhaps the mildest measurability concept one can use in stochastic analysis. The main reason why one should introduce this concept is the following much-used observation:

Proposition 1.20 *Assume that V is a right-regular, adapted process and assume that every trajectory of V has finite variation on every finite interval $[0,t]$.*

1. *If for every ω the trajectories $X(\omega)$ are integrable on any finite interval with respect to the measure generated by $V(\omega)$ then the parametric*

---

[13]If $P(N) = 0$ if N is countable otherwise $P(N) = 1$, then the probability space $(\Omega, \mathcal{A}, P)$ is complete.

[14]Specially, if $X(t, \omega)$ is measurable in ω and continuous in t then X is product measurable.

integral process

$$Y(t, \omega) \overset{\circ}{=} \int_0^t X(s, \omega) V(ds, \omega) \overset{\circ}{=} \tag{1.4}$$

$$\overset{\circ}{=} \int_{(0,t]} X(s, \omega) V(ds, \omega)$$

forms a right-regular process and $\Delta Y = X \cdot \Delta V$.
2. *If additionally* X *is progressively measurable then* Y *is adapted.*

Proof. The first statement of the proposition is a direct consequence of the Dominated Convergence Theorem. Observe that to prove the second statement one cannot directly apply Fubini's theorem, but one can easily adapt its usual proof: Let $\mathcal{H}$ denote the set of bounded processes for which $Y(t)$ in (1.4) is $\mathcal{F}_t$-measurable. As the measure of finite intervals is finite $\mathcal{H}$ is a linear space, it contains the constant process $X \equiv 1$, and if $0 \leq H_n \in \mathcal{H}$ and $H_n \nearrow H$ and H is bounded then by the Monotone Convergence Theorem $H \in \mathcal{H}$. This implies that $\mathcal{H}$ is a λ-system. If $C \in \mathcal{F}_t$ and $s_1, s_2 \leq t$, and $B \overset{\circ}{=} (s_1, s_2] \times C$ then as V is adapted the integral

$$\int_0^t \chi_B dV = \chi_C \left[V(s_2) - V(s_1) \right]$$

is $\mathcal{F}_t$-measurable. These processes form a π-system, hence by the Monotone Class Theorem $\mathcal{H}$ contains the processes which are measurable with respect to the σ-algebra generated by the processes $\chi_C \chi((s_1, s_2])$. As $C \in \mathcal{F}_t$ the π-system generates the σ-algebra of the product measurable sets $\mathcal{B}((0,t]) \times \mathcal{F}_t$. X is progressively measurable so its restriction to $(0, t]$ is $(\mathcal{B}((0,t]) \times \mathcal{F}_t)$-measurable. Hence the proposition is true if X is bounded. From this the general case follows from the Dominated Convergence Theorem. $\qquad \square$

What is the intuitive idea behind progressive measurability? Generally the filtration $\mathcal{F}$ is generated by some process X. Recall that if $Z \overset{\circ}{=} (\xi_\alpha)_{\alpha \in A}$ is a set of random variables and $\mathcal{X} \overset{\circ}{=} \sigma(\xi_\alpha : \alpha \in A)$ denotes the σ-algebra generated by them then $\mathcal{X} = \cup_{S \subseteq A} \mathcal{X}_S$ where the subsets S are arbitrary countable subsets of A and for any S the set $\mathcal{X}_S$ denotes the σ-algebra generated by the countably many variables $(\xi_{\alpha_i})_{\alpha_i \in S}$ of Z, that is $\mathcal{X}_S \overset{\circ}{=} \sigma(\xi_{\alpha_i} : \alpha_i \in S)$. By this structure of the generated σ-algebras, $\mathcal{F}_t^X$ contains all the information one can obtain observing X up to time t countably many times. If a process Y is adapted with respect to $\mathcal{F}^X$ then Y reflects the information one can obtain from countable many observations of X. But sometimes, like in (1.4), we want information which depends on uncountable number of observations of the underlying random source. In these cases one needs progressive measurability!

1.2.2 Stopping times

After filtration, stopping time is perhaps the most important concept of the theory of stochastic processes. As stopping times describe the moments when certain random events occur, it is not a great surprise that most of the relevant questions of the theory are somehow related to stopping times. It is important that not every random time is a stopping time. Stopping times are related to events described by the filtration of the stochastic base[15]. At every time t one can observe only the events of the probability space $(\Omega, \mathcal{F}_t, \mathbf{P})$. If τ is a random time then at time t one cannot observe the whole τ. One can observe only the random variable $\tau \wedge t$! By definition τ is a stopping time if $\tau \wedge t$ is an $(\Omega, \mathcal{F}_t, \mathbf{P})$-random variable for all t.

Definition 1.21 *Let Ω be the set of outcomes and let $\mathcal{F}$ be a filtration on Ω. Let $\tau : \Omega \to \Theta \cup \{\infty\}$.*

1. *The function τ is a* stopping time *if for every $t \in \Theta$*

$$\{\tau \le t\} \in \mathcal{F}_t.$$

We denote the set of stopping times by Υ.

2. *The function τ is a* weak stopping time *if for every $t \in \Theta$*

$$\{\tau < t\} \in \mathcal{F}_t.$$

Example 1.22 Almost-surely zero functions and stopping times.

Assume that the probability space $(\Omega, \mathcal{A}, \mathbf{P})$ is complete and for every t the σ-algebra $\mathcal{F}_t$ contains the measure-zero sets of $\mathcal{A}$. If $N \subseteq \Omega$ is a measure-zero set and the function $\tau \ge 0$ is zero on the complement of N, then τ is stopping time, as for all t $\{\tau \le t\} \subseteq N \in \mathcal{F}_t$, hence $\{\tau \le t\} \in \mathcal{F}_t$. In a similar way if $\sigma \ge 0$ is almost surely $+\infty$ then σ is a stopping time. These examples are special cases of the following: If $(\Omega, \mathcal{A}, \mathbf{P}, \mathcal{F})$ satisfies the usual conditions and τ is a stopping time and $\sigma \ge 0$ is almost surely equal to τ then σ is also a stopping time. $\square$

We shall see several times that in the theory of stochastic processes the time axis is not symmetric. The filtration defines an orientation on the real axis.

---

[15]If we travel from a city to the countryside then the moment when we arrive at the first pub after we leave the city is a stopping time, but the time when we arrive at the last pub before we leave the city is not a stopping time. In a similar way when X is a stochastic process the first time X is zero is a stopping time, but the last time it is zero is not a stopping time. One of the most important random times which is generally not a stopping time is the moment when X reaches its maximum on a certain interval. See: Example 1.110, page 73.

An elementary but very import consequence of this orientation is the following proposition:

Proposition 1.23 *Every stopping time is a weak stopping time. If the filtration $\mathcal{F}$ is right-continuous then every weak stopping time is a stopping time.*

Proof. As the filtration $\mathcal{F}$ is increasing, if τ is a stopping time then for all n

$$\left\{ \tau \le t - \frac{1}{n} \right\} \in \mathcal{F}_{t-1/n} \subseteq \mathcal{F}_t.$$

Therefore

$$\{\tau < t\} = \cup_n \left\{ \tau \le t - \frac{1}{n} \right\} \in \mathcal{F}_t.$$

On the other hand if $\mathcal{F}$ is right-continuous that is if $\mathcal{F}_{t+} = \mathcal{F}_t$ then

$$\{\tau \le t\} = \cap_n \left\{ \tau < t + \frac{1}{n} \right\} \in \cap_n \mathcal{F}_{t+1/n} \stackrel{\circ}{=} \mathcal{F}_{t+} = \mathcal{F}_t. \qquad \square$$

The right-continuity of the filtration is used in the next proposition as well.

Proposition 1.24 *If τ and σ are stopping times then $\tau \wedge \sigma$ and $\tau \vee \sigma$ are also stopping times. If (τ_n) is an increasing sequence of stopping times then*

$$\tau \stackrel{\circ}{=} \lim_{n \to \infty} \tau_n$$

is a stopping time. If the filtration $\mathcal{F}$ is right-continuous and (τ_n) is a decreasing sequence of stopping times then

$$\tau \stackrel{\circ}{=} \lim_{n \to \infty} \tau_n$$

is a stopping time.

Proof. If τ and σ are stopping times then

$$\{\tau \wedge \sigma \le t\} = \{\tau \le t\} \cup \{\sigma \le t\} \in \mathcal{F}_t,$$
$$\{\tau \vee \sigma \le t\} = \{\tau \le t\} \cap \{\sigma \le t\} \in \mathcal{F}_t.$$

If $\tau_n \nearrow \tau$ then for all t

$$\{\tau \le t\} = \cap_n \{\tau_n \le t\} \in \mathcal{F}_t.$$

If $\tau_n \searrow \tau$ then for all t

$$\{\tau \ge t\} = \cap_n \{\tau_n \ge t\} = \cap_n \{\tau_n < t\}^c \in \mathcal{F}_t$$

that is

$$\{\tau < t\} = \cup_n \{\tau_n < t\} \in \mathcal{F}_t.$$

Hence τ is a weak stopping time. If the filtration $\mathcal{F}$ is right-continuous then τ is a stopping time. $\qquad\square$

Corollary 1.25 *If the filtration $\mathcal{F}$ is right-continuous and (τ_n) is a sequence of stopping times then*

$$\sup_n \tau_n, \quad \inf_n \tau_n \quad \limsup_{n\to\infty} \tau_n, \quad \liminf_{n\to\infty} \tau_n$$

are stopping times.

The next definition concretizes the abstract definition of stopping times:

Definition 1.26 *If $\Gamma \subseteq \mathbb{R}_+ \times \Omega$ then the expression*

$$\tau_\Gamma(\omega) \overset{\circ}{=} \inf\{t : (t, \omega) \in \Gamma\} \tag{1.5}$$

is called the début *of the set Γ. If $B \subseteq \mathbb{R}^n$ and X is a vector valued stochastic process then*

$$\tau_B(\omega) \overset{\circ}{=} \inf\{t : X(t, \omega) \in B\} \tag{1.6}$$

is called the hitting time *of set B.*

If $B \subseteq \mathbb{R}$ and X is a stochastic process and if $\Gamma \overset{\circ}{=} \{X \in B\}$ then $\tau_\Gamma = \tau_B$ which means that every hitting time is a special début.

Example 1.27 The most important hitting times are the random functions

$$\tau_a(\omega) \overset{\circ}{=} \inf\{t : X(t, \omega)\mathcal{R}a\}$$

where $\mathcal{R}$ is one of the relations $\geq, >, \leq, <$. These type of hitting times are the so-called *first passage times.*

Theorem 1.28 (Construction of stopping times) *If the stochastic base $(\Omega, \mathcal{A}, \mathbf{P}, \mathcal{F})$ satisfies the usual conditions and Γ is progressively measurable then the début of Γ is a stopping time.*

Proof. Define the set $\Gamma_t \overset{\circ}{=} \Gamma \cap [0, t) \times \Omega$. If $\tau_\Gamma(\omega) < t$ then for some s obviously $(s, \omega) \in \Gamma_t$. Hence $\omega \in \text{proj}_\Omega(\Gamma_t)$. On the other hand if ω is in $\text{proj}_\Omega(\Gamma_t)$

then for some $s \in [0,t)$ obviously $(s, \omega) \in \Gamma$, hence $\tau_\Gamma(\omega) \leq s < t$ that is

$$\{\tau_\Gamma < t\} = \operatorname{proj}_\Omega (\Gamma_t).$$

Γ is progressively measurable, hence

$$\Gamma_t \overset{\circ}{=} (\Gamma \cap [0,t] \times \Omega) \cap [0,t) \times \Omega \in \mathcal{B}([0,t]) \times \mathcal{F}_t.$$

Recall that the projections of product measurable sets are not necessarily measurable. By the usual conditions $\mathcal{A}$ is complete, and also by the usual conditions $\mathcal{F}_t$ contains all the measure-zero sets, hence $\mathcal{F}_t$ is also complete and, therefore, by the Projection Theorem[16], the projection of the product measurable set Γ_t is $\mathcal{F}_t$-measurable, so

$$\{\tau_\Gamma < t\} = \operatorname{proj}_\Omega (\Gamma_t) \in \mathcal{F}_t.$$

As $\mathcal{F}$ is right-continuous[17] every weak stopping time is a stopping time so τ_Γ is a stopping time. $\square$

Corollary 1.29 *If the stochastic base $(\Omega, \mathcal{A}, \mathbf{P}, \mathcal{F})$ satisfies the usual conditions, the process X is progressively measurable and B is a Borel set then the hitting time of B is a stopping time.*

Let X be a progressively measurable process and let σ be a stopping time. Instead of (1.5) very often we are interested in variables of the type

$$\tau \overset{\circ}{=} \inf \{t > \sigma : X(t) \in B\}.$$

The set

$$\Gamma \overset{\circ}{=} \{(t, \omega) : X(t, \omega) \in B\} \cap \{(t, \omega) : t > \sigma(\omega)\}$$

is progressively measurable since by the progressive measurability of X the first set in the intersection is progressively measurable, and the characteristic function of the other set is adapted and left-continuous hence it is also progressively measurable. By the theorem above if $(\Omega, \mathcal{A}, \mathbf{P}, \mathcal{F})$ satisfies the usual conditions then the expression

$$\tau = \tau_\Gamma \overset{\circ}{=} \inf \{t : (t, \omega) \in \Gamma\}$$

is a stopping time.

---

[16] See: Theorem A.12, page 550.

[17] It can happen that $(s, \omega) \in \Gamma$ for all $s > t$, but $(t, \omega) \notin \Gamma$. In this case $\tau_\Gamma(\omega) = t$, but $\omega \notin \operatorname{proj}_\Omega (\Gamma \cap [0,t) \times \Omega)$, therefore in the proof we used the right-continuity of the filtration.

Corollary 1.30 *If the stochastic base* $(\Omega, \mathcal{A}, \mathbf{P}, \mathcal{F})$ *satisfies the usual conditions, the process* X *is progressively measurable and* B *is a Borel set then the hitting times*

$$\tau_0 \stackrel{\circ}{=} 0, \quad \tau_{n+1} \stackrel{\circ}{=} \inf\{t > \tau_n : X(t) \in B\}$$

are stopping times.

Example 1.31 If X is not progressively measurable then the hitting times of Borel sets are not necessarily stopping times.

Let $X \stackrel{\circ}{=} \chi_D$ be the adapted but not progressively measurable process in Example 1.17. The hitting time of the set $B \stackrel{\circ}{=} \{1\}$ is obviously not a stopping time as

$$\{\tau_B \leq 1/2\} = [0, 1/2] \notin \mathcal{A} \stackrel{\circ}{=} \mathcal{F}_{1/2}. \qquad \square$$

The main advantage of the above construction is its generality. An obvious disadvantage of the just proved theorem is that it builds on the Projection Theorem. Very often we do not need the generality of the above construction and we can construct stopping times without referring to the Projection Theorem.

Example 1.32 Construction of stopping times without the Projection Theorem.

---

1. If the set B is closed and X is a continuous, adapted process then one can easily proof that the hitting time (1.6) is a stopping time. As the trajectories are continuous the sets $K(t, \omega) \stackrel{\circ}{=} X([0, t], \omega)$ are compact for every outcome ω. As B is closed $K(t, \omega) \cap B = \emptyset$ if and only, if the distance between the two sets is positive. Therefore $K(t, \omega) \cap B = \emptyset$ if and only if $\tau_B(\omega) > t$. As the trajectories are continuous $X([0, t] \cap \mathbb{Q}, \omega)$ is dense in the set $K(t, \omega)$. As the metric is a continuous function

$$\{\tau_B \leq t\} = \{K(t) \cap B \neq \emptyset\} = \{d(K(t), B) = 0\} =$$
$$= \{\omega : \inf\{d(X(s, \omega), B) : s \leq t, s \in \mathbb{Q}\} = 0\}.$$

$X(s)$ is $\mathcal{F}_t$-measurable for a fixed $s \leq t$, hence as $x \mapsto d(x, B)$ is continuous $d(X(s), B)$ is also $\mathcal{F}_t$-measurable. The infimum of a countable number of measurable functions is measurable, hence $\{\tau_B \leq t\} \in \mathcal{F}_t$.

2. We prove that if B is open, the trajectories of X are right-continuous and adapted, and the filtration $\mathcal{F}$ is right-continuous then the hitting time (1.6) is a stopping time. It is sufficient to prove that $\{\tau_B < t\} \in \mathcal{F}_t$ for all t. As the trajectories are right-continuous and as B is open $X(s, \omega) \in B$, if and only if,

there is an $\varepsilon > 0$ such that whenever $u \in [s, s + \varepsilon)$ then $X(u, \omega) \in B$. From this

$$\{\tau_B < t\} = \cup_{s \in \mathbb{Q} \cap [0,t)} \{X(s) \in B\} \in \mathcal{F}_t.$$

3. In a similar way one can prove that if X is left-continuous and adapted, $\mathcal{F}$ is right-continuous, and B is open, then the hitting time τ_B is a stopping time.

4. If the filtration is right-continuous, and X is a right or left-continuous adapted process, then for any number c the first passage time

$$\tau \overset{\circ}{=} \inf \{t : X(t) > c\}$$

is a stopping time.

5. If B is open and the filtration is not right-continuous, then even for continuous processes the hitting time τ_B is not necessarily a stopping time[18]. If

$$X(t, \omega) \overset{\circ}{=} t \cdot \xi(\omega),$$

where ξ is a Gaussian random variable, and $\mathcal{F}_t$ is the filtration generated by X, then $\mathcal{F}_0 = \{0, \Omega\}$, and the hitting time τ_B of the set $B \overset{\circ}{=} \{x > 0\}$ is

$$\tau_B(\omega) \overset{\circ}{=} \begin{cases} 0 & \text{if} \quad \xi(\omega) > 0 \\ \infty & \text{if} \quad \xi(\omega) \leq 0 \end{cases}.$$

Obviously $\{\tau_B \leq 0\} \notin \mathcal{F}_0$, so τ_B is not a stopping time.

6. Finally we show that if σ is an arbitrary stopping time and X is a right-regular, adapted process and $c > 0$, then the first passage time

$$\tau(\omega) \overset{\circ}{=} \inf \{t > \sigma : |\Delta X(t, \omega)| \geq c\}$$

is stopping time. Let us fix an outcome ω and let assume that $\infty > t_n \searrow \tau(\omega)$, where $|\Delta X(t_n, \omega)| \geq c$. The trajectory $X(\omega)$ is right-regular, therefore the jumps which are bigger than c do not have an accumulation point. Hence for all indexes n large enough t_n is already constant, that is $\tau(\omega) = t_n > \sigma(\omega)$, so $|\Delta X(\tau(\omega))| = |\Delta X(t_n)| \geq c$ for some n. This means that $|\Delta X(\tau)| \geq c$ on the set $\{\tau < \infty\}$ and on the set $\{\sigma < \infty\}$ one has $\tau > \sigma$.

Let $A(t) \overset{\circ}{=} ([0, t] \cap \mathbb{Q}) \cup \{t\}$. We prove that $\tau(\omega) \leq t$ if and only if for all $n \in \mathbb{N}$ one can find a pair $q_n, p_n \in A(t)$ for which

$$\sigma(\omega) < p_n < q_n < p_n + \frac{1}{n}$$

---

[18]The reason for this is clear as the event $\{\tau_B = t\}$ can contain such outcomes ω that the trajectory will hit the set B just after t therefore one should investigate the events $\{\tau_B < t\}$.

and

$$|X(p_n, \omega) - X(q_n, \omega)| \geq c - \frac{1}{n}. \tag{1.7}$$

One implication is evident, that is if $\tau(\omega) \leq t$, then as the jumps bigger than c do not have accumulation points, $|\Delta X(s, \omega)| \geq c$ for some $\sigma(\omega) < s \leq t$. Hence by the regularity of the trajectories one can construct the necessary sequences. On the other hand, let us assume that the sequences (p_n), (q_n) exist. Without loss of generality one can assume that (p_n) and (q_n) are convergent. Let $\sigma(\omega) \leq s \leq t$ be the common limit point of these sequences. If for an infinite number of indexes $p_n \geq s$, then in any right neighbourhood of s there is an infinite number of intervals $[p_n, q_n]$, on which X changes more then $c/2 > 0$, which is impossible as X is right-continuous. Similarly, only for a finite number of indexes $q_n \leq s$ as otherwise for an infinite number of indexes $p_n < q_n \leq s$ which is impossible as $X(\omega)$ is left-continuous. This means that for indexes n big enough $\sigma(\omega) < p_n \leq s \leq q_n$. Taking the limit in the line (1.7) $|\Delta X(s, \omega)| \geq c$ and hence $\tau(\omega) \leq s \leq t$. Using this property one can easily proof that

$$\{\tau \leq t\} = \bigcap_{n \in \mathbb{N}} \bigcup_{\substack{p, q \in A(t) \\ p < q < p + 1/n}} \left(\{\sigma < q\} \cap \left\{ |X(p) - X(q)| \geq c - \frac{1}{n} \right\} \right).$$

$A(t)$ is countable, X is adapted therefore $\{\tau \leq t\} \in \mathcal{F}_t$, which means that τ is a stopping time.

7. If X is a regular process and $c > 0$ then the hitting time

$$\tau \overset{\circ}{=} \inf \{t : |\Delta X(t)| \geq c\}$$

is a stopping time. $\qquad \square$

1.2.3 Stopped variables, σ-algebras, and truncated processes

With stopping times one can define stopped variables, truncated processes, and the stopped σ-algebras:

Definition 1.33 *Let X be a stochastic process, and let τ be a stopping time.*

1. *By a* stopped *or* truncated process *we mean the process*

$$X^\tau(t, \omega) \overset{\circ}{=} X(\tau(\omega) \wedge t, \omega).$$

2. *We shall call the random variable*

$$X_\tau(\omega) \overset{\circ}{=} X(\tau(\omega), \omega)$$

a stopped variable. *Instead of X_τ we shall very often use the more readable notation $X(\tau)$. Observe that the definition of stopped variable is not entirely correct as X is generally not defined on the set $\{\tau = \infty\}$ and it is not clear what the definition of X_τ on this set is. If $\tau(\omega) \notin \Theta$ then one can use the definition*[19]

$$X_\tau(\omega) \overset{\circ}{=} 0.$$

If one uses the convention that the product of an undefined value with zero is zero, then one can write the definition of the stopped variable X_τ in the following way:

$$X_\tau(\omega) \overset{\circ}{=} X(\tau(\omega), \omega) \chi(\tau \in \Theta)(\omega).$$

3. *The stopped σ-algebra $\mathcal{F}_\tau$ is the set of events $A \in \mathcal{A}$ for which for all t*

$$A \cap \{\tau \leq t\} \in \mathcal{F}_t.$$

4. *$\mathcal{F}_{\tau+}$ is the set of events $A \in \mathcal{A}$ for which*

$$A \cap \{\tau \leq t\} \in \mathcal{F}_{t+}$$

for all t.

One can easily check that $\mathcal{F}_\tau$ and $\mathcal{F}_{\tau+}$ are really σ-algebras. For example, if $A \in \mathcal{F}_\tau$ then $A^c \in \mathcal{F}_\tau$ as for every t

$$A^c \cap \{\tau \leq t\} = \{\tau \leq t\} \setminus (A \cap \{\tau \leq t\}) \in \mathcal{F}_t,$$

and if $A_n \in \mathcal{F}_\tau$ then

$$(\cup_n A_n) \cap \{\tau \leq t\} = \cup_n (A_n \cap \{\tau \leq t\}) \in \mathcal{F}_t.$$

It is easy to see that if $\tau \equiv t$, then $\mathcal{F}_\tau = \mathcal{F}_t$ and $\mathcal{F}_{\tau+} = \mathcal{F}_{t+}$, hence the notation is unambiguous. If we assume that the usual conditions are satisfied then of course $\mathcal{F}_\tau = \mathcal{F}_{\tau+}$ hence there are not too many important theorems where the σ-algebra $\mathcal{F}_{\tau+}$ plays a role.

There are many simple observations related to the stopped processes, variables, and σ-algebras. Their proof is generally one or two lines. Let us show some of them:

Proposition 1.34 *Let $\mathcal{F}$ be a filtration, and let τ and σ be stopping times.*

1. *τ is $\mathcal{F}_\tau$-measurable.*
2. *If $\sigma \leq \tau$ then $\mathcal{F}_\sigma \subseteq \mathcal{F}_\tau$.*

---

[19] If X is the value process of some game and τ is an exit strategy then the present definition of X_τ is quite reasonable.

3. $\mathcal{F}_\sigma \cap \mathcal{F}_\tau = \mathcal{F}_{\sigma \wedge \tau}$.

4. $\{\sigma \leq \tau\}, \{\sigma < \tau\}, \{\sigma = \tau\}$ *are* $\mathcal{F}_{\sigma \wedge \tau}$-*measurable.*

Proof. The proofs are simple consequences of the definitions.

1. We prove that $\{\tau \leq s\} \in \mathcal{F}_\tau$ for all s. Let t be arbitrary. As τ is a stopping time

$$\{\tau \leq s\} \cap \{\tau \leq t\} = \{\tau \leq s \wedge t\} \in \mathcal{F}_{s \wedge t} \subseteq \mathcal{F}_t$$

so $\{\tau \leq s\} \in \mathcal{F}_\tau$ by the definition of $\mathcal{F}_\tau$. Hence τ is $\mathcal{F}_\tau$-measurable.

2. If $\sigma \leq \tau$ then $\{\tau \leq t\} \subseteq \{\sigma \leq t\}$. If $A \in \mathcal{F}_\sigma$ then

$$A \cap \{\tau \leq t\} = (A \cap \{\sigma \leq t\}) \cap \{\tau \leq t\} \in \mathcal{F}_t,$$

as both sets in the intersection are in $\mathcal{F}_t$. Hence $A \in \mathcal{F}_\tau$.

3. By the previous property $\mathcal{F}_{\tau \wedge \sigma} \subseteq \mathcal{F}_\tau \cap \mathcal{F}_\sigma$. On the other hand if $A \in \mathcal{F}_\tau \cap \mathcal{F}_\sigma$ then

$$A \cap \{\sigma \wedge \tau \leq t\} = A \cap (\{\sigma \leq t\} \cup \{\tau \leq t\}) =$$
$$= (A \cap \{\sigma \leq t\}) \cup (A \cap \{\tau \leq t\}) \in \mathcal{F}_t.$$

Hence $A \in \mathcal{F}_{\sigma \wedge \tau}$.

4. It is sufficient to prove that if σ and τ are stopping times then $\{\sigma \leq \tau\}, \{\tau \leq \sigma\} \in \mathcal{F}_\sigma$. From this by the symmetry $\{\sigma \leq \tau\} \in \mathcal{F}_\sigma \cap \mathcal{F}_\tau = \mathcal{F}_{\sigma \wedge \tau}$, and $\{\sigma = \tau\} = \{\sigma \leq \tau\} \cap \{\tau \leq \sigma\} \in \mathcal{F}_{\sigma \wedge \tau}$ and $\{\sigma < \tau\} = \{\sigma \leq \tau\} \setminus \{\sigma = \tau\} \in \mathcal{F}_{\sigma \wedge \tau}$.

From the definition of stopping times if $r \leq t$ then

$$\{\sigma > r > \tau\} = \{\sigma > r\} \cap \{\tau < r\} =$$
$$= \{\sigma \leq r\}^c \cap \{\tau < r\} \in \mathcal{F}_t.$$

From this

$$\{\sigma \leq \tau\}^c \cap \{\sigma \leq t\} = \{\sigma > \tau\} \cap \{\sigma \leq t\} =$$
$$= \cup_{r \in \mathbb{Q}} \{\sigma > r > \tau\} \cap \{\sigma \leq t\} =$$
$$= \cup_{r \in \mathbb{Q}, r \leq t} \{\sigma > r > \tau\} \cap \{\sigma \leq t\} \in \mathcal{F}_t.$$

Hence by the definition of $\mathcal{F}_\sigma$ one has $\{\sigma \leq \tau\} \in \mathcal{F}_\sigma$. On the other hand

$$\{\tau \leq \sigma\} \cap \{\sigma \leq t\} = \{\sigma \leq t\} \cap \{\tau \leq t\} \cap \{\tau \wedge t \leq \sigma \wedge t\} \in \mathcal{F}_t,$$

since the first two sets, by the definition of stopping times, are in $\mathcal{F}_t$ and the two random variables in the third set are $\mathcal{F}_t$-measurable. Hence $\{\tau \leq \sigma\} \in \mathcal{F}_\sigma$. $\square$

Proposition 1.35 *If X is progressively measurable and τ is an arbitrary stopping time then the stopped variable X_τ is $\mathcal{F}_\tau$-measurable, and the truncated process X^τ is progressively measurable.*

Proof. The first part of the proposition is an easy consequence of the second as, if B is a Borel measurable set and X^τ is adapted, then for all s

$$\{X_\tau \in B\} \cap \{\tau \leq s\} = \{X(\tau \wedge s) \in B\} \cap \{\tau \leq s\} =$$
$$= \{X^\tau(s) \in B\} \cap \{\tau \leq s\} \in \mathcal{F}_s,$$

that is, in this case the stopped variable X_τ is $\mathcal{F}_\tau$-measurable. Therefore it is sufficient to prove that if X is progressively measurable then X^τ is also progressively measurable. Let

$$Y(t,w) \triangleq \begin{cases} 1 & \text{if } t < \tau(w) \\ 0 & \text{if } t \geq \tau(w) \end{cases}.$$

Y is right-regular. τ is a stopping time so $\{Y(t) = 0\} = \{\tau \leq t\} \in \mathcal{F}_t$. Hence Y is adapted, therefore it is progressively measurable[20]. Obviously if $\tau(w) > 0$ then[21]

$$Z(t,w) \triangleq \int_{(0,t]} X(s,w) Y(ds,w) = \begin{cases} 0 & \text{if } t < \tau(w) \\ -X(\tau(w),w) & \text{if } t \geq \tau(w) \end{cases}.$$

As X is progressively measurable Z is adapted[22] and also right-regular so it is again progressively measurable. As

$$X^\tau = XY - Z + X(0)\chi(\tau = 0)$$

X^τ is obviously progressively measurable. $\square$

Corollary 1.36 *If $\mathcal{G} \triangleq \sigma(X(\tau) : X$ is right-regular and adapted) then $\mathcal{G} = \mathcal{F}_\tau$.*

Proof. As every right-regular and adapted process is progressively measurable $\mathcal{G} \subseteq \mathcal{F}_\tau$. If $A \in \mathcal{F}_\tau$ then the process $X(t) \triangleq \chi_A \chi(\tau \leq t)$ is right-regular and by

---

[20] See: Example 1.18, page 11.
[21] If $\tau(w) = 0$ then $Z(w) = 0$.
[22] See: Proposition 1.20, page 11.

the definition of $\mathcal{F}_\tau$

$$\{X(t) = 1\} = A \cap \{\tau \leq t\} \in \mathcal{F}_t.$$

Hence X is adapted. Obviously $X(\tau) = \chi_A$. Therefore $\mathcal{F}_\tau \subseteq \mathcal{G}$. □

1.2.4 Predictable processes

The class of progressively measurable processes is too large. As we have already remarked, the interesting stochastic processes have regular trajectories. There are two types of regular processes: some of them have left- and some of them have right-continuous trajectories. It is a bit surprising that there is a huge difference between these two classes. But one should recall that the trajectories are not just functions: the time parameter has an obvious orientation: the time line is not symmetric, the time flows from left to right.

Definition 1.37 *Let $(\Omega, \mathcal{A}, \mathbf{P}, \mathcal{F})$ be a stochastic base, and let us denote by $\mathcal{P}$ the σ-algebra of the subsets of $\Theta \times \Omega$ generated by the adapted, continuous processes. The sets in the σ-algebra $\mathcal{P}$ are called* predictable. *A process X is* predictable *if it is measurable with respect to $\mathcal{P}$.*

Example 1.38 A deterministic process is predictable if and only if its single trajectory is a Borel-measurable function.

Obviously we call a process X *deterministic* if it does not depend on the random parameter ω, more exactly a process X is called deterministic if it is a stochastic process on $(\Omega, \{\Omega, \emptyset\})$. If $\mathcal{A} \triangleq \{\Omega, \emptyset\}$ then the set of continuous stochastic processes is equivalent to the set of continuous functions, and the σ-algebra generated by the continuous functions is equivalent to the σ-algebra of the Borel measurable sets on Θ. □

The set of predictable processes is closed for the usual operations of analysis[23]. The most important and specific operation related to stochastic processes is the truncation:

Proposition 1.39 *If τ is an arbitrary stopping time and X is a predictable stochastic process then the truncated process X^τ is also predictable.*

Proof. Let $\mathcal{L}$ be the set of bounded stochastic processes X for which X^τ is predictable. It is obvious that $\mathcal{L}$ is a λ-system. If X is continuous then X^τ is also continuous hence the π-system of the bounded continuous processes is in $\mathcal{L}$. From the Monotone Class Theorem it is obvious that $\mathcal{L}$ contains the set of bounded predictable processes. If X is an arbitrary predictable process then

[23] Algebraic and lattice type operations, usual limits etc.

$X_n \doteq X\chi(|X| \leq n)$ is a predictable bounded process and therefore X_n^τ is also predictable. $X_n^\tau \to X^\tau$ therefore X^τ is obviously predictable. $\qquad\square$

To discuss the structure of the predictable processes let us introduce some notation:

Definition 1.40 *If τ and σ are stopping times then one can define the random intervals*

$$\{(t,\omega) \in [0,\infty) \times \Omega : \tau(\omega)\, \mathcal{R}_1 t \mathcal{R}_2 \sigma(\omega)\}$$

where $\mathcal{R}_1$ and $\mathcal{R}_2$ are one of the relations $<$ or $\leq$. One can define four random intervals $[\sigma,\tau]$, $[\sigma,\tau)$, $(\sigma,\tau]$ and (σ,τ) where the meaning of these notations is obvious.

One should emphasize that, in the definition of the stochastic intervals, the value of the time parameter t is always finite. Therefore if $\tau(\omega) = \infty$ for some ω then $(\infty,\omega) \notin [\tau,\tau]$. In measure theory we are used to the fact that the σ-algebras generated by the different types of intervals are the same. In $\mathbb{R}$ or in $\mathbb{R}^n$ one can construct every type of interval from any other type of interval with a countable number of set operations. For random intervals this is not true! For example, if we want to construct the semi-closed random interval $[0,\tau)$ with random closed segments $[0,\sigma]$ then we need a sequence of stopping times (σ_n) for which $\sigma_n \nearrow \tau$, and $\sigma_n < \tau$. If there is such a sequence[24] then of course $[0,\sigma_n] \nearrow [0,\tau)$, that is, in this case $[0,\tau)$ is in the σ-algebra generated by the closed random segments. But for an arbitrary stopping time τ such a sequence does not exist. If τ is a stopping time and $c > 0$ is a constant, then $\tau - c$ is generally not a stopping time! On the other hand if $c > 0$ then $\tau + c$ is always a stopping time, hence as $[0,\tau] = \cap_n [0,\tau+1/n)$ the closed random intervals $[0,\tau]$ are in the σ-algebra generated by the intervals $[0,\sigma)$. This shows again that in the theory of the stochastic processes the time line is not symmetric!

Definition 1.41 *Y is a* predictable simple *process if there is a sequence of stopping times*

$$0 = \tau_0 < \tau_1 < \ldots < \tau_n < \ldots$$

such that

$$Y = \eta_0 \chi(\{0\}) + \sum_i \eta_i \chi((\tau_i, \tau_{i+1}]) \tag{1.8}$$

---

[24] If for τ there is a sequence of stopping times $\sigma_n \nearrow \tau$, $\sigma_n \leq \tau$ and $\sigma_n < \tau$ on the set $\{\tau > 0\}$ then we shall say that τ is a *predictable* stopping time. Of course the main problem is that not every stopping time is predictable. See: Definition 3.5, page 182. The simplest examples are the jumps of the Poisson processes. See: Example 3.7, page 183.

where η_0 is $\mathcal{F}_0$-measurable and η_i are $\mathcal{F}_{\tau_i}$-measurable random variables. If the stopping times (τ_i) are constant then we say that Y is a predictable step processes.

Now we are ready to discuss the structure of predictable processes.

Proposition 1.42 *Let X be a stochastic process on $\Theta \triangleq [0, \infty)$. The following statements are equivalent[25]:*

1. *X is predictable.*
2. *X is measurable with respect to the σ-algebra generated by the adapted left-regular processes.*
3. *X is measurable with respect to the σ-algebra generated by the adapted left-continuous processes.*
4. *X is measurable with respect to the σ-algebra generated by the predictable step processes.*
5. *X is measurable with respect to the σ-algebra generated by the predictable simple processes.*

Proof. Let $\mathcal{P}_1, \mathcal{P}_2, \mathcal{P}_3, \mathcal{P}_4$ and $\mathcal{P}_5$ denote the σ-algebras in the proposition. Obviously it is sufficient to prove that these five σ-algebras are equal.

1. Obviously $\mathcal{P}_1 \subseteq \mathcal{P}_2 \subseteq \mathcal{P}_3$.

2. Let X be one of the processes generating $\mathcal{P}_3$, that is let X be a left-continuous, adapted process. As X is adapted

$$X_n(t) \triangleq X(0) \chi(\{0\}) + \sum_k X\left(\frac{k}{2^n}\right) \chi\left(\left(\frac{k}{2^n}, \frac{k+1}{2^n}\right]\right)$$

is a predictable step process. As X is left-continuous obviously $X_n \to X$ so X is $\mathcal{P}_4$-measurable hence $\mathcal{P}_3 \subseteq \mathcal{P}_4$.

3. Obviously $\mathcal{P}_4 \subseteq \mathcal{P}_5$.

4. Let $F \in \mathcal{F}_0$ and let f_n be such a continuous functions that $f_n(0) = 1$ and f_n is zero on the interval $[1/n, \infty)$. If $X_n \triangleq f_n \chi_F$ then X_n is obviously $\mathcal{P}_1$-measurable, therefore the process

$$\chi_F \chi(\{0\}) = \lim_{n \to \infty} X_n$$

[25]Let us recall that by definition $X(0-) \triangleq X(0)$. Therefore if ξ is an arbitrary $\mathcal{F}_0$-measurable random variable then the process $X \triangleq \xi \chi(\{0\})$ is adapted and left-regular, so if Z is predictable then $Z + X$ is also predictable. Hence we cannot generate $\mathcal{P}$ without the measurable rectangles $\{0\} \times F$, $F \in \mathcal{F}_0$. If one wants to avoid these sets then one should define the predictable processes on the open half line $(0, \infty)$. This is not necessarily a bad idea as the predictable processes are the integrands of stochastic integrals, and we shall always integrate only on the intervals $(0, t]$, so in the applications of the predictable processes the value of the these processes is entirely irrelevant at $t = 0$.

is also $\mathcal{P}_1$-measurable. If η_0 is an $\mathcal{F}_0$-measurable random variable then η_0 is a limit of $\mathcal{F}_0$-measurable step functions therefore the process $\eta_0 \chi(\{0\})$ is $\mathcal{P}_1$-measurable. This means that the first term in (1.8) is $\mathcal{P}_1$-measurable. Let us now discuss the second kind of term in (1.8). Let τ be an arbitrary stopping time. If

$$X_n(t, \omega) \overset{\circ}{=} \begin{cases} 1 & \text{if} & t \leq \tau(\omega) \\ 1 - n(t - \tau(\omega)) & \text{if} & \tau(\omega) < t < \tau(\omega) + 1/n \\ 0 & \text{if} & t \geq \tau(\omega) + 1/n \end{cases}$$

then X_n has continuous trajectories, and it is easy to see that X_n is adapted. Therefore

$$\chi([0, \tau]) = \lim_{n \to \infty} X_n \in \mathcal{P}_1.$$

If $\sigma \leq \tau$ is another stopping time then

$$\chi((\sigma, \tau]) = \chi([0, \tau] \setminus [0, \sigma]) = \chi([0, \tau]) - \chi([0, \sigma]) \in \mathcal{P}_1.$$

If $F \in \mathcal{F}_\sigma$ then

$$\sigma_F(\omega) \overset{\circ}{=} \begin{cases} \sigma(\omega) & \text{if} & \omega \in F \\ \infty & \text{if} & \omega \notin F \end{cases}$$

is also a stopping time as

$$\{\sigma_F \leq t\} = \{\sigma \leq t\} \cap F \in \mathcal{F}_t.$$

If $\sigma \leq \tau$ then $\mathcal{F}_\sigma \subseteq \mathcal{F}_\tau$, therefore not only σ_F but τ_F is also a stopping time.

$$\chi_F \chi((\sigma, \tau]) = \chi((\sigma_F, \tau_F]) \in \mathcal{P}_1.$$

If η is $\mathcal{F}_\sigma$-measurable, then η is a limit of step functions, hence if η is $\mathcal{F}_\sigma$-measurable and $\sigma \leq \tau$ then the process $\eta \chi((\sigma, \tau])$ is $\mathcal{P}_1$-measurable. By the definition of the predictable simple processes every predictable simple process is $\mathcal{P}_1$-measurable. Hence $\mathcal{P}_5 \subseteq \mathcal{P}_1$. $\qquad \square$

Corollary 1.43 *If* $\Theta = [0, \infty)$ *then the random intervals* $\{0\} \times F$, $F \in \mathcal{F}_0$ *and* $(\sigma, \tau]$ *generate the σ-algebra of the predictable sets.*

Corollary 1.44 *If* $\Theta = [0, \infty)$ *then the random intervals* $\{0\} \times F$, $F \in \mathcal{F}_0$ *and* $[0, \tau]$ *generate the σ-algebra of the predictable sets.*

Definition 1.45 *Let* $\mathcal{T}$ *denote the set of measurable rectangles*

$$\{0\} \times F, \quad F \in \mathcal{F}_0$$

and

$$\{(s,t] \times F, \quad F \in \mathcal{F}_s\}.$$

The sets in $\mathcal{T}$ *are called* predictable rectangles.

Corollary 1.46 *If* $\Theta = [0, \infty)$ *then the predictable rectangles generate the* σ-*algebra of predictable sets.*

It is quite natural to ask what the difference is between the σ-algebras generated by the right-regular and by the left-regular processes.

Definition 1.47 *The* σ-*algebra generated by the adapted, right-regular processes is called the* σ-*algebra of the optional sets. A process is called optional if it is measurable with respect to the* σ-*algebra of the optional sets.*

As every continuous process is right-regular so the σ-algebra of the optional sets is never smaller than the σ-algebra of the predictable sets $\mathcal{P}$.

Example 1.48 Adapted, right-regular process which is not predictable.

The simplest example of a right-regular process which is not predictable is the Poisson process. Unfortunately, at the present moment it is a bit difficult to prove[26]. The next example is 'elementary'. Let $\Omega \stackrel{\circ}{=} [0, 1]$ and for all t let

$$\mathcal{F}_t \stackrel{\circ}{=} \begin{cases} \sigma\left(\mathcal{B}\left([0, t]\right) \cup (t, 1]\right) & \text{if} \quad t < 1 \\ \mathcal{B}\left([0, 1]\right) & \text{if} \quad t \geq 1 \end{cases}.$$

If $s \leq t$ then $\mathcal{F}_s \subseteq \mathcal{F}_t$, and hence $\mathcal{F}$ is a filtration. It is easy to see that the random function $\tau(\omega) \stackrel{\circ}{=} \omega$ is a stopping time. Let $A \stackrel{\circ}{=} [\tau] \stackrel{\circ}{=} [\tau, \tau]$ be the graph of τ, which is the diagonal of the closed rectangle $[0, 1]^2$.

1. Let us show that A is optional. It is easy to see that the process

$$X_n \stackrel{\circ}{=} \chi\left([\tau, \tau + 1/n)\right)$$

is right-continuous. X_n is adapted as

$$\{X_n(t) = 1\} = \left\{\tau \leq t < \tau + \frac{1}{n}\right\} \in \mathcal{F}_t.$$

As $X_n \to \chi_A$, A is optional.

[26]See: Example 3.36, page 200, Example 3.56, page 219.

2. Now we show that $[\tau, \tau]$ is not predictable. We show that if $D \subseteq \mathbb{R}_+ \times \Omega$ and $D = P \cap [0, \tau]$ for some $P \in \mathcal{P}$ then there is a $B \in \mathcal{B}([0, 1])$ that

$$D = (B \times \Omega) \cap [0, \tau]. \tag{1.9}$$

Obviously $[\tau, \tau] = [\tau, \tau] \cap [0, \tau]$, therefore if $P \overset{\circ}{=} [\tau, \tau] \in \mathcal{P}$ then for some $B \in \mathcal{B}([0, 1])$

$$[\tau, \tau] = (B \times \Omega) \cap [0, \tau]$$

which is impossible and therefore $[\tau, \tau]$ cannot be predictable.

3. It remains to show the validity of the decomposition (1.9). Let $F \in \mathcal{F}_s$ and $R \overset{\circ}{=} (s, t] \times F$ with $s < t$. There are two possibilities[27]. If $F \cap (s, 1] = \emptyset$ then $F \subseteq [0, s]$, and hence

$$R \overset{\circ}{=} (s, t] \times F \subseteq (s, t] \times [0, s] \subseteq \{(x, y) : x > y\}$$

so, as

$$[0, \tau] = \{(t, \omega) : t \le \tau(\omega)\} \overset{\circ}{=} \{(t, \omega) : t \le \omega\} =$$
$$= \{(x, y) : x \le y\},$$

obviously

$$R \cap [0, \tau] = \emptyset = (\emptyset \times \Omega) \cap [0, \tau].$$

By the structure of $\mathcal{F}_s$ the interval $(s, 1]$ is an atom of $\mathcal{F}_s$. Hence if $F \cap (s, 1] \neq \emptyset$, then $(s, 1] \subseteq F$, hence for some $B \in \mathcal{B}([0, s])$

$$R \overset{\circ}{=} (s, t] \times F = (s, t] \times (B \cup (s, 1]).$$

So

$$R \cap [0, \tau] = (s, t] \times (B \cup (s, 1]) \cap \{(x, y) : x \le y\} =$$
$$= ((s, t] \times (s, 1]) \cap \{(x, y) : x \le y\} =$$
$$= ((s, t] \times \Omega) \cap [0, \tau]$$

and therefore in both cases the intersection has representation of type $B \times \Omega$. This remains true if we take the rectangles of type $\{0\} \times F$, $F \in \mathcal{F}_0$. As

---

[27]If we draw Ω on the y-axis and we draw on the time line the x-axis then $[\tau, \tau]$ is the line $y = x$, $[0, \tau]$ is the upper triangle. In the following argument F is under the diagonal hence the whole rectangle R is under the diagonal.

the generation and the restriction of the σ-algebras are interchangeable operations

$$\mathcal{P} \cap [0, \tau] = \sigma(\mathcal{T}) \cap [0, \tau] = \sigma(\mathcal{T} \cap [0, \tau]) =$$

$$= \sigma((B \times \Omega) \cap [0, \tau]) = \sigma(B \times \Omega) \cap [0, \tau] =$$

$$= (\mathcal{B}([0, 1]) \times \Omega) \cap [0, \tau],$$

which is exactly (1.9).

4. As the left-regular $\chi([0, \tau])$ is adapted and $\chi([\tau, \tau])$ is not predictable, the right-regular, adapted process

$$\chi([0, \tau)) = \chi([0, \tau]) - \chi([\tau, \tau])$$

is also not predictable. □

1.3 Martingales

In this section we introduce and discuss some important properties of continuous-time martingales. As martingales are stochastic processes one should fix the properties of their trajectories. We shall assume that the trajectories of the martingales are right-regular. The right-continuity of martingales is essential in the proof of the Optional Sampling Theorem, which describes one of the most important properties of martingales. There are a lot of good books on martingales, so we will not try to prove the theorems in their most general form. We shall present only those results from martingale theory which we shall use in the following. The presentation below is a bit redundant. We could have first proved the Downcrossing Inequality and from it we could have directly proved the Martingale Convergence Theorem. But I don't think that it is a waste of time and paper to show these theorems from different angles.

Definition 1.49 *Let us fix a filtration* $\mathcal{F}$. *The adapted process* X *is a* submartingale *if*

1. *the trajectories of* X *are right-regular,*
2. *for any time* t *the expected value of* $X^+(t)$ *is finite*[28],
3. *if* $s < t$, *then* $\mathbf{E}(X(t) \mid \mathcal{F}_s) \overset{a.s.}{\geq} X(s)$.

---

[28]Some authors, see: [53], assume that if X is a submartingale then $X(t)$ is integrable for all t. If we need this condition then we shall say that X is an *integrable submartingale*. The same remark holds for supermartingales as well. Of course martingales are always integrable.

We say that X is a supermartingale, *if* $-X$ *is a submartingale. X is a* martingale *if X is a supermartingale and a submartingale at the same time. This means that*

1. *the trajectories of X are right-regular,*
2. *for any time t the expected value of $X(t)$ is finite,*
3. *if $s < t$, then $\mathbf{E}(X(t) \mid \mathcal{F}_s) \overset{a.s.}{=} X(s)$.*

The conditional expectation is always a random variable—that is, the conditional expectation $\mathbf{E}(X(t) \mid \mathcal{F}_s)$ is always an equivalence class. As X is a stochastic process $X(s)$ is a function and not an equivalence class. Hence the two sides in the definition can be equal only in almost sure sense. Generally we shall not emphasize this, and we shall use the simpler $=$, $\geq$ and $\leq$ relations.

If X is a martingale, and g is a convex function[29] on $\mathbb{R}$ and $\mathbf{E}(g(X(t))^+) < \infty$ for all t, then the process $Y(t) \overset{\circ}{=} g(X(t))$ is a submartingale as by Jensen's inequality

$$g(X(s)) = g(\mathbf{E}(X(t) \mid \mathcal{F}_s)) \leq \mathbf{E}(g(X(t)) \mid \mathcal{F}_s).$$

In particular, if X is a martingale, $p \geq 1$, and $|X(t)|^p$ is integrable for all t, then the process $|X|^p$ is a submartingale. If X is a submartingale, g is convex and increasing, and $Y(t) \overset{\circ}{=} g(X(t))$ is integrable, then Y is a submartingale, as in this case

$$\mathbf{E}(g(X(t)) \mid \mathcal{F}_s) \geq g(\mathbf{E}(X(t) \mid \mathcal{F}_s)) \geq g(X(s)).$$

In particular, if X is a submartingale, then X^+ is also a submartingale.

1.3.1 Doob's inequalities

The most well-known inequalities of the theory of martingales are Doob's inequalities. First we prove the discrete-time versions, and then we discuss the continuous-time cases.

Proposition 1.50 (Doob's inequalities, discrete-time) *Let $X \overset{\circ}{=} (X_k, \mathcal{F}_k)_{k=1}^n$ be a non-negative submartingale.*

1. *If $\lambda \geq 0$, then*

$$\lambda \mathbf{P}\left(\max_{1 \leq k \leq n} X_k \geq \lambda\right) \leq \mathbf{E}(X_n). \tag{1.10}$$

2. *If $p > 1$, then[30]*

$$\|X_k\|_p \leq \left\|\max_{1 \leq k \leq n} X_k\right\|_p \leq \frac{p}{p-1}\|X_n\|_p \overset{\circ}{=} q\|X_n\|_p. \tag{1.11}$$

---

[29]Convex functions are continuous so $g(X)$ is adapted.
[30]Of course as usual $1/p + 1/q = 1$.

Proof. Let us remark that both inequalities estimate the size of the maximum of the non-negative submartingales.

1. Let $\lambda > 0$.

$$A_1 \overset{\circ}{=} \{X_1 \geq \lambda\}, \quad A_k \overset{\circ}{=} \left\{ \max_{1 \leq i < k} X_i < \lambda \leq X_k \right\}, \quad A \overset{\circ}{=} \left\{ \max_{1 \leq k \leq n} X_k \geq \lambda \right\}.$$

We show that

$$\lambda \mathbf{P}(A) \overset{\circ}{=} \lambda \mathbf{P}\left(\max_{1 \leq k \leq n} X_k \geq \lambda \right) \leq \int_A X_n d\mathbf{P}. \tag{1.12}$$

As $X_n \geq 0$

$$\int_A X_n d\mathbf{P} \leq \int_\Omega X_n d\mathbf{P} = \mathbf{E}(X_n),$$

therefore from (1.12) inequality (1.10) is obvious. The sets A_k are disjoint and $A = \cup_k A_k$. Observe that $A_k \in \mathcal{F}_k$. Hence by the submartingale property, by the definition of the sets A_k and by the definition of the conditional expectation

$$\lambda \mathbf{P}(A) = \lambda \sum_{k=1}^n \mathbf{P}(A_k) \leq \sum_{k=1}^n \int_{A_k} X_k d\mathbf{P} \leq \sum_{k=1}^n \int_{A_k} \mathbf{E}(X_n \mid \mathcal{F}_k) d\mathbf{P} =$$

$$= \sum_{k=1}^n \int_{A_k} X_n d\mathbf{P} = \int_A X_n d\mathbf{P}.$$

2. Let us prove inequality (1.11). Let us introduce the notation $X_n^* \overset{\circ}{=} \max_{1 \leq k \leq n} X_k$. As $0 \leq X_k \leq X_n^*$, the first inequality is trivial. If $\|X_n^*\|_p \geq \|X_n\|_p = \infty$, then the second inequality is also obvious. Let us assume that $\|X_n\|_p < \infty$. One cannot exclude that $\|X_n^*\|_p = \infty$, hence we should truncate the variable X_n^*. Fix a number N, and let $\eta \overset{\circ}{=} N \wedge X_n^*$. If

$$A(x) \overset{\circ}{=} \{X_n^* \geq x\}$$

then by (1.12)

$$x\mathbf{P}(A(x)) \leq \int_{A(x)} X_n d\mathbf{P}.$$

As $X_n \geq 0$ by Fubini's theorem and by the Fundamental Theorem of the Calculus

$$\mathbf{E}\left(\eta^p\right) = \mathbf{E}\left(\int_0^\eta px^{p-1}dx\right) \stackrel{\circ}{=} \mathbf{E}\left(\int_0^{X_n^* \wedge N} px^{p-1}dx\right) =$$

$$= \mathbf{E}\left(\int_0^N px^{p-1}\chi\left(X_n^* \geq x\right)dx\right) \stackrel{\circ}{=} \mathbf{E}\left(\int_0^N px^{p-1}\chi_{A(x)}dx\right) =$$

$$= p\int_0^N x^{p-1}\int_\Omega \chi_{A(x)}d\mathbf{P}dx = p\int_0^N x^{p-1}\mathbf{P}\left(A\left(x\right)\right)dx \leq$$

$$\leq p\int_0^N x^{p-2}\int_{A(x)}X_n d\mathbf{P}dx = p\int_0^N \int_\Omega X_n x^{p-2}\chi_{A(x)}d\mathbf{P}dx =$$

$$= p\int_\Omega X_n\int_0^\eta x^{p-2}dxd\mathbf{P} = \frac{p}{p-1}\int_\Omega X_n\eta^{p-1}d\mathbf{P}.$$

By Hölder's inequality

$$\mathbf{E}\left(\eta^p\right) \leq q\left\|X_n\right\|_p\left\|\eta^{p-1}\right\|_q = q\left\|X_n\right\|_p\mathbf{E}\left(\eta^p\right)^{1/q}.$$

Dividing[31] both sides by $\mathbf{E}\left(\eta^p\right)^{1/q}$ we get $\left\|\eta\right\|_p \leq q\left\|X_n\right\|_p$. By the Monotone Convergence Theorem

$$\lim_{N\to\infty}\left\|N \wedge X_n^*\right\|_p = \left\|X_n^*\right\|_p,$$

from which the inequality (1.11) follows. □

Corollary 1.51 *If $X \stackrel{\circ}{=} \left(X_k, \mathcal{F}_k\right)_{k=1}^n$ is a submartingale, then for arbitrary λ*

$$\lambda\mathbf{P}\left(\max_{1\leq k\leq n}X_k \geq \lambda\right) \leq \max_{1\leq k\leq n}\mathbf{E}\left(X_k^+\right) = \qquad (1.13)$$

$$= \mathbf{E}\left(X_n^+\right) \leq \mathbf{E}\left(|X_n|\right).$$

Proof. We can assume that $\lambda \geq 0$. As we remarked, if (X_k) is a submartingale, then $\left(X_k^+\right)$ is a non-negative submartingale. Hence the sequence of expected values of $\left(X_k^+\right)$ is not decreasing, so

$$\max_{1\leq k\leq n}\mathbf{E}\left(X_k^+\right) = \mathbf{E}\left(X_n^+\right).$$

[31] If $\mathbf{E}\left(\eta^p\right) = 0$ then the inequality holds.

As $\lambda \geq 0$ obviously

$$\mathbf{P}\left(\max_{1\leq k\leq n} X_k \geq \lambda\right) = \mathbf{P}\left(\max_{1\leq k\leq n} X_k^+ \geq \lambda\right),$$

so (1.13) follows from (1.10). □

Corollary 1.52 If $X \overset{\circ}{=} (X_k, \mathcal{F}_k)_{k=1}^n$ is a martingale or a non-negative submartingale, then for any $\lambda \geq 0$ and exponent $p \geq 1$

$$\lambda^p \mathbf{P}\left(\max_{1\leq k\leq n} |X_k| \geq \lambda\right) \leq \max_{1\leq k\leq n} \mathbf{E}\left(|X_k|^p\right) \leq \mathbf{E}\left(|X_n|^p\right). \qquad (1.14)$$

Proof. The function $|x|$ is convex, hence if X is a martingale then $(|X_k|, \mathcal{F}_k)_{k=1}^n$ is a non-negative submartingale, so we can assume that X is a non-negative submartingale. By Jensen's inequality and by the definition of submartingales

$$\mathbf{E}\left(X_n^p\right) = \mathbf{E}\left((\mathbf{E}\left(X_n^p \mid \mathcal{F}_k\right))\right) \geq \mathbf{E}\left((\mathbf{E}\left(X_n \mid \mathcal{F}_k\right))^p\right) \geq \mathbf{E}\left(X_k^p\right),$$

so the first inequality holds. If $\mathbf{E}\left(X_k^p\right) = \infty$ for some k, then (1.14) trivially holds. On the half-line $\mathbb{R}_+$ the function x^p is convex and increasing. If $\mathbf{E}(X_k^p) < \infty$ for all k, then (X_k^p) is a submartingale, therefore we can apply (1.13). As

$$\left\{\max_{1\leq k\leq n} |X_k| \geq \lambda\right\} = \left\{\max_{1\leq k\leq n} |X_k|^p \geq \lambda^p\right\}$$

from (1.13) the inequality (1.14) is obvious. □

If Θ is an interval and the process X has right-regular trajectories, then to calculate the supremum of the trajectories it is sufficient to calculate the supremum over the rational numbers in Θ. Hence, if X has right-regular trajectories, then $X^* \overset{\circ}{=} \sup_t X(t)$ is measurable. The submartingales by definition have right-regular trajectories, and, therefore, after applying the just proved Doob's inequalities for a finite number of rational numbers with the Monotone Convergence Theorem one can easily prove the next continuous-time inequalities:

Corollary 1.53 (Doob's inequalities, continuous-time) Let Θ be an interval.

1. If $p \geq 1$ and X is a martingale, or non-negative submartingale, then

$$\lambda^p \mathbf{P}\left(\sup_{t\in\Theta} |X(t)| \geq \lambda\right) \leq \sup_{t\in\Theta} \|X(t)\|_p^p. \qquad (1.15)$$

2. *If $p > 1$, then*

$$\left\| \sup_{t \in \Theta} |X(t)| \right\|_p \leq \frac{p}{p-1} \sup_{t \in \Theta} \|X(t)\|_p. \qquad (1.16)$$

3. *If Θ is closed and b is the finite or infinite right endpoint of Θ then under the conditions above*

$$\lambda \mathbf{P} \left(\sup_{t \in \Theta} X(t) \geq \lambda \right) \leq \|X^+(b)\|_1, \qquad (1.17)$$

$$\lambda^p \mathbf{P} \left(\sup_{t \in \Theta} |X(t)| \geq \lambda \right) \leq \|X(b)\|_p^p,$$

$$\left\| \sup_{t \in \Theta} |X(t)| \right\|_p \leq \frac{p}{p-1} \|X(b)\|_p. \qquad (1.18)$$

We shall very often use the following corollary of (1.16):

Corollary 1.54 *If X is a martingale and $p > 1$, then*

$$X^* \overset{\circ}{=} \sup_{t \in \Theta} |X_k| \in L^p(\Omega)$$

or

$$(X^*)^p \overset{\circ}{=} \left(\sup_{t \in \Theta} |X_k| \right)^p = \sup_{t \in \Theta} |X_k|^p \in L^1(\Omega)$$

if and only if X is bounded in $L^p(\Omega)$.

Definition 1.55 *If $p \geq 1$, then $\mathcal{H}^p$ will denote the space of martingales X for which*

$$\left\| \sup_{t} |X(t)| \right\|_p < \infty.$$

$\mathcal{H}^p$ *also denotes the equivalence classes of these martingales, where two martingales are equivalent whenever they are indistinguishable.*

Definition 1.56 *If $X \in \mathcal{H}^2$, then we shall say that X is a* square-integrable *martingale.*

If $\|\sup_t |X_n(t) - X(t)|\|_p \to 0$ then for a subsequence

$$\sup_t |X_{n_k}(t) - X(t)| \overset{a.s}{\to} 0,$$

hence if X_n is right-regular for every n, then X is almost surely right-regular. From the definition of the $\mathcal{H}^p$ spaces it is trivial that for all $p \geq 1$ the $\mathcal{H}^p$-martingales are uniformly integrable. From these the next observation is obvious:

Proposition 1.57 $\mathcal{H}^p$ *as a set of equivalence classes with the norm*

$$\|X\|_{\mathcal{H}^p} \overset{\circ}{=} \left\| \sup_t |X(t)| \right\|_p \tag{1.19}$$

is a Banach space.

If $p > 1$ then by Corollary 1.54 $X \in \mathcal{H}^p$ if and only if X is bounded in $L^p(\Omega)$.

1.3.2 The energy equality

An important elementary property of martingales is the following:

Proposition 1.58 (Energy equality) *Let X be a martingale and assume that $X(t)$ is square integrable for all t. If $s < t$ then*

$$\mathbf{E}\left((X(t) - X(s))^2\right) = \mathbf{E}\left(X^2(t)\right) - \mathbf{E}\left(X^2(s)\right).$$

Proof. The difference of the two sides is

$$d \overset{\circ}{=} 2 \cdot \mathbf{E}\left(X(s) \cdot (X(s) - X(t))\right).$$

As $s < t$, by the martingale property

$$
\begin{aligned}
d_n &\overset{\circ}{=} 2 \cdot \mathbf{E}\left(X(s)\chi(|X(s)| \leq n) \cdot (X(s) - X(t))\right) = \\
&= 2 \cdot \mathbf{E}\left(\mathbf{E}\left(X(s)\chi(|X(s)| \leq n) \cdot (X(s) - X(t)) \mid \mathcal{F}_s\right)\right) = \\
&= 2 \cdot \mathbf{E}\left(X(s)\chi(|X(s)| \leq n) \cdot \mathbf{E}\left(X(s) - X(t) \mid \mathcal{F}_s\right)\right) = \\
&= 2 \cdot \mathbf{E}\left(X(s)\chi(|X(s)| \leq n) \cdot 0\right) = 0.
\end{aligned}
$$

As $X(s), X(t) \in L^2(\Omega)$ obviously $|X(s) \cdot (X(s) - X(t))|$ is integrable. Hence one can use the Dominated Convergence Theorem on both sides so

$$d = \lim_{n \to \infty} d_n = 0. \qquad \square$$

Corollary 1.59 *If $X \in \mathcal{H}^2$ then there is a random variable, denoted by $X(\infty)$, such that $X(\infty) \in L^2(\Omega, \mathcal{F}_\infty, \mathbf{P})$ and*

$$X(t) \overset{a.s.}{=} \mathbf{E}(X(\infty) \mid \mathcal{F}_t) \tag{1.20}$$

for every t. In $L^2(\Omega)$-convergence

$$\lim_{t\to\infty} X(t) = X(\infty).$$

Proof. Let $t_n \nearrow \infty$ be arbitrary. By the energy equality the sequence $\left(\|X(t_n)\|_2^2\right)$ is increasing, and by the definition of $\mathcal{H}^2$ it is bounded from above. Also by the energy equality if $n > m$ then

$$\|X(t_n) - X(t_m)\|_2^2 = \|X(t_n)\|_2^2 - \|X(t_m)\|_2^2,$$

hence $(X(t_n))$ is a Cauchy sequence in $L^2(\Omega)$. As $L^2(\Omega)$ is complete the sequence $(X(t_n))$ is convergent in $L^2(\Omega)$. It is obvious from the construction that the limit $X(\infty)$ as an object in $L^2(\Omega)$ is unique, that is $X(\infty) \in L^2(\Omega)$ is independent of the sequence (t_n). X is a martingale, so if $s \geq 0$ then

$$X(t) \overset{a.s.}{=} \mathbf{E}\left(X(t+s) \mid \mathcal{F}_t\right).$$

In probability spaces L^1-convergence follows from L^2-convergence and as the conditional expectation is continuous in $L^1(\Omega)$, if $s \to \infty$ then

$$X(t) \overset{a.s.}{=} \mathbf{E}\left(\lim_{s\to\infty} X(t+s) \mid \mathcal{F}_t\right) \overset{\circ}{=} \mathbf{E}\left(X(\infty) \mid \mathcal{F}_t\right). \qquad \square$$

Example 1.60 Wiener processes and the structure of the square-integrable martingales.

Let $u < \infty$ and let w be a Wiener process on the interval $\Theta \overset{\circ}{=} [0, u]$. As w has independent increments, for every $t \leq u$

$$\mathbf{E}\left(w(u) \mid \mathcal{F}_t\right) = \mathbf{E}\left(w(u) - w(t) \mid \mathcal{F}_t\right) + \mathbf{E}\left(w(t) \mid \mathcal{F}_t\right) = w(t).$$

On the half-line $\mathbb{R}_+$ w is not bounded in $L^2(\Omega)$ that is, if $\Theta = \mathbb{R}_+$ then $w \notin \mathcal{H}^2$, and of course the representation (1.20) does not hold. $\qquad \square$

Proposition 1.61 *Let X be a martingale and let $p \geq 1$. If for some random variable $X(\infty)$*

$$X(t) \overset{L^p(\Omega)}{\to} X(\infty),$$

then

$$X(t) \overset{a.s.}{\to} X(\infty)$$

and

$$X(t) \stackrel{a.s.}{=} \mathbf{E}\left(X(\infty) \mid \mathcal{F}_t\right), \quad t \geq 0. \tag{1.21}$$

Proof. As the conditional expectation is continuous in $L^1(\Omega)$ if $s \to \infty$ then from the relation

$$X(t) \stackrel{a.s.}{=} \mathbf{E}\left(X(t+s) \mid \mathcal{F}_t\right), \quad t \geq 0$$

(1.21) follows. For an arbitrary s the increment $N(u) \stackrel{\circ}{=} X(u+s) - X(s)$ is a martingale with respect to the filtration $\mathcal{G}_u \stackrel{\circ}{=} \mathcal{F}_{s+u}$. Let

$$\beta(s) \stackrel{\circ}{=} \sup_{u \geq 0} |X(u+s) - X(\infty)| \leq \sup_u |N(u)| + |X(s) - X(\infty)|.$$

X is right-regular, so it is sufficient to take the supremum over the rational numbers, so $\beta(s)$ is measurable.

$$\sup_u \|N(u)\|_p \leq \|X(s) - X(\infty)\|_p + \sup_u \|X(u+s) - X(\infty)\|_p.$$

Let $\varepsilon > 0$ be arbitrary. As $X(s) \stackrel{L^p}{\to} X(\infty)$ if s is large enough then the right-hand side is less than $\varepsilon > 0$. By Doob's and by Markov's inequalities

$$\mathbf{P}\left(\beta(s) > 2\delta\right) \leq \mathbf{P}\left(|X(s) - X(\infty)| > \delta\right) + \mathbf{P}\left(\sup_u |N(u)| > \delta\right) \leq$$

$$\leq \frac{\|X(s) - X(\infty)\|_p^p}{\delta^p} + \left(\frac{\varepsilon}{\delta}\right)^p.$$

Therefore if $s \to \infty$ then $\beta(s) \stackrel{P}{\to} 0$. Every stochastically convergent sequence has an almost surely convergent subsequence. By the definition of $\beta(s)$ if $\beta(s_k) \stackrel{a.s.}{\to} 0$ then $X(t) \stackrel{a.s.}{\to} X(\infty)$. $\square$

Corollary 1.62 *If $X \in \mathcal{H}^2$ then there is a random variable $X(\infty) \in L^2(\Omega)$ such that $X(t) \to X(\infty)$, where the convergence holds in $L^2(\Omega)$ and almost surely.*

1.3.3 The quadratic variation of discrete time martingales

Our goal is to extend the result just proved to spaces $\mathcal{H}^p, p \geq 1$. The main tool of stochastic analysis is the so-called *quadratic variation*. Let us first investigate the quadratic variation of discrete-time martingales.

Proposition 1.63 (Austin) *Let $\mathbb{Z}$ denote the set of integers. Let $X = (X_n, \mathcal{F}_n)_{n \in \mathbb{Z}}$ be a martingale over $\mathbb{Z}$, that is let us assume that $\Theta = \mathbb{Z}$. If X*

is bounded in $L^1(\Omega)$ then the 'quadratic variation' of X is almost surely finite:

$$\sum_{n=-\infty}^{\infty} (X_{n+1} - X_n)^2 \overset{a.s.}{<} \infty. \tag{1.22}$$

Proof. As X is bounded in $L^1(\Omega)$ there is a $k < \infty$ such that $\|X_n\|_1 \leq k$ for all $n \in \mathbb{Z}$. Let $X^* \overset{\circ}{=} \sup_n |X_n|$. $|X|$ is a non-negative submartingale so by Doob's inequality

$$\mathbf{P}(X^* \geq p) \leq \frac{k}{p},$$

therefore X^* is almost surely finite. Fix a number p and define the continuously and differentiable, convex function

$$f(t) \overset{\circ}{=} \begin{cases} t^2 & \text{if } |t| \leq p \\ 2p|t| - p^2 & \text{if } |t| > p \end{cases}.$$

As f is convex the expression

$$g(s_1, s_2) \overset{\circ}{=} f(s_2) - f(s_1) - (s_2 - s_1) f'(s_1)$$

is non-negative. If $|s_1|, |s_2| \leq p$ then

$$g(s_1, s_2) = s_2^2 - s_1^2 - (s_2 - s_1) 2s_1 = (s_2 - s_1)^2.$$

By the definition of f obviously $f(t) \leq 2p|t|$. Therefore

$$\mathbf{E}(f(X_n)) \leq 2p\mathbf{E}(|X_n|) \leq 2pk. \tag{1.23}$$

By the elementary properties of the conditional expectation

$$\mathbf{E}((X_{n+1} - X_n) f'(X_n)) = \mathbf{E}(\mathbf{E}((X_{n+1} - X_n) f'(X_n)) \mid \mathcal{F}_n) =$$
$$= \mathbf{E}(f'(X_n) \mathbf{E}((X_{n+1} - X_n)) \mid \mathcal{F}_n) = 0$$

for all $n \in \mathbb{Z}$. From this and from (1.23), using the definition of g, for all n

$$2pk \geq \mathbf{E}(f(X_n)) \geq \mathbf{E}(f(X_n) - f(X_{-n})) =$$
$$= \sum_{i=-n}^{n-1} \mathbf{E}(f(X_{i+1}) - f(X_i)) =$$

$$= \sum_{i=-n}^{n-1} \mathbf{E} \left(f\left(X_{i+1}\right) - f\left(X_i\right) - \left(X_{i+1} - X_i\right) f'\left(X_i\right)\right) \stackrel{\circ}{=}$$

$$\stackrel{\circ}{=} \sum_{i=-n}^{n-1} \mathbf{E}\left(g\left(X_{i+1}, X_i\right)\right).$$

By the Monotone Convergence Theorem

$$\mathbf{E}\left(\sum_{n\in\mathbb{Z}}\left(X_{n+1} - X_n\right)^2 \chi\left(X^* \leq p\right)\right) = \mathbf{E}\left(\sum_{n\in\mathbb{Z}} g\left(X_{n+1}, X_n\right) \chi\left(X^* \leq p\right)\right) \leq$$

$$\leq \mathbf{E}\left(\sum_{n\in\mathbb{Z}} g\left(X_{n+1}, X_n\right)\right) =$$

$$= \sum_{n\in\mathbb{Z}} \mathbf{E}\left(g\left(X_{n+1}, X_n\right)\right) \leq 2pk.$$

As X^* is almost surely finite, $\sum_{n\in\mathbb{Z}}\left(X_{n+1} - X_n\right)^2$ is almost surely convergent. ◻

Corollary 1.64 Let $X \stackrel{\circ}{=} (X_n, \mathcal{F}_n)$ be a martingale over the natural numbers $\mathbb{N}$. If X is bounded in $L^1(\Omega)$ and (τ_n) is an increasing sequence of stopping times then almost surely

$$\sum_{n=1}^{\infty} \left(X(\tau_{n+1}) - X(\tau_n)\right)^2 < \infty. \tag{1.24}$$

Proof. For every m let us introduce the bounded stopping times $\tau_n^m \stackrel{\circ}{=} \tau_n \wedge m$. By the discrete-time version of the Optional Sampling Theorem[32]

$$X^m \stackrel{\circ}{=} \left(X\left(\tau_n^m\right), \mathcal{F}_{\tau_n^m}\right)_n$$

is a martingale, and therefore from the proof of the previous proposition

$$2pk \geq \mathbf{E}\left(\sum_{n=1}^{\infty}\left(X\left(\tau_{n+1}^m\right) - X\left(\tau_n^m\right)\right)^2 \chi\left(X^* \leq p\right)\right).$$

If $m \to \infty$ then by Fatou's lemma

$$\mathbf{E}\left(\sum_{n=1}^{\infty}\left(X\left(\tau_{n+1}\right) - X\left(\tau_n\right)\right)^2 \chi\left(X^* \leq p\right)\right) \leq 2pk,$$

from which (1.24) is obvious. ◻

[32]See: Lemma 1.83, page 49.

Corollary 1.65 *Let $X \doteq (X_n, \mathcal{F}_n)$ be a martingale over the natural numbers $\mathbb{N}$. If X is bounded in $L^1(\Omega)$ then there is a variable X_∞ such that $|X_\infty| < \infty$ and*

$$\lim_{n \to \infty} X_n \overset{a.s.}{=} X_\infty.$$

Proof. Assume that for some $\varepsilon > 0$ on a set of positive measure A

$$\limsup_{p,q \to \infty} |X_p - X_q| \geq 2\varepsilon. \tag{1.25}$$

Let $\tau_0 \doteq 1$, and let

$$\tau_{n+1} \doteq \inf \left\{ m \geq \tau_n : |X_m - X_{\tau_n}| \geq \varepsilon \right\}.$$

Obviously τ_n is a stopping time for all n and the sequence (τ_n) is increasing. On the set A

$$|X(\tau_{n+1}) - X(\tau_n)| \geq \varepsilon.$$

By (1.24) almost surely $\sum_{n=0}^{\infty} (X(\tau_{n+1}) - X(\tau_n))^2 < \infty$ which is impossible. $\square$

Corollary 1.66 *If $X = (X_n, \mathcal{F}_n)$ is a non-negative martingale then there exists a finite, non-negative variable X_∞ such that $X_\infty \in L^1(\Omega)$ and almost surely $X_n \to X_\infty$.*

Proof. X is non-negative and the expected value of X_n is the same for all n, hence X is obviously bounded in $L^1(\Omega)$. So $X_n \overset{a.s.}{\to} X_\infty$ exists. By Fatou's lemma

$$X(0) = \mathbf{E}(X_n \mid \mathcal{F}_0) = \lim_{n \to \infty} \mathbf{E}(X_n \mid \mathcal{F}_0) \geq \mathbf{E}\left(\liminf_{n \to \infty} X_n \mid \mathcal{F}_0\right) =$$

$$= \mathbf{E}(X_\infty \mid \mathcal{F}_0) \geq 0$$

and therefore $X_\infty \in L^1(\Omega)$. $\square$

Corollary 1.67 *Assume that $\Theta = \mathbb{R}_+$. If X is a uniformly integrable martingale then there is a variable $X(\infty) \in L^1(\Omega)$ such that $X(t) \to X(\infty)$, where the convergence holds in $L^1(\Omega)$ and almost surely. For all t*

$$X(t) \overset{a.s.}{=} \mathbf{E}(X(\infty) \mid \mathcal{F}_t). \tag{1.26}$$

Proof. Every uniformly integrable set is bounded in L^1, so if $t_n \nearrow \infty$, then there is an $X(\infty)$ such that $X(t_n) \overset{a.s.}{\to} X(\infty)$. By the uniform integrability the convergence holds in $L^1(\Omega)$ as well. Obviously $X(\infty)$ as an equivalence class is independent of (t_n). The relation (1.26) is an easy consequence of the $L^1(\Omega)$-continuity of the conditional expectation. $\square$

Corollary 1.68 *Assume that $p \geq 1$ and $\Theta = \mathbb{R}_+$. If $X \in \mathcal{H}^p$ then there is a variable $X(\infty) \in L^p(\Omega)$ such that $X(t) \to X(\infty)$, where the convergence holds in $L^p(\Omega)$ and almost surely. For all t*

$$X(t) \stackrel{a.s.}{=} \mathbf{E}(X(\infty) \mid \mathcal{F}_t). \tag{1.27}$$

Proof. If the measure is finite and $p \leq q$ then $L^q \subseteq L^p$. Hence if $p \geq 1$ and $X \in \mathcal{H}^p$ then $X \in \mathcal{H}^1$ so, if $t_n \nearrow \infty$, then there is a variable $X(\infty)$ such that $X(t_n) \stackrel{a.s.}{\to} X(\infty)$. As by the definition of $\mathcal{H}^p$ spaces $|X(t)|^p \leq \sup_s |X(s)|^p \in L^1(\Omega)$, so $X(\infty) \in L^p(\Omega)$ and by the Dominated Convergence Theorem the convergence holds in $L^p(\Omega)$ as well. Obviously $X(\infty)$, as an equivalence class, is independent of (t_n). The relation (1.27) is an easy consequence of the $L^1(\Omega)$ continuity of the conditional expectation. $\qquad\square$

Theorem 1.69 (Lévy's convergence theorem) *If $(\mathcal{F}_n)$ is an increasing sequence of σ-algebras, $\xi \in L^1(\Omega)$ and*

$$\mathcal{F}_\infty \stackrel{\circ}{=} \sigma(\cup_n \mathcal{F}_n),$$

then

$$X_n \stackrel{\circ}{=} \mathbf{E}(\xi \mid \mathcal{F}_n) \to \mathbf{E}(\xi \mid \mathcal{F}_\infty),$$

where the convergence holds in $L^1(\Omega)$ and almost surely.

Proof. Let $X_n \stackrel{\circ}{=} \mathbf{E}(\xi \mid \mathcal{F}_n)$. As

$$\mathbf{E}(|X_n|) \stackrel{\circ}{=} \mathbf{E}(|\mathbf{E}(\xi \mid \mathcal{F}_n)|) \leq \mathbf{E}(\mathbf{E}(|\xi| \mid \mathcal{F}_n))) = \mathbf{E}(|\xi|) < \infty,$$

$X = (X_n, \mathcal{F}_n)$ is an $L^1(\Omega)$ bounded martingale. Therefore $X_n \stackrel{a.s.}{\to} X_\infty$. After the proof we shall prove as a separate lemma that the sequence (X_n) is uniformly integrable, hence $X_n \stackrel{L^1}{\to} X_\infty$. If $A \in \mathcal{F}_n$, and $m \geq n$, then

$$\int_A X_m d\mathbf{P} = \int_A \xi d\mathbf{P},$$

hence as $X_m \stackrel{L^1}{\to} X_\infty$

$$\int_A X_\infty d\mathbf{P} = \int_A \xi d\mathbf{P}, \quad A \in \cup_n \mathcal{F}_n. \tag{1.28}$$

As X_∞ and ξ are integrable it is easy to see that the sets A for which (1.28) is true is a λ-system. As $(\mathcal{F}_n)$ is increasing $\cup_n \mathcal{F}_n$ is obviously a π-system. Therefore by the Monotone Class Theorem (1.28) is true if $A \in \mathcal{F}_\infty \stackrel{\circ}{=} \sigma(\cup \mathcal{F}_n)$. X_∞ is obviously $\mathcal{F}_\infty$-measurable, hence $X_\infty \stackrel{a.s.}{=} \mathbf{E}(\xi \mid \mathcal{F}_\infty)$. $\qquad\square$

Lemma 1.70 *If $\xi \in L^1$, and $(\mathcal{F}_\alpha)_{\alpha \in A}$ is an arbitrary set of σ-algebras then the set of random variables*

$$X_\alpha \overset{\circ}{=} \mathbf{E}\left(\xi \mid \mathcal{F}_\alpha\right), \quad \alpha \in A \tag{1.29}$$

is uniformly integrable.

Proof. By Markov's inequality for every α

$$\mathbf{P}\left(|X_\alpha| \geq n\right) \leq \frac{1}{n}\mathbf{E}\left(\mathbf{E}\left(|\xi| \mid \mathcal{F}_\alpha\right)\right) = \frac{1}{n}\mathbf{E}\left(|\xi|\right).$$

Therefore for any δ there is an n_0 that if $n \geq n_0$, then $\mathbf{P}\left(|X_\alpha| \geq n\right) < \delta$. As $X \in L^1(\Omega)$ the integral function $\int_A |\xi|\, d\mathbf{P}$ is absolutely continuous, that is for arbitrary $\varepsilon > 0$ there is a δ such that if $\mathbf{P}\left(A\right) < \delta$, then $\int_A |\xi|\, d\mathbf{P} < \varepsilon$. Hence if n is large enough, then

$$\int_{\{|X_\alpha|>n\}} |X_\alpha|\, d\mathbf{P} \leq \int_{\{|X_\alpha|>n\}} \mathbf{E}\left(|\xi| \mid \mathcal{F}_\alpha\right) d\mathbf{P} = \int_{\{|X_\alpha|>n\}} |\xi|\, d\mathbf{P} < \varepsilon,$$

which means that the set (1.29) is uniformly integrable. $\square$

1.3.4   The downcrossings inequality

Let X be an arbitrary adapted stochastic process and let $a < b$. Let us fix a point of time t, and let

$$S \overset{\circ}{=} \{s_0 < s_1 < \cdots < s_m\}$$

be a certain finite number of moments in the time interval $[0, t)$. Let[33]

$$\tau_0 \overset{\circ}{=} \inf\{s \in S : X\left(s\right) > b\} \wedge t.$$

With induction define

$$\tau_{2k+1} \overset{\circ}{=} \inf\{s \in S : s > \tau_{2k}, X\left(s\right) < a\} \wedge t,$$
$$\tau_{2k} \overset{\circ}{=} \inf\{s \in S : s > \tau_{2k-1}, X\left(s\right) > b\} \wedge t.$$

It is easy to check that τ_k is a stopping time for all k. It is easy to see that if X is an integrable submartingale then the inequality

$$\tau_{2k} \leq \tau_{2k+1} \overset{a.s.}{<} t$$

---

[33] If the set after inf is empty, then the infimum is by definition $+\infty$.

is impossible as in this case $X(\tau_{2k}) > b$, $X(\tau_{2k+1}) < a$ and by the submartingale property[34]

$$b < \mathbf{E}(X(\tau_{2k})) \le \mathbf{E}(X(\tau_{2k+1})) < a,$$

which is impossible. We say that function f *downcrosses* the interval $[a, b]$ if there are points $u < v$ with $f(u) > b$ and $f(v) < a$. By definition f has n downcrosses with thresholds a, b on the set S if there are points in S

$$u_1 < v_1 < u_2 < v_2 < \cdots < u_n < v_n$$

with $f(u_k) > b$, $f(v_k) < a$. Let us denote by $D_S^{a,b}$ the $a < b$ downcrossings of X in the set S. Obviously

$$\left\{ D_S^{a,b} \ge n \right\} = \{\tau_{2n-1} < t\} \in \mathcal{F}_t,$$

and hence $D_S^{a,b}$ is $\mathcal{F}_t$-measurable. We show that

$$\chi\left(D_S^{a,b} \ge n\right) \le \frac{\sum_{k=0}^{m}(X(\tau_{2k}) - X(\tau_{2k+1})) + (X(t) - b)^+}{n(b-a)}. \tag{1.30}$$

Recall that m is the number of points in S. Therefore the maximum number of possible downcrossings is obviously m. If we have more than n downcrossings then in the sum the first n term is bigger than $b - a$. For every trajectory all but the last non-zero terms of the sum are positive as they are all not smaller than $b - a > 0$. There are two possibilities: in the last non-zero term either $\tau_{2k+1} < t$ or $\tau_{2k+1} = t$. In the first case $X(\tau_{2k}) - X(\tau_{2k+1}) > b - a > 0$. In the second case still $X(\tau_{2k}) > b$, therefore in this case

$$X(\tau_{2k}) - X(\tau_{2k+1}) > b - X(t).$$

Of course $b - X(t)$ can be negative. This is the reason why we added to the sum the correction term $(X(t) - b)^+$. If $b - X(t) < 0$ then

$$
\begin{aligned}
X(\tau_{2k}) - X(\tau_{2k+1}) + (X(t) - b)^+ &= X(\tau_{2k}) - X(\tau_{2k+1}) + X(t) - b = \\
&= X(\tau_{2k}) - X(t) + X(t) - b = \\
&= X(\tau_{2k}) - b > 0,
\end{aligned}
$$

[34]See: Lemma 1.83, page 49.

which means that (1.30) always holds. Taking the expectation on both sides

$$\mathbf{P}\left(D_S^{a,b} \geq n\right) \leq \mathbf{E}\left(\frac{\sum_{k=0}^{m}(X(\tau_{2k}) - X(\tau_{2k+1})) + (X(t) - b)^+}{n(b-a)}\right) =$$

$$= \frac{1}{n(b-a)}\sum_{k=0}^{m}\mathbf{E}\left(X(\tau_{2k}) - X(\tau_{2k+1})\right) +$$

$$+ \frac{1}{n(b-a)}\mathbf{E}\left((X(t) - b)^+\right).$$

Now assume that X is an integrable submartingale. As $t \geq \tau_{2k+1} \geq \tau_{2k}$ by the discrete Optional Sampling Theorem[35]

$$\mathbf{E}\left(X(\tau_{2k}) - X(\tau_{2k+1})\right) \leq 0,$$

so

$$\mathbf{P}\left(D_S^{a,b} \geq n\right) \leq \frac{\mathbf{E}\left((X(t) - b)^+\right)}{n(b-a)}.$$

If the number of points of S increases by refining S then the number of downcrossings $D_S^{a,b}$ does not decrease. If S is an arbitrary countable set then the number of downcrossings in S is the supremum of the downcrossings of the finite subsets of S. With the Monotone Convergence Theorem we get the following important inequality:

Theorem 1.71 (Downcrossing inequality) *If X is an integrable submartingale and S is an arbitrary finite or countable subset of the time interval $[0, t)$ then*

$$\mathbf{P}\left(D_S^{a,b} \geq n\right) \leq \frac{\mathbf{E}\left((X(t) - b)^+\right)}{n(b-a)}.$$

In particular

$$\mathbf{P}\left(D_S^{a,b} = \infty\right) = 0.$$

There are many important consequences of this inequality. The first one is a generalization of the martingale convergence theorem.

Corollary 1.72 (Submartingale convergence theorem) *Let $X \overset{\circ}{=} (X_n, \mathcal{F}_n)$ be a submartingale over the natural numbers $\mathbb{N}$. If X is bounded in $L^1(\Omega)$ then*

---

[35]See: Lemma 1.83, page 49.

there is a variable $X_\infty \in L^1(\Omega)$ such that

$$\lim_{n \to \infty} X_n \overset{a.s.}{=} X_\infty. \tag{1.31}$$

Proof. If $X_n \overset{a.s.}{\to} X_\infty$ then by Fatou's lemma

$$\mathbf{E}\left(|X_\infty|\right) \le \liminf_{n \to \infty} \mathbf{E}\left(|X_n|\right) \le k < \infty$$

and $X_\infty \in L^1(\Omega)$. Let $a < b$ be rational thresholds, and let $S_m \overset{\circ}{=} \{1, 2, \ldots, m\}$. As $\mathbf{E}\left(|X_m|\right) \le k$ for all m

$$\mathbf{P}\left(D^{a,b}_{S_m} \ge n\right) \le \frac{\mathbf{E}\left((X_m - b)^+\right)}{n\left(b - a\right)} \le \frac{k}{n(b-a)}.$$

If $m \nearrow \infty$ then for all n

$$\mathbf{P}\left(D^{a,b}_{\mathbb{N}} = \infty\right) \le \mathbf{P}\left(D^{a,b}_{\mathbb{N}} \ge n\right) \le \frac{k}{n(b-a)},$$

which implies that $\mathbf{P}\left(D^{a,b}_{\mathbb{N}} = \infty\right) = 0$. The convergence in (1.31) easily follows from the next lemma: □

Lemma 1.73 *Let (c_n) be an arbitrary sequence of real numbers. If for every $a < b$ rational thresholds the number of downcrossings of the sequence (c_n) is finite then the (finite or infinite) limit $\lim_{n \to \infty} c_n$ exists.*

Proof. The $\limsup_n c_n$ and the $\liminf_n c_n$ extended real numbers always exist. If

$$\liminf_{n \to \infty} c_n < a < b < \limsup_{n \to \infty} c_n$$

then the number of the downcrossings of (c_n) is infinite. □

Definition 1.74 *Let $\xi \in L^1(\Omega)$ and let $X_n \overset{\circ}{=} \mathbf{E}\left(\xi \mid \mathcal{F}_n\right)$, $n \in \mathbb{N}$. Assume that the sequence of σ-algebras $(\mathcal{F}_n)$ is decreasing, that is $\mathcal{F}_{n+1} \subseteq \mathcal{F}_n$ for all $n \in \mathbb{N}$. These type of sequences are called* reversed martingales. *If $Y_{-n} \overset{\circ}{=} X_n$ for all $n \in \mathbb{N}$ and $\mathcal{G}_{-n} \overset{\circ}{=} \mathcal{F}_n$ then $Y = (Y_n, \mathcal{G}_n)$ is martingale over the parameter set $\Theta = \{-1, -2, \ldots\}$.*

If $(X_n, \mathcal{F}_n)$ is a reversed martingale then one can assume that $X_n = \mathbf{E}\left(X_0 \mid \mathcal{F}_n\right)$ for all n. If X is a continuous-time martingale and $t_n \searrow t_\infty$ then the sequence $(X(t_n), \mathcal{F}_{t_n})_n$ is a reversed martingale.

Theorem 1.75 (Lévy) *If $(\mathcal{F}_n)$ is a decreasing sequence of σ-algebras, $X_0 \in L^1(\Omega)$ and*

$$\mathcal{F}_\infty \overset{\circ}{=} \cap_n \mathcal{F}_n$$

then

$$X_n \overset{\circ}{=} \mathbf{E}\left(X_0 \mid \mathcal{F}_n\right) \to \mathbf{E}\left(X_0 \mid \mathcal{F}_\infty\right),$$

where the convergence holds in $L^1(\Omega)$ and almost surely.

Proof. As (X_n) is uniformly integrable[36], it is sufficient to prove that (X_n) is almost surely convergent. Let $a < b$ be rational thresholds. On the set

$$A \overset{\circ}{=} \left\{ \liminf_{n\to\infty} X_n < a < b < \limsup_{n\to\infty} X_n \right\}$$

the number of downcrossings is infinite. As $n \mapsto X_{-n}$ is a martingale on $\mathbb{Z}$, the probability of A is zero. Hence

$$\liminf_{n\to\infty} X_n \overset{a.s.}{=} \limsup_{n\to\infty} X_n. \qquad \square$$

1.3.5 Regularization of martingales

Recall that, by definition, every continuous-time martingale is right-regular. Let $\mathcal{F}$ be an arbitrary filtration, and let $\xi \in L^1(\Omega)$. In discrete-time the sequence $X_n \overset{\circ}{=} \mathbf{E}(\xi \mid \mathcal{F}_n)$ is a martingale as for every $s < t$

$$\mathbf{E}\left(X(t) \mid \mathcal{F}_s\right) \overset{\circ}{=} \mathbf{E}\left(\mathbf{E}(\xi \mid \mathcal{F}_t) \mid \mathcal{F}_s\right) \overset{a.s.}{=} \mathbf{E}\left(\xi \mid \mathcal{F}_s\right) \overset{\circ}{=}$$
$$\overset{\circ}{=} X(s).$$

In continuous-time X is not necessarily a martingale as the trajectories of X are not necessarily right-regular.

Definition 1.76 *A stochastic process X has* martingale structure *if $\mathbf{E}\left(X(t)\right)$ is finite for every t and*

$$\mathbf{E}\left(X(t) \mid \mathcal{F}_s\right) \overset{a.s.}{=} X(s)$$

for all $s < t$.

Our goal is to show that if the filtration $\mathcal{F}$ satisfies the usual conditions then every stochastic process with martingale structure has a modification which is a

---

[36]See: Lemma 1.70, page 42.

martingale. The proof depends on the following simple lemma:

Lemma 1.77 *If X has a martingale structure then there is an $\Omega_0 \subseteq \Omega$ with $\mathbf{P}(\Omega_0) = 1$, such that for every trajectory $X(\omega)$ with $\omega \in \Omega_0$ and for every rational threshold $a < b$ the number of downcrossings over the rational numbers $D_{\mathbb{Q}}^{a,b}$ is finite. In particular if $\omega \in \Omega_0$ then for every $t \in \Theta$ the (finite or infinite) limits*

$$\lim_{\substack{s \searrow t, \\ s \in \mathbb{Q}}} X(s, \omega), \quad \lim_{\substack{s \nearrow t, \\ s \in \mathbb{Q}}} X(s, \omega)$$

exist.

Proof. The first part of the lemma is a direct consequence of the downcrossings inequality. If $\lim_n X(s_n, \omega)$ does not exist for some $s_n \searrow t$ then for some rational thresholds $a < b$ the number of downcrossings of $(X(s_n, \omega))$ is infinite, so $\omega \notin \Omega_0$.

Assume that X has a martingale structure. Let $\Omega_0 \subseteq \Omega$ be the subset in the lemma above.

$$\widetilde{X}(t, \omega) \overset{\circ}{=} \begin{cases} 0 & \text{if } \omega \notin \Omega_0 \\ \lim_{s \searrow t, s \in \mathbb{Q}} X(s, \omega) & \text{if } \omega \in \Omega_0 \end{cases}. \tag{1.32}$$

We show that $\widetilde{X}$ is right-regular. Let $t < s$, $\varepsilon > 0$.

$$\left| \widetilde{X}(t, \omega) - \widetilde{X}(s, \omega) \right| \leq \left| \widetilde{X}(t, \omega) - X(t_n, \omega) \right| +$$

$$+ |X(t_n, \omega) - X(s_n, \omega)| + \left| X(s_n, \omega) - \widetilde{X}(s, \omega) \right|.$$

As for an arbitrary $\omega \in \Omega_0$ the number of $\varepsilon/3$ downcrossings of X over the $\mathbb{Q}$ is finite, so one can assume that in a right neighbourhood $(t.t + u)$ of t for every $t_n, s_n \in \mathbb{Q}$

$$|X(t_n, \omega) - X(s_n, \omega)| < \frac{\varepsilon}{3}.$$

From this, obviously, there is a δ such that if $t < s < t + \delta$, then

$$\left| \widetilde{X}(t, \omega) - \widetilde{X}(s, \omega) \right| < \varepsilon.$$

In a similar way one can prove that in Ω_0 the trajectories of $\widetilde{X}$ have left limits. Of course, without further assumptions we do not know that $\widetilde{X}$ is a modification of X. Assume that $\mathcal{F}$ satisfies the usual conditions. If $t_n \searrow t$ and $t_n \in \mathbb{Q}$ then by Lévy's theorem[37]

$$\widetilde{X}(t) \stackrel{a.s.}{=} \lim_{n \to \infty} X(t_n) \stackrel{a.s.}{=} \lim_{n \to \infty} \mathbf{E}\left(X(t_0) \mid \mathcal{F}_{t_n}\right) \stackrel{a.s}{=} \mathbf{E}\left(X(t_0) \mid \mathcal{F}_{t+}\right).$$

As $\mathcal{F}$ is right-continuous $\mathcal{F}_{t+} = \mathcal{F}_t$. As X has a martingale structure

$$\widetilde{X}(t) \stackrel{a.s.}{=} \mathbf{E}\left(X(t_0) \mid \mathcal{F}_t\right) \stackrel{a.s.}{=} X(t),$$

and therefore $\widetilde{X}$ is a modification of X. As $\mathcal{F}$ contains the measure-zero sets $\widetilde{X}$ is $\mathcal{F}$-adapted. This proves the following observation:

Theorem 1.78 *If X has a martingale structure and the filtration $\mathcal{F}$ satisfies the usual conditions then X has a modification which is a martingale.*

Corollary 1.79 *If $\mathcal{F}$ satisfies the usual conditions and X is a uniformly integrable martingale over $\Theta = [0, \infty)$ then there is a martingale $\widetilde{X}$ over $\widetilde{\Theta} \stackrel{\circ}{=} [0, \infty]$ which is indistinguishable from X over Θ.*

Proof. From the martingale convergence theorem[38] there is an $X(\infty) \in L^1(\Omega)$ such that $X(t) \stackrel{a.s.}{=} \mathbf{E}(X(\infty) \mid \mathcal{F}_t)$ for all t. On $[0, \infty]$ X has a martingale structure so it has a modification $\widetilde{X}$ which is a martingale over $[0, \infty]$. On $[0, \infty)$ X and $\widetilde{X}$ are right-regular so X and $\widetilde{X}$ are indistinguishable[39]. $\qquad \square$

From now on if X is a uniformly integrable martingale over $\Theta \stackrel{\circ}{=} \mathbb{R}_+$ then we shall always assume that X is a martingale over $[0, \infty]$.

Example 1.80 One cannot extend the theorem to submartingales.

For arbitrary Ω let $\mathcal{F}_t \stackrel{\circ}{=} \mathcal{A} \stackrel{\circ}{=} \{\emptyset, \Omega\}$. Every function f on $\mathbb{R}_+$ is an adapted stochastic process. $(\Omega, \mathcal{A}, \mathbf{P}, \mathcal{F})$ obviously satisfies the usual conditions. If f is increasing then f has a 'submartingale structure', but if f is not right-continuous then f is not a submartingale. If f is not right-continuous then it does not have a right-continuous modification. $\qquad \square$

Sometimes one cannot assume that the filtration satisfies the usual conditions. In this case one can use the following proposition:

Proposition 1.81 *Assume that X has a martingale structure and X is continuous in probability from the right. If $\mathcal{F}$ contains the measure-zero sets then X has a modification which is a martingale.*

---

[37]See: Theorem 1.75, page 46.
[38]See: (1.26), page 40.
[39]See: Proposition 1.9, page 7.

Proof. X is continuous from the right in probability. Therefore $X(s_n) \overset{a.s.}{\to} X(t)$ for some $s_n \searrow t$. So if $\widetilde{X}$ is the right-regular process in (1.32) then

$$\widetilde{X}(t) \overset{a.s.}{=} \lim_{n \to \infty} X(s_n) \overset{a.s.}{=} X(t).$$

Therefore $\widetilde{X}$ is a right-regular and adapted modification of X. □

The regularity of the trajectories is an essential condition.

Example 1.82 If the trajectories of martingales were not regular then most of the results of the continuous-time martingale theory would not be true.

Let $(\Omega, \mathcal{A}, \mathbf{P}) \overset{\circ}{=} ([0,1], \mathcal{B}([0,1]), \lambda)$ where λ denote the Lebesgue measure. If $\mathcal{F}_t \overset{\circ}{=} \mathcal{A} \overset{\circ}{=} \mathcal{B}([0,1])$ and Δ denote the diagonal of $[0,1]^2$ then $X \overset{\circ}{=} \chi_\Delta$ has a martingale structure, but if $a = 1$ then

$$1 = a\mathbf{P}\left(\sup_{t \in [0,1]} X(t) \geq a \right) > \sup_{t \in I} \left\| X^+(t) \right\|_1 = 0,$$

that is, without the regularity of the trajectories Doob's inequality does not hold. Of course $Y \equiv 0$ is regular modification of X, and for Y Doob's inequality holds. □

1.3.6 The Optional Sampling Theorem

As a first step let us prove the discrete-time version of the Optional Sampling Theorem[40].

Lemma 1.83 *Let $X = (X_n, \mathcal{F}_n)$ be a discrete-time, integrable submartingale. If τ_1 and τ_2 are stopping times and for some $p < \infty$*

$$\mathbf{P}(\tau_1 \leq \tau_2) = \mathbf{P}(\tau_2 \leq p) = 1,$$

then

$$X(\tau_1) \leq \mathbf{E}(X(\tau_2) \mid \mathcal{F}_{\tau_1})$$

and

$$\mathbf{E}(X_0) \leq \mathbf{E}(X(\tau_1)) \leq \mathbf{E}(X(\tau_2)) \leq \mathbf{E}(X_p).$$

If X is a martingale then in both lines above equality holds everywhere.

---

[40]The reader should observe that we have already used this lemma several times. Of course the proof of the lemma is independent of the results above.

Proof. Let $\tau_1 \leq \tau_2 \leq p$ and

$$\varphi_k \overset{\circ}{=} \chi\left(\tau_1 < k \leq \tau_2\right).$$

Observe that

$$\{\varphi_k = 1\} = \{\tau_1 < k, \tau_2 \geq k\} =$$
$$= \{\tau_1 \leq k - 1\} \cap \{\tau_2 \leq k - 1\}^c \in \mathcal{F}_{k-1}.$$

By the assumptions X_k is integrable for all k, so $X_k - X_{k-1}$ is also integrable, therefore the conditional expectation of the variable $X_k - X_{k-1}$ with respect to the σ-algebra $\mathcal{F}_{k-1}$ exists. φ_k is bounded, hence

$$\mathbf{E}\left(\eta\right) \overset{\circ}{=} \mathbf{E}\left(\sum_{k=1}^{p} \varphi_k \left[X_k - X_{k-1}\right]\right) =$$
$$= \sum_{k=1}^{p} \mathbf{E}\left(\mathbf{E}\left(\varphi_k \left[X_k - X_{k-1}\right] \mid \mathcal{F}_{k-1}\right)\right) =$$
$$= \sum_{k=1}^{p} \mathbf{E}\left(\varphi_k \mathbf{E}\left(X_k - X_{k-1} \mid \mathcal{F}_{k-1}\right)\right) \geq 0.$$

If $\tau_1\left(\omega\right) = \tau_2\left(\omega\right)$ for some outcome ω, then $\varphi_k\left(\omega\right) = 0$ for all k, hence $\eta\left(\omega\right) \overset{\circ}{=} 0$. If $\tau_1\left(\omega\right) < \tau_2\left(\omega\right)$, then

$$\eta\left(\omega\right) \overset{\circ}{=} X\left(\tau_1\left(\omega\right) + 1\right) - X\left(\tau_1\left(\omega\right)\right) + X\left(\tau_1\left(\omega\right) + 2\right) - X\left(\tau_1\left(\omega\right) + 1\right) + \ldots$$
$$+ X\left(\tau_2\left(\omega\right)\right) - X\left(\tau_2\left(\omega\right) - 1\right),$$

which is $X\left(\tau_2\left(\omega\right)\right) - X\left(\tau_1\left(\omega\right)\right)$. Therefore

$$\mathbf{E}\left(\eta\right) = \mathbf{E}\left(X\left(\tau_2\right) - X\left(\tau_1\right)\right) \geq 0.$$

X_k is integrable for all k, therefore $\mathbf{E}\left(X\left(\tau_1\right)\right)$ and $\mathbf{E}\left(X\left(\tau_2\right)\right)$ are finite. By the finiteness of these expected values

$$\mathbf{E}\left(X\left(\tau_2\right) - X\left(\tau_1\right)\right) = \mathbf{E}\left(X\left(\tau_2\right)\right) - \mathbf{E}\left(X\left(\tau_1\right)\right),$$

hence

$$\mathbf{E}\left(X\left(\tau_2\right)\right) \geq \mathbf{E}\left(X\left(\tau_1\right)\right). \tag{1.33}$$

Let $A \in \mathcal{F}_{\tau_1} \subseteq \mathcal{F}_{\tau_2}$, and let us define the variables

$$\tau_k^*\left(\omega\right) \overset{\circ}{=} \begin{cases} \tau_k\left(\omega\right) & \text{if} \quad \omega \in A \\ p + 1 & \text{if} \quad \omega \notin A \end{cases}.$$

τ_1^* and τ_2^* are stopping times since if $n \leq p$, then

$$\{\tau_k^* \leq n\} = A \cap \{\tau_k \leq n\} = A \cap \{\tau_k \leq n\} \in \mathcal{F}_n.$$

By (1.33)

$$\mathbf{E}\left(X\left(\tau_2^*\right)\right) = \int_A X\left(\tau_2\right) d\mathbf{P} + \int_{A^c} X\left(p+1\right) d\mathbf{P} \geq \mathbf{E}\left(X\left(\tau_1^*\right)\right) =$$

$$= \int_A X\left(\tau_1\right) d\mathbf{P} + \int_{A^c} X\left(p+1\right) d\mathbf{P}.$$

As X_{p+1} is integrable one can cancel $\int_{A^c} X\left(p+1\right) d\mathbf{P}$ from both sides of the inequality so

$$\int_A X\left(\tau_2\right) d\mathbf{P} \geq \int_A X\left(\tau_1\right) d\mathbf{P}.$$

$X\left(\tau_1\right)$ is $\mathcal{F}_{\tau_1}$-measurable and therefore

$$\mathbf{E}\left(X\left(\tau_2\right) \mid \mathcal{F}_{\tau_1}\right) \geq X\left(\tau_1\right). \qquad \square$$

To prove the continuous-time version of the Optional Sampling Theorem we need some technical lemmas:

Lemma 1.84 *If τ is a stopping time, then there is a sequence of stopping times (τ_n) such that τ_n has finite number of values*[41]*, $\tau < \tau_n$ for all n and $\tau_n \searrow \tau$.*

Proof. Divide the interval $[0, n)$ into $n2^n$ equal parts. $I_k^{(n)} \overset{\circ}{=} [(k-1)/2^n, k/2^n)$. Let

$$\tau_n(\omega) \overset{\circ}{=} \begin{cases} k/2^n & \text{if} \quad \omega \in \tau^{-1}(I_k^{(n)}) \\ +\infty & \text{otherwise} \end{cases}.$$

Obviously $\tau < \tau_n$. At every step the subintervals $I_k^{(n)}$ are divided equally, and the value of τ_n on $\tau^{-1}(I_k^{(n)})$ is always the right endpoint of the interval $I_k^{(n)}$. Therefore $\tau_n \searrow \tau$. τ is a stopping time, hence, using that, every stopping time is a weak stopping time

$$\tau^{-1}\left(I_k^{(n)}\right) = \left\{\tau < \frac{k}{2^n}\right\} \cap \left\{\tau < \frac{k-1}{2^n}\right\}^c \in \mathcal{F}_{k/2^n}.$$

[41]$\tau_n(\omega) = +\infty$ is possible.

Therefore

$$\left\{ \tau_n \le \frac{i}{2^n} \right\} = \bigcup_{k \le i} \left\{ \tau_n = \frac{k}{2^n} \right\} = \bigcup_{k \le i} \left\{ \tau^{-1} \left(I_k^{(n)} \right) \right\} \in \mathcal{F}_{i/2^n}.$$

The possible values of τ_n are among the dyadic numbers $i/2^n$ and therefore τ_n is a stopping time. $\qquad\square$

Lemma 1.85 *If (τ_n) is a sequence of stopping times and $\tau_n \searrow \tau$ then $\mathcal{F}_{\tau_n+} \searrow \mathcal{F}_{\tau+}$. If $\tau_n > \tau$ and $\tau_n \searrow \tau$ then $\mathcal{F}_{\tau_n} \searrow \mathcal{F}_{\tau+}$.*

Proof. Recall that by definition $A \in \mathcal{F}_{\rho+}$ if

$$A \cap \{\rho \le t\} \in \mathcal{F}_{t+}$$

for every t. If $A \in \mathcal{F}_{\rho+}$, then

$$A \cap \{\rho < t\} = \bigcup_n \left(A \cap \left\{ \rho \le t - \frac{1}{n} \right\} \right) \in \cup_n \mathcal{F}_{(t-1/n)+} \subseteq \mathcal{F}_t.$$

1. Let $A \in \mathcal{F}_{\rho+}$ and let $\rho \le \sigma$.

$$A \cap \{\sigma \le t\} = A \cap \{\rho \le t\} \cap \{\sigma \le t\} \in \mathcal{F}_{t+}$$

as $A \cap \{\rho \le t\} \in \mathcal{F}_{t+}$ and $\{\sigma \le t\} \in \mathcal{F}_t$. From this it is easy to see that $\mathcal{F}_{\tau+} \subseteq \cap_n \mathcal{F}_{\tau_n+}$. If $A \in \cap_n \mathcal{F}_{\tau_n+}$, then as $\tau_n \searrow \tau$

$$A \cap \{\tau < t\} = A \cap (\cup_n \{\tau_n < t\}) = \bigcup_n (A \cap \{\tau_n < t\}) \in \mathcal{F}_t.$$

So

$$A \cap \{\tau \le t\} = \bigcap_n \left(A \cap \left\{ \tau < t + \frac{1}{n} \right\} \right) \in \bigcap_n \mathcal{F}_{t+1/n} = \mathcal{F}_{t+}$$

that is $A \in \mathcal{F}_{\tau+}$.

2. If $A \in \mathcal{F}_{\rho+}$ and $\rho < \sigma$ then as $\{\sigma \le t\} \in \mathcal{F}_t$

$$A \cap \{\sigma \le t\} = A \cap \{\rho < t\} \cap \{\sigma \le t\} \in \mathcal{F}_t,$$

so $A \in \mathcal{F}_\sigma$. From this it is easy to see that if $\tau < \tau_n$ then $\mathcal{F}_{\tau+} \subseteq \cap_n \mathcal{F}_{\tau_n}$. On the other hand

$$\cap_n \mathcal{F}_{\tau_n} \subseteq \cap_n \mathcal{F}_{\tau_n+} = \mathcal{F}_{\tau+}$$

and so if $\tau < \tau_n$ then $\cap_n \mathcal{F}_{\tau_n} = \mathcal{F}_{\tau+}$. $\qquad\square$

Theorem 1.86 (Optional Sampling Theorem for martingales) *Let X be a martingale, and let $\tau_1 \leq \tau_2$ be stopping times. If X is uniformly integrable, or if τ_1 and τ_2 are bounded then*

$$X(\tau_1) \overset{a.s.}{=} \mathbf{E}\left(X(\tau_2) \mid \mathcal{F}_{\tau_1+}\right). \tag{1.34}$$

Proof. Let τ be a bounded stopping time and let $\tau^{(n)} \searrow \tau$, $\tau^{(n)} > \tau$ be a finite-valued approximating sequence[42]. As τ is bounded there is an N large enough that $\tau^{(n)} \leq N$. By the first lemma

$$X(\tau^{(n)}) = \mathbf{E}\left(X(N) \mid \mathcal{F}_{\tau^{(n)}}\right). \tag{1.35}$$

As $\tau^{(n)} > \tau$, by the last lemma $\cap_n \mathcal{F}_{\tau^{(n)}} = \mathcal{F}_{\tau+}$. So by the definition of the conditional expectation (1.35) means that

$$\int_A X(\tau^{(n)})d\mathbf{P} = \int_A X(N)d\mathbf{P}, \quad A \in \mathcal{F}_{\tau+}.$$

$X(N)$ is integrable therefore the sequence $\left(X(\tau^{(n)})\right)$ is uniformly integrable[43] by (1.35). By the right-continuity of the martingales $X(\tau) = \lim_{n \to \infty} X\left(\tau^{(n)}\right)$, so if $A \in \mathcal{F}_{\tau+}$ then

$$\int_A X(N)d\mathbf{P} = \lim_{n \to \infty} \int_A X(\tau^{(n)})d\mathbf{P} = \int_A \lim_{n \to \infty} X(\tau^{(n)})d\mathbf{P} =$$
$$= \int_A X(\tau)d\mathbf{P}.$$

As $X(\tau)$ is $\mathcal{F}_\tau$-measurable and $\mathcal{F}_\tau \subseteq \mathcal{F}_{\tau+}$,

$$X(\tau) = \mathbf{E}\left(X(N) \mid \mathcal{F}_{\tau+}\right).$$

If X is uniformly integrable then one can assume that X is a martingale on $[0, \infty]$. There is a continuous bijective time transformation f between the intervals $[0, \infty]$ and $[0, 1]$. During this transformation the properties of X and τ do not change, but $f(\tau)$ will be bounded, so using the same argument as above one can prove that

$$X(\tau) = \mathbf{E}\left(X(\infty) \mid \mathcal{F}_{\tau+}\right).$$

Finally if $\tau_1 \leq \tau_2$, then as $\mathcal{F}_{\tau_1+} \subseteq \mathcal{F}_{\tau_2+}$

$$\mathbf{E}\left(X(\tau_2) \mid \mathcal{F}_{\tau_1+}\right) = \mathbf{E}\left(\mathbf{E}\left(X(N) \mid \mathcal{F}_{\tau_2+}\right) \mid \mathcal{F}_{\tau_1+}\right) =$$
$$= \mathbf{E}\left(X(N) \mid \mathcal{F}_{\tau_1+}\right) = X(\tau_1),$$

where if X is uniformly integrable, then $N \overset{\circ}{=} \infty$. $\qquad\square$

[42]See: Lemma 1.84, page 51.
[43]See: Lemma 1.70, page 42.

Corollary 1.87 *If X is a non-negative martingale and $\tau_1 \leq \tau_2$, then*

$$X(\tau_1) \geq \mathbf{E}\left(X(\tau_2) \mid \mathcal{F}_{\tau_1+}\right). \tag{1.36}$$

Proof. First of all let us remark, that as X is a non-negative martingale $X(\infty)$ is meaningful[44], and if $n \nearrow \infty$ then $X(\tau \wedge n) \to X(\tau)$ for every stopping time τ. Let $\mathcal{G} \overset{\circ}{=} \sigma\left(\cup_n \mathcal{F}_{(\tau \wedge n)+}\right)$. Obviously $\mathcal{G} \subseteq \mathcal{F}_{\tau+}$. Let $A \in \mathcal{F}_{\tau+}$.

$$A \cap \{\tau \leq n\} \cap \{\tau \wedge n \leq t\} = A \cap \{\tau \leq t \wedge n\} \in \mathcal{F}_{t+},$$

therefore $A \cap \{\tau \leq n\} \in \mathcal{F}_{(\tau \wedge n)+}$. So $A \cap \{\tau < \infty\} \in \mathcal{G}$. Also

$$A \cap \{\tau > n\} \cap \{\tau \wedge n \leq t\} = A \cap \{t \geq \tau > n\} \in \mathcal{F}_{t+}$$

so $A \cap \{\tau > n\} \in \mathcal{F}_{(\tau \wedge n)+}$. Hence

$$A \cap \{\tau = \infty\} = A \cap \left(\cap_n \{\tau > n\}\right) \in \mathcal{G},$$

therefore $\mathcal{G} = \mathcal{F}_{\tau+}$. Let $n_1 \leq n_2$. By the Optional Sampling Theorem

$$X(\tau_1 \wedge n_1) = \mathbf{E}\left(X(\tau_2 \wedge n_2) \mid \mathcal{F}_{(\tau_1 \wedge n_1)+}\right).$$

$X(\tau_2 \wedge n_2) \in L^1(\Omega)$ and therefore by Lévy's theorem

$$X(\tau_1) = \mathbf{E}\left(X(\tau_2 \wedge n_2) \mid \mathcal{F}_{\tau_1+}\right).$$

By Fatou's lemma

$$X(\tau_1) = \lim_{n_2 \to \infty} \mathbf{E}\left(X(\tau_2 \wedge n_2) \mid \mathcal{F}_{\tau_1+}\right) \geq \mathbf{E}\left(\lim_{n_2 \to \infty} X(\tau_2 \wedge n_2) \mid \mathcal{F}_{\tau_1+}\right) =$$

$$= \mathbf{E}\left(X(\tau_2) \mid \mathcal{F}_{\tau_1+}\right). \qquad \square$$

Proposition 1.88 (Optional Sampling Theorem for submartingales) *Let $\tau_1 \leq \tau_2$ bounded stopping times. If X is an integrable submartingale then $X(\tau_1)$ and $X(\tau_2)$ are integrable and*

$$X(\tau_1) \leq \mathbf{E}\left(X(\tau_2) \mid \mathcal{F}_{\tau_1}\right). \tag{1.37}$$

The inequality also holds if $\tau_1 \leq \tau_2$ are arbitrary stopping times and X can be extended as an integrable submartingale to $[0, \infty]$.

Proof. The proof of the proposition is nearly the same as the proof in the martingale case. Again it is sufficient to prove the inequality in the bounded

---

[44]See: Corollary 1.66, page 40.

case. Assume that $\tau_1 \leq \tau_2 \leq K$ and let $(\tau_1^{(n)})_n$ and $(\tau_2^{(n)})_n$ be the finite-valued approximating sequences of τ_1 and τ_2. By the construction $\tau_1^{(n)} \leq \tau_2^{(n)}$, so by the first lemma of the subsection

$$\int_F X(\tau_1^{(n)}) d\mathbf{P} \leq \int_F X(\tau_2^{(n)}) d\mathbf{P}, \quad F \in \mathcal{F}_{\tau_1+}.$$

By the right-continuity of submartingales $X(\tau_k^{(n)}) \to X(\tau_k)$ and therefore one should prove that the convergence holds in $L^1(\Omega)$, that is, one should prove the uniform integrability of the sequences $(X(\tau_k^{(n)}))$. Since in this case one can take the limits under the integral signs therefore

$$\int_F X(\tau_1) d\mathbf{P} \leq \int_F X(\tau_2) d\mathbf{P}, \quad F \in \mathcal{F}_{\tau_1+}.$$

As $X(\tau_1)$ is $\mathcal{F}_{\tau_1+}$-measurable by the definition of the conditional expectation

$$X(\tau_1) = \mathbf{E}(X(\tau_1) \mid \mathcal{F}_{\tau_1}) \leq \mathbf{E}(X(\tau_2) \mid \mathcal{F}_{\tau_1+}).$$

This means that (1.37) holds. Let us prove that the sequence $\left(X(\tau_k^{(n)})\right)$ is uniformly integrable.

1. As X is submartingale, X^+ is also submartingale, therefore from the finite Optional Sampling Theorem

$$0 \leq X^+\left(\tau_k^{(n)}\right) \leq \mathbf{E}\left(X^+(K) \mid \mathcal{F}_{\tau_k^{(n)}}\right).$$

The right-hand side is uniformly integrable[45], so the left-hand side is also uniformly integrable.

2. Let $X_n \overset{\circ}{=} X(\tau_k^{(n)})$. By the finite Optional Sampling Theorem (X_n) is obviously an integrable reversed submartingale. Let $n > m$. As (X_n) is a reversed submartingale

$$0 \leq \int_{\{X_n^- \geq N\}} X_n^- d\mathbf{P} = -\int_{\{X_n^- \geq N\}} X_n d\mathbf{P} = \int_{\{X_n^- < N\}} X_n d\mathbf{P} - \mathbf{E}(X_n) \leq$$

$$\leq \int_{\{X_n^- < N\}} X_m d\mathbf{P} - \mathbf{E}(X_n) = \mathbf{E}(X_m) - \mathbf{E}(X_n) - \int_{\{X_n^- \geq N\}} X_m d\mathbf{P}.$$

As $n > m$ and as X_n is a reversed submartingale

$$\mathbf{E}(X(0)) \leq \mathbf{E}(X_n) \leq \mathbf{E}(X_m) \leq \mathbf{E}(X(K)).$$

[45]See: Lemma 1.70, page 42.

X is an integrable submartingale, so $\mathbf{E}(X(0))$ is finite. Hence the sequence $(\mathbf{E}(X_n))$ is convergent. If m is large enough then $0 \leq \mathbf{E}(X_m) - \mathbf{E}(X_n) \leq \varepsilon/2$. X_m is integrable so the integral function $\int_A X_m d\mathbf{P}$ is absolutely continuous. Therefore it is sufficient to prove that for an arbitrary δ there is an N such that for all n

$$\mathbf{P}\left(X_n^- \geq N\right) \leq \delta.$$

By Markov's inequality

$$\mathbf{P}\left(X_n^- \geq N\right) \leq \frac{\mathbf{E}(X_n^-)}{N} \leq \frac{\mathbf{E}(|X_n|)}{N} = \frac{\|X_n\|_1}{N}.$$

But since (X_n^+) is also a reversed submartingale

$$\|X_n\|_1 = \mathbf{E}\left(\left|X_n^+ - X_n^-\right|\right) = \mathbf{E}\left(X_n^+ + X_n^- - X_n\right) =$$
$$= 2\mathbf{E}\left(X_n^+\right) - \mathbf{E}(X_n) \leq 2\mathbf{E}\left(X^+(K)\right) - \mathbf{E}(X_n) \leq$$
$$\leq 2\mathbf{E}\left(X^+(K)\right) - \mathbf{E}(X(0)) \stackrel{\circ}{=} L.$$

Hence for all n

$$\mathbf{P}\left(X_n^- \geq N\right) \leq \frac{L}{N} \to 0$$

and therefore (X_n) is uniformly integrable. □

Example 1.89 The right-regularity and the assumption of uniform integrability of the martingales are important.

Let P be a Poisson process with parameter λ, and let $\pi(t) \stackrel{\circ}{=} P(t) - \lambda t$ be the so-called *compensated Poisson process*. Recall that the Poisson processes are, by definition, right-continuous, so it is easy to see that π is a martingale. Let τ denote the time of the first jump of P. If $N > 0$, then $\sigma \stackrel{\circ}{=} \tau \wedge N$ is a bounded stopping time. If π were not right but left-continuous then one could not apply the Optional Sampling Theorem: if P were left-continuous then $P(\sigma) = 0$, and

$$\mathbf{E}(\pi(0)) = 0 \neq \mathbf{E}(-\lambda\sigma) = \mathbf{E}(P(\sigma) - \lambda\sigma) = \mathbf{E}(\pi(\sigma)).$$

Let w be a Wiener process and let

$$\tau_a \stackrel{\circ}{=} \inf\{t : w(t) = a\}$$

be the first passage time of an $a \neq 0$. As w is not uniformly integrable and τ_a is unbounded, one cannot apply the Optional Sampling Theorem: almost surely[46] $\tau_a < \infty$, hence $w(\tau_a) \stackrel{a.s.}{=} a$. Therefore

$$\mathbf{E}(w(\tau_a)) = \mathbf{E}(a) = a \neq 0 = \mathbf{E}(w(0)).$$ $\square$

Example 1.90 The exponential martingales of Wiener processes are not uniformly integrable.

Let w be a Wiener process. If the so-called *exponential martingale*

$$X(t) \stackrel{\circ}{=} \exp(w(t) - t/2)$$

were uniformly integrable, then for every stopping time one could apply the Optional Sampling Theorem. X is a non-negative martingale, therefore there is[47] a random variable $X(\infty)$ such that almost surely $X(t) \to X(\infty)$. For almost all trajectories of w the set $\{w = 0\}$ is unbounded[48], therefore $w(\sigma_n) = 0$ for some sequence $\sigma_n \nearrow \infty$. Therefore

$$X(\infty) \stackrel{a.s.}{=} \lim_{n \to \infty} X(\sigma_n) \stackrel{\circ}{=} \lim_{n \to \infty} \exp\left(w(\sigma_n) - \frac{\sigma_n}{2}\right) = \lim_{n \to \infty} \exp\left(-\frac{\sigma_n}{2}\right) = 0.$$

Since $X(0) = 1$, $X(\infty) \stackrel{a.s.}{=} 0$ and X is continuous, if $a < 1$ then almost surely

$$\tau_a \stackrel{\circ}{=} \inf\{t : X(t) = a\} < \infty.$$

That is $X(\tau_a) \stackrel{a.s.}{=} a$. So if $a < 1$, then

$$\mathbf{E}(X(0)) = 1 > a = \mathbf{E}(X(\tau_a)).$$

Hence X is not uniformly integrable. $\square$

Proposition 1.91 (Martingales and conservation of the expected value)
Let X be an adapted and right-regular process. X is a martingale if and only if

$$X(\tau) \in L^1(\Omega) \quad and \quad \mathbf{E}(X(\tau)) = \mathbf{E}(X(0))$$

for all bounded stopping times τ. This property holds for every stopping time τ if and only if X is a uniformly integrable martingale.

[46]See: Proposition B.7, page 564.
[47]See: Corollary 1.66, page 40.
[48]See: Corollary B.8, page 565.

Proof. If X is a martingale, or uniformly integrable martingale, then by the Optional Sampling Theorem the proposition holds. Let $s < t$ and let $A \in \mathcal{F}_s$. It is easy to check that

$$\tau = t\chi_{A^c} + s\chi_A \tag{1.38}$$

is a bounded stopping time. By the assumption of the proposition

$$\mathbf{E}\left(X(0)\right) = \mathbf{E}\left(X(\tau)\right) = \mathbf{E}\left(X(t)\chi_{A^c}\right) + \mathbf{E}\left(X(s)\chi_A\right).$$

As $\tau \equiv t$ is also a stopping time,

$$\mathbf{E}\left(X(0)\right) = \mathbf{E}\left(X(t)\right) = \mathbf{E}\left(X(t)\chi_{A^c}\right) + \mathbf{E}\left(X(t)\chi_A\right).$$

Comparing the two equations $\mathbf{E}\left(X(s)\chi_A\right) = \mathbf{E}\left(X(t)\chi_A\right)$, that is

$$\mathbf{E}\left(X(s) \mid \mathcal{F}_s\right) = \mathbf{E}\left(X(t) \mid \mathcal{F}_s\right).$$

As X is adapted, $X(s)$ is $\mathcal{F}_s$-measurable so $X(s) = \mathbf{E}\left(X(t) \mid \mathcal{F}_s\right)$. If one can apply the property $\mathbf{E}\left(X(\tau)\right) = \mathbf{E}\left(X(0)\right)$ for every stopping time τ then one can apply it for the stopping time $\tau \equiv \infty$ as well. Hence $X\left(\infty\right)$ exists and in (1.38) $t = \infty$ is possible, hence

$$X(s) = \mathbf{E}\left(X(\infty) \mid \mathcal{F}_s\right),$$

so X is uniformly integrable[49]. $\qquad\square$

Corollary 1.92 (Conservation of the martingale property under truncation) *If X is a martingale and τ is a stopping time then the truncated process X^τ is also a martingale.*

Proof. If X is right-regular then the truncated process X^τ is also right-regular. By Proposition 1.35 X^τ is adapted. Let ϕ be a bounded stopping time. As $v \overset{\circ}{=} \phi \wedge \tau$ is a bounded stopping time by Proposition 1.91

$$\mathbf{E}\left(X^\tau(\phi)\right) = \mathbf{E}\left(X(v)\right) = \mathbf{E}\left(X(0)\right) = \mathbf{E}\left(X^\tau(0)\right)$$

and therefore X^τ is a martingale. $\qquad\square$

1.3.7 Application: elementary properties of Lévy processes

Lévy processes are natural generalizations of Wiener and Poisson processes. Let us fix a stochastic base space $(\Omega, \mathcal{A}, \mathbf{P}, \mathcal{F})$ and assume that $\Theta = [0, \infty)$.

Definition 1.93 *Let X be an adapted stochastic process. X is a process with independent increments with respect to the filtration $\mathcal{F}$ if*

---

[49] See: Lemma 1.70, page 42 .

1. $X(0) = 0$,
2. X is right-regular,
3. whenever $s < t$ then the increment $X(t) - X(s)$ is independent of the σ-algebra $\mathcal{F}_s$.

A process X with independent increments is a Lévy process, if it has stationary or homogeneous increments *that is for every t and for every* $h > 0$ *the distribution of the increment* $X(t+h) - X(t)$ *is the same as the distribution of* $X(h) - X(0)$.

By definition every Lévy process and every process with independent increments has right-regular trajectories. This topological assumption is very important as it is not implied by the other assumptions:

Example 1.94 Not every process starting from zero and having stationary and independent increments is a Lévy process.

Let Ω be arbitrary and $\mathcal{A} = \mathcal{F}_t = \{\emptyset, \Omega\}$ and let $(x_\alpha)_\alpha$ be a Hamel basis of $\mathbb{R}$ over the rational numbers. For every t let $X(t)$ be the sum of the coordinates of t in the Hamel basis. Obviously $X(t+s) = X(t) + X(s)$ so X has stationary and independent increments. But as X is highly discontinuous[50] it does not have a modification which is a Lévy process. $\qquad\qquad\square$

Example 1.95 The sum of two Lévy processes is not necessarily a Lévy process[51].

We show that even the sum of two Wiener processes is not a Wiener process. The present counter example is very important as it shows that, although the Lévy processes are the canonical and most important examples of semimartingales, they are not the right objects from the point of view of the theory. The sum of two semimartingales[52] is a semimartingale and the same is true for martingales or for local martingales. But it is not true for Lévy processes!

1. Let Ω be the set of two-dimensional continuous functions $\mathbb{R}_+ \to \mathbb{R}^2$ with the property $f(0) = (0, 0)$. Let $\mathbf{P}_1$ be a measure on the Borel σ-algebra of Ω for which the canonical stochastic process $X(\omega, t) = \omega(t)$ is a two-dimensional Wiener process with correlation coefficient 1. In the same way let $\mathbf{P}_2$ be the measure on Ω under which X is a Wiener process with correlation coefficient -1. Let $\mathbf{P} \overset{\circ}{=} (\mathbf{P}_1 + \mathbf{P}_2)/2$. It is easy to see that the coordinate processes $w_1(t)$ and

---

[50]The image space of X is the rational numbers!

[51]The example depends on results which we shall prove later. So the reader can skip the example at the first reading.

[52]We shall introduce the definitions of semimartingales and local martingales later.

$w_2(t)$ are Wiener processes. On the other hand, a simple calculation shows that the distribution of $Z \overset{\circ}{=} w_1 + w_2$ is not Gaussian. Z is continuous and every continuous Lévy process is a linear combination of a Wiener process and a linear trend[53], therefore, as Z is not a Gaussian process it cannot be a Lévy process.

2. The next example is bit more technical, but very similar: Let w be a Wiener process with respect to some filtration $\mathcal{F}$. Let $X(t) \overset{\circ}{=} \int_0^t \text{sign}(w) \, dw$, where the integral, of course, is an Itô integral. The quadratic variation of X is

$$[X](t) = \int_0^t (\text{sign}(w))^2 \, d[w] = \int_0^t 1 ds = t$$

so by Lévy's characterization theorem[54] the continuous local martingale X is also a Wiener process[55] with respect to $\mathcal{F}$. If

$$Z \overset{\circ}{=} w + X = 1 \bullet w + \text{sign}(w(s)) \bullet w = (1 + \text{sign}(w(s))) \bullet w$$

then Z is a continuous martingale with respect to $\mathcal{F}$ with zero expected value.

$$[Z](t) = \int_0^t (1 + \text{sign}(w))^2 \, d[w] = \int_0^t (1 + \text{sign}(w(s)))^2 \, ds$$

so Z is not a Wiener process. As in the first example, every continuous Lévy process is a linear combination of a Wiener process and a linear trend, therefore, as Z is not a Wiener process it cannot be a Lévy process. $\square$

During the proof of the next proposition, we shall need the next very useful simple observation:

Lemma 1.96 $\boldsymbol{\xi}_1$ *and* $\boldsymbol{\xi}_2$ *are independent vector-valued random variables if and only if*

$$\varphi = \varphi_1 \cdot \varphi_2,$$

where φ_1 *is the Fourier transform of* $\boldsymbol{\xi}_1$ *and* φ_2 *is the Fourier transform of* $\boldsymbol{\xi}_2$ *and* φ *is the Fourier transform of the joint distribution of* $(\boldsymbol{\xi}_1, \boldsymbol{\xi}_2)$.

Proof. If $\boldsymbol{\xi}_1$ and $\boldsymbol{\xi}_2$ are independent then the decomposition obviously holds. The other implication is an easy consequence of the Monotone Class Theorem:

---

[53]See: Theorem 6.11, page 367.
[54]See: Theorem 6.13, page 368.
[55]See: Example 6.14, page 370.

fix a vector $\mathbf{v}$ and let $\mathcal{L}$ be the set of bounded functions u for which

$$\mathbf{E}\left(u\left(\boldsymbol{\xi}_1\right)\cdot\exp\left(i\left(\mathbf{v},\boldsymbol{\xi}_2\right)\right)\right)=\mathbf{E}\left(u\left(\boldsymbol{\xi}_1\right)\right)\cdot\mathbf{E}\left(\exp\left(i\left(\mathbf{v},\boldsymbol{\xi}_2\right)\right)\right).$$

$\mathcal{L}$ is obviously a λ-system. Under the conditions of the lemma $\mathcal{L}$ contains the π-system of the functions $u\left(\mathbf{x}\right)=\exp\left(i\left(\mathbf{u},\mathbf{x}\right)\right)$, so it contains the characteristic functions of the sets of the σ-algebra generated by these exponential functions. Therefore it is easy to see that for every Borel measurable set B

$$\mathbf{E}\left(\chi_B\left(\boldsymbol{\xi}_1\right)\cdot\exp\left(i\left(\mathbf{v},\boldsymbol{\xi}_2\right)\right)\right)=\mathbf{P}\left(\boldsymbol{\xi}_1\in B\right)\cdot\mathbf{E}\left(\exp\left(i\left(\mathbf{v},\boldsymbol{\xi}_2\right)\right)\right).$$

Now let $\mathcal{L}$ be the set of bounded functions v for which

$$\mathbf{E}\left(\chi_B\left(\boldsymbol{\xi}_1\right)\cdot v\left(\boldsymbol{\xi}_2\right)\right)=\mathbf{P}\left(\boldsymbol{\xi}_1\in B\right)\cdot\mathbf{E}\left(v\left(\boldsymbol{\xi}_2\right)\right).$$

With the same argument as above, by the Monotone Class Theorem for any Borel measurable set D, one can choose $v=\chi_D$. So

$$\mathbf{P}\left(\boldsymbol{\xi}_1\in B,\boldsymbol{\xi}_2\in D\right)=\mathbf{E}\left(\chi_B\left(\boldsymbol{\xi}_1\right)\cdot\chi_D\left(\boldsymbol{\xi}_2\right)\right)=\mathbf{P}\left(\boldsymbol{\xi}_1\in B\right)\cdot\mathbf{P}\left(\boldsymbol{\xi}_2\in D\right)$$

therefore, by definition, the random vectors $\boldsymbol{\xi}_1$ and $\boldsymbol{\xi}_2$ are independent. $\square$

Proposition 1.97 *For an adapted process X the increments are independent if and only if the σ-algebra $\mathcal{G}_t$ generated by the increments*

$$X\left(u\right)-X\left(v\right),\quad u\geq v\geq t$$

is independent of $\mathcal{F}_t$ for every t.

Proof. To make the notation as simple as possible let $X\left(t_0\right)$ denote an arbitrary $\mathcal{F}_{t_0}$-measurable random variable. Let

$$0=t_{-1}\leq t=t_0\leq t_1\leq t_2\leq\ldots\leq t_n.$$

We show that if X has independent increments then the random variables

$$X(t_0),X(t_1)-X(t_0),X(t_2)-X(t_1),\ldots,X(t_n)-X(t_{n-1})\qquad(1.39)$$

are independent. To prove this one should prove that the Fourier transform of the joint distribution of the variables in (1.39) is the product of the Fourier transforms of the distributions of these increments:

$$\varphi(\mathbf{u}) \stackrel{\circ}{=} \mathbf{E}\left(\exp\left(i\sum_{j=0}^{n} u_j \left[X(t_j) - X(t_{j-1})\right]\right)\right) =$$

$$= \mathbf{E}\left(\mathbf{E}\left(\exp\left(i\sum_{j=0}^{n} u_j \Delta X(t_j)\right) \mid \mathcal{F}_{t_{n-1}}\right)\right) =$$

$$= \mathbf{E}\left(\exp\left(i\sum_{j=0}^{n-1} u_j \Delta X(t_j)\right) \mathbf{E}\left(\exp\left(iu_n \Delta X(t_n)\right) \mid \mathcal{F}_{t_{n-1}}\right)\right) =$$

$$= \mathbf{E}\left(\exp\left(i\sum_{j=0}^{n-1} u_j \Delta X(t_j)\right) \mathbf{E}\left(\exp\left(iu_n \Delta X(t_n)\right)\right)\right) =$$

$$= \mathbf{E}\left(\exp\left(i\sum_{j=0}^{n-1} u_j \Delta X(t_j)\right) \varphi_{t_n, t_{n-1}}(u_n)\right) =$$

$$= \mathbf{E}\left(\exp\left(i\sum_{j=0}^{n-1} u_j \Delta X(t_j)\right)\right) \varphi_{t_n, t_{n-1}}(u_n) = \cdots =$$

$$= \prod_{j=0}^{n} \varphi_{t_j, t_{j-1}}(u_j).$$

Of course this means that the σ-algebra generated by a finite number of increments is independent of $\mathcal{F}_t$ for any t. As the union of σ-algebras generated by finite number of increments is a π-system, with the uniqueness of the extension of the probability measures from π-systems one can prove that the σ-algebra generated by the increments is independent of $\mathcal{F}_t$. $\square$

Let us denote by φ_t the Fourier transform of $X(t)$. As X has stationary and independent increments, for every u

$$\varphi_{t+s}(u) \stackrel{\circ}{=} \mathbf{E}\left(\exp\left(iuX(t+s)\right)\right) = \tag{1.40}$$
$$= \mathbf{E}\left(\exp\left(iu\left(X(t+s) - X\left(t\right)\right)\right)\exp\left(iuX(t)\right)\right) =$$
$$= \mathbf{E}\left(\exp\left(iu\left(X(t+s) - X\left(t\right)\right)\right)\right) \cdot \mathbf{E}\left(\exp\left(iuX(t)\right)\right) =$$
$$= \mathbf{E}\left(\exp\left(iuX(s)\right)\right) \cdot \mathbf{E}\left(\exp\left(iuX(t)\right)\right) \stackrel{\circ}{=} \varphi_t(u) \cdot \varphi_s(u),$$

therefore

$$\left|\varphi_{t+s}(u)\right| = \left|\varphi_t(u)\right| \cdot \left|\varphi_s(u)\right|.$$

As $|\varphi_t(u)| \leq 1$ for all u and as $|\varphi_0(u)| = 1$ from Cauchy's functional equation

$$|\varphi_t(u)| = \exp\left(t \cdot c(u)\right).$$

This implies that $\varphi_t(u)$ is never zero. Let $h > 0$.

$$\left|\varphi_t(u) - \varphi_{t+h}(u)\right| = |\varphi_t(u)| \left|1 - \frac{\varphi_{t+h}(u)}{\varphi_t(u)}\right| \leq$$

$$\leq |1 - \varphi_h(u)|.$$

X is right-continuous so if $h \searrow 0$ then by the Dominated Convergence Theorem, using that $X(0) = 0$

$$\lim_{h \searrow 0} \varphi_h(u) = \varphi_0(u) = 1.$$

So $\varphi_t(u)$ is right-continuous. If $t > 0$ then

$$\left|\varphi_t(u) - \varphi_{t-h}(u)\right| = |\varphi_{t-h}(u)| \left|1 - \frac{\varphi_t(u)}{\varphi_{t-h}(u)}\right| \leq$$

$$\leq |1 - \varphi_h(u)| \to 0,$$

so $\varphi_t(u)$ is also left-continuous. Hence $\varphi_t(u)$ is continuous in t. Therefore

$$\mathbf{E}(\exp(iu\Delta X(t))) = \lim_{h \searrow 0} \mathbf{E}(\exp(iu(X(t) - X(t-h)))) =$$

$$= \lim_{h \searrow 0} \frac{\varphi_t(u)}{\varphi_{t-h}(u)} = 1,$$

so $\Delta X(t) = 0$ almost surely.

Hence for some subsequence $X(t_{n_k}) \overset{a.s.}{\to} X(t)$. This implies that $X(t-) \overset{a.s.}{=} X(t)$. Therefore one can make the next important observation:

Proposition 1.98 *If X is a Lévy process then $\varphi_t(u) \neq 0$ for every u and the probability of a jump at time t is zero for every t.*

This implies that every Lévy process is continuous in probability. We shall need the following generalization:

Proposition 1.99 *If X is a process with independent increments and X is continuous in probability then*

$$\varphi_t(u) \overset{\circ}{=} \varphi(u, t) \overset{\circ}{=} \mathbf{E}\left(\exp\left(iuX(t)\right)\right)$$

is never zero.

Proof. Let us fix the parameter u. As X is continuous in probability $\varphi(u,t)$ is continuous in t. Let

$$t_0(u) \doteq \inf\{t : \varphi(u,t) = 0\}.$$

One should prove that $t_0(u) = \infty$. By definition $X(0) = 0$ therefore $\varphi(u,0) = 1$ and as $\varphi(u,t)$ is continuous in t obviously $t_0(u) > 0$. Let

$$h(u,s,t) \doteq \mathbf{E}(\exp(iu(X(t) - X(s)))).$$

X has independent increments, so if $s < t$ then

$$\varphi(u,t) = \varphi(u,s) h(u,s,t). \tag{1.41}$$

By the right-regularity of X

$$\varphi(u, t_0(u)) = 0.$$

As $X(t)$ has limits from the left if $t_0(u) < \infty$ then $\varphi(u, t_0(u)-)$ is well-defined. We show that it is not zero. By (1.41) if $s < t_0(u) < \infty$ then

$$\varphi(u, t_0(u)-) = \varphi(u,s) h(u,s,t_0(u)-).$$

$\varphi(u,s) \neq 0$ by the definition of $t_0(u)$, so if $\varphi(u, t_0(u)-) = 0$ then

$$h(u,s,t_0(u)-) = 0$$

for every $s < t_0(u)$.

$$0 = \lim_{s \nearrow t_0(u)} h(u,s,t_0(u)-) =$$

$$= \lim_{s \nearrow t_0(u)} \mathbf{E}(\exp(iuX(t_0(u)-) - iuX(s))) =$$

$$= \mathbf{E}(\exp(0)) = 1,$$

which is impossible. Therefore

$$0 = \varphi(u, t_0(u)) = \varphi(u, t_0(u)-) \neq 0,$$

which is impossible since φ is continuous. $\qquad\square$

Let us recall the following simple observation:

Proposition 1.100 *Let ψ be a complex-valued, continuous curve defined on $\mathbb{R}$. If $\psi(t) \neq 0$ for every t then it has a logarithm that is there is a continuous curve ϕ with the property that $\psi = \exp(\phi)$. If $\phi_1(t_0) = \phi_2(t_0)$ for some point t_0 and $\psi = \exp(\phi_1) = \exp(\phi_2)$ for some continuous curves ϕ_1 and ϕ_2 then $\phi_1 = \phi_2$.*

Proof. The proposition and its proof is quite well-known, so we just sketch it:

1. $\psi \neq 0$, so if $\psi = \exp(\phi_1) = \exp(\phi_2)$ then

$$1 = \frac{\psi}{\psi} = \frac{\exp(\phi_1)}{\exp(\phi_2)} = \exp(\phi_1 - \phi_2).$$

Hence for all t

$$\phi_1(t) = \phi_2(t) + 2\pi in(t),$$

where $n(t)$ is a continuous integer-valued function. As $n(t_0) = 0$ obviously $n \equiv 0$, so $\phi_1 = \phi_2$.

2. The complex series

$$\ln(1+z) = \sum_{n=1}^{\infty} (-1)^{n+1} \frac{z^n}{n}$$

is convergent if $|z| < 1$. On the real line

$$\exp(\ln(1+z)) = 1 + z. \tag{1.42}$$

As $\ln(1+z)$ is analytic (1.42) holds for every $|z| < 1$. To simplify notation as much as possible let us assume that $t_0 = 0$ and $\varphi(t_0) = 1$ and let us assume that we are looking for a curve with $\phi(t_0) = 0$. From (1.42) there is an $r > 0$ that $\psi(t) \overset{\circ}{=} \ln(\varphi(t))$ is well-defined for $|t| < r$.

3. Let a be the infimum and let b be the supremum of the endpoints of closed intervals where one can define a ϕ. If $a_n \searrow a$ and $b_n \nearrow b$ and ϕ is defined on $[a_n, b_n]$ then by the first point of the proof $\phi(t)$ is well-defined on (a, b). Let assume that $b < \infty$. As $\psi(b) \neq 0$ we can define the curve $\theta(t) \overset{\circ}{=} \psi(b+t)/\psi(b)$. Applying the part of the proposition just proved for some $r > 0$

$$\frac{\psi(t)}{\psi(b)} = \exp(\varrho(t)), \quad |b - t| < r,$$

with $\varrho(b) = 0$. Let $t \in (b-r, b)$. As the range of the complex exponential function is $\mathbb{C} \backslash \{0\}$ there is a $z \in \mathbb{C}$ with $\psi(b) = \exp(z)$.

$$\exp(\phi(t)) = \psi(b)\exp(\varrho(t)) = \exp(z + \varrho(t)).$$

Hence $\phi(t) = z + \varrho(t) + 2n\pi i$. With $z + \varrho(t) + 2n\pi i$ one can easily continue ϕ to $(a, b+r)$. This contradiction shows that one can define ϕ for the whole $\mathbb{R}$. $\qquad\square$

$\varphi_1(u) \stackrel{\circ}{=} \mathbf{E}\left(\exp\left(iuX\left(1\right)\right)\right)$ is non-zero and by the Dominated Convergence Theorem it is obviously continuous in u. By the observation just proved

$$\varphi_1(u) = \exp\left(\log \varphi_1(u)\right) \stackrel{\circ}{=} \exp(\phi(u)),$$

where by definition $\phi(0) = 0$. From this by (1.40) $\varphi_n(u) = \exp(n\phi(u))$ and $\varphi_{1/n}(u) = \exp(n^{-1}\phi(u))$ for every $n \in \mathbb{N}$. Hence if r is a rational number then $\varphi_r(u) = \exp(r\phi(u))$. By the just proved continuity in t

$$\varphi_t(u) = \exp\left(t\phi(u)\right), \quad t \in \mathbb{R}_+. \tag{1.43}$$

Lévy processes are not martingales but we can use martingale theory to investigate their properties. The key tool is the so-called *exponential martingale* of X. Let us define the process

$$Z_t\left(u, \omega\right) \stackrel{\circ}{=} Z\left(t, u, \omega\right) \stackrel{\circ}{=} \frac{\exp\left(iuX(t, \omega)\right)}{\varphi_t\left(u\right)}. \tag{1.44}$$

$\varphi_t(u)$ is continuous in t for every fixed u, and therefore $Z_t\left(u, \omega\right)$ is a right-regular stochastic process. Let $t > s$.

$$
\begin{aligned}
\mathbf{E}\left(Z_t\left(u\right) \mid \mathcal{F}_s\right) &\stackrel{\circ}{=} \mathbf{E}\left(\frac{\exp\left(iuX\left(t\right)\right)}{\varphi_t\left(u\right)} \,\middle|\, \mathcal{F}_s\right) = \\
&= \mathbf{E}\left(\frac{\exp\left(iu\left(X\left(t\right) - X\left(s\right)\right)\right)\exp\left(iuX\left(s\right)\right)}{\varphi_{t-s}\left(u\right)\varphi_s\left(u\right)} \,\middle|\, \mathcal{F}_s\right) = \\
&= \frac{\exp\left(iuX\left(s\right)\right)}{\varphi_s\left(u\right)}\frac{\mathbf{E}\left(\exp\left(iu\left(X\left(t\right) - X\left(s\right)\right)\right)\right)}{\varphi_{t-s}\left(u\right)} = \\
&= Z_s\left(u\right)\frac{\mathbf{E}\left(\exp\left(iuX\left(t-s\right)\right)\right)}{\varphi_{t-s}\left(u\right)} = \\
&= Z_s\left(u\right) \cdot 1 \stackrel{\circ}{=} Z_s\left(u\right),
\end{aligned}
$$

therefore $Z_t\left(u\right)$ is a martingale in t for any fixed u.

Definition 1.101 $Z_t\left(u\right)$ *is called the exponential martingale of* X.

Example 1.102 The exponential martingale of a Wiener process.

If w is a Wiener process then

$$Z_t\left(u, \omega\right) \stackrel{\circ}{=} \frac{\exp\left(iuw(t)\right)}{\exp(-tu^2/2)} = \exp\left(iuw(t) + t\frac{u^2}{2}\right).$$

If instead of the Fourier transform we normalize with the Laplace transform, then[56]

$$\frac{\exp\left(uw(t)\right)}{\exp(tu^2/2)} = \exp\left(uw(t) - t\frac{u^2}{2}\right).$$

$\square$

Let X be a Lévy process and assume that the filtration is generated by X. Denote this filtration by $\mathcal{F}^X$. Obviously $\mathcal{F}^X$ does not necessarily contain the measure-zero sets[57], so $\mathcal{F}^X$ does not satisfy the usual conditions. Let $\mathcal{N}$ denotes the collection of measure-zero sets and let us introduce the so-called *augmented filtration*:

$$\mathcal{F}_t \overset{\circ}{=} \sigma\left(\sigma\left(X\left(s\right) : s \le t\right) \cup \mathcal{N}\right). \tag{1.45}$$

It is a bit surprising, but for every Lévy process the augmented filtration satisfies the usual conditions. That is, for Lévy processes the augmented filtration $\mathcal{F}$ is always right-continuous[58]:

Proposition 1.103 *If X is a Lévy process then (1.45) is right-continuous that is $\mathcal{F}_t = \mathcal{F}_{t+}$.*

Proof. Let us take the exponential martingale of X. If $t < w < s$ then

$$\frac{\exp\left(iuX\left(w\right)\right)}{\varphi_w\left(u\right)} \overset{\circ}{=} Z_w\left(u\right) = \mathbf{E}\left(Z_s\left(u\right) \mid \mathcal{F}_w\right) \overset{\circ}{=} \mathbf{E}\left(\frac{\exp\left(iuX\left(s\right)\right)}{\varphi_s\left(u\right)} \mid \mathcal{F}_w\right),$$

therefore

$$Z_w\left(u\right)\varphi_s\left(u\right) \overset{\circ}{=} \exp\left(iuX\left(w\right)\right)\frac{\varphi_s\left(u\right)}{\varphi_w\left(u\right)} = \mathbf{E}\left(\exp\left(iuX\left(s\right)\right) \mid \mathcal{F}_w\right).$$

If $w \searrow t$ then from the continuity of φ_t and from the right-continuity of X, with Lévy's theorem[59]

$$\exp\left(iuX\left(t\right)\right)\frac{\varphi_s\left(u\right)}{\varphi_t\left(u\right)} \overset{a.s.}{=} \mathbf{E}\left(\exp\left(iuX\left(s\right)\right) \mid \mathcal{F}_{t+}\right).$$

As $\exp\left(iuX\left(t\right)\right)$ is $\mathcal{F}_t$-measurable, and $Z_t\left(u\right)$ is a martingale

$$\exp\left(iuX\left(t\right)\right)\frac{\varphi_s\left(u\right)}{\varphi_t\left(u\right)} \overset{a.s.}{=} \mathbf{E}\left(\exp\left(iuX\left(s\right)\right) \mid \mathcal{F}_t\right).$$

---

[56] See: Example 1.118, page 82.
[57] See: Example 1.13, page 9.
[58] See: Example 1.13, page 9.
[59] See: Theorem 1.75, page 46.

Therefore

$$\mathbf{E}\left(\exp\left(iuX\left(s\right)\right)\mid\mathcal{F}_t\right)\overset{a.s.}{=}\mathbf{E}\left(\exp\left(iuX\left(s\right)\right)\mid\mathcal{F}_{t+}\right). \tag{1.46}$$

This equality can be extended to multidimensional trigonometric polynomials. For example, if $t < w \le s_1 \le s_2$ and $\eta \overset{\circ}{=} u_1 X\left(s_1\right) + u_2 X\left(s_2\right)$ then, as $X(s_2) - X\left(s_1\right)$ is independent of $\mathcal{F}_{s_1}$:

$$\mathbf{E}\left(\exp\left(i\eta\right)\mid\mathcal{F}_w\right) = \mathbf{E}\left(\exp\left(iu_1 X\left(s_1\right)\right)\cdot\exp\left(iu_2 X\left(s_2\right)\right)\mid\mathcal{F}_w\right) =$$
$$= \mathbf{E}\left(\exp\left(i(u_1 + u_2)X\left(s_1\right)\right)\cdot\mathbf{E}\left(\exp\left(iu_2\left(X(s_2) - X\left(s_1\right)\right)\right)\mid\mathcal{F}_{s_1}\right)\mid\mathcal{F}_w\right) =$$
$$= \mathbf{E}\left(\exp\left(i(u_1 + u_2)X\left(s_1\right)\right)\cdot\mathbf{E}\left(\exp\left(iu_2\left(X(s_2) - X\left(s_1\right)\right)\right)\right)\mid\mathcal{F}_w\right) =$$
$$= \mathbf{E}\left(\exp\left(i\left(u_1 + u_2\right)X\left(s_1\right)\right)\cdot\varphi_{s_2-s_1}\left(u_2\right)\mid\mathcal{F}_w\right) =$$
$$= \varphi_{s_2-s_1}\left(u_2\right)\cdot\mathbf{E}\left(\exp\left(i\left(u_1 + u_2\right)X\left(s_1\right)\right)\mid\mathcal{F}_w\right) =$$
$$= \varphi_{s_2-s_1}\left(u_2\right)\cdot\varphi_{s_1}\left(u_1 + u_2\right)\cdot Z_w\left(u_1 + u_2\right).$$

If $w \searrow t$ then by the right-continuity of Z_s and by Lévy's theorem[60]

$$\mathbf{E}\left(\exp\left(i\eta\right)\mid\mathcal{F}_{t+}\right)\overset{a.s.}{=}\varphi_{s_2-s_1}\left(u_2\right)\cdot\varphi_{s_1}\left(u_1 + u_2\right)\cdot Z_t\left(u_1 + u_2\right).$$

On the other hand with the same calculation if $w = t$

$$\mathbf{E}\left(\exp\left(i\eta\right)\mid\mathcal{F}_t\right)\overset{a.s.}{=}\varphi_{s_2-s_1}\left(u_2\right)\cdot\varphi_{s_1}\left(u_1 + u_2\right)\cdot Z_t\left(u_1 + u_2\right).$$

Therefore

$$\mathbf{E}\left(\exp\left(i\eta\right)\mid\mathcal{F}_t\right)\overset{a.s.}{=}\mathbf{E}\left(\exp\left(i\eta\right)\mid\mathcal{F}_{t+}\right).$$

That is if $s_k > t$ then

$$\mathbf{E}\left(\exp\left(i\sum_k u_k X(s_k)\right)\mid\mathcal{F}_{t+}\right)\overset{a.s.}{=}\mathbf{E}\left(\exp\left(i\sum_k u_k X(s_k)\right)\mid\mathcal{F}_t\right). \tag{1.47}$$

If $s_k \le t$ then equation (1.47) trivially holds. Hence if $\mathcal{L}$ is the set of bounded functions f for which

$$\mathbf{E}\left(f\left(X\left(s_1\right),\ldots,X\left(s_n\right)\right)\mid\mathcal{F}_{t+}\right)\overset{a.s.}{=}\mathbf{E}\left(f\left(X\left(s_1\right),\ldots,X\left(s_n\right)\right)\mid\mathcal{F}_t\right)$$

then $\mathcal{L}$ contains the π-system of the trigonometric polynomials. $\mathcal{L}$ is trivially a λ-system, therefore, by the Monotone Class Theorem, $\mathcal{L}$ contains the characteristic functions of the sets of the σ-algebra generated by the trigonometric polynomials.

---

[60]See: Theorem 1.75, page 46.

That is if $B \in \mathcal{B}(\mathbb{R}^n)$ then one can write in place of f the characteristic functions χ_B. Collection $\mathcal{Z}$ of sets A for which

$$\mathbf{E}(\chi_A \mid \mathcal{F}_{t+}) \stackrel{a.s.}{=} \mathbf{E}(\chi_A \mid \mathcal{F}_t)$$

is also a λ-system which contains the sets of the π-system

$$\cup_n \sigma\left((X(s_k))_{k=1}^n, s_k \geq 0\right).$$

Again, by the Monotone Class Theorem, $\mathcal{Z}$ contains the σ-algebra

$$\mathcal{F}_\infty^0 = \sigma(X(s) : s \geq 0).$$

If $A \in \mathcal{F}_{t+} \stackrel{\circ}{=} \cap_n \mathcal{F}_{t+1/n}$ then $A \in \mathcal{F}_\infty \stackrel{\circ}{=} \sigma\left(\mathcal{F}_\infty^0 \cup \mathcal{N}\right)$. Therefore there is an $\tilde{A} \in \mathcal{F}_\infty^0$, with $\chi_A \stackrel{a.s.}{=} \chi_{\tilde{A}}$. As $\tilde{A} \in \mathcal{F}_\infty^0 \subseteq \mathcal{Z}$

$$\chi_A \stackrel{a.s.}{=} \mathbf{E}(\chi_A \mid \mathcal{F}_{t+}) \stackrel{a.s.}{=} \mathbf{E}\left(\chi_{\tilde{A}} \mid \mathcal{F}_{t+}\right) \stackrel{a.s.}{=} \mathbf{E}\left(\chi_{\tilde{A}} \mid \mathcal{F}_t\right).$$

Hence up to a measure-zero set χ_A is almost surely equal to an $\mathcal{F}_t$-measurable function $\mathbf{E}\left(\chi_{\tilde{A}} \mid \mathcal{F}_t\right)$. As $\mathcal{F}_t$ contains all the measure-zero set χ_A is $\mathcal{F}_t$-measurable, that is $A \in \mathcal{F}_t$. □

In a similar way one can prove the next proposition:

Proposition 1.104 *If X is a process with independent increments and X is continuous in probability then (1.45) is right-continuous, that is $\mathcal{F}_t = \mathcal{F}_{t+}$.*

Example 1.105 One cannot drop the condition of independent increments.

If $\zeta \cong N(0,1)$ and $X(t,\omega) \stackrel{\circ}{=} t\zeta(\omega)$ then the trajectories of X are continuous and X has stationary increments. If $\mathcal{F}$ is the augmented filtration, then $\mathcal{F}_0 = \sigma(\mathcal{N})$, and if $t > 0$, then $\mathcal{F}_t = \sigma(\sigma(X), \mathcal{N})$, hence $\mathcal{F}_t$ is not right-continuous. □

Example 1.106 The augmentation is important: if w is a Wiener process then

$$\mathcal{F}_t^w \stackrel{\circ}{=} \sigma(w(s) : s \leq t)$$

is not necessarily right-continuous[61].

From now on we shall assume that the filtration of every Lévy process satisfies the usual assumptions.

---

[61]See: Example 1.13, page 9.

Proposition 1.107 *If the process X is left-continuous then the filtration*

$$\mathcal{F}_t^X \overset{\circ}{=} \sigma\left(X\left(s\right) : s \leq t\right)$$

is left-continuous. This remains true for the augmented filtration.

Proof. Let $\mathcal{F}_{t-}^X \overset{\circ}{=} \sigma\left(\cup_{s<t}\mathcal{F}_s^X\right)$. Evidently $\mathcal{F}_{t-}^X \subseteq \mathcal{F}_t^X$, and if $s < t$, then $X\left(s\right)$ is $\mathcal{F}_{t-}^X$-measurable. As the trajectories are left-continuous, $X\left(t\right) = \lim_{s \nearrow t} X\left(s\right)$, hence $X\left(t\right)$ is also $\mathcal{F}_{t-}^X$-measurable, that is $\mathcal{F}_t^X \supseteq \mathcal{F}_{t-}^X$. If $\mathcal{F}_t$ denotes the filtration augmented with the measure-zero sets, then $\mathcal{F}_{t-} \subseteq \mathcal{F}_t$ is evident again. If $F \in \mathcal{F}_t$, then there is an $F^0 \in \mathcal{F}_t^X$ such that $F \Delta F^0 = N \in \mathcal{N}$. As $F^0 \in \mathcal{F}_{t-}^X$ and $\mathcal{F}_{t-}$ contains the measure-zero sets, $F \in \mathcal{F}_{t-}$. $\qquad\square$

Corollary 1.108 *The augmented filtration of a Wiener process is continuous.*

A crucial property of Lévy processes is the so-called *strong Markov property*:

Proposition 1.109 (Strong Markov property for Lévy processes) *Let $\tau < \infty$ be a stopping time and let X be a Lévy process.*

1. *If*

$$X^*\left(t, \omega\right) \overset{\circ}{=} X\left(\tau\left(\omega\right) + t, \omega\right) - X\left(\tau\left(\omega\right), \omega\right), \quad t \geq 0,$$

   then X^* is a Lévy process with respect to the filtration[62] $\mathcal{F}_t^* \overset{\circ}{=} \mathcal{F}_{\tau+t}$.
2. *The distributions of X and X^* are the same[63].*
3. *In particular the set $\{X^*\left(t\right) : t \geq 0\}$ is independent of the stopped σ-algebra $\mathcal{F}_0^* = \mathcal{F}_\tau$.*

Proof. Since X is right-regular, X is progressively measurable, so X^* is obviously $\mathcal{F}_t^*$-adapted.

1. For every u (1.44) is a martingale. If τ is a bounded stopping time then by the Optional Sampling Theorem

$$\mathbf{E}\left(Z\left(\tau + t\right) \mid \mathcal{F}_\tau\right) = Z\left(\tau\right).$$

$Z\left(\tau\right)$ is $\mathcal{F}_\tau$-measurable and as $Z \neq 0$ and as τ is bounded $Z\left(\tau\right)^{-1}$ is bounded so

$$\mathbf{E}\left(\frac{Z_{\tau+t}}{Z_\tau} \mid \mathcal{F}_\tau\right) = 1. \tag{1.48}$$

---

[62]Observe that if $t \geq 0$ then $\sigma \overset{\circ}{=} \tau + t$ is a stopping time. Obviously $\mathcal{F}_t^* \overset{\circ}{=} \mathcal{F}_{\tau+t} \overset{\circ}{=} \mathcal{F}_\sigma$ is the stopped σ-algebra.

[63]That is the infinite dimensional distribution of $\{X\left(t\right) : t \geq 0\}$ and $\{X^*\left(t\right) : t \geq 0\}$ are equal.

By (1.40) for every ω

$$\frac{\varphi_{\tau(\omega)}(u)}{\varphi_{\tau(\omega)+t}(u)} = \frac{\varphi_{\tau(\omega)}(u)}{\varphi_{\tau(\omega)}(u)\,\varphi_t(u)} = \frac{1}{\varphi_t(u)}.$$

So for every $A \in \mathcal{F}_\tau$ using (1.48) by the definition of the conditional expectation

$$\varphi_t(u)\,\mathbf{P}(A) = \varphi_t(u)\int_A 1 d\mathbf{P} = \varphi_t(u)\int_A \frac{Z_{\tau+t}}{Z_\tau} d\mathbf{P} = \tag{1.49}$$

$$= \varphi_t(u)\int_A \frac{\exp(iu(X(\tau+t)-X(\tau)))}{\varphi_t(u)} d\mathbf{P} =$$

$$= \int_A \exp\left(iuX^*(t)\right) d\mathbf{P}.$$

If τ is arbitrary and $\tau_n \stackrel{\circ}{=} n \wedge \tau$ then, as $\tau < \infty$, for every outcome

$$X(\tau_n+t) - X(\tau_n) \to X(\tau+t) - X(\tau) \stackrel{\circ}{=} X^*(t). \tag{1.50}$$

If $A \in \mathcal{F}_\tau$ then for every $t \geq 0$

$$A \cap \{\tau \leq n\} \cap \{\tau_n \leq t\} \in \mathcal{F}_t, \tag{1.51}$$

since if $t \leq n$ then

$$\{\tau_n \leq t\} = \{\tau \leq t\} \subseteq \{\tau \leq n\},$$

and if $t > n$ then $\{\tau_n \leq t\} = \{\tau \leq n\}$. From (1.51) by the definition of the stopped σ-algebra

$$A_n \stackrel{\circ}{=} A \cap \{\tau \leq n\} \in \mathcal{F}_{\tau_n}. \tag{1.52}$$

As τ_n is bounded, by (1.49)

$$\int_{A_n} \exp\left(iu\left(X(\tau_n+t) - X(\tau_n)\right)\right) d\mathbf{P} = \mathbf{P}(A_n)\,\varphi_t(u).$$

From (1.50) and by the Dominated Convergence Theorem

$$\int_A \exp\left(iuX^*(t)\right) d\mathbf{P} =$$

$$= \int_A \lim_{n\to\infty} \chi\left(\tau \le n\right) \exp\left(iu\left(X\left(\tau_n + t\right) - X\left(\tau_n\right)\right)\right) d\mathbf{P} =$$

$$= \lim_{n\to\infty} \int_A \chi\left(\tau \le n\right) \exp\left(iu\left(X\left(\tau_n + t\right) - X\left(\tau_n\right)\right)\right) d\mathbf{P} =$$

$$= \lim_{n\to\infty} \int_{A_n} \exp\left(iu\left(X\left(\tau_n + t\right) - X\left(\tau_n\right)\right)\right) d\mathbf{P} =$$

$$= \lim_{n\to\infty} \mathbf{P}\left(A_n\right) \varphi_t\left(u\right) = \mathbf{P}\left(A\right) \varphi_t\left(u\right) = \mathbf{P}\left(A\right) \int_\Omega \exp\left(iuX(t)\right) d\mathbf{P}.$$

2. If $A \overset{\circ}{=} \Omega$ then the equation above means that the Fourier transform of $X^*(t)$ is φ_t. That is, the distribution of $X^*(t)$ and $X(t)$ is the same. Let $\mathcal{L}$ be the set of bounded functions f for which for all $A \in \mathcal{F}_\tau$

$$\int_A f\left(X^*(t)\right) d\mathbf{P} = \mathbf{P}\left(A\right) \int_\Omega f\left(X^*(t)\right) d\mathbf{P}.$$

Obviously $\mathcal{L}$ is a λ-system, and $\mathcal{L}$ contains the π-system of the trigonometric polynomials

$$x \mapsto \exp\left(iux\right), \quad u \in \mathbb{R}.$$

By the Monotone Class Theorem, $\mathcal{L}$ contains the functions $f \overset{\circ}{=} \chi_B$ with $B \in \mathcal{B}(\mathbb{R})$. Therefore for every $A \in \mathcal{F}_\tau$ and $B \in \mathcal{B}(\mathbb{R})$

$$\int_A \chi_B\left(X^*(t)\right) d\mathbf{P} = \mathbf{P}\left(A \cap \{X^*(t) \in B\}\right) =$$

$$= \mathbf{P}\left(A\right) \int_\Omega \chi_B\left(X^*(t)\right) d\mathbf{P} = \mathbf{P}\left(A\right) \cdot \mathbf{P}\left(X^*(t) \in B\right).$$

So $X^*(t)$ is independent of $\mathcal{F}_\tau$.

3. One should prove that X^* has stationary and independent increments. If $\sigma \overset{\circ}{=} \tau + t$ and

$$X^{**}(h) \overset{\circ}{=} X\left(\sigma + h\right) - X\left(\sigma\right),$$

then using the part of the proposition already proved for the stopping time σ

$$X^* (t + h) - X^* (t) \overset{\circ}{=} (X (\tau + t + h) - X (\tau)) - (X (\tau + t) - X (\tau)) =$$
$$= X (\sigma + h) - X (\sigma) = X^{**}(h) \cong X(h),$$

which is independent of t and therefore X^* has stationary increments. Also by the already proved part of the proposition

$$X^* (t + h) - X^* (t) = X^{**}(h)$$

is independent of $\mathcal{F}_\sigma \overset{\circ}{=} \mathcal{F}_t^*$. Obviously $X^* (0) = 0$ and X^* is right-regular therefore X^* is a process with independent increments.

4. Now we prove that X and X^* have the same distribution. Let

$$0 = t_0 < t_1 < \ldots < t_n$$

be arbitrary. As we proved

$$X^* (t_k) - X^* (t_{k-1}) \cong X^* (t_k - t_{k-1}) \cong X (t_k - t_{k-1}) \cong$$
$$\cong X (t_k) - X (t_k - 1).$$

As the increments are independent $(X^* (t_k) - X^* (t_{k-1}))_{k=1}^n$ has the same distribution as $(X (t_k) - X (t_{k-1}))_{k=1}^n$. This implies that $(X (t_k))_{k=1}^n$ has the same distribution as $(X^* (t_k))_{k=1}^n$. Which, by the Monotone Class Theorem, implies that X^* and X has the same distribution.

5. As we proved X^* is a process with independent increments so $\mathcal{F}_t^*$ is independent of the σ-algebra $\mathcal{G}_t^*$ generated by the increments[64]

$$X^* (u) - X^* (v), \quad u \geq v \geq t.$$

So as a special case the set $\{X^* (t) : t \geq 0\}$ is independent of $\mathcal{F}_0^* = \mathcal{F}_\tau$. □

Example 1.110 Random times which are not stopping times.

Let $a > 0$ and let w be a Wiener process.

1. Let

$$\gamma_a \overset{\circ}{=} \sup \{0 \leq s \leq a : w (s) = 0\} = \inf \{s \geq 0 : w (a - s) = 0\}.$$

[64]See: Proposition 1.97, page 61.

Obviously γ_a is $\mathcal{F}_a$-measurable, so it is a random time. As $\mathbf{P}\left(w\left(a\right)=0\right)=0$ almost surely $\gamma_a < a$. Assume that γ_a is a stopping time. In this case by the strong Markov property

$$w^*\left(t\right) \stackrel{\circ}{=} w\left(t+\gamma_a\right) - w\left(\gamma_a\right)$$

is also a Wiener process. It is easy to see that if w^* is a Wiener process then $\widetilde{w}\left(t\right) \stackrel{\circ}{=} tw^*\left(1/t\right)$ is also a Wiener process[65]. As every one-dimensional Wiener process almost surely returns to the origin[66], with the strong Markov property it is easy to prove that $\widetilde{w}$ returns to the origin almost surely after any time t. This means that there is a sequence $t_n \searrow 0$ with $t_n > 0$ that almost surely $w^*\left(t_n\right) = 0$. But this is impossible as almost surely w^* does not have a zero on the interval $\left(0, a - \gamma_a\right]$.

2. Let

$$\beta_a \stackrel{\circ}{=} \max\left\{w\left(s\right) : 0 \le s \le a\right\},$$
$$\rho_a \stackrel{\circ}{=} \inf\left\{0 \le s \le a : w\left(s\right) = \beta_a\right\}.$$

We show that ρ_a is not a stopping time. As $\mathbf{P}\left(w\left(a\right) - w\left(a/2\right) < 0\right) = 1/2$

$$\mathbf{P}\left(\rho_a < a\right) > 0.$$

If ρ_a were a stopping time, then by the strong Markov property

$$w^*\left(t\right) \stackrel{\circ}{=} w\left(t+\rho_a\right) - w\left(\rho_a\right)$$

would be a Wiener process. But this is impossible as with positive probability the interval $\left(0, a - \rho_a\right]$ is not empty and on this interval w^* cannot have a positive value. $\square$

An important consequence of the strong Markov property is the following:

Proposition 1.111 *If the size of the jumps of a Lévy process X are smaller than a constant $c > 0$, that is $|\Delta X| \le c$ then on any interval $[0, t]$ the moments of X are uniformly bounded. That is for each m there is a constant $K\left(m, t\right)$, that*

$$\mathbf{E}\left(|X^m\left(s\right)|\right) \le K\left(m, t\right), \quad s \in [0, t].$$

Proof. One may assume that the stopping time[67]

$$\tau_1 \stackrel{\circ}{=} \inf\left\{t : |X\left(t\right)| > c\right\}$$

---

[65] See: Corollary B.10, page 566.
[66] See: Corollary B.8, page 565.
[67] Recall that $\mathcal{F}$ satisfies the usual assumptions. See: Example 1.32, page 17.

is finite, as by the zero-one law the set of outcomes ω where $\tau_1(\omega) = \infty$ has probability 0 or 1. If with probability one $\tau_1(\omega) = \infty$ then X is uniformly bounded, hence in this case the proposition holds. Then define the stopping time

$$\tau_2 \overset{\circ}{=} \inf \{t : |X^*(t)| > c\} + \tau_1 \overset{\circ}{=} \inf \{t : |X(t + \tau_1) - X(\tau_1)| > c\} + \tau_1.$$

In a similar way let us define τ_3 etc. By the strong Markov property the variables $\{X^*(t) : t \geq 0\}$ are independent of the σ-algebra $\mathcal{F}_{\tau_1}$. The variable

$$\tau_2 - \tau_1 \overset{\circ}{=} \inf \{t \geq 0 : |X^*(t)| > c\}$$

is measurable with respect to the σ-algebra generated by the variables $\{X^*(t) : t \geq 0\}$ hence $\tau_2 - \tau_1$ is independent of $\mathcal{F}_{\tau_1}$. In general $\tau_n - \tau_{n-1}$ is independent of $\mathcal{F}_{\tau_{n-1}}$. Also by the strong Markov property for all n the distribution of $\tau_n - \tau_{n-1}$ is the same as the distribution of τ_1. Therefore if $\tau_0 \overset{\circ}{=} 0$, then using the independence of variables $(\tau_k - \tau_{k-1})$

$$\mathbf{E}\left(\exp\left(-\tau_n\right)\right) = \mathbf{E}\left(\exp\left(-\sum_{k=1}^{n}(\tau_k - \tau_{k-1})\right)\right) = \left(\mathbf{E}\left(\exp\left(-\tau_1\right)\right)\right)^n \overset{\circ}{=} q^n,$$

where $0 < q \leq 1$. If $q = 1$ then almost surely $\tau_1 = 0$, which by the right-continuity implies that $|X(0)| \geq c > 0$, which, by the definition of Lévy processes, is not the case, so $q < 1$. As the jumps are smaller than c

$$|X(\tau_1)| \leq |X(\tau_1-)| + |\Delta X(\tau_1)| \leq$$
$$\leq |X(\tau_1-)| + c \leq 2c.$$

In a same way it is easy to see that in general

$$\sup_t |X^{\tau_n}(t)| = \sup |\{X(t) : t \in [0, \tau_n]\}| \leq 2nc.$$

Therefore by Markov's inequality

$$\mathbf{P}\left(|X(t)| > 2nc\right) \leq \mathbf{P}\left(\tau_n < t\right) = \mathbf{P}\left(\exp\left(-\tau_n\right) > \exp\left(-t\right)\right) \leq$$
$$\leq \frac{\mathbf{E}\left(\exp\left(-\tau_n\right)\right)}{\exp\left(-t\right)} \leq \exp\left(t\right) q^n.$$

As $q < 1$

$$L(m) \overset{\circ}{=} \sum_{n=0}^{\infty} [2(n+1)c]^m q^n < \infty,$$

so

$$\mathbf{E}\left(|X(t)|^m\right) \le \sum_{n=0}^{\infty} \left[2(n+1)c\right]^m \cdot \mathbf{P}\left(|X(t)| > 2nc\right) \le$$

$$\le \exp(t) \sum_{n=0}^{\infty} \left[2(n+1)c\right]^m q^n \stackrel{\circ}{=} \exp(t) L(m),$$

from which the proposition is evident. □

One can generalize these observations.

Proposition 1.112 (Strong Markov property for processes with independent increments) *Let X be a process with independent increments and assume that X is continuous in probability. Let $D\left([0,\infty)\right)$ denote the space of right-regular functions over $[0,\infty)$ and let $\mathcal{H}$ be the σ-algebra over $D\left([0,\infty)\right)$ generated by the coordinate functionals. If f is a non-negative $\mathcal{H}$-measurable functional[68] over $D\left([0,\infty)\right)$, then for every stopping time $\tau < \infty$*

$$\mathbf{E}\left(f(X^*) \mid \mathcal{F}_\tau\right) = \mathbf{E}\left(f(X_s^*)\right)|_{s=\tau}$$

where

$$X_s^*(t) \stackrel{\circ}{=} X(s+t) - X(s).$$

Proof. Let $\varphi(u,t)$ be the Fourier transform of $X(t)$. As X is continuous in probability $\varphi(u,t) \ne 0$ and

$$Z(u,t) \stackrel{\circ}{=} \frac{\exp\left(iuX(t)\right)}{\varphi(u,t)}$$

is a martingale[69]. Let τ be a bounded stopping time. By the Optional Sampling Theorem

$$\mathbf{E}\left(Z(u,\tau+s) \mid \mathcal{F}_\tau\right) = Z(u,\tau).$$

$\varphi(u,\tau+t)$ is $\mathcal{F}_\tau$-measurable. Therefore

$$\mathbf{E}\left(\exp\left(iuX^*(t)\right) \mid \mathcal{F}_\tau\right) \stackrel{\circ}{=} \tag{1.53}$$

$$\stackrel{\circ}{=} \mathbf{E}\left(\exp\left(iu\left(X(\tau+t) - X(\tau)\right)\right) \mid \mathcal{F}_\tau\right) = \frac{\varphi(u,\tau+t)}{\varphi(u,\tau)} = \frac{\varphi(u,s+t)}{\varphi(u,s)}\bigg|_{s=\tau} =$$

---

[68]It is easy to see that $f(X) = g(X(t_1), X(t_2),\ldots)$ where g is an $\mathbb{R}^\infty \to \mathbb{R}$ Borel measurable function and (t_k) is a countable sequence in $\mathbb{R}_+$. The canonical example is $f(X) \stackrel{\circ}{=} \sup_{s \le t} |X(s)|$.

[69]See: Proposition 1.99, page 63.

$$= \frac{\varphi(u,s)\, \mathbf{E}\left(\exp\left(iu\left(X(t+s) - X(s)\right)\right)\right)}{\varphi(u,s)}\bigg|_{s=\tau} =$$

$$= \mathbf{E}\left(\exp\left(iu\left(X_s^*(t)\right)\right)\right)\big|_{s=\tau}.$$

If τ is not bounded then $\tau_n \overset{\circ}{=} \tau \wedge n$ is a bounded stopping time. Let

$$h(s) \overset{\circ}{=} \mathbf{E}\left(\exp\left(iu\left(X(s+t) - X(s)\right)\right)\right)$$

As $\tau < \infty$

$$X(\tau_n + t) - X(\tau_n) \to X(\tau + t) - X(\tau)$$

So by the Dominated Convergence Theorem $h(\tau_n) \to h(\tau)$. If $A \in \mathcal{F}_\tau$ then

$$A \cap \{\tau \leq n\} \in \mathcal{F}_{\tau_n}$$

therefore

$$\int_A \chi(\tau \leq n) \exp\left(iu\left(X(\tau_n + t) - X(\tau_n)\right)\right) = \int_A \chi(\tau \leq n)\, h(\tau_n)\, d\mathbf{P}.$$

By the Dominated Convergence Theorem one can take the limit $n \to \infty$. Hence in (1.53) we can drop the condition that τ is bounded. With the Monotone Class Theorem one can prove that for any Borel measurable set B

$$\mathbf{E}\left(\chi_B\left(X^*(t)\right) \mid \mathcal{F}_\tau\right) = \mathbf{E}\left(\chi_B\left(X_s^*(t)\right)\right)\big|_{s=\tau}$$

In the usual way, using multi-dimensional trigonometric polynomials and the Monotone Class Theorem several times, one can extend the relation to every $\mathcal{H}$-measurable and bounded function. Finally one can prove the proposition with the Monotone Convergence Theorem. □

Corollary 1.113 *Under the same conditions as above*

$$\mathbf{E}\left(f\left(X^*\right) \mid \tau = s\right) = \mathbf{E}\left(f\left(X_s^*\right)\right).$$

Let us remark, that if X is a Lévy process then the distribution of X_s^* is the same as the distribution of X for every s so

$$\mathbf{E}\left(f\left(X^*\right) \mid \mathcal{F}_\tau\right) = \mathbf{E}\left(f\left(X\right)\right)$$

for every $\tau < \infty$. If $f(X) \doteq \exp\left(i\sum_{k=1}^{n} u_k X(t_k)\right)$ then

$$\mathbf{E}\left(\exp\left(i\sum_{k=1}^{n} u_k X^*(t_k)\right) \mid \mathcal{F}_\tau\right) = \mathbf{E}\left(\exp\left(i\sum_{k=1}^{n} u_k X(t_k)\right)\right).$$

The right-hand side is deterministic which implies that $(X^*(t_1), X^*(t_2), \dots X^*(t_n))$ is independent of $\mathcal{F}_\tau$ and has the same distribution as $(X(t_1), X(t_2), \dots, X(t_n))$.

Proposition 1.114 *If X is a process with independent increments and X is continuous in probability, and the jumps of X are bounded by some constant c, then all the moments of X are uniformly bounded on any finite interval, that is, for every t*

$$\mathbf{E}(|X^m(s)|) \leq K(m,t) < \infty, \quad s \in [0,t].$$

Proof. Let us fix a t. X has right-regular trajectories so on any finite interval the trajectories are bounded. Therefore $\sup_{s \leq 2t} |X(s)| < \infty$. Hence if b is sufficiently large then

$$\mathbf{P}\left(\sup_{s \leq 2t} |X(s)| > \frac{b}{2}\right) < q < 1.$$

Let

$$\tau \doteq \inf\{s : |X(s)| > a\} \wedge 2t.$$

By the definition of τ

$$\{\tau < t\} \subseteq \left\{\sup_{s \leq t} |X(s)| > a\right\} \subseteq \{\tau \leq t\}.$$

If for some ω.

$$\omega \in \left\{\sup_{s \leq t} |X(s)| > a\right\} \setminus \{\tau < t\}$$

then

$$\sup_{s < t} |X(s, \omega)| \leq a$$

but $X(t, \omega) > a$, so process X has a jump at (t, ω), which by the stochastic continuity of X has probability zero. As the size of the jumps is bounded by the right-continuity

$$\sup_{s \leq \tau} |X(s)| \leq \sup_{s \leq \tau} |X(s-)| + \sup_{s \leq \tau} |\Delta X(s)| \leq a + c.$$

We show that this implies that

$$\left\{ \sup_{s \leq t} |X(s)| > a + b + c \right\} \subseteq \left\{ \sup_{s \leq t} |X(s)| > a, \ \sup_{s \leq t} |X(\tau + s) - X(\tau)| > b \right\}.$$

If

$$\sup_{s \leq t} |X(s)| > a + b + c$$

then obviously $\sup_{s \leq t} |X(s)| > a$, hence $\tau \leq t$, so if $\sup_{s \leq t} |X(\tau + s) - X(\tau)| \leq b$, then

$$\sup_{s \leq t} |X(s)| \leq \sup_{s \leq \tau} |X(s)| + \sup_{s \leq t} |X(\tau + s) - X(\tau)| \leq a + b + c.$$

Which is impossible. If $u \leq t$, then

$$\sup_{s \leq t} |X(u + s) - X(u)| \leq 2 \sup_{s \leq 2t} |X(s)|.$$

Therefore if $u \leq t$, then

$$\left\{ \sup_{s \leq t} |X(u + s) - X(u)| > b \right\} \subseteq \left\{ \sup_{s \leq 2t} |X(s)| > \frac{b}{2} \right\}.$$

Let F be the distribution function of τ. By the just proved strong Markov property

$$\mathbf{P} \left(\sup_{s \leq t} |X(s)| > a + b + c \right) \leq$$

$$\leq \mathbf{P} \left(\sup_{s \leq t} |X(s)| > a, \sup_{s \leq t} |X(\tau + s) - X(\tau)| > b \right) =$$

$$= \mathbf{P} \left(\tau < t, \sup_{s \leq t} |X(\tau + s) - X(\tau)| > b \right) =$$

$$= \int_{[0,t)} \mathbf{P} \left(\sup_{s \leq t} |X((\tau + s)) - X(\tau)| > b \mid \tau = u \right) dF(u) =$$

$$= \int_{[0,t)} \mathbf{P} \left(\sup_{s \leq t} |X(u + s) - X(u)| > b \right) dF(u) \leq$$

$$\leq \mathbf{P}\left(\sup_{s \leq 2t} |X(s)| > \frac{b}{2}\right) \cdot \mathbf{P}(\tau < t) =$$

$$= q \cdot \mathbf{P}(\tau < t) \leq q \cdot \mathbf{P}\left(\sup_{s \leq t} |X(s)| > a\right).$$

From this for an arbitrary n

$$\mathbf{P}\left(\sup_{s \leq t} |X(s)| > n(b+c)\right) \leq q^n.$$

Hence

$$\mathbf{E}\left(|X(t)|^m\right) \leq \mathbf{E}\left(\sup_{s \leq t} |X(s)|^m\right) \leq \sum_{n=1}^{\infty} (n(b+c))^m q^{n-1} < \infty. \qquad \square$$

We shall return to Lévy processes in section 7.1. If the reader is interested only in Lévy processes then they can continue the reading there.

1.3.8 Application: the first passage times of the Wiener processes

In this subsection we present some applications of the Optional Sampling Theorem. Let w be a Wiener process. We shall discuss some properties of the *first passage times*

$$\tau_a \overset{\circ}{=} \inf \{t : w(t) = a\}. \tag{1.54}$$

The set $\{a\}$ is closed and w is continuous, hence τ_a is a stopping time[70]. Recall that[71] almost surely

$$\limsup_{t \to \infty} w(t) = \infty, \quad \liminf_{t \to \infty} w(t) = -\infty. \tag{1.55}$$

Therefore as w is continuous τ_a is almost surely finite.

Example 1.115 The martingale convergence theorem does not hold in $L^1(\Omega)$.

Let w be a Wiener process and let $X \overset{\circ}{=} w + 1$. Let τ be the first passage time of zero for X, that is let

$$\tau \overset{\circ}{=} \inf \{t : X(t) = 0\} = \tau_{-1} \overset{\circ}{=} \inf \{t : w(t) = -1\}.$$

---

[70]See: Example 1.32, page 17.
[71]See: Proposition B.7, page 564.

As X is martingale X^τ is a non-negative martingale. By the martingale convergence theorem for non-negative martingales [72] if $t \nearrow \infty$ then $X^\tau(t)$ is almost surely convergent. As we remarked, τ is almost surely finite therefore obviously $X^\tau(\infty) = 0$. By the Optional Sampling Theorem

$$\|X^\tau(t)\|_1 = \|X(\tau \wedge t)\|_1 = \mathbf{E}\left(X(\tau \wedge t)\right) = \mathbf{E}\left(X(0)\right) = 1$$

for any t. Hence the convergence does not hold in $L^1(\Omega)$. □

Example 1.116 If $a < 0 < b$ and τ_a and τ_b are the respective first passage times of some Wiener process w, then

$$\mathbf{P}\left(\tau_a < \tau_b\right) = \frac{b}{b-a}, \quad \mathbf{P}\left(\tau_b < \tau_a\right) = \frac{-a}{b-a}.$$

By (1.55) with probability one, the trajectories of w are unbounded. Therefore as w starts from the origin the trajectories of w finally leave the interval $[a, b]$. So

$$\mathbf{P}\left(\tau_a < \tau_b\right) + \mathbf{P}\left(\tau_b < \tau_a\right) = 1.$$

If $\tau \overset{\circ}{=} \tau_a \wedge \tau_b$ then w^τ is a bounded martingale. Hence one can use the Optional Sampling Theorem. Obviously w_τ^τ is either a or b, hence

$$\mathbf{E}\left(w_\tau^\tau\right) = a\mathbf{P}\left(\tau_a < \tau_b\right) + b\mathbf{P}\left(\tau_b < \tau_a\right) = \mathbf{E}\left(w^\tau(0)\right) = 0.$$

We have two equations with two unknowns. Solving this system of linear equations, one can easily deduce the formulas above. □

Example 1.117 Let $a < 0 < b$ and let τ_a and τ_b be the respective first passage times of some Wiener process w. If $\tau \overset{\circ}{=} \tau_a \wedge \tau_b$, then $\mathbf{E}\left(\tau\right) = |ab|$.

With direct calculation it is easy to see that the process $w^2(t) - t$ is a martingale. From this it is easy to show that the process

$$X(t) \overset{\circ}{=} (w(t) - a)(b - w(t)) + t$$

is also a martingale. By the Optional Sampling Theorem

$$|ab| = -ab = \mathbf{E}\left(X(0)\right) = \mathbf{E}\left(X(\tau \wedge n)\right) =$$
$$= \mathbf{E}\left(w(\tau \wedge n) - a\right)(b - w(\tau \wedge n)) + \mathbf{E}\left(\tau \wedge n\right).$$

---

[72]See: Corollary 1.66, page 40.

If $n \nearrow \infty$ then by the Monotone and by the Dominated Convergence Theorems the limit of the right-hand side is $\mathbf{E}(\tau)$. $\qquad\qquad\square$

Example 1.118 Let w be a Wiener process. The Laplace transform of the first passage time τ_a is

$$L(s) \overset{\circ}{=} \mathbf{E}(\exp(-s\tau_a)) = \exp\left(-|a|\sqrt{2s}\right), \quad s \geq 0. \tag{1.56}$$

Let $a > 0$. For every u the process $X(t) \overset{\circ}{=} \exp\left(u \cdot w(t) - t \cdot u^2/2\right)$ is a martingale[73]. So the truncated process X^{τ_a} is also a martingale. If $u \geq 0$, then

$$0 \leq X^{\tau_a}(t) \leq \exp\left(ua - \frac{u^2 t}{2}\right) \leq \exp(au),$$

hence X^{τ_a} is a bounded martingale. Every bounded martingale is uniformly integrable, therefore one can apply the Optional Sampling Theorem. So

$$\mathbf{E}\left(X^{\tau_a}_{\tau_a}\right) = \mathbf{E}\left(\exp\left(ua - \frac{u^2 \tau_a}{2}\right)\right) = \mathbf{E}(X^{\tau_a}(0)) = 1.$$

Hence

$$\mathbf{E}\left(\exp\left(-\frac{u^2 \tau_a}{2}\right)\right) = \exp(-ua).$$

If $u \overset{\circ}{=} \sqrt{2s} \geq 0$ then

$$L(s) \overset{\circ}{=} \mathbf{E}(\exp(-s\tau_a)) = \exp\left(-a\sqrt{2s}\right).$$

If $a < 0$ then repeating the calculations for the Wiener process $-w$

$$L(s) = \exp\left(-|a|\sqrt{2s}\right). \qquad\qquad\square$$

Example 1.119 The Laplace transform of the first passage time of the *reflected Wiener process* $|w|$ is

$$\tilde{L}(s) \overset{\circ}{=} \mathbf{E}(\exp(-s\tilde{\tau}_a)) = \frac{1}{\cosh(a\sqrt{2s})}, \quad s \geq 0. \tag{1.57}$$

---

[73]See: (1.44), page 66.

By definition

$$\tilde{\tau}_a \overset{\circ}{=} \inf\{t : |w(t)| = a\}.$$

Let

$$X(t) \overset{\circ}{=} \frac{\exp(uw(t)) + \exp(-uw(t))}{2} \exp\left(-\frac{u^2 t}{2}\right) \overset{\circ}{=}$$

$$\overset{\circ}{=} \cosh(uw(t)) \exp\left(-\frac{u^2 t}{2}\right).$$

X is the sum of two martingales, hence it is a martingale. $|X^{\tilde{\tau}_a}| \leq \cosh(ua)$, therefore one can again apply the Optional Sampling Theorem.

$$\mathbf{E}\left(X_{\tilde{\tau}_a}^{\tilde{\tau}_a}\right) = \mathbf{E}\left(\cosh(ua)\exp\left(-\frac{u^2 \tilde{\tau}_a}{2}\right)\right) = 1,$$

therefore

$$\mathbf{E}\left(\exp\left(\frac{-u^2 \tilde{\tau}_a}{2}\right)\right) = \frac{1}{\cosh(ua)}.$$

If $u \overset{\circ}{=} \sqrt{2s}$ then

$$\mathbf{E}\left(\exp(-s\tilde{\tau}_a)\right) = \frac{1}{\cosh(a\sqrt{2s})}. \qquad \square$$

Example 1.120 The density function of the distribution of the first passage time τ_a of a Wiener process is

$$f(x) = |a| (2\pi x^3)^{-1/2} \exp\left(-\frac{a^2}{2x}\right). \tag{1.58}$$

By the uniqueness of the Laplace transform it is sufficient to prove that the Laplace transform of (1.58) is $\exp(-|a|\sqrt{2s})$. By the definition of the Laplace transform

$$L(s) \overset{\circ}{=} \int_0^\infty \exp(-sx) f(x) \, dx, \quad s \geq 0.$$

If F denotes the distribution function of (1.58) then

$$F(x) \overset{\circ}{=} \int_0^x f(t) \, dt = 2 \int_a^\infty \frac{1}{\sqrt{2\pi x}} \exp\left(-\frac{u^2}{2x}\right) du, \tag{1.59}$$

since if we substitute $t \overset{\circ}{=} xa^2/u^2$, then

$$F(x) = \int_\infty^a \frac{au^3}{a^3\sqrt{2\pi x^3}} \exp\left(-\frac{u^2}{2x}\right) xa^2(-2)u^{-3}du =$$

$$= 2\int_a^\infty \frac{1}{\sqrt{2\pi x}} \exp\left(-\frac{u^2}{2x}\right) du.$$

Integrating by parts and using that $F(0) = 0$, if $s > 0$ then

$$L(s) = [\exp(-sx)F(x)]_0^\infty + \int_0^\infty s\exp(-sx)F(x)\,dx =$$

$$= s\int_0^\infty \exp(-sx)F(x)\,dx.$$

By (1.59)

$$L(s) = 2s\int_0^\infty \exp(-sx)\int_a^\infty \frac{1}{\sqrt{2\pi x}}\exp\left(-\frac{u^2}{2x}\right)du\,dx.$$

Fix s and let us take $L(s)$ as a function of a. Let us denote this function by $g(a)$. We show that if $a > 0$ then $g(a)$ satisfies the differential equation

$$\frac{d^2}{da^2}g(a) = 2sg(a). \tag{1.60}$$

The integrand is non-negative, so by Fubini's theorem one can change the order of the integration, so

$$g(a) = 2s\int_a^\infty \int_0^\infty \exp(-sx)\frac{1}{\sqrt{2\pi x}}\exp\left(-\frac{u^2}{2x}\right)dx\,du.$$

As

$$\int_0^\infty \frac{1}{\sqrt{2\pi x}}\exp(-sx)\,dx = \frac{1}{\sqrt{2\pi s}}\Gamma\left(\frac{1}{2}\right) < \infty$$

$1/\sqrt{2\pi x}\exp(-sx)$ is integrable, and dominates the integrand. Hence the inner integral is a continuous function of u. Using this, one can differentiate with respect the integral sign

$$g'(a) = -2s\int_0^\infty \exp(-sx)\frac{1}{\sqrt{2\pi x}}\exp\left(-\frac{a^2}{2x}\right)dx.$$

We can differentiate under the integral sign as on the interval $a \in (b, c)$

$$\exp\left(-sx\right) \frac{c}{\sqrt{2\pi x^3}} \exp\left(-\frac{b^2}{2x}\right)$$

is integrable and dominates the partial derivatives

$$\frac{\partial}{\partial a}\left(\exp(-sx)\frac{1}{\sqrt{2\pi x}}\exp\left(-\frac{a^2}{2x}\right)\right) = \exp(-sx)\frac{-a}{\sqrt{2\pi x^3}}\exp\left(-\frac{a^2}{2x}\right).$$

Hence

$$g''\left(a\right) = 2s\int_0^\infty \exp\left(-sx\right)\frac{a}{\sqrt{2\pi x^3}}\exp\left(-\frac{a^2}{2x}\right)dx =$$

$$= 2s\int_0^\infty \exp\left(-sx\right)f\left(x\right)dx = 2s\cdot L\left(s\right) \stackrel{\circ}{=} 2s\cdot g\left(a\right).$$

The characteristic polynomial of this second-order linear differential equation is $\lambda^2 - 2s = 0$, which has the roots $\lambda_{1,2} = \pm\sqrt{2s}$. So the general solution of equation (1.60) is

$$A\exp\left(a\sqrt{2s}\right) + B\exp\left(-a\sqrt{2s}\right).$$

As $L\left(0\right) = A + B = 1$ and $L\left(\infty\right) = 0$

$$L\left(s\right) = \exp\left(-a\sqrt{2s}\right). \qquad \square$$

Example 1.121 The Fourier transform of τ_a is

$$\varphi\left(t\right) = \exp\left(-a\sqrt{|t|}\left(1 - i\cdot\operatorname{sgn}t\right)\right).$$

The formula $\exp\left(-a\sqrt{2s}\right)$ has an analytic extension to the half plane $\operatorname{Re}\left(z\right) > 0$. If $s > 0$ and $z \stackrel{\circ}{=} s + it$ then

$$z^{1/2} \stackrel{\circ}{=} \exp\left(\frac{1}{2}\log z\right) = \exp\left(\frac{1}{2}\ln\left(|z|\right)\right)\exp\left(\frac{1}{2}i\arg\left(\frac{z}{|z|}\right)\right) =$$

$$= \sqrt[4]{s^2 + t^2}\left(\cos\frac{\arctan\left(t/s\right)}{2} + i\sin\frac{\arctan\left(t/s\right)}{2}\right).$$

The complex Laplace transform is continuous so

$$\varphi(t) = L(-it) =$$

$$= \lim_{s \searrow 0} \exp\left(-a\sqrt{2}\sqrt[4]{s^2 + t^2}\left[\cos\frac{\arctan\left(\frac{-t}{s}\right)}{2} + i\sin\frac{\arctan\left(\frac{-t}{s}\right)}{2}\right]\right) =$$

$$= \exp\left(-a\sqrt{2|t|}\left(\cos\left(-\frac{\pi}{4}\operatorname{sgn}t\right) + i\sin\left(-\frac{\pi}{4}\operatorname{sgn}t\right)\right)\right) =$$

$$= \exp\left(-a\sqrt{|t|}(1 - i \cdot \operatorname{sgn}t)\right). \qquad \qquad \square$$

Example 1.122 The maximum process of a Wiener process.

Let w be a Wiener process, and let us introduce the maximum process

$$S(t) \overset{\circ}{=} \sup_{s \leq t} w(s) = \max_{s \leq t} w(s).$$

We show that for every $a \geq 0$ and $t \geq 0$

$$\mathbf{P}(S(t) \geq a) = \mathbf{P}(\tau_a \leq t) = 2 \cdot \mathbf{P}(w(t) \geq a) = \mathbf{P}(|w(t)| \geq a). \qquad (1.61)$$

The first and last equality are trivial. We prove the second one: recall that the density function of the distribution of τ_a is

$$\frac{d}{dt}\mathbf{P}(\tau_a \leq t) \overset{\circ}{=} \frac{d}{dt}F(t) \overset{\circ}{=} f(t) = a\frac{1}{\sqrt{2\pi t^3}}\exp\left(-\frac{a^2}{2t}\right).$$

$w(t) \cong N(0, \sqrt{t})$, so

$$U(t) \overset{\circ}{=} 2 \cdot \mathbf{P}(w(t) \geq a) = 2\left(1 - \Phi\left(\frac{a}{\sqrt{t}}\right)\right) =$$

$$= \frac{2}{\sqrt{2\pi}}\int_{a/\sqrt{t}}^{\infty}\exp\left(-\frac{u^2}{2}\right)du.$$

Differentiating with respect to t

$$\frac{d}{dt}U(t) = \frac{a}{\sqrt{2\pi}}\exp\left(-\frac{a^2}{2t}\right)t^{-3/2},$$

hence the derivatives of $\mathbf{P}\left(\tau_a \leq t\right)$ and $2 \cdot \mathbf{P}\left(w\left(t\right) \geq a\right)$ with respect to t are the same. The two functions are equal if $t = 0$, therefore

$$2 \cdot \mathbf{P}\left(w\left(t\right) \geq a\right) = \mathbf{P}\left(\tau_a \leq t\right)$$

for every t. $\qquad\qquad\qquad\qquad\qquad\qquad\qquad\qquad\qquad\qquad\qquad\qquad\square$

Example 1.123 The density function of $S\left(t\right) \overset{\circ}{=} \sup_{s \leq t} w\left(s\right)$ is

$$f\left(x\right) = \frac{2}{\sqrt{2\pi t}} \exp\left(-\frac{x^2}{2t}\right), \quad x > 0.$$

By (1.61) $\mathbf{P}\left(S\left(t\right) \geq x\right) = 2\left(1 - \Phi\left(x/\sqrt{t}\right)\right)$. Differentiating we get the formula. $\qquad\qquad\qquad\qquad\qquad\qquad\qquad\qquad\qquad\qquad\qquad\qquad\qquad\square$

Example 1.124 If w is a Wiener process then

$$\mathbf{E}\left(\sup_{s \leq 1}\left|w\left(s\right)\right|\right) = \sqrt{\frac{\pi}{2}}, \quad \mathbf{E}\left(\sup_{s \leq 1} w\left(s\right)\right) = \sqrt{\frac{2}{\pi}}.$$

Let

$$\widetilde{S}\left(t\right) \overset{\circ}{=} \sup_{s \leq t}\left|w\left(s\right)\right| = \max_{s \leq t}\left|w\left(s\right)\right|,$$

$$\widetilde{\tau}_a \overset{\circ}{=} \inf\left\{t : \left|w\left(t\right)\right| = a\right\}.$$

If $x > 0$, then[74]

$$\mathbf{P}\left(\widetilde{S}\left(t\right) \leq x\right) = \mathbf{P}\left(\max_{s \leq t}\left|xw\left(\frac{s}{x^2}\right)\right| \leq x\right) = \mathbf{P}\left(\max_{s \leq t}\left|w\left(\frac{s}{x^2}\right)\right| \leq 1\right) =$$

$$= \mathbf{P}\left(\max_{s \leq t/x^2}\left|w\left(s\right)\right| \leq 1\right) = \mathbf{P}\left(\widetilde{\tau}_1 \geq \frac{t}{x^2}\right) =$$

$$= \mathbf{P}\left(\frac{1}{\sqrt{\widetilde{\tau}_1}} \leq \frac{x}{\sqrt{t}}\right).$$

If $\sigma > 0$, then

$$\sqrt{\frac{2}{\pi}} \int_0^\infty \exp\left(-\frac{x^2}{2\sigma^2}\right) dx = \sigma.$$

[74]Recall that $s \mapsto xw\left(s/x^2\right)$ is also a Wiener process.

The expected value depends only on the distribution, so by Fubini's theorem and by (1.57)

$$\mathbf{E}\left(\tilde{S}(1)\right) = \mathbf{E}\left(\frac{1}{\sqrt{\tilde{\tau}_1}}\right) = \mathbf{E}\left(\sqrt{\frac{2}{\pi}}\int_0^\infty \exp\left(-\frac{x^2\tilde{\tau}_1}{2}\right)dx\right) =$$

$$= \sqrt{\frac{2}{\pi}}\int_0^\infty \mathbf{E}\left(\exp\left(-\frac{x^2\tilde{\tau}_1}{2}\right)\right)dx =$$

$$= \sqrt{\frac{2}{\pi}}\int_0^\infty \frac{1}{\cosh x}dx = 2\sqrt{\frac{2}{\pi}}\int_0^\infty \frac{\exp(x)}{\exp(2x)+1}dx =$$

$$= 2\sqrt{\frac{2}{\pi}}\int_1^\infty \frac{1}{y^2+1}dy = 2\sqrt{\frac{2}{\pi}}\cdot\frac{\pi}{4} = \sqrt{\frac{\pi}{2}}.$$

In a similar way, if S denotes the supremum of w then

$$\mathbf{E}\left(S(1)\right) = \mathbf{E}\left(\frac{1}{\sqrt{\tau_1}}\right) = \mathbf{E}\left(\sqrt{\frac{2}{\pi}}\int_0^\infty \exp\left(-\frac{x^2\tau_1}{2}\right)dx\right) =$$

$$= \sqrt{\frac{2}{\pi}}\int_0^\infty \mathbf{E}\left(\exp\left(-\frac{x^2\tau_1}{2}\right)\right)dx =$$

$$= \sqrt{\frac{2}{\pi}}\int_0^\infty \exp(-x)\,dx = \sqrt{\frac{2}{\pi}}.$$

One can prove the last relation with (1.61) as well:

$$\mathbf{E}\left(S(1)\right) = \mathbf{E}\left(|w(1)|\right) = \sqrt{\frac{2}{\pi}}\int_0^\infty x\exp\left(-\frac{x^2}{2}\right)dx = \sqrt{\frac{2}{\pi}} \qquad \square$$

Example 1.125 The intersection of a two-dimensional Wiener process with a line has Cauchy distribution.

Let w_1 and w_2 be independent Wiener processes, and let us consider the line[75] $L \stackrel{\circ}{=} \{x = a\}$ where $a > 0$. The two-dimensional process $\mathbf{w}(t) \stackrel{\circ}{=} (w_1(t), w_2(t))$ meets L the first time at

$$\tau_a \stackrel{\circ}{=} \inf\{t : w_1(t) = a\}.$$

---

[75]The Wiener processes are invariant under rotation so the result is true for an arbitrary line. One can generalize the result to an arbitrary dimension. In the general case, we are investigating the distribution of the intersection of the Wiener processes with hyperplanes.

What is the distribution of the y coordinate that is what is the distribution of $w_2 (\tau_a)$?

1. For an arbitrary u the process $t \mapsto u^{-1} w_1 (u^2 t)$ is also a Wiener process, hence the distribution of its maximum process is the same as the distribution of the maximum process of w_1. Let us denote this maximum process by S_1.

$$\mathbf{P} (\tau_a \geq x) = \mathbf{P} (S_1 (x) \leq a) = \mathbf{P} \left(\sqrt{x} S_1 \left(\frac{1}{(\sqrt{x})^2} x \right) \leq a \right) =$$

$$= \mathbf{P} (\sqrt{x} S_1 (1) \leq a) = \mathbf{P} \left(\frac{a^2}{S_1^2 (1)} \geq x \right).$$

$\mathbf{w}$ intersects L at $w_2 (\tau_a)$. τ_a is $\sigma (w_1)$-measurable, and as w_1 and w_2 are independent, that is the σ-algebras $\sigma (w_2)$ and $\sigma (w_1)$ are independent, τ_a is independent of w_2. We show that

$$w_2 (\tau_a) \cong \sqrt{\tau_a} \cdot w_2 (1) \tag{1.62}$$

that is, the distribution of $w_2 (\tau_a)$ is the same as the distribution of $\sqrt{\tau_a} \cdot w_2 (1)$. Using the independence of τ_a and w_2

$$\mathbf{P} (w_2 (\tau_a) \leq x \mid \tau_a = t) = \mathbf{P} (w_2 (t) \leq x) = \mathbf{P} \left(\sqrt{t} w_2 (1) \leq x \right),$$

and

$$\mathbf{P} (\sqrt{\tau_a} w_2 (1) \leq x \mid \tau_a = t) = \mathbf{P} \left(\sqrt{t} w_2 (1) \leq x \right).$$

Integrating both equations by the distribution of τ_a we get (1.62). Hence

$$w_2 (\tau_a) \cong \sqrt{\tau_a} \cdot w_2 (1) \cong \frac{a}{S_1} \cdot w_2 (1) \cong \frac{a}{|w_1 (1)|} \cdot w_2 (1).$$

$w_1 (1)$ and $w_2 (1)$ are independent with distribution $N (0, 1)$. Therefore $w_2 (\tau_a)$ has a Cauchy distribution.

2. One can also prove the relation with Fourier transforms. Let us calculate the Fourier transform of $w_2(\tau_a)$! The Fourier transform of $N (0, 1)$ is $\exp (-t^2/2)$. By the independence of τ_a and w_2 and by (1.56)

$$\varphi (t) \overset{\circ}{=} \mathbf{E} (\exp (itw_2 (\tau_a))) =$$

$$= \int_0^\infty \mathbf{E} (\exp (itw_2 (\tau_a)) \mid \tau_a = u) \, dG (u) =$$

$$= \int_0^\infty \mathbf{E} (\exp (itw_2 (u))) \, dG (u) =$$

$$= \int_0^\infty \exp\left(-\frac{t^2}{2}u\right) dG(u) =$$

$$= \mathbf{E}\left(\exp\left(-\frac{t^2}{2}\tau_a\right)\right) \stackrel{\circ}{=} L\left(\frac{t^2}{2}\right) =$$

$$= \exp\left(-a\sqrt{t^2}\right) = \exp\left(-a\,|t|\right),$$

which is the Fourier transform of a Cauchy distribution. $\qquad\qquad\Box$

Example 1.126 The process of first passage times of Wiener processes.

Let w be a Wiener process and let us define the hitting times

$$\tau_a \stackrel{\circ}{=} \inf\{t : w(t) = a\}, \quad \sigma_a \stackrel{\circ}{=} \inf\{t : w(t) > a\}.$$

w is continuous, the set $\{x > a\}$ is open, hence σ_a is a weak stopping time. As the augmented filtration of w is right-continuous σ_a is a stopping time[76]. w has continuous trajectories so obviously $\tau_a \le \sigma_a$. As the trajectories of w can contain 'peaks and flat segments' it can happen that for some outcomes τ_a is strictly smaller than σ_a. As we shall immediately see almost surely $\tau_a = \sigma_a$. One can define the stochastic processes

$$T(a, \omega) \stackrel{\circ}{=} \tau_a(\omega), \quad S(a, \omega) \stackrel{\circ}{=} \sigma_a(\omega)$$

with $a \in \mathbb{R}_+$. It is easy to see that T and S have strictly increasing trajectories. If $a_n \nearrow a$ then $w(\tau_{a_n}) = a_n \nearrow a$, hence obviously $\tau_{a_n} \nearrow \tau_a$, so T is left-continuous. On the other hand, it is easy to see that if $a_n \searrow a$, then $\sigma_{a_n} \searrow \sigma_a$, hence S is right-continuous. It is also easy to see, that $T(a+, \omega) = S(a, \omega)$ and $S(a-, \omega) = T(a, \omega)$ for all ω. Obviously τ_a and σ_a are almost surely finite. By the strong Markov property of w

$$w^*(t) \stackrel{\circ}{=} w(\tau_a + t) - w(\tau_a)$$

is also a Wiener process. $\{\tau_a < \sigma_a\}$ is in the set

$$\{w^*(t) \le 0 \text{ on some interval } [0, r], r \in \mathbb{Q}\}.$$

As w^* is a Wiener process it is not difficult to prove[77] that if $r > 0$ then

$$\mathbf{P}\left(w^*(t) \le 0, \forall t \in [0, r]\right) = 0.$$

---

[76]See: Example 1.32, page 17.
[77]See: Corollary B.12, page 566.

Hence

$$\mathbf{P}\left(\tau_a \neq \sigma_a\right) = \mathbf{P}\left(\tau_a < \sigma_a\right) = 0$$

for every a. Therefore S is a right-continuous modification of T. Obviously if $b > a$ and τ_{b-a}^* is the first passage time of w^* to $b - a$ then $\tau_b - \tau_a = \tau_{b-a}^*$. By the strong Markov property τ_{b-a}^* is independent of $\mathcal{F}_{\tau_a}$. Therefore $T(b) - T(a)$ is independent of $\mathcal{F}_{\tau_a}$. In general, one can easily prove that T and therefore S have independent increments with respect to the filtration $\mathcal{G}_a \triangleq \mathcal{F}_{\tau_a}$. Obviously $S(0) = 0$, hence S is a Lévy process with respect to the filtration $\mathcal{G}$. $\qquad\square$

1.3.9 Some remarks on the usual assumptions

The usual assumptions are crucial conditions of stochastic analysis. Without them very few statements of the theory would hold. The most important objects of stochastic analysis are related to stopping times, as these objects express the timing of events. The main tool of stochastic analysis is measure theory. In measure theory, objects are defined up to measure-zero sets. From a technical point of view, of course it is not a great surprise that we want to guarantee that every random time, which is almost surely equal to a stopping time, should also be a stopping time. The definition of a stopping time is very natural: at time t one can observe only $\tau \wedge t$ so we should assume $\tau \wedge t$ to be $\mathcal{F}_t$-measurable for every t. Hence if τ and τ' are almost surely equal and they differ on a set N, then every subset of N should be also $\mathcal{F}_t$-measurable. This implies that one should add all the measure-zero sets and all their subsets to the filtration[78].

The right-continuity of the filtration is more problematic; it assumes that somehow we can foresee the events of the near future. At first sight is seems natural; in our usual experience we always have some knowledge about the near future. Our basic experience is speed and momentum, and these objects are by definition the derivatives of the trajectories. By definition, differentiability means that the right-derivative is equal to the left-derivative and the left-derivative depends on the past and the present. So in our differentiable world we always know the right-derivative, hence—infinitesimally—we can always see the future. But in stochastic analysis we are interested in objects which are non-differentiable. Recall that for a continuous process the hitting time of a closed set is a stopping time[79]. At the moment that we hit a closed set we know that we are in the set. But what about the hitting times[80] of open sets? We hit an open set at its boundary and when we hit it we are generally still outside the set. Recall that the hitting time of an open set is a stopping time only when the filtration is right-continuous[81]. That is, when we hit the boundary of an open set—by the

---

[78]See: Example 6.37, page 386.
[79]See: Example 1.32, page 17.
[80]See: Definion 1.26, page 15.
[81]See: Example 1.32, page 17.

right-continuity of the filtration—we can ask for some extra information about the future which tells us whether we shall really enter the set or not. This is, of course, a very strong assumption. If we want to go to a restaurant and we are at the door, we know that we shall enter the restaurant. But a Wiener process can easily turn back at the door. One of the most surprising statements of the theory is that the augmented filtration of a Lévy process is right-continuous. This is true not only for Lévy processes, but under more general conditions[82]. It is important to understand the reason behind this phenomena. The probability that a one-dimensional Wiener process hits the boundary of an open set without actually entering the set itself has zero[83] probability! And in general the right-continuity of an augmented filtration means that all the events which need some insight into the future[84] have zero probability. We cannot see the future, we are just ignoring the irrelevant information!

1.4 Localization

Localization is one of the most frequently used concepts of mathematical analysis. For example, if f is a continuous function on $\mathbb{R}$, then of course generally f is not integrable on the whole real line. But this is not a problem at all. We can still talk about the integral function $F(x) \doteq \int_0^x f(t)dt$ of f. The functions of Calculus are not integrable, they are just locally integrable. In the real analysis we say that a certain property holds locally if it holds on every compact subset of the underlying topological space[85]. In the real line it is enough to ask that the property holds on any closed, bounded interval, in particular for any t the property should hold on any interval $[0, t]$. Very often, like in the case of local integrability, it is sufficient to ask that the property should hold on some intervals $[0, t_n]$ where $t_n \nearrow \infty$. In stochastic analysis we should choose the upper bounds t_n in a measurable way with respect to the underlying filtration. This explains the next definition:

Definition 1.127 *Let $\mathcal{X}$ be a family of processes. We say that process X is locally in $\mathcal{X}$ if there is a sequence of stopping times (τ_n) for which almost surely[86] $\tau_n \nearrow \infty$, and the truncated processes X^{τ_n} belong to $\mathcal{X}$ for every n. The sequence (τ_n) is called the* localizing sequence *of X. $\mathcal{X}_{loc}$ denotes the set of processes locally belonging to $\mathcal{X}$.*

A specific problem of the definition above, is that with localization one cannot modify the value of the variable $X(0)$, since every truncated process X^{τ_n} at the

---

[82]This is true e.g. for so called Feller processes, which form an important subclass of the Markov processes.

[83]See: Example 1.126, page 90, Corollary B.12, page 566. But see: Example 6.10, page 364.

[84]Like sudden jumps of the Poisson processes.

[85]Generally the topological space is locally compact.

[86]Almost surely and not everywhere! See: Proposition 1.130, page 94.

time $t = 0$ has the same value $X(0)$. To overcome this problem some authors[87] instead of using X^{τ_n} use the process $X^{\tau_n}\chi(\tau_n > 0)$ in the definition of the localization or instead of X they localize the process $X - X(0)$. In most cases it does not matter how we define the localization. First of all we shall use the localization procedure to define the different classes of local martingales. From the point of view of stochastic analysis, one can always assume that every local martingale is zero at time $t = 0$, as our final goal is to investigate the class of semimartingale, and the semimartingales have the representation $X(0) + L + V$, where L is a local martingale, zero at time $t = 0$. Just to fix the ideas we shall later explicitly concretize the definitions in the cases of local martingales and locally bounded processes. In both cases we localize the processes $X - X(0)$.

1.4.1 Stability under truncation

It is quite natural to ask for which type of processes $\mathcal{X}$ one has $(\mathcal{X}_{\text{loc}})_{\text{loc}} = \mathcal{X}_{\text{loc}}$.

Definition 1.128 *We say that space of processes $\mathcal{X}$ is closed or stable under truncation or closed under stopping if whenever $X \in \mathcal{X}$ then $X^\tau \in \mathcal{X}$ for arbitrary stopping time τ.*

It is an important consequence of this property that if $\mathcal{X}$ is closed under truncation and $X_k \in \mathcal{X}_{\text{loc}}$ and $\left(\tau_n^{(k)}\right)$ are the localizing sequences of the processes X_k, then $\tau_n \overset{\circ}{=} \wedge_{k=1}^m \tau_n^{(k)}$ for any finite m is a common localizing sequence of the first m processes. That is, if $\mathcal{X}$ is closed under the truncation, then for a finite number of processes we can always assume that they have a common localizing sequence. From the definition it is clear that if $\mathcal{X}$ is closed under the truncation, then $\mathcal{X}_{\text{loc}}$ is also closed under the truncation as, if (τ_n) is a localizing sequence of X and τ is an arbitrary stopping time, then (τ_n) is obviously a localizing sequence of the truncated process X^τ.

Example 1.129 $\mathcal{M}$, the space of uniformly integrable martingales, $\mathcal{H}^2$, the space of the square-integrable martingales and $\mathcal{K}$, the set of bounded processes are closed under truncation.

It is obvious from the definition that $\mathcal{K}$ is closed under truncation. By the Optional Sampling Theorem if $M \in \mathcal{M}$, then $M^\tau \in \mathcal{M}$. As

$$\mathbf{E}\left(\left(\sup_{t \geq 0} |X^\tau(t)|\right)^2\right) \leq \mathbf{E}\left(\left(\sup_{t \geq 0} |X(t)|\right)^2\right) < \infty$$

$\mathcal{H}^2$ is also closed under truncation. $\qquad\square$

---

[87]See: e.g. [78].

Proposition 1.130 *If the filtration $\mathcal{F}$ is right-continuous and if $\mathcal{X}$ is closed under truncation, then $(\mathcal{X}_{loc})_{loc} = \mathcal{X}_{loc}$.*

Proof. $\tau_n \equiv \infty$ is a localizing sequence, therefore $\mathcal{X}_{\text{loc}} \subseteq (\mathcal{X}_{\text{loc}})_{\text{loc}}$. Let $X \in (\mathcal{X}_{\text{loc}})_{\text{loc}}$ and let (τ_n) be a localizing sequence with $X^{\tau_n} \in \mathcal{X}_{\text{loc}}$. As $\mathcal{X}_{\text{loc}}$ is closed under truncation $X^{\tau_n \wedge n} \in \mathcal{X}_{\text{loc}}$ for an arbitrary n, so one can assume that τ_n is bounded. Let σ_k be such a stopping time that $(X^{\tau_k})^{\sigma_k} = X^{\tau_k \wedge \sigma_k} \in \mathcal{X}$, and $\mathbf{P}(\sigma_k \leq \tau_k) \leq 2^{-k}$. Let us define the sequence

$$\rho_n \overset{\circ}{=} \tau_n \wedge \inf_{k \geq n} \sigma_k.$$

Obviously

$$\left\{ \inf_{k \geq n} \sigma_k < t \right\} = \cup_{k \geq n} \{\sigma_k < t\} \in \mathcal{F}_t$$

so $\inf_{k \geq n} \sigma_k$ is a weak stopping time. The filtration $\mathcal{F}$ is right-continuous, hence $\inf_{k \geq n} \sigma_k$ is a stopping time and therefore ρ_n is a stopping time. The sequence (ρ_n) is increasing and

$$\mathbf{P}(\rho_n < \tau_n) \leq \sum_{k=n}^{\infty} \mathbf{P}(\sigma_k < \tau_n) \leq \sum_{k=n}^{\infty} \mathbf{P}(\sigma_k \leq \tau_k)$$

$$\leq \sum_{k=n}^{\infty} 2^{-k} = 2^{-n+1} \to 0.$$

Almost surely $\tau_n \nearrow \infty$ and therefore outside of a measure-zero set $\rho_n \nearrow \infty$, that is (ρ_n) is a localizing sequence[88] of $X \in (\mathcal{X}_{\text{loc}})_{\text{loc}}$ so $X \in \mathcal{X}_{\text{loc}}$. $\square$

1.4.2 Local martingales

Let us specify the definition of localization to local martingales.

Definition 1.131 *Let us denote by $\mathcal{M}$ the set of uniformly integrable martingales. Process X is a local martingale if $X - X(0) \in \mathcal{M}_{loc}$, that is a process X is a local martingale if there is a sequence of stopping times (τ_n), $\tau_n \overset{a.s.}{\nearrow} \infty$ such that $X^{\tau_n} - X(0) \in \mathcal{M}$ for all n. Let us denote[89] by $\mathcal{L}$ the set of local martingales which are zero at time $t = 0$. X is a local martingale, if there is an $L \in \mathcal{L}$ such that $X = X(0) + L$.*

---

[88] Let us observe that (ρ_n) converges to infinity just almost surely.
[89] See: Definition 3.1, page 179.

Example 1.132 Every martingale is a local martingale.

If M is a martingale, then $M(0)$ is integrable. The process $X \equiv M(0)$ is a martingale, so without loss of generality one can assume that $M(0) = 0$. If M is a martingale and $\tau_n \stackrel{\circ}{=} n$, then (τ_n) is a localizing sequence and

$$M^{\tau_n}(t) = M(t \wedge n) = \mathbf{F}(M(n) \mid \mathcal{F}_t),$$

therefore the set $(M^{\tau_n}(t))_{t \geq 0}$ is uniformly integrable[90] for all n. □

Example 1.133 Local martingale which is not a martingale.

Let $(\Omega, \mathcal{A}, \mathbf{P}) \stackrel{\circ}{=} ([0,1], \mathcal{B}([0,1]), \lambda)$, where λ denotes the Lebesgue measure. If $\mathcal{F}_t \stackrel{\circ}{=} \mathcal{A}$ and ξ is not integrable, then the constant process $X(t) \stackrel{\circ}{=} \xi$ is not a martingale, but as $X - X(0) \equiv 0$ by the definition of local martingales X is a local martingale. □

Proposition 1.134 *If X and Y are local martingales and ξ and η are $\mathcal{F}_0$-measurable random variables then $Z \stackrel{\circ}{=} \xi X + \eta Y$ is also a local martingale.*

Proof. If $L \in \mathcal{L}$ and ζ is an arbitrary $\mathcal{F}_0$-measurable variable then $\zeta + L$ is a local martingale, therefore one can assume that $X, Y \in \mathcal{L}$. Let (τ_n) be the localizing sequence of X and (σ_n) be the localizing sequence of Y. As ξ and η are $\mathcal{F}_0$-measurable the variables

$$\alpha_n \stackrel{\circ}{=} \begin{cases} \infty & \text{if } |\xi| \leq n \\ 0 & \text{if } |\xi| > n \end{cases}, \quad \beta_n \stackrel{\circ}{=} \begin{cases} \infty & \text{if } |\eta| \leq n \\ 0 & \text{if } |\eta| > n \end{cases}$$

are stopping times. Obviously

$$\rho_n \stackrel{\circ}{=} \tau_n \wedge \sigma_n \wedge \alpha_n \wedge \beta_n$$

is a stopping time and $\rho_n \nearrow \infty$ so (ρ_n) is a localizing sequence.

$$Z^{\rho_n} \stackrel{\circ}{=} (\xi X + \eta Y)^{\rho_n} = \chi(|\xi| \leq n)\xi X^{\rho_n} + \chi(|\eta| \leq n)\eta Y^{\rho_n}. \qquad (1.63)$$

As $X^{\rho_n}, Y^{\rho_n} \in \mathcal{M}$ and as $\chi(|\xi| \leq n)\xi$ and $\chi(|\eta| \leq n)\eta$ are bounded $\mathcal{F}_0$-measurable variables, obviously $Z^{\rho_n} \in \mathcal{M}$ and therefore Z is a local martingale. Let us observe that in line (1.63) we used that $X, Y \in \mathcal{L}$ that is $X(0) = Y(0) = 0$. If in the definition of local martingales one had used the simpler $X \in \mathcal{M}_{\text{loc}}$ definition, then in this proposition one should have assumed the ξ and η to be bounded. □

---

[90]See: Lemma 1.70, page 42.

One can observe that in the definition of local martingales we used the class of uniformly integrable martingales and not the class of martingales. If L^{τ_n} is a martingale for some τ_n, then $L^{\tau_n \wedge n} \in \mathcal{M}$, so the class of local martingales is the same as the class of 'locally uniformly integrable martingales'. Very often we prove different theorems first for uniformly integrable martingales and then with localization we extend the proofs to local martingales. In most cases one should use the same method if one wants to extend the result from uniformly integrable martingales just to martingales.

An important subclass of local martingales is the space of locally square-integrable martingales:

Definition 1.135 *X is a locally square-integrable martingale if $X - X(0) \in \mathcal{H}_{loc}^2$.*

Example 1.136 Every martingale which has square-integrable values is a locally square-integrable martingale.

By definition a martingale X is square-integrable in ω if $X(t) \in L^2(\Omega)$ for every t. In this case $X(0) \in L^2(\Omega)$, therefore for all t $X(t) - X(0) \in L^2(\Omega)$, so again one can assume that $X(0) = 0$. If $\tau_n \stackrel{\circ}{=} n$ then (τ_n) is a localizing sequence. By Doob's inequality

$$\left\| \sup_t |X^{\tau_n}(t)| \right\|_2 = \left\| \sup_{t \leq n} |X(t)| \right\|_2 \leq 2 \cdot \|X(n)\|_2 < \infty,$$

so $X^{\tau_n} \in \mathcal{H}^2$ and therefore $X \in \mathcal{H}_{loc}^2$. □

Example 1.137 Every continuous local martingale is locally square-integrable[91].

Let X be a continuous local martingale and let (τ_n) be a localizing sequence of X. As X is continuous

$$\sigma_n \stackrel{\circ}{=} \inf \{t : |X(t)| \geq n\}$$

is a stopping time. If $\rho_n \stackrel{\circ}{=} \tau_n \wedge \sigma_n$ then $\rho_n \nearrow \infty$ and $|X^{\rho_n}| \leq n$ by the continuity of X, so X^{ρ_n} is a bounded, hence it is a square-integrable martingale. Therefore $M \in \mathcal{H}_{loc}^2$. □

Example 1.138 Martingales which are not in $\mathcal{H}_{loc}^2$.

---

[91]One can easily generalize this example. If the jumps of X are bounded then X is in $\mathcal{H}_{loc}^2$. See: Proposition 1.152, page 107.

Let us denote by $\sigma(\mathcal{N})$ the σ-algebra generated by the measure-zero sets. Let

$$\mathcal{F}_t \stackrel{\circ}{=} \begin{cases} \sigma(\mathcal{N}) & \text{if } t < 1 \\ \mathcal{A} & \text{if } t \geq 1 \end{cases},$$

and let $\xi \in L^1(\Omega)$, but $\xi \notin L^2(\Omega)$. Let us also assume that $\mathbf{E}(\xi) = 0$. $\mathcal{F}$ satisfies the usual conditions, hence $X(t) \stackrel{\circ}{=} \mathbf{E}(\xi \mid \mathcal{F}_t)$ is martingale. $X(0) \stackrel{a.s.}{=} 0$, hence not only $X \in \mathcal{M}_{\mathrm{loc}}$, but also $X \in \mathcal{L}$. On the other hand $X \notin \mathcal{H}^2_{\mathrm{loc}}$ as, if the stopping time τ is not almost surely constant, then almost surely $\tau \geq 1$, hence for all $t \geq 1$ $X^\tau(t) = \xi \notin L^2(\Omega)$. $\square$

It is a quite natural, but wrong, guess that local martingales are badly integrable martingales. The local martingales are far more mysterious objects.

Example 1.139 Integrable local martingale which is not a martingale.

Let $\Omega \stackrel{\circ}{=} C[0, \infty)$, that is let Ω be the set of continuous functions defined on the half-line $\mathbb{R}_+$. Let X be the canonical coordinate process, that is if $\omega \in \Omega$, then let $X(t, \omega) \stackrel{\circ}{=} \omega(t)$, and let the filtration $\mathcal{F}$ be the filtration generated by X. Let $\mathbf{P}$ be the probability measure defined on Ω for which X is a Wiener process starting from point 1. Let

$$\tau_0 \stackrel{\circ}{=} \inf\{t : X(t) = 0\}.$$

Let us define the measure $\mathbf{Q}(t)$ on the σ-algebra $\mathcal{F}_t$ with the Radon–Nikodym derivative

$$\frac{d\mathbf{Q}(t)}{d\mathbf{P}} \stackrel{\circ}{=} X(t \wedge \tau_0) = X(t)\chi(t < \tau_0) + X(\tau_0)\chi(t \geq \tau_0) =$$
$$= X(t)\chi(t < \tau_0).$$

As the truncated martingales are martingales, X^{τ_0} is a martingale under the measure $\mathbf{P}$. Hence

$$\mathbf{E}(X(t \wedge \tau_0) \mid \mathcal{F}_s) = X(s \wedge \tau_0).$$

The measures $(\mathbf{Q}(t))_{t \geq 0}$ are consistent: if $s < t$ and $F \in \mathcal{F}_s \subseteq \mathcal{F}_t$, then

$$\mathbf{Q}(s)(F) = \int_F \frac{d\mathbf{Q}(s)}{d\mathbf{P}} d\mathbf{P} \stackrel{\circ}{=} \int_F X(s \wedge \tau_0) d\mathbf{P} = \int_F X(t \wedge \tau_0) d\mathbf{P} \stackrel{\circ}{=}$$
$$\stackrel{\circ}{=} \int_F \frac{d\mathbf{Q}(t)}{d\mathbf{P}} d\mathbf{P} = \mathbf{Q}(t)(F).$$

In particular

$$\mathbf{Q}\,(t)\,(\Omega) \doteq \int_{\Omega} X\,(t \wedge \tau_0)\,d\mathbf{P} = \int_{\Omega} X\,(0)\,d\mathbf{P} = 1,$$

so $\mathbf{Q}\,(t)$ is a probability measure for every t. The space $C\,[0, \infty)$ is a Kolmogorov type measure space, so on the Borel sets of $C\,[0, \infty)$ there is a probability measure $\mathbf{Q}$, which, restricted to $\mathcal{F}_t$ is $\mathbf{Q}\,(t)$. $\{\tau_0 \le t\} \in \mathcal{F}_t$ for every t so

$$\mathbf{Q}\,(\tau_0 \le t) = \mathbf{Q}\,(t)\,(\tau_0 \le t) \doteq \int_{\Omega} \chi\,(\tau_0 \le t)\,X\,(\tau_0 \wedge t)\,d\mathbf{P} =$$

$$= \int_{\Omega} \chi\,(\tau_0 \le t)\,X\,(\tau_0)\,d\mathbf{P} = 0,$$

so $\mathbf{Q}\,(\tau_0 = \infty) = 1$, that is X is almost surely never zero under $\mathbf{Q}$. Hence $X \overset{a.s.}{>} 0$ under $\mathbf{Q}$, so under $\mathbf{Q}$ the process $Y \doteq 1/X$ is almost surely well-defined.

1. As a first step let us show that Y is not a martingale under $\mathbf{Q}$. To show this it is sufficient to prove that the $\mathbf{Q}$-expected value of Y is decreasing to zero. As $\mathbf{P}(\tau_0 < \infty)$ if $t \nearrow \infty$

$$\mathbf{E}^{\mathbf{Q}}(Y\,(t)) \doteq \int_{\Omega} Y\,(t)\,d\mathbf{Q} = \int_{\Omega} \frac{1}{X\,(t)}\,d\mathbf{Q}\,(t) =$$

$$= \int_{\Omega} \frac{1}{X\,(t)} \chi\,(t < \tau_0)\,X\,(t)\,d\mathbf{P} =$$

$$= \int_{\Omega} \chi\,(t < \tau_0)\,d\mathbf{P} = \mathbf{P}\,(t < \tau_0) \to 0.$$

2. Now we prove that Y is a local martingale under $\mathbf{Q}$. Let $\varepsilon > 0$ and let

$$\tau_\varepsilon \doteq \inf\,\{t : X(t) = \varepsilon\}\,.$$

X is continuous, therefore if $\varepsilon \searrow 0$ then $\tau_\varepsilon\,(\omega) \nearrow \tau_0\,(\omega)$ for every outcome ω. Since $\mathbf{Q}(\tau_0 = \infty) = 1$ obviously $\mathbf{Q}$-almost surely[92] $\tau_\varepsilon \nearrow \infty$. Let us show, that under $\mathbf{Q}$ the truncated process Y^{τ_ε} is a martingale. Almost surely $0 < Y^{\tau_\varepsilon} \le 1/\varepsilon$ hence Y^{τ_ε} is almost surely bounded, hence it is uniformly integrable. One should

---

[92]Let us recall that by the definition of the localizing sequence, it is sufficient if the localizing sequence converges just almost surely to infinity.

only prove that Y^{τ_ε} is a martingale under $\mathbf{Q}$. If $s < t$ and $F \in \mathcal{F}_s$, then as $\tau_\varepsilon < \tau_0$

$$\int_F Y^{\tau_\varepsilon}(t)\, d\mathbf{Q} \stackrel{\circ}{=} \int_F \frac{1}{X(t \wedge \tau_\varepsilon)}\, d\mathbf{Q}(t) = \tag{1.64}$$

$$= \int_F \frac{1}{X(t \wedge \tau_\varepsilon)} X(t \wedge \tau_0)\, d\mathbf{P} =$$

$$= \int_F \left(\frac{\chi(t < \tau_\varepsilon)}{X(t)} + \frac{\chi(t \geq \tau_\varepsilon)}{X(\tau_\varepsilon)} \right) X(t)\, \chi(t < \tau_0)\, d\mathbf{P} =$$

$$= \int_F \chi(t < \tau_\varepsilon) + \frac{X(t)}{\varepsilon} \chi(\tau_0 > t \geq \tau_\varepsilon)\, d\mathbf{P} =$$

$$= \frac{1}{\varepsilon} \int_F \varepsilon + (X^{\tau_0}(t) - \varepsilon)\, \chi(t \geq \tau_\varepsilon)\, d\mathbf{P}.$$

Let us prove that

$$M(t) \stackrel{\circ}{=} (X^{\tau_0}(t) - \varepsilon)\, \chi(t \geq \tau_\varepsilon)$$

is a martingale under $\mathbf{P}$. If σ is a bounded stopping time, then as $\tau_\varepsilon < \tau_0$ by the elementary properties of the conditional expectation[93] and by the Optional Sampling Theorem

$$\mathbf{E}(M(\sigma)) \stackrel{\circ}{=} \mathbf{E}((X^{\tau_0}(\sigma) - \varepsilon)\, \chi(\sigma \geq \tau_\varepsilon)) =$$

$$= \mathbf{E}(\mathbf{E}((X^{\tau_0}(\sigma) - \varepsilon)\, \chi(\sigma \geq \tau_\varepsilon) \mid \mathcal{F}_{\sigma \wedge \tau_\varepsilon})) =$$

$$= \mathbf{E}(\mathbf{E}(X^{\tau_0}(\sigma) - \varepsilon \mid \mathcal{F}_{\sigma \wedge \tau_\varepsilon})\, \chi(\sigma \geq \tau_\varepsilon)) =$$

$$= \mathbf{E}(\mathbf{E}(X(\tau_0 \wedge \sigma) - \varepsilon \mid \mathcal{F}_{\sigma \wedge \tau_\varepsilon})\, \chi(\sigma \geq \tau_\varepsilon)) =$$

$$= \mathbf{E}((X(\sigma \wedge \tau_\varepsilon) - \varepsilon)\, \chi(\sigma \geq \tau_\varepsilon)) =$$

$$= \mathbf{E}((X(\tau_\varepsilon) - \varepsilon)\, \chi(\sigma \geq \tau_\varepsilon)) = 0,$$

which means that M is really a martingale[94]. As M is a martingale under $\mathbf{P}$ in the last integral of (1.64) one can substitute s on the place of t, so calculating backwards

$$\int_F Y^{\tau_\varepsilon}(t)\, d\mathbf{Q} \stackrel{\circ}{=} \int_F \frac{1}{X(t \wedge \tau_\varepsilon)}\, d\mathbf{Q} = \int_F \frac{1}{X(s \wedge \tau_\varepsilon)}\, d\mathbf{Q} \stackrel{\circ}{=}$$

$$\stackrel{\circ}{=} \int_F Y^{\tau_\varepsilon}(s)\, d\mathbf{Q},$$

that is Y^{τ_ε} is a martingale under $\mathbf{Q}$. Therefore $(\tau_{1/n})$ localizes Y under $\mathbf{Q}$. $\square$

---

[93]See: Proposition 1.34, page 20.
[94]See: Proposition 1.91, page 57.

Example 1.140 $L^2(\Omega)$ bounded local martingale, which is not a martingale [95].

Let $\mathbf{w}$ be a standard Wiener process in $\mathbb{R}^3$, and let $\mathbf{X}(t) \overset{\circ}{=} \mathbf{w}(t)+\mathbf{u}$ where $\mathbf{u} \neq \mathbf{0}$ is a fixed vector. By the elementary properties of Wiener processes[96] if $t \to \infty$ then

$$R(t) \overset{\circ}{=} \|\mathbf{X}(t)\|_2 \to \infty. \tag{1.65}$$

With direct calculation it is easy to check that on $\mathbb{R}^3 \setminus \{0\}$ the function

$$g(\mathbf{x}) \overset{\circ}{=} \frac{1}{\|\mathbf{x}\|_2} = \frac{1}{\sqrt{x_1^2 + x_2^2 + x_3^2}}$$

is harmonic, that is[97]

$$\Delta g \overset{\circ}{=} \frac{\partial^2}{\partial x_1^2} g + \frac{\partial^2}{\partial x_2^2} g + \frac{\partial^2}{\partial x_3^2} g = 0.$$

Hence by Itô's formula[98] $M \overset{\circ}{=} 1/R$ is a local martingale. The density function of the $\mathbf{X}(t)$ is

$$f_t(\mathbf{x}) \overset{\circ}{=} \frac{1}{\left(\sqrt{2\pi t}\right)^3} \exp\left(-\frac{1}{2t} \|\mathbf{x} - \mathbf{u}\|_2^2\right).$$

If $t \geq 1$ then f_t is uniformly bounded so if $t \geq 1$ then obviously

$$\mathbf{E}\left(M^2(t)\right) = \int_{\mathbb{R}^3} \frac{1}{\|\mathbf{x}\|_2^2} f_t(\mathbf{x}) \, d\lambda_3(\mathbf{x}) \leq$$

$$\leq \int_{\mathbb{R}^3} \frac{1}{\|\mathbf{x}\|_2^2} d\lambda_3(\mathbf{x}).$$

Evidently the last integral can diverge only around $\mathbf{x} = \mathbf{0}$.

$$I \overset{\circ}{=} \int_{\|\mathbf{x}\| \leq 1} \frac{1}{\|\mathbf{x}\|_2^2} d\lambda_3(\mathbf{x}) = \sum_k \int_{G(k)} \frac{1}{\|\mathbf{x}\|_2^2} d\lambda_3(\mathbf{x})$$

---

[95] We shall use several results which we shall prove later, so one can skip this example during the first reading.

[96] See: Proposition B.7, page 564, Corollary 6.9, page 363.

[97] Now Δ denotes the Laplace operator.

[98] See: Theorem 6.2, page 353. As $n = 3$ almost surely $\mathbf{X}(t) \neq 0$ hence we can use the formula. See: Theorem 6.7, page 359.

where

$$G\left(k\right) = \left\{\frac{1}{2^{k+1}} < \|\mathbf{x}\|_2 \leq \frac{1}{2^k}\right\}.$$

As $2^k G\left(k\right) = G\left(0\right)$ using the transformation $T\left(\mathbf{x}\right) \overset{\circ}{=} 2^k \mathbf{x}$

$$\int_{G(0)} \frac{1}{\|\mathbf{x}\|_2^2} d\lambda_3\left(\mathbf{x}\right) = \int_{G(k)} \frac{1}{\|2^k\mathbf{x}\|_2^2} 2^{3k} d\lambda\left(\mathbf{x}\right) =$$

$$= 2^k \int_{G(k)} \frac{1}{\|\mathbf{x}\|_2^2} d\lambda_3\left(\mathbf{x}\right).$$

Hence

$$I = \sum_{k=0}^{\infty} 2^{-k} \int_{G(0)} \frac{1}{\|\mathbf{x}\|_2^2} d\lambda_3\left(\mathbf{x}\right) < \infty.$$

It is easy to show that $\mathbf{E}\left(M^2\left(t\right)\right)$ is continuous in t. Therefore it is bounded on $[0,t]$. Hence $\mathbf{E}\left(M^2\left(t\right)\right)$ is bounded on $\mathbb{R}_+$. By (1.65) $M(t) \to 0$. M is bounded in $L^2(\Omega)$ therefore it is uniformly integrable, so $M(t) \overset{L^1}{\to} 0$. If M were a martingale then

$$0 \neq M\left(t\right) = \mathbf{E}\left(M\left(\infty\right) \mid \mathcal{F}_t\right) = \mathbf{E}\left(0 \mid \mathcal{F}_t\right) = 0,$$

which is impossible. $\square$

As the uniformly integrable local martingales are not necessarily martingales even the next, nearly trivial observation is very useful:

Proposition 1.141 *Every non-negative local martingale is a supermartingale.*

Proof: Let $M = M(0) + L$ be a non-negative local martingale. Observe that by the definition of supermartingales, $M(t) \geq 0$ is not necessarily integrable so one cannot assume that $M(0)$ is integrable. As $L \in \mathcal{L}$ there is a localizing sequence (τ_n) that $L^{\rho_n} \in \mathcal{M}$ for all n. If $t > s$, then as $M \geq 0$, by Fatou's lemma

$$\mathbf{E}\left(M\left(t\right) \mid \mathcal{F}_s\right) = \mathbf{E}\left(\liminf_{n\to\infty} M^{\tau_n}\left(t\right) \mid \mathcal{F}_s\right) \leq \liminf_{n\to\infty} \mathbf{E}\left(M^{\tau_n}\left(t\right) \mid \mathcal{F}_s\right) =$$

$$= M(0) + \liminf_{n\to\infty} \mathbf{E}\left(L^{\tau_n}\left(t\right) \mid \mathcal{F}_s\right) =$$

$$= M(0) + \liminf_{n\to\infty} L^{\tau_n}\left(s\right) = M\left(s\right). \quad \square$$

Corollary 1.142 *If $M \in \mathcal{L}$ and $M \geq 0$ then $M = 0$.*

Proof: As M is a supermartingale $0 \leq \mathbf{E}\left(M(t)\right) \leq \mathbf{E}\left(M(0)\right) = 0$ for all $t \geq 0$, so $M(t) \overset{a.s.}{=} 0$. $\square$

The most striking and puzzling feature of local martingales is that even uniform integrability is not sufficient to guarantee that local martingales are proper martingales. The reason for it is the following: If Γ is a set of stopping times, then the uniform integrability of the family $(X(t))_{t \in \Theta}$ does not guarantee the uniform integrability of the stopped family $(X(\tau))_{\tau \in \Gamma}$. This cannot happen if the local martingale belongs to the so-called class $\mathcal{D}$.

Definition 1.143 *Process X belongs to the Dirichlet–Doob class[99], shortly X is in class $\mathcal{D}$, if the set*

$$\{X(\tau) : \tau < \infty \text{ is an arbitrary finite-valued stopping time}\}$$

is uniformly integrable. We shall also denote by $\mathcal{D}$ the set of processes in class $\mathcal{D}$.

Proposition 1.144 *Let L be a local martingale. L is in class $\mathcal{D}$ if and only if $L \in \mathcal{M}$ that is if L is a uniformly integrable martingale.*

Proof: Recall that we constructed a non-negative $L^2(\Omega)$-bounded local martingales which is not a proper martingale.

1. Let $L \in \mathcal{D}$ and let L be a local martingale. As $\tau = 0$ is a stopping time, by the definition of $\mathcal{D}$, $L(0)$ is integrable, so one can assume that $L \in \mathcal{L}$. If (τ_n) is a localizing sequence of L then

$$L(\tau_n \wedge s) = L^{\tau_n}(s) = \mathbf{E}\left(L^{\tau_n}(t) \mid \mathcal{F}_s\right) =$$
$$= \mathbf{E}\left(L(\tau_n \wedge t) \mid \mathcal{F}_s\right).$$

$\tau_n \nearrow \infty$, hence the sequences $(L(\tau_n \wedge s))_n$ and $(L(\tau_n \wedge t))_n$ converge to $L(s)$ and $L(t)$. By uniform integrability the convergence $L(\tau_n \wedge t) \to L(t)$ holds in $L^1(\Omega)$ as well. By the L^1-continuity of the conditional expectation

$$L(s) = \mathbf{E}\left(L(t) \mid \mathcal{F}_s\right),$$

hence L is a martingale[100]. Obviously the set $\{L(t)\}_t \subseteq \{L(\tau)\}_\tau$ is uniformly integrable so $L \in \mathcal{M}$.

2. The reverse implication is obvious: If L is a uniformly integrable martingale then by the Optional Sampling Theorem $L(\tau) = \mathbf{E}\left(L(\infty) \mid \mathcal{F}_\tau\right)$ for every stopping time τ, hence the family $(L(\tau))_\tau$ is uniformly integrable[101]. $\square$

---

[99] In [77] on page 244 class $\mathcal{D}$ is called Dirchlet class. [74] on page 107 remarks that class $\mathcal{D}$ is for Doob's class and the definition was introduced by P.A. Meyer in 1963.

[100] Observe that it is enough to asssume that $\{L(\tau)\}_\tau$ is uniformly integrable for the set of bounded stopping times τ.

[101] See: Lemma, 1.70, page 42.

Corollary 1.145 *If a process X is dominated by an integrable variable then $X \in \mathcal{D}$, hence if X is a local martingale and X is dominated by an integrable variable[102] then $X \in \mathcal{M}$.*

Example 1.146 Let us assume that L has independent increments. If $X \overset{\circ}{=} \exp(L)$ then X is a local martingale if and only if X is a martingale.

One should only prove that if X is a local martingale, then X is a martingale. By the definition of processes with independent increments, $L(0) = 0$, hence $X(0) = 1$. X is a non-negative local martingale, so it is a supermartingale[103]. If $m(t)$ denotes the expected value of $X(t)$ then by the supermartingale property $1 \geq m(t) > 0$. Let us prove that $M(t) \overset{\circ}{=} X(t)/m(t)$ is a martingale. As L has independent increments, if $t > s$, then

$$m(t) \overset{\circ}{=} \mathbf{E}(X(t)) = \mathbf{E}(X(s))\mathbf{E}(\exp(L(t) - L(s))) \overset{\circ}{=}$$
$$\overset{\circ}{=} m(s)\mathbf{E}(\exp(L(t) - L(s))).$$

From this

$$\mathbf{E}(M(t) \mid \mathcal{F}_s) \overset{\circ}{=} \mathbf{E}\left(\frac{\exp(L(t))}{m(t)} \mid \mathcal{F}_s\right) =$$
$$= \mathbf{E}\left(\frac{\exp(L(t) - L(s) + L(s))}{m(t)} \mid \mathcal{F}_s\right) =$$
$$= \frac{\exp(L(s))}{m(t)}\mathbf{E}(\exp(L(t) - L(s)) \mid \mathcal{F}_s) =$$
$$= \frac{\exp(L(s))}{m(t)}\mathbf{E}(\exp(L(t) - L(s))) =$$
$$= \frac{\exp(L(s))}{m(s)} \overset{\circ}{=} M(s),$$

hence M is martingale. For arbitrary $T < \infty$ on the interval $[0, T]$ M is uniformly integrable, that is, M is in class $\mathcal{D}$. As on interval $[0, T]$

$$0 \leq X = Mm \leq M,$$

hence X is also in class $\mathcal{D}$. Therefore $X \in \mathcal{D}$ and X is a local martingale on $[0, T]$. This means that X is a martingale on $[0, T]$ for every T, hence X is a martingale on $\mathbb{R}_+$. □

---

[102] See: Davis' inequality. Theorem 4.62, page 277.
[103] See: Proposition 1.141, page 101.

If a process has independent increments and the expected value of the process is zero, then it is obviously a martingale. Therefore martingales are the generalization of random walks. From an intuitive point of view one can also think about local martingales as generalized random walks as we shall later prove the next — somewhat striking— theorem:

Theorem 1.147 *Assume that the stochastic base satisfies the usual conditions. If a local martingale has independent increments then it is a true martingale*[104].

1.4.3 Convergence of local martingales: uniform convergence on compacts in probability

Let $\mathcal{X}$ be an arbitrary space. In $\mathcal{X}_{\mathrm{loc}}$ it is very natural to define the topology with localization; $X_m \to X$, if X and the elements of the sequence (X_m) have a common localizing sequence (τ_n) and for every n in the topology of $\mathcal{X}$

$$\lim_{m \to \infty} X_m^{\tau_n} = X^{\tau_n}.$$

Let us assume[105] that (X_m) and X are in $\mathcal{H}_{\mathrm{loc}}^p$. In $\mathcal{H}^p$ one should define the topology with the norm

$$\|X\|_{\mathcal{H}^p} \doteq \left\| \sup_s |X(s)| \right\|_p.$$

If $\tau_n \nearrow \infty$ and $t < \infty$, then for every $\delta > 0$ one can find an n, that $\mathbf{P}(\tau_n \le t) < \delta$. Let $\varepsilon > 0$ be arbitrary. If

$$A \doteq \left\{ \sup_{s \le t} |X_m(s) - X(s)| > \varepsilon \right\},$$

then

$$\mathbf{P}(A) = \mathbf{P}((\tau_n \le t) \cap A) + \mathbf{P}((\tau_n > t) \cap A) \le$$
$$\le \mathbf{P}(\tau_n \le t) + \mathbf{P}((\tau_n > t) \cap A) \le \delta + \mathbf{P}((\tau_n > t) \cap A) \le$$
$$\le \delta + \mathbf{P}\left(\sup_{s \le t} |X_m^{\tau_n}(s) - X^{\tau_n}(s)| > \varepsilon \right) \le$$
$$\le \delta + \mathbf{P}\left(\sup_s |X_m^{\tau_n}(s) - X^{\tau_n}(s)| > \varepsilon \right).$$

---

[104] Of course the main point is that a local martingale with independent increments has finite expected value. See: Theorem 7.97, page 545.

[105] It is an important consequence of the Fundamental Theorem of Local Martingales that every local martingale is in $\mathcal{H}_{\mathrm{loc}}^1$. See Corollary 3.59, page 221.

By Markov's inequality the stochastic convergence follows from the convergence in $L^p(\Omega)$. Therefore if $\lim_{m\to\infty} X_m^{\tau_n} = X^{\tau_n}$ in $\mathcal{H}^p$ then

$$\lim_{m\to\infty} \mathbf{P}\left(\sup_s |X_m^{\tau_n}(s) - X^{\tau_n}(s)| > \varepsilon\right) = 0.$$

This implies that for every $\varepsilon > 0$ and for every t

$$\lim_{m\to\infty} \mathbf{P}\left(\sup_{s\le t} |X_m(s) - X(s)| > \varepsilon\right) = 0.$$

Hence one should expect that the next definition is very useful[106]:

Definition 1.148 *We say that the sequence of stochastic processes (X_n) converges uniformly on compacts in probability to process X if for arbitrary[107] $t < \infty$*

$$\sup_{s\le t} |X_n(s) - X(s)| \xrightarrow{P} 0.$$

We shall denote this type of convergence by $\xrightarrow{ucp}$.

Every stochastically convergent sequence has an almost surely convergent subsequence. Hence if $X_n \xrightarrow{ucp} X_\infty$ and $X_n \xrightarrow{ucp} Y_\infty$, then for every $t < \infty$ for some subsequence

$$\sup_{s\le t} |X_{n_k}(s) - X(s)| \xrightarrow{a.s.} 0,$$

and

$$\sup_{s\le t} |X_{n_k}(s) - Y(s)| \xrightarrow{a.s.} 0$$

therefore X_∞ and Y_∞ are indistinguishable. One can also easily prove the next observations:

Proposition 1.149 *If for all n the processes X_n are right-regular and $X_n \xrightarrow{ucp} X_\infty$ then X_∞ is indistinguishable from a process which has right-regular trajectories. If X_n is continuous for all n and $X_n \xrightarrow{ucp} X_\infty$ then X_∞ is indistinguishable from a process which has continuous trajectories.*

---

[106] Later, as a consequence of the Fundamental Theorem of Local Martingales, we shall prove that every local martingale is in $\mathcal{H}^1_{\text{loc}}$, see: Corollary 3.59, page 221, so the natural topology for the space of local martingales is the uniform convergence on compacts in probability.

[107] If X and X_n has regular trajectories then the supremums are measurable.

1.4.4 Locally bounded processes

One can use localization to define the class of locally bounded processes:

Definition 1.150 *Denote by* $\mathcal{K}$ *the set of bounded process.* $X \in \mathcal{K}$ *if and only if there is a real number* k *such that* $|X(t,\omega)| \leq k$ *for all* (t,ω)*. A process* X *is locally bounded if* $X - X(0) \in \mathcal{K}_{loc}$*.*

As we have seen[108] every regular function is bounded on every compact interval. Let us consider the stopping times

$$\tau_a \stackrel{\circ}{=} \inf \{t : |X(t)| > a\}.$$

If X is right-regular then $|X(\tau_a)| \geq a$, but as X can reach the level a with a jump, it can happen that for certain outcomes $|X(\tau_a)| > a$. For right-continuous processes one can only use the estimation

$$|X(\tau_a)| \leq a + |\Delta X(\tau_a)|.$$

As the jump $|\Delta X(\tau_a)|$ can be arbitrarily large X is not necessarily bounded on the random interval

$$[0, \tau_a] \stackrel{\circ}{=} \{(t, \omega) : 0 \leq t \leq \tau_a(\omega) < \infty\}. \tag{1.66}$$

On the other hand, let us assume that X is left-continuous. If $\tau_a(\omega) > 0$ and

$$|X(\tau_a(\omega)), \omega| > a$$

for some outcome ω then by the left-continuity one can decrease the value of $\tau_a(\omega)$, which by definition is impossible. Hence $|X(\tau_a)| \leq a$ on the set $\{\tau_a > 0\}$. This means that if X is left-continuous and $X(0) = 0$ then X is bounded on the random interval (1.66). These observations are the core of the next two propositions:

Proposition 1.151 *If the filtration is right-continuous then every left-regular process is locally bounded.*

Proof: Let X be left-regular. The process $X - X(0)$ is also left-regular so one can assume that $X(0) = 0$. Define the random times

$$\tau_n \stackrel{\circ}{=} \inf \{t : |X(t)| > n\}.$$

The filtration is right-continuous, X is left-regular so τ_n is a stopping time[109]. As $X(0) = 0$, if $\tau_n(\omega) = 0$ then $|X(\tau_n)| \leq n$. If $\tau_n(\omega) > 0$ then $|X(\tau_n(\omega), \omega)| > n$ is impossible as in this case, by the left-continuity of X one could decrease $\tau_n(\omega)$.

---

[108] See: Proposition 1.6, page 5.
[109] See: Example 1.32, page 17.

Hence the truncated process X^{τ_n} is bounded. Let us show that $\tau_n \nearrow \infty$, that is, let us show that the sequence (τ_n) is a localizing sequence. Obviously (τ_n) is never decreasing. If for some outcome ω the sequence $(\tau_n(\omega))$ were bounded then one would find a bounded sequence (t_n) for which $|X(t_n, \omega)| > n$. Let $(t_{n_k})_k$ be a monotone, convergent subsequence of (t_n). If $t_{n_k} \to t^*$, then $|X(t_n, \omega)| \to \infty$, which is impossible as X has finite left and right limits. ☐

Proposition 1.152 *If the filtration is right-continuous and the jumps of the right-regular process X are bounded then X is locally bounded.*

Proof: We can again assume that $X(0) = 0$. Assume that $|\Delta X| \le a$. As in the previous proposition if

$$\tau_n \overset{\circ}{=} \inf \{t : |X(t)| > n\}$$

then (τ_n) is a localizing sequence, $|X(\tau_n-)| \le n$, therefore

$$|X^{\tau_n}| \le n + |\Delta X(\tau_n)| \le n + a. \qquad \square$$

Example 1.153 In the previous propositions one cannot drop the condition of regularity.

The process

$$X(t) \overset{\circ}{=} \begin{cases} 1/t & \text{if} \quad t > 0 \\ 0 & \text{if} \quad t = 0 \end{cases}$$

is continuous from the left but not regular, and it is obviously not locally bounded. The

$$X(t) \overset{\circ}{=} \begin{cases} 1/(1-t) & \text{if} \quad t < 1 \\ 0 & \text{if} \quad t \ge 1 \end{cases}$$

is continuous from the right but it is also not locally bounded. ☐

2

STOCHASTIC INTEGRATION WITH LOCALLY SQUARE-INTEGRABLE MARTINGALES

In this chapter we shall present a relatively simple introduction to stochastic integration theory. Our main simplifying assumption is that we assume that the integrators are locally square-integrable martingales. Every continuous process is locally bounded, hence the space $\mathcal{H}^2_{\mathrm{loc}}$ contains the continuous local martingales. In most of the applications the integrator is continuous, therefore in this chapter we shall mainly concentrate on the continuous case. As we shall see, the slightly more general case, when the integrator is in $\mathcal{H}^2_{\mathrm{loc}}$ is nearly the same as the continuous one. The central concept of this chapter is the quadratic variation $[X]$. We shall show that if X is a continuous local martingale then $[X]$ is continuous, increasing and $X^2 - [X]$ is also a local martingale. It is a crucial observation that in the continuous case these properties characterize the quadratic variation. When the integrator X is discontinuous then the quadratic variation $[X]$ is also discontinuous. As in the continuous case, $X^2 - [X]$ is still a local martingale, but this property does not characterize the quadratic variation for local martingales in general. The jump process of the quadratic variation $\Delta[X]$ satisfies the identity $\Delta[X] = (\Delta X)^2$, and $[X]$ is the only right-continuous, increasing process for which $X^2 - [X]$ is a local martingale and the identity $\Delta[X] = (\Delta X)^2$ holds. When the integrators are continuous one can define the stochastic integral for progressively measurable integrands. The main difference between the continuous and the $\mathcal{H}^2_{\mathrm{loc}}$ case is that in the discontinuous case we should take into account the jumps of the integral. Because of this extra burden in the discontinuous case one can define the stochastic integral only when the integrands are predictable.

In the first part of the chapter we shall introduce the so-called Itô–Stieltjes integral. We shall use the existence theorem of the Itô–Stieltjes integral to prove the existence of the quadratic variation. After this, we present the construction of stochastic integral when the integrators are continuous local martingales. At the end of the chapter we briefly discuss the difference between the continuous and the $\mathcal{H}_{\mathrm{loc}}^2$ case.

In the present chapter we assume that the filtration is right-continuous and if $N \in \mathcal{A}$ has probability zero, then $N \in \mathcal{F}_s$ for all s. But we shall not need the assumption that $(\Omega, \mathcal{A}, \mathbf{P})$ is complete.

2.1 The Itô–Stieltjes Integrals

In this section we introduce the simplest concept of stochastic integration, which I prefer to call *Itô–Stieltjes integration*. Every integral is basically a limit of certain approximating sums. The meaning of the integral is generally obvious for the finite approximations and by definition the integral operator extends the meaning of the finite sums to some more complicated infinite objects. In stochastic integration theory we have two stochastic processes: the integrator X and the integrand Y. As in elementary analysis, let us fix an interval $[a, b]$ and let

$$\Delta_n : a = t_0^{(n)} < t_1^{(n)} < \cdots < t_{m_n}^{(n)} = b \qquad (2.1)$$

be a partition of $[a, b]$. For a fixed partition Δ_n let us define the finite approximating sum

$$S_n \stackrel{\circ}{=} \sum_{k=1}^{m_n} Y\left(\tau_k^{(n)}\right) \left(X\left(t_k^{(n)}\right) - X\left(t_{k-1}^{(n)}\right)\right),$$

where the test points $\tau_k^{(n)}$ have been chosen in some way from the time subintervals $[t_{k-1}^{(n)}, t_k^{(n)}]$. If the integrator X is the price of some risky asset then $X(t_k^{(n)}) - X(t_{k-1}^{(n)})$ is the change of the price during the time interval $[t_{k-1}^{(n)}, t_k^{(n)}]$ and if $Y(\tau_k^{(n)})$ is the number of assets one holds during this time period then S_n is the net change of the value of the portfolio during the whole time period $[a, b]$. If

$$\lim_{n \to \infty} \max_k \left| t_k^{(n)} - t_{k-1}^{(n)} \right| = 0$$

then the sequence of partitions (Δ_n) is called *infinitesimal*. In this section we say that the integral $\int_a^b Y \, dX$ exists if for any infinitesimal sequence of partitions of $[a, b]$ the sequence of approximating sums (S_n) is convergent and the limit is independent of the partition (Δ_n). The main problem is the following: under which conditions and in which sense does the limit $\lim_{n \to \infty} S_n$ exist? Generally we can only guarantee that the approximating sequence (S_n) is convergent in probability and for the existence of the integral we should assume that the test points $\tau_k^{(n)}$ have been chosen in a very restricted way. That is, we should assume, that $\tau_k^{(n)} = t_{k-1}^{(n)}$. This type of integral we shall call the Itô–Stieltjes integral of Y against X. Perhaps the most important and most unusual point in the theory is that we should restrict the choice of the test points $\tau_k^{(n)}$. The simplest example showing why it is necessary follows:

Example 2.1 Let w be a Wiener process. Try to define the integral $\int_a^b w\,dw$!

Consider the approximating sums

$$S_n \overset{\circ}{=} \sum_k w(t_k^{(n)}) \left(w(t_k^{(n)}) - w(t_{k-1}^{(n)}) \right),$$

and

$$I_n \overset{\circ}{=} \sum_k w(t_{k-1}^{(n)}) \left(w(t_k^{(n)}) - w(t_{k-1}^{(n)}) \right).$$

In the first case $\tau_k^{(n)} \overset{\circ}{=} t_k^{(n)}$ and in the second case $\tau_k^{(n)} \overset{\circ}{=} t_{k-1}^{(n)}$. Obviously

$$S_n - I_n = \sum_k \left(w(t_k^{(n)}) - w(t_{k-1}^{(n)}) \right)^2,$$

which is the approximating sum for the quadratic variation of the Wiener process. As we will prove[1] if $n \to \infty$ then in $L^2(\Omega)$-norm

$$\lim_{n \to \infty} (S_n - I_n) = b - a \neq 0,$$

that is the limit of the approximating sums is dependent on the choice of the test points $\tau_k^{(n)}$. As the interpretation of the stochastic integral is basically the net gain of some gambling process, it is quite reasonable to choose $\tau_k^{(n)}$ as $t_{k-1}^{(n)}$ as one should decide about the size of a portfolio before the prices change, since it is quite unrealistic to assume that one can decide about the size of an investment after the new prices have already been announced. It is very simple to see that

$$I_n = \frac{1}{2} \left(\sum_k \left(w^2(t_k^{(n)}) - w^2(t_{k-1}^{(n)}) \right) - \sum_k \left(w(t_k^{(n)}) - w(t_{k-1}^{(n)}) \right)^2 \right) =$$

$$= \frac{1}{2} \left(w^2(b) - w^2(a) \right) - \frac{1}{2} \sum_k \left(w(t_k^{(n)}) - w(t_{k-1}^{(n)}) \right)^2,$$

---

[1]See: Example 2.27, page 129, Theorem B.17, page 571.

hence

$$\lim_{n\to\infty} I_n = \frac{1}{2} \left[w^2(t) \right]_a^b - \frac{1}{2}(b-a) =$$

$$= \frac{1}{2}\left(w^2(b) - w^2(a)\right) - \frac{1}{2}(b-a),$$

and similarly

$$\lim_{n\to\infty} S_n = \frac{1}{2} \left[w^2(t) \right]_a^b - \frac{1}{2}(b-a) + (b-a) =$$

$$= \frac{1}{2}\left(w^2(b) - w^2(a)\right) + \frac{1}{2}(b-a). \qquad \square$$

2.1.1 Itô–Stieltjes integrals when the integrators have finite variation

Integration theory is quite simple when the trajectories of the integrator X have finite variation on any finite interval. As a point of departure it is worth recalling a classical theorem from elementary analysis. The following simple proposition is well-known and it is just a parametrized version of one of the most important existence theorems of the calculus.

Proposition 2.2 (Existence of Riemann–Stieltjes integrals) *Let us fix a finite time interval $[a,b]$. If the trajectories of the integrator X have finite variation and the integrand Y is continuous, then for all outcomes ω the limit of the integrating sums*

$$S_n \doteq \sum_{k=1}^{m_n} Y(\tau_k^{(n)}) \left(X(t_k^{(n)}) - X(t_{k-1}^{(n)}) \right), \qquad (2.2)$$

exists and it is independent of the choice of the infinitesimal sequence of partitions (2.1) and of the choice of the test points $\tau_k^{(n)} \in \left[t_{k-1}^{(n)}, t_k^{(n)} \right]$.

Proof. As the trajectories $Y(\omega)$ are continuous on $[a,b]$ they are uniformly continuous and therefore for any $\varepsilon > 0$ there is a $\delta(\omega) > 0$, such that if $|t' - t''| < \delta(\omega)$, then[2]

$$|Y(t',\omega) - Y(t'',\omega)| < \frac{\varepsilon}{\mathrm{Var}\,(X(\omega),a,b)}. \qquad (2.3)$$

---

[2]We can assume that $\mathrm{Var}\,(X(\omega),a,b) > 0$, otherwise $X(\omega)$ is constant on $[a,b]$ and the integral trivially exists.

If all partitions of $[a, b]$ are finer than $\delta(\omega)/2$, that is, if for all n

$$\max_k \left| t_k^{(n)} - t_{k-1}^{(n)} \right| < \frac{\delta(\omega)}{2}$$

then by (2.3)

$$0 \le |S_i - S_j| \overset{\circ}{=}$$

$$\overset{\circ}{=} \left| \sum_k Y(\tau_k^{(i)}) \left(X(t_k^{(i)}) - X(t_{k-1}^{(i)}) \right) - \sum_l Y(\tau_l^{(j)}) \left(X(t_l^{(j)}) - X(t_{l-1}^{(j)}) \right) \right| \overset{\circ}{=}$$

$$\overset{\circ}{=} \left| \sum_r \left(Y(\theta_r^{(i)}) - Y(\theta_r^{(j)}) \right) (X(s_r) - X(s_{r-1})) \right| \le$$

$$\le \max_r \left| Y(\theta_r^{(i)}) - Y(\theta_r^{(j)}) \right| \sum_r |X(s_r) - X(s_{r-1})| \le$$

$$\le \max_r \left| Y(\theta_r^{(i)}) - Y(\theta_r^{(j)}) \right| \operatorname{Var}(X, a, b) \le \varepsilon,$$

where (s_r) is any partition containing the points $(t_k^{(i)})$ and $(t_l^{(j)})$ and the $\theta_r^{(i)}$ and $\theta_r^{(j)}$ are the original test points $\tau_k^{(i)}$ and $\tau_k^{(j)}$ corresponding to $[s_{r-1}, s_r]$ respectively. So for any ω, $(S_n(\omega))$ is a Cauchy sequence. so for all ω the limit

$$\left(\int_a^b Y \, dX \right)(\omega) \overset{\circ}{=} \lim_{n \to \infty} S_n(\omega)$$

exists. If (S_p) and (S_q) are two different approximating sequences generated by different infinitesimal sequences of partitions of $[a, b]$ or they belong to different choices of test points and

$$I_n \overset{\circ}{=} \begin{cases} S_p & \text{if } n = 2p \\ S_q & \text{if } n = 2q - 1 \end{cases}$$

then by the argument just presented (I_n) also has a limit, which is of course the common limit of (S_p) and (S_q). Hence the limit does not depend on the infinitesimal sequence of partitions $(t_k^{(n)})$ and does not depend on the way of choosing the test points $(\tau_k^{(n)})$. $\qquad \square$

Definition 2.3 *If the value of the integral is independent of the choice of test points $(\tau_k^{(n)})$ then the integral is called the Riemann–Stieltjes integral of Y against X. Of course the integral is denoted by $\int_a^b Y \, dX$.*

Example 2.4 If Y and X have common points of discontinuity then the Riemann–Stieltjes integral $\int_a^b Y dX$ does not exist.

If

$$Y(t) \overset{\circ}{=} \begin{cases} 0 & \text{if } t \leq 0 \\ 1 & \text{if } t > 0 \end{cases} \quad \text{and} \quad X(t) \overset{\circ}{=} \begin{cases} 0 & \text{if } t < 0 \\ 1 & \text{if } t \geq 0 \end{cases}$$

then the Riemann–Stieltjes integral $\int_{-1}^1 X dY$ does not exist. If $\tau_k^{(n)} \leq 0$ for the subinterval containing $t = 0$ then $S_n = 0$, otherwise $S_n = 1$. Observe that if the test point $\tau_k^{(n)}$ is the left endpoint of the subinterval, then $S_n = 0$, hence the so-called Itô–Stieltjes integral[3] is zero. $\qquad\square$

Our goal is to extend the integral to discontinuous integrands. As a first step, we extend the integral to regular integrands. As we saw in the previous example even for left-regular integrands we cannot choose the test points $\tau_k^{(n)}$ arbitrarily.

Definition 2.5 *If the value of the test point* $\tau_k^{(n)}$ *is always the left endpoint of the subinterval* $[t_{k-1}^{(n)}, t_k^{(n)}]$, *that is if* $\tau_k^{(n)} = t_{k-1}^{(n)}$ *for all* k, *then the integral is called the Itô–Stieltjes integral of* Y *against* X. *Of course the* Itô–Stieltjes *integrals are also denoted by* $\int_a^b Y dX$.

Example 2.6 If f is a simple predictable jump that is

$$f(t) \overset{\circ}{=} \begin{cases} c_1 & \text{if } t \leq t_0 \\ c_2 & \text{if } t > t_0 \end{cases}$$

then for any regular function g the Itô–Stieltjes integral is

$$\int_a^b f dg = c_1 \left(g(t_0+) - g(a) \right) + c_2 \left(g(b) - g(t_0+) \right). \tag{2.4}$$

If f is a simple jump that is

$$f(t) \overset{\circ}{=} \begin{cases} c_1 & \text{if } t < t_0 \\ c_3 & \text{if } t = t_0 \\ c_2 & \text{if } t > t_0 \end{cases}$$

then for any right-regular function g the Itô–Stieltjes integral is again (2.4).

---

[3] See the definition below.

If $t_0 = b$ then by definition $g(t_0+) = g(b+) = g(b)$ so in this case (2.4) is obvious. Let $(t_k^{(n)})$ be an infinitesimal sequence of partitions. By the definition of the integral

$$S_n \overset{\circ}{=} \sum_k f(t_{k-1}^{(n)}) \left(g(t_k^{(n)}) - g(t_{k-1}^{(n)})\right) =$$

$$= c_1 \left(g(t_j^{(n)}) - g(a)\right) + c_2 \left(g(b) - g(t_j^{(n)})\right),$$

where $t_0 \in [t_{j-1}^{(n)}, t_j^{(n)})$. If $n \to \infty$, then $t_j^{(n)} \searrow t_0+$ and as g is regular the limit $\lim_n S_n$ exists and it is equal to the formula given. Assume that g is right-regular. If $t_0 \neq t_{j-1}^{(n)}$ then the approximating sums do not change. If $t_0 = t_{j-1}^{(n)}$ then

$$S_n = c_1 \left(g\left(t_0\right) - g\left(a\right)\right) + c_3 \left(g(t_0) - g(t_j^{(n)})\right) +$$

$$+ c_2 \left(g(b) - g(t_j^{(n)})\right).$$

g is right-continuous at t_0 so $g\left(t_0\right) - g(t_j^{(n)}) \to 0$, hence the limit is again the same as in the previous case. □

One can easily generalize the example above[4]:

Lemma 2.7 *If every trajectory of the integrand Y is a finite number of jumps and X is a right-continuous process, then for arbitrary $a < b$ the Itô–Stieltjes integral $\int_a^b Y \, dX$ exists and the approximating sums converge for every outcome ω.*

Example 2.8 If f is a simple spike, that is if

$$f(t) \overset{\circ}{=} \begin{cases} c & \text{if} \quad t = t_0 \\ 0 & \text{if} \quad t \neq t_0 \end{cases},$$

then for any right-continuous integrator the Itô–Stieltjes integral of f is zero.

The approximating sum is

$$S_n = \begin{cases} 0 & \text{if} \quad t_0 \neq t_j^{(n)} \\ c \cdot \left(g(t_{j+1}^{(n)}) - g(t_j^{(n)})\right) & \text{if} \quad t_0 = t_j^{(n)} \end{cases}.$$

[4]Let us observe that the Itô–Stieltjes integral is, trivially, additive.

In the first case of course $\lim_n S_n = 0$, in the second case as g is right-continuous

$$\lim_{n \to \infty} S_n = c \lim_{n \to \infty} \left(g(t_{j+1}^{(n)}) - g(t_j^{(n)}) \right) = c \lim_{n \to \infty} \left(g(t_{j+1}^{(n)}) - g(t_0) \right) = 0.$$

Observe that if g has bounded variation, then g defines a signed measure on $\mathbb{R}$. The Lebesgue–Stieltjes integral is $\int_a^b f \, dg = f(t_0) \Delta g(t_0)$ which is different from the Itô–Stieltjes integral. Later[5] we shall show that for left-regular processes the Lebesgue–Stieltjes and the Itô–Stieltjes integrals are equal but, as in this case f is not left-regular, the theorem is not applicable[6]. □

We shall very often use the following simple observation:

Proposition 2.9 (The existence of the Itô–Stieltjes integral) *If the integrator X is right-continuous[7] and it has finite variation and the integrand Y is regular then for any time interval $[a, b]$ the Itô–Stieltjes integral $\int_a^b Y \, dX$ exists and for all outcome ω the approximating sequences*

$$I_n(\omega) \stackrel{\circ}{=} \sum_k Y(t_{k-1}^{(n)}, \omega) \left(X(t_k^{(n)}, \omega) - X(t_{k-1}^{(n)}, \omega) \right)$$

are convergent.

Proof. The proof is similar to the proof of the existence of Riemann–Stieltjes integrals. Fix an outcome ω and let (I_n) be the sequence of the approximating sums. Fix an $\varepsilon > 0$ and an outcome ω. By the regularity of $Y(\omega)$ there are only a finite number of jumps bigger than[8]

$$c \stackrel{\circ}{=} \frac{\varepsilon}{4 \cdot \mathrm{Var}(X)(a, b, \omega)}.$$

Let $J \stackrel{\circ}{=} \sum \Delta Y \cdot \chi(|\Delta Y| \geq c)$ and $Z \stackrel{\circ}{=} Y - J$.

1. Let us denote by $(I_n^{(J)})$ the approximating sums formed with J. As Y is regular the number of 'big jumps' on every trajectory is finite. X is right-continuous, hence by the previous lemma the integral $\int_a^b J(\omega) \, dX(\omega)$ exists for any ω. Hence if i and j are big enough, then

$$\left| I_i^{(J)}(\omega) - I_j^{(J)}(\omega) \right| \leq \frac{\varepsilon}{2}.$$

[5] It is an easy consequence of the Dominated Convergence Theorem. See: Theorem 2.88, page 174. See also the properties of the stochastic integral on page 434.

[6] Recall that the Riemann–Stieltjes integral $\int_a^b f \, dg$ does not exist.

[7] If X is not right-continuous then we should assume that Y is left-regular.

[8] See: Proposition 1.5, page 5. We can assume that $\mathrm{Var}(X(\omega), a, b) > 0$ otherwise $X(\omega)$ is constant on $[a, b]$ and the proposition is trivially satisfied.

2. Finally let us define the approximating sums

$$I_n^{(Z)} \doteq \sum_k Z(t_{k-1}^{(n)}, \omega) X\left((t_k^{(n)}, \omega) - X(t_{k-1}^{(n)}, \omega)\right).$$

The jumps of Z are smaller than c and Z is regular, hence[9] there is a $\delta(\omega)$ such that if $|s - t| \leq \delta(\omega)$ then

$$|Z(s, \omega) - Z(t, \omega)| \leq 2c.$$

If $\max_k \left| t_k^{(n)} - t_{k-1}^{(n)} \right| \leq \delta(\omega)/2$ for all $n \geq N$ then as in the case of the ordinary Riemann–Stieltjes integral

$$\left| I_i^{(Z)}(\omega) - I_j^{(Z)}(\omega) \right| \leq 2c \cdot \text{Var}\left(X(\omega), a, b\right) \leq \frac{\varepsilon}{2}.$$

3. Adding up the two inequalities above if i and j are sufficiently large then

$$|I_i(\omega) - I_j(\omega)| \leq \left| I_i^{(J)}(\omega) - I_j^{(J)}(\omega) \right| + \qquad (2.5)$$

$$+ \left| I_i^{(Z)}(\omega) - I_j^{(Z)}(\omega) \right| \leq \varepsilon.$$

This means that $(I_n(\omega))$ is a Cauchy sequence for any ω. The rest of the proof is the same as the last part of the proof of the previous proposition. $\qquad \square$

Example 2.10 The Itô–Stieltjes and the Lebesgue–Stieltjes integrals are not equal.

One should emphasize that as X has bounded variation one can also define the pathwise Lebesgue–Stieltjes integral of Y with respect to the measures generated by the trajectories of X. If Y is left-continuous then

$$Y = \lim_{n \to \infty} \sum_k Y\left(t_{k-1}^{(n)}\right) \chi\left(\left(t_{k-1}^{(n)}, t_k^{(n)}\right]\right)$$

so by the Dominated Convergence Theorem the two integrals are equal. But in general the Itô–Stieltjes and the Lebesgue–Stieltjes integrals are not equal. If

$$Y(t) = X(t) \doteq \begin{cases} 0 & \text{if } t < 1/2 \\ 1 & \text{if } t \geq 1/2 \end{cases}$$

---

[9]See: Proposition 1.7, page 6.

then the measure generated by X is the Dirac measure $\delta_{1/2}$ so the Lebesgue–Stieltjes integral over $(0, 1]$ is one, while the Itô–Stieltjes integral is zero[10]. □

2.1.2 Itô–Stieltjes integrals when the integrators are locally square-integrable martingales

Perhaps the most important stochastic processes are the Wiener processes. As the trajectories of Wiener processes almost surely do not have finite variation[11], we cannot apply the previous construction when the integrator is a Wiener process.

Theorem 2.11 (Fisk) *Let L be a continuous local martingale. If the trajectories of L have finite variation then for almost all outcomes ω the trajectories of L are constant functions.*

Proof. Consider the local martingale $M \doteq L - L(0)$. It is sufficient to prove that $M = 0$. Let $V \doteq \mathrm{Var}(M)$ and let (ρ_n) be a localizing sequence of M. As the variation of a continuous function is continuous

$$\upsilon_n(\omega) \doteq \inf\{t : |M(t, \omega)| \geq n\}$$

and

$$\kappa_n(\omega) \doteq \inf\{t : V(t, \omega) \geq n\}$$

are stopping times. Hence $\tau_n \doteq \upsilon_n \wedge \kappa_n \wedge \rho_n$ is also a stopping time. Obviously $\tau_n \nearrow \infty$, hence if $M^{\tau_n} = 0$ for all n then M is zero on $[0, \tau_n]$ for all n and therefore M will be zero on $\cup_n [0, \tau_n] = \mathbb{R}_+ \times \Omega$, so $M = 0$. As the trajectories of M^{τ_n} and V^{τ_n} are bounded one can assume that M and $V \doteq \mathrm{Var}(M)$ are bounded. Let $(t_k^{(n)})$ be an arbitrary infinitesimal sequence of partitions of $[0, t]$. By the energy identity[12] if $u > v$ then

$$\mathbf{E}\left((M(u) - M(v))^2\right) = \mathbf{E}\left(M^2(u) - M^2(v)\right), \qquad (2.6)$$

hence as $M(0) = 0$

$$\mathbf{E}\left(M^2(t)\right) = \mathbf{E}\left(M^2(t)\right) - \mathbf{E}\left(M^2(0)\right) =$$

$$= \mathbf{E}\left(\sum_k \left(M^2\left(t_k^{(n)}\right) - M^2\left(t_{k-1}^{(n)}\right)\right)\right) =$$

$$= \mathbf{E}\left(\sum_k \left(M\left(t_k^{(n)}\right) - M\left(t_{k-1}^{(n)}\right)\right)^2\right).$$

[10]See: Example 2.6, page 113.
[11]See: Theorem B.17, page 571.
[12]See: Proposition 1.58, page 35.

V is bounded hence $V \stackrel{\circ}{=} \operatorname{Var}(M) \le c$.

$$\mathbf{E}\left(M^2(t)\right) \le$$

$$\le \mathbf{E}\left(\sum_k \left|M\left(t_k^{(n)}\right) - M\left(t_{k-1}^{(n)}\right)\right| \cdot \max_k \left|M\left(t_k^{(n)}\right) - M\left(t_{k-1}^{(n)}\right)\right|\right) \le$$

$$\le \mathbf{E}\left(V(t) \cdot \max_k \left|M\left(t_k^{(n)}\right) - M\left(t_{k-1}^{(n)}\right)\right|\right)$$

$$\le c \cdot \mathbf{E}\left(\max_k \left|M\left(t_k^{(n)}\right) - M\left(t_{k-1}^{(n)}\right)\right|\right).$$

The trajectories of M are continuous hence they are uniformly continuous on $[0, t]$ so

$$\lim_{n \to \infty} \max_k \left|M\left(t_k^{(n)}\right) - M\left(t_{k-1}^{(n)}\right)\right| = 0.$$

On the other hand

$$\max_k \left|M\left(t_k^{(n)}\right) - M\left(t_{k-1}^{(n)}\right)\right| \le V(t) \le c,$$

so we can use the Dominated Convergence Theorem:

$$\lim_{n \to \infty} \mathbf{E}\left(\max_k \left|M\left(t_k^{(n)}\right) - M\left(t_{k-1}^{(n)}\right)\right|\right) = 0.$$

Hence $M(t) \stackrel{a.s.}{=} 0$ for every t. The trajectories of M are continuous and therefore[13] for almost all outcomes ω one has that $M(t, \omega) = 0$ for all t. □

This means that when the integrators are continuous local martingales we need another approach. First we prove two very simple lemmata:

Lemma 2.12 Let $(M_k, \mathcal{F}_k)$ be a discrete-time martingale and let (N_k) be an $\mathcal{F} \stackrel{\circ}{=} (\mathcal{F}_k)$ adapted process. If the variables

$$N_{k-1} \cdot (M_k - M_{k-1})$$

are integrable then the sequence

$$Z_0 \stackrel{\circ}{=} 0, \quad Z_n \stackrel{\circ}{=} \sum_{k=1}^n N_{k-1} \cdot (M_k - M_{k-1})$$

[13] See: Proposition 1.9, page 7.

is an $\mathcal{F}$-martingale. Specifically, if N is uniformly bounded and M is an arbitrary discrete-time martingale then Z is a martingale.

Proof. By the assumptions $N_{k-1} \cdot (M_k - M_{k-1})$ is integrable, hence if $k-1 \geq m$ then

$$\mathbf{E}\left(N_{k-1}\left(M_k - M_{k-1}\right) \mid \mathcal{F}_m\right) = \mathbf{E}\left(\mathbf{E}\left(N_{k-1}\left(M_k - M_{k-1}\right) \mid \mathcal{F}_{k-1}\right) \mid \mathcal{F}_m\right) =$$
$$= \mathbf{E}\left(N_{k-1}\mathbf{E}\left(M_k - M_{k-1} \mid \mathcal{F}_{k-1}\right) \mid \mathcal{F}_m\right) =$$
$$= \mathbf{E}\left(N_{k-1} \cdot 0 \mid \mathcal{F}_m\right) = 0,$$

from which the lemma is evident. $\qquad\square$

Lemma 2.13 *Let $(M_k, \mathcal{F}_k)$ be a discrete-time $L^2(\Omega)$-valued martingale. If $|N_k| \leq c$ is an $\mathcal{F}$-adapted sequence and*

$$Z_0 \overset{\circ}{=} 0, \quad Z_n \overset{\circ}{=} \sum_{k=1}^{n} N_{k-1} \cdot (M_k - M_{k-1})$$

then

$$\|Z_n\|_2 \leq c\sqrt{\|M_n\|_2^2 - \|M_0\|_2^2}.$$

Proof. By the previous lemma (Z_n) is a martingale, so by the energy equality

$$\|Z_n\|_2^2 = \sum_{k=1}^{n} \|N_{k-1}\left(M_k - M_{k-1}\right)\|_2^2.$$

Using the energy equality again

$$\|Z_n\|_2^2 \leq c^2 \sum_{k=1}^{n} \|M_k - M_{k-1}\|_2^2 =$$

$$= c^2 \sum_{k=1}^{n} \left(\|M_k\|_2^2 - \|M_{k-1}\|_2^2\right) =$$

$$= c^2 \left(\|M_n\|_2^2 - \|M_0\|_2^2\right).$$

$\qquad\square$

First we prove the existence of the integral for continuous integrands.

Proposition 2.14 (Existence of Itô–Stieltjes integrals for continuous integrands) *If $X \in \mathcal{H}^2$ and Y is adapted and continuous on a finite interval* [a, b] *then the Itô –Stieltjes integral* $\int_a^b Y\,dX$ *exists and the approximating sums*

$$I_n \triangleq \sum_k Y(t_{k-1}^{(n)})\left(X(t_k^{(n)}) - X(t_{k-1}^{(n)})\right)$$

converge in probability.

Proof. The proof is similar to the proof of the existence of the integral when the integrator has finite variation.

1. The basic, but not entirely correct trick is that as Y is continuous it is uniformly continuous, hence if I_n and I_m are two approximating sums of the integral then by the previous lemma

$$\|I_n - I_m\|_2 \triangleq$$

$$\triangleq \left\| \sum_k Y(t_{k-1}^{(n)})\left(X(t_k^{(n)}) - X(t_{k-1}^{(n)})\right) - \sum_k Y(t_{k-1}^{(m)})\left(X(t_k^{(m)}) - X(t_{k-1}^{(m)})\right) \right\|_2 =$$

$$= \left\| \sum_k \left(Y(t_{k-1}') - Y(t_{k-1}'')\right)\left(X(t_k''') - X(t_{k-1}''')\right) \right\|_2 \leq$$

$$\leq c\sqrt{\|X(b)\|_2^2 - \|X(a)\|_2^2}.$$

Of course the main problem with this estimation is that one cannot guarantee that for any fixed partition

$$\left| Y(t_{k-1}', \omega) - Y(t_{k-1}'', \omega) \right| \leq c \tag{2.7}$$

for every ω. What one can show is that if the partitions $(t_k^{(n)})$ and $(t_k^{(m)})$ are sufficiently fine then outside of an event with small probability the estimation (2.7) is valid. That is the reason why one can prove only that the integrating sums converge in probability and not in $L^2(\Omega)$.

2. To show the correct proof fix an α and a β and let

$$c \triangleq \sqrt{\frac{\beta\alpha^2}{2\left(\|X(b)\|_2^2 - \|X(a)\|_2^2\right)}}.$$

For every $\delta > 0$ let us define the modulus of continuity of Y:

$$M_\delta(\omega, u) \triangleq \sup\left\{ |Y(t, \omega) - Y(s, \omega)| : |t - s| \leq \delta,\ t, s \in [a, u] \right\}.$$

As Y is continuous one can calculate the supremum when s and t are rational numbers so M_δ is adapted and as Y is continuous obviously M_δ is also continuous.

Y is continuous, so every trajectory of Y is uniformly continuous on $[a, b]$, hence for every ω

$$\lim_{\delta \searrow 0} M_\delta(\omega, b) = 0.$$

This means that if δ is sufficiently small then

$$\mathbf{P}(M_\delta(b) \geq c) \leq \frac{\beta}{2}.$$

Fix this δ and let us define the stopping time

$$\tau \stackrel{\circ}{=} \inf \{u : M_\delta(u) \geq c\} \wedge b.$$

As τ is a stopping time, $Z \stackrel{\circ}{=} Y^\tau$ is adapted and if $|x - y| \leq \delta$ then $|Z(x) - Z(y)| \leq c$. Let

$$I_n^{(Z)} \stackrel{\circ}{=} \sum_k Z(t_{k-1}^{(n)}) \left(X(t_k^{(n)}) - X(t_{k-1}^{(n)}) \right).$$

If the partitions $\left(t_k^{(i)} \right)$ and $\left(t_k^{(j)} \right)$ are finer than $\delta/2$ then by the previous lemma

$$\left\| I_i^{(Z)} - I_j^{(Z)} \right\|_2^2 \leq c^2 \left(\|X(b)\|_2^2 - \|X(a)\|_2^2 \right) = \frac{\beta \alpha^2}{2}.$$

Let $A \stackrel{\circ}{=} \{M_\delta(b) \geq c\}$. It is easy to see that $Z = Y$ on A^c. By Chebyshev's inequality

$$\mathbf{P}\left(|I_i - I_j| > \alpha \right) =$$
$$= \mathbf{P}\left(\{|I_i - I_j| > \alpha\} \cap A \right) + \mathbf{P}\left(\{|I_i - I_j| > \alpha\} \cap A^c \right) \leq$$
$$\leq \mathbf{P}(A) + \mathbf{P}\left(\{|I_i - I_j| > \alpha\} \cap A^c \right) =$$
$$= \mathbf{P}(A) + \mathbf{P}\left(\left\{ \left| I_i^{(Z)} - I_j^{(Z)} \right| > \alpha \right\} \cap A^c \right) \leq$$
$$\leq \frac{\beta}{2} + \mathbf{P}\left(\left| I_i^{(Z)} - I_j^{(Z)} \right| > \alpha \right) \leq \frac{\beta}{2} + \frac{\left\| I_i^{(Z)} - I_j^{(Z)} \right\|_2^2}{\alpha^2} \leq$$
$$\leq \frac{\beta}{2} + \frac{c^2 \left(\|X(b)\|_2^2 - \|X(a)\|_2^2 \right)}{\alpha^2} = \frac{\beta}{2} + \frac{\beta}{2} = \beta.$$

Hence (I_n) is convergent in probability. □

Now we generalize the theorem for regular integrands.

Proposition 2.15 (The existence of the Itô–Stieltjes integral for $\mathcal{H}^2$ integrators) *If on a finite interval $[a,b]$ the adapted stochastic process Y is regular and $X \in \mathcal{H}^2$ then the Itô–Stieltjes integral $\int_a^b Y\,dX$ exists and the Itô-type approximating sums converge in probability.*

Proof. The proof is similar to the proof of the existence of the integral when the integrator has finite variation. Let (I_n) be an approximating sequence of the integral $\int_a^b Y\,dX$. Fix an ε and a β.

$$c \doteq \sqrt{\frac{\beta\varepsilon^2}{48\left(\|X(b)\|_2^2 - \|X(a)\|_2^2\right)}}$$

Let again $J \doteq \sum \Delta Y \chi\left(|\Delta Y| \geq c\right)$, $Z \doteq Y - J$.

1. As the trajectories of Y are regular for any ω the trajectory $Y(\omega)$ has a finite number of jumps which are larger than c. $X \in \mathcal{H}^2$ and by definition X is right-continuous, hence the integral $\int_a^b J\,dX$ exists. As it converges for every outcome ω it converges stochastically as well, so if i and j are big enough, then

$$\mathbf{P}\left(\left|I_i^{(J)} - I_j^{(J)}\right| > \frac{\varepsilon}{2}\right) \leq \frac{\beta}{3}.$$

2. The jumps of Z are smaller than c. As in the continuous case[14] if $\delta > 0$ is small enough then there is a stopping time τ such that

$$\mathbf{P}\left(\tau < b\right) \doteq \mathbf{P}\left(A\right) \leq \frac{\beta}{3}$$

and if $|x - y| \leq \delta$ then $|Z(x) - Z(y)| \leq 2c$ on the random interval $[a, \tau]$. If $V \doteq Z^\tau$ then $|V(x) - V(y)| \leq 2c$ whenever $|x - y| \leq \delta$. If the partitions $(t_k^{(i)})$ and $(t_k^{(j)})$ are finer than $\delta/2$ then again as in the continuous case

$$\left\|I_i^{(V)} - I_j^{(V)}\right\|_2^2 \leq (2c)^2 \left(\|X(b)\|_2^2 - \|X(a)\|_2^2\right).$$

By Chebyshev's inequality

$$\mathbf{P}\left(\left|I_i^{(V)} - I_j^{(V)}\right| > \frac{\varepsilon}{2}\right) \leq \frac{(2c)^2 \left(\|X(b)\|_2^2 - \|X(a)\|_2^2\right)}{(\varepsilon/2)^2} = \frac{\beta}{3}.$$

[14]See: Proposition 1.7, page 6.

3. If i and j are big enough, then

$$\mathbf{P}\left(|I_i - I_j| > \varepsilon\right) \leq \mathbf{P}\left(\left|I_i^{(J)} - I_j^{(J)}\right| > \frac{\varepsilon}{2}\right) + \mathbf{P}\left(\left|I_i^{(Z)} - I_j^{(Z)}\right| > \frac{\varepsilon}{2}\right) \leq$$

$$\leq \frac{\beta}{3} + \mathbf{P}\left(\left|I_i^{(Z)} - I_j^{(Z)}\right| > \frac{\varepsilon}{2}\right) \leq$$

$$\leq \frac{\beta}{3} + \mathbf{P}\left(A\right) + \mathbf{P}\left(A^c \cap \left|I_i^{(Z)} - I_j^{(Z)}\right| > \frac{\varepsilon}{2}\right) \leq$$

$$\leq \frac{2\beta}{3} + \mathbf{P}\left(\left|I_i^{(V)} - I_j^{(V)}\right| > \frac{\varepsilon}{2}\right) \leq \beta.$$

This means that (I_n) is a Cauchy sequence in probability and hence it converges in probability. $\qquad\square$

Corollary 2.16 *Let Y be an adapted, regular process on a finite interval $[a, b]$. If $X \in \mathcal{H}_{loc}^2$ then the Itô–Stieltjes integral $\int_a^b Y\, dX$ exists and the approximating sums converge in probability.*

Proof. Assume that $X \in \mathcal{H}_{\mathrm{loc}}^2$ and let (τ_n) be a localizing sequence of X. As $\tau_n \nearrow \infty$ for any $\beta > 0$ if s is big enough then $\mathbf{P}\left(\tau_s \leq b\right) < \beta/2$. Let

$$I_n \doteq \sum_k Y(t_{k-1}^{(n)})\left(X(t_k^{(n)}) - X(t_{k-1}^{(n)})\right),$$

$$S_n \doteq \sum_k Y(t_{k-1}^{(n)})\left(X^{\tau_s}(t_k^{(n)}) - X^{\tau_s}(t_{k-1}^{(n)})\right).$$

For any $\alpha > 0$

$$\mathbf{P}\left(|I_n - I_m| > \alpha\right) \leq \mathbf{P}\left(\tau_s \leq b\right) + \mathbf{P}\left(|I_n - I_m| > \alpha, \tau_s \geq b\right) \leq$$

$$\leq \frac{\beta}{2} + \mathbf{P}\left(|I_n - I_m| > \alpha, \tau_s \geq b\right) \leq$$

$$\leq \frac{\beta}{2} + \mathbf{P}\left(|S_n - S_m| > \alpha\right).$$

As $X^{\tau_s} \in \mathcal{H}^2$ by the previous proposition $\mathbf{P}\left(|S_n - S_m| > \alpha\right) \to 0$. Hence (I_n) is a stochastic Cauchy sequence, so it is convergent in probability. $\qquad\square$

2.1.3 Itô–Stieltjes integrals when the integrators are semimartingales

As we can integrate with respect to processes with finite variation and with respect to locally square-integrable martingales, the next definition is very natural:

Definition 2.17 *An adapted process X is called a* semimartingale *if X has a decomposition*

$$X = X(0) + V + H \tag{2.8}$$

where V is a right-continuous, adapted process with finite variation and $H \in \mathcal{H}^2_{loc}$ and $V(0) = H(0) = 0$.

It is important to emphasize that at the moment we do not know too much about the class of semimartingales. As there are martingales which are not locally square-integrable it is not even evident from the definition that every martingale is a semimartingale. Later we shall prove that every local martingale is a semimartingale in the above sense[15]. We shall later also prove that every integrable sub- and supermartingale is a semimartingale[16]. Therefore the class of semimartingales is a very broad one.

Every continuous local martingale is locally square-integrable [17], therefore in the continuous case we can use the following definition:

Definition 2.18 *An adapted continuous stochastic process X is called a continuous semimartingale if X has a decomposition (2.8) where H is a continuous local martingale and V is a continuous, adapted process with finite variation.*

Proposition 2.19 *If X is a continuous semimartingale then the decomposition (2.8) is unique.*

Proof. If $X = X(0) + H_1 + V_1$ and $X = X(0) + H_2 + V_2$ then $H_1 - H_2 = V_2 - V_1$ is a continuous local martingale having finite variation. Hence by Fisk's theorem[18] $H_1 - H_2 = V_1 - V_2 = 0$. $\qquad\square$

Example 2.20 For discontinuous semimartingales the decomposition (2.8) is not necessarily unique.

---

[15]This is the so called Fundamental Theorem of Local Martingales. See: Theorem 3.57, page 220.

[16]This is a direct consequence of the so called Doob–Meyer decomposition. See: Proposition 5.11, page 303.

[17]See: Example 1.137, page 96.

[18]See: Theorem 2.11, page 117.

The simplest example is the compensated Poisson process. If π is a Poisson process with parameter λ then the *compensated Poisson* process $X(t) \overset{\circ}{=} \pi(t) - \lambda t$ is in $\mathcal{H}^2_{\text{loc}}$ and the trajectories of X on any finite interval have finite variation. So $H \overset{\circ}{=} X$, $V \overset{\circ}{=} 0$ and $H \overset{\circ}{=} 0$, $V \overset{\circ}{=} X$ are both proper decompositions of X. □

Almost surely convergent sequences are convergent in probability, therefore one can easily prove the following theorem:

Theorem 2.21 (Existence of Itô–Stieltjes integrals) *If X is a semimartingale and Y is a regular and adapted process then for any finite interval $[a, b]$ the Itô–Stieltjes integral $\int_a^b Y\,dX$ exists and it is convergent in probability. The value of the integral is independent of the value of the jumps of Y, that is for any regular Y*

$$\int_a^b Y\,dX = \int_a^b Y_-\,dX = \int_a^b Y_+\,dX.$$

Proof. We have already proved the first part of the theorem. Let (I_n) be the sequence of the approximating sums for $\int_a^b Y\,dX$ and let (S_n) be the sequence of approximating sums when the integrand is Y_-. We need to prove that

$$I_n - S_n = \sum_k \left(Y\left(t_{k-1}^{(n)}\right) - Y_-\left(t_{k-1}^{(n)}\right) \right) \left[X\left(t_k^{(n)}\right) - X\left(t_{k-1}^{(n)}\right) \right] \overset{P}{\to} 0. \quad (2.9)$$

Observe that the situation is very similar to that in the proof of Theorem 2.15. We can separate the big jumps and the small jumps and apply the same argument as above[19]. □

Example 2.22 Wiener integrals.

The simplest case of stochastic integration is the so-called Wiener integral: the integrator is a Wiener process w, the integrand is a deterministic function f. If f is regular, then f, as a stochastic process, is adapted and regular, hence by the above theorem the expression $\int_a^b f(s)\,dw(s)$ is meaningful. The increments of a Wiener process are independent. As the sum of independent normally distributed variables is again normally distributed

$$\sum_i f(t_{i-1}^{(n)}) \left(w(t_i^{(n)}) - w(t_{i-1}^{(n)}) \right) \cong N\left(0, \sum_i f^2\left(t_{i-1}^{(n)}\right) \left(t_i^{(n)} - t_{i-1}^{(n)}\right) \right).$$

[19]See: Example 2.8, page 114.

Stochastic convergence implies convergence in distribution, hence

$$\int_a^b f\,dw \cong N\left(0, \int_a^b f^2(t)\,dt\right),$$

where $N(\mu, \sigma^2)$ denotes the normal distribution with expected value μ and variance σ^2. □

2.1.4 Properties of the Itô–Stieltjes integral

The next properties of the Itô–Stieltjes integral are obvious:

Proposition 2.23 *If X_1, X_2 and X are semimartingales, Y_1, Y_2 and Y are adapted regular processes, α and β are constants then*

1. $\alpha \int_a^b Y_1 dX + \beta \int_a^b Y_2 dX \overset{a.s.}{=} \int_a^b (\alpha Y_1 + \beta Y_2)\, dX,$
2. $\int_a^b Y d\,(\alpha X_1 + \beta X_2) \overset{a.s}{=} \alpha \int_a^b Y dX_1 + \beta \int_a^b Y dX_2.$
3. *If $a < c < b$, then* $\int_a^b Y dX \overset{a.s.}{=} \int_a^c Y dX + \int_c^b Y dX.$
4. *If $Y_1 \chi_A$ is an equivalent modification of $Y_2 \chi_A$ for some $A \subseteq \Omega$ then the integrals $\int_a^b Y_1 dX$ and $\int_a^b Y_2 dX$ are almost surely equal on A.*

Since the approximating sums are convergent in probability it is important to note that the Itô–Stieltjes integral is defined only as an equivalence class. In the following we shall not distinguish between functions and equivalence classes so when it is not important to emphasize this difference instead of $\overset{a.s.}{=}$ we shall use the simpler sign $=$.

2.1.5 The integral process

Let us briefly investigate the *integral process*

$$(Y \bullet X)\,(t) \overset{\circ}{=} \int_a^t Y dX.$$

We have defined the stochastic integral only for fixed time intervals. On every time interval the definition determines the value of the stochastic integral up to a measure-zero set, hence the properties of the integral process $t \mapsto (Y \bullet X)\,(t)$ are unclear. It is not a stochastic process, just an indexed set of random variables! When does it have a version which is a martingale? Assume that $X \in \mathcal{H}^2$ and that Y is adapted. Assume also that Y is uniformly bounded that is $|Y| \leq c$ for some constant c. As the filtration $\mathcal{F}$ is right-continuous, the right-regular process

Y_+ is also adapted. As we have seen[20] for every $t \in [a, b]$

$$\|I_n(t)\|_2^2 \doteq \mathbf{E}\left(\left(\sum_k Y_+\left(t_{k-1}^{(n)} \wedge t\right)\left(X\left(t_k^{(n)} \wedge t\right) - X\left(t_{k-1}^{(n)} \wedge t\right)\right)\right)^2\right) \leq$$

$$\leq c^2\left(\mathbf{E}\left(X^2(b)\right) - \mathbf{E}\left(X^2(a)\right)\right) \doteq K,$$

hence the sequence

$$I_n(t) \doteq \sum_k Y_+\left(t_{k-1}^{(n)} \wedge t\right)\left(X\left(t_k^{(n)} \wedge t\right) - X\left(t_{k-1}^{(n)} \wedge t\right)\right)$$

is bounded in $L^2(\Omega)$ so the sequence of the approximating sums is uniformly integrable hence not only

$$I_n(t) \xrightarrow{p} (Y \bullet X)(t)$$

but also

$$I_n(t) \xrightarrow{L^1} (Y \bullet X)(t).$$

It is easy to see[21] that if $s < t$ then

$$\mathbf{E}(I_n(t) \mid \mathcal{F}_s) = I_n(s).$$

As $I_n(t) \xrightarrow{L^1} \int_a^t Y dX$ using the $L^1(\Omega)$-continuity of the conditional expectation operator

$$\mathbf{E}\left(\int_a^t Y dX \mid \mathcal{F}_s\right) = \int_a^s Y dX.$$

Observe that $I_n(t)$ is right-regular so $I_n(t)$ is a martingale for every n. As $I_m - I_n$ is a martingale by Doob's inequality, for any $\lambda > 0$

$$\lambda \mathbf{P}\left(\sup_t |I_n(t) - I_m(t)| \geq \lambda\right) \leq \|I_n(b) - I_m(b)\|_1.$$

$(I_n(b))$ is convergent in $L^1(\Omega)$ so

$$\sup_t |I_n(t) - I_m(t)| \xrightarrow{P} 0,$$

[20]See: Lemma 2.13, page 119.
[21]See: Lemma 2.12, page 118.

hence for a subsequence

$$\sup_t |I_{n_k}(t) - I_{m_k}(t)| \overset{a.s.}{\to} 0, \tag{2.10}$$

so except for a measure-zero set the continuity-type properties of trajectories of (I_n) are preserved, so we get the following proposition:

Proposition 2.24 *If Y is an adapted, regular, and uniformly bounded process, $X \in \mathcal{H}^2$ then the integral process*

$$(Y \bullet X)(t) \overset{\circ}{=} \int_a^t Y \, dX, \quad t \geq a$$

has a version which is a martingale. If (I_n) is the sequence of approximating sums then for every t

$$\sup_{a \leq s \leq t} |I_n(s) - (X \bullet M)(s)| \overset{P}{\to} 0. \tag{2.11}$$

If X is continuous and bounded then $Y \bullet X$ has a continuous version.

Let us emphasize that in the argument above the set of exceptional points N in (2.10) is in $\mathcal{F}_b$. Of course we should define the integral process on N as well, and of course we should guarantee that the integral process is adapted. We can do this only when we assume that for all $s \leq b$, $N \in \mathcal{F}_s$. This assumption is part of the usual conditions. Observe that in the continuous case we do not explicitly use the right-continuity of the filtration. On the other hand, this is a very uninteresting remark since, in most cases[22], if we add the measure-zero sets to the filtration then the augmented filtration is right-continuous.

2.1.6 Integration by parts and the existence of the quadratic variation

One of the most important concepts of stochastic analysis is the quadratic variation. The main reason to introduce the Itô–Stieltjes integral is that from the existence theorem of the Itô–Stieltjes integral one can easily deduce the existence of the quadratic variation of semimartingales.

Definition 2.25 *Let U and V be stochastic processes on $[a, b]$. If for every infinitesimal sequence of partitions $\left(t_k^{(n)}\right)$ of $[a, b]$ the sequence*

$$Q_n \overset{\circ}{=} \sum_k \left(U\left(t_k^{(n)}\right) - U\left(t_{k-1}^{(n)}\right)\right)\left(V\left(t_k^{(n)}\right) - V\left(t_{k-1}^{(n)}\right)\right)$$

---

[22]E.g. if the filtration is generated by a Lévy process. See: Proposition 1.103, page 67.

is convergent in probability then the limit $\lim_{n \to \infty} Q_n$ *is called the* quadratic
co-variation *of U and V. The quadratic co-variation of U and V on $[a, b]$ is*
denoted by $[U, V]_a^b$. *If $V = U$ then $[U, U]_a^b \overset{\circ}{=} [U]_a^b$ is called the* quadratic variation
of U. Of course in stochastic convergence

$$[U]_a^b \overset{\circ}{=} \lim_{n \to \infty} \sum_k \left(U\left(t_k^{(n)}\right) - U\left(t_{k-1}^{(n)}\right) \right)^2.$$

Example 2.26 *If the trajectories of X are continuous and the trajectories of V have*
finite variation then $[X, V]_a^b \overset{a.s.}{=} 0$ for any interval $[a, b]$.

By the continuity assumption, the trajectories of X are uniformly continuous on
the compact interval $[a, b]$. Hence if $\max_k \left| t_k^{(n)} - t_{k-1}^{(n)} \right| \to 0$ then for every ω

$$\lim_{n \to \infty} \max_k \left| X(t_k^{(n)}, \omega) - X(t_{k-1}^{(n)}, \omega) \right| \to 0.$$

Therefore, as $\operatorname{Var}(V, a, b) < \infty$

$$|Q_n| \overset{\circ}{=} \left| \sum_k \left(X(t_k^{(n)}) - X(t_{k-1}^{(n)}) \right) \left(V(t_k^{(n)}) - V(t_{k-1}^{(n)}) \right) \right| \le$$

$$\le \max_k \left| X(t_k^{(n)}) - X(t_{k-1}^{(n)}) \right| \operatorname{Var}(V, a, b) \to 0. \qquad \square$$

Example 2.27 *If w is a Wiener process[23] then $[w]_0^t \overset{a.s.}{=} t$. If π is a Poisson process*
then $[\pi]_0^t \overset{a.s.}{=} \pi(t)$.

If π is a Poisson process then for any ω the number of the jumps on any finite
interval $[0, t]$ is finite, so for any ω one can assume that every subinterval contains
just one jump, hence $Q_n(t, \omega)$ is the number of jumps of the trajectory $\pi(\omega)$
during the time interval $[0, t]$. So evidently $Q_n(t, \omega) = \pi(t, \omega)$. $\qquad \square$

Proposition 2.28 (Integration By Parts Formula) *If M and N are*
semimartingales then:

1. *For any finite interval $[a, b]$ the quadratic co-variation $[M, N]_a^b$ exists.*
2. *The following integration by parts formula holds:*

$$(MN)(b) - (MN)(a) = \int_a^b M_- dN + \int_a^b N_- dM + [M, N]_a^b. \qquad (2.12)$$

---

[23]See: Theorem B.17, page 571.

Proof. By definition semimartingales are right-regular processes so the processes M_- and N_- are well-defined left-regular processes. For any partition $(t_k^{(n)})$ of $[a, b]$ let us define the approximating sums

$$\sum_k M\left(t_{k-1}^{(n)}\right) \Delta N\left(t_k^{(n)}\right) + \sum_k N\left(t_{k-1}^{(n)}\right) \Delta M\left(t_k^{(n)}\right) +$$

$$+ \sum_k \Delta M\left(t_k^{(n)}\right) \Delta N\left(t_k^{(n)}\right).$$

With elementary calculation for all k

$$M(t_k^{(n)})N(t_k^{(n)}) - M(t_{k-1}^{(n)})(Nt_{k-1}^{(n)}) =$$

$$= M\left(t_{k-1}^{(n)}\right)\left(N(t_k^{(n)}) - N(t_{k-1}^{(n)})\right) +$$

$$+ N\left(t_{k-1}^{(n)}\right)\left(M(t_k^{(n)}) - M(t_{k-1}^{(n)})\right) +$$

$$+ \left(M(t_k^{(n)}) - M(t_{k-1}^{(n)})\right)\left(N(t_k^{(n)}) - N(t_{k-1}^{(n)})\right).$$

Adding up by k, on the left side one gets a telescopic sum which adds up to

$$M\left(b\right) N\left(b\right) - M\left(a\right) N\left(a\right),$$

which is the expression on the left-hand side of (2.12). The integrating sums on the right-hand side converge to the Itô–Stieltjes integrals

$$\int_a^b M dN = \int_a^b M_- dN \quad \text{and} \quad \int_a^b N dM = \int_a^b N_- dM$$

so $[M, N]_a^b$ exits and the formula (2.12) holds. $\qquad\square$

Example 2.29 The jumps of independent Poisson processes.

Let N_1 and N_2 be two Poisson processes with respect to the same filtration[24] $\mathcal{F}$. For $s \geq 0$ let

$$U_i\left(s, t\right) \doteq \frac{\exp\left(-sN_i\left(t\right)\right)}{\mathbf{E}\left(\exp\left(-sN_i\left(t\right)\right)\right)}, \quad i = 1, 2$$

[24]That is N_1 and N_2 are counting Lévy processes with respect to the same filtration.

be the exponential martingales defined by the Laplace transforms of the Poisson processes. By the Integration By Parts Formula

$$U_1(s_1, t) U_2(s_2, t) - 1 = \int_0^t U_1(s_1, r-) U_2(s_2, dr) +$$
$$+ \int_0^t U_2(s_2, r-) U_1(s_1, dr) +$$
$$+ [U_1(s_1), U_2(s_2)](t).$$

It is easy to see that U_1 and U_2 are bounded martingales, with respect to $\mathcal{F}$ for any $s \geq 0$ on any finite interval $[0, t]$. As they are also $\mathcal{F}$-adapted the stochastic integrals are martingales[25]. Therefore the expected value of the stochastic integrals are zero. So

$$\mathbf{E}(U_1(s_1, t) U_2(s_2, t)) - 1 = \mathbf{E}([U_1(s_1), U_2(s_2)](t)).$$

By the definition of U_1 and U_2

$$\mathbf{E}\left(\exp\left(-\left(\sum_{i=1}^2 s_i N_i(t)\right)\right)\right) = \prod_{i=1}^2 \mathbf{E}(\exp(-s_i N_i(t)))$$

if and only if

$$\mathbf{E}([U_1(s_1), U_2(s_2)](t)) = 0. \tag{2.13}$$

That is $N_1(t)$ and $N_2(t)$ are independent if and only if (2.8) holds[26]. As Laplace transform is continuous in time

$$\Delta U_i(s, r) = \frac{\exp(-s N_i(r)) - \exp(-s N_i(r-))}{\mathbf{E}(\exp(-s N_i(r)))} \leq 0$$

it is easy to see that

$$[U_1(s_1), U_2(s_2)](t) = \sum_{r \leq t} \Delta U_1(s_1, r) \Delta U_2(s_2, r) \geq 0.$$

Therefore its expected value is zero if and only if it is almost surely zero. Hence $N_1(t)$ and $N_2(t)$ are independent if and only if with probability one N_1 and N_2 do not have common jumps on the interval $[0, t]$. □

[25]See: Proposition 2.24, page 128.
[26]One can easily modify the proof of Lemma 1.96 on page 60.

The next property of the quadratic co-variation is obvious:

Proposition 2.30 *If M, N and U are arbitrary semimartingales, ξ and η are $\mathcal{F}_0$-measurable random variables then for any interval $[a, b]$*

$$[\xi M + \eta N, U]_a^b \stackrel{a.s.}{=} \xi [M, U]_a^b + \eta [N, U]_a^b .$$

Specifically

$$[M + N] = [M] + 2 [M, N] + [N] .$$

Example 2.31 If $X = X(0) + L + V$ is a continuous semimartingale then $[X]_a^b \stackrel{a.s.}{=} [L]_a^b$ for any interval $[a, b]$, where L is the continuous local martingale part of X .

As V and L are continuous and the trajectories of V have finite variation $[V]_a^b \stackrel{a.s.}{=} 0$ and $[V, L] \stackrel{a.s.}{=} 0$. By the additivity:

$$[X]_a^b \stackrel{\circ}{=} [X(0) + L + V]_a^b \stackrel{a.s.}{=} [L + V]_a^b \stackrel{a.s.}{=}$$

$$\stackrel{a.s.}{=} [L]_a^b + 2 [L, V]_a^b + [V]_a^b \stackrel{a.s.}{=} [L]_a^b .$$ □

Example 2.32 Assume that F is a deterministic, right-regular function with finite variation. If w is a Wiener process then

$$\int_0^t w(s)\, dF(s) \cong N\left(0, \int_0^t (F(t) - F(s))^2\, ds\right) .$$

---

w is continuous and F has finite variation therefore $[w, F] = 0$. By the integration by parts formula

$$w(t) F(t) = \int_0^t w\, dF + \int_0^t F_-\, dw,$$

hence

$$\int_0^t w\, dF = w(t) F(t) - \int_0^t F_-\, dw =$$

$$= \int_0^t F(t)\, dw - \int_0^t F_-\, dw =$$

$$= \int_0^t (F(t) - F(s-))\, dw(s) .$$

The last integral is a Wiener integral, so

$$\int_0^t w \, dF \cong N\left(0, \int_0^t (F(t) - F(s-))^2 \, ds\right) =$$
$$= N\left(0, \int_0^t (F(t) - F(s))^2 \, ds\right). \qquad \square$$

As we have remarked, if X has finite variation and Y is continuous then[27] $[X, Y] = 0$. Hence in this case the integration by parts formula is

$$XY - X(0)Y(0) = Y \bullet X + X_- \bullet Y.$$

For this formula we do not in fact need the continuity of Y. Observe that as X has finite variation every trajectory of X defines a measure on $\mathbb{R}_+$. Let Y be an arbitrary semimartingale, and let ΔY denote the jumps of Y. We show, that in this case

$$[Y, X] = \Delta Y \bullet X,$$

where the integral is the Lebesgue–Stieltjes integral defined by the trajectories of X. If $U \stackrel{\circ}{=} \sum \Delta Y \chi(|\Delta Y| \geq \varepsilon)$ are the jumps of Y which are bigger than ε then as the number of such jumps on every finite interval is finite

$$[Y, X] = [Y - U, X] + [U, X] =$$
$$= [Y - U, X] + \sum \Delta Y \chi(|\Delta Y| \geq \varepsilon) \Delta X =$$
$$= [Y - U, X] + \Delta Y \chi(|\Delta Y| \geq \varepsilon) \bullet X.$$

The jumps of the regular process $Z \stackrel{\circ}{=} Y - U$ are smaller than ε, hence if the partition of the interval $[a, b]$ is fine enough, then[28]

$$\left| Z(t_k^{(n)}, \omega) - Z(t_{k-1}^{(n)}, \omega) \right| \leq 2\varepsilon$$

for any ω. Therefore if $n \to \infty$

$$\left| \sum_k \left(Z(t_k^{(n)}) - Z(t_{k-1}^{(n)}) \right) \left(X(t_k^{(n)}) - X(t_{k-1}^{(n)}) \right) \right| \leq 2\varepsilon \mathrm{Var}(X, a, b) \to 0.$$

As X has finite variation and the integral is a Lebesgue–Stieltjes integral one can use the Dominated Convergence Theorem. From this theorem for every

---

[27]See: Example 2.26, page 129.
[28]See: Proposition 1.7, page 6.

trajectory

$$\Delta Y \chi(|\Delta Y| \geq \varepsilon) \bullet X \to \Delta Y \bullet X = \sum \Delta Y \Delta X,$$

assuming of course that for every trajectory, on every finite interval, $|\Delta Y|$ is integrable. But this has to be true as the trajectories of Y are regular so on every finite interval every trajectory of Y will be bounded[29].

Proposition 2.33 *If X is right-continuous and has finite variation, Y is an arbitrary semimartingale then*

$$[X, Y] = \sum \Delta Y \Delta X = \Delta Y \bullet X \qquad (2.14)$$

therefore[30]

$$XY - X(0)Y(0) = Y_- \bullet X + X_- \bullet Y + [X, Y] =$$
$$= Y_- \bullet X + X_- \bullet Y + \Delta Y \bullet X =$$
$$= Y \bullet X + X_- \bullet Y$$

where the integral with respect to X is a Lebesgue–Stieltjes integral and the integral with respect to Y is an Itô–Stieltjes integral.

2.1.7 The Kunita–Watanabe inequality

In the construction of the stochastic integral below we shall use the following simple inequality:

Proposition 2.34 (Kunita–Watanabe inequality) *If X, Y are product measurable processes, and M, N are semimartingales, $a \leq b \leq \infty$ and $V \doteq \mathrm{Var}([M, N])$ then*

$$\int_a^b |XY| \, dV \overset{a.s.}{\leq} \sqrt{\int_a^b X^2 d[M]} \sqrt{\int_a^b Y^2 d[N]}. \qquad (2.15)$$

Remark first that the meaning of the proposition is not really clear as it is not clear what is the meaning of $[M]$, $[N]$ and $[M, N]$. So far we have defined the quadratic variation only for fixed time intervals, and the quadratic variation for every time interval is defined as a limit in stochastic convergence, and hence the quadratic variation on any interval is defined just up to a measure-zero set. If X is a semimartingale then for every t one can define $[X](t) \doteq [X]_0^t$, but this $[X]$ is not a stochastic process since for a fixed ω and t the value of $[X](t, \omega)$

---

[29]See: Proposition 1.6, page 5.

[30]Observe that the Lebesgue–Stieltjes integral $Y \bullet X$ exists: The trajectories of Y are regular, hence they are bounded on every finite interval.

is undefined. Of course, if t is restricted to the set of the rational numbers then we can collect the corresponding measure-zero sets in just one measure-zero set, but it is unclear how one can extend this process to the irrational values of t as at the moment we have not proved any continuity property of the quadratic variation. Observe, that we do not know anything about integral processes. In particular we do not know when they will be martingales. If the integral process is a semimartingale then, by definition, it has a right-continuous version, so by (2.12) the quadratic variation also has a right-continuous version. One of the goals of the later developments will be to provide a right-continuous version for the quadratic variation process or, which is the same, to prove some martingale-type properties for the stochastic integral.

So, to prove the inequality up to the end of the section we assume that there are processes $[M]$, $[N]$ and $[M, N]$ which are right-continuous, and that for any t they provide a version of the related quadratic variation. In this case $[M](\omega)$, $[N](\omega)$ and $\text{Var}([M, N], \omega)$ are increasing, right-continuous functions for every ω, hence they define a measure and for every ω the integrals in (2.15) are defined as Lebesgue–Stieltjes integrals.

Proof. It is sufficient to prove the proposition for finite a and b. One can prove the case $b = \infty$ by the Monotone Convergence Theorem. Also by the Monotone Convergence Theorem one can assume that X any Y are bounded. We should prove the inequality when on the left-hand side we have $\left| \int_a^b XY \, d[M, N] \right|$ since to prove (2.15) one can replace Y by

$$\widetilde{Y} \overset{\circ}{=} Y \cdot \text{sgn}(XY) \frac{dV}{d[M, N]}.$$

1. First assume that $X = 1$ and $Y = 1$. In this case, the inequality is

$$\left| [M, N]_a^b \right| \overset{a.s.}{\leq} \sqrt{[M]_a^b} \sqrt{[N]_a^b}. \tag{2.16}$$

Fix a u and a v. The proof of (2.16) is nearly the same as the proof of the classical Cauchy–Schwarz inequality. It is easy to see that for all rational numbers r

$$0 \overset{a.s.}{\leq} [M + rN]_u^v \overset{a.s.}{=} [M, M]_u^v + 2r \cdot [M, N]_u^v + r^2 \cdot [N, N]_u^v \overset{\circ}{=} Ar^2 + Br + C.$$

Hence there is a measure-zero set Z such that on the complement of Z the inequality above is true for all rational, and therefore all real, r. Hence, as in the proof of the Cauchy–Schwarz inequality $B^2 - 4AC \overset{a.s.}{\leq} 0$ so (2.16) holds with $a = u$ and $b = v$. Unifying the measure-zero sets one can easily prove (2.16) for every rational numbers u and v. By the assumption above the quadratic variation is right-continuous, so the relation (2.16) holds for every real $a = u$ and $b = v$.

2. Let (t_k) be a partition of $[a, b]$ and assume that X and Y are constant on every subinterval $(t_{k-1}, t_k]$. We are integrating by trajectory so

$$\left| \int_a^b XY\, d\,[M, N] \right| \leq \sum_k |X(t_k)\, Y(t_k)| \left| [M, N]_{t_k}^{t_{k+1}} \right| \leq$$

$$\leq \sum_k |X(t_k)\, Y(t_k)| \sqrt{[M]_{t_k}^{t_{k+1}}} \sqrt{[N]_{t_k}^{t_{k+1}}}.$$

Using the Cauchy–Schwarz inequality we can continue

$$\left| \int_a^b XY\, d\,[M, N] \right| \leq \sqrt{\sum_k |X(t_k)|^2 [M]_{t_k}^{t_{k+1}}} \sqrt{\sum_k Y^2(t_k) [N]_{t_k}^{t_{k+1}}} =$$

$$= \sqrt{\int_a^b X^2 d\,[M]} \sqrt{\int_a^b Y^2 d\,[N]}.$$

3. Using standard measure theory one can easily prove[31] that if μ is a finite, regular measure on the real line, and g is a bounded Borel measurable function, then there is a sequence of step functions

$$s_n \stackrel{\circ}{=} \sum_i c_i \chi\left(\left(t_i^{(n)}, t_{i+1}^{(n)} \right] \right)$$

that $s_n \to g$ almost surely in μ. As μ is finite and g is bounded $s_n \to g$ in $L^2(\mu)$.

4. We prove that Kunita–Watanabe inequality holds for every outcome where (2.16) holds for every real a and b. Fix the process Y and an outcome ω, and consider the set of processes X for which the inequality (2.16) holds for this ω. Let $s_n \to X(\omega)$ be a set of step functions. By (2.16) the measure generated by $[M, N](\omega)$ is absolutely continuous with respect to the measure generated by $[M](\omega)$. Hence $s_n \to X(\omega)$ almost surely in $[M, N](\omega)$. Therefore by the Dominated Convergence theorem, using that X and Y are bounded, a and b are finite and that the convergence holds almost everywhere in $[M, N](\omega)$ and in $L^2([M](\omega))$

$$\left| \int_a^b XY\, d\,[M, N] \right| \leq \sqrt{\int_a^b X^2 d\,[M]} \sqrt{\int_a^b Y^2 d\,[N]}$$

[31] Use Lusin's theorem [80], page 56, and the uniform continuity of continuous functions on compact sets.

for outcome ω. If X is product measurable then by Fubini's theorem every trajectory of X is Borel measurable. Hence if X is product measurable then inequality (2.15) holds for almost all outcome ω.

5. Now we fix X and repeat the argument for Y. $\qquad\Box$

Corollary 2.35 *If $q, p \geq 1$ and $1/p + 1/q$ then*

$$\mathbf{E}\left(\int_0^\infty |XY| \, d\,[M, N]\right) \leq \left\|\sqrt{\int_0^\infty X^2 d\,[M]}\right\|_p \left\|\sqrt{\int_0^\infty Y^2 d\,[N]}\right\|_q.$$

Proof. By Hölder's inequality and by (2.15)

$$\mathbf{E}\left(\int_0^\infty |XY| \, d\,[M, N]\right) \leq \mathbf{E}\left(\sqrt{\int_0^\infty X^2 d\,[M]} \sqrt{\int_0^\infty Y^2 d\,[N]}\right) \leq$$

$$\leq \left\|\sqrt{\int_0^\infty X^2 d\,[M]}\right\|_p \left\|\sqrt{\int_0^\infty Y^2 d\,[N]}\right\|_q.$$

$\qquad\Box$

Corollary 2.36 *If M and N are semimartingales then*

$$|[M, N]| \leq \sqrt{[M]\,[N]} \qquad\qquad (2.17)$$

and

$$[M + N]^{1/2} \leq [M]^{1/2} + [N]^{1/2}$$

and

$$[M + N] \leq 2\,([M] + [N]).$$

Proof. The first inequality is just the Kunita–Watanabe inequality when $X = Y = 1$.

$$[M + N] = [M] + 2\,[M, N] + [N] \leq$$

$$\leq [M] + 2\sqrt{[M]\,[N]} + [N] =$$

$$= \left([M]^{1/2} + [N]^{1/2}\right)^2$$

from which the second inequality is obvious. In a similar way

$$[M + N] \leq [M] + 2\sqrt{[M][N]} + [N] \leq$$
$$\leq [M] + ([M] + [N]) + [N] =$$
$$= 2([M] + [N]).\qquad \Box$$

2.2 The Quadratic Variation of Continuous Local Martingales

The following proposition is the starting point in our construction of the stochastic integral process.

Proposition 2.37 (Simple Doob–Meyer decomposition) *If M is a uniformly bounded, continuous martingale, then:*

1. *The quadratic variation $P(t) \overset{\circ}{=} [M](t) \overset{\circ}{=} [M]_0^t$ exists.*
2. *$[M]$ has a version which is increasing and continuous.*
3. *For this version $M^2 - [M]$ is a martingale.*
4. *$[M]$ is indistinguishable from any increasing, continuous process P for which $P(0) = 0$ and $M^2 - P$ is a martingale.*

If $(t_k^{(n)})$ is an infinitesimal sequence of partitions of $[0, t]$ then

$$\sup_{s \leq t} |Q_n(s) - [M](s)| \overset{p}{\to} 0 \qquad (2.18)$$

for any t, where

$$Q_n(s) \overset{\circ}{=} \sum_k \left(M\left(t_k^{(n)} \wedge s\right) - M\left(t_{k-1}^{(n)} \wedge s\right) \right)^2.$$

Proof. By the Integration By Parts Formula for any t

$$M^2(t) - M^2(0) = 2\int_0^t M dM + [M](t) = 2 \cdot (M \bullet M)(t) + [M](t).$$

As M is continuous and uniformly bounded the integral process $M \bullet M$ has a version which is a continuous martingale[32], therefore as M^2 is continuous

$$[M] \overset{\circ}{=} M^2 - M^2(0) - 2 \cdot M \bullet M$$

is continuous, and by Proposition 2.24

$$M^2 - [M] = M^2(0) + 2 \cdot (M \bullet M)$$

---

[32]See: Proposition 2.24, page 128.

is a martingale. $[M](t)$ is a version of the quadratic variation $[M]_0^t$ for any t. For any rational numbers $p \leq q$ we have $[M]_0^p \stackrel{a.s.}{\leq} [M]_0^q$. Taking the union the measure-zero sets and using the continuity of $[M]$ we can construct a version which is increasing. If P is another continuous, increasing process for which $P(0) = 0$ and $M^2 - P$ is a martingale, then $N \stackrel{\circ}{=} P - [M]$ is also a continuous martingale and $N(0) = 0$. As N is the difference of two increasing processes the trajectories of N have finite variation. By Fisk's theorem[33] $N = 0$, so P is indistinguishable from $[M]$. The convergence (2.18) is a simple consequence of (2.11). $\qquad\square$

First we extend the proposition to continuous local martingales. In order to do it we need the following rule:

Proposition 2.38 *Under the assumptions of the previous proposition if τ is an arbitrary stopping time then $[M^\tau] = [M]^\tau$.*

Proof. As $(M^\tau)^2 = (M^2)^\tau$

$$(M^\tau)^2 - [M]^\tau = (M^2)^\tau - [M]^\tau = (M^2 - [M])^\tau.$$

Stopped martingales are martingales hence $(M^2 - [M])^\tau$ is a martingale. $[M]^\tau$ is increasing, so by the uniqueness of the quadratic variation $[M^\tau] = [M]^\tau$. $\qquad\square$

Proposition 2.39 *If M is a continuous local martingale then there is one and only one continuous, increasing process $[M]$ such that:*

1. *$[M](0) = 0$ and*
2. *$M^2 - [M]$ is a continuous local martingale.*

For any t if $\left(t_k^{(n)}\right)$ is an infinitesimal sequence of partitions of $[0, t]$ then

$$\sup_{s \leq t} |Q_n(s) - [M](s)| \stackrel{p}{\to} 0 \qquad (2.19)$$

where

$$Q_n(s) \stackrel{\circ}{=} \sum_k \left(M\left(t_k^{(n)} \wedge s\right) - M\left(t_{k-1}^{(n)} \wedge s\right) \right)^2.$$

Proof. Let M be a continuous local martingale and let (σ_n) be a localizing sequence of M. As M is continuous the hitting times

$$\upsilon_n \stackrel{\circ}{=} \inf \{t : |M(t)| \geq n\}$$

[33]See: Theorem 2.11, page 117.

are stopping times. Stopped martingales are martingales, so if instead of σ_n we take the localizing sequence $\tau_n \stackrel{\circ}{=} \sigma_n \wedge \upsilon_n$ then the processes $M_n \stackrel{\circ}{=} M^{\tau_n}$ are bounded martingales.

1. As M_n is a bounded, continuous martingale $[M_n]$ is an increasing processes and $M_n^2 - [M_n]$ is a continuous martingale. By the previous proposition

$$[M_{n+1}]^{\tau_n} = [M_{n+1}^{\tau_n}] = [M_n],$$

hence $[M_n] = [M_{n+1}]$ on the interval $[0, \tau_n]$. As $\tau_n \nearrow \infty$ one can define the process $[M]$ as the 'union' of the processes $[M_n]$, that is

$$[M](t, \omega) \stackrel{\circ}{=} [M_n](t, \omega), \quad t \leq \tau_n(\omega).$$

Evidently $[M]$ is continuous, increasing and $[M](0) = 0$. Of course

$$\left(M^2 - [M]\right)^{\tau_n} = (M^{\tau_n})^2 - [M]^{\tau_n} \stackrel{\circ}{=} M_n^2 - [M_n],$$

which is a martingale, hence $M^2 - [M]$ is a local martingale.

2. Assume that $A(0) = 0$ and $M^2 - A$ is a continuous local martingale for some continuous, increasing process A.

$$Z \stackrel{\circ}{=} \left(M^2 - [M]\right) - \left(M^2 - A\right) = A - [M]$$

is a continuous local martingale and Z, as the difference of two increasing processes, has finite variation. So by Fisk's theorem Z is constant. As $Z(0) = A(0) - [M](0) = 0$, obviously $Z \equiv 0$.

3. Finally, let us prove (2.19). Fix $\varepsilon, \delta, t > 0$ and $(t_k^{(n)})_k$. Let Q_n be the approximating sum for $[M]$ and let $Q_n^{(m)}$ be the approximating sum for $[M_m]$.

$$A \stackrel{\circ}{=} \left\{ \sup_{s \leq t} |Q_n(s) - [M](s)| > \varepsilon \right\},$$

$$A^{(m)} \stackrel{\circ}{=} \left\{ \sup_{s \leq t} \left| Q_n^{(m)}(s) - [M_m](s) \right| > \varepsilon \right\}.$$

As $\tau_m \nearrow \infty$, for m large enough $\mathbf{P}(\tau_m \leq t) \leq \delta/2$ and $\mathbf{P}(A^{(m)}) \leq \delta/2$. Obviously

$$\mathbf{P}(A) = \mathbf{P}(A \cap (\tau_m \leq t)) + \mathbf{P}(A \cap (\tau_m > t)) \leq$$
$$\leq \mathbf{P}((\tau_m \leq t)) + \mathbf{P}(A \cap (\tau_m > t)) \leq$$

$$\leq \frac{\delta}{2} + \mathbf{P}\left(A \cap (\tau_m > t)\right) =$$

$$= \frac{\delta}{2} + \mathbf{P}\left(A^{(m)} \cap (\tau_m > t)\right) \leq \frac{\delta}{2} + \mathbf{P}\left(A^{(m)}\right) \leq \frac{\delta}{2} + \frac{\delta}{2},$$

hence (2.19) holds. □

Proposition 2.40 *If M and N are continuous local martingales then $[M, N]$ is the only continuous process with finite variation on finite intervals for which:*

1. $[M, N](0) = 0$ *and*
2. $MN - [M, N]$ *is a continuous local martingale.*

For any infinitesimal sequence of partitions $(t_k^{(n)})$ of $[0, t]$

$$\sup_{s \leq t} |Q_n(s) - [M, N](s)| \xrightarrow{p} 0$$

where

$$Q_n(s) \stackrel{\circ}{=} \sum_k \left(M(t_k \wedge s) - M(t_{k-1} \wedge s)\right)\left(N(t_k \wedge s) - N(t_{k-1} \wedge s)\right). \quad (2.20)$$

Proof. From Fisk's theorem the uniqueness of $[M, N]$ is again trivial, as $MN - A$ and $MN - B$ are continuous local martingales for some A and B, then $A - B$ is a continuous local martingale with finite variation, so $A - B$ is a constant. As $A(0) = B(0) = 0$ obviously $A = B$.

$$MN = \frac{1}{4}\left((M + N)^2 - (M - N)^2\right),$$

so it is easy to see that Proposition 2.39 can be applied to

$$[M, N] \stackrel{\circ}{=} \frac{1}{4}\left([M + N] - [M - N]\right) \quad (2.21)$$

in order to show that $MN - [M, N]$ is a continuous local martingale and that (2.21) holds. □

Definition 2.41 *If for some process X there is a process P such that $X - P$ is a local martingale, then we say that P is a compensator of X. If P is continuous then we say that P is a continuous compensator of X. If P is predictable then we say that P is a predictable compensator of X etc.*

So far we have proved that if M is a continuous local martingale then $[M]$ is the only increasing, continuous compensator of M^2. It is important to emphasize that this property of $[M]$ holds only for continuous local martingales.

Example 2.42 Quadratic variation of the compensated Poisson processes.

Let π be a Poisson process with parameter λ. The increments of π are independent and the expected value of $\pi(t)$ is λt, hence the compensated process $\nu(t) \doteq \pi(t) - \lambda t$ is a martingale. We show that $\nu^2(t) - \lambda t$ is also a martingale, that is: λt is a continuous, increasing compensator for ν^2.

$$\mathbf{E}\left(\nu^2(t) - \lambda t \mid \mathcal{F}_s\right) = \nu(s)^2 + 2\nu(s)\mathbf{E}\left(\nu(t) - \nu(s) \mid \mathcal{F}_s\right) +$$
$$+ \mathbf{E}\left((\nu(t) - \nu(s))^2 \mid \mathcal{F}_s\right) - \lambda t.$$

The increments of π are independent, hence the conditional expectation is a real expectation. Given that the increments are stationary

$$2\nu(s)\mathbf{E}\left(\nu(t) - \nu(s) \mid \mathcal{F}_s\right) = 2\nu(s)\mathbf{E}\left(\nu(t-s)\right) = 0$$
$$\mathbf{E}\left((\nu(t) - \nu(s))^2 \mid \mathcal{F}_s\right) = \mathbf{E}\left((\nu(t-s))^2\right) = \lambda(t-s),$$

hence

$$\mathbf{E}\left(\nu^2(t) - \lambda t \mid \mathcal{F}_s\right) = \nu^2(s) + \lambda(t-s) - \lambda t =$$
$$= \nu^2(s) - \lambda s.$$

If we partition the interval $[0, t]$ then if $Q_n^{(\nu)}$ is the sequence of the approximating sum for $[\nu]$ and $Q_n^{(\pi)}$ is for $[\pi]$, then

$$Q_n^{(\nu)} = Q_n^{(\pi)} - 2\lambda \sum_k \left(\pi\left(t_k^{(n)}\right) - \pi\left(t_{k-1}^{(n)}\right)\right)\left(t_k^{(n)} - t_{k-1}^{(n)}\right) +$$
$$+ \lambda^2 \sum_k \left(t_k^{(n)} - t_{k-1}^{(n)}\right)^2.$$

It is easy to see that if $\max_k \left(t_k^{(n)} - t_{k-1}^{(n)}\right) \to 0$ then the limit of $Q_n^{(\pi)}$ is the process π. The limits of the other expressions are zero. Hence $[\nu] = \pi$.   □

Proposition 2.43 If M, N and U are continuous local martingales; ξ and η are $\mathcal{F}_0$-measurable random variables then

$$[\xi M + \eta N, U] = \xi[M, U] + \eta[N, U].$$

Proof. $MU - [M, U]$ and $NU - [N, U]$ are local martingales hence $(M + N)U - ([M, U] + [N, U])$ is also a local martingale, and by the uniqueness property of the quadratic co-variation

$$[M + N, U] = [M, U] + [N, U].$$

In a similar way: $MU - [M, U]$ is a local martingale, ξ is $\mathcal{F}_0$-measurable, hence $\xi (MU - [M, U])$ is also a local martingale, hence again by the uniqueness property of the quadratic co-variation $[\xi M, N] = \xi [M, N]$. $\square$

Proposition 2.44 *If M and N are continuous local martingales then*

$$[M, N] = [M - M(0), N - N(0)] = [M - M(0), N].$$

Proof. Obviously $[M - M(0), N] = [M, N] - [M(0), N]$. As $M(0)$ is $\mathcal{F}_0$-measurable $M(0) N$ is a continuous local martingale. Hence $[M(0), N] = 0$. $\square$

Proposition 2.45 (Stopping rule for quadratic variation) *Let τ be an arbitrary stopping time.*

1. *If M is a continuous local martingale then $[M^\tau] = [M]^\tau$.*
2. *If M and N are continuous local martingales then $[M^\tau, N^\tau] = [M, N]^\tau = [M^\tau, N]$.*

Proof. $[M^\tau]$ is the only continuous, increasing process A for which $A(0) = 0$ and $(M^\tau)^2 - A$ is a continuous local martingale. $M^2 - [M]$ is a continuous local martingale, hence

$$\left(M^2 - [M]\right)^\tau = \left(M^2\right)^\tau - [M]^\tau = (M^\tau)^2 - [M]^\tau$$

is a continuous local martingale, hence by the uniqueness $[M]^\tau = [M^\tau]$. From (2.21) and from the first part of the proof

$$[M^\tau, N^\tau] \overset{\circ}{=} \frac{1}{4} \left([(M + N)^\tau] - [(M - N)^\tau]\right) =$$

$$= \frac{1}{4} \left([M + N]^\tau - [M - N]^\tau\right) = [M, N]^\tau.$$

If U and V are martingales and τ is a stopping time, then for any bounded stopping time σ by the Optional Sampling Theorem

$$\mathbf{E}\left((U^\tau \cdot (V - V^\tau))(\sigma)\right) = \mathbf{E}\left(U(\tau \wedge \sigma) \cdot \mathbf{E}\left(V(\sigma) - V(\tau \wedge \sigma) \mid \mathcal{F}_{\tau \wedge \sigma}\right)\right) =$$

$$= \mathbf{E}\left(U(\tau \wedge \sigma) \cdot 0\right) = 0,$$

hence $U^\tau (V - V^\tau)$ is a martingale. From this it is easy to prove with localization that $M^\tau (N - N^\tau)$ is a local martingale, hence

$$M^\tau N - [M, N]^\tau = M^\tau N - M^\tau N^\tau + M^\tau N^\tau - [M, N]^\tau =$$

$$= M^\tau (N - N^\tau) + ((MN)^\tau - [M, N]^\tau)$$

is also a local martingale. From the uniqueness of the quadratic co-variation

$$[M^\tau, N] = [M, N]^\tau = [M^\tau, N^\tau].$$ $\qquad\square$

Example 2.46 If M and N are independent and they are continuous local martingales with respect to their own filtration then $[M, N] = 0$.

Let $\mathcal{F}^M$ and $\mathcal{F}^N$ be the filtrations generated by M and N. Let $\mathcal{F}_s$ be the σ-algebra generated by the sets

$$A \cap B, \quad A \in \mathcal{F}_s^M, B \in \mathcal{F}_s^N.$$

We shall prove that if M and N are independent martingales then MN is a martingale under the filtration $\mathcal{F}$. As M and N are martingales, $M(t)$ and $N(t)$ are integrable. $M(t)$ and $N(t)$ are independent for any t. Hence the product $M(t) N(t)$ is also integrable. If $F \doteq A \cap B$, $A \in \mathcal{F}_s^M$ and $B \in \mathcal{F}_s^N$ then

$$\mathbf{E} (MN(t) \chi_F) = \mathbf{E} (M(t) \chi_A N(t) \chi_B) = \mathbf{E} (M(t) \chi_A) \mathbf{E} (N(t) \chi_B) =$$

$$= \mathbf{E} (M(s) \chi_A) \mathbf{E} (N(s) \chi_B) = \mathbf{E} (MN(s) \chi_F),$$

which by the uniqueness of the extension of finite measures can be extended for every $F \in \mathcal{F}_s$. Hence MN is an $\mathcal{F}$-martingale so $[M, N] = 0$. The quadratic co-variation is independent of the filtration[34] so $[M, N] = 0$ under the original filtration. If M and N are local martingales with respect to their own filtration, then the localized processes are independent martingales. Hence if (τ_n) is a common localizing sequence then $[M, N]^{\tau_n} = [M^{\tau_n}, N^{\tau_n}] = 0$. Hence $[M, N] = 0$. $\qquad\square$

Proposition 2.47 *Let M be a continuous local martingale. M is indistinguishable from a constant if and only if the quadratic variation $[M]$ is zero.*

---

[34]Here we directly used the definition of the quadratic variation as the limit of the approximating sums.

Proof. If M is a constant then M^2 is also a constant, hence M^2 is a local martingale[35] so $[M] = 0$. On the other hand if $[M] = 0$ then $M^2 - [M] = M^2$ is a local martingale. The proposition follows from the next proposition. $\square$

Proposition 2.48 M and M^2 are continuous local martingales, if and only if M is a constant.

Proof. If M is constant then M and M^2 are local martingales. On the other hand

$$(M - M(0))^2 = M^2 - 2 \cdot M \cdot M(0) + M^2(0).$$

Since M and M^2 are local martingales and $M(0)$ is $\mathcal{F}_0$-measurable, $(M - M(0))^2$ is also a local martingale. Let (τ_n) be a localizing sequence for $(M - M(0))^2$. By the martingale property

$$\mathbf{E}\left((M^{\tau_n}(t) - M^{\tau_n}(0))^2\right) = \mathbf{E}\left((M^{\tau_n}(0) - M^{\tau_n}(0))^2\right) = 0,$$

hence for any t

$$M(t \wedge \tau_n) \overset{a.s.}{=} M(0).$$

Therefore for any t

$$M(t) = \lim_{n \to \infty} M(t \wedge \tau_n) \overset{a.s.}{=} M(0).$$

The local martingales are right-regular therefore M is indistinguishable from $M(0)$. $\square$

Corollary 2.49 Let $a \leq b < \infty$. A continuous local martingale M is constant on $[a, b]$ if and only if $[M]$ is constant on $[a, b]$.

Proof. If $\tau_n \nearrow \infty$ then a process X is constant on an interval $[a, b]$ if and only if X^{τ_n} is constant on $[a, b]$ for all n. Using this fact and that $[M^{\tau_n}] = [M]^{\tau_n}$ one can assume that M is a martingale.

1. Define the stochastic process

$$N(t) \overset{\circ}{=} M(t + a) - M(a).$$

N is trivially a martingale for the filtration $\mathcal{G}_t \overset{\circ}{=} \mathcal{F}_{t+a}$, $t \geq 0$.

$$N^2(t) - ([M](t + a) - [M](a)) = M^2(t + a) - ([M](t + a) - [M](a)) -$$
$$- 2M(t + a)M(a) + M^2(a).$$

---

[35]See: Definition 1.131, page 94.

Obviously

$$M^2 (t + a) - ([M] (t + a) - [M] (a))$$

is a $\mathcal{G}$-martingale. $M (t + a)$ is also a $\mathcal{G}$-martingale hence

$$M (t + a) M (a) + M^2 (a)$$

is obviously a $\mathcal{G}$-local martingale, hence by the uniqueness of the quadratic variation

$$[N] (t) = [M] (t + a) - [M] (a) .$$

2. M is constant on the interval $[a, b]$ if and only if N is zero on the interval $[0, b - a]$. As we proved N is constant on $[0, b - a]$ if and only if $[N] = 0$ on $[0, b - a]$. Hence M is constant on $[a, b]$ if and only if $[M]$ is constant on $[a, b]$.□

We summarize the statements above in the following proposition:

Proposition 2.50 $[M, N]$ *is a symmetric bilinear form and* $[M] \geq 0$. $[M] = 0$ *if and only if M is constant. This is also true on any half-line $[a, \infty)$ if instead of $[M, N]$ we use the increments $[M, N] - [M, N] (a)$.*

2.3   Integration when Integrators are Continuous Semimartingales

In this section we introduce a simple construction of the stochastic integral when the integrator X is a continuous semimartingale and the integrand Y is progressively measurable[36]. Every continuous semimartingale has a unique decomposition of type $X = X (0) + L + V$, where V is continuous and has finite variation and L is a continuous local martingale. The integration with respect to V is a simple measure theoretic exercise: $V (\omega)$ generates a σ-finite measure on $\mathbb{R}_+$ for every ω. Every progressively measurable process is product measurable, hence all trajectories $Y (\omega)$ are measurable. For every ω and for every t one can define the pathwise integral

$$(Y \bullet V) (t, \omega) \stackrel{\circ}{=} \int_0^t Y (s, \omega) V (ds, \omega) ,$$

where the integrals are simple Lebesgue integrals[37]

The main problem is how to define the stochastic integral with respect to the local martingale part L!

---

[36]See: [78]

[37]See: Proposition 1.20, page 11.

2.3.1 The space of square-integrable continuous local martingales

Recall the definition and some elementary properties of square-integrable martingales:

Definition 2.51 *As before $\mathcal{H}^2$ is the space of $L^2(\Omega)$ bounded martingales*[38] *on $\mathbb{R}_+$. Let $\mathcal{G}^2 \stackrel{\circ}{=} \mathcal{H}_c^2$ denote the space of $L^2(\Omega)$-bounded, continuous martingales.*

$$\mathcal{H}_0^2 \stackrel{\circ}{=} \left\{ M \in \mathcal{H}^2 : M(0) = 0 \right\}, \quad \mathcal{G}_0^2 \stackrel{\circ}{=} \left\{ M \in \mathcal{G}^2 : M(0) = 0 \right\}.$$

The elements of $\mathcal{H}^2, \mathcal{G}^2, \mathcal{H}_0^2$ and $\mathcal{G}_0^2$ are equivalence classes: M_1 and M_2 are in the same equivalence class if they are indistinguishable.

Proposition 2.52 *$M \in \mathcal{H}^2$ if and only if*

$$\sup_t M^2(t) \in L^1(\Omega).$$

$\left(\mathcal{H}^2, \|\cdot\|_{\mathcal{H}^2} \right)$ is a Hilbert space where

$$\|M\|_{\mathcal{H}^2} \stackrel{\circ}{=} \|M(\infty)\|_2 = \lim_{t \to \infty} \|M(t)\|_2.$$

the set of continuous square-integrable martingales $\mathcal{G}^2$ is a closed subspace of $\mathcal{H}^2$.

Proof. The first statement follows from Doob's inequality[39]. The relation

$$\|M(\infty)\|_2 = \lim_{t \to \infty} \|M(t)\|_2$$

is obviously true as $M(t)$ converges[40] to $M(\infty)$ in $L^2(\Omega)$, and the norm is a continuous function. In order to show that $\mathcal{G}^2$ is closed, let (M_n) be a sequence of continuous square-integrable martingales and assume that $M_n \xrightarrow{\mathcal{H}^2} M$. By Doob's inequality[41]

$$\mathbf{E}\left(\left(\sup_t |M_n(t) - M(t)| \right)^2 \right) \leq 4 \|M_n(\infty) - M(\infty)\|_2^2 \stackrel{\circ}{=}$$

$$\stackrel{\circ}{=} 4 \|M_n - M\|_{\mathcal{H}^2}^2 \to 0.$$

---

[38]That is if M is a martingale then $M \in \mathcal{H}^2$, that is M is square-integrable, if and only if $\sup_t \|M(t)\|_2 < \infty$.

[39]See: Corollary 1.54, page 34.

[40]See: Corollary 1.59, page 35.

[41]See: (1.18) line, page 34.

From the L^2-convergence one has a subsequence for which

$$\sup_t |M_{n_k}(t) - M(t)| \overset{a.s.}{\to} 0,$$

hence $M_{n_k}(t, \omega) \to M(t, \omega)$ uniformly in t for almost all ω. Hence $M(t, \omega)$ is continuous in t for almost all ω. So the trajectories of M are almost surely continuous, therefore $\mathcal{G}^2$ is closed. □

Our direct goal is to prove that if M is a square-integrable martingale and $M(0) = 0$ then

$$\|M\|_{\mathcal{H}^2}^2 \overset{\circ}{=} \|M(\infty)\|_2^2 = \mathbf{E}\left(M^2(\infty)\right) = \mathbf{E}\left([M](\infty)\right).$$

To do this one should prove that $M^2 - [M]$ is not only a local martingale but it is a uniformly integrable martingale.

Proposition 2.53 (Characterization of square-integrable martingales)
Let M be a continuous local martingale. The following statements are equivalent:

1. *M is square integrable,*
2. *$M(0) \in L^2(\Omega)$ and $\mathbf{E}\left([M](\infty)\right) < \infty$.*

In both cases $M^2 - [M]$ is a uniformly integrable martingale.

Proof. The proof of the equivalence of the statements is the following:

1. Let (τ_n) be a localizing sequence of the local martingale $M^2 - [M]$ and let $\sigma_n \overset{\circ}{=} \tau_n \wedge n$. By the martingale property of $(M^{\tau_n})^2 - [M^{\tau_n}]$

$$\mathbf{E}\left(M^2(\sigma_n) - [M](\sigma_n)\right) = \mathbf{E}\left(M^2(0)\right). \tag{2.22}$$

As M is square-integrable

$$M^2(\sigma_n) \leq \sup_t M^2(t) \in L^1(\Omega),$$

so by the Dominated Convergence Theorem

$$\lim_{n \to \infty} \mathbf{E}\left(M^2(\sigma_n)\right) = \mathbf{E}\left(\lim_{n \to \infty} M^2(\sigma_n)\right) = \mathbf{E}\left(M^2(\infty)\right) < \infty.$$

$[M]$ is increasing therefore by the Monotone Convergence Theorem and by (2.22)

$$\mathbf{E}\left([M](\infty)\right) = \lim_{n \to \infty} \mathbf{E}\left([M](\sigma_n)\right) = \lim_{n \to \infty}\left(\mathbf{E}\left(M^2(\sigma_n) - \mathbf{E}\left(M^2(0)\right)\right)\right) < \infty,$$

that is $[M]\,(\infty) \in L^1(\Omega)$ and 1. implies 2. For every stopping time τ

$$\left| \left(M^2 - [M]\right)(\tau) \right| \leq \sup_t M^2\,(t) + \sup_t [M]\,(t) =$$

$$= \sup_t M^2\,(t) + [M]\,(\infty) \in L^1\,(\Omega)\,,$$

hence the set $\left\{ M^2\,(\tau) - [M]\,(\tau) \right\}_\tau$ is dominated by an integrable variable and therefore it is uniformly integrable. By this $M^2 - [M]$ is a class $\mathcal{D}$ local martingale hence it is a uniformly integrable martingale[42].

2. Let τ be an arbitrary stopping time. Let (σ_n) be a localizing sequence of M. One can assume that $M^{\sigma_n} - M\,(0)$ is bounded[43]. Let $N \overset{\circ}{=} M^{\tau \wedge \sigma_n} - M\,(0)$. By the definition of the quadratic variation

$$N^2\,(t) = 2 \int_0^t N_- \, dN + [N]\,(t)\,.$$

As N_- is bounded the Itô–Stieltjes integral defines a martingale[44]. So

$$\mathbf{E}\left(N^2\,(t)\right) = \mathbf{E}\left([N]\,(t)\right) = \mathbf{E}\left([M^{\tau \wedge \sigma_n}]\,(t)\right) \leq \mathbf{E}\left([M]\,(\infty)\right)\,.$$

Applying Fatou's lemma

$$\mathbf{E}\left((M - M\,(0))^2\,(\tau)\right) \leq \mathbf{E}\left([M]\,(\infty)\right)\,. \tag{2.23}$$

By the second assumption of 2. the expected value on the right-hand side is finite so the set of variables $\mathcal{S}$ of type $(M - M\,(0))\,(\tau)$ is bounded in $L^2\,(\Omega)$. Hence $\mathcal{S}$ is a uniformly integrable set and therefore $M - M\,(0)$ is a class $\mathcal{D}$ local martingale and hence it is a martingale[45]. By (2.23) $M - M\,(0)$ is trivially bounded in $L^2\,(\Omega)$, that is $M - M\,(0) \in \mathcal{G}^2$. As $M\,(0) \in L^2\,(\Omega)$ by the first assumption of 2. obviously $M \in \mathcal{G}^2$. □

Corollary 2.54 *If $M \in \mathcal{G}^2$ and $\sigma \leq \tau$ are stopping times then*

$$\mathbf{E}\left(M^2\,(\tau) - M^2\,(\sigma) \mid \mathcal{F}_\sigma\right) = \mathbf{E}\left([M]\,(\tau) - [M]\,(\sigma) \mid \mathcal{F}_\sigma\right) =$$

$$= \mathbf{E}\left((M\,(\tau) - M\,(\sigma))^2 \mid \mathcal{F}_\sigma\right)\,,$$

specifically

$$\mathbf{E}\left(M^2\,(\tau)\right) - \mathbf{E}\left(M^2\,(0)\right) = \mathbf{E}\left([M]\,(\tau)\right)\,. \tag{2.24}$$

[42]See: Proposition 1.144, page 102.

[43]In the general case when M is not necessarily continuous one can assume that $M^{\sigma_n}_- - M\,(0)$ is bounded.

[44]See: Proposition 2.24, page 128.

[45]See: Proposition 1.144, page 102.

Proof. By the previous proposition $M^2 - [M]$ is a uniformly integrable martingale, hence if $\sigma \leq \tau$ then by the Optional Sampling Theorem

$$\mathbf{E}\left(M^2\left(\tau\right) - [M]\left(\tau\right) \mid \mathcal{F}_\sigma\right) = M^2\left(\sigma\right) - [M]\left(\sigma\right)$$

from which the first equation follows. M is also uniformly integrable hence again by the Optional Sampling Theorem $M\left(\sigma\right) = \mathbf{E}\left(M\left(\tau\right) \mid \mathcal{F}_\sigma\right).$

$$\mathbf{E}\left(\left(M\left(\tau\right) - M\left(\sigma\right)\right)^2 \mid \mathcal{F}_\sigma\right) =$$
$$= \mathbf{E}\left(M^2\left(\tau\right) + M^2\left(\sigma\right) - 2M\left(\sigma\right)M\left(\tau\right) \mid \mathcal{F}_\sigma\right) =$$
$$= \mathbf{E}\left(M^2\left(\tau\right) + M^2\left(\sigma\right) - 2M\left(\sigma\right)^2 \mid \mathcal{F}_\sigma\right) =$$
$$= \mathbf{E}\left(M^2\left(\tau\right) - M^2\left(\sigma\right) \mid \mathcal{F}_\sigma\right). \qquad \square$$

Let M be a semimartingale. Let us define

$$\alpha_M\left(C\right) \stackrel{\circ}{=} \mathbf{E}\left(\int_0^\infty \chi_C d\left[M\right]\right)$$

where the integral with respect $[M]$ is the pathwise Lebesgue–Stieltjes integral generated by the increasing, right-regular[46] process $[M]$. It is not entirely trivial that α_M is well-defined, that is the expression under the expected value is measurable. By the Monotone Convergence Theorem

$$\mathbf{E}\left(\int_0^\infty \chi_C d\left[M\right]\right) = \mathbf{E}\left(\lim_{n\to\infty}\int_0^n \chi_C d\left[M\right]\right).$$

As $\int_0^n \chi_C d\left[M\right]$ is measurable[47] for every n the parametric integral under the expected value is measurable. Obviously α_M is a measure on $\mathcal{B}\left(\mathbb{R}_+\right) \times \mathcal{A}$.

Example 2.55 If $M \in \mathcal{G}^2$ and τ is a stopping time then

$$\alpha_M\left([0,\tau]\right) = \mathbf{E}\left(M^2\left(\tau\right)\right) - \mathbf{E}\left(M^2\left(0\right)\right).$$

If $M \in \mathcal{G}_0^2$ then

$$\|M\|_{\mathcal{H}^2} \stackrel{\circ}{=} \sqrt{\mathbf{E}\left(M^2\left(\infty\right)\right)} = \sqrt{\mathbf{E}\left(\left[M\right]\left(\infty\right)\right)} \stackrel{\circ}{=} \qquad (2.25)$$
$$\stackrel{\circ}{=} \left\|\sqrt{\left[M\right]\left(\infty\right)}\right\|_2 = \sqrt{\alpha_M\left(\mathbb{R}_+ \times \Omega\right)}.$$

---

[46]Of course tacitly we again assume that $[M]$ has a right-regular version.
[47]See: Proposition 1.20, page 11.

If τ is an arbitrary random time then

$$\alpha_M\left([0,\tau]\right) \stackrel{\circ}{=} \mathbf{E}\left(\int_0^\infty \chi\left([0,\tau]\right)d\left[M\right]\right) = \mathbf{E}\left([M]\left(\tau\right) - [M]\left(0\right)\right) =$$
$$= \mathbf{E}\left([M]\left(\tau\right)\right).$$

By (2.24) for every stopping time

$$\mathbf{E}\left([M]\left(\tau\right)\right) = \mathbf{E}\left(M^2\left(\tau\right)\right) - \mathbf{E}\left(M^2\left(0\right)\right),$$

hence

$$\alpha_M\left([0,\tau]\right) = \mathbf{E}\left([M]\left(\tau\right)\right) - \mathbf{E}\left([M]\left(0\right)\right) = \mathbf{E}\left([M]\left(\tau\right)\right) =$$
$$= \mathbf{E}\left(M^2\left(\tau\right)\right) - \mathbf{E}\left(M^2\left(0\right)\right).$$

If $M \in \mathcal{G}_0^2$ then $M\left(0\right) \stackrel{\circ}{=} 0$ hence by (2.24)

$$\mathbf{E}\left(M^2\left(\infty\right)\right) = \mathbf{E}\left(M^2\left(\infty\right)\right) - \mathbf{E}\left(M^2\left(0\right)\right) = \mathbf{E}\left([M]\left(\infty\right)\right).$$

The other relations are consequences of the definitions. $\qquad\square$

Definition 2.56 α_M *is called the* Doléans measure[48] *generated by the quadratic variation of* M.

2.3.2 Integration with respect to continuous local martingales

Let us start with the simplest case:

Definition 2.57 *Let* M *be a continuous local martingale. Let* $\mathcal{L}^2\left(M\right)$ *denote the space of equivalence classes of square-integrable and progressively measurable functions on the measure space* $\left(\mathbb{R}_+ \times \Omega, \mathcal{R}, \alpha_M\right)$ *that is let*

$$\mathcal{L}^2\left(M\right) \stackrel{\circ}{=} L^2\left(\mathbb{R}_+ \times \Omega, \mathcal{R}, \alpha_M\right)$$

where $\mathcal{R}$ *, as before, denote the* σ*-algebra of progressively measurable sets. Let* $\|\cdot\|_M$ *denote the norm of the Hilbert space* $\mathcal{L}^2\left(M\right)$:

$$\|X\|_M \stackrel{\circ}{=} \sqrt{\int_{\mathbb{R}_+ \times \Omega} X^2 d\alpha_M} \stackrel{\circ}{=} \sqrt{\mathbf{E}\left(\int_0^\infty X^2 d\left[M\right]\right)}.$$

Example 2.58 The space $\mathcal{L}^2(w)$.

---

[48]See: Definition 5.4, page 295.

The quadratic variation of a Wiener process on an interval $[0, s]$ is s. Hence $\|X\|_w^2 = \mathbf{E} \left(\int_0^t X^2(s) \, ds \right)$ on the interval $[0, t]$. If $t < \infty$ then $w \in \mathcal{L}^2(w)$, since by Fubini's theorem

$$\|w\|_w^2 \overset{\circ}{=} \mathbf{E} \left(\int_0^t w^2(s) \, ds \right) = \int_0^t \mathbf{E} \left(w^2(s) \right) \, ds = \int_0^t s \, ds < \infty. \qquad \square$$

The main result of this section is the following:

Proposition 2.59 (Stochastic integration and quadratic variation) *If M is a continuous local martingale and $X \in \mathcal{L}^2(M)$ then there is a unique process in $\mathcal{G}_0^2$ denoted by $X \bullet M$ such that for every $N \in \mathcal{G}^2$*

$$[X \bullet M, N] = X \bullet [M, N]. \tag{2.26}$$

If we denote $X \bullet M$ by $\int_0^t X \, dM$ then (2.26) can be written as

$$\left[\int_0^t X \, dM, N \right] = \int_0^t X \, d[M, N].$$

Proof. We divide the proof into several steps. We prove that $X \bullet M$ exists, and the definition of $X \bullet M$ is correct—that is, the process $X \bullet M$ is unique.

1. The proof of uniqueness is easy. If I_1 and I_2 are two processes in $\mathcal{G}_0^2$ satisfying (2.26) then $[I_1, N] = [I_2, N]$ for all $N \in \mathcal{G}_0^2$. Hence $[I_1 - I_2, N] = 0$ for all $N \in \mathcal{G}_0^2$. As $I_1 - I_2 \in \mathcal{G}_0^2$

$$[I_1 - I_2, I_1 - I_2] \overset{\circ}{=} [I_1 - I_2] = 0,$$

hence $I_1 - I_2$ is constant[49]. As $I_1 - I_2 \in \mathcal{G}_0^2$, $I_1 - I_2 = 0$, so $I_1 = I_2$.

2. Now we prove the existence of $X \bullet M$. Assume first that $N \in \mathcal{G}_0^2$. By the Kunita–Watanabe inequality[50] and by the formula (2.25)

$$\left| \mathbf{E} \left(\int_0^\infty X \, d[M, N] \right) \right| \leq \left\| \sqrt{\int_0^\infty X^2 d[M]} \right\|_2 \left\| \sqrt{\int_0^\infty d[N]} \right\|_2 \overset{\circ}{=} \tag{2.27}$$

$$\overset{\circ}{=} \|X\|_M \sqrt{\mathbf{E} \left(\int_0^\infty d[N] \right)} =$$

$$= \|X\|_M \sqrt{\mathbf{E} \left([N](\infty) \right)} = \|X\|_M \|N\|_{\mathcal{H}^2}.$$

---

[49]See: Proposition 2.47, page 144.
[50]See: Corollary 2.35, page 137.

Observe that $\|X\|_M \|N\|_{\mathcal{H}^2} < \infty$, hence $\int_0^\infty X d\,[M, N]$ is almost surely finite. So the right-hand side of (2.26) is well-defined. By the bilinearity of the quadratic co-variation

$$N \mapsto \mathbf{E}\left(\int_0^\infty X d\,[M, N]\right)$$

is a continuous linear functional on the Hilbert space $\mathcal{G}_0^2$. As every continuous linear functional on a Hilbert space has a scalar product representation there is an $X \bullet M \in \mathcal{G}_0^2$ such that for every $N \in \mathcal{G}_0^2$

$$\mathbf{E}\left(\int_0^\infty X d\,[M, N]\right) = (X \bullet M, N) \stackrel{\circ}{=} \mathbf{E}\left((X \bullet M)\,(\infty)\,N\,(\infty)\right). \qquad (2.28)$$

3. The main part of the proof is to show that for $X \bullet M$ the identity (2.26) holds. Define the process

$$S \stackrel{\circ}{=} (X \bullet M)\,N - X \bullet [M, N].$$

To prove (2.26) we show that S is a continuous martingale, hence by the uniqueness of the quadratic co-variation

$$[X \bullet M, N] = X \bullet [M, N]!$$

First observe that S is adapted: $(X \bullet M)\,N$ is a product of two martingales, that is the product of two adapted processes. X is progressively measurable, by the definition of $\mathcal{L}^2(M)$, so the integral $\int_0^t X d\,[M, N]$ is also adapted[51]. S is continuous as by the construction $(X \bullet M)\,N$ is a product of two continuous functions so it is continuous, and since M and N are continuous the quadratic variation $[M, N]$ is also continuous. Therefore the integral $\int_0^t X d\,[M, N]$ as a function of t is continuous. Finally to show that S is a martingale one should prove that[52]

$$\mathbf{E}\,(S\,(\tau)) = \mathbf{E}\,(S\,(0)) = 0 \qquad (2.29)$$

for every bounded stopping time τ. By definition $X \bullet M$ is a uniformly integrable martingale. Therefore by the Optional Sampling Theorem

$$(X \bullet M)\,(\tau) = \mathbf{E}\,((X \bullet M)\,(\infty) \mid \mathcal{F}_\tau).$$

---

[51]See: Proposition 1.20, page 11.
[52]See: Proposition 1.91, page 57.

Using that $N^\tau \in \mathcal{G}_0^2$ and (2.28)

$$\mathbf{E}\left(S\left(\tau\right)\right) \overset{\circ}{=} \mathbf{E}\left(\left(X \bullet M\right)\left(\tau\right) N\left(\tau\right) - \int_0^\tau X\left[M, N\right]\right) =$$

$$= \mathbf{E}\left(\left(X \bullet M\right)\left(\tau\right) N\left(\tau\right)\right) - \mathbf{E}\left(\int_0^\tau X\left[M, N\right]\right) =$$

$$= \mathbf{E}\left(\mathbf{E}\left(\left(X \bullet M\right)\left(\infty\right) \mid \mathcal{F}_\tau\right) N\left(\tau\right)\right) - \mathbf{E}\left(\int_0^\infty X\left[M, N\right]^\tau\right) =$$

$$= \mathbf{E}\left(\mathbf{E}\left(\left(X \bullet M\right)\left(\infty\right) N\left(\tau\right) \mid \mathcal{F}_\tau\right)\right) - \mathbf{E}\left(\int_0^\infty X\left[M, N^\tau\right]\right) =$$

$$= \mathbf{E}\left(X \bullet M\left(\infty\right) N\left(\tau\right)\right) - \mathbf{E}\left(\int_0^\infty X\left[M, N^\tau\right]\right) =$$

$$= \mathbf{E}\left(\left(X \bullet M\left(\infty\right) N\left(\tau\right)\right) - \int_0^\infty X\left[M, N^\tau\right]\right) = 0.$$

Therefore (2.29) holds.

4. Finally if $N \in \mathcal{G}^2$ then $N - N\left(0\right) \in \mathcal{G}_0^2$, hence

$$\left[X \bullet M, N\right] = \left[X \bullet M, N - N\left(0\right)\right] =$$

$$= X \bullet \left[M, N - N\left(0\right)\right] = X \bullet \left[M, N\right]. \qquad \square$$

Proposition 2.60 (Stopping rule for stochastic integrals) *If M is an arbitrary continuous local martingale, $X \in \mathcal{L}^2\left(M\right)$ and τ is an arbitrary stopping time then*

$$X \bullet M^\tau = \left(\chi\left(\left[0, \tau\right]\right) X\right) \bullet M = \left(X \bullet M\right)^\tau = X^\tau \bullet M^\tau. \qquad (2.30)$$

Proof. By (2.26) and by the stopping rule for the quadratic variation, if $N \in \mathcal{G}^2$

$$\left[\left(X \bullet M\right)^\tau, N\right] = \left[\left(X \bullet M\right), N\right]^\tau = \left(X \bullet \left[M, N\right]\right)^\tau = X \bullet \left[M, N\right]^\tau =$$

$$= X \bullet \left[M^\tau, N\right] = \left[X \bullet M^\tau, N\right].$$

By the bilinearity of the quadratic variation

$$\left[\left(X \bullet M\right)^\tau - X \bullet M^\tau, N\right] = 0, \ N \in \mathcal{G}^2,$$

from which $\left[\left(X \bullet M\right)^\tau - X \bullet M^\tau\right] = 0$ that is

$$\left(X \bullet M\right)^\tau = X \bullet M^\tau.$$

If $X \in \mathcal{L}^2(M)$ then trivially $\chi([0,\tau])X \in \mathcal{L}^2(M)$. For every $N \in \mathcal{G}^2$

$$[X \bullet M^\tau, N] = X \bullet [M^\tau, N] = X \bullet [M, N]^\tau =$$
$$= (\chi([0,\tau])X) \bullet [M, N] =$$
$$= [(\chi([0,\tau])X) \bullet M, N],$$

hence again

$$X \bullet M^\tau = (\chi([0,\tau])X) \bullet M. \qquad \square$$

Using stopping rule (2.30) we can extend the stochastic integral to the space $\mathcal{L}^2_{\mathrm{loc}}(M)$.

Definition 2.61 *Let M be a continuous local martingale. The space $\mathcal{L}^2_{\mathrm{loc}}(M)$ is the set of progressively measurable processes X for which there is a localizing sequence of stopping times (τ_n) such that*

$$\mathbf{E}\left(\int_0^\infty X^2 d[M^{\tau_n}]\right) = \mathbf{E}\left(\int_0^\infty X^2 d[M]^{\tau_n}\right) =$$
$$= \mathbf{E}\left(\int_0^{\tau_n} X^2 d[M]\right) = \mathbf{E}\left(\int_0^\infty \chi([0,\tau_n])X^2 d[M]\right) \overset{\circ}{=}$$
$$\overset{\circ}{=} \int_{(0,\infty)\times\Omega} \chi([0,\tau_n])X^2 d\alpha_M < \infty.$$

Example 2.62 If M is a continuous local martingale and X is locally bounded then $X \in \mathcal{L}^2_{\mathrm{loc}}(M)$.

One can assume that $X(0) = 0$ as obviously every $\mathcal{F}_0$-measurable constant process is in $\mathcal{L}^2_{\mathrm{loc}}$. As M is continuous $M \in \mathcal{H}^2_{\mathrm{loc}}$. Let (τ_n) be a common localizing sequence of X and M. $M^{\tau_n} \in \mathcal{H}^2$ so[53] $[M^{\tau_n}](\infty) \in L^1(\Omega)$. Therefore

$$\mathbf{E}\left(\int_0^\infty X^2 d[M^{\tau_n}]\right) \leq \sup_{t \leq \tau_n} \left|X^2(t)\right| \mathbf{E}([M^{\tau_n}](\infty)) < \infty. \qquad \square$$

Proposition 2.63 *If M is a continuous local martingale then for every $X \in \mathcal{L}^2_{loc}(M)$ there is a process denoted by $X \bullet M$ such that*

1. *$(X \bullet M)(0) = 0$ and $X \bullet M$ is a continuous local martingale,*
2. *for every continuous local martingale N*

$$[X \bullet M, N] = X \bullet [M, N]. \tag{2.31}$$

[53]See: Proposition 2.53, page 148.

$X \bullet M$ *is unambiguously defined by (2.31), that is* $X \bullet M$ *is the only continuous local martingale for which for every continuous local martingale N (2.31) holds.*

Proof. M is a continuous local martingale so it is locally bounded hence $M \in \mathcal{H}_{\text{loc}}^2$. Assume that $X \in \mathcal{L}_{\text{loc}}^2(M)$ and let (τ_n) be such a localizing sequence of X for which $\mathbf{E}\left(\int_0^\infty X^2 d\left[M^{\tau_n}\right]\right) < \infty$ that is let $X \in \mathcal{L}^2(M^{\tau_n})$. Consider the integrals $I_n \stackrel{\circ}{=} X \bullet M^{\tau_n}$.

$$I_{n+1}^{\tau_n} \stackrel{\circ}{=} \left(X \bullet M^{\tau_{n+1}}\right)^{\tau_n} = X \bullet \left(M^{\tau_{n+1}}\right)^{\tau_n} = X \bullet M^{\tau_n} = I_n,$$

hence I_{n+1} and I_n are equal on $[0, \tau_n]$. One can define the integral process $X \bullet M$ unambiguously if for all n the value of $X \bullet M$ is by definition is I_n on the interval $[0, \tau_n]$. By the stopping rule for stochastic integrals it is obvious from the construction that $X \bullet M$ is independent of the localizing sequence (τ_n). Obviously $(X \bullet M)(0) = 0$ and $X \bullet M$ is continuous. Trivially

$$(X \bullet M)^{\tau_n} \stackrel{\circ}{=} \left(X \bullet M^{\tau_n}\right)^{\tau_n} = X \bullet M^{\tau_n}$$

and $X \bullet M^{\tau_n} \in \mathcal{G}_0^2$, hence $(X \bullet M)^{\tau_n}$ is a uniformly integrable martingale so $X \bullet M$ is a local martingale. We should prove (2.31). Let (τ_n) be such a localizing sequence that $X \in \mathcal{L}^2(M^{\tau_n})$ and $N^{\tau_n} \in \mathcal{G}^2$. As $X \in \mathcal{L}^2(M^{\tau_n})$ and $N^{\tau_n} \in \mathcal{G}^2$ by the stopping rule for the quadratic variation[54]

$$[X \bullet M, N]^{\tau_n} = \left[(X \bullet M)^{\tau_n}, N^{\tau_n}\right] \stackrel{\circ}{=}$$

$$\stackrel{\circ}{=} \left[X \bullet M^{\tau_n}, N^{\tau_n}\right] = X \bullet \left[M^{\tau_n}, N^{\tau_n}\right] =$$

$$= X \bullet [M, N]^{\tau_n} = (X \bullet [M, N])^{\tau_n},$$

hence (2.31) is valid. $\qquad \qquad \square$

Let us prove some elementary properties of the stochastic integral. The most important properties are simple consequences of (2.31), the basic properties of the quadratic variation and the analogous properties of the pathwise integration.

Proposition 2.64 (Itô's isometry) *If M is a continuous local martingale then the mapping $X \mapsto X \bullet M$ is an $\mathcal{L}^2(M) \to \mathcal{G}_0^2$ isometry. That is if $X \in \mathcal{L}^2(M)$ then*

$$\mathbf{E}\left((X \bullet M)^2(\infty)\right) \stackrel{\circ}{=} \|X \bullet M\|_{\mathcal{H}^2}^2 = \|X\|_M^2 \stackrel{\circ}{=} \mathbf{E}\left(\int_0^\infty X^2 d[M]\right).$$

---

[54]See: Proposition 2.45, page 143.

Proof. Using the definition of the norm in $\mathcal{H}^2$ and (2.25), by (2.31)

$$\|X \bullet M\|_{\mathcal{H}^2}^2 \overset{\circ}{=} \mathbf{E}\left((X \bullet M)^2(\infty)\right) = \mathbf{E}\left([X \bullet M](\infty)\right) \overset{\circ}{=}$$

$$\overset{\circ}{=} \mathbf{E}\left([X \bullet M, X \bullet M](\infty)\right) =$$

$$= \mathbf{E}\left(\int_0^\infty Xd[X \bullet M, M]\right) = \mathbf{E}\left(\int_0^\infty Xd(X \bullet [M])\right).$$

In the right-hand side of the identity $[X \bullet M, M] = X \bullet [M, M]$. The integral is taken pathwise, hence

$$\|X \bullet M\|_{\mathcal{H}^2}^2 = \mathbf{E}\left(\int_0^\infty Xd(X \bullet [M])\right) =$$

$$= \mathbf{E}\left(\int_0^\infty X^2 d[M]\right) \overset{\circ}{=} \|X\|_M^2,$$

and hence the mapping $X \mapsto X \bullet M$ is an isometry. $\qquad\square$

Example 2.65 The standard deviation of $\int_0^1 wdw$ is $1/\sqrt{2}$.

The integral is meaningful and as on finite intervals $w \in \mathcal{L}^2(w)$ the integral process $w \bullet w$ is a martingale. Hence the expected value of the integral $\int_0^1 wdw$ is zero. By Itô's isometry and by Fubini's theorem

$$\mathbf{E}\left(\left(\int_0^1 wdw\right)^2\right) = \mathbf{E}\left(\int_0^1 w^2(s)\,ds\right) = \int_0^1 \mathbf{E}\left(w^2(s)\right)ds =$$

$$= \int_0^1 sds = \frac{1}{2}.$$

Hence the standard deviation is $1/\sqrt{2}$. We can calculate the standard deviation in the following way as well:

$$\left(\int_0^t wdw\right)^2 - \left[\int_0^t wdw\right]$$

is a martingale, hence

$$\mathbf{E}\left(\left(\int_0^1 wdw\right)^2\right) = \mathbf{E}\left(\left[\int_0^1 wdw\right]\right) \overset{\circ}{=} \mathbf{E}\left(\left[\int_0^1 wdw, \int_0^1 wdw\right]\right) =$$

$$= \mathbf{E}\left(\int_0^1 w^2 d[w]\right) = \int_0^1 \mathbf{E}\left(w^2(s)\right)ds = \frac{1}{2},$$

using (2.26) directly. $\qquad\square$

Proposition 2.66 *If M is a continuous local martingale and $X \in \mathcal{L}_{loc}^2(M)$ then*

$$[X \bullet M] = X^2 \bullet [M]. \tag{2.32}$$

Proof. By simple calculation using (2.31), and that on the right-hand side of (2.31), we have a pathwise integral

$$[X \bullet M] \overset{\circ}{=} [X \bullet M, X \bullet M] = X \bullet [M, X \bullet M] =$$
$$= X \bullet (X \bullet [M, M]) = X^2 \bullet [M]. \qquad \square$$

Corollary 2.67 *If M is a continuous local martingale and X is a progressively measurable process then $X \in \mathcal{L}_{loc}^2(M)$ if and only if for all t almost surely*

$$\int_0^t X^2 d[M] \overset{\circ}{=} \left(X^2 \bullet [M]\right)(t) < \infty. \tag{2.33}$$

Proof. The quadratic variation $[X \bullet M]$, like every quadratic variation, is almost surely finite, hence if $X \in \mathcal{L}_{loc}^2(M)$ then by (2.32), (2.33) holds. On the other hand, assume that (2.33) holds. For all n let us define the stopping times

$$\tau_n \overset{\circ}{=} \inf\left\{t : \left|\int_0^t X^2 d[M]\right| \geq n\right\}.$$

As $[M]$ is continuous, $X^2 \bullet [M]$ is also continuous, hence

$$\int_0^{\tau_n} X^2 d[M] \leq n,$$

that is $X \in \mathcal{L}^2(M^{\tau_n})$, hence $X \in \mathcal{L}_{loc}^2(M)$, so the space $\mathcal{L}_{loc}^2(M)$ contains all the $\mathcal{R}$-measurable processes, for which (2.33) holds for all t. $\qquad \square$

Corollary 2.68 *Assume that M is a local martingale and $X \in \mathcal{L}_{loc}^2(M)$. If on an interval $[a, b]$*

1. $X(t, \omega) = 0$ for all ω or
2. $M(t, \omega) = M(a, \omega)$,

then $X \bullet M$ is constant on $[a, b]$.

Proof. The integral $X^2 \bullet [M, M]$ is a pathwise integral, hence under the assumptions $X^2 \bullet [M, M]$ is constant on $[a, b]$. As $[X \bullet M] = X^2 \bullet [M]$, the local martingale $X \bullet M$ is constant on[55] $[a, b]$. $\qquad \square$

---

[55]See: Proposition 2.47, page 144.

Proposition 2.69 (Stopping rule for stochastic integrals) *If* M *is a continuous local martingale,* $X \in \mathcal{L}^2_{loc}(M)$ *and* τ *is an arbitrary stopping time then*

$$(X \bullet M)^\tau = \chi\left([0, \tau]\right) X \bullet M = X^\tau \bullet M^\tau = X \bullet M^\tau. \tag{2.34}$$

Proof. Let τ be an arbitrary stopping time. If $X \in \mathcal{L}^2_{loc}(M)$, then as $|\chi\left([0, \tau]\right) X| \leq |X|$ trivially $\chi\left([0, \tau]\right) X \in \mathcal{L}^2_{loc}(M)$. Using the analogous properties of the $\mathcal{L}^2(M)$ integrals

$$((X \bullet M)^\tau)^{\tau_n} = ((X \bullet M)^{\tau_n})^\tau \stackrel{\circ}{=} (X \bullet M^{\tau_n})^\tau =$$
$$= \chi\left([0, \tau]\right) X \bullet M^{\tau_n} \stackrel{\circ}{=}$$
$$\stackrel{\circ}{=} (\chi\left([0, \tau]\right) X \bullet M)^{\tau_n}.$$

The proof of the other parts of (2.34) are analogous. □

Proposition 2.70 (Linearity) $X \bullet M$ *is bilinear, that is if* α_1 *and* α_2 *are constants then*

$$X \bullet (\alpha_1 M_1 + \alpha_2 M_2) = \alpha_1 (X \bullet M_1) + \alpha_2 (X \bullet M_2)$$

and

$$(\alpha_1 X_1 + \alpha_2 X_2) \bullet M = \alpha_1 (X_1 \bullet M) + \alpha_2 (X_2 \bullet M)$$

when all the expressions are meaningful. In these relations if two integrals are meaningful then the third one is meaningful.

Proof. If $X \in \mathcal{L}^2_{loc}(M_1) \cap \mathcal{L}^2_{loc}(M_2)$ then for all t

$$\int_0^t X^2 d[M_1] < \infty \quad \text{and} \quad \int_0^t X^2 d[M_2] < \infty.$$

Obviously, by the Kunita–Watanabe inequality[56]

$$[M_1 + M_2] \leq 2\left([M_1] + [M_2]\right)$$

hence

$$\int_0^t X^2 d[M_1 + M_2] \leq 2\left(\int_0^t X^2 d[M_1] + \int_0^t X^2 d[M_2]\right) < \infty,$$

[56]See: Corollary 2.36, page 137.

therefore $X \in \mathcal{L}^2_{\text{loc}}(M_1 + M_2)$. From the linearity of the pathwise integration and from the bilinearity of the quadratic variation

$$
[X \bullet (\alpha_1 M_1 + \alpha_2 M_2), N] = X \bullet [(\alpha_1 M_1 + \alpha_2 M_2), N] =
$$
$$
= X \bullet (\alpha_1 [M_1, N] + \alpha_2 [M_2, N]) =
$$
$$
= \alpha_1 X \bullet [M_1, N] + \alpha_2 X \bullet [M_2, N] =
$$
$$
= [\alpha_1 X \bullet M_1 + \alpha_2 X \bullet M_2, N],
$$

from which the linearity of the integral in the integrand is evident. The linearity in the integrator is also evident as

$$
[(\alpha_1 X_1 + \alpha_2 X_2) \bullet M, N] = (\alpha_1 X_1 + \alpha_2 X_2) \bullet [M, N] =
$$
$$
= \alpha_1 X_1 \bullet [M, N] + \alpha_2 X_2 \bullet [M, N] =
$$
$$
= [\alpha_1 X_1 \bullet M, N] + [\alpha_2 X_2 \bullet M, N] =
$$
$$
= [\alpha_1 X_1 \bullet M + \alpha_2 X_2 \bullet M, N].
$$

The remark about the integrability is evident from the trivial linearity of the space $\mathcal{L}^2_{\text{loc}}(M)$. $\qquad\square$

Proposition 2.71 (Associativity) *If $X \in \mathcal{L}^2(M)$ then $Y \in \mathcal{L}^2(X \bullet M)$ if and only if $XY \in \mathcal{L}^2(M)$. If $X \in \mathcal{L}^2_{loc}(M)$ then $Y \in \mathcal{L}^2_{loc}(X \bullet M)$, if and only if $XY \in \mathcal{L}^2_{loc}(M)$. In both cases*

$$
(YX) \bullet M = Y \bullet (X \bullet M). \tag{2.35}
$$

Proof. Using the construction of the stochastic integral and given that the associativity formula (2.35) is valid for pathwise integration

$$
[X \bullet M] = [X \bullet M, X \bullet M] = X \bullet [M, X \bullet M] =
$$
$$
= X \bullet (X \bullet [M, M]) = X^2 \bullet [M, M].
$$

By the associativity of the pathwise integration for non-negative integrands

$$
\mathbf{E}\left(\int_0^\infty Y^2 d[X \bullet M] \right) = \mathbf{E}\left(\int_0^\infty Y^2 d \int_0^s X^2 d[M] \right) =
$$
$$
= \mathbf{E}\left(\int_0^\infty Y^2 X^2 d[M] \right),
$$

hence $YX \in \mathcal{L}^2(M)$ if and only if $Y \in \mathcal{L}^2(X \bullet M)$. If $X \in \mathcal{L}^2(M)$, then by the Kunita–Watanabe inequality for almost all ω the trajectory $X(\omega)$ is integrable with respect to $[M, N](\omega)$. If $XY \in \mathcal{L}^2(M)$ then using (2.26) again

$$[(YX) \bullet M, N] = (YX) \bullet [M, N] \overset{\circ}{=} \tag{2.36}$$

$$\overset{\circ}{=} \int_0^t YX d[M, N] = \int_0^t Y d \int_0^s X d[M, N] \overset{\circ}{=} Y \bullet (X \bullet [M, N]),$$

Using (2.26) and that $Y \in \mathcal{L}^2(X \bullet M)$,

$$Y \bullet (X \bullet [M, N]) = Y \bullet [X \bullet M, N] = [Y \bullet (X \bullet M), N].$$

Comparing it with line (2.36),

$$[(YX) \bullet M, N] = [Y \bullet (X \bullet M), N].$$

Hence by the uniqueness of the stochastic integral

$$(YX) \bullet M = Y \bullet (X \bullet M).$$

To prove the general case, observe that $XY \in \mathcal{L}^2_{\mathrm{loc}}(M)$ if and only if for some localizing sequence (τ_n)

$$\mathbf{E}\left(\chi([0, \tau_n]) X^2 Y^2 \bullet [M]\right) < \infty.$$

As

$$\chi([0, \tau_n]) Y^2 \bullet \left(X^2 \bullet [M]\right) = \chi([0, \tau_n]) Y^2 X^2 \bullet [M]$$

$XY \in \mathcal{L}^2_{\mathrm{loc}}(M)$ if and only if $Y \in \mathcal{L}^2_{\mathrm{loc}}(X \bullet M)$. Let (τ_n) be a common localizing sequence for M and $X \bullet M$. If $Y \in \mathcal{L}^2_{\mathrm{loc}}(X \bullet M)$ then evidently

$$Y \in \mathcal{L}^2((X \bullet M)^{\tau_n}) = \mathcal{L}^2((X \bullet M^{\tau_n})).$$

So

$$(Y \bullet (X \bullet M))^{\tau_n} \overset{\circ}{=} Y \bullet (X \bullet M)^{\tau_n} = Y \bullet (X \bullet M^{\tau_n}) =$$

$$= (YX \bullet M^{\tau_n}) \overset{\circ}{=} ((YX \bullet M))^{\tau_n},$$

from which the associativity is evident. $\qquad\square$

2.3.3 Integration with respect to semimartingales

We can extend again the definition of the stochastic integration to semimartingales:

Definition 2.72 *Let $X = X(0) + L + V$ be a continuous semimartingale. If for some process Y the integrals $Y \bullet L$ and $Y \bullet V$ are meaningful then the stochastic integral $Y \bullet X$ of Y with respect to X by definition is the sum*

$$Y \bullet X \overset{\circ}{=} Y \bullet L + Y \bullet V.$$

Remember that by Fisk's theorem the decomposition $X = X(0) + L + V$ is unique, hence the integral is well-defined.

Proposition 2.73 *The most important properties of the stochastic integral $Y \bullet X$ are the following:*

1. *$Y \bullet X$ is bilinear, that is*

$$Y \bullet (\alpha_1 X_1 + \alpha_2 X_2) = \alpha_1 (Y \bullet X_1) + \alpha_2 (Y \bullet X_2)$$

and

$$(\alpha_1 Y_1 + \alpha_2 Y_2) \bullet X = \alpha_1 (Y_1 \bullet X) + \alpha_2 (Y_2 \bullet X)$$

assuming that all the expressions are meaningful. If two integrals are meaningful then the third is meaningful.

2. *For all locally bounded processes Y, Z*

$$Z \bullet (Y \bullet X) = (ZY) \bullet X.$$

3. *For every stopping time τ*

$$(Y \bullet X)^{\tau} = (Y\chi([0,\tau]) \bullet X) = Y \bullet X^{\tau}.$$

4. *If the integrator X is a local martingale or if X has bounded variation on finite intervals then the same is true for the integral process $Y \bullet X$.*
5. *$Y \bullet X$ is constant on any interval where either $Y = 0$, or X is constant.*
6. *$[Y \bullet X, Z] = Y \bullet [X, Z]$ for any continuous semimartingale Z.*

2.3.4 The Dominated Convergence Theorem for stochastic integrals

A crucial property of every integral is that under some conditions one can swap the order of taking limit and the integration:

Proposition 2.74 (Dominated Convergence Theorem for stochastic integrals) *Let X be a continuous semimartingale, and let (Y_n) be a sequence of* *progressively measurable processes. Assume that $(Y_n(t,\omega))$ converges to $Y_\infty(t,\omega)$ in every point (t,ω). If there is an integrable process Y such that[57] $|Y_n| \leq Y$ for all n, then $Y_n \bullet X \to Y_\infty \bullet X$, where the convergence is uniform in probability on every compact interval, that is*

$$\sup_{s \leq t} |(Y_n \bullet X)(s) - (Y_\infty \bullet X)(s)| \xrightarrow{P} 0, \quad \text{for all } t \geq 0.$$

Proof. One can prove the proposition separately when X has finite variation and when X is a local martingale. It is sufficient to prove the proposition when $Y_\infty \equiv 0$.

1. First, assume that X has finite variation. In this case the integrability of Y means that for every t

$$\int_0^t |Y| \, d\mathrm{Var}(X) < \infty.$$

As $|Y_n| \leq Y$, for every ω the trajectory $Y_n(\omega)$ is also integrable on every interval $[0, t]$. Applying the classical Dominated Convergence Theorem for every trajectory individually, for all $s \leq t$

$$\left| \int_0^s Y_n \, dX \right| \leq \int_0^t |Y_n| \, d\mathrm{Var}(X) \to 0.$$

Hence the integral, as a function of the upper bound uniformly converges to zero. Pointwise convergence on a finite measure space implies convergence in measure, so when the integrator has finite variation then the proposition holds.

2. Let X be a local martingale. Y is integrable with respect to X, hence by definition $Y \in \mathcal{L}^2_{\mathrm{loc}}(X)$. Let $\varepsilon, \delta > 0$ be arbitrary, and let (τ_n) be a localizing sequence of Y. To make the notation simpler, let us denote by σ a τ_n for which $\mathbf{P}(\tau_n < t) \leq \delta/2$. By the stopping rule $(Y_n \bullet X)^\sigma = Y_n \bullet X^\sigma$, that is if $s \leq \sigma(\omega)$ then

$$(Y_n \bullet X)(s, \omega) = (Y_n \bullet X^\sigma)(s, \omega).$$

If

$$A \doteq \left\{ \sup_{s \leq t} |Y_n \bullet X|(s) > \varepsilon \right\}, \quad A_\sigma \doteq \left\{ \sup_{s \leq t} |Y_n \bullet X^\sigma|(s) > \varepsilon \right\},$$

---

[57]The integrability of Y depends on the integrator X. If X is a local martingale, then by definition this means that $Y \in \mathcal{L}^2_{\mathrm{loc}}(X)$.

then

$$\mathbf{P}(A) = \mathbf{P}((\sigma < t) \cap A) + \mathbf{P}((\sigma \geq t) \cap A) \leq$$

$$\leq \mathbf{P}(\sigma < t) + \mathbf{P}((t \leq \sigma) \cap A) \leq \frac{\delta}{2} + \mathbf{P}(A_\sigma).$$

Since $Y \in \mathcal{L}^2(X^\sigma)$, obviously $Y_n \in \mathcal{L}^2(X^\sigma)$. Hence by the classical Dominated Convergence Theorem as $Y_n \to 0$ and $|Y_n| \leq Y$

$$\|Y_n\|_{X^\sigma}^2 \stackrel{\circ}{=} \mathbf{E}\left(\int_0^\infty Y_n^2 d[X^\sigma]\right) = \mathbf{E}\left(\int_0^\infty Y_n^2 d[X]^\sigma\right) =$$

$$= \mathbf{E}\left(\int_0^\infty \chi([0,\sigma]) Y_n^2 d[X]\right) \to 0,$$

that is $Y_n \to 0$ in $\mathcal{L}^2(X^\sigma)$. By Itô's isometry the correspondence $Z \mapsto Z \bullet X^\sigma$ is an $\mathcal{L}^2(X^\sigma) \to \mathcal{H}^2$ isometry[58]. Hence $Y_n \bullet X^\sigma \stackrel{\mathcal{H}^2}{\to} 0$. By Doob's inequality[59]

$$\mathbf{E}\left(\left(\sup_{s \leq \infty} |Y_n \bullet X^\sigma|(s)\right)^2\right) \leq 4\mathbf{E}\left(((Y_n \bullet X^\sigma)(\infty))^2\right) \stackrel{\circ}{=}$$

$$\stackrel{\circ}{=} 4\|Y_n \bullet X^\sigma\|_{\mathcal{H}^2}^2 \to 0.$$

By Markov's inequality, stochastic convergence follows from the $L^2(\Omega)$-convergence, hence

$$\mathbf{P}(A_\sigma) \stackrel{\circ}{=} \mathbf{P}\left(\sup_{s \leq t} |Y_n \bullet X^\sigma|(s) > \varepsilon\right) \to 0.$$

Hence for n large enough

$$\mathbf{P}(A) \stackrel{\circ}{=} \mathbf{P}\left(\sup_{s \leq t} |Y_n \bullet X|(s) > \varepsilon\right) \leq \delta. \qquad \square$$

2.3.5 Stochastic integration and the Itô–Stieltjes integral

As we mentioned, every integral is in some sense the limits of certain approximating sums. From the construction above it is not clear in which sense the integral $X \bullet M$ is a limit of the approximating sums.

[58] See: Itô's isometry, Proposition 2.64, page 156.
[59] See: line (1.17) page 34. Proposition 2.52 page 147.

Lemma 2.75 *If X is a continuous semimartingale and*

$$Y \overset{\circ}{=} \sum_i \eta_i \cdot \chi \left((\tau_i, \tau_{i+1}] \right)$$

is an integrable, non-negative predictable simple process[60] *then*

$$(Y \bullet X)(t) = \sum_i \eta_i \cdot \left(X(\tau_{i+1} \wedge t) - X(\tau_i \wedge t) \right).$$

Proof. If $\sigma \le \tau$ are stopping times, then using the linearity and the stopping rule

$$\chi \left((\sigma, \tau] \right) \bullet X = \left(\chi \left([0, \tau] \right) - \chi \left([0, \sigma] \right) \right) \bullet X =$$
$$= (1 \bullet X)^{\tau} - (1 \bullet X)^{\sigma} = X^{\tau} - X^{\sigma}.$$

Hence the formula holds with $\eta \equiv 1$. It is easy to check that if $F \in \mathcal{F}_\sigma \subseteq \mathcal{F}_\tau$ then

$$\sigma_F (\omega) \overset{\circ}{=} \begin{cases} \sigma (\omega) & \text{if} \quad \omega \in F \\ \infty & \text{if} \quad \omega \notin F \end{cases}, \qquad \tau_F (\omega) \overset{\circ}{=} \begin{cases} \tau (\omega) & \text{if} \quad \omega \in F \\ \infty & \text{if} \quad \omega \notin F \end{cases}$$

are also stopping times, hence

$$\left(\chi_F \chi \left((\sigma, \tau] \right) \right) \bullet X = \chi \left((\sigma_F, \tau_F] \right) \bullet X = X^{\tau_F} - X^{\sigma_F} = \chi_F \left(X^{\tau} - X^{\sigma} \right),$$

hence the formula is valid if $\eta = \chi_F$, $F \in \mathcal{F}_\sigma$. If η is an $\mathcal{F}_\sigma$-measurable step function, then since the integral is linear one can write η in the place of χ_F. It is easy to show that for any $\mathcal{F}_\sigma$-measurable function η the process $\eta \chi \left((\sigma, \tau] \right)$ is integrable with respect to X, hence using the Dominated Convergence Theorem one can prove the formula when η is an arbitrary $\mathcal{F}_\sigma$-measurable function. As $Y \ge 0$

$$0 \le Y_n \overset{\circ}{=} \sum_{i=1}^{n} \eta_i \chi \left((\tau_i, \tau_{i+1}] \right) \le Y.$$

The general case follows from the Dominated Convergence Theorem and from the linearity of the integral. $\qquad \square$

Corollary 2.76 *If X is a continuous semimartingale, $\tau_n \nearrow \infty$ and $Y \overset{\circ}{=} \sum_i \eta_i \cdot \chi \left((\tau_i, \tau_{i+1}] \right)$ is a predictable simple process then*

$$\int_0^t Y \, dX \overset{\circ}{=} (Y \bullet X)(t) = \sum_i \eta_i \cdot \left(X(\tau_{i+1} \wedge t) - X(\tau_i \wedge t) \right).$$

---

[60]See: Definition 1.41, page 24.

Proof. As $\tau_n \nearrow \infty$, Y is left-continuous and has right-hand side limits. So Y is locally bounded on $[0, \infty)$ and therefore $Y^{\pm}$ are integrable. $\qquad\square$

Proposition 2.77 *If X is a continuous semimartingale, Y is a left-continuous, adapted and locally bounded process, then $(Y \bullet X)(t)$ is the Itô–Stieltjes integral for every t. The convergence of the approximating sums is uniform in probability on every compact interval. The partitions of the intervals can be random as well.*

Proof. More precisely, let $\tau_k^{(n)} \leq \tau_{k+1}^{(n)} \nearrow \infty$ be a sequence of stopping times. For each t let

$$\sum_k Y(\tau_k^{(n)}) \left(X(\tau_{k+1}^{(n)} \wedge t) - X(\tau_k^{(n)} \wedge t) \right)$$

be the sequence of Itô-type approximating processes. Assume that for each ω

$$\lim_{n \to \infty} \max_k \left| \tau_{k+1}^{(n)}(\omega) - \tau_k^{(n)}(\omega) \right| = 0.$$

Define the locally bounded simple predictable processes

$$Y^{(n)} \overset{\circ}{=} \sum_k Y\left(\tau_k^{(n)}\right) \chi\left(\tau_k^{(n)}, \tau_{k+1}^{(n)}\right].$$

As we saw

$$\left(Y^{(n)} \bullet X\right)(t) = \sum_k Y(\tau_k^{(n)}) \left(X(\tau_{k+1}^{(n)} \wedge t) - X(\tau_k^{(n)} \wedge t) \right).$$

Y is continuous from the left, hence in every point $Y^{(n)} \to Y$. Let

$$K(t) \overset{\circ}{=} \sup_{s < t} |Y(s)|.$$

Y is continuous from the left, so one can take the supremum over the rational points only, hence K is progressively measurable. If Y is locally bounded then K is also locally bounded, and as $\left|Y^{(n)}\right| \leq K$, by the Dominated Convergence Theorem

$$Y^{(n)} \bullet X \to Y \bullet X,$$

where the convergence is uniform in probability on every compact interval. $\qquad\square$

2.4 Integration when Integrators are Locally Square-Integrable Martingales

So far, we have assumed that the integrator processes are continuous. It is obvious from the construction that during the discussion the continuity of the integrator was rarely explicitly used. In fact the only place where the continuity is used is the construction and the characterization of the quadratic variation process. The main point in the continuous case is that if M is a continuous local martingale then the quadratic variation is continuous, hence by Fisk's theorem it is the only increasing process P for which $M^2 - P$ is a local martingale. In this section we briefly discuss the case when the integrators are in $\mathcal{H}^2_{\mathrm{loc}}$. If $M \in \mathcal{H}^2_{\mathrm{loc}}$ then the quadratic variation is generally not continuous hence one cannot use Fisk's theorem. On the other hand as we shall show the jumps $\Delta[M]$ of $[M]$ are the squares of the jumps of M and $[M] - \Delta[M] = [M] - (\Delta M)^2$ is continuous so $[M]$ is the only right-continuous, increasing process P for which $M^2 - P$ is a local martingale and $\Delta P = (\Delta M)^2$. If we use this observation then the rest of the construction is nearly the same as for continuous local martingales and hence we can make the discussion of the case $\mathcal{H}^2_{\mathrm{loc}}$ very short.

2.4.1 The quadratic variation of locally square-integrable martingales

Recall that we have defined the quadratic variation only for finite intervals. The first step of the discussion is to construct the quadratic variation process for $\mathcal{H}^2_{\mathrm{loc}}$ local martingales.

Proposition 2.78 *If $M \in \mathcal{H}^2$ and M_- is uniformly bounded then there is an increasing and right-continuous process $[M]$ such that $[M](t) \overset{a.s.}{=} [M]_0^t$ for any t and*

$$\Delta[M] = (\Delta M)^2.$$

This version is indistinguishable from any increasing, right-continuous process P for which $P(0) = 0$, $\Delta P = (\Delta M)^2$ and $M^2 - P$ is a martingale. If $(t_k^{(n)})$ is an infinitesimal sequence of partitions of $[0, t]$, then

$$\sup_{s \le t} |Q_n(s) - [M](s)| \overset{p}{\to} 0, \tag{2.37}$$

where

$$Q_n(s) \overset{\circ}{=} \sum_k \left(M(t_k^{(n)} \wedge s) - M(t_{k-1}^{(n)} \wedge s) \right)^2. \tag{2.38}$$

Proof. By the integration by parts formula for any t

$$M^2(t) - [M](t) = M^2(0) + 2\int_0^t M_- dM = M^2(0) + 2(M_- \bullet M)(t).$$

As $M \in \mathcal{H}^2$ and M_- is uniformly bounded, the integral process $M_- \bullet M$ has a version which is a martingale[61] and

$$\sup_{s \leq t} |I_n(s) - (M_- \bullet M)(s)| = \sup_{s \leq t} |I_n(s) - (M \bullet M)(s)| \xrightarrow{P} 0, \qquad (2.39)$$

where

$$I_n(s) \stackrel{\circ}{=} \sum_k M(t_{k-1}^{(n)} \wedge s)\left(M(t_k^{(n)} \wedge s) - M(t_{k-1}^{(n)} \wedge s)\right).$$

As M^2 and $M_- \bullet M$ are right-continuous

$$[M] \stackrel{\circ}{=} M^2 - M^2(0) - 2M_- \bullet M$$

is also right-continuous. $[M](t)$ is a version of the quadratic variation $[M]_0^t$ for any t. $[M]_0^p \stackrel{a.s.}{\leq} [M]_0^q$ for rational numbers $p \leq q$. Unifying the measure-zero sets and using the right-continuity of $[M]$ we can construct an increasing version. (2.37) follows from (2.39). By (2.38) there is a subsequence such that

$$\sup_{s \leq t} |Q_{n_k}(s) - [M](s)| \stackrel{a.s.}{\to} 0.$$

From the uniform convergence

$$\Delta[M](s) \stackrel{\circ}{=} [M](s) - [M](s-) = \lim_{n \to \infty} Q_n(s) - \lim_{n \to \infty} Q_n(s-) =$$

$$= \lim_{n \to \infty} (Q_n(s) - Q_n(s-)) = \lim_{n \to \infty} (\Delta Q_n(s)) =$$

$$= \lim_{n \to \infty} \left(\left(M(s) - M\left(t_{k-1}^{(n)}\right)\right)^2 - \left(M(s-) - M\left(t_{k-1}^{(n)}\right)\right)^2\right) =$$

$$= (\Delta M(s))^2 - 0^2 = (\Delta M(s))^2.$$

If P is another right-continuous, increasing process for which $P(0) = 0$, $\Delta P = (\Delta M)^2$ and $M^2 - P$ is martingale, then $N \stackrel{\circ}{=} P - [M]$ is a continuous martingale, $N(0) = 0$ and the trajectories of N have finite variation. By Fisk's theorem $N = 0$ so P is indistinguishable from $[M]$. $\qquad\square$

---

[61]See: Proposition 2.24, page 128.

Proposition 2.79 *Under the assumption of the previous proposition, if τ is an arbitrary stopping time then $[M^\tau] = [M]^\tau$.*

Proof. As $(M^\tau)^2 = (M^2)^\tau$

$$(M^\tau)^2 - [M]^\tau = (M^2)^\tau - [M]^\tau = (M^2 - [M])^\tau$$

Since stopped martingales are martingales, stopping the martingale $M^2 - [M]$ at τ we get a martingale again. $[M]^\tau$ is increasing and

$$\Delta([M]^\tau) = (\Delta[M])^\tau = ((\Delta M)^2)^\tau = ((\Delta M^\tau))^2$$

so by the uniqueness property of the quadratic variation $[M^\tau] = [M]^\tau$.
$\square$

Proposition 2.80 *If $M \in \mathcal{H}_{loc}^2$ then there is one and only one right-continuous increasing process $[M]$ such that $[M](0) = 0$, $\Delta[M] = (\Delta M)^2$ and $M^2 - [M]$ is a local martingale. For any t if $\left(t_k^{(n)}\right)$ is an infinitesimal sequence of partitions of $[0, t]$ then*

$$\sup_{s \le t} |Q_n(s) - [M](s)| \xrightarrow{p} 0$$

where

$$Q_n(s) \stackrel{\circ}{=} \sum_k \left(M(t_k^{(n)} \wedge s) - M(t_{k-1}^{(n)} \wedge s)\right)^2.$$

Proof. If $M \in \mathcal{H}_{loc}^2$ then by definition there is a localizing sequence such that $M^{\tau_n} \in \mathcal{H}^2$. As M_- is left-regular M_- is locally bounded[62], hence we can assume that $M_-^{\tau_n}$ is bounded. If $M_n \stackrel{\circ}{=} M^{\tau_n}$ then by the previous proposition

$$[M_{n+1}]^{\tau_n} = \left[M_{n+1}^{\tau_n}\right] = [M_n^{\tau_n}] = [M_n]$$

hence $[M_n] = [M_{n+1}]$ on the interval $[0, \tau_n]$ and $[M_n]$ is constant on $[\tau_n, \infty)$. As $\tau_n \nearrow \infty$ one can define the process $[M]$ as the 'union' of processes $[M_n]$ that is

$$[M](t, \omega) \stackrel{\circ}{=} [M_n](t, \omega), \quad t \le \tau_n(\omega).$$

Evidently $[M]$ is right-continuous, increasing and $[M](0) = 0$ and $\Delta[M] = (\Delta M)^2$. Of course

$$(M^2 - [M])^{\tau_n} = (M^{\tau_n})^2 - [M]^{\tau_n} \stackrel{\circ}{=} M_n^2 - [M_n],$$

---

[62]See: Proposition 1.151, page 106.

is a martingale, hence $M^2 - [M]$ is a local martingale. The proof of the uniqueness is the same as above. $\qquad\square$

Corollary 2.81 *If* $M, N \in \mathcal{H}^2_{loc}$ *then there is one and only one right-continuous process with bounded variation* $[M, N]$ *such that*

1. $[M, N](0) = 0$,
2. $\Delta[M, N] = \Delta M \Delta N$ *and*
3. $MN - [M, N]$ *is a local martingale.*

For any t *if* $(t_k^{(n)})_k$ *is an infinitesimal sequence of partitions of* $[0, t]$ *then*

$$\sup_{s \leq t} |Q_n(s) - [M](s)| \xrightarrow{p} 0$$

where

$$Q_n(s) \doteq \sum_k \left(M(t_k^{(n)} \wedge s) - M(t_{k-1}^{(n)} \wedge s) \right) \left(N(t_k^{(n)} \wedge s) - N(t_{k-1}^{(n)} \wedge s) \right).$$

One can easily prove the next propositions with a trivial modification of proofs of the corresponding theorems in the continuous case[63].

Proposition 2.82 *If* M *is a local martingale then the quadratic variation* $[M]$ *is zero if and only if* M *is indistinguishable from a constant[64].*

Proposition 2.83 (Stopping rule for quadratic variation) *Let* τ *be an arbitrary stopping time[65].*

1. *If* M *is a local martingale then* $[M^\tau] = [M]^\tau$.
2. *If* M *and* N *are local martingales then* $[M^\tau, N^\tau] = [M, N]^\tau = [M, N^\tau]$.

Proposition 2.84 (Characterization of $\mathcal{H}^2$ martingales) *Let* M *be a local martingale. The following statements are equivalent:*

1. $M \in \mathcal{H}^2$.
2. $M(0) \in L^2(\Omega)$ *and* $\mathbf{E}([M](\infty)) < \infty$.

In both cases $M^2 - [M]$ *is a uniformly integrable martingale[66].*

---

[63] Let us emphasize that we have not proved yet that for an arbitrary local martingale L the difference $L^2 - [L]$ is also a local martingale. To prove this we shall need the Fundamental Theorem of Local Martingales. See: 3.62, page 222. At the moment we have proved only for locally square-integrable martingales that $L^2 - [L]$ is a local martingale, so at the moment one can use the results below only for locally square-integrable martingales.

[64] See: Proposition 2.47, page 144.

[65] See: Proposition 2.45, page 143.

[66] See: Proposition 2.53, page 148.

Proposition 2.85 *If $M \in \mathcal{H}_0^2$ then*

$$\|M\|_{\mathcal{H}^2} \overset{\circ}{=} \sqrt{\mathbf{E}\left(M^2\left(\infty\right)\right)} = \sqrt{\mathbf{E}\left(\left[M\right]\left(\infty\right)\right)} \overset{\circ}{=} \left\|\sqrt{\left[M\right]\left(\infty\right)}\right\|_2 = \sqrt{\alpha_M\left(\mathbb{R}_+ \times \Omega\right)}.$$

2.4.2 Integration when the integrators are locally square-integrable martingales

In the discontinuous case, we can define the stochastic integral only when the integrand X is predictable, therefore when M is not continuous then by definition we assume that the members of $\mathcal{L}^2\left(M\right)$, and of course the members of $\mathcal{L}_{\text{loc}}^2\left(M\right)$, are predictable.

Proposition 2.86 *If $M \in \mathcal{H}^2$ and $X \in \mathcal{L}^2\left(M\right)$, then there is a unique process in $\mathcal{H}_0^2$ denoted by $X \bullet M$ such that for every $N \in \mathcal{H}^2$*

$$[X \bullet M, N] = X \bullet [M, N]. \qquad (2.40)$$

If we denote $X \bullet M$ as $\int_0^t X dM$ then (2.40) can be written as

$$\left[\int_0^t X dM, N\right] = \int_0^t X d[M, N].$$

Proof. Most of the proof is the same as in the continuous case.

1. One can prove the existence of $X \bullet M$ as in the continuous case. One can also define the process

$$S \overset{\circ}{=} \left(X \bullet M\right) N - X \bullet [M, N].$$

As in the continuous case one can show that S is a martingale.

2. As $X \bullet [M, N]$ is a pathwise integral it is easy to show[67] that the jumps of $X \bullet [M, N]$ are

$$X \cdot \Delta[M, N] = X \cdot \Delta M \Delta N.$$

Using the predictability of the integrands we prove that the jumps of $X \bullet M$ are $X \Delta M$, that is

$$\Delta\left(X \bullet M\right) = X \cdot \Delta M. \qquad (2.41)$$

From this the jumps of $\left(X \bullet M\right) N$ are $\left(X \Delta M\right) \Delta N$. By the characterization of the quadratic co-variation this implies that

$$[X \bullet M, N] = X \bullet [M, N].$$

That is if $N \in \mathcal{H}_0^2$ then (2.40) holds!

---

[67]See: Proposition 1.20, page 11.

3. So let us prove (2.41)! Assume that $M \in \mathcal{H}^2$. First let

$$X \overset{\circ}{=} \xi \chi \left((a, b] \right),$$

where the ξ is bounded and $\mathcal{F}_a$-measurable. Observe that X is left-regular, hence it is predictable. We prove that $X \bullet M = \xi \left(M^b - M^a \right)$.

$$\mathbf{E} \left(\int_0^\infty X d \left[M, N \right] \right) \overset{\circ}{=} \mathbf{E} \left(\xi \int_0^\infty \chi \left((a, b] \right) d \left[M, N \right] \right) =$$

$$= \mathbf{E} \left(\xi \left(\left[M, N \right] (b) - \left[M, N \right] (a) \right) \right) =$$

$$= \mathbf{E} \left(\xi \left(\left[M, N \right]^b (\infty) - \left[M, N \right]^a (\infty) \right) \right) =$$

$$= \mathbf{E} \left(\xi \left[M^b - M^a, N \right] (\infty) \right).$$

As $M^b - M^a \in \mathcal{H}^2$ if $N \in \mathcal{H}^2$

$$\left(M^b - M^a \right) N - \left[M^b - M^a, N \right]$$

is uniformly integrable. As ξ is bounded and $\mathcal{F}_a$-measurable

$$\xi \left(M^b - M^a \right) N - \xi \left[M^b - M^a, N \right]$$

is also uniformly integrable. Hence

$$\mathbf{E} \left(\xi \left[M^b - M^a, N \right] (\infty) \right) = \mathbf{E} \left(\xi \left(M^b - M^a \right) (\infty) N (\infty) \right) \overset{\circ}{=}$$

$$\overset{\circ}{=} \left(\xi \left(M^b - M^a \right), N \right).$$

By the definition of the integral for all $N \in \mathcal{H}^2$

$$\left(\xi \left(M^b - M^a \right), N \right) = \mathbf{E} \left(\int_0^\infty X d \left[M, N \right] \right) = \left(X \bullet M, N \right).$$

This means that, as we said,

$$X \bullet M = \xi \left(M^b - M^a \right).$$

4. The mapping $X \to \mathbf{E} \left(\int_0^\infty X d \left[M, N \right] \right)$ is linear, the mapping $X \to X \bullet M$ is obviously also linear, hence if

$$X \overset{\circ}{=} \sum_{i=1}^n \xi_i \chi \left((t_i, t_{i+1}] \right), \tag{2.42}$$

where the ξ_i are bounded and $\mathcal{F}_{t_i}$-measurable then

$$(X \bullet M)(t) = \sum_{i=1}^{n} \xi_i (M(t \wedge t_{i+1}) - M(t \wedge t_i)) =$$

$$= (X \bullet M)(t).$$

For processes (2.42) relations (2.40) and the jump condition (2.41) obviously hold. As elements of $\mathcal{L}^2(M)$ are predictable the bounded predictable step processes[68] are dense in $\mathcal{L}^2(M)$. Let $X \in \mathcal{L}^2(M)$ and let $X_n \to X$ where (X_n) are step processes. By Doob's inequality and by (2.27) and (2.28)

$$\mathbf{E}\left(\left(\sup_t |(X_n \bullet M)(t) - (X \bullet M)(t)|\right)^2\right) \leq$$

$$\leq 4 \|(X_n - X) \bullet M(\infty)\|_2^2 =$$

$$= 4((X_n - X) \bullet M, (X_n - X) \bullet M) =$$

$$= 4\mathbf{E}\left(\int_0^\infty (X_n - X) d[M, (X_n - X) \bullet M]\right) \leq$$

$$\leq 4 \|X_n - X\|_M \|(X_n - X) \bullet M\|_{\mathcal{H}^2} \leq$$

$$\leq 4 \|X_n - X\|_M (\|X_n \bullet M\|_{\mathcal{H}^2} + \|X \bullet M\|_{\mathcal{H}^2}).$$

(2.40) holds for step processes so

$$\|X_n \bullet M\|_{\mathcal{H}^2} = \sqrt{\mathbf{E}([X_n \bullet M](\infty))} = \sqrt{\mathbf{E}\left(\int_0^\infty X_n^2 d[M]\right)} \stackrel{\circ}{=} \|X_n\|_M$$

is a bounded sequence. As $\|X_n - X\|_M \to 0$

$$\mathbf{E}\left(\left(\sup_t |(X_n \bullet M)(t) - (X \bullet M)(t)|\right)^2\right) \to 0.$$

Hence for a subsequence $(X_{n_k} \bullet M)$ almost surely

$$\sup_t |(X_{n_k} \bullet M)(t) - (X \bullet M)(t)| \to 0.$$

Therefore for the jumps almost surely

$$\Delta(X_{n_k} \bullet M) \to \Delta(X \bullet M).$$

[68]See: Proposition 1.42, page 25.

As we have proved $\Delta\left(X_{n_k} \bullet M\right) = X_{n_k}\Delta M$, therefore

$$\Delta\left(X_{n_k} \bullet M\right) = X_{n_k}\Delta M \to X\Delta M.$$

This means that if $M \in \mathcal{H}^2$ then

$$\Delta\left(X \bullet M\right) = X\Delta M,$$

hence the proposition holds. $\qquad\qquad\qquad\qquad\qquad\qquad\qquad\qquad\square$

We can extend the stochastic integral $X \bullet M$ to processes $M \in \mathcal{H}^2_{\mathrm{loc}}$ and $X \in \mathcal{L}^2_{\mathrm{loc}}(M)$ exactly as we did it for continuous local martingales. It is easy to show that if X is locally bounded then $X \in \mathcal{L}^2_{\mathrm{loc}}(M)$. If $M \in \mathcal{H}^2_{\mathrm{loc}}$ and (τ_n) is a localizing sequence of M then, as in the continuous case, one can prove that

$$X \bullet M^{\tau_n} = X \bullet \left(M^{\tau_{n+1}}\right)^{\tau_n} = \left(X \bullet M^{\tau_{n+1}}\right)^{\tau_n}$$

so one can 'paste together' $X \bullet M$. Let us observe that

$$\Delta\left(X \bullet M\right)^{\tau_n} = \Delta\left(X \bullet M^{\tau_n}\right) = X\Delta M^{\tau_n}$$

which implies that $\Delta\left(X \bullet M\right) = X\Delta M$. Using that the members of $\mathcal{L}^2(M)$ are predictable we showed that $\Delta\left(X \bullet M\right) = X\Delta M$. With localization one can easily prove the following important observation:

Corollary 2.87 *If $M \in \mathcal{H}^2_{loc}$ and $X \in \mathcal{L}^2_{loc}(M)$ then $\Delta\left(X \bullet M\right) = X\Delta M$.*

Let us summarize the properties of the stochastic integration when the integrator is in $\mathcal{H}^2_{\mathrm{loc}}$. The proofs of these properties are direct modifications of the proofs of the corresponding properties presented in the continuous case.

Theorem 2.88 (Properties of stochastic integration) *If the integrators are in $\mathcal{H}^2_{loc}$ then stochastic integration has the following properties:*

1. *We defined the stochastic integral only for predictable integrands.*
2. *If $M \in \mathcal{H}^2_{loc}$ and $X \in \mathcal{L}^2_{loc}(M)$ then $\left(X \bullet M\right)(0) = 0$ and $X \bullet M \in \mathcal{H}^2_{loc}$.*
3. *If $M \in \mathcal{H}^2_{loc}$ and X is locally bounded and predictable then $X \bullet M$ exists.*
4. *Observe that when M is continuous then $X \in \mathcal{L}^2_{loc}(M)$ if and only if $X^2 \bullet [M] < \infty$. In the general case this characterization is not true.*
5. *If $M \in \mathcal{H}^2_{loc}$ then $X \mapsto X \bullet M$ is an $\mathcal{L}^2(M) \to \mathcal{H}^2_0$ isometry.*
6. *If $M, N \in \mathcal{H}^2_{loc}$ and $X \in \mathcal{L}^2_{loc}(M)$ then $[X \bullet M, N] = X \bullet [M, N]$.*
7. *Assume that $M \in \mathcal{H}^2_{loc}$ and $X \in \mathcal{L}^2_{loc}(M)$. If on an interval $[a, b]$ one has $X(t, \omega) = 0$ or $M(t, \omega) = M(a, \omega)$ for all ω then $X \bullet M$ is constant on $[a, b]$.*

8. *If $M \in \mathcal{H}_{loc}^2$, $X \in \mathcal{L}_{loc}^2(M)$ and τ is an arbitrary stopping time then*

$$(X \bullet M)^\tau = \chi([0, \tau]) X \bullet M = X^\tau \bullet M^\tau = X \bullet M^\tau.$$

9. *$X \bullet M$ is bilinear.*

10. *Assume that $M \in \mathcal{H}_{loc}^2$. If $X \in \mathcal{L}^2(M)$ then $Y \in \mathcal{L}^2(X \bullet M)$ if and only if $XY \in \mathcal{L}^2(M)$. If $X \in \mathcal{L}_{loc}^2(M)$ then $Y \in \mathcal{L}_{loc}^2(X \bullet M)$, if and only if $XY \in \mathcal{L}_{loc}^2(M)$. In both cases*

$$(YX) \bullet M = Y \bullet (X \bullet M).$$

11. *If $M \in \mathcal{H}_{loc}^2$ and $X \in \mathcal{L}_{loc}^2(M)$ then $\Delta(X \bullet M) = X\Delta M$.*

12. *If $M \in \mathcal{H}_{loc}^2$ and (X_n) is a sequence of predictable processes, $X_n \to X_\infty$ in every point and there is an $X \in \mathcal{L}_{loc}^2(M)$ such that $|X_n| \leq X$ then $X_n \bullet M \to X_\infty \bullet M$, where the convergence is uniform on every compact interval in probability, that is*

$$\sup_{s \leq t} |(X_n \bullet M)(s) - (X_\infty \bullet M)(s)| \xrightarrow{p} 0, \quad \text{for every } t \geq 0.$$

13. *If $M \in \mathcal{H}_{loc}^2$, $\tau_n \nearrow \infty$ and $X \overset{\circ}{=} \sum_i \xi_i \chi((\tau_i, \tau_{i+1}])$ is a predictable simple process then $X \bullet M$ exists and*

$$\int_0^t X \, dM \overset{\circ}{=} (X \bullet M)(t) = \sum_i \xi_i (M(\tau_{i+1} \wedge t) - M(\tau_i \wedge t)).$$

14. *If $M \in \mathcal{H}_{loc}^2$, X is left-regular then $(X \bullet M)(t)$ is an Itô–Stieltjes integral for every t where the convergence of the approximating sums is uniform in probability on every compact interval. The approximating partitions can be random as well*[69].

Remark that if $M \in \mathcal{H}_{loc}^2$ it is possible that the trajectories of M have finite variation on finite intervals[70]. In this case we potentially might have two different definitions for the stochastic integral. Fortunately this not the case.

Proposition 2.89 *Let us assume that for some process M we can define two different concepts of integration. Assume that*

1. *both concepts of integration is linear over the bounded processes,*
2. *for both concepts of integration bounded predictable processes are integrable,*

---

[69]If X is not left-regular then the property does not hold. See: Example 2.8, page 114.
[70]See: Example 2.20, page 124.

3. *the integral of the bounded simple processes*

$$X \overset{\circ}{=} \sum_{i=1}^{n} \xi_i \chi\left(\left(t_i, t_{i+1}\right]\right)$$

is

$$\left(X \bullet M\right)(t) = \sum_{i=1}^{n} \xi_i \left(M\left(t_{i+1} \wedge t\right) - M\left(t_i \wedge t\right)\right),$$

4. *in both cases the Theorem on Dominated Convergence is true.*

If for some predictable process X both integrals exist then they are indistinguishable.

Proof. Let us denote by $\mathcal{L}$ the set of bounded processes where the two concepts of integration coincide. $\mathcal{L}$ is obviously a linear space and $1 \in \mathcal{L}$. By the dominated convergence property it is obvious that $\mathcal{L}$ is a λ-system. The set of bounded elementary processes is a π-system, hence by the Monotone Class Theorem $\mathcal{L}$ contains all the bounded predictable processes[71]. If X predictable then for all n the integrals of $X_n \overset{\circ}{=} X\chi\left(|X| \leq n\right)$ are equal. If X is integrable for both concepts then by the Dominated Convergence Theorem the two integrals should be equal. $\square$

2.4.3 Stochastic integration when the integrators are semimartingales

The decomposition of continuous semimartingales is unique. In the discontinuous case this is not true. Hence we need the following new definition:

Definition 2.90 *Let X be a semimartingale. We say that the predictable process Y is integrable with respect to X if there is a decomposition*

$$X = X(0) + H + V \tag{2.43}$$

where $H \in \mathcal{H}_{loc}^2$ and V has finite variation and the integrals $Y \bullet H$ and $Y \bullet V$ exist. In this case

$$Y \bullet X \overset{\circ}{=} Y \bullet H + Y \bullet V.$$

The next example is very important.

---

[71] See: Proposition 1.42, page 25.

Example 2.91 If the integrand is not locally bounded then it is possible that for some decomposition of a semimartingale the two integrals in the above definition exist but for some other decomposition they do not exist.

If X Poisson process with parameter λ, then X has two different decompositions. One can write

$$X\left(t\right) = \left(X\left(t\right) - \lambda t\right) + \lambda t \overset{\circ}{=} H\left(t\right) + V\left(t\right),$$

where H is the compensated Poisson process and we can decompose X as

$$X = 0 + X = H + V$$

where $H = 0$ and $V \overset{\circ}{=} X$ has finite variation. Let τ be the time of the first jump of X, and let $Y\left(t, \omega\right) \overset{\circ}{=} \chi\left(0, \tau\left(\omega\right)\right] / t$. Y is predictable as it is the limit of the predictable processes $Y_n \overset{\circ}{=} \chi\left(1/n, \tau\right] / t$, but not locally bounded. If we use the decomposition $X = X + 0$ then $Y \bullet 0 = 0$, $Y \bullet X = 1/\tau$, hence the integral exists. On the other hand if $V\left(t\right) = \lambda t$ then for all ω $\int_0^s Y dV = \lambda \int_0^t Y dt$ is ∞, hence in the other decomposition the integral does not exist. $\qquad\Box$

If Y is locally bounded and X is a semimartingale then obviously for any decomposition (2.43) of X the integrals $Y \bullet H$ and $Y \bullet V$ exist. If we restrict the set of possible integrands to locally bounded processes then the stochastic integration with respect to semimartingales is very simple. One can easily show that the integral has all the usual properties of the stochastic integration[72]. If Y is integrable then there is a decomposition (2.43) that $Y \bullet H$ and $Y \bullet V$ exist. If $|Y_n| \leq Y$ then $Y_n \bullet H$ and $Y_n \bullet V$ exist for every n. For stochastic integration with respect to a $\mathcal{H}_{\text{loc}}^2$ integrator and for the classical pathwise integration the Dominated Convergence Theorem holds. For bounded predictable processes the stochastic integral exists for any decomposition of the semimartingale integrand and it is linear for any fixed decomposition over the bounded integrands. Therefore, by Proposition 2.89, if for two different decompositions of a semimartingale the integral exists for some predictable process, then the two possible integrals are equal. Hence for predictable processes the definition of the integral is independent of the decomposition of the semimartingale.

Every left-regular process is locally bounded, hence one can easily prove the following very important observation:

[72]See: Theorem 2.88, page 174,

Proposition 2.92 (Existence of quadratic variation) *If X and Y are arbitrary semimartingales then the quadratic co-variation $[X, Y]$ has a right-continuous version and*

$$XY - X(0) Y(0) = X_- \bullet Y + Y_- \bullet X + [X, Y]$$

where the integrals are stochastic integrals. The jumps of the quadratic co-variation are $\Delta [X, Y] = \Delta X \Delta Y$.

Proof. It is sufficient to prove the relation $\Delta [X, Y] = \Delta X \Delta Y$. From the formula for the jumps of stochastic integrals

$$
\begin{aligned}
\Delta [X, Y] &= \Delta (XY) - X_- \Delta Y - Y_- \Delta X = \\
&= XY - X_- Y_- - X_- \Delta Y - Y_- \Delta X = \\
&= XY - X_- (Y_- + \Delta Y) - Y_- (\Delta X + X_-) + Y_- X_- = \\
&= XY - X_- Y - Y_- X + Y_- X_- = \\
&= (X - X_-) (Y - Y_-) = \Delta X \Delta Y.
\end{aligned}
$$

$\square$

Recall that we have defined semimartingales as the sums of locally square-integrable martingales and processes with finite variation. At this stage there are three things we do not know about semimartingales.

1. We have not proved that the local martingales are semimartingales[73].

2. We also have not proved that if M is a local martingale then $M^2 - [M]$ is a local martingale.

3. Let X be a semimartingale and let us assume that $Y_1 \bullet X$ and $Y_2 \bullet X$ exist. Can we prove the existence of $(Y_1 + Y_2) \bullet X$? At the moment of course not: If Y_1 and Y_2 are not locally bounded then it is possible that $Y_1 \bullet X$ exists in one decomposition and $Y_2 \bullet X$ exists in some different decomposition of X.

These are serious problems, to overcome them in the next chapters will take a concerted effort!

---

[73]To prove that every local martingale is semimartingale one needs the Fundamental Theorem of Local Martingales. See: Theorem 3.57, page 220.

3

THE STRUCTURE OF LOCAL MARTINGALES

The main result of this chapter is the theorem which we shall call the *Fundamental Theorem of Local Martingales*, which states that every local martingale can be decomposed as the sum of a locally bounded local martingale and a local martingale which has locally integrable variation[1]. The Fundamental Theorem of Local Martingales has many important consequences. Perhaps the most important one is that for any local martingale L the quadratic variation process $[L]$ exists and the difference $L^2 - [L]$ is a local martingale.

From now on we assume that the space $(X, \mathcal{A}, \mathbf{P})$ is complete, that is if $N \subseteq A \in \mathcal{A}$ and $\mathbf{P}(A) = 0$ then $N \in \mathcal{A}$. We also assume that the filtration $\mathcal{F}$ satisfies the usual conditions. Of course the usual conditions is not a big surprise, but it is remarkable that we need the completeness of the base space $(X, \mathcal{A}, \mathbf{P})$. The reader can take this fact as an indicator of the forthcoming measure-theoretic difficulties.

Let us introduce some useful notation.

Definition 3.1 *Fix a stochastic base* $(\Omega, \mathcal{A}, \mathbf{P}, \mathcal{F})$.

1. $\mathcal{L}$ *will denote the set of local martingales* L *for which* $L(0) = 0$.
2. $\mathcal{V}$ *will denote the set of right-regular, adapted processes* V *which have finite variation on every finite interval and for which* $V(0) = 0$.
3. $\mathcal{A}$ *will denote the set of processes* $A \in \mathcal{V}$, *for which*

$$\mathbf{E}(\text{Var}(A)(\infty)) < \infty.$$

$\mathcal{A}$ *is called the space of processes with* integrable variation[2].

---

[1]See: Definition 3.1 below.

[2]The careful reader would notice that the symbol $\mathcal{A}$ denotes two objects. $\mathcal{A}$ is the set of possible events in the probability space $(\Omega, \mathcal{A}, \mathbf{P})$ and $\mathcal{A}$ is also the set of processes with integrable variation. In the theory of stochastic processes the events of $(\Omega, \mathcal{A}, \mathbf{P})$ play a very minor role, so generally $\mathcal{A}$ will denote the set of processes with integrable variation. Nevertheless we apologize for this inconvenience.

4. *If $X \in \mathcal{A}_{\text{loc}}$ that is there is a localizing sequence (τ_n) that $X^{\tau_n} \in \mathcal{A}$ for all n then we shall say that X has* locally integrable variation.

5. *$\mathcal{V}^+$ will denote the processes in $\mathcal{V}$ which have increasing trajectories.*

6. *$\mathcal{A}^+$ will denote the processes in $\mathcal{A}$ which have increasing trajectories.*

7. *The adapted process S is a* semimartingale *if it has a decomposition*

$$S = S(0) + L + V \tag{3.1}$$

where $V \in \mathcal{V}$ and $L \in \mathcal{L}$. $\mathcal{S}$ will denote the set of semimartingales[3].

8. *A semimartingale S is a* special semimartingale *if there is a decomposition (3.1) where V is predictable. $\mathcal{S}_p$ will denote the space of special semimartingales.*

Observe that we have changed the definition of semimartingales[4]. An important message of the Fundamental Theorem is that the present definition is the same as the old one.

The sequence of jumps of a Poisson process does not have an accumulation point, hence every compound Poisson process is in $\mathcal{V}$.

Example 3.2 If X is a compound Poisson process then the distribution of the jumps of X has finite expected value if and only if $X \in \mathcal{A}_{\text{loc}}$.

Let (τ_n) be the jump-times of X and let (ξ_n) be the size of the jumps. If for the common distribution of the jumps the expected value $m \stackrel{\circ}{=} \mathbf{E}(|\xi_n|)$ is finite then

$$\mathbf{E}\left(\text{Var}\left(X^{\tau_n}\right)(\infty)\right) = \mathbf{E}\left(\sum_{k=1}^{n} |\xi_k|\right) = nm < \infty,$$

hence $X \in \mathcal{A}_{\text{loc}}$. On the other hand assume that $X \in \mathcal{A}_{\text{loc}}$, but $m = \infty$. Let (σ_n) be an $\mathcal{A}$-localizing sequence of X. Define the stopping times $\rho_n \stackrel{\circ}{=} \sigma_n \wedge \tau_1$. $\rho_n \leq \tau_1$, hence $\mathcal{F}_{\rho_n} \subseteq \mathcal{F}_{\tau_1}$ so ρ_n is $\mathcal{F}_{\tau_1}$-measurable. $\mathbf{P}(\tau_1 < \infty) = 1$, hence if n is large enough then $\mathbf{P}(\rho_n = \tau_1) > 0$. The compound Poisson processes are Lévy processes, hence by the strong Markov property of Lévy processes $\xi_1 = \Delta X(\tau_1)$ is independent of $\mathcal{F}_{\tau_1}$. Hence

$$\infty > \mathbf{E}\left(\text{Var}\left(X^{\sigma_n}\right)(\infty)\right) \geq \mathbf{E}\left(\text{Var}\left(X^{\rho_n}\right)(\infty)\right) = \mathbf{E}\left(|\xi_1| \chi\left(\tau_1 = \rho_n\right)\right) =$$
$$= \mathbf{E}\left(|\xi_1|\right) \mathbf{P}\left(\tau_1 = \rho_n\right) = \infty,$$

which is impossible. Of course, in the argument we did not use the fact that the distributions of the jumps were the same. If (τ_n) is a strictly increasing sequence

---

[3]The decomposition is not necessarily unique.
[4]See: Definition 2.17, page 124.

of stopping times and at each τ_k we have a jump ξ_k, which is independent of $\mathcal{F}_{\tau_k}$ then $\mathbf{E}\left(|\xi_k|\right)$ is finite for all k if and only if the process $X\left(t\right) \overset{\circ}{=} \sum_k \xi_k \chi\left(\tau_k \leq t\right)$ is in $\mathcal{A}_{\mathrm{loc}}$. $\qquad\square$

Example 3.3 If $L \in \mathcal{L}$ and $L^*\left(t\right) \overset{\circ}{=} \sup_{s \leq t} |L\left(s\right)|$ then $L^* \in \mathcal{A}_{\mathrm{loc}}^+$.

Since L is right-continuous

$$\sup_{s \leq t} |L\left(s\right)| = \sup_{s \leq t, s \in \mathbb{Q}} |L\left(s\right)|,$$

hence L^* is adapted, and increasing. As L is right-continuous, L^* is also right-continuous. Let $\left(\tau_n\right)$ be a localizing sequence of L. Let

$$\sigma_n \overset{\circ}{=} \inf\left\{t : |L\left(t\right)| > n\right\} \wedge \tau_n.$$

Since L is right-continuous, σ_n is a stopping time[5] for all n. If

$$\sigma_\infty\left(\omega\right) \overset{\circ}{=} \sup_n \sigma_n\left(\omega\right) < \infty$$

for some outcome ω then $|L\left(\sigma_\infty\left(\omega\right)-\right)| = \infty$ which is impossible as for every outcome L has finite limits from the left. Therefore obviously $\sigma_n \nearrow \infty$.

$$\left(L^*\right)^{\sigma_n}\left(t\right) \leq \left(L^*\right)^{\sigma_n}\left(\infty\right) = L^*\left(\sigma_n\right) \leq L^*\left(\sigma_n-\right) + |\Delta L^*\left(\sigma_n\right)| \leq$$
$$\leq n + |L\left(\sigma_n\right)| = n + |L^{\tau_n}\left(\sigma_n\right)|.$$

$L^{\tau_n} \in \mathcal{M}$ therefore by the Optional Sampling Theorem $|L^{\tau_n}\left(\sigma_n\right)|$ is integrable, hence $\left(L^*\right)^{\sigma_n} \in \mathcal{A}^+$ and so $L^* \in \mathcal{A}_{\mathrm{loc}}^+$. $\qquad\square$

Proposition 3.4 *If* $V \in \mathcal{V} \cap \mathcal{L}$ *then* $V \in \mathcal{A}_{\mathrm{loc}}$.

Proof. Let $\left(\tau_n\right)$ be a localizing sequence of a $V \in \mathcal{L}$.

$$\sigma_n \overset{\circ}{=} \inf\left\{t : \mathrm{Var}\left(V\right)\left(t\right) > n\right\} \wedge \tau_n.$$

V is right-continuous and adapted, hence $\mathrm{Var}\left(V\right)$ is also right-continuous and adapted. The filtration is right-continuous hence σ_n is a stopping time and by the right-regularity of $\mathrm{Var}\left(V\right)$ again $\sigma_n \nearrow \infty$. If ΔV denotes the jumps of V,

---

[5]See: Example 1.32, page 17.

then

$$\text{Var}\,(V)\,(\sigma_n) = \text{Var}\,(V)\,(\sigma_n-) + \Delta\text{Var}\,(V)\,(\sigma_n) =$$
$$= \text{Var}\,(V)\,(\sigma_n-) + |\Delta V\,(\sigma_n)| \le$$
$$\le n + |V\,(\sigma_n)| + |V\,(\sigma_n-)| \le 2n + |V\,(\sigma_n)| \,.$$

V^{τ_n} is a uniformly integrable martingale, hence $V\,(\sigma_n)$ is integrable so $\text{Var}\,(V)\,(\sigma_n)$ is also integrable, hence by the definition of $\mathcal{A}$ obviously $V \in \mathcal{A}_{\text{loc}}$. $\square$

3.1 Predictable Projection

Our main tool in analysing the structure of local martingales is the so-called *predictable projection*. It is very natural to ask that how 'far' are the predictable processes from the other classes of measurable processes. If X is a product measurable process, then with the predictable projection one can find a predictable process denoted by ^pX which is in some sense 'close' to X. This closeness means that for every so-called *predictable stopping time* τ the expected value of the stopped variables $X(\tau)$ and $(^pX)\,(\tau)$ are equal. If X is the gain process of some game and τ is an exit strategy, then the stopped variable $X(\tau)$ is the value of the game if one plays the exit strategy τ. If a stopping time is 'predictable' then somehow we can foresee, predict it. As $X(\tau)$ and $(^pX)\,(\tau)$ have the same expected value for predictable exit rules it will be irrelevant, on average, whether we play the game X or the predictable game ^pX. So, as an interpretation one can say that ^pX is the predictable part[6] of X. If L is a local martingale then the 'unpredictable' part of L are the jumps ΔL of L. The most important examples of the predictable projection are the rules $^pL = L_-$ and $^p\,(\Delta L) = 0$. This means that one cannot 'predict' the size of the jumps of a local martingale[7].

3.1.1 Predictable stopping times

Let us first define when a stopping time is predictable.

Definition 3.5 *We say that a stopping time σ announces τ if $\sigma\,(\omega) \le \tau\,(\omega)$ for all outcomes ω and $\sigma\,(\omega) < \tau\,(\omega)$ whenever $\tau\,(\omega) > 0$. We say that the stopping time τ is predictable if there is a sequence of stopping times (σ_n) such that $\sigma_n \nearrow \tau$ and σ_n announces τ for all n. The sequence (σ_n) is called the announcing or predicting sequence of τ.*

---

[6]More exactly, of course, ^pX by definition is the predictable part of X, since ^pX is mathematically well-defined, but the expression 'predictable part' is not a mathematical concept.

[7]It is very natural to ask how one can predict the jumps of stock prices. As in mathematical finance we assume that the stock prices are basically driven by some local martingale, the answer is that nobody can predict the jumps of the price processes. So the humble relation $^p\,(\Delta L) = 0$ just mentioned has extraordinarily important theoretical and applied implications.

Definition 3.6 *We say that a stopping time σ is* totally inaccessible *if*

$$\mathbf{P}\left(\tau = \sigma < \infty\right) = 0$$

for every predictable stopping time τ.

Example 3.7 The jump-times of Poisson processes are totally inaccessible[8].

Let N be a Poisson process with parameter $\lambda > 0$. Let τ be the time of the first jump of N that is let

$$\tau \overset{\circ}{=} \inf\left\{t : N\left(t\right) = 1\right\}.$$

Obviously τ is a stopping time and almost surely $0 < \tau < \infty$. It is well-known[9] that τ has an exponential distribution with parameter λ. We show that τ is not predictable. Suppose (τ_n) to be an announcing sequence of τ. $\tau_n < \tau < \infty$, so trivially $N\left(\tau_n\right) = 0$ a.s. and by the strong Markov property of Lévy processes

$$N^{(n)}(t) \overset{\circ}{=} N\left(t + \tau_n\right) - N\left(\tau_n\right) = N\left(t + \tau_n\right)$$

is also a Poisson process with parameter λ. If σ_n is the time of the first jump of $N^{(n)}$, then $\sigma_n = \tau - \tau_n$, and σ_n also has an exponential distribution with parameter λ.

$$1 = \mathbf{P}\left(\tau_n \nearrow \tau\right) = \mathbf{P}\left(\sigma_n \searrow 0\right),$$

which is impossible since the convergence of distributions follows from the almost sure convergence.

In the same way one can prove that no part of τ is predictable, that there is no predictable stopping time ρ for which on a set B with positive probability $\rho = \tau$: Assume that there is such a ρ and let (ρ_n) be a sequence announcing ρ. For the stopping times $\tau_n \overset{\circ}{=} \rho_n \wedge \tau$

$$N^{(n)}(t) \overset{\circ}{=} N\left(t + \tau_n\right) - N\left(\tau_n\right)$$

is again a Poisson process with parameter λ. Again, if σ_n denotes the time of the first jump of $N^{(n)}$ then σ_n has exponential distribution with parameter λ. (σ_n) is almost surely convergent, hence it converges in distribution, which is impossible since if σ_∞ is the limit of (σ_n) then σ_∞ must be zero on B where B has positive probability. $\square$

[8]See: Example 7.5, page 465 and Example 7.75, page 517.
[9]See: page 461.

Example 3.8 If w is a Wiener process then for all a the first passage time $\tau_a \overset{\circ}{=} \inf \{t: w(t) = a\}$ is a predictable stopping time.

By the continuity of the trajectories of w the sequence $(\tau_{a-1/n})$ obviously announces τ. $\square$

Proposition 3.9 *If τ and σ are predictable stopping times then $\tau \wedge \sigma$ and $\tau \vee \sigma$ are predictable stopping times. If $\tau_n \nearrow \tau$ and τ_n are predictable stopping times then τ is a predictable stopping time.*

Proof. If (τ_n) announces τ and (σ_n) announces σ then $(\tau_n \wedge \sigma_n)$ will be an announcing sequence for $\tau \wedge \sigma$ and $(\tau_n \vee \sigma_n)$ will be an announcing sequence for $\tau \vee \sigma$. If $(\tau_k^{(n)})$ announces τ_n, then obviously

$$\sigma_n \overset{\circ}{=} \max \left\{ \tau_n^{(k)}: k = 1, \ldots, n \right\}$$

announces τ and

$$\sigma_n \leq \max \left\{ \tau_{n+1}^{(k)}: k = 1, \ldots, n \right\} \leq \sigma_{n+1}.$$

For any $\varepsilon > 0$ and for any outcome ω, there is an N, depending on the outcome, such that for any outcome $\tau_n(\omega) \geq \tau(\omega) - \varepsilon$ for every $n \geq N(\omega)$. $\left(\tau_k^{(N)}\right)$ announces τ_N hence $\tau_k^{(N)} \geq \tau_N - \varepsilon$ for all $k \geq M \geq N$.

$$\sigma_M \overset{\circ}{=} \max \left\{ \tau_M^{(k)} : k = 1, \ldots, M \right\} \geq \tau_M^{(M)} \geq \tau_M^{(N)} \geq \tau_N - \varepsilon \geq \tau - 2\varepsilon$$

so $\sigma_n \nearrow \tau$. $\square$

Example 3.10 If the stopping times τ_n are predictable and $\tau_n \searrow \tau$ then τ is not necessarily predictable.

Let τ be an arbitrary non-predictable stopping time. Obviously $\tau_n \overset{\circ}{=} \tau + 1/n$ is a predictable stopping time for every n and $\tau_n \searrow \tau$. $\square$

Proposition 3.11 *If $\tau \overset{a.s.}{=} 0$ then τ is a predictable stopping time. If τ is a predictable stopping time and $\sigma \overset{a.s.}{=} \tau$ then σ is also a predictable stopping time.*

Proof. $(\Omega, \mathcal{A}, \mathbf{P})$ is complete and the filtration contains the measure-zero sets

$$\{\tau \leq t\} \overset{a.s.}{=} \{\tau = 0\} \in \mathcal{F}_t$$

for any t. Hence τ is a stopping time. As τ is announced by

$$\tau_n \stackrel{\circ}{=} \begin{cases} \tau - 1/n & \text{if} \quad \tau > 1/n \\ 0 & \text{if} \quad \tau \leq 1/n \end{cases}$$

τ is predictable. To prove the second statement let (τ_n) be an announcing sequence of τ. By the usual conditions

$$\sigma_n \stackrel{\circ}{=} \begin{cases} \tau_n & \text{if} & \tau = \sigma \\ \sigma - 1/n & \text{if} & \tau \neq \sigma \text{ and } \sigma > 1/n \\ 0 & \text{otherwise} \end{cases}$$

then (σ_n) is an announcing sequence for σ. $\square$

Definition 3.12 *If τ is a predictable stopping time and (τ_n) is an announcing sequence of τ then let*

$$\mathcal{F}_{\tau-} \stackrel{\circ}{=} \sigma\left(\cup_n \mathcal{F}_{\tau_n}\right).$$

Of course one should now prove that the definition of $\mathcal{F}_{\tau-}$ is independent of the announcing sequence (τ_n).

Proposition 3.13 *If τ is a predictable stopping time then*

$$\mathcal{F}_{\tau-} = \sigma\left(\mathcal{F}_0, \{A \cap \{t < \tau\}, A \in \mathcal{F}_t\}\right), \tag{3.2}$$

hence the definition of $\mathcal{F}_{\tau-}$ is independent of the announcing sequence (τ_n).

Proof. For an arbitrary stopping time ρ let

$$\widetilde{\mathcal{F}}_\rho \stackrel{\circ}{=} \sigma\left(\mathcal{F}_0, \{A \cap \{t < \rho\}, A \in \mathcal{F}_t\}\right).$$

Assume that (τ_n) announces τ.

1. We show that if σ announces τ then $\mathcal{F}_\sigma \subseteq \widetilde{\mathcal{F}}_\tau$, hence

$$\sigma\left(\cup_n \mathcal{F}_{\tau_n}\right) \stackrel{\circ}{=} \mathcal{F}_{\tau-} \subseteq \widetilde{\mathcal{F}}_\tau.$$

Since σ announces τ, for every A

$$A = \left(\cup_{r \in \mathbb{Q}} A \cap \{\sigma < r < \tau\}\right) \cup \left(A \cap \{\tau = 0\}\right).$$

If $A \in \mathcal{F}_\sigma$ then $B \stackrel{\circ}{=} A \cap \{\sigma < r\} \in \mathcal{F}_r$, hence by the definition of $\widetilde{\mathcal{F}}_\tau$

$$A \cap \{\sigma < r < \tau\} \stackrel{\circ}{=} B \cap \{r < \tau\} \in \widetilde{\mathcal{F}}_\tau.$$

As $\sigma \leq \tau$

$$A \cap \{\tau = 0\} = A \cap \{\sigma = 0\} \cap \{\tau = 0\}.$$

As τ is a stopping time $\{\tau = 0\} \in \mathcal{F}_0$. $A \in \mathcal{F}_\sigma$ so $A \cap \{\sigma = 0\} \in \mathcal{F}_0$, hence again by the definition of $\tilde{\mathcal{F}}_\tau$

$$A \cap \{\tau = 0\} = A \cap \{\sigma = 0\} \cap \{\tau = 0\} \in \tilde{\mathcal{F}}_\tau,$$

so $\mathcal{F}_\sigma \subseteq \tilde{\mathcal{F}}_\tau$.

2. We show that $\tilde{\mathcal{F}}_\sigma \subseteq \mathcal{F}_\sigma$ for every stopping time σ. If $A \in \mathcal{F}_t$ for some t then for all possible u

$$\{t < \sigma\} \cap A \cap \{\sigma \leq u\} = A \cap \{t < \sigma \leq u\} \in \mathcal{F}_u,$$

hence $\{t < \sigma\} \cap A \in \mathcal{F}_\sigma$. $0 \leq \sigma$. Therefore $\mathcal{F}_0 \subseteq \mathcal{F}_\sigma$, so by the definition of $\tilde{\mathcal{F}}_\sigma$ obviously $\tilde{\mathcal{F}}_\sigma \subseteq \mathcal{F}_\sigma$.

3. Let $A \in \mathcal{F}_t$. Since $\tau_n \nearrow \tau$ and as $\tilde{\mathcal{F}}_{\tau_n} \subseteq \mathcal{F}_{\tau_n}$

$$A \cap \{t < \tau\} = \cup_n (A \cap \{t < \tau_n\}) \in \cup_n \tilde{\mathcal{F}}_{\tau_n} \subseteq \sigma(\cup_n \mathcal{F}_{\tau_n}) \stackrel{\circ}{=} \mathcal{F}_{\tau-}.$$

Trivially $\mathcal{F}_0 \subseteq \mathcal{F}_{\tau_n} \subseteq \mathcal{F}_{\tau-}$. Hence $\tilde{\mathcal{F}}_\tau \subseteq \mathcal{F}_{\tau-}$. So $\mathcal{F}_{\tau-} = \tilde{\mathcal{F}}_\tau$. That is the proposition holds. $\qquad \square$

Proposition 3.14 *If τ is predictable then τ is measurable with respect to $\mathcal{F}_{\tau-}$.*

Proof. If (τ_n) announces τ then τ_n is measurable with respect to $\mathcal{F}_{\tau_n} \subseteq \mathcal{F}_{\tau-}$ for all n. $\tau_n \nearrow \tau$ and therefore τ is also measurable with respect to $\mathcal{F}_{\tau-}$. $\qquad \square$

Proposition 3.15 *If X is a predictable process and τ is a predictable stopping time, then the stopped variable $X_\tau \stackrel{\circ}{=} X(\tau) \chi(\tau < \infty)$ is measurable with respect to $\mathcal{F}_{\tau-}$.*

Proof. If X is left-continuous and adapted then it is progressively measurable. Hence if (τ_n) announces τ then

$$X_{\tau_n} \stackrel{\circ}{=} \chi(\tau_n < \infty) X(\tau_n)$$

is measurable[10] with respect to $\mathcal{F}_{\tau_n} \subseteq \mathcal{F}_{\tau-}$. As X is left-continuous $X_{\tau_n} \to X_\tau$. Hence X_τ is measurable with respect to $\mathcal{F}_{\tau-}$. The set of bounded processes X for which the proposition holds is a λ-system. The left-continuous, adapted processes form a π-system and the predictable processes are measurable with

[10]See: Proposition 1.35, page 22.

respect to the σ-algebra generated by the adapted left-continuous processes[11].
Hence by the Monotone Class Theorem one can prove the $\mathcal{F}_{\tau-}$-measurability of
X_τ in the usual way for every predictable process X. $\qquad\square$

Proposition 3.16 *If τ is a predictable stopping time and $B \in F_{\tau-}$ then*

$$\tau_B(\omega) \overset{\circ}{=} \begin{cases} \tau(\omega) & \text{if } \omega \in B \\ \infty & \text{if } \omega \notin B \end{cases}$$

is also a predictable stopping time.

Proof. Let $\mathcal{B}$ be the collection of such sets B that the proposition holds both
for B and for B^c. As the maximum and the minimum of a finite number of
predictable stopping times are again predictable stopping times, $\mathcal{B}$ is an algebra.
As the limit of an increasing sequence of predictable stopping times is again a
predictable stopping time $\mathcal{B}$ is a σ-algebra. Let (τ_n) be an announcing sequence
of τ. Fix a $C \in \mathcal{F}_{\tau_n}$. If $m \geq n$ then $\mathcal{F}_{\tau_n} \subseteq \mathcal{F}_{\tau_m}$ and therefore

$$\tau_C^{(m)} \overset{\circ}{=} \begin{cases} \tau_m(\omega) & \text{if } \omega \in C \\ \infty & \text{if } \omega \notin C \end{cases}$$

is again a stopping time. Hence $\tau_C^{(m)} \wedge m$ is also a stopping time, and
$\left(\tau_C^{(m)} \wedge m\right)_{m \geq n}$ announces τ_C. Hence τ_C is predictable. By the definition of
$\mathcal{F}_{\tau-}$, using the fact that $\mathcal{B}$ is a σ-algebra, $\mathcal{F}_{\tau-} \overset{\circ}{=} \sigma(\cup \mathcal{F}_{\tau_n}) \subseteq \mathcal{B}$, hence the
proposition holds. $\qquad\square$

Proposition 3.17 *If δ and ρ are predictable stopping times then*

$$\{\rho = \delta\}, \{\rho \leq \delta\}, \{\delta \leq \rho\}, \{\delta < \rho\}, \{\rho < \delta\} \in \mathcal{F}_{\rho-}.$$

Proof. It is enough to prove that

$$\{\rho \leq \delta\}, \{\delta \leq \rho\} = \{\rho < \delta\}^c \in \mathcal{F}_{\rho-}.$$

If (ρ_n) announces ρ and (δ_n) announces δ, then[12]

$$\{\rho \leq \delta\} = \cap_n \{\rho_n \leq \delta\} =$$
$$= (\cup_n \{\rho_n \leq \delta\}^c)^c \in \sigma(\cup_n \mathcal{F}_{\rho_n}) \overset{\circ}{=} \mathcal{F}_{\rho-},$$
$$\{\rho < \delta\} = (\cup_n \{\rho \leq \delta_n\}) \setminus \{\rho = \delta = 0\} =$$
$$= (\cup_n \cap_m \{\rho_m \leq \delta_n\}) \setminus \{\rho = \delta = 0\} \in \sigma(\cup \mathcal{F}_{\rho_m}) \overset{\circ}{=} \mathcal{F}_{\rho-}.$$
$$\qquad\square$$

---

[11] See: Proposition 1.42, page 25.
[12] See: Proposition 1.34, page 20.

3.1.2 Decomposition of thin sets

Assume that the trajectories of some process X are regular. This implies that for every outcome ω the set $\{t : \Delta X(t, \omega) \neq 0\}$ is at most countable and the jumps, which are larger than a given $c > 0$ cannot have an accumulation point. If

$$\tau_0 = 0, \quad \tau_{n+1} = \inf \{t > \tau_n : |\Delta X| \geq c\}$$

then the union of the graphs of the stopping times[13] (τ_n) covers the set $\{|\Delta X| \geq c\}$. This implies that the set $\{\Delta X \neq 0\}$ can be covered by the graphs of at most a countable number of stopping times.

Definition 3.18 *A set $A \subseteq \mathbb{R}_+ \times \Omega$ is called* thin *if $A \subseteq \cup_n [\rho_n]$ where (ρ_n) is a sequence of stopping times. If $[\rho_n] \cap [\rho_m] = \emptyset$ for all $m \neq n$ then we say that (ρ_n) is an* exhausting sequence *of A.*

Proposition 3.19 (Accessible and inaccessible part of stopping times) *For every stopping time τ there are, at most, countably many, predictable stopping times (σ_k) and a set $X \in \mathcal{F}_\tau$, for which τ_X is totally inaccessible and the graph of τ_{X^c} is covered by the disjoint union of the graphs of the predictable stopping times (σ_k).*

Proof. If τ is totally inaccessible then there is nothing to prove. If $\tau^{(0)} \stackrel{\circ}{=} \tau$ is not totally inaccessible then $\mathbf{P}\left(\tau^{(0)} = \sigma_1 < \infty\right) > 0$ for some predictable stopping time σ_1. Let

$$B_1 \stackrel{\circ}{=} \left\{\tau^{(0)} = \sigma_1 < \infty\right\} = \{\tau = \sigma_1 < \infty\}.$$

Let us delete from τ the set B_1: As[14] $B_1 \in \mathcal{F}_{\tau^{(0)}} = \mathcal{F}_\tau$

$$\tau^{(1)}(\omega) \stackrel{\circ}{=} \tau^{(0)}_{B_1^c}(\omega) \stackrel{\circ}{=} \begin{cases} \tau(\omega) & \text{if } \omega \notin B_1 \\ \infty & \text{if } \omega \in B_1 \end{cases}$$

is a stopping time. If $\tau^{(1)}$ is totally inaccessible then we stop. If $\tau^{(1)}$ is not totally inaccessible then $\mathbf{P}\left(\tau^{(1)} = \sigma_2 < \infty\right) > 0$ for some predictable stopping time σ_2. Let

$$B_2 \stackrel{\circ}{=} \left\{\tau^{(1)} = \sigma_2 < \infty\right\} = \{\tau = \sigma_2 < \infty\} \cap \{\tau = \sigma_1 < \infty\}^c \in \mathcal{F}_\tau$$

etc. The only problem is that we should finish the process in at most countably many steps. Let

$$b_n \stackrel{\circ}{=} \sup \left\{\mathbf{P}\left(\tau^{(n)} = \sigma < \infty, \ \sigma \text{ is predictable}\right)\right\}$$

---

[13]See: Example 1.32, page 17.
[14]See: Proposition 1.34, page 20.

and for any n let σ_{n+1} be such a predictable stopping time that

$$\mathbf{P}\left(B_{n+1}\right) \overset{\circ}{=} \mathbf{P}\left(\tau^{(n)} = \sigma_{n+1} < \infty\right) \geq \frac{b_n}{2}.$$

If $b_n \neq 0$ then as above

$$B_{n+1} \overset{\circ}{=} \left\{\tau^{(n)} = \sigma_{n+1} < \infty\right\} = \{\tau = \sigma_{n+1} < \infty\} \setminus \left(\cup_{k=1}^{n} B_k\right) \in \mathcal{F}_\tau$$

and

$$\tau^{(n+1)}\left(\omega\right) = \tau^{(n)}_{B^c_{n+1}}\left(\omega\right) \overset{\circ}{=} \begin{cases} \tau\left(\omega\right) & \text{if} \quad \omega \notin \left(\cup_{k=1}^{n+1} B_k\right) \\ \infty & \text{if} \quad \omega \in \cup_{k=1}^{n+1} B_k \end{cases}.$$

As the sets B_n are disjoint

$$\sum_{n=1}^{\infty} b_n \leq 2 \sum_{n=1}^{\infty} \mathbf{P}\left(B_n\right) = \mathbf{P}\left(\cup_n B_n\right) < \infty.$$

Therefore $b_n \to 0$. $B_n \in \mathcal{F}_\tau$ for every n so

$$X \overset{\circ}{=} \Omega \setminus \left(\cup_n B_n\right) \in \mathcal{F}_\tau.$$

Let σ be a predictable stopping time. For an arbitrary n

$$\mathbf{P}\left(\tau_X = \sigma < \infty\right) \leq \mathbf{P}\left(\tau^{(n)} = \sigma < \infty\right) \leq b_n \to 0,$$

so $\mathbf{P}\left(\tau_X = \sigma < \infty\right) = 0$. That is

$$\tau_X\left(\omega\right) \overset{\circ}{=} \begin{cases} \tau\left(\omega\right) & \text{if} \quad \omega \notin \cup_n B_n \\ \infty & \text{if} \quad \omega \in \cup_n B_n \end{cases}.$$

is totally inaccessible. $\qquad \square$

Definition 3.20 *A stopping time σ is called* accessible *if there is a sequence of predictable stopping times (τ_n) such that*

$$\mathbf{P}\left(\cup_n \{\sigma = \tau_n < \infty\}\right) = \mathbf{P}\left(\sigma < \infty\right).$$

Corollary 3.21 *If τ is a stopping time then there are disjoint events $X, Y \in \mathcal{F}_\tau$ such that $X \cup Y \overset{a.s.}{=} \{\tau < \infty\}$ and τ_X is totally inaccessible and τ_Y is accessible and*

$$\tau \overset{a.s.}{=} \tau_X \wedge \tau_Y.$$

Proposition 3.22 *Every thin set X has an exhausting sequence. We may assume that the members of the exhausting sequence are either predictable or totally inaccessible.*

Proof. By the previous proposition every stopping time can be covered by a countable number of predictable and one totally inaccessible stopping time. Hence one can cover the set X with the sets $\cup_n [\sigma_n]$ and $\cup_m [\tau_m]$ where σ_n is totally inaccessible for all n and τ_m is predictable for all m. We show that one can find such a sequence of stopping times which are either predictable or totally inaccessible and the graphs of the stopping times being disjoint. As σ_n is predictable and τ_m is totally inaccessible $\{\tau_m = \sigma_n < \infty\}$ has zero measure, so it is sufficient to make the sequences (τ_n) and (σ_m) disjoint[15].

1. One can easily make the sequence (σ_n) disjoint. Since[16] $\{\sigma_1 = \sigma_2\} \in \mathcal{F}_{\sigma_2}$,

$$\sigma_2' (\omega) \stackrel{\circ}{=} \left\{ \begin{array}{ll} \sigma_2 (\omega) & \text{if} \quad \omega \notin \{\sigma_1 = \sigma_2\} \\ \infty & \text{if} \quad \omega \in \{\sigma_1 = \sigma_2\} \end{array} \right.$$

is also a totally inaccessible stopping time and σ_1 and σ_2' have disjoint graphs.

2. To do the same with the sequence (τ_n) it is sufficient to observe that if τ_1 and τ_2 are predictable stopping times then[17] $\{\tau_1 = \tau_2\} \in \mathcal{F}_{\tau_2-}$, hence[18]

$$\tau_2' (\omega) \stackrel{\circ}{=} \left\{ \begin{array}{ll} \tau_2 (\omega) & \text{if} \quad \omega \notin \{\tau_1 = \tau_2\} \\ \infty & \text{if} \quad \omega \in \{\tau_1 = \tau_2\} \end{array} \right.$$

is a predictable stopping time.   □

3.1.3   The extended conditional expectation

In this subsection we introduce a generalization of the conditional expectation. Let us first recall the definition of the conditional expectation:

Definition 3.23 (Conditional expectation) *Let η be an arbitrary random variable and let $\mathcal{F}$ be a σ-algebra. We say that the random variable ξ is the conditional expectation of η with respect to the σ-algebra $\mathcal{F}$ if*

1. ξ *is $\mathcal{F}$-measurable and*
2. $\int_F \xi \, d\mathbf{P} = \int_F \eta \, d\mathbf{P}$ *for any $F \in \mathcal{F}$.*

The usual notation of the conditional expectation is $\mathbf{E}(\eta \mid \mathcal{F})$.

---

[15]Observe that we have used the completeness of $(\Omega, \mathcal{A}, \mathbf{P})$ implicitly as we tacitly assumed that if a function is almost surely zero, then it is a predictable stopping time.

[16]See: Proposition 1.34, page 20.

[17]See: Proposition 3.17, page 187.

[18]See: Proposition 3.16, page 187.

Let us recall that if $\eta \geq 0$ then almost surely $\mathbf{E}(\eta \mid \mathcal{F}) \geq 0$, but it is possible that $\mathbf{E}(\eta \mid \mathcal{F})$ has infinite values. The following theorem is well-known and it is a direct consequence of the Radon–Nikodym theorem:

Theorem 3.24 *Let $\mathcal{F}$ be an arbitrary σ-algebra.*

1. *If η is non-negative then $\mathbf{E}(\eta \mid \mathcal{F})$ exists.*
2. *If η is quasi-integrable[19] then $\mathbf{E}(\eta \mid \mathcal{F})$ exists and $\mathbf{E}(\eta \mid \mathcal{F}) = \mathbf{E}(\eta^+ \mid \mathcal{F}) - \mathbf{E}(\eta^- \mid \mathcal{F})$.*
3. *$\mathbf{E}(\eta \mid \mathcal{F})$ is unique up to a measure-zero set.*

Definition 3.25 (Generalized conditional expectation) *Let η be an arbitrary random variable and let $\mathcal{F}$ be a σ-algebra. If the conditional expectations $\mathbf{E}\left(\eta^+ \mid \mathcal{F}\right)$ and $\mathbf{E}\left(\eta^- \mid \mathcal{F}\right)$ are finite then by definition the generalized conditional expectation of η with respect to the σ-algebra $\mathcal{F}$ will be the difference*

$$\mathbf{E}\left(\eta^+ \mid \mathcal{F}\right) - \mathbf{E}\left(\eta^- \mid \mathcal{F}\right).$$

We shall also denote the generalized conditional expectation by $\mathbf{E}\left(\eta \mid \mathcal{F}\right)$.

One can easily reformulate most of the properties of the conditional expectation for the generalized conditional expectation:

Proposition 3.26 *The generalized conditional expectation has the following properties:*

1. *The generalized conditional expectation $\mathbf{E}\left(\xi \mid \mathcal{F}\right)$ exists if and only if the conditional expectation of $|\xi|$ is almost surely finite that is $\mathbf{E}\left(|\xi| \mid \mathcal{F}\right) \overset{a.s.}{<} \infty$.*
2. *The generalized conditional expectation $\mathbf{E}\left(\xi \mid \mathcal{F}\right)$ is unique up to a measure-zero set.*
3. *If ξ is $\mathcal{F}$-measurable then $\mathbf{E}\left(\xi \mid \mathcal{F}\right) \overset{a.s.}{=} \xi$.*
4. *If the variables ξ and η have generalized conditional expectation then for arbitrary numbers a and b the variable $a\xi + b\eta$ also has generalized conditional expectation and*

$$\mathbf{E}\left(a\xi + b\eta \mid \mathcal{F}\right) \overset{a.s.}{=} a\mathbf{E}\left(\xi \mid \mathcal{F}\right) + b\mathbf{E}\left(\eta \mid \mathcal{F}\right).$$

5. *If the generalized conditional expectation $\mathbf{E}\left(\xi \mid \mathcal{F}\right)$ exists and η is $\mathcal{F}$-measurable then the generalized conditional expectation $\mathbf{E}\left(\eta\xi \mid \mathcal{F}\right)$ also exists and*

$$\mathbf{E}\left(\eta\xi \mid \mathcal{F}\right) \overset{a.s.}{=} \eta\mathbf{E}\left(\xi \mid \mathcal{F}\right).$$

[19]That is the expectation $\mathbf{E}(\eta) \overset{\circ}{=} \mathbf{E}(\eta^+) - \mathbf{E}(\eta^-)$ exists. Recall that by definition the integral $\mathbf{E}(\eta)$ exists if the difference $\mathbf{E}(\eta^+) - \mathbf{E}(\eta^-)$ is not of the form $\infty - \infty$. See: [71].

6. *If $\mathcal{G} \subseteq \mathcal{F}$ and the generalized conditional expectation $\mathbf{E}(\xi \mid \mathcal{G})$ exits then the generalized conditional expectation $\mathbf{E}(\xi \mid \mathcal{F})$ also exists and*

$$\mathbf{E}(\mathbf{E}(\xi \mid \mathcal{G}) \mid \mathcal{F}) \overset{a.s.}{=} \mathbf{E}(\xi \mid \mathcal{G}),$$

$$\mathbf{E}(\mathbf{E}(\xi \mid \mathcal{F}) \mid \mathcal{G}) \overset{a.s.}{=} \mathbf{E}(\xi \mid \mathcal{G}).$$

Recall that the difference $\mathbf{E}(\eta^+ \mid \mathcal{F}) - \mathbf{E}(\eta^- \mid \mathcal{F})$ is not necessarily meaningful for every η. To make the definition of the predictable projection as simple as possible we introduce the operator

$$x \ominus y \overset{\circ}{=} \begin{cases} x - y & \text{if } x, y \geq 0 \text{ are not both infinite,} \\ \infty & \text{otherwise.} \end{cases}$$

Definition 3.27 (Extended conditional expectation) *Let η be an arbitrary random variable and let $\mathcal{F}$ be an arbitrary σ-algebra.*

1. *The extended conditional expectation $\widehat{\mathbf{E}}(\eta \mid \mathcal{F})$ is by definition*

$$\widehat{\mathbf{E}}(\eta \mid \mathcal{F}) \overset{\circ}{=} \mathbf{E}(\eta^+ \mid \mathcal{F}) \ominus \mathbf{E}(\eta^- \mid \mathcal{F}). \tag{3.3}$$

2. *We say that the extended conditional expectation $\widehat{\mathbf{E}}(\eta \mid \mathcal{F})$ is well-defined if except on a measure-zero set one need not use the convention $\infty \ominus \infty \overset{\circ}{=} \infty$.*

Let us remark that $\widehat{\mathbf{E}}(\eta \mid \mathcal{F})$ is $\mathcal{F}$-measurable, but it is different from the conditional expectation and from the generalized conditional expectation. Observe that $\widehat{\mathbf{E}}(\eta \mid \mathcal{F})$ is finite if and only if $\widehat{\mathbf{E}}(\eta \mid \mathcal{F})$ is the generalized conditional expectation. Observe also that in general the extended conditional expectation operator $\widehat{\mathbf{E}}$ is not linear.

3.1.4   Definition of the predictable projection

Using the extended conditional expectation, let us define the predictable projection. Before the definition let us discuss the next, very important example.

Example 3.28 The predictable projection of a local martingale.

Let $L \in \mathcal{L}$ be a local martingale and let L be the net gain of some game. Let $\tau < \infty$ be some exit strategy from L. Let assume that we want to predict, foresee, at least infinitesimally, the value of $L(\tau)$. Of course we should assume that τ is a predictable stopping time otherwise there is no hope to predict the value of $L(\tau)$. If (τ_n) announces τ and L is a uniformly integrable martingale then at τ_n our reasonable prediction of $L(\tau)$ is $\mathbf{E}(L(\tau) \mid \mathcal{F}_{\tau_n}) = L(\tau_n)$. By Lévy's theorem

$$\lim_{n \to \infty} \mathbf{E}\left(L\left(\tau\right) \mid \mathcal{F}_{\tau_n}\right) = \mathbf{E}\left(L\left(\tau\right) \mid \mathcal{F}_{\tau-}\right)$$

which is obviously

$$\lim_{n \to \infty} L\left(\tau_n\right) = L\left(\tau-\right).$$

If L is a local martingale and (σ_m) is localizing L then

$$\lim_{m \to \infty} L^{\sigma_m}\left(\tau-\right) = L\left(\tau-\right)$$

is a reasonable infinitesimal prediction of $L\left(\tau\right)$. Therefore we have got that the 'predictable part' of L is L_-. Which is not a great surprise after all. Perhaps it is more interesting to study the limit of the estimating formula $\mathbf{E}\left(L^{\sigma_m}\left(\tau\right) \mid \mathcal{F}_{\tau-}\right)$.

$$L^{\sigma_m}\left(\tau\right) = L\left(\tau\right) \chi\left(\{\tau > \sigma_m\}^c\right) + L\left(\sigma_m\right) \chi\left(\tau > \sigma_m\right).$$

Obviously[20]

$$\{\tau > \sigma_m\} = \cup_n \{\tau_n > \sigma_m\} \in \sigma\left(\cup_n \mathcal{F}_{\tau_n}\right) = \mathcal{F}_{\tau-}$$

therefore

$$\mathbf{E}\left(L^{\sigma_m}\left(\tau\right) \mid \mathcal{F}_{\tau-}\right) \overset{\circ}{=} \mathbf{E}\left(\left(L^+\right)^{\sigma_m}\left(\tau\right) \mid \mathcal{F}_{\tau-}\right) - \mathbf{E}\left(\left(L^-\right)^{\sigma_m}\left(\tau\right) \mid \mathcal{F}_{\tau-}\right) =$$

$$= \chi\left(\tau \leq \sigma_m\right) \mathbf{E}\left(L^+\left(\tau\right) \mid \mathcal{F}_{\tau-}\right) +$$

$$+ \chi\left(\tau > \sigma_m\right) \mathbf{E}\left(L^+\left(\sigma_m\right) \mid \mathcal{F}_{\tau-}\right) -$$

$$- \chi\left(\tau \leq \sigma_m\right) \mathbf{E}\left(L^-\left(\tau\right) \mid \mathcal{F}_{\tau-}\right) -$$

$$- \chi\left(\tau > \sigma_m\right) \mathbf{E}\left(L^-\left(\sigma_m\right) \mid \mathcal{F}_{\tau-}\right).$$

If $m \to \infty$ then as $\tau < \infty$ the limit is almost surely

$$\mathbf{E}\left(L^+\left(\tau\right) \mid \mathcal{F}_{\tau-}\right) - \mathbf{E}\left(L^-\left(\tau\right) \mid \mathcal{F}_{\tau-}\right).$$

Almost surely

$$\chi\left(\tau \leq \sigma_m\right) \mathbf{E}\left(L^+\left(\tau\right) \mid \mathcal{F}_{\tau-}\right) = \mathbf{E}\left(\chi\left(\tau \leq \sigma_m\right) L^+\left(\tau\right) \mid \mathcal{F}_{\tau-}\right) < \infty$$

for all m and almost surely $\cup_m \{\tau \leq \sigma_m\} = \Omega$. Therefore from the calculation it is clear that almost surely $\mathbf{E}\left(L^{\pm}\left(\tau\right) \mid \mathcal{F}_{\tau-}\right) < \infty$. Hence

$$L_-\left(\tau\right) = \mathbf{E}\left(L^+\left(\tau\right) \mid \mathcal{F}_{\tau-}\right) - \mathbf{E}\left(L^-\left(\tau\right) \mid \mathcal{F}_{\tau-}\right) \overset{\circ}{=} \mathbf{E}\left(L\left(\tau\right) \mid \mathcal{F}_{\tau-}\right)$$

where $\mathbf{E}$ denotes the extended conditional expectation with respect to $\mathcal{F}_{\tau-}$. $\square$

---

[20]See: Proposition 1.34, page 20.

Definition 3.29 (Predictable projection) *We say that the predictable process ^PX is the* predictable projection *of a process X if*

$$^PX(\tau) \stackrel{a.s.}{=} \widehat{\mathbf{E}}(X(\tau) \mid \mathcal{F}_{\tau-}) \quad \text{on the set } \{\tau < \infty\} \tag{3.4}$$

that is[21]

$$\chi(\tau < \infty)\,\widehat{\mathbf{E}}(X(\tau) \mid \mathcal{F}_{\tau-}) \stackrel{a.s.}{=} \widehat{\mathbf{E}}(X(\tau)\chi(\tau < \infty) \mid \mathcal{F}_{\tau-}) \stackrel{a.s.}{=} {}^PX(\tau)\chi(\tau < \infty)$$

for every predictable stopping time τ.

1. *^PX is well-defined if for every predictable stopping time τ the extended conditional expectation in (3.4) is well defined.*
2. *^PX is finite if for every predictable stopping time τ the extended conditional expectation in (3.4) is a.s. finite. In this case the definition of the predictable projection is the following: for every predictable stopping time τ on the set $\{\tau < \infty\}$*

$$^PX(\tau) \stackrel{a.s.}{=} \widehat{\mathbf{E}}(X(\tau) \mid \mathcal{F}_{\tau-}) = \mathbf{E}(X(\tau) \mid \mathcal{F}_{\tau-}) =$$
$$= \mathbf{E}(X^+(\tau) \mid \mathcal{F}_{\tau-}) - \mathbf{E}(X^-(\tau) \mid \mathcal{F}_{\tau-}),$$

   where $\mathbf{E}$ denotes the generalized conditional expectation.

Example 3.30 For non-negative processes, if the predictable projection exists then it is well-defined; for bounded processes, if it exists then it is finite.

3.1.5 The uniqueness of the predictable projection, the predictable section theorem

Our direct goal in this subsection is to prove the next observation[22]:

Theorem 3.31 (Uniqueness of the predictable projection) *If $X_1 \geq X_2$ and $^PX_1$ is a predictable projection of X_1 and $^PX_2$ is a predictable projection of X_2 then $^PX_1(\omega) \geq {}^PX_2(\omega)$ for almost all outcomes ω. Specifically, if $(^PX)_1$ and $(^PX)_2$ are both predictable projections of some process X then $(^PX)_1$ and $(^PX)_2$ are indistinguishable.*

Before we start to prove the theorem, let us remark that from the definition of the predictable projection it is obvious that the result of the operation ^PX is unique up to modification. But stochastic processes are not just indexed sets of random

---

[21] See: Proposition 3.14, page 186. Observe that $X(\tau)$ can have a meaning even on $\{\tau = \infty\}$.

[22] At moment we do not know when the predictable projection exists. Later we shall prove that every product measurable process has a predictable projection. See: Proposition 3.44, page 206.

variables, they are functions of two variables! The main point of the theorem is that we can select from every equivalence class $^PX(t)$ a representative in such a way that the resulting process, as a function of two variables, is predictable. From this it is not really surprising that the proof of the theorem depends on some deep and difficult properties of measurable sets. The theorem is an easy and direct consequence of the next result:

Proposition 3.32 *If $B \subseteq \mathbb{R}_+ \times \Omega$ is a predictable set and $\mathbf{P}(\mathrm{proj}_\Omega B) > 0$ then there is a predictable stopping time σ such that*

$$[\sigma] = \mathrm{Graph}(\sigma) \triangleq \{(\sigma(\omega), \omega) : \sigma(\omega) < \infty\} \subseteq B,$$

and $0 < \mathbf{P}(\sigma < \infty)$.

Proof of the Theorem: If the Theorem were not true then the projection on Ω of the set

$$B \triangleq \{(t, \omega) : \ ^PX_1(t, \omega) < \ ^PX_2(t, \omega)\}$$

would have positive probability. The set B is predictable, since $^PX_1$ and $^PX_2$ are predictable by definition. Hence, by the proposition above, there is a predictable stopping time σ such that with positive probability the graph of σ is in B. It is easy to see that the extended conditional expectation is a monotone operation. Hence as $X_1 \geq X_2$ by the definition of the predictable projection on the set $\{\sigma < \infty\}$

$$^PX_1(\sigma) \overset{a.s.}{=} \widehat{\mathbf{E}}(X_1(\sigma) \mid \mathcal{F}_{\sigma-}) \overset{a.s.}{\geq} \widehat{\mathbf{E}}(X_2(\sigma) \mid \mathcal{F}_{\sigma-}) \overset{a.s.}{=} \ ^PX_2(\sigma)$$

which is impossible as on a set of size $\mathbf{P}(\sigma < \infty) > 0$

$$^PX_1(\sigma) < \ ^PX_2(\sigma).$$

If $(^PX)_1$ and $(^PX)_2$ are two different predictable projections of some X then for almost all outcomes $(^PX)_1(\omega) \geq (^PX)_2(\omega)$ and $(^PX)_2(\omega) \geq (^PX)_1(\omega)$. Hence the trajectories of $(^PX)_1$ and $(^PX)_2$ are equal almost surely. $\qquad \square$

Proposition 3.32 is a direct consequence of the next important 'technical' theorem. This theorem is the 'hard part' of the whole stochastic analysis, and during the proof of it we shall use the completeness of $(\Omega, \mathcal{A}, \mathbf{P})$ several times.

Theorem 3.33 (Predictable Section Theorem) *If $B \subseteq \mathbb{R}_+ \times \Omega$ is a predictable set then for every $\varepsilon > 0$ there is a predictable stopping time σ such that*

$$[\sigma] = \mathrm{Graph}(\sigma) \triangleq \{(t, \omega) : t = \sigma(\omega) < \infty\} \subseteq B,$$

and

$$\mathbf{P}(\mathrm{proj}_\Omega B) \leq \mathbf{P}(\sigma < \infty) + \varepsilon.$$

Proof. The proof of the theorem contains several steps.

1. Observe that if τ is an arbitrary stopping time then the process $\chi\left([0,\tau]\right)$ is adapted and left-continuous. Recall that the random intervals of type $[0,\tau]$ together with the products $\{0\} \times F$, $F \in \mathcal{F}_0$, generate the predictable sets[23] $\mathcal{P}$. As $(\tau,\infty) = [0,\tau]^c$, the random intervals (τ,∞) together with the sets $\{0\} \times F$, $F \in \mathcal{F}_0$, also generate the predictable sets. If σ is a predictable stopping time and (σ_n) announces σ then $[\sigma,\infty) = \cap_n (\sigma_n,\infty)$. Hence if σ is a predictable stopping time then the random interval $[\sigma,\infty)$ is a predictable set. For an arbitrary stopping time τ

$$(\tau,\infty) = \cup_n \left[\tau + 1/n,\infty\right).$$

$\tau + 1/n$ is trivially a predictable stopping time, hence the intervals $[\sigma,\infty)$, where σ is a predictable stopping time, together with the sets $\{0\} \times F$, $F \in \mathcal{F}_0$, generate the predictable sets $\mathcal{P}$. Let $\mathcal{I}$ denote the set of random intervals $[\sigma,\tau)$, where σ and τ are predictable stopping times. Using that the minima and the maxima of predictable stopping times are predictable stopping times, it is easy to see that $\mathcal{I}$ is a semi-algebra. Observe that if $F \in \mathcal{F}_0$ then

$$\sigma\left(\omega\right) \stackrel{\circ}{=} \begin{cases} 0 & \text{if } \omega \in F \\ \infty & \text{if } \omega \notin F \end{cases}$$

is a predictable stopping time. If $\tau \stackrel{\circ}{=} 0$ then $\{0\} \times F = [\sigma,\tau] = \cap_n [\sigma,\tau + 1/n)$ that is $\{0\} \times F \in \sigma(\mathcal{I})$. This implies that $\mathcal{P} = \sigma(\mathcal{I})$.

2. Since $(\Omega, \mathcal{A}, \mathbf{P})$ is complete, by the Measurable Selection Theorem[24] there exists a measurable function $f \colon \Omega \to [0,\infty]$ such that

$$\text{Graph}\left(f\right) \subseteq B \quad \text{and} \quad \{f < \infty\} = \text{proj}_\Omega B.$$

On the product σ-algebra

$$\left(\mathbb{R}_+ \times \Omega, \mathcal{B}\left(\mathbb{R}_+\right) \times \mathcal{A}\right)$$

let us define the set-function

$$\mu\left(E\right) \stackrel{\circ}{=} \mathbf{P}\left(\text{proj}_\Omega\left(E \cap \text{Graph}\left(f\right)\right)\right).$$

Since f is measurable the set $\text{Graph}\left(f\right)$ is product measurable. Hence the set after the operator proj_Ω is product measurable. As $(\Omega, \mathcal{A}, \mathbf{P})$ is complete, by the Projection Theorem[25]

$$\text{proj}_\Omega\left(E \cap \text{Graph}\left(f\right)\right) \in \mathcal{A}.$$

---

[23] See: Corollary 1.44, page 26.
[24] See: Theorem A.13, page 551.
[25] See: Theorem A.12, page 550.

This implies that the definition of μ is correct. One should prove that μ is a measure. To show this it is sufficient to observe that if $E_1 \cap E_2 = \emptyset$ then[26]

$$\text{proj}_\Omega \left(E_1 \cap \text{Graph}\,(f)\right) \bigcap \text{proj}_\Omega \left(E_2 \cap \text{Graph}\,(f)\right) = \emptyset. \qquad (3.5)$$

3. Consider the algebra $\mathcal{H}$ generated by $\mathcal{I}$. As we have remarked $\mathcal{P} = \sigma\,(\mathcal{H})$. By the assumption of the theorem $B \in \mathcal{P}$ and, as $\mathcal{P} = \sigma\,(\mathcal{H})$, of course $B \in \sigma\,(\mathcal{H})$. Obviously μ is a finite measure. Hence from the uniqueness of the extension of finite measures from an algebra to the generated σ-algebra and from the construction of the extended measure, $\mu\,(B) \overset{\circ}{=} \mu_*\,(B)$, there is a $C \subseteq B$, where $C = \cap_n H_n$, $H_n \in \mathcal{H}$ such that

$$\mathbf{P}\,(\text{proj}_\Omega B) \overset{\circ}{=} \mu\,(B) \le \mu\,(C) + \varepsilon.$$

Therefore it is sufficient to prove that there is a predictable stopping time σ for which almost surely

$$\text{Graph}\,(\sigma) \subseteq C.$$

4. Consider the début of C:

$$\tau_C\,(\omega) \overset{\circ}{=} \inf\,\{t : (t,\omega) \in C\}\,.$$

Random intervals are progressively measurable and therefore C is progressively measurable. Hence τ_C is a stopping time[27]. We should show that τ_C is a predictable stopping time and for almost all ω if $\tau_C\,(\omega) < \infty$ then $(\tau_C\,(\omega),\omega) \in C$.

5. Assume first that $C \overset{\circ}{=} [\sigma, \rho) \in \mathcal{I}$. If $D \overset{\circ}{=} \{\sigma < \rho\}$ then

$$\tau_C\,(\omega) = \sigma_D\,(\omega) \overset{\circ}{=} \begin{cases} \sigma\,(\omega) & \text{if} \quad \omega \in D \\ \infty & \text{if} \quad \omega \notin D \end{cases} = \begin{cases} \sigma\,(\omega) & \text{if} \quad \sigma\,(\omega) < \rho\,(\omega) \\ \infty & \text{if} \quad \rho\,(\omega) \le \sigma\,(\omega) \end{cases}.$$

Obviously $\tau_C\,(\omega) < \infty$ if and only if $(\tau_C\,(\omega),\omega) \in C$. By the definition of $\mathcal{I}$ the stopping times σ and ρ are predictable. Therefore[28]

$$D \overset{\circ}{=} \{\sigma < \rho\} \in \mathcal{F}_{\sigma-}.$$

As σ is predictable σ_D is also a predictable stopping time. Therefore if $C \in \mathcal{I}$ the theorem holds.

---

[26]If the intersection were not empty and x was in the intersection then $(x,y) \in \text{Graph}\,(f)$ if and only if $y = f\,(x)$ so $(x, f\,(x)) \in E_1 \cap E_2 \ne \emptyset$. The σ-additivity of μ is obvious as for an arbitrary mapping $F\,(\cup_n A_n) = \cup_n F\,(A_n)$.

[27]See: Theorem 1.28, page 15.

[28]See: Proposition 3.17, page 187.

6. Assume that the theorem is true for $C_1, C_2, \ldots, C_n$ and let $C \overset{\circ}{=} \cup_{i=1}^n C_i$. Obviously

$$\tau_C = \min_i \tau_{C_i} \in C.$$

As the minimum of a finite number of predictable stopping times is a predictable stopping time, the theorem holds for C as well. This means that the theorem is valid if $C \in \mathcal{H}$.

7. If $C_1, C_2, \ldots, C_n \in \mathcal{H}$ and $C \overset{\circ}{=} \cap_{i=1}^n C_i$ then

$$\tau_C = \max_i \tau_{C_i} \in C$$

and, as the maximum of a finite number of predictable stopping times is a predictable stopping time, the theorem holds again.

8. Finally let $C_n \searrow C$ and let assume that τ_{C_n} is predictable and $[\tau_{C_n}] \subseteq C_n$ for all n. Let

$$\tau \overset{\circ}{=} \operatorname{ess\,sup} \{\sigma : \sigma \leq \tau_C, \ \sigma \text{ is a predictable stopping time}\}.$$

By the usual construction of the essential supremum, given that the increasing limit of predictable stopping times is again a predictable stopping time, one can easily prove that τ is a predictable stopping time. Let $D_n \overset{\circ}{=} C_n \cap [\tau, \infty)$ and let τ_{D_n} be the début of D_n. Obviously

$$\tau_{D_n} = \tau_{C_n} \vee \tau.$$

τ is predictable, hence τ_{D_n} is also predictable. As $[\tau_{C_n}] \subseteq C_n$ it is obvious that $[\tau_{D_n}] \subseteq D_n$. As $\tau \leq \tau_C$

$$C \subseteq C_n \cap [\tau_C, \infty) \subseteq C_n \cap [\tau, \infty) \overset{\circ}{=} D_n \subseteq C_n \searrow C.$$

Hence $C = \cap_n D_n$. Obviously $\tau_{C_n} \leq \tau_C$ so

$$\tau \leq \tau_{C_n} \vee \tau = \tau_{D_n} \leq \tau_C.$$

So, by the definition of τ

$$\tau \overset{a.s.}{=} \tau_{D_n} \quad \text{for all } n.$$

As $[\tau_{D_n}] \subseteq D_n$ if $\tau(\omega) < \infty$ then

$$\{(\tau(\omega), \omega)\} \in \cap_n D_n = C, \tag{3.6}$$

for almost all ω. This implies that $\tau \overset{a.s.}{\geq} \tau_C$, so $\tau \overset{a.s.}{=} \tau_C$. The filtration is complete and τ is predictable so τ_C is a predictable stopping time. By (3.6) $(\tau_C(\omega), \omega) \in C$ if $\tau_C(\omega) < \infty$ for almost all ω. $\square$

Corollary 3.34 *A random variable τ is a predictable stopping time if and only if*

$$[\tau] \stackrel{\circ}{=} [\tau, \tau] = \mathrm{Graph}\,(\tau)$$

is a predictable set.

Proof. If τ is a predictable stopping time and (τ_n) announces τ then

$$\mathrm{Graph}\,(\tau) = (\cap_n (\tau_n, \tau]) \cup (\{0\} \times \{\tau = 0\})\,,$$

hence $\mathrm{Graph}\,(\tau)$ is a predictable set.

1. On the other hand let us assume that $\mathrm{Graph}\,(\tau)$ is predictable. Applying the predictable section theorem for $\mathrm{Graph}\,(\tau)$ one can find a predictable stopping time σ_n for all n such that $\tau = \sigma_n$ for the finite values of τ outside an event with probability smaller than $1/n$. If $\tau_n \stackrel{\circ}{=} \min_{k \le n} \sigma_k$ then τ_n is a predictable stopping time and except on a measure-zero set $\tau_n \searrow \tau$. Almost surely zero functions are stopping times, hence τ is a stopping time. But we should prove that τ is a *predictable* stopping time[29].

2. Let $(\tau_k^{(n)})$ be a finite announcing sequence for τ_n. If d denotes the metric generating the topology of $[0, \infty]$ then we can assume that for all k

$$\mathbf{P}\left(d\left(\tau_k^{(n)}, \tau_n\right) > 2^{-k}\right) < 2^{-(k+n)}.$$

Introduce the stopping times $\rho_k \stackrel{\circ}{=} \inf_n \tau_k^{(n)}$. Obviously $\rho_k \le \rho_{k+1}$ for every k and $\rho_k \le \tau_k^{(n)} \le \tau_n$ announces τ for every k. Let $\rho \stackrel{\circ}{=} \lim_k \rho_k$. If for some ω and for some k

$$d\left(\tau_k^{(n)}\,(\omega), \tau\,(\omega)\right) \le 2^{-k}, \quad \text{for all } n$$

then $d\,(\rho_k\,(\omega), \tau\,(\omega)) \le 2^{-k}$. Therefore $d\,(\rho\,(\omega), \tau\,(\omega)) \le 2^{-k}$. Therefore

$$\mathbf{P}\left(d\,(\rho, \tau) > 2^{-k}\right) \le \sum_n \mathbf{P}\left(d\left(\tau_k^{(n)}, \tau\right) > 2^{-k}\right).$$

If τ_n is finite then $\tau_n = \tau$, so

$$\left\{d\left(\tau_k^{(n)}, \tau\right) > 2^{-k}\right\} \subseteq \left\{d\left(\tau_k^{(n)}, \tau_n\right) > 2^{-k}\right\},$$

---

[29]Remember that in general if $\tau_n \searrow \tau$ and τ_n are predictable stopping times for all n then τ is not necessarily a predictable stopping time.

hence

$$\mathbf{P}\left(d\left(\rho,\tau\right) > 2^{-k}\right) \leq \sum_{n} \mathbf{P}\left(d\left(\tau_k^{(n)}, \tau_n\right) > 2^{-k}\right) \leq 2^{-k}.$$

Therefore $\tau \overset{a.s.}{=} \rho$ so τ is a predictable stopping time[30]. $\qquad\square$

We shall often use the following observation:

Proposition 3.35 *Every right-regular, predictable process is locally bounded.*

Proof. To simplify the notation we assume, that $X\left(0\right) = 0$.

$$\tau_n \overset{\circ}{=} \inf\left\{t : |X\left(t\right)| > n\right\},$$

is a stopping time. X is regular, hence on every finite time-interval every trajectory of X is bounded so $\tau_n \nearrow \infty$. Observe that X can have a jump at time τ_n, so X is not necessarily bounded on the interval $[0, \tau_n]$. As X is right-continuous $|X\left(\tau_n\right)| \geq n$. Hence

$$[\![\tau_n, \tau_n]\!] \overset{\circ}{=} \left\{(t, \omega) : t = \tau_n\left(\omega\right)\right\} = [0, \tau_n] \cap \left\{(t, \omega) : |X\left(t, \omega\right)| \geq n\right\}.$$

X is predictable hence the set $\{|X| \geq n\}$ is predictable. So the graph of τ_n is predictable. Hence by the just proved corollary τ_n is a predictable stopping time. If $(\sigma_m^{(n)})$ announces τ_n then $|X^{\sigma_m^{(n)}}| \leq n$. For any n let us choose such a $\sigma_n \overset{\circ}{=} \sigma_m^{(n)}$ for which

$$\mathbf{P}\left(\tau_n - \sigma_n \geq 2^{-n}\right) \leq 2^{-n}.$$

By the Borel–Cantelli lemma, outside of a measure-zero set, for all ω there exists an $n_0\left(\omega\right)$ such that if $n \geq n_0\left(\omega\right)$ then

$$\sigma_n < \tau_n < \sigma_n + 2^{-n}.$$

This implies that $\sigma_n \overset{a.s.}{\to} \infty$. If $\rho_m \overset{\circ}{=} \max_{1 \leq k \leq m} \sigma_k$ then almost surely $\rho_m \nearrow \infty$. It is easy to see that X^{ρ_m} is bounded for all m, so X is locally bounded. $\qquad\square$

Example 3.36 The Poisson processes are not predictable [31].

---

[30] See: Proposition 3.11, page 184.
[31] See: Example 3.56, page 219.

Let X be a Poisson process. Recall[32] that

$$\tau_1 = \inf\{t : X(t) = 1\}$$

is not predictable, so $[\tau_1]$ is not a predictable set. As $[\tau_1] = [0, \tau_1] \cap \{X \geq 1\}$ and $[0, \tau_1]$ is predictable $\{X \geq 1\}$ is not predictable. So X is not a predictable process. $\qquad\square$

3.1.6 Properties of the predictable projection

Later we shall prove that every product measurable process has a predictable projection. During the proof of the existence we shall need some properties of the operation, so we first summarize them.

Proposition 3.37 (Properties of the predictable projection) *Up to indistinguishability the predictable projection has the following properties:*

1. *If Y is predictable then ^PY exists and $^PY = Y$.*

2. *If Y is a finite-valued predictable process and X has a finite predictable projection ^PX, then (YX) also has a finite valued predictable projection $^P(YX)$ and*

$$^P(Y \cdot X) = Y \cdot (^PX). \tag{3.7}$$

   The identity (3.7) also holds if Y is non-negative and predictable, X is non-negative and ^PX exists, or when Y is non-negative, predictable and finite, and ^PX exists.

3. *The correspondence $X \mapsto {}^PX$ is increasing, that is if $0 \leq X \leq Y$ and ^PX and ^PY are meaningful then $0 \leq {}^PX \leq {}^PY$.*

4. *If processes X and Y have finite predictable projection, then $X + Y$ also has finite predictable projection and*

$$^P(X + Y) = {}^PX + {}^PY.$$

   The additivity property also holds when X and Y are non-negative.

5. *The correspondence $X \mapsto {}^PX$ is homogeneous, that is, if process X has a predictable projection and a is an arbitrary real number, then aX also has a predictable projection. If $a \geq 0$ then $^P(aX) = a({}^PX)$.*

6. *The predictable projection satisfies the Monotone Convergence Theorem, that is, if $0 \leq X_n \nearrow X_\infty$ and the processes X_n have predictable projections then the process X_∞ also has a predictable projection and*

$$^PX_n(t, \omega) \nearrow {}^PX_\infty(t, \omega) \tag{3.8}$$

   for all t and for almost all ω.

---

[32]See: Example 3.7, page 183.

7. *If σ is an arbitrary stopping time then $^p(X^\sigma) = (^pX)^\sigma$ on the random interval $[0, \sigma]$.*

8. *The predictable projection is localizable. If (σ_n) is a localizing sequence, $^p(X^{\sigma_n})$ exists for all n and $^p(X^{\sigma_n}) = Y^{\sigma_n}$ on the random intervals $[0, \sigma_n]$, then ^pX exists and $^pX = Y$. If $^p(X^{\sigma_n})$ is well-defined or finite for all n then ^pX is also well-defined or finite.*

Proof. The proof is built on the analogous properties of the conditional expectation and on the uniqueness of the predictable projection.

1. If Y is predictable then $Y^+(\tau)$ and $Y^-(\tau)$ are $\mathcal{F}_{\tau-}$-measurable[33]. If ^pY exists then for every predictable stopping time τ, on the set $\{\tau < \infty\}$

$$^pY(\tau) = \widehat{\mathbf{E}}\left(Y(\tau) \mid \mathcal{F}_{\tau-}\right) \overset{\circ}{=} \mathbf{E}\left(Y^+(\tau) \mid \mathcal{F}_{\tau-}\right) \ominus \mathbf{E}\left(Y^-(\tau) \mid \mathcal{F}_{\tau-}\right) =$$
$$= Y^+(\tau) \ominus Y^-(\tau) = Y(\tau).$$

As the predictable projection is unique, ^pY and Y are indistinguishable. Reading the line above in reverse order one can see that ^pY exists.

2. If Y is predictable, then the stopped variable Y_τ is $\mathcal{F}_{\tau-}$-measurable[34] for every predictable stopping time τ. By the first assumption ^pX is finite hence on the set $\{\tau < \infty\}$

$$^pX(\tau) = \widehat{\mathbf{E}}\left(X(\tau) \mid \mathcal{F}_{\tau-}\right) = \mathbf{E}\left(X(\tau) \mid \mathcal{F}_{\tau-}\right).$$

Multiplying the equation by $Y(\tau)$ and using the analogous properties of the generalized conditional expectation

$$Y(\tau)\left(^pX(\tau)\right) = Y(\tau)\mathbf{E}\left(X(\tau) \mid \mathcal{F}_{\tau-}\right) = \mathbf{E}\left(Y(\tau)X(\tau) \mid \mathcal{F}_{\tau-}\right).$$

As the predictable projection is unique, $(^pX)Y = {}^p(XY)$. If X and Y are non-negative, one can use the same argument, but instead of the generalized conditional expectation one should use the similar properties of the conditional expectation. To prove the last property it is sufficient to remark that if Y is non-negative and finite we can multiply the relation

$$^pX(\tau) = \widehat{\mathbf{E}}\left(X(\tau) \mid \mathcal{F}_{\tau-}\right) = \mathbf{E}\left(X^+(\tau) \mid \mathcal{F}_{\tau-}\right) \ominus \mathbf{E}\left(X^-(\tau) \mid \mathcal{F}_{\tau-}\right)$$

by Y.

3. We have already proved this property[35].

---

[33] See: Proposition 3.15, page 186.
[34] See: Proposition 3.15, page 186.
[35] See: Theorem 3.31, page 194.

4. By the definition of the predictable projection, for an arbitrary predictable stopping time τ since on the set $\{\tau < \infty\}$

$$^P X \left(\tau\right) \stackrel{a.s.}{=} \widehat{\mathbf{E}} \left(X \left(\tau\right) \mid \mathcal{F}_{\tau -}\right), \quad ^P Y \left(\tau\right) \stackrel{a.s.}{=} \widehat{\mathbf{E}} \left(Y \left(\tau\right) \mid \mathcal{F}_{\tau -}\right)$$

by the assumptions, predictable projections are finite or non-negative one can write $\mathbf{E}$ instead of $\widehat{\mathbf{E}}$. Using the additivity of the generalized conditional expectation

$$\left(^P X \left(\tau\right) + {}^P Y \left(\tau\right)\right) \chi \left(\tau < \infty\right) =$$
$$= \left(\mathbf{E} \left(X \left(\tau\right) \mid \mathcal{F}_{\tau -}\right) + \mathbf{E} \left(Y \left(\tau\right) \mid \mathcal{F}_{\tau -}\right)\right) \chi \left(\tau < \infty\right) =$$
$$= \left(\mathbf{E} \left(X \left(\tau\right) \chi \left(\tau < \infty\right) \mid \mathcal{F}_{\tau -}\right) + \mathbf{E} \left(Y \left(\tau\right) \chi \left(\tau < \infty\right) \mid \mathcal{F}_{\tau -}\right)\right) =$$
$$= \mathbf{E} \left(\left(X \left(\tau\right) + Y \left(\tau\right)\right) \chi \left(\tau < \infty\right) \mid \mathcal{F}_{\tau -}\right) =$$
$$= \widehat{\mathbf{E}} \left(X \left(\tau\right) + Y \left(\tau\right) \mid \mathcal{F}_{\tau -}\right) \chi \left(\tau < \infty\right),$$

hence $^P \left(X + Y\right)$ exists and $^P \left(X + Y\right) = {}^P X + {}^P Y$.

5. The proof is analogous.

6. Assume that $0 \leq X_n \nearrow X_\infty$, and that $^P X_n$ exists for all n. For an arbitrary predictable stopping time τ on the set $\{\tau < \infty\}$

$$^P X_n \left(\tau\right) \stackrel{a.s.}{=} \mathbf{E} \left(X_n \left(\tau\right) \mid \mathcal{F}_{\tau -}\right).$$

The limit $Z \stackrel{\circ}{=} \liminf_{n \to \infty} {}^P X_n$ is predictable. By the monotonicity of the predictable projection $\lim_{n \to \infty} {}^P X_n \left(\tau\right) \stackrel{\circ}{=} Z \left(\tau\right)$ exists for every trajectory except on a measure-zero set. The Monotone Convergence Theorem is true for the conditional expectation, hence if $n \to \infty$, then

$$Z \left(\tau\right) = \liminf_{n \to \infty} {}^P X_n \left(\tau\right) \stackrel{a.s.}{=} \lim_{n \to \infty} {}^P X_n \left(\tau\right) \stackrel{a.s.}{=} \lim_{n \to \infty} \mathbf{E} \left(X_n \left(\tau\right) \mid \mathcal{F}_{\tau -}\right) =$$

$$= \mathbf{E} \left(\lim_{n \to \infty} X_n \left(\tau\right) \mid \mathcal{F}_{\tau -}\right) \stackrel{\circ}{=} \mathbf{E} \left(X_\infty \left(\tau\right) \mid \mathcal{F}_{\tau -}\right),$$

so $Z \left(\tau\right) \stackrel{a.s.}{=} \mathbf{E} \left(X_\infty \left(\tau\right) \mid \mathcal{F}_{\tau -}\right)$, hence Z is the predictable projection of X_∞.

7. First of all let us remark that in general the relation $^P \left(X^\sigma\right) = \left(^P X\right)^\sigma$ is not true. As we shall prove[36], if $L \in \mathcal{L}$ then $^P L = L_-$ and obviously $\left(L_-\right)^\sigma \neq \left(L^\sigma\right)_-$. To prove the property let us notice that $\chi \left(\left[0, \sigma\right]\right)$ is a predictable process so by

---

[36]See: Proposition 3.39, page 205.

the second property of the predictable projection just proved

$$\chi\left([0,\sigma]\right)\cdot {}^{P}\left(X^{\sigma}\right) = {}^{P}\left(\chi\left([0,\sigma]\right)\cdot X^{\sigma}\right) =$$
$$= {}^{P}\left(\chi\left([0,\sigma]\right)\cdot X\right) =$$
$$= \chi\left([0,\sigma]\right)\cdot {}^{P}X = \chi\left([0,\sigma]\right)\cdot \left({}^{P}X\right)^{\sigma}.$$

that is ${}^{P}\left(X^{\sigma}\right) = \left({}^{P}X\right)^{\sigma}$ on the random interval $[0,\sigma]$.

8. Let τ be an arbitrary predictable stopping time. If (τ_{k}) announces τ then

$$\{\tau \leq \sigma_{n}\} = \cap_{k}\{\tau_{k} \leq \sigma_{n}\} \in \sigma\left(\cup_{k}\mathcal{F}_{\tau_{k}}\right) \subseteq \mathcal{F}_{\tau-}.$$

In a similar way as above

$$\chi\left(\tau \leq \sigma_{n}\right)\cdot \widehat{\mathbf{E}}\left(X\left(\tau\right) \mid \mathcal{F}_{\tau-}\right) = \tag{3.9}$$
$$= \chi\left(\tau \leq \sigma_{n}\right)^{2}\cdot \widehat{\mathbf{E}}\left(X\left(\tau\right) \mid \mathcal{F}_{\tau-}\right) =$$
$$= \chi\left(\tau \leq \sigma_{n}\right)\cdot \widehat{\mathbf{E}}\left(\chi\left(\tau \leq \sigma_{n}\right)X\left(\tau\right) \mid \mathcal{F}_{\tau-}\right) =$$
$$= \chi\left(\tau \leq \sigma_{n}\right)\cdot \widehat{\mathbf{E}}\left(\chi\left(\tau \leq \sigma_{n}\right)X^{\sigma_{n}}\left(\tau\right) \mid \mathcal{F}_{\tau-}\right) =$$
$$= \chi\left(\tau \leq \sigma_{n}\right)\cdot \widehat{\mathbf{E}}\left(X^{\sigma_{n}}\left(\tau\right) \mid \mathcal{F}_{\tau-}\right) =$$
$$= \chi\left(\tau \leq \sigma_{n}\right)\cdot \left({}^{P}\left(X^{\sigma_{n}}\right)\right)\left(\tau\right) =$$
$$= \chi\left(\tau \leq \sigma_{n}\right)\cdot Y^{\sigma_{n}}\left(\tau\right).$$

As $\sigma_{n} \nearrow \infty$, obviously $\widehat{\mathbf{E}}\left(X\left(\tau\right) \mid \mathcal{F}_{\tau-}\right) \stackrel{a.s.}{=} Y\left(\tau\right)$ on the set $\{\tau < \infty\}$. Y is a limit of predictable processes so it is predictable, hence the first part of the property holds. The second part of the property follows from (3.9). □

3.1.7   Predictable projection of local martingales

Let us first prove some interesting results.

Proposition 3.38 (Predictable Optional Sampling) *If X is a uniformly integrable martingale and τ is a predictable stopping time then*

$$X\left(\tau-\right) \stackrel{a.s.}{=} \mathbf{E}\left(X\left(\tau\right) \mid \mathcal{F}_{\tau-}\right) = \mathbf{E}\left(X\left(\infty\right) \mid \mathcal{F}_{\tau-}\right). \tag{3.10}$$

Proof. If (τ_{n}) announces τ then by the Optional Sampling Theorem

$$X\left(\tau_{n}\right) \stackrel{a.s.}{=} \mathbf{E}\left(X\left(\tau\right) \mid \mathcal{F}_{\tau_{n}}\right) = \mathbf{E}\left(X\left(\infty\right) \mid \mathcal{F}_{\tau_{n}}\right).$$

By the uniform integrability $X(\infty) \in L^1(\Omega)$. Hence $X(\tau) \in L^1(\Omega)$. Every martingale has left-limits so

$$\lim_{n \to \infty} X(\tau_n) = X(\tau-).$$

As $\mathcal{F}_{\tau_n} \nearrow \mathcal{F}_{\tau-}$ by Lévy's martingale convergence theorem[37]

$$\lim_{n \to \infty} \mathbf{E}\left(X(\infty) \mid \mathcal{F}_{\tau_n}\right) = \mathbf{E}\left(X(\tau) \mid \mathcal{F}_{\tau-}\right) = \mathbf{E}\left(X(\infty) \mid \mathcal{F}_{\tau-}\right).$$

Therefore (3.10) holds. □

Proposition 3.39 *If L is a local martingale then L has a finite predictable projection ^pL and*

$$^pL(t) = L_-(t) \overset{\circ}{=} L(t-).$$

Proof. L_- is left-continuous therefore it is predictable. If L is a uniformly integrable martingale and τ is a predictable stopping time then by (3.10)

$$L_-(\tau) = L(\tau-) \overset{a.s.}{=} \mathbf{E}\left(L(\tau) \mid \mathcal{F}_{\tau-}\right)$$

that is by the definition of the predictable projection $^pL = L_-$. The general case is evident from the last property of the predictable projection. □

Corollary 3.40 *A local martingale is predictable if and only if it is continuous.*

Proof. If L is continuous then L is predictable. The reverse implication follows from $L_+ = L = {}^pL = L_-$. □

Corollary 3.41 *If X is a special semimartingale then there is just one decomposition of the kind*

$$X = X(0) + V + L, \ V \in \mathcal{V}, \ L \in \mathcal{L}$$

where V is predictable. This decomposition is called the canonical decomposition of X.

Proof. If $V_1 + L_1 = V_2 + L_2$ then $M \overset{\circ}{=} V_1 - V_2 = L_2 - L_1$ is a predictable local martingale, hence M is a continuous local martingale. On the other hand the trajectories of M have finite variation, so by Fisk's theorem[38] $M \overset{a.s.}{=} 0$. □

The most important example of the predictable projection is the following.

[37]See: Theorem 1.69, page 41.
[38]See: Theorem 2.11, page 117.

Corollary 3.42 *If L is a local martingale and ΔL is the jump process of L then*

$$^p(\Delta L) = 0.$$

Proof. It is evident from the additivity of the predictable projection:

$$^p(\Delta L) = {}^p(L - L_-) = {}^pL - {}^p(L_-) =$$
$$= L_- - L_- = 0. \qquad \square$$

3.1.8 Existence of the predictable projection

Now we are ready to discuss the question of the existence of the predictable projection.

Corollary 3.43 *Let η be an integrable random variable and let $M(t) \overset{\circ}{=} \mathbf{E}(\eta \mid \mathcal{F}_t)$ be the martingale generated by the random variable η. If $X \equiv \eta$ then X has a predictable projection and $^pX = M_-$.*

Proof. As the usual conditions hold $M(t) \overset{\circ}{=} \mathbf{E}(\eta \mid \mathcal{F}_t)$ has a right-regular version, hence M is a uniformly integrable martingale. If τ is a predictable stopping time then by the Predictable Optional Sampling Theorem on the set $\{\tau < \infty\}$

$$M(\tau-) = \mathbf{E}(M(\infty) \mid \mathcal{F}_{\tau-}) =$$
$$= \mathbf{E}(\eta \mid \mathcal{F}_{\tau-}) \overset{\circ}{=} \mathbf{E}(X(\tau) \mid \mathcal{F}_{\tau-}),$$

that is, X has a predictable projection and $^pX = M_-$. $\qquad \square$

Proposition 3.44 (Existence of the predictable projection) *Every product-measurable process X has a predictable projection.*

Proof. If η is an integrable random variable then as we have seen $X \equiv \eta$ has a predictable projection. If $I \in \mathcal{B}(\mathbb{R}_+)$ then χ_I is predictable, hence

$$^p(\chi_I \cdot X) = \chi_I \cdot {}^pX,$$

that is, the processes of type $X \overset{\circ}{=} \chi_I \cdot \eta$ also have predictable projection. The processes of type $\chi_I \cdot \eta$ form a π-system, which generate the set of product-measurable processes. Denote by $\mathcal{H}$ the set of bounded processes X which have predictable projection. Observe that by the properties of the predictable projection, $\mathcal{H}$ is a λ-system:

1. The constant process 1 is predictable, hence trivially $1 \in \mathcal{H}$.

2. By the linearity of the predictable projection among the bounded processes $\mathcal{H}$ is a linear space.

3. By the Monotone Convergence Theorem for the predictable projection, $\mathcal{H}$ is a monotone class.

By the Monotone Class Theorem $\mathcal{H}$ contains the set of processes generated by the π-system of the processes $\chi_I \cdot \eta$ that is $\mathcal{H}$ contains the product-measurable bounded process. Again by the Monotone Convergence Theorem every non-negative product-measurable process has a predictable projection. If X is a product-measurable process then the process

$$^pX \doteq {}^pX^+ \ominus {}^pX^-$$

satisfies (3.4) and therefore X has a predictable projection. $\qquad\qquad\square$

3.2 Predictable Compensators

Assume that X is some non-negative increasing process. That is assume that X describes some 'unpredictable' cumulative losses[39]. Although the losses are 'unpredictable' one can still ask whether they are 'insurable', that is, whether there is an insurance contract which compensates for the losses of X. Under compensation we mean that there is some payments process P for which the net risk $X - P$ is a local martingale[40]. Of course, since the process X is 'risky', one cannot forecast the jumps of X, so to make the insurance contract fair one should pay the price of the compensator 'before' the jumps, so we are looking for a 'predictable' payment process P. The main result of this section is that if a right-continuous, increasing process X has locally integrable variation, then X has an increasing, right-continuous and locally integrable predictable compensator. This theorem is the 'home edition' of the 'professional' Doob–Meyer decomposition[41].

3.2.1 Predictable Radon–Nikodym Theorem

Before the proof of the existence of the compensator, we introduce some tools which we shall use during the proof. Let $V \in \mathcal{V}^+$. Since V is right-continuous and increasing, every trajectory $V(\omega)$ generates a σ-finite measure $\mu(\omega)$ on $\mathbb{R}_+$. Let us denote by $\mathcal{G}$ the set of possible events[42] in Ω. If the process Y is measurable with respect to the product σ-algebra $\widetilde{\mathcal{G}} \doteq \mathcal{B}(\mathbb{R}_+) \times \mathcal{G}$ then the trajectories $Y(\omega)$ are $\mathcal{B}(\mathbb{R}_+)$-measurable. Hence if $Y(\omega) \geq 0$ then the integrals

$$\int_0^\infty Y(s,\omega)\,d\mu(s,\omega) \doteq \int_0^\infty Y(s,\omega)\,dV(s,\omega) \qquad (3.11)$$

are meaningful. Of course the value of the integral depends on ω. Let $Y \doteq \chi_{(s,t]}\chi_F$ for some $F \in \mathcal{G}$. V is adapted hence $V(s,\omega)$, as a function of ω, is

---

[39]Of course in this paragraph we use the word 'predictable' in a colloquial sense.

[40]See: Definition 2.41, page 141.

[41]See: Theorem 5.1, page 292.

[42]Recall that symbol $\mathcal{A}$ is ambiguous. It denotes the set of events and the set of processes with integrable variation. To avoid confusion in the present subsection we shall use the symbol $\mathcal{G}$ for the events of Ω.

$\mathcal{G}$-measurable for all s and therefore the integral

$$\int_0^\infty Y(s,\omega)\, dV(s,\omega) = \chi_F \cdot (V(t,\omega) - V(s,\omega))$$

is also $\mathcal{G}$-measurable. With the Monotone Class Theorem it is now easy to prove that (3.11) is $\mathcal{G}$-measurable[43] for every non-negative product-measurable processes Y. Hence the expression

$$\mathbf{E}\left(\int_0^\infty Y\, dV\right) \doteq \int_\Omega \int_0^\infty Y(s,\omega)\, dV(s,\omega)\, d\mathbf{P}(\omega)$$

is meaningful for all non-negative product-measurable Y.

Definition 3.45 *For $V \in \mathcal{V}^+$ let us define*

$$\mu_V(B) \doteq \mathbf{E}\left(\int_0^\infty \chi_B\, dV\right) \doteq \mathbf{E}\left(\int_0^\infty \chi_B(s,\omega)\, dV(s,\omega)\right), \quad B \in \widetilde{\mathcal{G}}.$$

It is easy to see that μ_V is a measure on the product σ-algebra $\widetilde{\mathcal{G}}$. If $V \in \mathcal{A}^+$, that is if $V \in \mathcal{V}^+$ and $\mathbf{E}(V(\infty)) < \infty$ then μ_V is finite. If $V \in \mathcal{A}^+_{\mathrm{loc}}$ then μ_V is σ-finite.

Definition 3.46 *If μ is a measure on the product σ-algebra $\widetilde{\mathcal{G}}$ and for some $V \in \mathcal{V}^+$*

$$\mu_V(B) = \mu(B), \quad B \in \widetilde{\mathcal{G}}$$

then we say that measure μ is generated *by V.*

Definition 3.47 *We say that the product-measurable set N is* evanescent *if its characteristic function is indistinguishable from the characteristic function of the empty set, that is outside an event with zero probability the trajectories of χ_N are zero[44].*

Definition 3.48 *Let μ be a measure on the product σ-algebra[45] $\widetilde{\mathcal{G}}$. We say that μ is* absolutely continuous *if $\mu(N) = 0$ for every evanescent set N.*

Proposition 3.49 (Generalized Radon–Nikodym Theorem) *Let μ be a measure on*

$$(\mathbb{R}_+ \times \Omega, \mathcal{B}(\mathbb{R}_+) \times \mathcal{G}) \doteq (\widetilde{\Omega}, \widetilde{\mathcal{G}}).$$

---

[43]Observe that we cannot directly apply the classical Radon–Nikodym theorem. The present argument is exactly that which one should use during the proof of the Radon–Nikodym theorem.

[44]It is contained in $\mathbb{R}_+ \times N$ for some N with $\mathbf{P}(N) = 0$.

[45]Recall that $\mathcal{G}$ denotes the set of events in Ω.

There exists an $A \in \mathcal{A}_{\mathrm{loc}}^+$ for which

$$\mu(B) = \mu_A(B) \overset{\circ}{=} \mathbf{E}\left(\int_0^\infty \chi_B \, dA\right), \quad \text{whenever } B \in \tilde{\mathcal{G}}$$

if and only if

1. $\mu([0]) \overset{\circ}{=} \mu(\{0\} \times \Omega) = 0$.
2. $\mu([0,t] \times F)$ *is σ-finite for every t.*
3. μ *is absolutely continuous.*

If $A, B \in \mathcal{V}$ are two processes representing μ then A and B are indistinguishable.

Proof. It is easy to see that the conditions are necessary. For example, if B is evanescent then, by definition, outside an event of zero probability the trajectories of χ_B are zero so, trivially, $\mu_A(B) = 0$.

Let us prove the sufficiency of the conditions. Let us define the measures

$$\rho_t(H) \overset{\circ}{=} \mu([0,t] \times H) = \mu([0] \times H) + \mu((0,t] \times H) =$$
$$= \mu((0,t] \times H), \quad H \in \mathcal{G}.$$

By the third assumption ρ_t is absolutely continuous with respect to $\mathbf{P}$ for every t. Hence the classical Radon–Nikodym derivative $d\rho_t/d\mathbf{P}$ exists for all t. By the second condition $d\rho_t/d\mathbf{P}$ is finite. Let A_t' be a non-negative version of the derivative. Trivially $A_0' \overset{a.s.}{=} 0$ and if $s \le t$ then $A_s' \overset{a.s.}{\le} A_t'$. By this monotonicity if $t_n \searrow t_\infty$ then $A_{t_n}' \overset{a.s.}{\searrow} A_*'$. By the second assumption and by the Dominated Convergence Theorem

$$\mu([0,t_\infty] \times F) = \lim_{n \to \infty} \mu([0,t_n] \times F) =$$

$$= \lim_{n \to \infty} \int_F A_{t_n}' \, d\mathbf{P} = \int_F \lim_{n \to \infty} A_{t_n}' \, d\mathbf{P} \overset{\circ}{=} \int_F A_*' \, d\mathbf{P}.$$

Hence by the uniqueness of the Radon–Nikodym derivative $A_*' \overset{a.s.}{=} A_{t_\infty}'$. Of course we have the usual problem: since the variables A_t' are defined up to a measure-zero set we can unambiguously define the process $t \mapsto A_t'$ only on the rational numbers. To define A for all real numbers $t \ge 0$ we use the definition

$$A_t \overset{\circ}{=} \inf\{A_r' : t < r, \, r \in \mathbb{Q}\}.$$

Since the filtration is right-continuous, A_t will be $\mathcal{G}$-measurable for all t and it is easy to see that the trajectories of A are increasing and right-continuous. From the first condition it is obvious that $A_0 = 0$ is possible, that is $A \in \mathcal{V}^+$. By the argument above A_t is a version of $d\rho_t/d\mathbf{P}$ for all t. The processes of type $\chi([0,t] \times H)$ form a π-system generating the σ-algebra of product measurable sets $\widetilde{\mathcal{G}}$. This π-system is contained in the λ-system of bounded functions for which

$$\mathbf{E}\left(\int_0^\infty f dA\right) = \int_{\mathbb{R}_+ \times \Omega} f d\mu \overset{\circ}{=} \int_{\widetilde{\Omega}} f d\mu.$$

By the Monotone Class Theorem and by the Monotone Convergence Theorem for every $B \in \widetilde{\mathcal{G}}$

$$\mathbf{E}\left(\int_0^\infty \chi_B dA\right) = \int_{\mathbb{R}_+ \times \Omega} \chi_B d\mu = \mu(B).$$

As μ is σ-finite, for every A there is just one μ associated with A. From the construction it is clear, that if $B \in \mathcal{V}^+$ is another right-continuous representation of μ then B is a modification of A. As both A and B are right-continuous they are indistinguishable.   □

The main non-trivial question is the following: when is the generalized Radon–Nikodym derivative A predictable? To answer the question we need the following simple observation:

Lemma 3.50 *Let $\mathcal{F} \subseteq \mathcal{G}$ be two σ-algebras and let $\xi \in L^1(\Omega, \mathcal{G}, \mathbf{P})$. Assume that $\mathcal{F}$ contains the measure-zero sets of $\mathcal{G}$. If for all $F \in \mathcal{G}$*

$$\mathbf{E}(\xi \cdot \chi_F) = \mathbf{E}(\xi \cdot \mathbf{E}(\chi_F \mid \mathcal{F})) \tag{3.12}$$

then ξ is $\mathcal{F}$-measurable.

Proof. We prove that

$$\widehat{\xi} \overset{\circ}{=} \mathbf{E}(\xi \mid \mathcal{F}) \overset{a.s}{=} \xi.$$

If $F \in \mathcal{G}$ and $h \overset{\circ}{=} \chi_F$ then by (3.12)

$$\mathbf{E}(\xi \cdot h) \overset{\circ}{=} \mathbf{E}(\xi \cdot \chi_F) = \mathbf{E}(\xi \cdot \mathbf{E}(\chi_F \mid \mathcal{F})) =$$
$$= \mathbf{E}(\mathbf{E}(\xi \cdot \mathbf{E}(\chi_F \mid \mathcal{F}) \mid \mathcal{F})) =$$
$$= \mathbf{E}(\mathbf{E}(\xi \mid \mathcal{F}) \cdot \mathbf{E}(\chi_F \mid \mathcal{F})) \overset{\circ}{=}$$
$$\overset{\circ}{=} \mathbf{E}\left(\widehat{\xi} \cdot \mathbf{E}(\chi_F \mid \mathcal{F})\right) = \mathbf{E}\left(\mathbf{E}\left(\widehat{\xi} \cdot \chi_F \mid \mathcal{F}\right)\right) =$$
$$= \mathbf{E}\left(\widehat{\xi} \cdot \chi_F\right) \overset{\circ}{=} \mathbf{E}\left(\widehat{\xi} \cdot h\right).$$

As $\xi \in L^1(\Omega)$ by the usual simple density argument, using that every $h \in L^\infty(\Omega, \mathcal{G}, \mathbf{P})$ is almost surely a uniform limit of step functions the equation

$$\mathbf{E}(\xi \cdot h) = \mathbf{E}\left(\widehat{\xi} \cdot h\right)$$

can be extended to all $h \in L^\infty (\Omega, \mathcal{G}, \mathbf{P})$. Therefore if $h \stackrel{\circ}{=} \text{sign}(\xi - \widehat{\xi})$ then

$$\mathbf{E} \left(\left| \xi - \widehat{\xi} \right| \right) = \mathbf{E} \left(\left(\xi - \widehat{\xi} \right) \text{sign} \left(\xi - \widehat{\xi} \right) \right) = 0.$$

Hence, as we said $\xi \stackrel{a.s.}{=} \widehat{\xi}$. By the definition of the conditional expectation $\widehat{\xi}$ is $\mathcal{F}$-measurable. $\mathcal{F}$ contains all the measure-zero sets of $\mathcal{G}$ and therefore ξ is also $\mathcal{F}$-measurable. $\qquad \square$

Proposition 3.51 *Assume that the measure μ_V is generated by a process $V \in \mathcal{V}^+$. Assume also that whenever*

$$^p X = {}^p Y$$

then

$$\mathbf{E} \left(\int_0^\infty X \, dV \right) \stackrel{\circ}{=} \mu_V (X) = \mu_V (Y) \stackrel{\circ}{=} \mathbf{E} \left(\int_0^\infty Y \, dV \right).$$

Then V is predictable.

Proof. We divide the proof into several steps.

1. First we prove that $V (\tau)$ is $\mathcal{F}_{\tau-}$-measurable for every predictable stopping time τ. By the lemma it is sufficient to prove that

$$\mathbf{E} (\chi_H \cdot V (\tau)) = \mathbf{E} (\mathbf{E} (\chi_H \mid \mathcal{F}_{\tau-}) \cdot V (\tau)), \quad \text{whenever } H \in \mathcal{G},$$

which is the same as

$$\mu_V (\chi_H \cdot \chi ([0, \tau])) = \mu_V (\mathbf{E} (\chi_H \mid \mathcal{F}_{\tau-}) \cdot \chi ([0, \tau])).$$

To prove this equation one should prove, by the assumption of the proposition, that the predictable projections of the two associated processes in the argument of μ_V are indistinguishable:

$$^p (\chi_H \cdot \chi ([0, \tau])) = {}^p (\mathbf{E} (\chi_H \mid \mathcal{F}_{\tau-}) \cdot \chi ([0, \tau])). \tag{3.13}$$

Let $M (t) \stackrel{\circ}{=} \mathbf{E} (\chi_H \mid \mathcal{F}_t)$. By the Predictable Optional Sampling Theorem

$$M_-^\tau (t) = M ((\tau \wedge t) -) = \mathbf{E} \left(M (\infty) \mid \mathcal{F}_{(\tau \wedge t)-} \right) =$$
$$= \mathbf{E} \left(\chi_H \mid \mathcal{F}_{(\tau \wedge t)-} \right) = \mathbf{E} \left(\mathbf{E} (\chi_H \mid \mathcal{F}_{\tau-}) \mid \mathcal{F}_{(t \wedge \tau)-} \right) \stackrel{\circ}{=}$$
$$\stackrel{\circ}{=} {}^p (\mathbf{E} (\chi_H \mid \mathcal{F}_{\tau-})) (t \wedge \tau),$$

that is

$$M_-^\tau = ({}^p (\mathbf{E} (\chi_H \mid \mathcal{F}_{\tau-})))^\tau.$$

$\chi\left([0,\tau]\right)$ is predictable and[46] $^{p}\left(\chi_{H}\right)=M_{-}$ hence[47]

$$
\begin{aligned}
{}^{p}\left(\chi_{H}\cdot\chi\left([0,\tau]\right)\right)={}^{p}\left(\chi_{H}\right)\cdot\chi\left([0,\tau]\right)=M_{-}\cdot\chi\left([0,\tau]\right)= \\
=M_{-}^{\tau}\cdot\chi\left([0,\tau]\right)=\left({}^{p}\left(\mathbf{E}\left(\chi_{H}\mid\mathcal{F}_{\tau-}\right)\right)\right)^{\tau}\cdot\chi\left([0,\tau]\right)= \\
={}^{p}\left(\mathbf{E}\left(\chi_{H}\mid\mathcal{F}_{\tau-}\right)\right)\cdot\chi\left([0,\tau]\right)= \\
={}^{p}\left(\mathbf{E}\left(\chi_{H}\mid\mathcal{F}_{\tau-}\right)\cdot\chi\left([0,\tau]\right)\right),
\end{aligned}
$$

which is what we wanted in (3.13). Therefore $V\left(\tau\right)$ is $\mathcal{F}_{\tau-}$-measurable.

2. Since $V\left(\tau\right)$ is $\mathcal{F}_{\tau-}$-measurable, from the definition of the predictable projection on $\{\tau<\infty\}$

$$
{}^{p}V\left(\tau\right)\stackrel{a.s.}{=}\mathbf{E}\left(V\left(\tau\right)\mid\mathcal{F}_{\tau-}\right)\stackrel{a.s.}{=}V\left(\tau\right) \tag{3.14}
$$

for every predictable stopping time τ.

3. As a special case we get that ${}^{p}V$ is a modification of V. As a next step we prove that V and ${}^{p}V$ are indistinguishable. Let $V=V^{c}+V^{d}$, where V^{c} is continuous and $V^{d}\stackrel{\circ}{=}\sum\Delta V$ is the jump part of V. As V^{c} and V^{d} are non-negative and as the predictable projection for non-negative processes is additive:

$$
{}^{p}V={}^{p}\left(V^{c}+V^{d}\right)={}^{p}\left(V^{c}\right)+{}^{p}\left(V^{d}\right)=V^{c}+{}^{p}\left(V^{d}\right).
$$

Since V is regular, the jumps of V form a thin set and as V is right-regular one may write

$$
V^{d}=\sum\Delta V=\sum_{k}\Delta V\left(\sigma_{k}\right)\chi\left([\sigma_{k},\infty)\right),
$$

where the σ_{k} are either predictable or totally inaccessible[48]. The terms in the sum are non-negative so by the Monotone Convergence Theorem for the predictable projection[49]

$$
{}^{p}\left(V^{d}\right)=\sum_{k}{}^{p}\left(\Delta V\left(\sigma_{k}\right)\chi\left([\sigma_{k},\infty)\right)\right).
$$

If σ_{k} is predictable, then $\chi\left([\sigma_{k},\infty)\right)$ is predictable and by (3.14) in the previous point of the proof[50]

$$
\begin{aligned}
{}^{p}\left(\Delta V\left(\sigma_{k}\right)\cdot\chi\left([\sigma_{k},\infty)\right)\right)={}^{p}\left(\Delta V\left(\sigma_{k}\right)\right)\cdot\chi\left([\sigma_{k},\infty)\right)= \\
={}^{p}\left(\left(V-V_{-}\right)\left(\sigma_{k}\right)\right)\cdot\chi\left([\sigma_{k},\infty)\right)= \\
=\Delta V\left(\sigma_{k}\right)\cdot\chi\left([\sigma_{k},\infty)\right)
\end{aligned}
$$

[46] See: Corollary 3.43, page 206.
[47] See: (3.7), page 201.
[48] See: Proposition 3.19, page 188.
[49] See: (3.8), page 201.
[50] See: (3.7), page 201.

almost surely. If σ_k is totally inaccessible, then

$$\mathbf{E}\left(\Delta V\left(\sigma_k\right)\right) = \int_\Omega \int_0^\infty \chi\left([\sigma_k]\right) \, dV \, d\mathbf{P} \stackrel{\circ}{=} \mu_V\left(\chi\left([\sigma_k]\right)\right).$$

Since σ_k is totally inaccessible $\chi\left([\sigma_k]\right)\left(\tau\right) \stackrel{a.s.}{=} 0$ for any predictable stopping time τ. So $^P\chi\left([\sigma_k]\right) = 0 = {}^P 0$. Hence by the assumption of the proposition

$$\mathbf{E}\left(\Delta V\left(\sigma_k\right)\right) = 0.$$

As V is increasing, $V\left(\sigma_k\right) \stackrel{a.s.}{=} V\left(\sigma_k-\right)$, that is if σ_k is totally inaccessible, then $\Delta V\left(\sigma_k\right) \stackrel{a.s.}{=} 0$. From this it is now obvious that V and the predictable process $^P V$ are indistinguishable.

4. As a final step it will be sufficient to prove that if X is a product measurable and indistinguishable from the zero process, then X is predictable. Let $\mathbf{P}\left(N\right) = 0$ and let $X\left(\omega\right) = 0$ if $\omega \notin N$. Let us first assume that $X = \chi_C$, where $C \stackrel{\circ}{=} (t_1, t_2] \times A$ with $A \subseteq N$. As $\mathcal{F}_t$ is complete

$$\sigma_t\left(\omega\right) \stackrel{\circ}{=} \begin{cases} t & \text{if } \omega \in A \\ 0 & \text{if } \omega \notin A \end{cases}$$

is a stopping time for all t. This implies that

$$C \stackrel{\circ}{=} \{(t, \omega) : t_1 < t \leq t_2, \, \omega \in A\} =$$
$$= \{(t, \omega) : \sigma_{t_1}\left(\omega\right) < t \leq \sigma_{t_2}\left(\omega\right)\} \stackrel{\circ}{=}$$
$$\stackrel{\circ}{=} (\sigma_{t_1}, \sigma_{t_2}] \in \mathcal{P}.$$

From this, using that X is product measurable, the predictability of X is already immediate. $\qquad\square$

3.2.2 Predictable Compensator of locally integrable processes

Now we are ready to prove the 'home edition' of the Doob–Meyer decomposition:

Theorem 3.52 (Existence of Predictable Compensators) *If $A \in \mathcal{A}_{\text{loc}}^+$ then there is a predictable process $A^p \in \mathcal{A}_{\text{loc}}^+$ which is unique up to indistinguishability and which satisfies each of the following three equivalent properties*[51]:

1. $A - A^p \in \mathcal{L}$.

---

[51]The interpretation of these equivalent properties is the following. If A is a process of some cumulative losses then A^p is the cumulative insurance fee which one would have to pay to an insurance company to cover the risk in A. $A - A^p \in \mathcal{L}$ means that to make the contract fair the net gain of the insurance company should contain no systematic trend. By the second condition, whatever stopping strategy one of the parties follows, on average nobody gains. By the third condition, if H is the number of insurance contracts then on average the net payouts of the company is the same as the amount of fees paid by the clients. Of course one can decide about the number of contracts and about the size of the fee before the losses represented by A occur, so H and A^p should be predictable.

2. $\mathbf{E}(A(\tau)) = \mathbf{E}(A^p(\tau))$ *for any stopping time*[52] τ.

3. *For all non-negative, predictable processes* H

$$\mathbf{E}\left(\int_0^\infty H\,dA\right) = \mathbf{E}\left(\int_0^\infty H\,dA^p\right), \tag{3.15}$$

where the integrals on both sides are are pathwise Lebesgue–Stieltjes integrals.

Proof. We divide the proof into several steps.

1. First we prove that the three conditions above are equivalent. Let $A-A^p \in \mathcal{L}$ and let (τ_n) be a joint localizing sequence of A, A^p and $A-A^p$. As the spaces $\mathcal{A}^+_{\text{loc}}$ and $\mathcal{L}$ are closed under stopping one can find such a sequence. $(A - A^p)^{\tau_n} \in \mathcal{M}$, hence by the Optional Sampling Theorem

$$\mathbf{E}(A(\tau \wedge \tau_n)) = \mathbf{E}(A^p(\tau \wedge \tau_n)).$$

A and A^p are increasing processes by the definition of $\mathcal{A}^+_{\text{loc}}$, hence by the Monotone Convergence Theorem

$$\mathbf{E}(A(\tau)) = \mathbf{E}(A^p(\tau)).$$

If $H \overset{\circ}{=} \chi([0,\tau])$ and $X \in \mathcal{V}^+$ arbitrary, then $\int_0^\infty H\,dX = X(\tau)$, hence if $H \overset{\circ}{=} \chi([0,\tau])$ then from the second condition the third one follows. If H is the characteristic function of some set $\{0\} \times F$, $F \in \mathcal{F}_0$ then (3.15) obviously holds. The σ-algebra generated by the intervals $[0,\tau]$ and by the sets $\{0\} \times F$ is the set of predictable processes, hence the general case follows from the Monotone Class Theorem and from the Monotone Convergence Theorem, since the processes of type $\chi([0,\tau])$ form a π-system[53], and the set of processes H satisfying (3.15) is trivially a λ-system. We can reverse the last argument so if the third condition holds, then

$$\mathbf{E}(A(\theta)) = \mathbf{E}\left(\int_0^\infty \chi([0,\theta])\,dA\right) = \mathbf{E}\left(\int_0^\infty \chi([0,\theta])\,dA^p\right) = \mathbf{E}(A^p(\theta))$$

for any stopping time θ. Of course, it can happen that both expected values are infinite. If (τ_n) is a joint localizing sequence for A and A^p, then $\mathbf{E}(A_{\tau \wedge \tau_n}) = \mathbf{E}(A^p_{\tau \wedge \tau_n})$ for every stopping time τ, where by the localization the two expected values are finite. Hence $(A - A^p)^{\tau_n} \in \mathcal{M}$, that is $A - A^p$ is a local martingale.

---

[52] Observe that since A and A^p are increasing $A(\infty)$ and $A^p(\infty)$ are well-defined although they can be $+\infty$.

[53] $\chi([0,\tau])\,\chi([0,\sigma]) = \chi([0,\sigma \wedge \tau])$ and $\sigma \wedge \tau$ is also stopping time.

2. We prove that A^p is unique. If $A_1^p, A_2^p \in \mathcal{A}_{\text{loc}}$ are predictable processes and $A - A_i^p$ are local martingales for indexes $i = 1, 2$, then $A_1^p - A_2^p \in \mathcal{V}$ is a predictable local martingale. As every predictable local martingale is continuous[54] $A_1^p - A_2^p \in \mathcal{V}$ is a continuous local martingale. Hence by Fisk's theorem[55] $A_1^p - A_2^p = A_1^p(0) - A_2^p(0) = 0$.

3. Finally we prove the existence of A^p. Let μ_A be the measure generated by A, that is if X is a product measurable set then let $\mu_A(X) \doteq \mathbf{E}\left(\int_0^\infty \chi(X)\, dA\right)$. On the product measurable sets let us define the set function

$$\mu(X) \doteq \mu_A\left({}^p\chi(X)\right) \doteq \mathbf{E}\left(\int_0^\infty {}^p\chi(X)\, dA\right).$$

Observe that since ${}^p\chi(X)$ is well-defined the set function μ is also well-defined. If X_1 and X_2 are disjoint then by the additivity of the predictable projection

$$\mu(X_1 \cup X_2) \doteq \mu_A\left({}^p\chi(X_1 \cup X_2)\right) \doteq \mathbf{E}\left(\int_0^\infty {}^p\chi(X_1 \cup X_2)\, dA\right) =$$

$$= \mathbf{E}\left(\int_0^\infty {}^p\left(\chi(X_1) + \chi(X_2)\right)\, dA\right) =$$

$$= \mathbf{E}\left(\int_0^\infty {}^p\chi(X_1)\, dA\right) + \mathbf{E}\left(\int_0^\infty {}^p\chi(X_2)\, dA\right) \doteq$$

$$\doteq \mu_A\left({}^p\chi(X_1)\right) + \mu_A\left({}^p\chi(X_2)\right) \doteq \mu(X_1) + \mu(X_2),$$

so μ is additive. It is clear from the Monotone Convergence Theorem for the predictable projection that μ is σ-additive. Hence μ is a measure. $A \in \mathcal{A}_{\text{loc}}^+$ therefore μ_A, hence μ is σ-finite. If X is evanescent, it is predictable[56], hence

$$\mu(X) \doteq \mu_A\left({}^p\chi(X)\right) = \mu_A\left(\chi(X)\right) \doteq \mathbf{E}\left(\int_0^\infty \chi(X)\, dA\right)$$

$$= \mathbf{E}\left(\int_0^\infty 0\, dA\right) = 0.$$

Hence μ is absolutely continuous. Therefore by the generalized Radon–Nikodym theorem there is an $A^p \in \mathcal{A}_{\text{loc}}^+$ for which $\mu = \mu_{A^p}$. That is for all predictable

---

[54]See: Corollary 3.40, page 205.

[55]See: Theorem 2.11. page 117.

[56]See: step 4 in the proof of Proposition 3.51, page 211.

sets X

$$\mathbf{E}\left(\int_0^\infty \chi(X)\, dA\right) \stackrel{\circ}{=} \mu(X) = \mu_{A^p}(X) \stackrel{\circ}{=} \mathbf{E}\left(\int_0^\infty \chi(X)\, dA^p\right).$$

From Proposition 3.51 it is clear that A^p is predictable. Hence A^p is the increasing, predictable compensator of A. $\qquad\square$

Corollary 3.53 *If $A \in \mathcal{A}_{\mathrm{loc}}$ then there is a predictable process $A^p \in \mathcal{A}_{\mathrm{loc}}$ for which $A - A^p$ is a local martingale. If A_1^p and A_2^p are two such processes then A_1^p and A_2^p are indistinguishable.*

Proof. As $A \in \mathcal{A}_{\mathrm{loc}} \subseteq \mathcal{V}$ the process $\mathrm{Var}(A)$ is well-defined and $\mathrm{Var}(A) \in \mathcal{A}_{\mathrm{loc}}^+$. By Jordan's decomposition $B \stackrel{\circ}{=} (A + \mathrm{Var}(A))/2 \in \mathcal{A}_{\mathrm{loc}}^+$ and $C \stackrel{\circ}{=} (\mathrm{Var}(A) - A)/2 \in \mathcal{A}_{\mathrm{loc}}^+$. The process $A^p \stackrel{\circ}{=} B^p - C^p$ is predictable and $A - A^p \in \mathcal{L}$. The proof of the uniqueness of A^p is the same as in the previous statement. $\qquad\square$

Let us remark, that the condition $A \in \mathcal{A}_{\mathrm{loc}}$ is in some sense necessary. If $A \in \mathcal{V}$ and there is an $A^p \in \mathcal{V}$ such that $A - A^p \in \mathcal{L}$, then $A - A^p \in \mathcal{L} \cap \mathcal{V}$, hence[57] $A - A^p \in \mathcal{A}_{\mathrm{loc}}$. As A^p is predictable and right-regular it is locally bounded[58], so $A \in \mathcal{A}_{\mathrm{loc}}$.

Example 3.54 Predictable compensator of compound Poisson processes.

Let X be a compound Poisson process. Assume that the expected value of the distribution of the jumps M is finite. If N is the Poisson process with parameter λ describing the number of the jumps of X then

$$\mathbf{E}(X(t)) = \sum_{k=0}^\infty \mathbf{E}(X(t) \mid N(t) = k)\, \mathbf{P}(N(t) = k) =$$

$$= \sum_{k=0}^\infty kM \frac{(\lambda t)^k}{k!} \exp(-\lambda t) = \lambda t M.$$

X has independent increments, so it is very easy to see that $X(t) - \lambda M t$ is a martingale[59]. The process $\lambda M t$ is continuous, hence it is predictable, so $X^p(t) = \lambda M t$. If the distribution of the jumps does not have an expected value then $X \notin \mathcal{A}_{\mathrm{loc}}$, hence X does not have a compensator. $\qquad\square$

---

[57] See: Proposition 3.4, page 181.

[58] See: Proposition 3.35, page 200.

[59] Every process with independent increments and zero expected value is martingale.

3.2.3 Properties of the Predictable Compensator

Let us summarize the most important properties of A^p. Let us remark that if $A \in \mathcal{A}_{\mathrm{loc}}$ and H is locally bounded then the frequently used condition $H \bullet A \in \mathcal{A}_{\mathrm{loc}}$ holds.

1. If $A \in \mathcal{A}_{\mathrm{loc}}$ and A is predictable then[60] $A^p = A$.
2. If $A \in \mathcal{A}_{\mathrm{loc}}$ then A is a local martingale if and only if $A^p = 0$.
3. $(A + B)^p = A^p + B^p$, $(cA)^p = cA^p$.
4. If $A \in \mathcal{A}$ then $A^p \in \mathcal{A}$ and $A - A^p \in \mathcal{M}$.

Let $A^{\pm} \stackrel{\circ}{=} \frac{1}{2}(A \pm \mathrm{Var}(A))$. $A^p \stackrel{\circ}{=} (A^+)^p - (A^-)^p$. Obviously $A^{\pm} \in \mathcal{A}^+$. By definition $(A^{\pm})^p \in \mathcal{A}_{\mathrm{loc}}^+$. But as for $\tau = \infty$ $\mathbf{E}(A^{\pm}(\tau)) = \mathbf{E}((A^{\pm})^p(\tau))$ so $(A^{\pm})^p \in \mathcal{A}^+$. Hence if $A \in \mathcal{A}$ then $A^p \in \mathcal{A}$. Therefore $A - A^p$ is in $\mathcal{A}$. Hence $A - A^p$ is a class $\mathcal{D}$ local martingale so $A - A^p \in \mathcal{M}$.

5. If $A \in \mathcal{A}_{\mathrm{loc}}$ and τ is a stopping time then $(A^{\tau})^p = (A^p)^{\tau}$.

$A - A^p \in \mathcal{L}$, hence $(A - A^p)^{\tau} = A^{\tau} - (A^p)^{\tau} \in \mathcal{L}$. Truncated predictable processes are predictable[61] hence, as $(A^{\tau})^p$ is unique, $(A^{\tau})^p = (A^p)^{\tau}$.

6. If $A \in \mathcal{A}_{\mathrm{loc}}$ then

$$\Delta(A^p) = {}^p(\Delta A). \tag{3.16}$$

$\Delta(A^p) \stackrel{\circ}{=} A^p - A^p_-$ is predictable hence[62]

$$^p(\Delta(A^p)) = \Delta(A^p).$$

If $L \in \mathcal{L}$ then $^p(\Delta L) = 0$ and as $^p(\Delta L)$ is a finite predictable projection by the linearity of the predictable projection[63]

$$0 = {}^p(\Delta L) = {}^p(\Delta(A - A^p)) = {}^p(\Delta A - \Delta(A^p)) =$$
$$= {}^p(\Delta A) - {}^p(\Delta(A^p)) = {}^p(\Delta A) - \Delta(A^p),$$

which is exactly (3.16).

7. If H is predictable, $A \in \mathcal{A}_{\mathrm{loc}}$ and $(H \bullet A)(t) \stackrel{\circ}{=} \int_0^t H dA \in \mathcal{A}_{\mathrm{loc}}$ then

$$(H \bullet A)^p = H \bullet A^p. \tag{3.17}$$

The predictable compensator by definition is member of the space $\mathcal{A}_{\mathrm{loc}}$, hence under the present conditions

$$H \bullet A^p \in \mathcal{A}_{\mathrm{loc}}.$$

---

[60]$0 \in \mathcal{L}$.
[61]See: Proposition 1.39, page 23.
[62]See: Proposition 3.37, page 201.
[63]See: Corollary 3.42, page 206.

Let $B - C$ be the Jordan decomposition of the integrator A. By the definition of the integral with respect to a signed measure $H^+ \bullet B \in \mathcal{A}_{\text{loc}}^+$. The integrator $B^p \in \mathcal{A}_{\text{loc}}^+$ is predictable and if $H^+ \stackrel{\circ}{=} \chi([0, \tau])$, then the integral process

$$\left(H^+ \bullet B^p \right)(t) \stackrel{\circ}{=} \int_0^t H^+ dB^p = B^p(\tau \wedge t) - B^p(0) =$$
$$= B^p(\tau \wedge t) = (B^p)^\tau(t)$$

is predictable. The processes of type $\chi([0, \tau])$ form a π-system which generates the predictable sets. The set of bounded processes H for which $H \bullet B^p$ is predictable is a λ-system. By the Monotone Class Theorem $H \bullet B^p$ is predictable for every bounded predictable process H. By the Monotone Convergence Theorem $H^+ \bullet B^p$ is predictable for every predictable, non-negative process H^+. By (3.15) if $H^+ \bullet B \in \mathcal{A}_{\text{loc}}^+$, then $H^+ \bullet B^p \in \mathcal{A}_{\text{loc}}^+$. If G is a non-negative, predictable process then by (3.15)

$$\mathbf{E}\left(\int_0^\infty G d\left(H^+ \bullet B\right) \right) = \mathbf{E}\left(\int_0^\infty G H^+ dB \right) =$$
$$= \mathbf{E}\left(\int_0^\infty G H^+ dB^p \right) =$$
$$= \mathbf{E}\left(\int_0^\infty G d\left(H^+ \bullet B^p\right) \right),$$

hence as $H^+ \bullet B^p \in \mathcal{A}_{\text{loc}}^+$ and the process is predictable

$$\left(H^+ \bullet B \right)^P = H^+ \bullet B^p.$$

The general case is evident from the definition of the integration with respect to signed measure and from the additivity of the compensator operator[64].

8. *If H predictable, $A \in \mathcal{A}_{\text{loc}}$ and $H \bullet A \in \mathcal{A}_{\text{loc}}$ then the integral process $H \bullet (A - A^p)$ is a local martingale.*
Indeed

$$\left(H \bullet (A - A^p) \right)^P = H \bullet (A - A^p)^P =$$
$$= H \bullet \left(A^p - (A^p)^P \right) =$$
$$= H \bullet 0 = 0.$$

hence by *2.* above $H \bullet (A - A^p) \in \mathcal{L}$.

---

[64] Let us remark that by this property if H and $A \in \mathcal{A}_{\text{loc}}$ are predictable and $\int_0^t H dA \in \mathcal{A}_{\text{loc}}$, then the integral process $\int_0^t H dA$ is predictable. For example, if $A \in \mathcal{A}_{\text{loc}}$, H is locally bounded and A and H are predictable then the integral process $\int_0^t H dA$ is predictable.

9. *If H is predictable, $V \in \mathcal{V} \cap \mathcal{L}$ and $H \bullet V \in \mathcal{A}_{\mathrm{loc}}$ then the integral process $H \bullet V$ is a local martingale[65].*

By the assumptions[66] $V \in \mathcal{A}_{\mathrm{loc}}$. By *7.* and using that $V \in \mathcal{L}$ and hence $V^p = 0$

$$(H \bullet V)^p = H \bullet V^p = H \bullet 0 = 0,$$

hence by *2.* $H \bullet V \in \mathcal{L}$. $\qquad\square$

Example 3.55 If the integrand is not predictable and the integrator is a discontinuous local martingale then it is possible that the stochastic integral is not a local martingale.

In the theory of stochastic integration it is very important, that if the integrand is predictable and locally bounded and the integrator is a local martingale then the integral is also a local martingale. By *9.* it holds if the integrator is in $\mathcal{V} \cap \mathcal{L}$. If the integrand is not predictable then the integral process is not necessarily a local martingale[67]. Let $L(t) \overset{\circ}{=} \pi(t) - t$ be a compensated Poisson process. Let us denote by τ the time of the first jump of X. If $H \overset{\circ}{=} -\chi(t < \tau)$ then the trajectories of H are right-regular. The process

$$(H \bullet L)(t) = - \int_0^t \chi(s < \tau) \, d(\pi(s) - s) = t \wedge \tau$$

is not a local martingale: The trajectories are continuous and increasing, hence if it is a local martingale then by Fisk's theorem it is a constant, which is impossible as τ has an exponential distribution. $\qquad\square$

Example 3.56 Right-regular, adapted process which is not predictable.

In the previous example $\chi(t < \tau) = \chi([0, \tau))$ is right-continuous and adapted, but it cannot be predictable by Property 8. $\qquad\square$

3.3 The Fundamental Theorem of Local Martingales

In the previous chapter we defined the semimartingales as processes $X = X(0) + H + V$ where $V \in \mathcal{V}$ and $H \in \mathcal{H}^2_{\mathrm{loc}}$. In this chapter we said that X was a semimartingale if $X = X(0) + L + V$, where $V \in \mathcal{V}$ and $L \in \mathcal{L}$. Now we are ready to prove that the two definitions are equivalent.

---

[65]See: Example 3.55, page 219.

[66]See: Proposition 3.4, page 181.

[67]Intuitive if in a game one should not decide before the jumps of the gain process, then it is possible to make a systematic profit from the jumps.

Theorem 3.57 (Fundamental Theorem of Local Martingales) *Every local martingale L has a decomposition*

$$L = L(0) + L' + L'', \quad L', L'' \in \mathcal{L},$$

where $L' \in \mathcal{A}_{\mathrm{loc}}$ and L'' is locally bounded, hence $L'' \in \mathcal{H}^2_{\mathrm{loc}}$.

Proof. To make the notation simple let us assume that $L(0) = 0$. Fix a $b > 0$ and let (ρ_n) be a localizing sequence of L. The trajectories of L are regular, hence the number of jumps which have absolute value larger than b, the number of the 'big jumps', is finite on every finite interval. Hence one can define the process A consisting of the 'big jumps' of L:

$$A \stackrel{\circ}{=} \sum \Delta L \chi \left(|\Delta L| > b \right).$$

Evidently the trajectories of A have finite variation on every finite interval. In the definition of A the jumps at t are in $A(t)$, hence A is right-regular and adapted. Therefore $A \in \mathcal{V}$. Let us introduce the stopping times

$$\tau_n \stackrel{\circ}{=} \inf \left\{ t : \mathrm{Var}(A)(t) > n \right\} \wedge \inf \left\{ t : |L(t)| > n \right\} \wedge \rho_n.$$

Obviously

$$\mathrm{Var}(A)(\tau_n) \leq n + |\Delta L(\tau_n)| \leq$$

$$\leq n + |L(\tau_n)| + |L(\tau_n-)| \leq$$

$$\leq 2n + |L(\tau_n)|.$$

As L^{ρ_n} is a uniformly integrable martingale $L(\tau_n)$ is integrable. Hence $\mathrm{Var}(A)(\tau_n)$ is integrable, so $A \in \mathcal{A}_{\mathrm{loc}}$. As $A \in \mathcal{A}_{\mathrm{loc}}$ we can take the compensator A^p of A. Let us define $L' \stackrel{\circ}{=} A - A^p \in \mathcal{A}_{\mathrm{loc}}$ and let $L'' \stackrel{\circ}{=} L - L'$. By the definition of A^p the processes L' and L'' are local martingales. We are going to show that the process $U \stackrel{\circ}{=} L - A$ is locally bounded. As U is right-continuous

$$\sigma_n \stackrel{\circ}{=} \inf \left\{ t : |U(t)| > n \right\},$$

is a stopping time. By the definition of σ_n obviously $|U(\sigma_n-)| \leq n$. The size of the jumps of U are bounded by b, so

$$|U(\sigma_n)| \leq |U(\sigma_n-)| + |\Delta U(\sigma_n)| \leq n + b,$$

hence U is really locally bounded. A^p is right-continuous and predictable, hence it is locally bounded[68], so

$$L'' \overset{\circ}{=} L - (A - A^p) = U - A^p$$

is also locally bounded. □

Example 3.58 The decomposition is not unique.

Let ξ be an integrable but not square-integrable random variable. If

$$\mathcal{F}_t \overset{\circ}{=} \begin{cases} \{\emptyset, \Omega\} & \text{if } t < 1 \\ \mathcal{A} & \text{if } t \geq 1 \end{cases}$$

then

$$L(t) \overset{\circ}{=} \mathbf{E}(\xi \mid \mathcal{F}_t) = \begin{cases} \mathbf{E}(\xi) & \text{if } t < 1 \\ \xi & \text{if } t \geq 1 \end{cases}$$

is a martingale, which is not in $\mathcal{H}^2_{\mathrm{loc}}$. If ξ is symmetric then $L \in \mathcal{L}$. If $\eta \overset{\circ}{=} \xi \chi(|\xi| \leq 1)$ then

$$L''(t) \overset{\circ}{=} \mathbf{E}(\eta \mid \mathcal{F}_t) \overset{\circ}{=} \begin{cases} \mathbf{E}(\eta) = 0 & \text{if } t < 1 \\ \eta & \text{if } t \geq 1 \end{cases}$$

is in $\mathcal{L}$ and it is bounded and $L' \overset{\circ}{=} L - L''$ has integrable variation. Observe that $L'' \overset{\circ}{=} 0$, $L' \overset{\circ}{=} L$ is also a good decomposition which shows that the decomposition is not unique. □

Corollary 3.59 *Every local martingale L has an $\mathcal{H}^1$-localization.*

Proof. Let $L = L(0) + L' + L''$ be a decomposition guaranteed by the Fundamental Theorem. L'' is locally bounded, hence $L'' \in \mathcal{H}^1_{\mathrm{loc}}$. $L' \in \mathcal{A}_{\mathrm{loc}} \cap \mathcal{L}$, and if (τ_n) is a localizing sequence then as

$$\left| (L')^{\tau_n} \right| \leq \mathrm{Var}\left((L')^{\tau_n} \right)(\infty) \in L^1(\Omega),$$

trivially $L' \in \mathcal{H}^1_{\mathrm{loc}}$. □

Corollary 3.60 *If X is a local martingale and Y is a locally bounded predictable process, then the stochastic integral $Y \bullet X$ is well-defined and $Y \bullet X \in \mathcal{L}$. If Y is left-regular, then for any t the random variable $(Y \bullet X)(t)$ is the Itô–Stieltjes integral $\int_0^t Y \, dX$ of Y with respect to X.*

[68]See: Proposition 3.35, page 200.

Proof. By the Fundamental Theorem X is a semimartingale in the sense of the previous chapter. Y is locally bounded hence $Y \bullet X$ is well-defined[69]. One should only prove the relation $Y \bullet X \in \mathcal{L}$. If $X = X(0) + V + H$, where $H \in \mathcal{H}^2_{\mathrm{loc}}$ then $Y \bullet H \in \mathcal{H}^2_{\mathrm{loc}} \subseteq \mathcal{L}$. $V \in \mathcal{L} \cap \mathcal{V}$, hence by the last property of the predictable compensator $Y \bullet V \in \mathcal{L}$. By definition $Y \bullet X = Y \bullet H + Y \bullet V$ so $Y \bullet X \in \mathcal{L}$. $\square$

3.4 Quadratic Variation

Perhaps the most important consequence of the Fundamental Theorem is the following:

Corollary 3.61 *If X and Y are arbitrary semimartingales[70] then $[X, Y]$ exits and*

$$XY - X(0)Y(0) = X_- \bullet Y + Y_- \bullet X + [X, Y]. \tag{3.18}$$

The jump process of the quadratic co-variation $[X, Y]$ is

$$\Delta [X, Y] = \Delta X \Delta Y. \tag{3.19}$$

If X and Y are local martingales then $XY - [X, Y]$ is a local martingale and $[X, Y]$ is the only process from $\mathcal{V}$ for which $XY - X(0)Y(0) - [X, Y]$ is a local martingale and $\Delta [X, Y] = \Delta X \Delta Y$.

Proof. By the Fundamental Theorem X and Y are semimartingales in the sense of the previous chapter. Recall that if X and Y are semimartingales in the sense of the previous chapter then the stochastic integrals $X_- \bullet Y$ and $Y_- \bullet X$ exist and (3.18) and (3.19) hold[71]. If X and Y are local martingales then, as X_- and Y_- are locally bounded, $X_- \bullet Y$ and $Y_- \bullet X$ are local martingales. If $XY - X(0)Y(0) - A \in \mathcal{L}$ for some process $A \in \mathcal{V}$ and $\Delta [X, Y] = \Delta A$ then $[X, Y] - A \in \mathcal{L} \cap \mathcal{V}$ is continuous so by Fisk's theorem $[X, Y] - A$ is constant. $\square$

Theorem 3.62 (Fundamental Properties of the Quadratic Variation)
If L is a local martingale then:

1. *the quadratic variation $[L]$ exits,*
2. *$L^2 - [L]$ is a local martingale,*
3. *it is a right-regular increasing process with*

$$[L]^{1/2} \in \mathcal{A}^+_{\mathrm{loc}}. \tag{3.20}$$

---

[69]See the discussion in 2.4.3.
[70]Of course by the new definition.
[71]See: Proposition 2.92, page 178

Proof. Let (τ_n) be a localizing sequence of L. The existence and the right-regularity of $[L]$ is obvious from the previous statements.

For every n

$$\sigma_n \overset{\circ}{=} \inf\{t : [L](t) > n\} \wedge \inf\{t : |L|(t) > n\} \wedge \tau_n$$

is a stopping time. As $[L]$ and $|L|$ are right-regular obviously $\sigma_n \nearrow \infty$. As L has limit from the left

$$[L]^{1/2}(\sigma_n) \leq \sqrt{n + \Delta[L](\sigma_n)} \leq \sqrt{n} + \sqrt{\Delta[L]}(\sigma_n) =$$
$$= \sqrt{n} + |\Delta L|(\sigma_n) \leq$$
$$\leq \sqrt{n} + |L(\sigma_n)| + |L(\sigma_n-)| \leq$$
$$\leq \sqrt{n} + |L(\sigma_n)| + n.$$

By the Optional Sampling Theorem the right-hand side is integrable hence $[L]^{1/2} \in \mathcal{A}_{\mathrm{loc}}^+$. $\qquad\square$

Example 3.63 Process with finite quadratic variation which is not a semimartingale.

Let X be the right-regular step process over $[0,1)$ with $X(1 - 1/n) \overset{\circ}{=} (-1)^n/n$. Obviously $\mathrm{Var}(X) = \sum_{n=1}^{\infty} 1/n = \infty$, but $[X] = \sum_{n=1}^{\infty} 1/n^2 < \infty$. As every deterministic semimartingale has finite variation[72] X cannot be a semimartingale. $\qquad\square$

Proposition 3.64 (characterization of the locally square-integrable martingales) Let $L \in \mathcal{L}$. The following statements are equivalent:

1. $L \in \mathcal{H}_{\mathrm{loc}}^2$.
2. $[L] \in \mathcal{A}_{\mathrm{loc}}^+$.
3. There is a predictable process in $\mathcal{A}_{\mathrm{loc}}^+$, denoted by $\langle L \rangle$, for which $L^2 - \langle L \rangle$ is a local martingale.
4. $\sup_{s \leq t} |L(s)|^2 \in \mathcal{A}_{\mathrm{loc}}^+$.

Proof. We show that each statement implies the following one:

1. $L(0) = 0$, therefore $L \in \mathcal{H}^2$ if and only if $\mathbf{E}([L](\infty))$ is finite[73], hence $L \in \mathcal{H}_{\mathrm{loc}}^2$ if and only if there is a localizing sequence (τ_n) for which $\mathbf{E}([L^{\tau_n}](\infty)) < \infty$ for all n.

2. By the elementary properties of the quadratic variation $L^2 - [L] \in \mathcal{L}$. As $[L] \in \mathcal{A}_{\mathrm{loc}}^+$ one can define $\langle L \rangle \overset{\circ}{=} [L]^p$ and of course $[L] - [L]^p \overset{\circ}{=} [L] - \langle L \rangle \in \mathcal{L}$. Hence $L^2 - \langle L \rangle$ is a local martingale.

---

[72]See: Theorem 7.83, page 524. Step 4. of the proof.
[73]See: Proposition 2.84, page 170.

3. By 3. in the proposition $L^2 - \langle L \rangle \in \mathcal{L}$ hence[74] $Y(t) \stackrel{\circ}{=} \sup_{s \leq t} |L^2 - \langle L \rangle|$ $\in \mathcal{A}^+_{\mathrm{loc}}$. $\langle L \rangle \in \mathcal{A}^+_{\mathrm{loc}}$ and it is increasing so $Z(t) \stackrel{\circ}{=} \sup_{s \leq t} |\langle L \rangle| \in \mathcal{A}^+_{\mathrm{loc}}$. Obviously $\sup_{s \leq t} |L(s)|^2 \leq Y(t) + Z(t)$ so 4. follows from 3.

4. If (τ_n) is a localizing sequence of $U(t) \stackrel{\circ}{=} \sup_{s \leq t} |L(s)|^2$ then

$$\mathbf{E}\left(L^{\tau_n}(t)^2 \right) \leq \mathbf{E}(U(\tau_n)) < \infty,$$

hence $L \in \mathcal{H}^2_{\mathrm{loc}}$. $\qquad\qquad\qquad\qquad\qquad\qquad\qquad\qquad\qquad\qquad\qquad\qquad$ □

Corollary 3.65 *If $H \in \mathcal{H}^2_{\mathrm{loc}}$ then $\langle H \rangle = [H]^p$.*

Corollary 3.66 *If $M, N \in \mathcal{H}^2_{\mathrm{loc}}$ then there is a predictable process with finite variation on finite intervals denoted by $\langle M, N \rangle$ such that $MN - \langle M, N \rangle$ is a local martingale.*

$$\langle M, N \rangle = [M, N]^p.$$

Proof. As $M + N, M - N \in \mathcal{H}^2_{\mathrm{loc}}$

$$\langle M, N \rangle = \frac{1}{4}\left(\langle M + N \rangle - \langle M - N \rangle \right).$$

$\qquad\qquad\qquad\qquad\qquad\qquad\qquad\qquad\qquad\qquad\qquad\qquad\qquad\qquad$ □

Definition 3.67 *If $M, N \in \mathcal{H}^2_{\mathrm{loc}}$ then $\langle M, N \rangle$ is called the* predictable quadratic co-variation *of M and N. If $M \in \mathcal{H}^2_{\mathrm{loc}}$ then $\langle M \rangle \stackrel{\circ}{=} \langle M, M \rangle$ is called the* predictable quadratic variation *of M.*

---

[74]See: Example 3.3, page 181.

4

GENERAL THEORY OF STOCHASTIC INTEGRATION

In this chapter we discuss the general theory of stochastic integration. We shall assume that the integrators are semimartingales, but we shall not assume that the integrands are locally bounded. In the first part of the chapter we shall prove that every local martingale is the sum of a continuous and a purely discontinuous local martingale[1]. One can think about purely discontinuous local martingales as sums of continuously compensated single jumps[2]. The quadratic variation of a purely discontinuous local martingale is the sum of the squares of the jumps of the local martingale[3]. This decomposition is unique, that is for every local martingale L there is just one continuous $L^c \in \mathcal{L}$ and one purely discontinuous local martingale $L^d \in \mathcal{L}$ for which $L = L(0) + L^c + L^d$. In the second part of the chapter we shall present the integration theory for general local martingales. Our starting point is the decomposition $L = L(0) + L^c + L^d$. By the second, chapter the integral is well-defined when the integrator is a continuous local martingale, so we need to present the integration theory only when the integrator is purely discontinuous.

4.1 Purely Discontinuous Local Martingales

If some process V has finite variation then for every outcome the process $V^d \overset{\circ}{=} \sum \Delta V$ is absolutely convergent. Therefore it is finite, so one can easily define the decomposition

$$V = \left(V - \sum \Delta V\right) + \left(\sum \Delta V\right) \overset{\circ}{=} V^c + V^d$$

where V^c is a continuous process with finite variation and V^d is a pure jump process also with finite variation. The trajectories of local martingales are

---

[1] See: Definition 4.5, page 228.
[2] See: Proposition 4.30, page 243.
[3] See: Theorem 4.33, page 244.

right-regular, hence one can easily define the jump process ΔL of a local martingale L. Unfortunately if L is a local martingale then the process

$$\left(\sum \Delta L\right)(t) \stackrel{\circ}{=} \sum_{s \leq t} \Delta L(s)$$

formed from the jumps of L is not necessarily finite[4] and, which is more important, generally it is not a local martingale. For example, if L is a compensated Poisson process then the $A \stackrel{\circ}{=} \sum \Delta L$ is increasing. Hence it is not a local martingale.

In which sense can one define the continuous and the discontinuous part of a local martingale?

Assume that a sequence of stopping times (σ_n) covers the jumps of L. First let us define the process

$$A_1 \stackrel{\circ}{=} \Delta L(\sigma_1) \chi([\sigma_1, \infty)).$$

A_1 is the 'first' jump of L. Obviously[5]

$$|A_1| \leq [L]^{1/2} \in \mathcal{A}_{\text{loc}}.$$

Let $L_1 \stackrel{\circ}{=} A_1 - A_1^p$ be the compensated first jump of L. As we know $\Delta A_1^p = {}^p(\Delta A_1)$. Arguing a bit heuristically[6] as ${}^p(\Delta L) = 0$ the jumps of the local martingales are unpredictable, hence ${}^p(\Delta A_1) = 0$ so $\Delta A_1^p = 0$, that is A_1^p is continuous. Let $\overline{L}_1 \stackrel{\circ}{=} L - L_1$. As L and L_1 are local martingales $\overline{L}_1$ is also a local martingale. Since A_1^p is continuous we have deleted from L just the 'first' jump of L. Then let

$$A_2 \stackrel{\circ}{=} \Delta L(\sigma_2) \chi([\sigma_2, \infty))$$

and $L_2 \stackrel{\circ}{=} A_2 - A_2^p$, $\overline{L}_2 \stackrel{\circ}{=} \overline{L}_1 - L_2$, etc. Of course at the moment we do not know whether the sum of the compensated jumps $\sum_n L_n$ is convergent or not. If $L^d \stackrel{\circ}{=} \sum_n L_n$ exists then it is reasonable to call $L^c \stackrel{\circ}{=} L - L^d$ the continuous part of L, and L^d the discontinuous part of L.

If U is an arbitrary continuous local martingale then as the trajectories of $L_n \stackrel{\circ}{=} A_n - A_n^p$ have finite variation $[U, L_n] = 0$ for all n. Hence by the integration by parts formula[7] $U L_n \in \mathcal{L}$. Assume that the compensators A_n^p are continuous and

---

[4]But $\sum(\Delta L(s))^2$ is finite as $[L] < \infty$. In some sense this is the executive summary of the theory of discontinuous local martingales. See also: (1.22) on page 38.

[5]See: Proposition 3.62, page 222.

[6]See: Proposition 4.28, page 240.

[7]See: Corollary 3.61, page 222, Corollary 3.60, page 221.

the jump times $[\sigma_i]$ and $[\sigma_j]$ are disjoint if $i \neq j$. In this case[8]

$$[L_i, L_j] \stackrel{\circ}{=} \left[A_i - A_i^p, A_j - A_j^p\right] = [A_i, A_j] = 0.$$

Definition 4.1 (Strong orthogonality) *We say that the local martingales M and N are* strongly orthogonal *if $[M, N] = 0$.*

4.1.1 Orthogonality of local martingales

Recall that if

$$N, M \in \mathcal{H}_0^2 \stackrel{\circ}{=} \left\{X \in \mathcal{H}^2 : X(0) = 0\right\}$$

and M and N are strongly orthogonal then $NM = NM - [N, M]$ is a uniformly integrable martingale[9]. Therefore

$$(N, M)_{\mathcal{H}^2} \stackrel{\circ}{=} \mathbf{E}\left(N(\infty)M(\infty)\right) = \mathbf{E}\left(N(0)M(0)\right) = 0.$$

This means that if $N, M \in \mathcal{H}_0^2$ and $[N, M] = 0$ then N and M are orthogonal in the Hilbert space $\mathcal{H}_0^2$. In view of these observations the next definition looks very promising:

Definition 4.2 (Orthogonality) *We say that the local martingales M and N are* orthogonal *if the product MN is a local martingale.*

Example 4.3 It is possible for some local martingales[10] M and N to be orthogonal nevertheless $[N, M] \neq 0$.

Let π be a Poisson process and let (τ_n) be the sequence of stopping times describing the jumps. Let (ξ_n) be a sequence of independent and identically distributed random variables which are independent of (τ_n). If $\mathbf{E}(\xi_n) = 0$ for all n then the compound Poisson process

$$M(t) \stackrel{\circ}{=} \sum_n \xi_n \chi(\tau_n \leq t)$$

is a martingale. If (η_n) is a similar sequence then

$$N(t) \stackrel{\circ}{=} \sum_n \eta_n \chi(\tau_n \leq t)$$

---

[8]See: Example 2.26, page 129.

[9]See: Proposition 2.84, page 170.

[10]Observe that M and N in the example are purely discontinuous.

is also a martingale. If the sequences (ξ_n) and (η_n) are independent and $\zeta_n \overset{\circ}{=} \xi_n \eta_n$ then

$$\mathbf{E}(\zeta_n) \overset{\circ}{=} \mathbf{E}(\eta_n \xi_n) = \mathbf{E}(\eta_n)\mathbf{E}(\xi_n) = 0$$

therefore the compound Poisson process

$$MN(t) = \sum_n \xi_n \eta_n \chi(\tau_n \leq t) \overset{\circ}{=} \sum_n \zeta_n \chi(\tau_n \leq t)$$

is also a martingale so M and N are orthogonal. Obviously $[M, N] = MN \neq 0$.
$\square$

Proposition 4.4 *The local martingale L is orthogonal to itself if and only if up to indistinguishability L is a constant*[11].

Proof. By Proposition 2.48, L and L^2 are local martingales if and only if L is a constant. $\square$

Definition 4.5 *A local martingale L is* purely discontinuous *if $L(0) = 0$, and L is orthogonal to every continuous local martingale.*

It is obvious from the definition that the purely discontinuous local martingales form a linear subspace of the local martingales. We show that in some sense it is the 'orthocomplement' of the subspace of the continuous local martingales.

Corollary 4.6 *If L is a continuous, purely discontinuous local martingale then $L = 0$.*

Proof. By the assumption of the proposition L is orthogonal to itself, hence L is constant. For every purely discontinuous local martingale by definition $L(0) = 0$, so $L \equiv 0$. $\square$

Corollary 4.7 *If M and N are purely discontinuous local martingales and $\Delta M = \Delta N$ then M and N are indistinguishable*

Proof. $L \overset{\circ}{=} M - N$ is a purely discontinuous, continuous local martingale. Therefore $L = 0$. $\square$

Proposition 4.8 *Local martingales M and N are orthogonal if and only if $[M, N]$ is a local martingale.*

Proof. As $MN - [M, N]$ is always a local martingale[12] MN is a local martingale if and only if $[M, N] \in \mathcal{L}$. $\square$

---

[11]Of course L is constant by time, that is $L(t) = L(0)$ for all t.
[12]See: Corollary 3.61, page 222.

Example 4.9 Orthogonality in $\mathcal{H}^2_{\text{loc}}$.

Recall[13] that if $M, N \in \mathcal{H}^2_{\text{loc}}$ then $[M, N] \in \mathcal{A}_{\text{loc}}$ and in this case one can define the predictable quadratic co-variation $\langle M, N \rangle \stackrel{\circ}{=} [M, N]^p$. If M and N are orthogonal then

$$\langle M, N \rangle = ((\langle M, N \rangle - [M, N]) + [M, N] \in \mathcal{L}.$$

Hence $\langle M, N \rangle \in \mathcal{V} \cap \mathcal{L}$. As $\langle M, N \rangle$ is a predictable local martingale it is continuous[14], so by Fisk's theorem

$$\langle M, N \rangle = \langle M, N \rangle^p = 0.$$

On the other hand if $\langle M, N \rangle = 0$ then

$$[M, N] = [M, N] - \langle M, N \rangle = [M, N] - [M, N]^p \in \mathcal{L}$$

so M and N are orthogonal. Hence we proved that if $N, M \in \mathcal{H}^2_{\text{loc}}$ then M and N are orthogonal if and only if $\langle M, N \rangle = 0$. □

Corollary 4.10 *If M is continuous and N is purely discontinuous then $[N, M] = 0$. $N \in \mathcal{L}$ is a purely discontinuous local martingale if and only if $[M, N] = 0$ for every continuous local martingale M. Therefore $N \in \mathcal{L}$ is a purely discontinuous local martingale if and only if N is strongly orthogonal to every continuous local martingale M.*

Proof. As M is continuous

$$\Delta [N, M] = \Delta M \Delta N = 0.$$

As M and N are orthogonal $[N, M]$ is a continuous local martingale which has finite variation so by Fisk's theorem it is zero. The rest follows from this. □

Theorem 4.11 (Generalized Fisk's theorem) *If $L \in \mathcal{V} \cap \mathcal{L}$ then L is purely discontinuous.*

Proof. If $L \in \mathcal{V}$ then $[M, L] = 0$ for every continuous local martingale M. Hence $0 = [M, L] \in \mathcal{L}$, which means that M and L are strongly orthogonal. Therefore L is purely discontinuous. □

Example 4.12 If $A \in \mathcal{A}_{\text{loc}}$, then $L \stackrel{\circ}{=} A - A^p \in \mathcal{L} \cap \mathcal{V}$ is a purely discontinuous local martingale.

---

[13]See: Corollary 3.66, page 224.
[14]See: Corollary 3.40, page 205.

Proposition 4.13 (Orthogonality and localization) *The orthogonality of local martingales has the following properties:*

1. *If τ and σ are stopping times, M and N are orthogonal local martingales, then the stopped processes M^τ and N^σ are also orthogonal.*
2. *If M and N are local martingales and (τ_n) is a localizing sequence of N, then M and N are orthogonal if and only if M and the stopped processes N^{τ_n} are orthogonal for all n.*

Proof. If N and M are orthogonal, then $[M, N]$ is a local martingale. $[M^\tau, N^\sigma] = [M, N]^{\sigma \wedge \tau}$ is also a local martingale, hence M^τ and N^σ are orthogonal. If M and N^{τ_n} are orthogonal, then $[M, N^{\tau_n}] = [M, N]^{\tau_n} \in \mathcal{L}$ for all n. Obviously $\mathcal{L}_{\mathrm{loc}} = \mathcal{L}$ so $[M, N] \in \mathcal{L}$. Hence M and N are orthogonal. $\square$

Every continuous local martingale is locally bounded hence:

Corollary 4.14 *If M is a local martingale then M is purely discontinuous if and only if one of the next statements holds:*

1. *M is orthogonal to every square-integrable continuous martingale.*
2. *M is orthogonal to every bounded continuous martingale.*

Recall that the space of square-integrable martingales $\mathcal{H}^2$, is a Hilbert space with the scalar product

$$(M, N) \doteq (M, N)_{\mathcal{H}^2} \doteq \mathbf{E}\left(M\left(\infty\right), N\left(\infty\right)\right), \quad M, N \in \mathcal{H}^2.$$

By Doob's inequality if $M \in \mathcal{H}^2$, then $\sup_t |M(t)| \in L^2(\Omega)$, hence

$$|MN(t)| \le \sup_t |M(t)| \sup_t |N(t)| \in L^1(\Omega),$$

so MN is in class $\mathcal{D}$ as it is dominated by an integrable variable.

Proposition 4.15 ($\mathcal{H}^2$-orthogonality) *If $M \in \mathcal{H}^2$ and $N \in \mathcal{H}^2_0$, then the following statements are equivalent:*

1. *M and N as local martingales are orthogonal.*
2. *$[M, N]$ is a uniformly integrable martingale.*
3. *$(M^\tau, N)_{\mathcal{H}^2} = 0$ that is M^τ and N as elements of the Hilbert space $\mathcal{H}^2$, are orthogonal for every stopping time τ.*

Proof. Recall that by definition $N \in \mathcal{H}^2_0$, if $N \in \mathcal{H}^2$ and $N(0) = 0$.

1. By the Kunita–Watanabe inequality[15]

$$|[M, N]| \leq \sqrt{[M][N]}.$$

As $M, N \in \mathcal{H}^2$ both[16] $[M] = [M - M(0)], [N] \in \mathcal{A}^+$, therefore $[M, N] \in \mathcal{A}$, so $[M, N]$ has integrable variation. Hence it is a uniformly integrable local martingale, that is $[M, N]$ is a uniformly integrable martingale.

2. Assume that $[M, N]$ is a uniformly integrable martingale. In this case $[M^\tau, N] = [M, N]^\tau$ is also a uniformly integrable martingale, so it is sufficient to prove, that if $M \in \mathcal{H}^2$, $N \in \mathcal{H}_0^2$ and $[M, N]$ is a uniformly integrable martingale then $(M, N)_{\mathcal{H}^2} = 0$. As $M, N \in \mathcal{H}^2$

$$|MN(t)| \leq \sup_s |M(s)| \sup_s |N(s)| \in L^1(\Omega).$$

This implies that $MN - [M, N]$ is a uniformly integrable martingale. As $N(0) = 0$

$$
\begin{aligned}
(M, N)_{\mathcal{H}^2} &\stackrel{\circ}{=} \mathbf{E}(MN(\infty)) = \mathbf{E}(MN(\infty)) - \mathbf{E}([M, N](0)) = \\
&= \mathbf{E}(MN(\infty)) - \mathbf{E}([M, N](\infty)) = \\
&= \mathbf{E}((MN - [M, N])(\infty)) = \\
&= \mathbf{E}((MN - [M, N])(0)) = \mathbf{E}(M(0)N(0)) = 0,
\end{aligned}
$$

hence $(M, N)_{\mathcal{H}^2} = 0$.

3. Assume that the third condition holds. If τ is an arbitrary stopping time then by the Optional Sampling Theorem

$$
\begin{aligned}
\mathbf{E}(M(\tau)N(\tau)) &= \mathbf{E}(M(\tau)\mathbf{E}(N(\infty) \mid \mathcal{F}_\tau)) = \\
&= \mathbf{E}(\mathbf{E}(M(\tau)N(\infty) \mid \mathcal{F}_\tau)) = \\
&= \mathbf{E}(M(\tau)N(\infty)) = \mathbf{E}(M^\tau(\infty)N(\infty)) = \\
&= (M^\tau, N)_{\mathcal{H}^2} = 0
\end{aligned}
$$

hence MN is a martingale. $\qquad\qquad\qquad\qquad\qquad\qquad\qquad\qquad\Box$

Example 4.16 The assumption in the third property about all possible stoppings of M is important.

From the proof it is obvious that if $M, N \in \mathcal{H}_0^2$ are orthogonal as local martingales, then they are always orthogonal in the Hilbert space $\mathcal{H}_0^2$. The reverse is not true. It is possible that $M, N \in \mathcal{H}_0^2$, they are orthogonal in Hilbert space

[15] See: (2.17), page 137.
[16] See: Proposition 2.84, page 170.

sense, but they are not orthogonal as local martingales. Perhaps the simplest counterexample is the following: Let $M \in \mathcal{H}_0^2$ and let ξ be an $\mathcal{F}_0$-measurable random variable with $\mathbf{P}(\xi = 1) = \mathbf{P}(\xi = -1) = 1/2$. Let us assume that ξ is independent of M. As ξ is $\mathcal{F}_0$-measurable $N \overset{\circ}{=} \xi M$ is also in $\mathcal{H}_0^2$. Since ξ and M are independent

$$(M, N)_{\mathcal{H}^2} \overset{\circ}{=} \mathbf{E}(M(\infty)N(\infty)) \overset{\circ}{=} \mathbf{E}(\xi M^2(\infty)) = \mathbf{E}(\xi)\mathbf{E}(M^2(\infty)) = 0,$$

hence M and N are orthogonal in $\mathcal{H}_0^2$. On the other hand, unless $M = 0$, MN is not a martingale. As ξ is $\mathcal{F}_0$-measurable

$$\mathbf{E}(MN(t) \mid \mathcal{F}_0) = \xi \mathbf{E}(M^2(t) \mid \mathcal{F}_0) \neq 0 = MN(0). \qquad \square$$

4.1.2 Decomposition of local martingales

If $\mathcal{H}_c^2$ denotes the continuous elements of $\mathcal{H}^2$ then by Doob's inequality $\mathcal{H}_c^2$ is a closed subspace of $\mathcal{H}^2$. It is not too surprising that the following proposition holds:

Proposition 4.17 *If $\mathcal{H}_d^2$ denotes the set of purely discontinuous elements of $\mathcal{H}^2$ then $\mathcal{H}_d^2$ is the orthogonal complement of $\mathcal{H}_c^2$.*

Proof. Let $N \in \mathcal{H}_d^2$ and $M \in \mathcal{H}_c^2$. By the definition of purely discontinuous local martingales $N(0) = 0$. M and N are orthogonal local martingales hence by the previous proposition $(M, N)_{\mathcal{H}^2} = 0$.

On the other hand let us assume that $N \in \mathcal{H}^2$ is orthogonal to the subspace $\mathcal{H}_c^2$. The constant process $M(t) \overset{\circ}{=} N(0)$ is a continuous martingale, hence

$$\mathbf{E}\left(N^2(0)\right) = \mathbf{E}\left(M(\infty)N(0)\right) = \mathbf{E}\left(M(\infty)\mathbf{E}\left(N(\infty) \mid \mathcal{F}_0\right)\right) =$$
$$= \mathbf{E}\left(M(\infty)N(\infty)\right) \overset{\circ}{=} (M, N)_{\mathcal{H}^2} = 0,$$

hence $N(0) = 0$. If $M \in \mathcal{H}_c^2$, then $M^\tau \in \mathcal{H}_c^2$ for any stopping time τ, hence $(M^\tau, N) = 0$. Again by the previous proposition N and M are orthogonal local martingales. By Corollary 4.14 N is purely discontinuous. $\qquad \square$

Corollary 4.18 *Every $M \in \mathcal{H}^2$ has a unique decomposition*

$$M = M(0) + M^c + M^d, \quad \text{where} \quad M^c \in \mathcal{H}_c^2, M^d \in \mathcal{H}_d^2. \qquad (4.1)$$

Theorem 4.19 (Continuous and purely discontinuous parts of local martingales) *Every local martingale L has a decomposition*

$$L = L(0) + L^c + L^d, \quad L^c, L^d \in \mathcal{L},$$

where L^c is continuous, and L^d is purely discontinuous. If $L = L(0) + L_1^c + L_1^d$ and $L = L(0) + L_2^c + L_2^d$ are two such decompositions then L_1^c, L_2^c and also L_1^d, L_2^d are indistinguishable.

Proof. If L_i^c, L_i^d, $i = 1, 2$ are two decompositions of L then $L_1^c - L_2^c = L_2^d - L_1^d$, which means that $L_1^c - L_2^c$ is purely discontinuous and continuous. Hence[17] $L_1^c - L_2^c = 0$, so $L_2^d - L_1^d = 0$ as well.

1. Let us take the decomposition $L = L(0) + L' + L''$ of the Fundamental Theorem. $L'' \in \mathcal{H}_{\mathrm{loc}}^2$. $L' \in \mathcal{V}$, hence by the generalized Fisk's theorem[18] L' is purely discontinuous. We may therefore assume that $L \in \mathcal{H}_{\mathrm{loc}}^2$.

2. Let (τ_n) be a $\mathcal{H}^2$-localizing sequence of L, and let L_k^c and L_k^d be the (4.1) decomposition of L^{τ_k} in $\mathcal{H}^2$. Of course

$$L^{\tau_k} = \left((L)^{\tau_{k+1}}\right)^{\tau_k} = \left(L_{k+1}^c + L_{k+1}^d\right)^{\tau_k} = \left(L_{k+1}^c\right)^{\tau_k} + \left(L_{k+1}^d\right)^{\tau_k}.$$

Obviously $\left(L_{k+1}^c\right)^{\tau_k} \in \mathcal{H}_c^2$. L_{k+1}^d is orthogonal to every continuous local martingale, hence $\left(L_{k+1}^d\right)^{\tau_k}$ is also orthogonal to every continuous local martingale[19], hence $\left(L_{k+1}^d\right)^{\tau_k}$ is purely discontinuous. As the decomposition (4.1) is unique

$$\left(L_{k+1}^d\right)^{\tau_k} = L_k^d, \quad \left(L_{k+1}^c\right)^{\tau_k} = L_k^c,$$

so

$$\left(L^c\right)^{\tau_k} \stackrel{\circ}{=} L_k^c \quad \text{and} \quad \left(L^d\right)^{\tau_k} \stackrel{\circ}{=} L_k^d.$$

unambiguously defines the local martingales L^c and L^d

3. L^c is trivially a continuous local martingale. We show that L^d is purely discontinuous. To prove it, it is sufficient to show that L^d is orthogonal to every continuous martingale[20] $U \in \mathcal{H}^2$. $\left(L^d\right)^{\tau_n} \in \mathcal{H}_d^2$, hence $\left(L^d\right)^{\tau_n}$ and U are orthogonal. Hence U and L^d are also orthogonal[21]. □

Example 4.20 Purely discontinuous local martingale[22] which is not in $\mathcal{V}$.

---

[17]See: Corollary 4.6, page 228.

[18]See: Corollary 4.11, page 229.

[19]See: Proposition 4.13, page 229.

[20]See: Corollary 4.14, page 230.

[21]See: Corollary 4.14, page 230.

[22]See: Example 7.35, page 484.

Let (N_i) be a sequence of independent Poisson processes with $\lambda = 1$. For any t the compensated Poisson processes

$$M_i(t) \stackrel{\circ}{=} N_i(t) - \lambda t = N_i(t) - t$$

on the finite time horizon $[0, t]$ are in $\mathcal{H}_0^2$. As they are independent they almost surely do not have common jumps[23]. Obviously $[M_i, M_j] = \sum \Delta M_i \Delta M_j$, therefore if $i \neq j$ then M_i and M_j are orthogonal local martingales so they are orthogonal in $\mathcal{H}_0^2$. As $\sum_i 1/i^2 < \infty$ the sequence

$$M \stackrel{\circ}{=} \sum_{i=1}^{\infty} \frac{1}{i} M_i$$

is convergent in the Hilbert space $\mathcal{H}_0^2$. Every M_i is in $\mathcal{V}$ so they are purely discontinuous. Therefore M is also purely discontinuous. The variables $N_i(t) - t$ are independent, they have zero expected value. So for any t the sequence

$$R_n \stackrel{\circ}{=} \sum_{i=1}^{n} \frac{N_i(t) - t}{i}$$

is a discrete-time martingale. Obviously (R_n) is bounded in $L^2(\Omega)$ so by the Martingale Convergence Theorem it is convergent almost surely. As $\sum_i 1/i = \infty$ obviously

$$\mathrm{Var}(M)(t) \geq \sum_{s \leq t} \Delta M(s) = \sum_{i=1}^{\infty} \frac{N_i(t)}{i} = \infty. \qquad \square$$

4.1.3 Decomposition of semimartingales

The decomposition theorem just proved can be transferred to semimartingales.

Theorem 4.21 (Continuous part of semimartingales) *If $S \in \mathcal{S}$ then there is a continuous local martingale L^c for which for any decomposition of S*

$$S = S(0) + V + L, \quad V \in \mathcal{V}, \ L \in \mathcal{L}$$

L^c is the continuous part of L. If L_1^c and L_2^c are two such local martingales then L_1^c and L_2^c are indistinguishable.

Proof. One should prove only the uniqueness of L^c, the other part of the theorem is trivial. If

$$S(0) + V_1 + L_1^c + L_1^d = S(0) + V_2 + L_2^c + L_2^d$$

---

[23]See: Proposition 7.13, page 471.

then

$$L_1^c - L_2^c = V_2 - V_1 + L_2^d - L_1^d. \tag{4.2}$$

$V_2 - V_1 \in \mathcal{V} \cap \mathcal{L}$, hence $V_2 - V_1$ is purely discontinuous[24], hence the right side of (4.2) is purely discontinuous, the left side is continuous, hence $L_1^c - L_2^c = 0$. $\square$

Example 4.22 If S is a semimartingale then the continuous part of S as a 'true semimartingale', is not 'well-defined'.

Let us take our usual counter-example, the Poisson process. If π is a Poisson process and $\pi(t) = \lambda t + (\pi(t) - \lambda t)$ then the continuous part of the local martingale part $L(t) \stackrel{\circ}{=} (\pi(t) - \lambda t)$ is zero. If $\pi(t) = \pi(t) + 0$ then the continuous part of the local martingale part $L = 0$ is again zero. In the first case the continuous part of the finite variation part $V(t) = \lambda t \in \mathcal{V}$ is λt, but in the second case the continuous part of the finite variation part $\pi \in \mathcal{V}$ is zero. What is the continuous part of the semimartingale π? $\square$

In Itô's formula[25] we shall use the notation S^c. Let us fix the definition of S^c in the following way:

Definition 4.23 *If S is a semimartingale then S^c denote the continuous part of the local martingale part of S.*

Example 4.24 If $S \stackrel{\circ}{=} \pi$ is a Poisson process then $\pi^c = 0$.

4.2 Purely Discontinuous Local Martingales and Compensated Jumps

During the construction of stochastic integrals with respect to local martingales, we shall need the next inequality:

Theorem 4.25 (Davis' inequality) *There are positive constants c and C such that for every local martingale $L \in \mathcal{L}$ and for any stopping time τ*

$$c \cdot \mathbf{E}\left(\sqrt{[L]}\,(\tau)\right) \leq \mathbf{E}\left(\sup_{t \leq \tau} |L(t)|\right) \leq C \cdot \mathbf{E}\left(\sqrt{[L]}\,(\tau)\right).$$

---

[24]See: Theorem 4.11, page 229.

[25]More precisely: In Itô's formula one uses only the quadratic variation of the continuous part of the semimartingales, which is independent of the decomposition. See: Corollary 4.36, page 246.

The proof of the inequality is a lengthy calculation which we shall present at the end of this chapter as a separate section.

The most important application of Davis' inequality is the following theorem.

Theorem 4.26 (Convergence of strongly orthogonal series) *Let us assume that $(L_n) \subseteq \mathcal{L}$ and if $i \neq j$ then L_i and L_j are strongly orthogonal, that is*

$$[L_i, L_j] = 0, \quad i \neq j.$$

If $\sqrt{\sum_{n=1}^{\infty} [L_n]} \in \mathcal{A}^+_{loc}$ then there is an $L \in \mathcal{L}$ such that on every compact interval in the topology of uniform convergence in probability

$$L = \sum_{n=1}^{\infty} L_n.$$

If $\sqrt{\sum_{n=1}^{\infty} [L_n]} \in \mathcal{A}^+$ then L is a uniformly integrable martingale and the convergence holds in the topology of uniform convergence in $L^1(\Omega)$ that is

$$\lim_{m \to \infty} \mathbf{E} \left(\sup_t \left| L(t) - \sum_{n=1}^{m} L_n(t) \right| \right) = 0.$$

Proof. First let us assume that $\sqrt{\sum_{n=1}^{\infty} [L_n]} \in \mathcal{A}^+$. By Davis' inequality and by the assumption $[L_i, L_j] = 0$ if $m > n$ then

$$\mathbf{E} \left(\sup_t \left| \sum_{i=n}^{m} L_i(t) \right| \right) \leq C \cdot \mathbf{E} \left(\sqrt{\left[\sum_{i=n}^{m} L_i \right](\infty)} \right) =$$

$$= C \cdot \mathbf{E} \left(\sqrt{\sum_{i=n}^{m} [L_i](\infty)} \right).$$

As $\sqrt{\sum_{n=1}^{\infty} [L_n]} \in \mathcal{A}^+$ by the Dominated Convergence Theorem

$$\lim_{n,m \to \infty} \mathbf{E} \left(\sqrt{\sum_{i=n}^{m} [L_i](\infty)} \right) = 0,$$

which implies that

$$\lim_{n,m \to \infty} \mathbf{E} \left(\sup_t \left| \sum_{i=n}^{m} L_i(t) \right| \right) = 0.$$

As $L^1(\Omega)$ is complete $\sup_t \left| \sum_{i=1}^m L_i(t) \right|$ is convergent in $L^1(\Omega)$. From the convergence in $L^1(\Omega)$ one has a subsequence which is almost surely convergent, therefore there is a process L such that for almost all ω

$$\lim_{k \to \infty} \sup_t \left| \sum_{i=1}^{n_k} L_i(t, \omega) - L(t, \omega) \right| = 0.$$

L is obviously right-regular and of course $\sum_{i=1}^n L_i$ converges to L uniformly in $L^1(\Omega)$, that is

$$\lim_{n \to \infty} \mathbf{E} \left(\sup_t \left| \sum_{i=1}^n L_i(t) - L(t) \right| \right) = 0.$$

Again by Davis' inequality

$$\mathbf{E} \left(\sup_t |L_i(t)| \right) \le C \cdot \mathbf{E} \left([L_i]^{1/2}(\infty) \right) < \infty,$$

hence L_i is a class $\mathcal{D}$ local martingale hence it is a martingale. From the convergence in $L^1(\Omega)$ it follows that $L \overset{\circ}{=} \sum_{i=1}^\infty L_i$ is also a martingale.

$$\mathbf{E} \left(\sup_t |L(t)| \right) \le \mathbf{E} \left(\sup_t \left| \sum_{i=1}^n L_i(t) \right| \right) + \mathbf{E} \left(\sup_t \left| L - \sum_{i=1}^n L_i(t) \right| \right) < \infty$$

hence the limit L is in $\mathcal{D}$ that is L a uniformly integrable martingale.

Now let us assume that $\sqrt{\sum_{n=1}^\infty [L_n]} \in \mathcal{A}_{\text{loc}}^+$. In this case there is a localizing sequence (τ_k) for which

$$\sqrt{\sum_{n=1}^\infty [L_n^{\tau_k}]} = \sqrt{\sum_{n=1}^\infty [L_n]^{\tau_k}} = \left(\sqrt{\sum_{n=1}^\infty [L_n]} \right)^{\tau_k} \in \mathcal{A}^+.$$

Observe that (τ_k) is a common localizing sequence for all L_n, that is $\sqrt{[L_n^{\tau_k}]} \in \mathcal{A}$ for all n. Observe also, that by Davis' inequality $L_n^{\tau_k} \in \mathcal{M}$ for every n and k. By the first part of the proof for every k there is an $L^{(k)} \in \mathcal{M}$ such that $\sum_{n=1}^\infty L_n^{\tau_k} = L^{(k)}$. Obviously $\left(L^{(k+1)} \right)^{\tau_k} = L^{(k)}$, so one can define an $L \in \mathcal{L}$ for which $L^{\tau_k} = L^{(k)}$. Let us fix an ε and a δ. As $\tau_k \nearrow \infty$ for every $t < \infty$ there is an n such that $\mathbf{P}\left(\tau_k \leq t\right) \leq \delta/2$ whenever $k \geq n$. In the usual way, for $k \geq n$

$$\mathbf{P}\left(\sup_{s \leq t}\left|L(s) - \sum_{k=1}^{n} L_k(s)\right| > \varepsilon\right) \leq$$

$$\leq \mathbf{P}\left(\tau_k \leq t\right) + \mathbf{P}\left(\sup_{s \leq t}\left|L(s) - \sum_{k=1}^{n} L_k(s)\right| > \varepsilon, \tau_k > t\right).$$

The first probability is smaller than $\delta/2$, the second probability is

$$\mathbf{P}\left(\sup_{s \leq t}\left|L^{\tau_k}(s) - \sum_{k=1}^{n} L_k^{\tau_k}(s)\right| > \varepsilon, \tau_k > t\right)$$

which is smaller than

$$\mathbf{P}\left(\sup_{s}\left|L^{\tau_k}(s) - \sum_{k=1}^{n} L_k^{\tau_k}(s)\right| > \varepsilon\right).$$

As $L_n^{\tau_k} \to L^{\tau_k}$ uniformly in $L^1(\Omega)$, by Markov's inequality

$$\mathbf{P}\left(\sup_{s}\left|L^{\tau_k}(s) - \sum_{k=1}^{n} L_k^{\tau_k}(s)\right| > \varepsilon\right) \to 0,$$

from which one can easily show that for n large enough

$$\mathbf{P}\left(\sup_{s \leq t}\left|L(s) - \sum_{k=1}^{n} L_k(s)\right| > \varepsilon\right) < \delta,$$

that is $\sum_{k=1}^{n} L_k \overset{\text{ucp}}{\Rightarrow} L$, which means that on every compact interval in the topology of uniform convergence in probability

$$\lim_{n \to \infty} \sum_{k=1}^{n} L_k \overset{\circ}{=} \sum_{k=1}^{\infty} L_k = L. \qquad \square$$

Theorem 4.27 (Parseval's identity) *Under the conditions of the theorem above for every t*

$$\lim_{n \to \infty} \left[L - \sum_{k=1}^{n} L_k\right](t) = 0 \qquad (4.3)$$

and

$$[L](t) \overset{a.s.}{=} \sum_{k=1}^{\infty} [L_k](t) \tag{4.4}$$

where in both cases the convergence holds in probability.

Proof. By Davis' inequality

$$\mathbf{E} \left(\sqrt{\left[L - \sum_{k=1}^{n} L_k \right](t)} \right) \leq \frac{1}{c} \cdot \mathbf{E} \left(\sup_{s \leq t} \left| L(s) - \sum_{n=1}^{m} L_n(s) \right| \right).$$

If $\sqrt{\sum_{n=1}^{\infty} [L_n]} \in \mathcal{A}^+$ then by the theorem just proved

$$\lim_{m \to \infty} \mathbf{E} \left(\sup_{s \leq t} \left| L(s) - \sum_{n=1}^{m} L_n(s) \right| \right) = 0.$$

By Markov's inequality convergence in $L^1(\Omega)$ implies convergence in probability, therefore if $\sqrt{\sum_{n=1}^{\infty} [L_n]} \in \mathcal{A}^+$ then (4.3) holds. Let $\sqrt{\sum_{n=1}^{\infty} [L_n]} \in \mathcal{A}_{loc}^+$ and let (τ_k) be a localizing sequence of $\sqrt{\sum_{n=1}^{\infty} [L_n]}$. Let us fix an ε and a δ. As $\tau_k \nearrow \infty$ for every $t < \infty$ there is a q such that $\mathbf{P}(\tau_k \leq t) \leq \delta/2$ whenever $k \geq q$. In the usual way, for $k \geq q$

$$\mathbf{P} \left(\sup_{s \leq t} \left| L(s) - \sum_{k=1}^{n} L_k(s) \right| > \varepsilon \right) \leq$$

$$\leq \mathbf{P}(\tau_k \leq t) + \mathbf{P} \left(\sup_{s \leq t} \left| L(s) - \sum_{k=1}^{n} L_k(s) \right| > \varepsilon, \tau_k > t \right).$$

Obviously

$$\mathbf{P} \left(\sup_{s \leq t} \left| L(s) - \sum_{k=1}^{n} L_k(s) \right| > \varepsilon, \tau_k > t \right) =$$

$$= \mathbf{P} \left(\sup_{s \leq t} \left| L^{\tau_k}(s) - \sum_{k=1}^{n} L_k^{\tau_k}(s) \right| > \varepsilon, \tau_k > t \right) \leq$$

$$\leq \mathbf{P} \left(\sup_{s \leq t} \left| L^{\tau_k}(s) - \sum_{k=1}^{n} L_k^{\tau_k}(s) \right| > \varepsilon \right).$$

By the stopping rule of the quadratic variation

$$\sqrt{\sum_{n=1}^{\infty}[L_n^{\tau_k}]} = \sqrt{\sum_{n=1}^{\infty}[L_n]^{\tau_k}} = \left(\sqrt{\sum_{n=1}^{\infty}[L_n]}\right)^{\tau_k} \in \mathcal{A}^+,$$

so by the first part of the proof if n is large enough

$$\mathbf{P}\left(\sup_{s\le t}\left|L(s) - \sum_{k=1}^{n}L_k(s)\right| > \varepsilon\right) \le \frac{\delta}{2} + \frac{\delta}{2}$$

that is (4.3) holds in the general case. By Kunita–Watanabe inequality[26]

$$\left|\sqrt{[L](t)} - \sqrt{\left[\sum_{k=1}^{n}L_k\right](t)}\right| \le \sqrt{\left[L - \sum_{k=1}^{n}L_k\right](t)}.$$

This implies that

$$[L](t) = \lim_{n\to\infty}\left[\sum_{k=1}^{n}L_k\right](t) = \lim_{n\to\infty}\sum_{k=1}^{n}[L_k](t) \overset{\circ}{=}$$

$$\overset{\circ}{=} \sum_{k=1}^{\infty}[L_k](t)$$

where convergences hold in probability. $\qquad\square$

4.2.1 Construction of purely discontinuous local martingales

The cornerstone of the construction of the general stochastic integral is the next proposition:

Proposition 4.28 *Let H be a progressively measurable process. There is one and only one purely discontinuous local martingale $L \in \mathcal{L}$ for which $\Delta L = H$ if and only if*

1. *the set $\{H \ne 0\}$ is thin,*
2. *$^{p}H = 0$ and*
3. *$\sqrt{\sum H^2} \in \mathcal{A}_{loc}^+$.*

Proof. By the definition of the thin sets, for every ω there exists just a countable number of points where the trajectory $H(\omega)$ is not zero. Hence the sum $\left(\sum H^2\right)(t) \overset{\circ}{=} \sum_{s\le t}H^2(s)$ is meaningful. Observe that from the condition $\sqrt{\sum H^2} \in \mathcal{A}_{loc}^+$ it implicitly follows that $H(0) = 0$.

---

[26]See: Corollary 2.36, page 137.

1. The uniqueness of L is obvious, as if purely discontinuous local martingales have the same jumps then they are indistinguishable[27].

2. If $H \stackrel{\circ}{=} \Delta L$ for some $L \in \mathcal{L}$ then ${}^p H \stackrel{\circ}{=} {}^p (\Delta L) = 0$, and as $(\Delta L)^2 = \Delta [L]$ and $[L]$ is increasing

$$\sum H^2 = \sum (\Delta L)^2 \leq \sum (\Delta L)^2 + [L]^c =$$
$$= [L].$$

Since[28] $\sqrt{[L]} \in \mathcal{A}_{\text{loc}}^+$ obviously $\sqrt{\sum H^2} \in \mathcal{A}_{\text{loc}}^+$, so the conditions are necessary.

3. Let us assume that $\sqrt{\sum H^2} \in \mathcal{A}_{\text{loc}}^+$ and let us assume that the sequence of stopping times (ρ_m) exhausting[29] for the thin set $\{H \neq 0\}$. We can assume that ρ_m is either totally inaccessible or predictable. For every stopping time ρ_m let us define a simple jump processes which jumps at ρ_m and for which the value of the jump is $H(\rho_m)$:

$$N_m \stackrel{\circ}{=} H(\rho_m) \chi([\rho_m, \infty)).$$

It is worth emphasizing that it is possible that $\cup_m [\rho_m] \neq \{H \neq 0\}$. That is, the inclusion

$$\{H \neq 0\} \subseteq \cup_m [\rho_m]$$

can be proper, but

$$\cup_m \{\Delta N_m \neq 0\} = \{H \neq 0\}.$$

N_m is right-regular, H is progressively measurable, hence the stopped variables $H(\rho_m)$ are $\mathcal{F}_{\rho_m}$-measurable and so N_m is adapted. As $\sqrt{\sum H^2} \in \mathcal{A}_{\text{loc}}^+$

$$|N_m| \leq \sqrt{\sum H^2} \in \mathcal{A}_{\text{loc}}^+$$

for every m, hence N_m has locally integrable variation, so it has a compensator N_m^p.

4. We show that N_m^p is continuous. If ρ_m is predictable then the graph $[\rho_m]$ of ρ_m is a predictable set[30] so using property 6. of the predictable

---

[27] See: Corollary 4.7, page 228.
[28] See: (3.20) line, page 222.
[29] See: Proposition 3.22, page 189.
[30] See: Corollary 3.34, page 199.

compensator[31] up to indistinguishability

$$\Delta \left(N_m^p \right) = \; ^p \left(\Delta N_m \right) \overset{\circ}{=} \; ^p \left(H \left(\rho_m \right) \chi \left([\rho_m] \right) \right) = \; ^p \left(H \chi \left([\rho_m] \right) \right) =$$
$$= \left(^p H \right) \chi \left([\rho_m] \right) = 0 \cdot \chi \left([\rho_m] \right) = 0.$$

Hence N_m^p is continuous. Let ρ_m be totally inaccessible. As above

$$\Delta \left(N_m^p \right) = \; ^p \left(\Delta N_m \right) = \; ^p \left(H \chi \left([\rho_m] \right) \right).$$

ρ_m is totally inaccessible and therefore $\mathbf{P} \left(\rho_m = \sigma \right) = 0$ for every predictable stopping time σ, hence if σ is predictable then

$$^p \left(H \chi \left([\rho_m] \right) \right) (\sigma) \overset{\circ}{=} \widehat{\mathbf{E}} \left(H \chi \left([\rho_m] \right) (\sigma) \mid \mathcal{F}_{\sigma-} \right) =$$
$$= \widehat{\mathbf{E}} \left(0 \mid \mathcal{F}_{\sigma-} \right) = 0.$$

By the definition of the predictable projection $\Delta \left(N_m^p \right) = 0$.

5. Let $L_m \overset{\circ}{=} N_m - N_m^p \in \mathcal{L}$ be the compensated jumps. As the compensators are continuous and have finite variation if $i \neq j$ then $[L_i, L_j] = [N_i, N_j] = 0$, and

$$\sqrt{\sum [L_k]} = \sqrt{\sum [N_k]} = \sqrt{\sum H^2} \in \mathcal{A}_{\mathrm{loc}}^+.$$

Hence[32] there is an $L \in \mathcal{L}$ for which $L = \sum_k L_k$. As the convergence is uniform in probability there is a sequence for which the convergence is almost surely uniform. Hence up to indistinguishability

$$\Delta L = \Delta \left(\sum L_k \right) = \sum \Delta L_k = H.$$

Observe that in the last step we have used the fact that

$$\{ H \neq 0 \} = \cup_m \{ \Delta N_m \neq 0 \} = \cup_m \{ \Delta L_m \neq 0 \}.$$

6. Let us prove that L is purely discontinuous. Let M be a continuous local martingale. Obviously $[L_k, M] = 0$. Therefore by the inequality of Kunita and

---

[31] See: page 217.
[32] See: Theorem 4.26, page 236.

Watanabe[33] and by (4.3)

$$|[M, L]| \leq \left\|\left[M, L - \sum_{k=1}^{n} L_k\right]\right\| + \left\|\left[M, \sum_{k=1}^{n} L_k\right]\right\| =$$

$$= \left\|\left[M, L - \sum_{k=1}^{n} L_k\right]\right\| \leq \sqrt{[M]} \sqrt{\left[L - \sum_{k=1}^{n} L_k\right]} \to 0$$

which implies that $[M, L] = 0$, that is M and L are orthogonal. Hence L is purely discontinuous. □

Definition 4.29 *The following definitions are useful:*

1. *We say that process X is a* single jump *if there is a stopping time ρ and an $\mathcal{F}_\rho$-measurable random variable ξ such that $X = \xi \chi([\rho, \infty))$.*
2. *We say that process X is a* compensated single jump *if there is a single jump Y for which $X = Y - Y^p$.*
3. *We say that the X is a* continuously compensated single jump *if Y^p in 2. is continuous.*

Proposition 4.30 (The structure of purely discontinuous local martingales) *If $L \in \mathcal{L}$ is a purely discontinuous local martingale then in the topology of uniform convergence in probability on compact intervals*

$$L \overset{\circ}{=} \sum_{k=1}^{\infty} L_k,$$

where for all k:

1. *$L_k \in \mathcal{L}$ is a continuously compensated single jump,*
2. *the jumps of L_k are jumps of L.*
3. *If $i \neq j$ then $[L_i, L_j] = 0$ that is L_i and L_j are strongly orthogonal,*
4. *$[L_k] = (\Delta L(\rho_k))^2 \chi([\rho_k, \infty))$, where ρ_k denotes the stopping time of L_k.*
5. *If $i \neq j$ then the graphs $[\rho_i]$ and $[\rho_j]$ are disjoint.*

If $\sqrt{[L]} \in \mathcal{A}^+$ then the convergence holds in the topology of uniform convergence in $L^1(\Omega)$.

Proof. It is sufficient to remark, that if $L \in \mathcal{L}$ is purely discontinuous then the jump process of L satisfies the conditions of the above proposition[34]. □

[33]See: Corollary 2.36, page 137.
[34]See: Proposition 4.28, page 240.

4.2.2 Quadratic variation of purely discontinuous local martingales

In this subsection we return to the investigation of the quadratic variation.

Definition 4.31 *We say that M is a pure quadratic jump process if*

$$[M] = \sum (\Delta M)^2. \qquad (4.5)$$

Example 4.32 Every $V \in \mathcal{V}$ is a pure quadratic jump process[35].

By (2.14) $[V, V] = \sum \Delta V \Delta V = \sum (\Delta V)^2$. $\qquad\qquad\qquad \square$

Theorem 4.33 (Quadratic variation of purely discontinuous local martingales) *A local martingale $L \in \mathcal{L}$ is a pure quadratic jump process if and only if it is purely discontinuous.*

Proof. Let $L \in \mathcal{L}$.

1. If L is purely discontinuous, then by the structure of purely discontinuous local martingales[36] $L = \sum_k L_k$, where

$$[L_k, L_j] = \begin{cases} 0 & \text{if } k \neq j \\ (\Delta L (\rho_k))^2 \chi ([\rho_k, \infty)) & \text{if } k = j \end{cases}.$$

By Parseval's identity (4.4) for every t

$$[L](t) \overset{a.s}{=} \sum_{k=1}^{\infty} [L_k](t) = \sum_{s \leq t} (\Delta L)^2 (s).$$

As both sides of the equation are right-regular $[L]$ and $\sum_{s \leq t} (\Delta L)^2$ are indistinguishable.

2. If L is a pure quadratic jump process, then

$$[L] = \sum (\Delta L)^2.$$

[35] See: Proposition 2.33, page 134.
[36] See: Proposition 4.30, page 243.

Let $L = L^c + L^d$ be the decomposition of $L \in \mathcal{L}$. As L^c is continuous[37]

$$[L] = [L^c + L^d] = [L^c] + 2[L^c, L^d] + [L^d] =$$
$$= [L^c] + [L^d].$$

By the part of the theorem already proved

$$[L^d] = \sum (\Delta L^d)^2 = \sum (\Delta L^d + \Delta L^c)^2 = \sum (\Delta L)^2.$$

Hence $[L^c] = 0$, therefore $L^c = 0$ and so $L = L^d$. $\qquad\square$

Corollary 4.34 *If X is a purely discontinuous local martingale then for every local martingale Y*

$$[X, Y] = \sum \Delta X \Delta Y. \tag{4.6}$$

Proof. Obviously

$$[X, Y] = [X, Y^c + Y^d] = [X, Y^c] + [X, Y^d].$$

By the definition of the orthogonality $[X, Y^c]$ is a local martingale. $\Delta [X, Y^c] = \Delta X \Delta Y^c = 0$, hence $[X, Y^c]$ is continuous. $[X, Y^c] \in \mathcal{V} \cap \mathcal{L}$ so by Fisk's theorem $[X, Y^c] = 0$. As the purely discontinuous local martingales form a linear space

$$[X, Y^d] = \frac{1}{4} \left([X + Y^d] - [X - Y^d] \right) =$$
$$= \frac{1}{4} \left(\sum (\Delta X + \Delta Y^d)^2 - \sum (\Delta X - \Delta Y^d)^2 \right) =$$
$$= \sum \Delta X \Delta Y^d = \sum \Delta X \left(\Delta Y^d + \Delta Y^c \right) = \sum \Delta X \Delta Y.$$
$\qquad\square$

Proposition 4.35 (Quadratic variation of semimartingales) *For every semimartingale X*

$$[X] = [X^c] + \sum (\Delta X)^2, \tag{4.7}$$

where, as before[38], X^c denotes the continuous part of the local martingale part of X. More generally if X and Y are semimartingales then

$$[X, Y] = [X^c, Y^c] + \sum \Delta X \Delta Y. \tag{4.8}$$

[37]See: Corollary 4.10, page 229.
[38]See: Definition 4.23, page 235.

Proof. Recall[39] that every semimartingale X has a decomposition,

$$X = X(0) + X^c + H + V,$$

where X^c is a continuous local martingale, $V \in \mathcal{V}$ and H is a purely discontinuous local martingale. By simple calculation

$$[X] = [X^c] + [V] + [H] +$$
$$+ 2[X^c, H] + 2[X^c, V] + 2[H, V].$$

As X^c is continuous and V has finite variation so $[X^c, V] = 0$. H is purely discontinuous and X^c is continuous, hence by (4.6) $[X^c, H] = 0$. Therefore

$$[X] = [X^c] + [V] + [H] + 2[H, V].$$

Every process with finite variation is a pure quadratic jump process so

$$[V] = \sum (\Delta V)^2.$$

H is purely discontinuous, hence it is also a pure quadratic jump process, so

$$[H] = \sum (\Delta H)^2.$$

As V has finite variation so by (2.14)

$$[H, V] = \sum \Delta H \Delta V.$$

Therefore

$$[V] + [H] + 2[H, V] = \sum (\Delta H + \Delta V)^2 = \sum (\Delta X)^2,$$

so (4.7) holds. The proof of the general case is similar. $\qquad\square$

Corollary 4.36 *If X is a semimartingale then $[X^c] = [X]^c$. More generally if X and Y are semimartingales then $[X^c, Y^c] = [X, Y]^c$.*

4.3 Stochastic Integration With Respect To Local Martingales

Recall that so far we have defined the stochastic integral with respect to local martingales only when the integrator Y was locally square-integrable. In fact, in this case the construction of the stochastic integral is nearly the same as the construction when the integrator is a continuous local martingale. The only

---

[39]See: Theorem 4.19, page 232.

difference is that when $Y \in \mathcal{H}^2_{\text{loc}}$ then one can integrate only predictable processes and one has to consider the condition for the jumps of the integral $\Delta(X \bullet Y) = X \Delta Y$ as well. Recall that if $Y \in \mathcal{H}^2_{\text{loc}}$ then a predictable process X is integrable if and only if

$$X \in \mathcal{L}^2_{\text{loc}}(Y) \overset{\circ}{=} \left\{ Z : Z^2 \bullet [Y] \in \mathcal{A}^+_{\text{loc}} \right\}.$$

In this case $X \bullet Y \in \mathcal{H}^2_{\text{loc}}$. Observe that the condition $X \in \mathcal{L}^2_{\text{loc}}(Y)$ is very natural. If M is a local martingale then $M \in \mathcal{H}^2_{\text{loc}}$ if and only if[40] $[M] \in \mathcal{A}^+_{\text{loc}}$. As $[X \bullet Y] = X^2 \bullet [Y]$, obviously $X \bullet Y \in \mathcal{H}^2_{\text{loc}}$ if and only if $X \in \mathcal{L}^2_{\text{loc}}(Y)$. As $\Delta(X \bullet Y) = X \Delta Y$, if Y is continuous then $X \bullet Y$ is also continuous. Let $Y = Y(0) + Y^c + Y^d$ be the decomposition of Y into continuous and purely discontinuous local martingales. As $[Y] \in \mathcal{A}^+_{\text{loc}}$ and as

$$[Y] = [Y^c] + [Y^d] \tag{4.9}$$

it is obvious that $[Y^c], [Y^d] \in \mathcal{A}^+_{\text{loc}}$. This immediately implies that Y^c and Y^d are in $\mathcal{H}^2_{\text{loc}}$. From (4.9) it is also clear that $X \in \mathcal{L}^2_{\text{loc}}(Y)$ if and only if $X \in \mathcal{L}^2_{\text{loc}}(Y^c)$ and $X \in \mathcal{L}^2_{\text{loc}}(Y^d)$. This implies that $X \bullet Y^c$ and $X \bullet Y^d$ exist and obviously

$$X \bullet Y = X \bullet Y^c + X \bullet Y^d.$$

By the construction $X \bullet Y^c$ is continuous. Observe that $X \bullet Y^d$ is a purely discontinuous local martingale as for any continuous local martingale L

$$\left[X \bullet Y^d, L \right] = X \bullet \left[Y^d, L \right] = X \bullet 0 = 0,$$

that is $X \bullet Y^d$ is strongly orthogonal to every continuous local martingale.

The goal of this section is to extend the integration to the case when the integrator is an arbitrary local martingale. To do this one should define the stochastic integral for every purely discontinuous local martingale. Extending the integration to purely discontinuous local martingales from the integration procedure we expect the following properties:

1. If $L \in \mathcal{L}$ is purely discontinuous then $X \bullet L \in \mathcal{L}$ should be also purely discontinuous.

2. Purely discontinuous local martingales are uniquely determined by their jumps[41], hence it is sufficient to prescribe the jumps of $X \bullet L$: it is very natural to ask that the formula $\Delta(X \bullet L) = X \Delta L$ should hold.

---

[40] See: Proposition 3.64, page 223.
[41] See: Corollary 4.7, page 228.

3. We have proved[42] $[L]^{1/2} \in \mathcal{A}_{\text{loc}}^{+}$ for any local martingale L, therefore if $X \bullet L$ is a purely discontinuous local martingale then the expression $\sqrt{[X \bullet L]} = \sqrt{\sum (X \Delta L)^2}$ should have locally integrable variation.

4. If $L \in \mathcal{L}$ then $^{p}(\Delta L) = 0$. By the jump condition, if X is predictable then

$$^{p}(\Delta (X \bullet L)) = \ ^{p}(X \cdot \Delta L) = X \cdot (^{p}(\Delta L)) = X \cdot 0 = 0$$

from which one can expect that one can guarantee only for predictable integrands X that $X \bullet L \in \mathcal{L}$ and $\Delta (X \bullet L) = X \Delta L$.

4.3.1 Definition of stochastic integration

Assume, that $L \in \mathcal{L}$ is a purely discontinuous local martingale. As L is a local martingale $^{p}(\Delta L)$ is finite and $^{p}(\Delta L) = 0$. If H is a predictable real valued process then as $^{p}(\Delta L)$ is finite[43]

$$^{p}(H \Delta L) = H (^{p}(\Delta L)) = 0,$$

hence if

$$\sqrt{\sum H^2 (\Delta L)^2} \in \mathcal{A}_{\text{loc}}^{+},$$

then there is one and only one purely discontinuous local martingale[44], denoted by $H \bullet L$, for which

$$\Delta (H \bullet L) = H \Delta L.$$

If one expects the properties

$$H \Delta L = \Delta (H \bullet L) \quad \text{and} \quad (H \bullet L)^d = H \bullet L^d$$

from the stochastic integral $H \bullet L$ then this definition is the only possible one for $H \bullet L$.

Definition 4.37 *If L is a purely discontinuous local martingale then $H \bullet L$ is the stochastic integral of H with respect to L.*

Definition 4.38 *If $L = L(0) + L^c + L^d$ is a local martingale and H is a predictable process for which $\sqrt{H^2 \bullet [L]} \in \mathcal{A}_{\text{loc}}^{+}$ then*

$$H \bullet L \stackrel{\circ}{=} H \bullet L^c + H \bullet L^d.$$

$H \bullet L$ is the stochastic integral of H with respect to L.

---

[42]See: (3.20), page 222.
[43]See: Proposition 3.37. page 201.
[44]See: Proposition 4.28, page 240.

Example 4.39 If $X \in \mathcal{V}$ is predictable[45] and L is a local martingale then $\Delta X \bullet L = \sum \Delta X \Delta L$.

1. The trajectories of L are right-regular, therefore they are bounded on finite intervals[46]. As $X \in \mathcal{V}$ obviously $\Delta L \bullet X$ exists and

$$\sum \Delta X \Delta L = \Delta L \bullet X.$$

X is predictable and right-regular, therefore it is locally bounded[47]. As $\text{Var}(X)$ is also predictable and it is also right-regular it is also locally bounded.

2. $|\Delta X| \leq \text{Var}(X)$, which implies that $\Delta X \bullet L$ is well-defined. Let $L = L(0) + L^c + L^d$ be the decomposition of L. For any local martingale N

$$\Delta X \bullet [L^c, N] = 0$$

hence $\Delta X \bullet L^c = 0$. Therefore one can assume that L is purely discontinuous.

$$\left| \sum \Delta X \Delta L \right| \leq \sum |\Delta X \Delta L| \leq \sqrt{\sum (\Delta X)^2} \sqrt{\sum (\Delta L)^2} \leq$$
$$\leq \sqrt{[X]} \sqrt{[L]} < \infty.$$

Obviously $\Delta \left(\sum \Delta X \Delta L \right) = \Delta X \Delta L$. As $\sum \Delta X \Delta L$ has finite variation, so if it is a local martingale then it is a purely discontinuous local martingale. Therefore we should prove that $\sum \Delta X \Delta L$ is a local martingale. Hence we should prove that $\Delta L \bullet X$ is a local martingale.

3. With localization one can assume that X and $\text{Var}(X)$ are bounded. As X and $\text{Var}(X)$ are bounded

$$|\Delta L| \bullet \text{Var}(X) = \sum |\Delta X| |\Delta L| \leq \sqrt{\sum (\Delta X)^2} \sqrt{[L]} \leq$$
$$\leq \sqrt{\sup |X| \cdot \text{Var}(X)} \sqrt{[L]} \in \mathcal{A}_{\text{loc}}^+.$$

Hence with further localization we can assume that $\Delta L \bullet X \in \mathcal{A}$. If τ is a stopping time then

$$\mathbf{E}\left((\Delta L \bullet X)(\tau)\right) = \mathbf{E}\left((\Delta L \bullet X^\tau)(\infty)\right).$$

As X^τ is also predictable[48] one should prove that if $\Delta L \bullet X \in \mathcal{A}$ and X is predictable, then $\mathbf{E}\left((\Delta L \bullet X)(\infty)\right) = 0$. By Dellacherie's formula[49], using that

[45]If X is not predictable then ΔX is also not predictable so $\Delta X \bullet L$ is undefined.
[46]See: Proposition 1.6, page 5.
[47]See: Proposition 3.35, page 200.
[48]See: Proposition 1.39, page 23.
[49]See: Proposition 5.9, page 301.

L is a local martingale hence $^p(\Delta L) = 0$,

$$\mathbf{E}\left((\Delta L \bullet X)(\infty)\right) = \mathbf{E}\left((\left(^p(\Delta L)\right) \bullet X)(\infty)\right) = 0.$$

That is $\Delta L \bullet X = \sum \Delta X \Delta L$ is a local martingale. □

4.3.2 Properties of stochastic integration

Let us discuss the properties of stochastic integration with respect to local martingales:

1. *If* $\sqrt{H^2 \bullet [L]} \in \mathcal{A}_{\text{loc}}^+$ *then the definition is meaningful and* $H \bullet L \in \mathcal{L}$. *Specifically every locally bounded predictable process is integrable[50].*
For any local martingale L

$$[L] = [L^c] + \sum (\Delta L)^2. \tag{4.10}$$

The integral $H^2 \bullet [L^c]$ is finite, hence the integral $H \bullet L^c$ exists[51]. By (4.10)

$$\sqrt{H^2 \bullet [L^d]} = \sqrt{H^2 \bullet \sum (\Delta L)^2} = \sqrt{\sum (H \Delta L)^2} \in \mathcal{A}_{\text{loc}},$$

hence $H \bullet L^d$ is also meaningful. Both integrals are local martingales, hence the sum $H \bullet L \stackrel{\circ}{=} H \bullet L^c + H \bullet L^d$ is also a local martingale. The second observation easily follows from the relation $\sqrt{[L]} \in \mathcal{A}_{\text{loc}}^+$.

2. $H\Delta L = \Delta(H \bullet L)$.
3. $(H \bullet L)^c = H \bullet L^c$ *and* $(H \bullet L)^d = H \bullet L^d$.
4. $[H \bullet L] = H^2 \bullet [L]$.

$$[H \bullet L] = [(H \bullet L)^c] + \sum (\Delta(H \bullet L))^2 =$$
$$= H^2 \bullet [L^c] + \sum (H \Delta L)^2 = H^2 \bullet [L^c] + H^2 \bullet [L^d] =$$
$$= H^2 \bullet [L].$$

5. $H \bullet L$ *is the only process in* $\mathcal{L}$ *for which*

$$[H \bullet L, N] = H \bullet [L, N]$$

holds for every $N \in \mathcal{L}$.
By the inequality of Kunita and Watanabe

$$|H| \bullet \text{Var}([L, N]) \leq \sqrt{H^2 \bullet [L]} \sqrt{[N]}$$

---

[50] $\sqrt{[M]} \in \mathcal{A}_{\text{loc}}^+$ for any local martingale M, hence the present construction of $H \bullet L$ is maximal in H, that is if one wants to extend the definition of the stochastic integral to a broader class of integrands H, then $H \bullet L$ will not necessarily be a local martingale.

[51] See: Corollary 2.67, page 158.

hence the integral $H \bullet [L, N]$ is meaningful. Therefore

$$
\begin{aligned}
[H \bullet L, N] &= [(H \bullet L)^c, N^c] + \left[(H \bullet L)^d, N^d \right] = \\
&= [H \bullet L^c, N^c] + \sum H \Delta L \Delta N \\
&= H \bullet [L^c, N^c] + H \bullet \left[L^d, N^d \right] = \\
&= H \bullet \left([L^c, N^c] + \left[L^d, N^d \right] \right) = \\
&= H \bullet [L, N].
\end{aligned}
$$

If $H \bullet [L, N] = [Y, N]$ for some local martingale Y, then $[Y - H \bullet L, N] = 0$. Hence if $N \overset{\circ}{=} Y - H \bullet L$ then $[Y - H \bullet L] = 0$. $Y - H \bullet L$ is a local martingale therefore[52] $Y - H \bullet L = 0$.

6. *If τ is an arbitrary stopping time, and $H \bullet L$ exists then*

$$
H \bullet L^\tau = (H \bullet L)^\tau = (\chi([0, \tau]) H) \bullet L.
$$

If $\sqrt{H^2 \bullet [L]} \in \mathcal{A}_{\text{loc}}$, then trivially

$$
\sqrt{H^2 \bullet [L^\tau]} = \sqrt{\chi([0, \tau]) H^2 \bullet [L]} \in \mathcal{A}_{\text{loc}}
$$

so the integrals above exists. By the stopping rule of the quadratic variation if $N \in \mathcal{L}$

$$
\begin{aligned}
[(H \bullet L)^\tau, N] &= [(H \bullet L), N]^\tau = (H \bullet [L, N])^\tau = H \bullet [L, N]^\tau = \\
&= H \bullet [L^\tau, N] = [H \bullet L^\tau, N],
\end{aligned}
$$

hence by the bilinearity of the quadratic variation

$$
[(H \bullet L)^\tau - H \bullet L^\tau, N] = 0, \ N \in \mathcal{L},
$$

from which

$$
(H \bullet L)^\tau = H \bullet L^\tau.
$$

For arbitrary $N \in \mathcal{L}$

$$
\begin{aligned}
[H \bullet L^\tau, N] &= H \bullet [L^\tau, N] = H \bullet [L, N]^\tau = \\
&= (\chi([0, \tau]) H) \bullet [L, N] = \\
&= [(\chi([0, \tau]) H) \bullet L, N],
\end{aligned}
$$

hence again $H \bullet L^\tau = (\chi([0, \tau]) H) \bullet L$ from Property 5.

[52]See: Proposition 2.82, page 170.

7. *The integral is linear in the integrand.*

By elementary calculation

$$\sqrt{(H_1 + H_2)^2 \bullet [L]} \le \sqrt{H_1^2 \bullet [L]} + \sqrt{H_2^2 \bullet [L]},$$

hence if $H_1 \bullet L$ and $H_2 \bullet L$ exist then the integral $(H_1 + H_2) \bullet L$ also exists. When the integrator is continuous the integral is linear. The linearity of the purely discontinuous part is a simple consequence of the relation.

$$(H_1 + H_2) \Delta L = H_1 \Delta L + H_2 \Delta L.$$

The proof of the homogeneity is analogous.

8. *The integral is linear in the integrator.*

By the inequality of Kunita and Watanabe[53]

$$[L_1 + L_2] \le 2 \left([L_1] + [L_2] \right),$$

hence if the integrals $H \bullet L_1$ and $H \bullet L_2$ exist then $H \bullet (L_1 + L_2)$ also exists. The decomposition of the local martingales into continuous and purely discontinuous martingales is unique so $(L_1 + L_2)^c = L_1^c + L_2^c$, and $(L_1 + L_2)^d = L_1^d + L_2^d$. For continuous local martingales we have already proved the linearity, the linearity of the purely discontinuous part is evident from the relation $\Delta (L_1 + L_2) = \Delta L_1 + \Delta L_2$.

9. *If $H \overset{\circ}{=} \sum_i \xi_i \chi \left((\tau_i, \tau_{i+1}] \right)$ is an adapted simple process then*

$$(H \bullet L)(t) = \sum_i \xi_i \left(L \left(\tau_{i+1} \wedge t \right) - L \left(\tau_i \wedge t \right) \right). \qquad (4.11)$$

By the linearity it is sufficient to calculate the integral just for one jump. For the continuous part we have already deduced the formula. For the discontinuous part it is sufficient to remark that if ξ_i is $\mathcal{F}_{\tau_i}$-measurable and L is a purely discontinuous local martingale then $\xi_i \left(L \left(\tau_{i+1} \wedge t \right) - L \left(\tau_i \wedge t \right) \right)$ is a purely discontinuous local martingale[54], with jumps $\xi_i \chi \left((\tau_i, \tau_{i+1}] \right) \Delta L$.

10. *Assume that the integral $H \bullet L$ exists. The integral $K \bullet (H \bullet L)$ exists if and only if the integral $(KH) \bullet L$ exists. In this case*

$$(KH) \bullet L = K \bullet (H \bullet L).$$

Let us remark that as the integrals are pathwise integrals with respect to processes with finite variation

$$\sqrt{K^2 \bullet (H^2 \bullet [L])} = \sqrt{(KH)^2 \bullet [L]}.$$

---

[53]See: Corollary 2.36, page 137.

[54]The space of purely discontinuous local martingales is closed under stopping.

$K \bullet (H \bullet L)$ exists if and only if

$$\sqrt{K^2 \bullet [H \bullet L]} = \sqrt{K^2 \bullet (H^2 \bullet [L])} = \sqrt{(KH)^2 \bullet [L]} \in \mathcal{A}_{\mathrm{loc}}^+,$$

from which the first part is evident. If N is an arbitrary local martingale then

$$\begin{aligned}
[K \bullet (H \bullet L), N] = K \bullet [H \bullet L, N] = KH \bullet [L, N] = \\
= [KH \bullet L, N],
\end{aligned}$$

from which the second part is evident.

11. *If τ is an arbitrary stopping time then*

$$H \bullet L^\tau = (\chi([0, \tau]) H) \bullet L = (H \bullet L)^\tau.$$

If N is an arbitrary local martingale, then

$$\begin{aligned}
[H \bullet L^\tau, N] = H \bullet [L^\tau, N] = H \bullet [L, N]^\tau = \\
= H\chi([0, \tau]) \bullet [L, N] = \\
= [H\chi([0, \tau]) \bullet L, N] = \\
= (H \bullet [L, N])^\tau = [H \bullet L, N]^\tau = \\
= [(H \bullet L)^\tau, N],
\end{aligned}$$

from which the property is evident.

12. *The Dominated Convergence Theorem is valid, that is if (H_n) is a sequence of predictable processes, $H_n \to H_\infty$ and there is a predictable process H, for which the integral $H \bullet L$ exists and $|H_n| \le H$ then the integrals $H_n \bullet L$ also exist and $H_n \bullet L \to H_\infty \bullet L$, where the convergence is uniform in probability on the compact time-intervals.*

As $H_n^2 \bullet [L] \le H^2 \bullet [L]$ for all $n \le \infty$ the integrals $H_n \bullet L$ exist. By Davis' inequality, for every stopping time τ

$$\mathbf{E}\left(\sup_t |((H_n - H_\infty) \bullet L^\tau)(t)| \right) \le C \cdot \mathbf{E}\left(\sqrt{\left((H_n - H_\infty)^2 \bullet [L]^\tau\right)(\infty)} \right).$$

There is a localizing sequence (τ_m), that $\mathbf{E}\left(\sqrt{H^2 \bullet [L]^{\tau_m}(\infty)} \right) < \infty$, hence by the classical Dominated Convergence Theorem

$$\mathbf{E}\left(\sqrt{(H_n - H_\infty)^2 \bullet [L]^{\tau_m}(\infty)} \right) \to 0$$

hence

$$\sup_t |((H_n - H_\infty) \bullet L^{\tau_m})(t)| \overset{L_1}{\to} 0,$$

from which as in the continuous case[55] one can guarantee on every compact interval the uniform convergence in probability.

13. *The definition of the integral is unambiguous that is if $L \in \mathcal{V} \cap \mathcal{L}$ then the two possible concepts of integration give the same result.*

It is trivial from Proposition 2.89.

14. *If X is left-continuous and locally bounded then $(X \bullet L)(t)$ is an Itô–Stieltjes integral for every t where the convergence of the approximating sums is uniform in probability on every compact interval. The approximating partitions can be random as well.*

The proof is the same as in the continuous case[56]. $\qquad\qquad\square$

4.4 Stochastic Integration With Respect To Semimartingales

Recall the definition of stochastic integration with respect to semimartingales:

Definition 4.40 *If semimartingale X has a decomposition*

$$X = X(0) + L + V, \quad V \in \mathcal{V}, \, L \in \mathcal{L}$$

for which the integrals $H \bullet L$ and $H \bullet V$ exist then

$$H \bullet X \overset{\circ}{=} H \bullet L + H \bullet V.$$

By Proposition 2.89 the next statement is trivial[57]:

Proposition 4.41 *For predictable integrands the definition is unambiguous, that is the integral is independent of the decomposition of the integrator.*

Proposition 4.42 *If X and Y are arbitrary semimartingales and the integrals $U \bullet X$ and $V \bullet Y$ exist, then*

$$[U \bullet X, V \bullet Y] = UV \bullet [X, Y].$$

Proof. Let $X_L + X_V$, and $Y_L + Y_V$ be the decomposition of X and Y.

$$[U \bullet X, V \bullet Y] = [U \bullet X_L, V \bullet Y_L] + [U \bullet X_L, V \bullet Y_V] +$$
$$+ [U \bullet X_V, V \bullet Y_L] + [U \bullet X_V, V \bullet Y_V].$$

---

[55]See: Proposition 2.74. page 162.
[56]See: Proposition 2.77, page 166.
[57]See: Subsection 2.4.3, page 176.

For integrals with respect to local martingales

$$[U \bullet X_L, V \bullet Y_L] = UV \bullet [X_L, Y_L].$$

In the three other expressions one factor has finite variation, hence the quadratic variation is the sum of the products of the jumps[58]. For example

$$[U \bullet X_L, V \bullet Y_V] = \sum \Delta (U \bullet X_L) \Delta (V \bullet Y_V) = \sum (U\Delta X_L) (V\Delta Y_V).$$

On the other hand for the same reason

$$UV \bullet [X_L, Y_V] = UV \bullet \left(\sum \Delta X_L \Delta Y_V\right) = \sum UV\Delta X_L \Delta Y_V,$$

hence

$$[U \bullet X_L, V \bullet Y_V] = UV \bullet [X_L, Y_V].$$

One can finish the proof with the same calculation for the other tags. $\square$

Observe that the existence of the integral $H \bullet X$ means that for some decomposition $X = X(0) + L + V$ one can define the integral and the existence of the integral does not mean that in every decomposition of X the two integrals are meaningful. Observe also that with the definition we extended the class of integrable processes even for local martingales. It is possible that the integral $H \bullet L$ as an integral with respect to the local martingale L does not exist, but L has a decomposition $L = L(0) + M + V$, $M \in \mathcal{L}$, $V \in \mathcal{V}$ for which H is integrable with respect to M and V. Of course in this general case we cannot guarantee that[59] $H \bullet L \in \mathcal{L}$.

Example 4.43 If the integrand is not locally bounded then the stochastic integral with respect to a local martingales is not necessarily a local martingale.

Let M be a compound Poisson process, where $\mathbf{P}(\xi_k = \pm 1) = 1/2$ for the jumps ξ_k. M is a martingale and the trajectories of M are not continuous. Let τ_1 be the time of the first jump of M and let

$$X(t, \omega) \stackrel{\circ}{=} \frac{1}{t} \cdot \chi((0, \tau_1(\omega)]).$$

---

[58]See: line (2.14), page 134.
[59]See: Example 4.43, page 255.

X is predictable but it is not locally bounded. As the trajectories of M have finite variation the pathwise stochastic integral

$$L\left(t,\omega\right) \overset{\circ}{=} \left(X \bullet M\right)\left(t,\omega\right) = \int_{(0,t]} \frac{1}{s}\chi\left(\left(0,\tau_1\left(\omega\right)\right]\right) dM\left(s,\omega\right) =$$

$$= \begin{cases} 0 & \text{if} \quad t < \tau_1\left(\omega\right) \\ \xi_1\left(\omega\right)/\tau_1\left(\omega\right) & \text{if} \quad \tau_1\left(\omega\right) \leq t \end{cases}$$

is meaningful. We prove that L is not a local martingale. If (ρ_k) would be a localization of L then L^{ρ_1} was a uniformly integrable martingale. Hence for the stopping time $\sigma \overset{\circ}{=} \rho_1 \wedge t$

$$\mathbf{E}\left(L\left(\sigma\right)\right) \overset{\circ}{=} \mathbf{E}\left(L\left(\rho_1 \wedge t\right)\right) = \mathbf{E}\left(L^{\rho_1}\left(t\right)\right) = \mathbf{E}\left(L\left(0\right)\right) = 0.$$

Therefore it is sufficient to prove that for any finite stopping time $\sigma \neq 0$

$$\mathbf{E}\left(\left|L\left(\sigma\right)\right|\right) = \infty. \tag{4.12}$$

Let σ be a finite stopping time with respect to the filtration $\mathcal{F}$ generated by M.

$$\mathbf{E}\left(\left|L\left(\sigma\right)\right|\right) = \int_{\Omega} \frac{1}{\tau_1}\chi\left(\tau_1 \leq \sigma\right) d\mathbf{P} \geq \int_{\Omega} \frac{1}{\tau_1}\chi\left(\tau_1 \leq \sigma \wedge \tau_1\right) d\mathbf{P}.$$

Hence to prove (4.12) one can assume that $\sigma \leq \tau_1$. In this case σ is $\mathcal{F}_{\tau_1}$-measurable. Hence it is independent of the variables (ξ_n). So one can assume that σ is a stopping time for the filtration generated by the point process part of M. By the formula of the representation of stopping times of point processes[60]

$$\sigma = \varphi_0 \chi\left(\sigma < \tau_1\right) + \sum_{n=1}^{\infty} \chi\left(\tau_n \leq \sigma < \tau_{n+1}\right) \varphi_n\left(\tau_0, \ldots, \tau_n\right) \overset{\circ}{=}$$

$$\overset{\circ}{=} \sum_{n=0}^{\infty} \chi\left(\tau_n \leq \sigma < \tau_{n+1}\right) \varphi_n\left(\tau_0, \ldots, \tau_n\right) =$$

$$= \varphi_0 \chi\left(\sigma < \tau_1\right) + \chi\left(\sigma \geq \tau_1\right) \varphi_1\left(\tau_1\right).$$

From this $\{\tau_1 \leq \varphi_0\} \subseteq \{\tau_1 \leq \sigma\}$. If $\varphi_0 > 0$ then using that τ_1 has an exponential distribution

$$\mathbf{E}\left(\left|L\left(\sigma\right)\right|\right) = \int_{\Omega} \frac{1}{\tau_1}\chi\left(\tau_1 \leq \sigma\right) d\mathbf{P} \geq \int_{\Omega} \frac{1}{\tau_1}\chi\left(\tau_1 \leq \varphi_0\right) d\mathbf{P} =$$

$$= \int_0^{\varphi_0} \frac{1}{x}\lambda \exp\left(-\lambda x\right) dx = \infty.$$

[60]See: Proposition C.6, page 581.

$\sigma \neq 0$ and $\mathcal{F}_0 = \{\emptyset, \Omega\}$, therefore $\{\sigma \leq 0\} = \emptyset$. Hence $\sigma > 0$, so if $\varphi_0 = 0$ then $\sigma \geq \tau_1$. Hence again

$$\mathbf{E}\left(|L\left(\sigma\right)|\right) = \int_\Omega \frac{1}{\tau_1} \chi\left(\tau_1 \leq \sigma\right) d\mathbf{P} = \int_\Omega \frac{1}{\tau_1} d\mathbf{P} = \infty. \qquad \square$$

By the definition of the integral it is clear that if a process H is integrable with respect to semimartingales X_1 and X_2 then H is integrable with respect to $aX_1 + bX_2$ for every constants a, b and

$$H \bullet (aX_1 + bX_2) = a\left(H \bullet X_1\right) + b\left(H \bullet X_2\right).$$

Observe that by the above definitions the other additivity of the integral, that is the relation

$$(H_1 + H_2) \bullet X = H_1 \bullet X + H_2 \bullet X$$

is not clear. Our direct goal in the following two subsections is to prove this additivity property of the integral.

4.4.1 Integration with respect to special semimartingales

Recall that by definition S is a special semimartingale if it has a decomposition

$$S = S\left(0\right) + V + L, \quad V \in \mathcal{V}, L \in \mathcal{L} \tag{4.13}$$

where V is predictable.

Theorem 4.44 (Characterization of special semimartingales) *Let S be a semimartingale. The next statements are equivalent:*

1. S is a special semimartingale, i.e. there is a decomposition *(4.13)* where V is predictable.
2. There is a decomposition *(4.13)*, where $V \in \mathcal{A}_{loc}$.
3. For all decompositions *(4.13)* $V \in \mathcal{A}_{loc}$.
4. $S^*\left(t\right) \overset{\circ}{=} \sup_{s \leq t} |S\left(s\right) - S\left(0\right)| \in \mathcal{A}_{loc}^+$.

Proof. We prove the equivalence of the statements backwards.

1. Let us assume that the last statement holds, and let $S = S\left(0\right) + V + L$ be a decomposition of S. Let $L^*\left(t\right) \overset{\circ}{=} \sup_{s \leq t} |L\left(s\right)|$. L^* is in[61] $\mathcal{A}_{loc}^+$, hence from the assumption of the fourth statement

$$V^*\left(t\right) \overset{\circ}{=} \sup_{s \leq t} |V\left(s\right)| \leq S^*\left(t\right) + L^*\left(t\right) \in \mathcal{A}_{loc}^+.$$

---

[61]See: Example 3.3, page 181.

The process $\text{Var}(V)_-$ is increasing and continuous from the left, hence it is locally bounded, hence $\text{Var}(A)_- \in \mathcal{A}^+_{\text{loc}}$. As

$$\text{Var}(V) \leq \text{Var}(V)_- + \Delta(\text{Var}(V)) \leq \text{Var}(V)_- + 2V^*$$

$\text{Var}(V) \in \mathcal{A}^+_{\text{loc}}$, hence the third condition holds.

2. From the third condition the second one follows trivially.

3. If $V \in \mathcal{A}_{\text{loc}}$ in the decomposition $S = S(0) + V + L$, then V^p, the predictable compensator of V, exists. $V - V^p$ is a local martingale, hence

$$S = S(0) + V^p + (V - V^p + L)$$

is a decomposition where $V^p \in \mathcal{V}$ is predictable, so S is a special semimartingale.

4. Let us assume that $S(0) = 0$ so $S = V + L$. If $V^*(t) \overset{\circ}{=} \sup_{s \leq t} |V(s)|$, then as $V^* \leq \text{Var}(V)$

$$S^* \leq V^* + L^* \leq \text{Var}(V) + L^*.$$

$L^* \in \mathcal{A}^+_{\text{loc}}$, so it is sufficient to prove that if $V \in \mathcal{V}$ is predictable then $\text{Var}(V) \in \mathcal{A}^+_{\text{loc}}$. It is sufficient to prove that $\text{Var}(V)$ is locally bounded. V is continuous from the right, hence when one calculates $\text{Var}(V)$ it suffices to use the partitions with dyadic rationals and hence if V is predictable then $\text{Var}(V)$ is also predictable. $\text{Var}(V)$ is right-continuous and predictable hence it is locally bounded[62]. $\square$

Example 4.45 $X \in \mathcal{V}$ is a special semimartingale if and only if $X \in \mathcal{A}_{\text{loc}}$. A compound Poisson process is a special semimartingale if and only if the expected value of the distribution of the jumps is finite.

The first remark is evident from the theorem. Recall, that a compound Poisson process has locally integrable variation if and only if the distribution of the jumps has finite expected value[63]. $\square$

Example 4.46 If a semimartingale S is locally bounded then S is a special semimartingale.

Example 4.47 If a semimartingale S has bounded jumps then S is a special semimartingale[64].

---

[62]See: Proposition 3.35, page 200.
[63]See: Example 3.2, page 180.
[64]See: Proposition 1.152, page 107.

Example 4.48 Decomposition of continuous semimartingales.

Recall that by definition S is a continuous semimartingale if S has a decomposition $S = S(0) + V + L$, where $V \in \mathcal{V}$, $L \in \mathcal{L}$ and V and L are continuous[65]. Let S now be a semimartingale and let us assume that S is continuous. As S is continuous it is locally bounded, so S is a special semimartingale. By the just proved proposition S has a decomposition $S(0) + V + L$, where $V \in \mathcal{V}$ is predictable and $L \in \mathcal{L}$. As S is continuous L is also predictable, hence it is continuous[66]. This implies that V is also continuous. This means that S is a continuous semimartingale.   □

The stochastic integral $X \bullet Y$ is always a semimartingale. One can ask: when is it a special semimartingale?

Theorem 4.49 (Integration with respect to special semimartingales)
Let X be a special semimartingale. Assume that for a predictable process H the integral $H \bullet X$ exists. Let $X \overset{\circ}{=} X(0) + A + L$ be the canonical decomposition of X. $H \bullet X$ is a special semimartingale if and only if the integrals $H \bullet A$ and $H \bullet L$ exist and $H \bullet L$ is a local martingale. In this case the canonical decomposition of $H \bullet X$ is exactly $H \bullet A + H \bullet L$.

Proof. Let us first remark that if U and W are predictable and $W \in \mathcal{V}$ and the integral $U \bullet W$ exists then it is predictable. This is obviously true if

$$U \overset{\circ}{=} \chi((s,t]) \chi_F, \quad F \in \mathcal{F}_s$$

as[67]

$$U \bullet W = \chi_F \left(W^t - W^s \right) = \left(\chi_F \chi((s,\infty)) \right) \left(W^t - W^s \right).$$

The general case follows from the Monotone Class Theorem. Assume that the integral[68]

$$Z \overset{\circ}{=} H \bullet X \overset{\circ}{=} H \bullet V + H \bullet M$$

exists and it is a special semimartingale. Let $Z \overset{\circ}{=} B + N$ be the canonical decomposition of Z. $B \in \mathcal{A}_{\text{loc}}$ and B is predictable. $\chi(|H| \leq n)$ is bounded and predictable, hence the integral

$$\chi(|H| \leq n) \bullet Z \overset{\circ}{=} \chi(|H| \leq n) \bullet B + \chi(|H| \leq n) \bullet N$$

[65]See: Definition 2.18, page 124.
[66]See: 3.40, page 205.
[67]See: Proposition 1.39, page 23.
[68]With some decomposition $X = X(0) + V + M$.

exists. $\chi\left(|H| \leq n\right)$ is bounded, $B \in \mathcal{A}_{\text{loc}}$ hence

$$\chi\left(|H| \leq n\right) \bullet B \in \mathcal{A}_{\text{loc}}.$$

As $\chi\left(|H| \leq n\right)$ and B are predictable $\chi\left(|H| \leq n\right) \bullet B$ is also predictable. Let $H_n \overset{\circ}{=} H\chi\left(|H| \leq n\right)$. H_n is bounded and predictable hence the integral

$$H_n \bullet X \overset{\circ}{=} H_n \bullet A + H_n \bullet L$$

is meaningful. $H_n \bullet A \in \mathcal{A}_{\text{loc}}$ and $H_n \bullet A$ is predictable and $H_n \bullet L \in \mathcal{L}$ so $H_n \bullet X$ is a special semimartingale and $H_n \bullet A + H_n \bullet L$ its canonical decomposition. By the associativity rule of the integration with respect to local martingales and processes with finite variation, and by the linearity in the integrator

$$\begin{aligned}
\chi\left(|H| \leq n\right) \bullet Z &\overset{\circ}{=} \chi\left(|H| \leq n\right) \bullet (H \bullet X) \overset{\circ}{=} \\
&\overset{\circ}{=} \chi\left(|H| \leq n\right) \bullet (H \bullet V + H \bullet M) = \\
&= \chi\left(|H| \leq n\right) \bullet (H \bullet V) + \chi\left(|H| \leq n\right) \bullet (H \bullet M) = \\
&= \left(\chi\left(|H| \leq n\right) H\right) \bullet V + \left(\chi\left(|H| \leq n\right) H\right) \bullet M \overset{\circ}{=} \\
&\overset{\circ}{=} \left(\chi\left(|H| \leq n\right) H\right) \bullet X \overset{\circ}{=} \\
&\overset{\circ}{=} H_n \bullet X = H_n \bullet A + H_n \bullet L.
\end{aligned}$$

The canonical decomposition of special semimartingales is unique, hence

$$\chi\left(|H| \leq n\right) \bullet B = H_n \bullet A, \quad \chi\left(|H| \leq n\right) \bullet N = H_n \bullet L.$$

As we have seen

$$\begin{aligned}
\chi\left(|H| \leq n\right) H^2 \bullet [L] &\overset{\circ}{=} H_n^2 \bullet [L] = [H_n \bullet L] = [\chi\left(|H| \leq n\right) \bullet N] = \\
&= \chi\left(|H| \leq n\right) \bullet [N] \leq [N].
\end{aligned}$$

$\sqrt{[N]} \in \mathcal{A}_{\text{loc}}^+$, so by the Monotone Convergence Theorem $\sqrt{H^2 \bullet [L]} \in \mathcal{A}_{\text{loc}}^+$ and therefore the integral $H \bullet L \in \mathcal{L}$ exists, and by the Dominated Convergence Theorem $N = H \bullet L$. Similarly, $H \bullet A$ exists, it is in $\mathcal{A}_{\text{loc}}$ and $H \bullet A = B$.

If H and A are predictable then $H \bullet A$ is predictable hence the other implication is evident. $\qquad\square$

Corollary 4.50 *Let L be a local martingale and let us assume that the integral $H \bullet L$ exists. $H \bullet L$ is a local martingale if and only if $\sup_{s \leq t} |(H \bullet L)(s)|$ is locally integrable, that is*

$$\sup_{s \leq t} |(H \bullet L)(s)| \in \mathcal{A}_{loc}^+.$$

Proof. As $\sup_{s \leq t} |M(s)|$ is locally integrable[69] for every local martingale $M \in \mathcal{L}$ one should only prove that if $\sup_{s \leq t} |(H \bullet L)(s)|$ is locally integrable then $H \bullet L$ is a local martingale. $X \overset{\circ}{=} L$ is a special semimartingale with canonical decomposition $X = L + 0$. Hence $H \bullet L$ is a local martingale if and only if $Y \overset{\circ}{=} H \bullet L$ is a special semimartingale. But as $Y(0) = 0$, the process Y is a special semimartingale[70] if and only if $\sup_{s \leq t} |Y(s)| \in \mathcal{A}_{\text{loc}}^+$. $\qquad \square$

4.4.2   Linearity of the stochastic integral

The most important property of every integral is the linearity in the integrand. Now we are ready to prove this important property:

Theorem 4.51 (Additivity of stochastic integration) *Let X be an arbitrary semimartingale. If H_1 and H_2 are predictable processes and the integrals $H_1 \bullet X$ and $H_2 \bullet X$ exist, then for arbitrary constants a and b the integral $(aH_1 + bH_2) \bullet X$ exists and*

$$(aH_1 + bH_2) \bullet X = a\,(H_1 \bullet X) + b\,(H_2 \bullet X). \tag{4.14}$$

Proof. Let

$$B \overset{\circ}{=} \{|\Delta X| > 1, |\Delta\,(H_1 \bullet X)| > 1, |\Delta\,(H_2 \bullet X)| > 1\}$$

be the set of the 'big jumps'. Observe that

$$\Delta\,(H_i \bullet X) \overset{\circ}{=} \Delta\,(H_i \bullet V_i + H_i \bullet L_i) =$$

$$= \Delta\,(H_i \bullet V_i) + \Delta\,(H_i \bullet L_i) =$$

$$= H_i \Delta V_i + H_i \Delta L_i = H_i \Delta X,$$

so

$$B = \{|\Delta X| > 1, |H_1 \Delta X| > 1, |H_2 \Delta X| > 1\}.$$

Obviously for an arbitrary ω the section $B\,(\omega)$ does not have an accumulation point. Let us separate the 'big jumps' from X. That is let

$$\widetilde{X} \overset{\circ}{=} \sum \Delta X \chi_B, \quad \overline{X} \overset{\circ}{=} X - \widetilde{X}.$$

Observe that, by the simple structure of B, $\widetilde{X} \in \mathcal{V}$ and the integrals $H_k \bullet \widetilde{X}$ are simple sums, so they exist. By the construction of the stochastic integral

---

[69]See: Example 3.3, page 181.

[70]See: Theorem 4.44, page 257.

$H_k \bullet \overline{X}$ also exists[71]. As the jumps of the $\overline{X}$ are bounded, $\overline{X}$ is a special semimartingale[72].

$$\Delta \left(H_k \bullet \overline{X} \right) = H_k \Delta \overline{X} = H_k \Delta \left(X - \widetilde{X} \right) =$$
$$= H_k \Delta X \chi_{B^c},$$

hence the jumps of $H_k \bullet \overline{X}$ are also bounded and therefore the processes $H_k \bullet \overline{X}$ are also special semimartingales. Let

$$\overline{X} = X(0) + A + L$$

be the canonical decomposition of $\overline{X}$. By the previous theorem integrals $H_k \bullet A$ and $H_k \bullet L$ also exist. The integration with respect to local martingales and with respect to processes with finite variation is additive, hence

$$(H_1 + H_2) \bullet A = H_1 \bullet A + H_2 \bullet A,$$
$$(H_1 + H_2) \bullet L = H_1 \bullet L + H_2 \bullet L,$$

which of course means that the integrals on the left-hand side exist. The integrals $H_k \bullet \widetilde{X}$ are ordinary sums, hence

$$(H_1 + H_2) \bullet \widetilde{X} = H_1 \bullet \widetilde{X} + H_2 \bullet \widetilde{X}.$$

Adding up these three lines above and using that the integral is additive in the integrator we get (4.14). The homogeneity of the integral is obvious by the definition of the integral. $\qquad\square$

4.4.3 The associativity rule

Like additivity, the associativity rule is also not directly evident from the definition of the stochastic integral.

Theorem 4.52 (Associativity rule) *Let X be an arbitrary semimartingale and let us assume that the integral $H \bullet X$ exists. The integral $K \bullet (H \bullet X)$ exists if and only if the integral $(KH) \bullet X$ exists. In this case*

$$K \bullet (H \bullet X) = (KH) \bullet X.$$

---

[71] $H^2 \bullet \left[\overline{L} \right] \leq H^2 \bullet [L]$ and $\mathrm{Var} \left(\overline{V} \right) \leq \mathrm{Var}(V)$!
[72] See: Example 4.47, page 258.

Proof. Assume that K is integrable with respect to the semimartingale $Y \overset{\circ}{=} H \bullet X$. Let B be again the set of the 'big jumps', that is

$$B \overset{\circ}{=} \{|\Delta X| > 1, |\Delta Y| > 1, |\Delta (K \bullet Y)| > 1\}.$$

As in the previous subsection for every ω the section $B(\omega)$ is a discrete set. Let us define the processes

$$\tilde{X} \overset{\circ}{=} \sum \chi_B \Delta X, \quad \overline{X} \overset{\circ}{=} X - \tilde{X},$$
$$\tilde{Y} \overset{\circ}{=} \sum \chi_B \Delta Y, \quad \overline{Y} \overset{\circ}{=} Y - \tilde{Y}.$$

Using the formula for the jumps of the integrals and the additivity of the integral in the integrator

$$\overline{Y} \overset{\circ}{=} Y - \tilde{Y} = H \bullet X - H \bullet \tilde{X} = H \bullet \overline{X}.$$

As the jumps of $\overline{X}$ are bounded, $\overline{X}$ is a special semimartingale. Let

$$\overline{X} = X(0) + A + L$$

be the canonical decomposition of $\overline{X}$. By the same reason $\overline{Y}$ is also a special semimartingale and as we saw above the canonical decomposition of $\overline{Y}$ is

$$\overline{Y} = H \bullet \overline{X} = H \bullet A + H \bullet L.$$

The integral $K \bullet \tilde{Y}$ on any finite interval is a finite sum, hence if $K \bullet Y$ exists then $K \bullet \overline{Y}$ also exists.

$$\Delta \left(K \bullet \overline{Y} \right) = K \Delta \overline{Y} = K \Delta Y \chi_{B^c}.$$

The jumps of $K \bullet \overline{Y}$ are bounded so $K \bullet \overline{Y}$ is also a special semimartingale. Therefore the integrals $K \bullet (H \bullet A)$ and $K \bullet (H \bullet L)$ exist and $K \bullet (H \bullet L)$ is a local martingale. By the associativity rule for local martingales and for processes with finite variation

$$K \bullet (H \bullet A) = (KH) \bullet A,$$
$$K \bullet (H \bullet L) = (KH) \bullet L.$$

Adding up the corresponding lines

$$K \bullet Y = K \bullet \overline{Y} + K \bullet \widetilde{Y} =$$
$$= K \bullet (H \bullet A + H \bullet L) + K \bullet \left(H \bullet \widetilde{X} \right) =$$
$$= (KH) \bullet A + (HL) \bullet L + (KH) \bullet \widetilde{X} =$$
$$= (KH) \bullet \overline{X} + (KH) \bullet \widetilde{X} = (KH) \bullet X.$$

The proof of the reverse implication is similar. Assume that the integrals $Y \triangleq H \bullet X$ and $(KH) \bullet X$ exist, and let

$$B \triangleq \{|\Delta X| > 1, |\Delta Y| > 1, |\Delta ((KH) \bullet X)| > 1\}.$$

In this case

$$H \bullet \overline{X} = H \bullet A + H \bullet L$$
$$(KH) \bullet \overline{X} = (KH) \bullet A + (KH) \bullet L =$$
$$= K \bullet (H \bullet A) + K \bullet (H \bullet L),$$

where of course the integrals exist. $(KH) \bullet \widetilde{X}$ is again a simple sum, therefore

$$(KH) \bullet X = (KH) \bullet \overline{X} + (KH) \bullet \widetilde{X} =$$
$$= K \bullet (H \bullet A) + K \bullet (H \bullet L) + K \bullet \left(H \bullet \widetilde{X} \right) =$$
$$= K \bullet \left(H \bullet A + H \bullet L + H \bullet \widetilde{X} \right) =$$
$$= K \bullet \left(H \bullet \left(A + L + \widetilde{X} \right) \right) = K \bullet (H \bullet X).$$

$\square$

4.4.4   Change of measure

In this subsection we discuss the behaviour of the stochastic integral when we change the measure on the underlying probability space.

Definition 4.53 *Let* $\mathbf{P}$ *and* $\mathbf{Q}$ *be two probability measures on a measure space* $(\Omega, \mathcal{A})$. *Let us fix a filtration* $\mathcal{F}$. *If* $\mathbf{Q}$ *is absolutely continuous with respect to* $\mathbf{P}$ *on the measure space* $(\Omega, \mathcal{F}_t)$ *for every* t *then we say that* $\mathbf{Q}$ *is locally absolutely continuous with respect to* $\mathbf{P}$. *In this case we shall use the notation* $\mathbf{Q} \overset{loc}{\ll} \mathbf{P}$.

If $\mathbf{Q} \overset{loc}{\ll} \mathbf{P}$ then one can define the Radon–Nikodym derivatives

$$\Lambda(t) \overset{\circ}{=} \frac{d\mathbf{Q}(t)}{d\mathbf{P}(t)}$$

where $\mathbf{Q}(t)$ is the restriction of $\mathbf{Q}$ and $\mathbf{P}(t)$ is the restriction of $\mathbf{P}$ to $\mathcal{F}_t$. If $s < t$ and $F \in \mathcal{F}_s$ then

$$\int_F \Lambda(t)\, d\mathbf{P} \overset{\circ}{=} \int_F \frac{d\mathbf{Q}(t)}{d\mathbf{P}(t)} d\mathbf{P} = \mathbf{Q}(t)(F) =$$

$$= \mathbf{Q}(s)(F) = \int_F \frac{d\mathbf{Q}(s)}{d\mathbf{P}(s)} d\mathbf{P} \overset{\circ}{=} \int_F \Lambda(s)\, d\mathbf{P}.$$

If filtration $\mathcal{F}$ satisfies the usual conditions then process Λ has a modification which is a martingale. As $\Lambda(t)$ is defined up to a set with measure-zero one can assume that the *Radon–Nikodym process* Λ is a martingale.

Lemma 4.54 *If $\mathbf{Q} \overset{loc}{\ll} \mathbf{P}$ and σ is a bounded stopping time then $\Lambda(\sigma)$ is the Radon–Nikodym derivative $d\mathbf{Q}/d\mathbf{P}$ on the σ-algebra $\mathcal{F}_\sigma$. If Λ is uniformly integrable then this is true for any stopping time σ.*

Proof. If σ is a bounded stopping time and $\sigma \leq t$ then by the Optional Sampling Theorem, since Λ is a martingale

$$\Lambda(\sigma) = \mathbf{E}(\Lambda(t) \mid \mathcal{F}_\sigma).$$

That is if $F \in \mathcal{F}_\sigma \subseteq \mathcal{F}_t$ then

$$\int_F \Lambda(\sigma)\, d\mathbf{P} = \int_F \Lambda(t)\, d\mathbf{P} = \mathbf{Q}(t)(F) = \mathbf{Q}(F).$$

$\square$

As Λ is not always a uniformly integrable martingale[73] the lemma is not valid for arbitrary stopping time σ. Since Λ is non-negative $\Lambda(t) \overset{a.s.}{\to} \Lambda(\infty)$, where $\Lambda(\infty) \geq 0$ is an integrable[74] variable. By Fatou's lemma

$$\Lambda(t) = \mathbf{E}(\Lambda(N) \mid \mathcal{F}_t) = \liminf_{N \to \infty} \mathbf{E}(\Lambda(N) \mid \mathcal{F}_t) \geq$$

$$\geq \mathbf{E}\left(\liminf_{N \to \infty} \Lambda(N) \mid \mathcal{F}_t\right) = \mathbf{E}(\Lambda(\infty) \mid \mathcal{F}_t).$$

Hence the extended process is a non-negative, integrable supermartingale on $[0, \infty]$. By the Optional Sampling Theorem for Submartingales[75] if $\sigma \leq \tau$ are

---

[73]See: Example 6.34, page 384.

[74]See: Corollary 1.66, page 40.

[75]See: Proposition 1.88, page 54.

arbitrary stopping times then

$$\Lambda\left(\sigma\right) \geq \mathbf{E}\left(\Lambda\left(\tau\right) \mid \mathcal{F}_{\sigma}\right). \tag{4.15}$$

Let us introduce the stopping time

$$\tau \overset{\circ}{=} \inf\left\{t : \Lambda\left(t\right) = 0\right\}.$$

Let L be a local martingale and let

$$U \overset{\circ}{=} \Delta L\left(\tau\right)\chi\left(\left[\tau, \infty\right)\right).$$

As L is a local martingale $U \in \mathcal{A}_{\mathrm{loc}}$. So U has a compensator U^p. With this notation we have the following theorem:

Proposition 4.55 *Let* $\mathbf{Q} \overset{\mathrm{loc}}{\ll} \mathbf{P}$. *If*

$$\Lambda\left(t\right) \overset{\circ}{=} \frac{d\mathbf{Q}\left(t\right)}{d\mathbf{P}\left(t\right)}$$

then Λ^{-1} *is meaningful and right-regular*[76] *under* $\mathbf{Q}$. *If* L *is a local martingale under measure* $\mathbf{P}$ *then the integral* $\Lambda^{-1} \bullet [L, \Lambda]$ *has finite variation on compact intervals under* $\mathbf{Q}$ *and*

$$\widehat{L} \overset{\circ}{=} L - \Lambda^{-1} \bullet [L, \Lambda] + U^p$$

is a local martingale[77] *under measure* $\mathbf{Q}$.

Proof. We divide the proof into several steps.

1. First we show that $\Lambda > 0$ almost surely under $\mathbf{Q}$. Let

$$\tau \overset{\circ}{=} \inf\left\{t : \Lambda\left(t\right) = 0\right\}.$$

Λ is right-continuous so if $\tau\left(\omega\right) < \infty$ then $\Lambda\left(\tau\left(\omega\right), \omega\right) = 0$. If $0 \leq q \in \mathbb{Q}$ then $\tau + q \geq \tau$. Hence by (4.15)

$$\Lambda\left(\tau\right)\chi\left(\tau < \infty\right) \geq \chi\left(\tau < \infty\right) \cdot \mathbf{E}\left(\Lambda\left(\tau + q\right) \mid \mathcal{F}_{\tau}\right) =$$
$$= \mathbf{E}\left(\Lambda\left(\tau + q\right)\chi\left(\tau < \infty\right) \mid \mathcal{F}_{\tau}\right).$$

Taking expected value

$$0 \geq \mathbf{E}\left(\Lambda\left(\tau + q\right)\chi\left(\tau < \infty\right)\right) \geq 0.$$

---

[76]That is Λ^{-1} is almost surely finite and right-regular with respect to $\mathbf{Q}$, that is $\Lambda \overset{a.s.}{>} 0$ with respect to $\mathbf{Q}$. In this case $\Lambda^{-1} \overset{a.s}{=} \Lambda^{\ominus}$ under $\mathbf{Q}$. See: (4.18).

[77]More precisely $\widehat{L}$ is indistinguishable from a local martingale under $\mathbf{Q}$.

Hence $\Lambda\,(\tau+q)\stackrel{a.s.}{=}0$ on the set $\{\tau<\infty\}$ for any $q\in\mathbb{Q}$. As Λ is right-continuous, outside a set with $\mathbf{P}$-measure-zero if $\tau\,(\omega)\le t<\infty$ then $\Lambda\,(t,\omega)=0$.

$$\mathbf{Q}\,(t)\,(\{\Lambda\,(t)=0\})=\int_{\{\Lambda(t)=0\}}\frac{d\mathbf{Q}}{d\mathbf{P}}\,(t)\,d\mathbf{P}=\int_{\{\Lambda(t)=0\}}\Lambda\,(t)\,d\mathbf{P}=0,$$

so $\Lambda\,(t)>0$ almost surely with respect to $\mathbf{Q}\,(t)$.

$$\mathbf{Q}\,(\Lambda\,(t)=0\text{ for some }t)=\mathbf{Q}\,(\tau<\infty)=\mathbf{Q}\,(\cup_n\Lambda\,(n)=0)\le$$

$$\le\sum_{n=1}^{\infty}\mathbf{Q}\,(\Lambda\,(n)=0)=\sum_{n=1}^{\infty}\mathbf{Q}\,(n)\,(\Lambda\,(n)=0)=0.$$

Hence Λ^{-1} is meaningful and $\Lambda^{-1}>0$ almost surely under $\mathbf{Q}$. We prove that Λ_{-} is also almost surely positive with respect to $\mathbf{Q}$. Let

$$\rho\stackrel{\circ}{=}\inf\{t:\Lambda_{-}\,(t)=0\},\qquad\qquad(4.16)$$

$$\rho_n\stackrel{\circ}{=}\inf\left\{t:\Lambda\,(t)\le\frac{1}{n}\right\}.$$

As Λ is right-regular $\Lambda\,(\rho_n)\le 1/n$. Obviously on the set $\{\rho<\infty\}$

$$\lim_{n\to\infty}\Lambda\,(\rho_n)=\Lambda\,(\rho-)=0.$$

By (4.15) for any positive rational number q

$$\Lambda\,(\rho_n)\,\chi\,(\rho_n<\infty)\ge\mathbf{E}\,\big(\Lambda\,(\rho_n+q)\,\chi\,(\rho_n<\infty)\mid\mathcal{F}_{\rho_n}\big).$$

Taking expected value

$$\frac{1}{n}\ge\mathbf{E}\,(\Lambda\,(\rho_n+q)\,\chi\,(\rho_n<\infty))\ge 0.$$

By Fatou's lemma

$$\mathbf{E}\,(\Lambda\,((\rho+q)-)\,\chi\,(\rho<\infty))=0.$$

Hence for every $q\ge 0$

$$\Lambda\,((\rho+q)-)\,\chi\,(\rho<\infty)\stackrel{a.s}{=}0.$$

Hence outside a set with **P**-measure-zero if $\rho(\omega) \leq t < \infty$ then $\Lambda_-(t, \omega) = 0$. Hence if $\rho(\omega) < t < \infty$ then $\Lambda(t, \omega) = 0$. Therefore $\tau(\omega) \leq \rho(\omega)$.

$$\mathbf{Q}(t)(\{\Lambda_-(t) = 0\}) \leq \mathbf{Q}(t)(\{\rho \leq t\}) = \int_{\{\rho \leq t\}} \Lambda(t) \, d\mathbf{P} \leq$$

$$\leq \int_{\{\tau \leq t\}} \Lambda(t) \, d\mathbf{P} = 0.$$

With the same argument as above one can easily prove that

$$\mathbf{Q}(\Lambda_-(t) = 0 \text{ for some } t) = 0.$$

If for some ω the trajectory $\Lambda(\omega)$ and $\Lambda_-(\omega)$ are positive then as $\Lambda(\omega)$ is right-regular $\Lambda^{-1}(\omega)$ is also right-regular. Therefore it is bounded on any finite interval[78]. Hence if $V \in \mathcal{V}$ then $\Lambda^{-1} \bullet V$ is well-defined and $\Lambda^{-1} \bullet V \in \mathcal{V}$ under **Q**.

2. Assume that for some right-regular, adapted process N the product $N\Lambda$ is a local martingale under **P**. We show that N is a local martingale under **Q**. Let σ be a stopping time and let us assume that the truncated process $(\Lambda N)^\sigma$ is a martingale under **P**. If $F \in \mathcal{F}_{\sigma \wedge t}$, and $r \geq t$, then[79]

$$\int_F N^\sigma(t) \, d\mathbf{Q} = \int_F N^\sigma(t) \Lambda^\sigma(t) \, d\mathbf{P} =$$

$$= \int_F N^\sigma(r) \Lambda^\sigma(r) \, d\mathbf{P} = \int_F N^\sigma(r) \, d\mathbf{Q}.$$

Hence N^σ is a martingale under **Q** with respect to the filtration $(\mathcal{F}_{\sigma \wedge t})_t$. We show that it is a martingale under **Q** with respect to the filtration $\mathcal{F}$. Let ρ be a bounded stopping time under $\mathcal{F}$. We show that $\tau \overset{\circ}{=} \rho \wedge \sigma$ is a stopping time under $(\mathcal{F}_{\sigma \wedge t})_t$. One should show that $\{\rho \wedge \sigma \leq t\} \in \mathcal{F}_{\sigma \wedge t}$. By definition this means that

$$\{\rho \wedge \sigma \leq t\} \cap \{\sigma \wedge t \leq r\} \in \mathcal{F}_r.$$

If $t \leq r$ then this is true as $\rho \wedge \sigma$ and $\sigma \wedge t$ are stopping times. If $t > r$ then the set above is $\{\sigma \leq r\} \in \mathcal{F}_r$. By the Optional Sampling Theorem, using that $\tau \overset{\circ}{=} \rho \wedge \sigma$ is a stopping time under $(\mathcal{F}_{\sigma \wedge t})_t$ and N^σ is a **Q**-martingale under this filtration

$$\int_\Omega N^\sigma(0) \, d\mathbf{Q} = \int_\Omega N^\sigma(\tau) \, d\mathbf{Q} = \int_\Omega N^\sigma(\rho) \, d\mathbf{Q}.$$

[78] See: Proposition 1.6, page 5.
[79] See: Lemma 4.54, page 265.

This implies that N^σ is a martingale under $\mathbf{Q}$. Hence N is a local martingale under $\mathbf{Q}$.

3. To simplify the notation let $L(0) = 0$, from which $\widehat{L}(0) = 0$. Integrating by parts

$$L\Lambda = L_- \bullet \Lambda + \Lambda_- \bullet L + [L, \Lambda]. \tag{4.17}$$

Λ and L are local martingales under $\mathbf{P}$ so the stochastic integrals on the right-hand side are local martingales under $\mathbf{P}$. Let

$$a^\ominus \overset{\circ}{=} \begin{cases} a^{-1} & \text{if} \quad a > 0 \\ 0 & \text{if} \quad a = 0 \end{cases}. \tag{4.18}$$

and let

$$A \overset{\circ}{=} \Lambda^\ominus \bullet [L, \Lambda]. \tag{4.19}$$

A is almost surely finite under $\mathbf{Q}$ as $\Lambda > 0$ and Λ_- are almost surely finite under $\mathbf{Q}$. But we are now defining A under $\mathbf{P}$ and with positive probability $\Lambda^\ominus$ can be unbounded on some finite intervals under $\mathbf{P}$. Hence we do not know that A is well-defined under $\mathbf{P}$. To solve this problem let us observe that (ρ_n) in (4.16) is a localizing sequence under $\mathbf{Q}$ and one can localize $\widehat{L}$. So it is sufficient to prove that $\widehat{(L^{\rho_n})} = (\widehat{L})^{\rho_n}$ is a local martingale under $\mathbf{Q}$ for every n. For L^{ρ_n} (4.19) is well-defined. So one can assume that A is finite. Again integrating by parts, noting that Λ is right-continuous

$$\begin{aligned} \Lambda A &= A_- \bullet \Lambda + \Lambda_- \bullet A + [A, \Lambda] = \\ &= A_- \bullet \Lambda + \Lambda_- \bullet A + \sum \Delta A \Delta \Lambda = \\ &= A_- \bullet \Lambda + \Lambda_- \bullet A + \Delta \Lambda \bullet A = \\ &= A_- \bullet \Lambda + \Lambda \bullet A = \\ &= A_- \bullet \Lambda + \Lambda \Lambda^\ominus \bullet [L, \Lambda] = \\ &= A_- \bullet \Lambda + \chi(\Lambda > 0) \bullet [L, \Lambda]. \end{aligned}$$

Finally[80]

$$\begin{aligned} \Lambda U^p &= U^p_- \bullet \Lambda + \Lambda_- \bullet U^p + [U^p, \Lambda] = \\ &= U^p_- \bullet \Lambda + \Lambda_- \bullet U^p + \sum \Delta U^p \Delta \Lambda = \\ &= U^p_- \bullet \Lambda + \Lambda_- \bullet U^p + \Delta U^p \bullet \Lambda = \end{aligned}$$

[80]See: Example 4.39, page 249.

$$= U^p \bullet \Lambda + \Lambda_- \bullet U^p =$$
$$= U^p \bullet \Lambda + \Lambda_- \bullet U^p \pm \Lambda_- \bullet U =$$
$$= U^p \bullet \Lambda + \Lambda_- \bullet (U - U^p) + \Lambda_- \bullet U$$

The stochastic integrals with respect to local martingales are local martingales, the sum of local martingales is a local martingale so

$$\Lambda \widehat{L} \stackrel{\circ}{=} \Lambda L - \Lambda A + \Lambda U^p =$$
$$= \text{local martingale} + [L, \Lambda] - \chi (\Lambda > 0) \bullet [L, \Lambda] + \Lambda_- \bullet U.$$

Observe that the last line is

$$\chi (\Lambda = 0) \bullet [L, \Lambda] + \Lambda_- \bullet U =$$
$$= \chi (t \geq \tau) \bullet [L, \Lambda] + \Lambda_- (\tau) \Delta L (\tau) \chi (t \geq \tau) =$$
$$= \chi (t \geq \tau) \Delta L (\tau) \Delta \Lambda (\tau) + \Lambda_- (\tau) \Delta L (\tau) \chi (t \geq \tau) = 0$$

where we have used that $[L, \Lambda]$ is constant[81] on $\{t \geq \tau\}$. Hence $\Lambda \widehat{L}$ is a local martingale under $\mathbf{P}$. So by the second part of the proof $\widehat{L}$ is a local martingale under $\mathbf{Q}$. $\qquad \square$

Corollary 4.56 *Let* $\mathbf{Q} \overset{loc}{\ll} \mathbf{P}$ *and let* $\mathbf{P} \overset{loc}{\ll} \mathbf{Q}$ *that is let assume that* $\mathbf{Q} \overset{loc}{\sim} \mathbf{P}$. *If*

$$\Lambda (t) \stackrel{\circ}{=} \frac{d\mathbf{Q} (t)}{d\mathbf{P} (t)}$$

then $\Lambda > 0$. *If* L *is a local martingale under measure* $\mathbf{P}$ *then the integral* $\Lambda^{-1} \bullet [L, \Lambda]$ *has finite variation on compact intervals under* $\mathbf{Q}$ *and*

$$\widehat{L} \stackrel{\circ}{=} L - \Lambda^{-1} \bullet [L, \Lambda]$$

is a local martingale under the measure $\mathbf{Q}$.

Corollary 4.57 *Let* $\mathbf{Q} \overset{loc}{\ll} \mathbf{P}$. *If*

$$\Lambda (t) \stackrel{\circ}{=} \frac{d\mathbf{Q} (t)}{d\mathbf{P} (t)}$$

and L *is a continuous local martingale under measure* $\mathbf{P}$ *then the integral* $\Lambda^{-1} \bullet [L, \Lambda]$ *has finite variation on compact intervals under measure* $\mathbf{Q}$ *and*

$$\widehat{L} \stackrel{\circ}{=} L - \Lambda^{-1} \bullet [L, \Lambda]$$

is a local martingale under the measure $\mathbf{Q}$.

---

[81]See: Corollary 2.49, page 145.

If $V \in \mathcal{V}$ under $\mathbf{P}$ and $\mathbf{Q} \overset{\text{loc}}{\ll} \mathbf{P}$ then obviously $V \in \mathcal{V}$ under $\mathbf{Q}$. Hence the proof of the following observation is trivial:

Corollary 4.58 *If X is a semimartingale under $\mathbf{P}$ and $\mathbf{Q} \overset{\text{loc}}{\ll} \mathbf{P}$ then X is a semimartingale under $\mathbf{Q}$.*

Let $V \in \mathcal{V}$ and assume that the integral $H \bullet V$ exists under measure $\mathbf{P}$. By definition this means that the pathwise integrals $(H \bullet V)(\omega)$ exist almost surely under $\mathbf{P}$. If $\mathbf{Q} \overset{\text{loc}}{\ll} \mathbf{P}$ then the integral $H \bullet V$ exists under the measure $\mathbf{Q}$ as well, and the value of the two processes are almost surely the same under $\mathbf{Q}$. It is not too surprising that it is true for any semimartingale.

Proposition 4.59 *Let X be an arbitrary semimartingale and let H be a predictable process. Assume that the integral $H \bullet X$ exists under measure $\mathbf{P}$. If $\mathbf{Q} \overset{\text{loc}}{\ll} \mathbf{P}$ then the integral $H \bullet X$ exists under measure $\mathbf{Q}$ as well, and the two integral processes are indistinguishable under measure $\mathbf{Q}$.*

Proof. By the remark above it is obviously sufficient to prove the proposition if $X \in \mathcal{L}$ under $\mathbf{P}$. It is also sufficient to prove that for every $T > 0$ the two integrals exist on the interval $[0, T]$ and they are almost surely equal.

1. Let $X = X^c + X^d$ be the decomposition of X into continuous and purely discontinuous local martingales. As the time horizon is finite, Λ is a uniformly integrable martingale. Recall that if L is a local martingale under the measure $\mathbf{P}$ then

$$\widetilde{L} \overset{\circ}{=} L - \Lambda^{-1} \bullet [L, \Lambda] + U^p \tag{4.20}$$

is a local martingale under measure $\mathbf{Q}$ and if L is continuous then U^p can be dropped.

$$\widetilde{X} \overset{\circ}{=} X - \Lambda^{-1} \bullet [X, \Lambda] + U^p =$$
$$= X^c + X^d - \Lambda^{-1} \bullet [X^c + X^d, \Lambda] + U^p =$$
$$= \left(X^c - \Lambda^{-1} \bullet [X^c, \Lambda]\right) + \left(X^d - \Lambda^{-1} \bullet [X^d, \Lambda] + U^p\right).$$

By (4.20) the processes

$$\widetilde{X^c} \overset{\circ}{=} X^c - \Lambda^{-1} \bullet [X^c, \Lambda] \quad \text{and} \quad \widetilde{X^d} \overset{\circ}{=} X^d - \Lambda^{-1} \bullet [X^d, \Lambda] + U^p$$

are local martingales under measure $\mathbf{Q}$. X^c is continuous, hence the quadratic co-variation $[X^c, \Lambda]$ is also continuous[82]. Hence $\widetilde{X^c}$ is continuous. If W and V

---

[82]See: line (3.19), page 222.

are pure quadratic jump processes then

$$[W + V] = [W] + 2[W, V] + [V] =$$
$$= \sum (\Delta W)^2 + 2 \sum \Delta W \Delta V + \sum (\Delta V)^2 =$$
$$= \sum (\Delta (W + V))^2$$

hence $W + V$ is also a pure quadratic jump process. Processes with finite variation are pure quadratic jump processes[83], hence $\widetilde{X^d}$ is a pure quadratic jump process under $\mathbf{P}$. Under the change of measure the quadratic variation does not change, hence $\widetilde{X^d}$ is a pure quadratic jump process under $\mathbf{Q}$. Hence $\widetilde{X^d}$ is a purely discontinuous local martingale under $\mathbf{Q}$. We want to show that $H \bullet \widetilde{X}$ exists under $\mathbf{Q}$. This means that $H \bullet \widetilde{X}$ exist on $(0, t]$ for every t. To prove this one need only prove that the integrals $H \bullet \widetilde{X^c}$ and $H \bullet \widetilde{X^d}$ exist under $\mathbf{Q}$.

2. $\widetilde{X^c}$ is a continuous local martingale, hence $H \bullet \widetilde{X^c}$ exists under $\mathbf{Q}$ if and only if $\mathbf{Q}\left(H^2 \bullet \left[\widetilde{X^c}\right] < \infty\right) = 1$. $H \bullet X$ exists under $\mathbf{P}$ therefore $\mathbf{P}\left(H^2 \bullet [X^c] < \infty\right) = 1$. As $\Lambda^{-1} \bullet [X^c, \Lambda]$ is continuous by quick calculation

$$[\widetilde{X^c}] \overset{\circ}{=} [X^c - \Lambda^{-1} \bullet [X^c, \Lambda]] = [X^c].$$

Therefore

$$\mathbf{P}\left(H^2 \bullet \left[\widetilde{X^c}\right] < \infty\right) = \mathbf{P}\left(H^2 \bullet [X^c] < \infty\right) = 1,$$

that is $\mathbf{P}(H^2 \bullet [\widetilde{X^c}] = \infty) = 0$. $\mathbf{Q} \ll \mathbf{P}$ on $\mathcal{F}_t$ so $\mathbf{Q}((H^2 \bullet [\widetilde{X^c}])(t) = \infty) = 0$ for every t, that is $\mathbf{Q}((H^2 \bullet [\widetilde{X^c}]) = \infty) = 0$. So $H \bullet \widetilde{X^c}$ exists under $\mathbf{Q}$.

3. $\widetilde{X^d}$ is purely discontinuous, hence $H \bullet \widetilde{X^d}$ exists under $\mathbf{Q}$ if and only if

$$Z \overset{\circ}{=} \sqrt{H^2 \bullet \left[\widetilde{X^d}\right]} \in \mathcal{A}_{\text{loc}}^+$$

under measure $\mathbf{Q}$. Z is obviously increasing, so we need only prove that $Z \in \mathcal{A}_{\text{loc}}^+$.

4. Let us prove the following general observation: if Λ is a non-negative martingale, τ is an arbitrary stopping time and $\Lambda \leq c$ on $[0, \tau)$ then $\mathbf{E}\left(\chi(\tau > 0)\Lambda(\tau)\right) \leq c$.

---

[83]See: line (2.14), page 134.

Let $M \overset{\circ}{=} \Lambda^{\tau}$. By Lévy's theorem

$$M(t-) = \mathbf{E}(M(t) \mid \mathcal{F}_{t-}).$$

Hence as $\{\tau < t\} \in \mathcal{F}_{t-}$

$$
\begin{aligned}
\mathbf{E}(\Lambda(t \wedge \tau)) \overset{\circ}{=} \mathbf{E}(M(t)) &= \mathbf{E}(M(t-)) = \\
&= \mathbf{E}(M(t-)\chi(\tau \geq t)) + \mathbf{E}(M(t-)\chi(\tau < t)) = \\
&= \mathbf{E}(\Lambda(t-)\chi(\tau \geq t)) + \mathbf{E}(M(t)\chi(\tau < t)) = \\
&= \mathbf{E}(\Lambda(t-)\chi(\tau \geq t)) + \mathbf{E}(\Lambda(\tau)\chi(\tau < t)).
\end{aligned}
$$

Hence as $\Lambda \geq 0$

$$\mathbf{E}(\Lambda(\tau)\chi(\tau < t)) \leq \mathbf{E}(\Lambda(t \wedge \tau)) < \infty.$$

So by the Optional Sampling Theorem for non-negative martingales[84]

$$
\begin{aligned}
\mathbf{E}(\Lambda(\tau)\chi(\tau \geq t)) = \mathbf{E}(\Lambda(\tau)) - \mathbf{E}(\Lambda(\tau)\chi(\tau < t)) &\leq \\
= \mathbf{E}(\Lambda(\tau \wedge t)) - \mathbf{E}(\Lambda(\tau)\chi(\tau < t)) &= \\
= \mathbf{E}(\Lambda(t-)\chi(\tau \geq t)) &\leq c.
\end{aligned}
$$

If $t \searrow 0$ then by Fatou's lemma $\mathbf{E}(\chi(\tau > 0)\Lambda(\tau)) \leq c$.

5. Z_- is locally bounded. Let (ρ_n) be a localizing sequence of Z_-. Let

$$\tau_n \overset{\circ}{=} \inf\{s : \Lambda(s) > n\} \wedge \rho_n \wedge n. \tag{4.21}$$

τ_n is a bounded stopping time and if $s < \tau_n(\omega)$ then $\Lambda(s, \omega) \leq n$. Hence using the estimate just proved

$$
\begin{aligned}
\mathbf{E}^{\mathbf{Q}}(Z(\tau_n-)) = \mathbf{E}\left(Z(\tau_n-)\frac{d\mathbf{Q}}{d\mathbf{P}}\right) &= \\
= \mathbf{E}\left(\mathbf{E}\left(Z(\tau_n-)\frac{d\mathbf{Q}}{d\mathbf{P}} \mid \mathcal{F}_{\tau_n}\right)\right) &= \\
= \mathbf{E}\left(Z(\tau_n-)\mathbf{E}\left(\frac{d\mathbf{Q}}{d\mathbf{P}} \mid \mathcal{F}_{\tau_n}\right)\right) &= \\
= \mathbf{E}(Z(\tau_n-)\Lambda(\tau_n)) \leq k_n \cdot \mathbf{E}(\Lambda(\tau_n)) &= \\
= k_n \cdot \mathbf{E}(\{\tau_n > 0\}\Lambda(\tau_n) + \{\tau_n = 0\}\Lambda(\tau_n)) &\leq \\
\leq k_n \cdot (n + \mathbf{E}(\Lambda(0))) &< \infty.
\end{aligned}
$$

[84]See: Corollary 1.87, page 54.

6. We show that $\Delta U^p = 0$. The stopping time τ can be covered by its predictable and totally inaccessible parts so one can assume that τ is either totally inaccessible or predictable. If τ is predictable then $\chi([\tau])$ is predictable therefore

$$\Delta(U^p) = {}^p(\Delta U) \stackrel{\circ}{=} {}^p(\Delta X(\tau)\chi([\tau])) = {}^p(\Delta X \cdot \chi([\tau])) =$$
$$= ({}^p\Delta X) \cdot \chi([\tau]) = 0 \cdot \chi([\tau]) = 0.$$

If τ is totally inaccessible then $\mathbf{P}(\tau = \sigma) = 0$ for every predictable stopping time σ, hence

$$ {}^p(\Delta X\chi([\tau]))(\sigma) \stackrel{\circ}{=} \widehat{\mathbf{E}}((\Delta X\chi([\tau]))(\sigma) \mid \mathcal{F}_{\sigma-}) = \widehat{\mathbf{E}}(0 \mid \mathcal{F}_{\sigma-}) = 0, $$

so $\Delta U^p = {}^p(\Delta X\chi([\tau])) = 0$. Therefore in both cases $\Delta U^p = 0$.

7. $\widetilde{X^d}$ is purely discontinuous, hence $\left[\widetilde{X^d}\right] = \sum \left(\Delta \widetilde{X^d}\right)^2$ and

$$\Delta \widetilde{X^d} = \Delta X^d - \Lambda^\ominus \Delta\left[X^d, \Lambda\right] + \Delta U^p.$$

Since $\Delta U^p = 0$

$$\left|\Delta \widetilde{X^d}\right| = \left|\Delta X^d - \Lambda^\ominus \cdot \left(\Delta X^d \Delta\Lambda\right)\right| =$$
$$= \left|\Delta X^d\left(1 - \Lambda^\ominus \cdot \Delta\Lambda\right)\right| =$$
$$= \left|\Delta X^d\left(\chi(\Lambda = 0) + \Lambda^\ominus \cdot \Lambda_-\right)\right|.$$

$\sqrt{H^2 \bullet [X^d]} \in \mathcal{A}_{\mathrm{loc}}$ under $\mathbf{P}$. One can assume that τ_n localizes $\sqrt{H^2 \bullet [X^d]}$ in (4.21). Therefore one may assume that

$$\mathbf{E}\left(\left|\Delta X^d(\tau_n) H(\tau_n)\right|\right) \leq \mathbf{E}\left(\sqrt{H^2 \bullet [X^d]}(\tau_n)\right) < \infty.$$

Using this

$$\mathbf{E}^Q\left(\sqrt{\Delta\left(H^2 \bullet \left[\widetilde{X^d}\right]\right)(\tau_n)}\right) =$$
$$= \mathbf{E}\left(\sqrt{H^2\left(\Delta\widetilde{X^d}\right)^2(\tau_n)}\frac{d\mathbf{Q}}{d\mathbf{P}}\right) = \mathbf{E}\left(\left|H\Delta\widetilde{X^d}(\tau_n)\right|\Lambda(\tau_n)\right) =$$
$$= \mathbf{E}\left(\left|H\Delta X^d(\tau_n)\right|\left|\chi(\Lambda = 0) + \Lambda_-\Lambda^\ominus\right|(\tau_n)\Lambda(\tau_n)\right) =$$
$$= \mathbf{E}\left(\left|H\Delta X^d(\tau_n)\right|\left|\Lambda_-\Lambda^\ominus\right|(\tau_n)\Lambda(\tau_n)\right) \leq$$
$$\leq \mathbf{E}\left(\left|H(\tau_n)\Delta X^d(\tau_n)\right|\Lambda(\tau_n-)\right) \leq n \cdot \mathbf{E}\left(\left|\Delta X^d(\tau_n)H(\tau_n)\right|\right) < \infty.$$

8. As $\sqrt{x + y} \leq \sqrt{x} + \sqrt{y}$

$$\mathbf{E}^{\mathbf{Q}}\left(Z\left(\tau_n\right)\right) \stackrel{\circ}{=} \mathbf{E}^{\mathbf{Q}}\left(\sqrt{H^2 \bullet \left[\widetilde{X^d}\right]\left(\tau_n\right)}\right) \leq$$

$$\leq \mathbf{E}^{\mathbf{Q}}\left(Z\left(\tau_n-\right)\right) + \mathbf{E}^{\mathbf{Q}}\left(\sqrt{\Delta\left(H^2 \bullet \left[\widetilde{X^d}\right]\right)\left(\tau_n\right)}\right) < \infty.$$

Therefore $Z \in \mathcal{A}_{\mathrm{loc}}$ under measure $\mathbf{Q}$.

9. Let us consider the decomposition

$$X = \widetilde{X} + \Lambda^{-1} \bullet [X, \Lambda] - U^p \stackrel{\circ}{=} \widetilde{X} + A - U^p$$

and let us assume that the integral $H \bullet X$ exists under measure $\mathbf{P}$. As the integral $H \bullet \widetilde{X}$ exists under $\mathbf{Q}$ one should prove that the Lebesgue–Stieltjes integrals $H \bullet A$ and $H \bullet U^p$ also exist. By the inequality of Kunita and Watanabe

$$\int_0^T |H| \, d\mathrm{Var}\left(A\right) = \int_0^T |H| \, \Lambda^{\ominus} d\mathrm{Var}\left([X, \Lambda]\right) \leq$$

$$\leq \sqrt{\int_0^T |H|^2 \, \Lambda^{\ominus} d\left[X\right]} \sqrt{\int_0^T \Lambda^{\ominus} d\left[\Lambda\right]} =$$

$$= \sqrt{\int_0^T \Lambda^{\ominus} d\left(|H|^2 \bullet [X]\right)} \sqrt{\int_0^T \Lambda^{\ominus} d\left[\Lambda\right]}.$$

$\Lambda > 0$ and $\Lambda_- > 0$ almost surely under $\mathbf{Q}$, that is almost all trajectories of Λ and Λ_- are positive[85] hence $\Lambda^{\ominus}$ has regular trajectories almost surely under $\mathbf{Q}$. Hence almost surely the trajectories of $\Lambda^{\ominus}$ are bounded on every finite interval, therefore the expression $\sqrt{\int_0^T \Lambda^{\ominus} d\left[\Lambda\right]}$ is finite. Similarly as $H \bullet X$ exists $R \stackrel{\circ}{=} |H|^2 \bullet [X] \in \mathcal{V}$, hence $\Lambda^{\ominus} \bullet R$ is finite under $\mathbf{Q}$. That is for every trajectory $\int_0^T |H| \, d\mathrm{Var}\left(A\right) < \infty$, hence $H \bullet A$ exists under $\mathbf{Q}$. Let σ be a stopping time in a localizing sequence of $\sqrt{H^2 \bullet [X]}$.

$$\mathbf{E}\left(\left(|H| \bullet U^p\right)\left(\sigma\right)\right) = \mathbf{E}\left(\left(|H| \bullet U\right)\left(\sigma\right)\right) \leq \mathbf{E}\left(\sqrt{|H|^2 \bullet [X]}\left(\sigma\right)\right) < \infty.$$

Hence $H \bullet U^p$ is almost surely finite under $\mathbf{P}$ so it is almost surely finite under $\mathbf{Q}$. Therefore the integral $H \bullet X$ exists under $\mathbf{Q}$.

10. Let us denote by $(\mathbf{P}) \, H \bullet X$ and by $(\mathbf{Q}) \, H \bullet X$ the value of $H \bullet X$ under $\mathbf{P}$ and under $\mathbf{Q}$ respectively. Let us denote by $\mathcal{H}$ the set of processes H for

---

[85]See: Proposition 4.55, page 266.

which $(\mathbf{P})\, H \bullet X$ and $(\mathbf{Q})\, H \bullet X$ are indistinguishable under $\mathbf{Q}$. From the Dominated Convergence Theorem and from the linearity of the stochastic integral it is obvious that $\mathcal{H}$ is a λ-system, which contains the π-system of the elementary processes. From the Monotone Class Theorem it is clear the $\mathcal{H}$ contains all the bounded predictable processes.

11. If $H_n \overset{\circ}{=} H\chi\,(|H| \leq n)$ then H_n is bounded. Hence the value of the integral $(\mathbf{P})\, H_n \bullet X$ is $\mathbf{Q}$ almost surely equal to the integral $(\mathbf{Q})\, H_n \bullet X$. As $H \bullet X$ exists under $\mathbf{P}$ and under $\mathbf{Q}$ by the Dominated Convergence Theorem uniformly in probability on compact intervals $(\mathbf{P})\, H_n \bullet X \overset{\mathrm{ucp}}{\rightarrow} (\mathbf{P})\, H \bullet X$ and $(\mathbf{Q})\, H_n \bullet X \overset{\mathrm{ucp}}{\rightarrow} (\mathbf{Q})\, H \bullet X$. The stochastic convergence under $\mathbf{P}$ implies[86] the stochastic convergence under $\mathbf{Q}$, hence $(\mathbf{P})\, H \bullet X = (\mathbf{Q})\, H \bullet X$ almost surely under $\mathbf{Q}$. $\square$

Let us prove some consequences of the proposition. During the construction of the stochastic integral we emphasized that we cannot define the integral pathwise. But it does not mean that the integral is not determined by the trajectories of the integrator and the integrand.

Corollary 4.60 *Let X and $\overline{X}$ be semimartingales. Assume that for the predictable processes H and $\overline{H}$ the integrals $H \bullet X$ and $\overline{H} \bullet \overline{X}$ exist. If*

$$A \overset{\circ}{=} \left\{ \omega : H\,(\omega) = \overline{H}\,(\omega) \right\} \cap \left\{ \omega : X\,(\omega) = \overline{X}\,(\omega) \right\}$$

then the processes $H \bullet X$ and $\overline{H} \bullet \overline{X}$ are indistinguishable on A.

Proof. One may assume that $\mathbf{P}\,(A) > 0$. Define the measure

$$\mathbf{Q}\,(B) \overset{\circ}{=} \frac{\mathbf{P}\,(A \cap B)}{\mathbf{P}\,(A)}.$$

Obviously $\mathbf{Q} \ll \mathbf{P}$. The processes $H, \overline{H}$ and $X, \overline{X}$ are indistinguishable under $\mathbf{Q}$. Hence processes $(\mathbf{Q})\overline{H} \bullet \overline{X}$ and $(\mathbf{Q})\, H \bullet X$ are indistinguishable under $\mathbf{Q}$. By the proposition above under $\mathbf{Q}$ up to indistinguishability

$$(\mathbf{P})\, H \bullet X = (\mathbf{Q})\, H \bullet X = (\mathbf{Q})\, \overline{H} \bullet \overline{X} = (\mathbf{P})\, \overline{H} \bullet \overline{X}$$

which means that $(\mathbf{P})\, H \bullet X = (\mathbf{P})\, \overline{H} \bullet \overline{X}$ on A. $\square$

The proof of the following corollary is similar:

Corollary 4.61 *Let X be a semimartingale and let assume that the integral $H \bullet X$ exists. If on a set B the trajectories of X have finite variation then almost surely on B the trajectories of $H \bullet X$ are equal to the pathwise integrals of H with respect to X.*

---

[86] A sequence is stochastically convergent if and only if every subsequence of the sequence has another subsequence which is almost surely convergent to the same, fixed random variable.

4.5 The Proof of Davis' Inequality

In this section we prove the following inequality:

Theorem 4.62 (Davis' inequality) *There are positive constants c and C such that for any local martingale $L \in \mathcal{L}$ and for any stopping time τ*

$$c \cdot \mathbf{E}\left(\sqrt{[L]}\,(\tau)\right) \leq \mathbf{E}\left(\sup_{t \leq \tau} |L(t)|\right) \leq C \cdot \mathbf{E}\left(\sqrt{[L]}\,(\tau)\right).$$

Example 4.63 In the inequality one cannot write $|L|(\tau)$ in the place of $\sup_{t \leq \tau} |L|$.

If w is a Wiener process and $\tau \stackrel{\circ}{=} \inf\{t : w(t) = 1\}$ then $L \stackrel{\circ}{=} w^\tau$ is a martingale. $\mathbf{E}(L(t)) = 0$ for every t, hence

$$\|L(t)\|_1 = \mathbf{E}(|L(t)|) = 2\mathbf{E}\left(L^+(t)\right) \leq 2.$$

On the other hand if $t \to \infty$

$$\left\|\sqrt{[L]}\,(t)\right\|_1 = \mathbf{E}\left(\sqrt{\tau \wedge t}\right) \to \mathbf{E}\left(\sqrt{\tau}\right).$$

The density function[87] of τ is

$$f(x) = \frac{1}{\sqrt{2x^3\pi}} \exp\left(-\frac{1}{2x}\right), \quad x > 0,$$

hence the expected value of $\sqrt{\tau}$ is

$$\mathbf{E}\left(\sqrt{\tau}\right) = \int_0^\infty \sqrt{x} \frac{1}{\sqrt{2x^3\pi}} \exp\left(-\frac{1}{2x}\right) dx =$$

$$= \int_0^\infty \frac{1}{\sqrt{2\pi}} \frac{1}{x} \exp\left(-\frac{1}{2x}\right) dx =$$

$$= \frac{1}{\sqrt{2\pi}} \int_0^\infty \frac{1}{u} \exp\left(-\frac{u}{2}\right) du = \infty. \qquad \square$$

If σ is an arbitrary stopping time then in place of L one can write L^σ in the inequality. On the other hand if for some localizing sequence $\sigma_n \nearrow \infty$ the inequality is true for all L^{σ_n} then by the Monotone Convergence Theorem it is true for L as well. By the Fundamental Theorem of Local Martingales $L \in \mathcal{L}$ has a decomposition $L = H + A$ where $H \in \mathcal{H}^2_{\mathrm{loc}}$ and $A \in \mathcal{A}_{\mathrm{loc}}$. With localization one can assume that $H \in \mathcal{H}^2$ and $A \in \mathcal{A}$. L_- is left-regular, hence it is locally

---

[87]See: (1.58) on page 83.

bounded, so with further localization of the inequality one can assume that L_- is bounded.

It suffices to prove the inequality on any finite time horizon $[0, T]$. It is sufficient to prove the inequality for finite, discrete-time horizons: If $\left(t_k^{(n)}\right)$ is an infinitesimal sequence of partitions of $[0, T]$ then trivially

$$\mathbf{E}\left(\sup_{t_k^{(n)} \leq T}\left|L\left(t_k^{(n)}\right)\right|\right) \nearrow \mathbf{E}\left(\sup_{t \leq T}|L(t)|\right).$$

Recall that as $L(0) = 0$ at any time t the quadratic variation $[L]$ is the limit in probability of the sequence

$$[L]^{(n)}(t) \doteq \sum_k \left[L\left(t_k^{(n)} \wedge t\right) - L\left(t_{k-1}^{(n)} \wedge t\right)\right]^2 =$$

$$= L^2(t) - 2\sum_k L\left(t_{k-1}^{(n)} \wedge t\right)\left[L\left(t_k^{(n)} \wedge t\right) - L\left(t_{k-1}^{(n)} \wedge t\right)\right].$$

If

$$Y_n(t) \doteq \sum_k L\left(t_{k-1}^{(n)} \wedge t\right) \chi\left(\left(t_{k-1}^{(n)} \wedge t, t_k^{(n)} \wedge t\right]\right),$$

then the sum in the above expression is $(Y_n \bullet L)(t)$. Obviously $Y_n \to L_-$ and

$$|Y_n(t)| \leq \sup_{s \leq t}|L_-(s)| \leq k.$$

Repeating the proof of the Dominated Convergence Theorem we prove that for all t

$$(Y_n \bullet L)(t) \to (L_- \bullet L)(t)$$

in $L^1(\Omega)$. As (Y_n) is uniformly bounded, by Itô's isometry the convergence $Y_n \bullet H \to L_- \bullet H$ holds in $\mathcal{H}^2$ and therefore

$$(Y_n \bullet H)(t) \xrightarrow{L^2} (L_- \bullet H)(t).$$

Obviously

$$|(Y_n \bullet A)(t) - (L_- \bullet A)(t)| \leq 2k \cdot \text{Var}(A)(t).$$

As $A \in \mathcal{A}$ by the classical Dominated Convergence Theorem

$$(Y_n \bullet A)(t) \xrightarrow{L^1} (L_- \bullet A)(t).$$

Therefore, as we said,

$$(Y_n \bullet A)(T) \xrightarrow{L^1} (L_- \bullet A)(T).$$

Hence $[L]^{(n)}(T) \xrightarrow{L^1} [L](T)$, so by Jensen's inequality

$$\left| \mathbf{E}\left(\sqrt{[L]^{(n)}(T)} \right) - \mathbf{E}\left(\sqrt{[L](T)} \right) \right| \leq \mathbf{E}\left(\left| \sqrt{[L]^{(n)}(T)} - \sqrt{[L](T)} \right| \right) \leq$$

$$\leq \mathbf{E}\left(\sqrt{\left| [L]^{(n)}(T) - [L](T) \right|} \right) \leq$$

$$\leq \sqrt{\mathbf{E}\left(\left| [L]^{(n)}(T) - [L](T) \right| \right)} \to 0.$$

This means that if the inequality holds in discrete-time then it is true in continuous-time.

4.5.1 Discrete-time Davis' inequality

Up to the end of this section we assume that if M is a martingale then $M(0) = 0$.

Definition 4.64 *Let us first introduce some notation. For any sequence* $M \stackrel{\circ}{=} (M_n)$

$$\Delta M_n \stackrel{\circ}{=} M_n - M_{n-1}.$$

If $M \stackrel{\circ}{=} (M_n)$ *is a discrete-time martingale then* (ΔM_n) *is the martingale difference of* M.

$$[M]_n \stackrel{\circ}{=} \sum_{k=1}^{n} (\Delta M_k)^2 = \sum_{k=1}^{n} (M_k - M_{k-1})^2$$

$$M_n^* \stackrel{\circ}{=} \sup_{k \leq n} |M_k|$$

for any n. *If* n *is the maximal element in the parameter set or* $n = \infty$ *then we drop the subscript* n.

With this notation the discrete-time Davis' inequality has the following form:

Theorem 4.65 (Discrete-time Davis' inequality) *There are positive constants* c *and* C *such that for every discrete-time martingale* M *for which* $M(0) = 0$

$$c \cdot \mathbf{E}\left(\sqrt{[M]} \right) \leq \mathbf{E}(M^*) \leq C \cdot \mathbf{E}\left(\sqrt{[M]} \right).$$

The proof of the discrete-time Davis' inequality is a simple but lengthy[88] calculation. Let us first prove two lemmas:

Lemma 4.66 *Let $M \stackrel{\circ}{=} (M_n, \mathcal{F}_n)$ be a martingale and let $V \stackrel{\circ}{=} (V_n, \mathcal{F}_{n-1})$ be a predictable sequence[89], for which*

$$|\Delta M_n| \stackrel{\circ}{=} |M_n - M_{n-1}| \leq V_n.$$

If $\lambda > 0$ and $0 < \delta < \beta - 1$ then

$$\mathbf{P}\left(M^* > \beta\lambda, \sqrt{[M]} \vee V^* \leq \delta\lambda\right) \leq \frac{2\delta^2}{(\beta - \delta - 1)^2}\mathbf{P}\left(M^* > \lambda\right),$$

$$\mathbf{P}\left(\sqrt{[M]} > \beta\lambda, M^* \vee V^* \leq \delta\lambda\right) \leq \frac{9\delta^2}{\beta^2 - \delta^2 - 1}\mathbf{P}\left(\sqrt{[M]} > \lambda\right).$$

Proof. The proof of the two inequalities are similar.

1. Let us introduce the stopping times

$$\mu \stackrel{\circ}{=} \inf\{n : |M_n| > \lambda\}, \quad \nu \stackrel{\circ}{=} \inf\{n : |M_n| > \beta\lambda\},$$

$$\sigma \stackrel{\circ}{=} \inf\left\{n : \sqrt{[M]_n} \vee V_{n+1} > \delta\lambda\right\}.$$

For every j

$$F_j \stackrel{\circ}{=} \{\mu < j \leq \nu \wedge \sigma\} = \{\mu < j\} \cap \{\nu \wedge \sigma < j\}^c \in \mathcal{F}_{j-1},$$

hence if

$$H_n \stackrel{\circ}{=} \sum_{j=1}^{n} \Delta M_j \chi_{F_j},$$

then

$$\mathbf{E}\left(H_n \mid \mathcal{F}_{n-1}\right) \stackrel{\circ}{=} \mathbf{E}(\sum_{j=1}^{n} \Delta M_j \chi_{F_j} \mid \mathcal{F}_{n-1}) =$$

$$= \sum_{j=1}^{n-1} \Delta M_j \chi_{F_j} + \mathbf{E}(\Delta M_n \chi_{F_n} \mid \mathcal{F}_{n-1}) =$$

---

[88] And boring.
[89] That is V_n is $\mathcal{F}_{n-1}$-measurable.

$$= \sum_{j=1}^{n-1} \Delta M_j \chi_{F_j} + \chi_{F_n} \mathbf{E}(\Delta M_n \mid \mathcal{F}_{n-1}) =$$

$$= \sum_{j=1}^{n-1} \Delta M_j \chi_{F_j} \overset{\circ}{=} H_{n-1},$$

therefore (H_n) is a martingale. By the assumptions of the lemma $|\Delta M_j| \leq V_j$, hence by the definition of σ

$$[H]_n \leq [M]_\sigma = \left([M]_{\sigma-1} + (\Delta M_\sigma)^2\right) \chi\,(\sigma < \infty) + [M]_\sigma\,\chi\,(\sigma = \infty) \leq$$
$$\leq \left([M]_{\sigma-1} + V_\sigma^2\right) \chi\,(\sigma < \infty) + [M]_\sigma\,\chi\,(\sigma = \infty) \leq$$
$$\leq 2\delta^2\lambda^2.$$

$\{M^* \leq \lambda\} = \{\mu = \infty\}$ hence on this set $H = 0$ so $[H] = 0$. Therefore

$$\mathbf{E}\,([H]) = \mathbf{E}\,([H]\,\chi\,(M^* > \lambda) + [H]\,\chi\,(M^* \leq \lambda)) =$$
$$= \mathbf{E}\,([H]\,\chi\,(M^* > \lambda)) \leq 2\delta^2\lambda^2\mathbf{P}\,(M^* > \lambda)\,.$$

Observe that

$$F_j \cap \{\nu < \infty, \sigma = \infty\} = \{\mu < j \leq \nu\} \cap \{\nu < \infty, \sigma = \infty\}$$

hence on the set $\{\nu < \infty, \sigma = \infty\}$

$$H_n = M_{\nu \wedge n} - M_{\mu \wedge n}.$$

On $\{\nu < \infty\}$ obviously $\sup_n |M_{\nu \wedge n}| \geq \lambda\beta$. On $\{\sigma = \infty\}$ by definition $V^* \leq \delta\lambda$, hence

$$|M_\mu| = |M_{\mu-1} + \Delta M_\mu| \leq \lambda + \delta\lambda.$$

This implies that on the set $\{\nu < \infty, \sigma = \infty\}$

$$H^* = \sup_n |M_{\nu \wedge n} - M_{\mu \wedge n}| > \lambda\beta - \lambda\,(\delta + 1) = \lambda\,(\beta - (1 + \delta))\,.$$

By Doob's inequality[90] using the definition of ν and σ

$$P_1 \overset{\circ}{=} \mathbf{P}\left(M^* > \beta\lambda, \sqrt{[M]} \vee V^* \leq \delta\lambda\right) =$$
$$= \mathbf{P}\left(\nu < \infty, \sigma = \infty\right) \leq$$
$$\leq \mathbf{P}\left(H^* > \lambda\left(\beta - (1+\delta)\right)\right) \leq \frac{\mathbf{E}\left(H_\infty^2\right)}{\lambda^2\left(\beta - 1 - \delta\right)^2} \leq$$
$$\leq \frac{\mathbf{E}\left([H]\right)}{\lambda^2\left(\beta - 1 - \delta\right)^2} \leq$$
$$\leq \frac{2\delta^2\lambda^2\mathbf{P}\left(M^* > \lambda\right)}{\lambda^2\left(\beta - (1+\delta)\right)^2} = \frac{2\delta^2}{\left(\beta - 1 - \delta\right)^2}\mathbf{P}\left(M^* > \lambda\right),$$

which is the first inequality.

2. Analogously, let us introduce the stopping times

$$\mu' \overset{\circ}{=} \inf\left\{n : \sqrt{[M]_n} > \lambda\right\}, \quad \nu' \overset{\circ}{=} \inf\left\{n : \sqrt{[M]_n} > \beta\lambda\right\},$$
$$\sigma' \overset{\circ}{=} \inf\left\{n : M_n^* \vee V_{n+1} > \delta\lambda\right\}.$$

Again for all j let

$$F_j' \overset{\circ}{=} \{\mu' < j \leq \nu' \wedge \sigma'\}.$$

As $F_j' \in \mathcal{F}_{j-1}$

$$G_n \overset{\circ}{=} \sum_{j=1}^n \Delta M_j \chi_{F_j'}$$

is again a martingale. If $\mu' \geq \sigma'$ then $G^* = 0$. Hence if $\sigma' < \infty$ then

$$G^* = G^* \chi\left(\mu' < \sigma'\right) \leq$$
$$\leq \left(M_{\mu'}^* + M_{\sigma'}^*\right)\chi\left(\mu' < \sigma'\right) \leq$$
$$\leq \left(M_{\sigma'-1}^* + M_{\sigma'}^*\right)\chi\left(\mu' < \sigma'\right) =$$
$$= \left(M_{\sigma'-1}^* + M_{\sigma'-1}^* + \Delta M_{\sigma'}^*\right)\chi\left(\mu' < \sigma'\right) \leq$$
$$\leq \left(M_{\sigma'-1}^* + M_{\sigma'-1}^* + V_{\sigma'}\right)\chi\left(\mu' < \sigma'\right) \leq$$
$$\leq \delta\lambda + \delta\lambda + \delta\lambda = 3\delta\lambda.$$

[90]See: line (1.14), page 33.

If $\sigma' = \infty$ then of course $\sigma' - 1$ is meaningless, but in this case obviously

$$\left(M^*_{\mu'} + M^*_{\sigma'}\right) \chi \left(\mu' < \sigma'\right) \le 2\delta\lambda,$$

so in this case the inequality $G^* \le 3\delta\lambda$ still holds. On the set $\left\{\sqrt{[M]} \le \lambda\right\} = \{\mu' = \infty\}$ obviously $G^* = 0$.

$$\mathbf{E}\left((G^*)^2\right) = \mathbf{E}\left((G^*)^2 \chi \left(\sqrt{[M]} > \lambda\right) + (G^*)^2 \chi \left(\sqrt{[M]} \le \lambda\right)\right) =$$
$$= \mathbf{E}\left((G^*)^2 \chi \left(\sqrt{[M]} > \lambda\right)\right) \le 9\delta^2\lambda^2 \mathbf{P}\left(\sqrt{[M]} > \lambda\right).$$

On the set $\{\nu' < \infty, \sigma' = \infty\}$

$$[G]_n = [M]_{\nu' \wedge n} - [M]_{\mu' \wedge n}.$$

By this using that $\nu' < \infty$ and $\sigma' = \infty$

$$[G] > (\beta\lambda)^2 - [M]_{\mu'-1} - (\Delta M_{\mu'})^2 \ge$$
$$\ge (\beta\lambda)^2 - \lambda^2 - (V_{\mu'})^2 \ge$$
$$\ge (\beta\lambda)^2 - \left(1 + \delta^2\right)\lambda^2.$$

By Markov's inequality and by the energy identity[91]

$$P_2 \stackrel{\circ}{=} \mathbf{P}\left(\sqrt{[M]} > \beta\lambda, M^* \vee V^* \le \delta\lambda\right) =$$
$$= \mathbf{P}\left(\nu' < \infty, \sigma' = \infty\right) \le$$
$$\le \mathbf{P}\left([G] > \lambda^2 \left(\beta^2 - (1 + \delta^2)\right)\right) \le \frac{\mathbf{E}\left([G]\right)}{\lambda^2 \left(\beta^2 - (1 + \delta^2)\right)} =$$
$$= \frac{\mathbf{E}\left(G^2\right)}{\lambda^2 \left(\beta^2 - (1 + \delta^2)\right)} \le \frac{\mathbf{E}\left((G^*)^2\right)}{\lambda^2 \left(\beta^2 - (1 + \delta^2)\right)} \le$$
$$\le \frac{9\delta^2}{\beta^2 - (1 + \delta)^2} \mathbf{P}\left(\sqrt{[M]} > \lambda\right).$$

$\square$

Lemma 4.67 *Let $M \stackrel{\circ}{=} (M_n, \mathcal{F}_n)$ be a martingale and let assume that $M_0 = 0$. If*

$$d_j \stackrel{\circ}{=} \Delta M_j \stackrel{\circ}{=} M_j - M_{j-1},$$
$$a_j \stackrel{\circ}{=} d_j \chi \left(|d_j| \le 2d^*_{j-1}\right) - \mathbf{E}\left(d_j \chi \left(|d_j| \le 2d^*_{j-1}\right) \mid \mathcal{F}_{j-1}\right),$$
$$b_j \stackrel{\circ}{=} d_j \chi \left(|d_j| > 2d^*_{j-1}\right) - \mathbf{E}\left(d_j \chi \left(|d_j| > 2d^*_{j-1}\right) \mid \mathcal{F}_{j-1}\right)$$

[91]If $\mathbf{E}\left(G^2\right) = \infty$ then the inequality is true, otherwise one can use Proposition 1.58 on page 35.

then the sequences

$$G_n \overset{\circ}{=} \sum_{j=1}^{n} a_j \quad and \quad H_n \overset{\circ}{=} \sum_{j=1}^{n} b_j,$$

are $\mathcal{F}$-martingales, $M = G + H$ and

$$|a_j| \leq 4d_{j-1}^*, \tag{4.22}$$

$$\sum_{j=1}^{\infty} \left| d_j \chi \left(|d_j| > 2d_{j-1}^* \right) \right| \leq 2d^*, \tag{4.23}$$

$$\sum_{j=1}^{\infty} \mathbf{E} \left(|b_j| \right) \leq 4 \mathbf{E} \left(d^* \right). \tag{4.24}$$

Proof. As $M_0 = 0$

$$\sum_{j=1}^{n} d_j \overset{\circ}{=} \sum_{j=1}^{n} \Delta M_j = M_n - M_0 = M_n.$$

One should only prove the three inequalities, since from this identity the other parts of the lemma are obvious[92].

1. (4.22) is evident.
2. $|d_j| + 2d_{j-1}^* \leq 2 |d_j|$ on $\left\{ |d_j| > 2d_{j-1}^* \right\}$, hence

$$\sum_{j=1}^{\infty} \left| d_j \chi \left(|d_j| > 2d_{j-1}^* \right) \right| \leq \sum_{j=1}^{\infty} \left(2 |d_j| - 2d_{j-1}^* \right) \chi \left(|d_j| > 2d_{j-1}^* \right) \leq$$

$$\leq 2 \sum_{j=1}^{\infty} \left(d_j^* - d_{j-1}^* \right) = 2d^*,$$

which is exactly (4.23).

3.

$$\sum_{j=1}^{\infty} \mathbf{E} \left(|b_j| \right) \leq \sum_{j=1}^{\infty} \mathbf{E} \left(|d_j| \chi \left(|d_j| > 2d_{j-1}^* \right) \right) +$$

$$+ \sum_{j=1}^{\infty} \mathbf{E} \left(\left| \mathbf{E} \left(d_j \chi \left(|d_j| > 2d_{j-1}^* \right) \mid \mathcal{F}_{j-1} \right) \right| \right).$$

---

[92] For any sequence $(\xi_n, \mathcal{F}_n)$ $\mathbf{E} \left(\xi_n \mid \mathcal{F}_{n-1} \right) = 0$ if and only if $(\xi_n, \mathcal{F}_n)$ is a martingale difference sequence.

If in the second sum we bring the absolute value into the conditional expectation, then

$$\sum_{j=1}^{\infty} \mathbf{E}\left(|b_j|\right) \le 2\mathbf{E}\left(\sum_{j=1}^{\infty} |d_j| \chi\left(|d_j| > 2d_{j-1}^*\right)\right).$$

By (4.23) the expression in the conditional expectation is not larger than $2d^*$, from which (4.24) is evident. $\qquad\square$

The proof of the discrete-time Davis' inequality: Let $M = H + G$ be the decomposition of the previous lemma. $G_n \stackrel{\circ}{=} \sum_{j=1}^{n} a_j$ is a martingale, $|a_j| \le 4d_{j-1}^*$, hence by the first lemma, if $\lambda > 0$ and $0 < \delta < \beta - 1$, then

$$\mathbf{P}\left(G^* > \beta\lambda, \sqrt{[G]} \vee 4d^* \le \delta\lambda\right) \le \frac{2\delta^2}{(\beta - \delta - 1)^2} \mathbf{P}\left(G^* > \lambda\right),$$

$$\mathbf{P}\left(\sqrt{[G]} > \beta\lambda, G^* \vee 4d^* \le \delta\lambda\right) \le \frac{9\delta^2}{\beta^2 - \delta^2 - 1} \mathbf{P}\left(\sqrt{[G]} > \lambda\right).$$

Hence for any $\lambda > 0$

$$\mathbf{P}\left(G^* > \beta\lambda\right) \le \mathbf{P}\left(\sqrt{[G]} > \delta\lambda\right) + \mathbf{P}\left(4d^* > \delta\lambda\right) +$$

$$+ \frac{2\delta^2}{(\beta - \delta - 1)^2} \mathbf{P}\left(G^* > \lambda\right),$$

and

$$\mathbf{P}\left(\sqrt{[G]} > \beta\lambda\right) \le \mathbf{P}\left(G^* > \delta\lambda\right) + \mathbf{P}\left(4d^* > \delta\lambda\right) +$$

$$+ \frac{9\delta^2}{\beta^2 - \delta^2 - 1} \mathbf{P}\left(\sqrt{[G]} > \lambda\right).$$

Integrating w.r.t. λ and using that if $\xi \ge 0$ then

$$\mathbf{E}\left(\xi\right) = \int_0^{\infty} 1 - F(x)dx = \int_0^{\infty} \mathbf{P}(\xi > x)dx,$$

one has that

$$\frac{\mathbf{E}\left(G^*\right)}{\beta} \le \frac{\mathbf{E}\left(\sqrt{[G]}\right)}{\delta} + \frac{4\mathbf{E}\left(d^*\right)}{\delta} +$$

$$+ \frac{2\delta^2}{(\beta - \delta - 1)^2} \mathbf{E}\left(G^*\right),$$

and

$$\frac{\mathbf{E}\left(\sqrt{[G]}\right)}{\beta} \leq \frac{\mathbf{E}\left(G^*\right)}{\delta} + \frac{4\mathbf{E}\left(d^*\right)}{\delta} +$$

$$+ \frac{9\delta^2}{\beta^2 - \delta^2 - 1} \mathbf{E}\left(\sqrt{[G]}\right).$$

For the stopped martingale G^n the expected values in the inequalities are finite, hence one can reorder the inequalities

$$\left(\frac{1}{\beta} - \frac{2\delta^2}{(\beta - \delta - 1)^2}\right) \mathbf{E}\left(G_n^*\right) \leq \frac{\mathbf{E}\left(\sqrt{[G]}_n\right)}{\delta} + \frac{4\mathbf{E}\left(\Delta M_n^*\right)}{\delta}.$$

and

$$\left(\frac{1}{\beta} - \frac{9\delta^2}{\beta^2 - \delta^2 - 1}\right) \mathbf{E}\left(\sqrt{[G]}_n\right) < \frac{\mathbf{E}\left(G_n^*\right)}{\delta} + \frac{4\mathbf{E}\left(\Delta M_n^*\right)}{\delta}.$$

If δ is small enough then the constants on the left-hand side are positive, hence we can divide by them. Hence if $n \nearrow \infty$ then by the Monotone Convergence Theorem

$$\mathbf{E}\left(G^*\right) \leq A_1 \mathbf{E}\left(\sqrt{[G]}\right) + A_2 \mathbf{E}\left((\Delta M)^*\right),$$

$$\mathbf{E}\left(\sqrt{[G]}\right) \leq B_1 \mathbf{E}\left(G^*\right) + B_2 \mathbf{E}\left((\Delta M)^*\right).$$

By the second lemma

$$\mathbf{E}\left(M^*\right) \leq \mathbf{E}\left(G^* + H^*\right) \leq$$

$$\leq \mathbf{E}\left(G^*\right) + \sum_j \mathbf{E}\left(|b_j|\right) \leq \mathbf{E}\left(G^*\right) + 4\mathbf{E}\left(d^*\right) \leq$$

$$\leq A_1 \mathbf{E}\left(\sqrt{[G]}\right) + A_2 \mathbf{E}\left((\Delta M)^*\right) + 4\mathbf{E}\left((\Delta M)^*\right),$$

$$\mathbf{E}\left(\sqrt{[M]}\right) \leq \mathbf{E}\left(\sqrt{[G]} + \sqrt{[H]}\right) \leq$$

$$\leq \mathbf{E}\left(\sqrt{[G]}\right) + \sum_j \mathbf{E}\left(|b_j|\right) \leq \mathbf{E}\left(\sqrt{[G]}\right) + 4\mathbf{E}\left(d^*\right) \leq$$

$$\leq B_1 \mathbf{E}\left(G^*\right) + B_2 \mathbf{E}\left((\Delta M)^*\right) + 4\mathbf{E}\left((\Delta M)^*\right).$$

As $G = M - H$ by the second lemma again

$$\mathbf{E}\left(G^*\right) \le \mathbf{E}\left(M^*\right) + \mathbf{E}\left(H^*\right) \le \mathbf{E}\left(M^*\right) + \sum_{j=1}^{\infty} \mathbf{E}\left(|b_j|\right) \le$$

$$\le \mathbf{E}\left(M^*\right) + 4\mathbf{E}\left((\Delta M)^*\right)$$

and

$$\mathbf{E}\left(\sqrt{[G]}\right) \le \mathbf{E}\left(\sqrt{[M]}\right) + \mathbf{E}\left(\sqrt{[H]}\right) \le \mathbf{E}\left(\sqrt{[M]}\right) + \sum_{j=1}^{\infty} \mathbf{E}\left(|b_j|\right) \le$$

$$\le \mathbf{E}\left(\sqrt{[M]}\right) + 4\mathbf{E}\left((\Delta M)^*\right).$$

From this with simple calculation

$$\mathbf{E}\left(M^*\right) \le A_1 \mathbf{E}\left(\sqrt{[M]}\right) + A_3 \mathbf{E}\left((\Delta M)^*\right) \le A \cdot \mathbf{E}\left(\sqrt{[M]}\right),$$

and

$$\mathbf{E}\left(\sqrt{[M]}\right) \le B_1 \mathbf{E}\left(M^*\right) + B_3 \mathbf{E}\left((\Delta M)^*\right) \le B \cdot \mathbf{E}\left(M^*\right),$$

from which Davis' inequality already follows, trivially. □

4.5.2 Burkholder's inequality

One can extend Davis' inequality in such a way that instead of the $L^1(\Omega)$-norm one can write the $L^p(\Omega)$-norm for every $p \ge 1$.

Theorem 4.68 (Burkholder's inequality) *For any $p > 1$ there are constants c_p and C_p, such that for every local martingale $L \in \mathcal{L}$ and for every stopping time τ*

$$c_p \left\| \sqrt{[L]}\,(\tau) \right\|_p \le \left\| \sup_{t \le \tau} |L(t)| \right\|_p \le C_p \left\| \sqrt{[L]}\,(\tau) \right\|_p.$$

During the proof of the inequality we shall use the next result:

Lemma 4.69 *Let A be a right-regular, non-negative, increasing, adapted process and let ξ be a non-negative random variable. Assume that almost surely for every t*

$$\mathbf{E}\left(A\left(\infty\right) - A\left(t\right) \mid \mathcal{F}_t\right) \le \mathbf{E}\left(\xi \mid \mathcal{F}_t\right) \tag{4.25}$$

and

$$\Delta A\left(t\right) \leq \xi.$$

Then for every $p \geq 1$

$$\left\|A\left(\infty\right)\right\|_p \leq 2p\left\|\xi\right\|_p. \tag{4.26}$$

Proof. A is increasing, so for every n

$$\chi\left(A\left(t\right) \geq n\right)\left(A\left(\infty\right) - A\left(t\right)\right) = \left(A \wedge n\right)\left(\infty\right) - \left(A \wedge n\right)\left(t\right).$$

So if (4.26) holds for some A then it holds for $A \wedge n$. Hence one can assume that A is bounded, since otherwise we can replace A with $A \wedge n$ and in (4.26) one can take $n \nearrow \infty$. If ξ is not integrable then the inequality trivially holds. Hence one can assume that ξ is integrable.

1. As ξ is integrable $\mathbf{E}\left(\xi \mid \mathcal{F}_t\right)$ is a uniformly integrable martingale. As A is bounded

$$\mathbf{E}\left(A\left(\infty\right) - A\left(t\right) - \xi \mid \mathcal{F}_t\right) = \mathbf{E}\left(A\left(\infty\right) \mid \mathcal{F}_t\right) - \mathbf{E}\left(\xi \mid \mathcal{F}_t\right) - A\left(t\right)$$

is a uniformly integrable, non-positive supermartingale. By the Optional Sampling Theorem for every stopping time τ

$$\mathbf{E}\left(A\left(\infty\right) - A\left(\tau\right) \mid \mathcal{F}_\tau\right) \leq \mathbf{E}\left(\xi \mid \mathcal{F}_\tau\right). \tag{4.27}$$

Let $x > 0$ and let

$$\tau_x \stackrel{\circ}{=} \inf\left\{t : A\left(t\right) \geq x\right\}.$$

Obviously $A\left(\tau_x-\right) \leq x$. By (4.27)

$$\begin{aligned}
\mathbf{E}\left(\left(A\left(\infty\right) - x\right)\chi\left(x < A\left(\infty\right)\right)\right) &\leq \mathbf{E}\left(\left(A\left(\infty\right) - x\right)\chi\left(\tau_x < \infty\right)\right) = \\
&\leq \mathbf{E}\left(\left(A\left(\infty\right) - A\left(\tau_x-\right)\right)\chi\left(\tau_x < \infty\right)\right) = \\
&= \mathbf{E}\left(\left(A\left(\infty\right) - A\left(\tau_x\right)\right)\chi\left(\tau_x < \infty\right)\right) \\
&\quad + \mathbf{E}\left(\Delta A\left(\tau_x\right)\chi\left(\tau_x < \infty\right)\right) \leq \\
&\leq \mathbf{E}\left(\xi\chi\left(\tau_x < \infty\right)\right) + \mathbf{E}\left(\xi\chi\left(\tau_x < \infty\right)\right) \leq \\
&\leq 2\mathbf{E}\left(\xi\chi\left(x \leq A\left(\infty\right)\right)\right).
\end{aligned}$$

2. With simple calculation using Fubini's theorem and Hölder's inequality

$$\|A\left(\infty\right)\|_p^p \doteq \mathbf{E}\left(A^p\left(\infty\right)\right) = p\mathbf{E}\left(A^p\left(\infty\right)\right) - \left(p-1\right)\mathbf{E}\left(A^p\left(\infty\right)\right) =$$

$$= p\left(p-1\right)\mathbf{E}\left(A\left(\infty\right)\int_0^{A\left(\infty\right)} x^{p-2}dx\right)$$

$$- p\left(p-1\right)\mathbf{E}\left(\int_0^{A\left(\infty\right)} x^{p-1}dx\right) =$$

$$= p\left(p-1\right)\mathbf{E}\left(\int_0^{A\left(\infty\right)}\left(A\left(\infty\right)-x\right)x^{p-2}dx\right) =$$

$$= p\left(p-1\right)\int_0^\infty \mathbf{E}\left(\left(A\left(\infty\right)-x\right)\chi\left(x<A\left(\infty\right)\right)\right)x^{p-2}dx \le$$

$$\le 2p\left(p-1\right)\int_0^\infty \mathbf{E}\left(\xi\chi\left(x\le A\left(\infty\right)\right)\right)x^{p-2}dx =$$

$$= 2p\left(p-1\right)\mathbf{E}\left(\int_0^{A\left(\infty\right)}\xi x^{p-2}dx\right) = 2p\cdot\mathbf{E}\left(\xi A^{p-1}\left(\infty\right)\right) \le$$

$$\le 2p\cdot\|\xi\|_p\|A\left(\infty\right)\|_p^{p-1}.$$

If $\|A\left(\infty\right)\|_p > 0$ then we can divide both sides by $\|A\left(\infty\right)\|_p^{p-1}$, otherwise the inequality trivially holds. $\square$

Proof of Burkholder's inequality: Let L be a local martingale. Let $B \in \mathcal{F}_t$ and let $N \doteq \chi_B\left(L-L^t\right)$. N is a local martingale so by Davis' inequality

$$c\cdot\mathbf{E}\left(\sqrt{[N]}\left(\infty\right)\right) \le \mathbf{E}\left(\sup_s|N\left(s\right)|\right) \le C\cdot\mathbf{E}\left(\sqrt{[N]}\left(\infty\right)\right),$$

which immediately implies that

$$c\cdot\mathbf{E}\left(\sqrt{[L-L^t]}\left(\infty\right)\mid\mathcal{F}_t\right) \le \mathbf{E}\left(\sup_s|L-L^s|\mid\mathcal{F}_t\right) \le$$

$$\le C\cdot\mathbf{E}\left(\sqrt{[L-L^t]}\left(\infty\right)\mid\mathcal{F}_t\right).$$

Let $L^*\left(t\right) \doteq \sup_{s\le t}|L\left(s\right)|$. Since

$$\sqrt{[L]\left(\infty\right)} - \sqrt{[L]\left(t\right)} \le \sqrt{[L]\left(\infty\right)-[L]\left(t\right)} = \sqrt{[L-L^t]\left(\infty\right)} \le \sqrt{[L]\left(\infty\right)}$$

and

$$L^* (\infty) - L^* (s) \leq \sup_s \left| L - L^t (s) \right| \leq 2L^* (\infty)$$

if

$$A (t) \overset{\circ}{=} \sqrt{[L] (t)} \quad \text{and} \quad \xi \overset{\circ}{=} c^{-1} 2 \cdot L^{\bullet} (\infty)$$

or if

$$A (t) \overset{\circ}{=} L^* (t) \quad \text{and} \quad \xi \overset{\circ}{=} C \sqrt{[L] (\infty)}$$

then estimation (4.25) in the lemma holds. Without loss of generality one can assume that the constants in the definition of ξ are larger than one. Since for every constant $k \geq 1$

$$\Delta L^* \leq |\Delta L| = \sqrt{\Delta [L]} \leq k \cdot \sqrt{[L] (\infty)}$$

$$\Delta \sqrt{[L]} \leq \sqrt{\Delta [L]} = |\Delta L| \leq k \cdot 2L^* (\infty)$$

in both cases we get that

$$\Delta A \leq \xi.$$

Hence $\|A (\infty)\|_p \leq 2p \|\xi\|_p$ which is just the two sides of Burkholder's inequality.

$\square$

Corollary 4.70 *If $L \in \mathcal{L}$ and $p \geq 1$ then $L \in \mathcal{H}_{loc}^p$ if and only if $[L]^{p/2} \in \mathcal{A}_{loc}$.*

Corollary 4.71 *If M is a local martingale and for some $p \geq 1$ for every sequence of infinitesimal partitions of the interval $[0, t]$*

$$[M]^{(n)} (t) \overset{L^p}{\to} [M] (t),$$

then

$$M^* (t) \overset{\circ}{=} \sup_{s \leq t} |M (s)| \in L^p (\Omega)$$

that is $M \in \mathcal{H}^p$ on the interval $[0, t]$.

Proof. Let (M_n) be a discrete-time approximation of M. If $[M]^{(n)} (t)$ is convergent in $L^p(\Omega)$, then $K \overset{\circ}{=} \sup_n \left\| [M]^{(n)} (t) \right\|_p < \infty$. By the Davis–Burkholder inequality and by Jensen's inequality

$$\left\| \sup_{s \leq t} |M_n| (s) \right\|_p \leq C_p \left\| \sqrt{[M]^{(n)} (t)} \right\|_p \leq C_p \sqrt{\left\| [M]^{(n)} (t) \right\|_p} \leq L < \infty.$$

For a subsequence $\sup |M_n| \nearrow \sup |M|$, hence by the Monotone Convergence Theorem

$$\|M^* (t)\|_p \leq L < \infty.$$

$\square$

Corollary 4.72 *If $q \geq 1$ and $L \in \mathcal{H}^q$ is purely discontinuous then L is the $\mathcal{H}^q$-sum of its compensated jumps.*

Proof. Let us denote by (ρ_k) the stopping times exhausting the jumps of L. Let $L \in \mathcal{H}^q$ be purely discontinuous and let $L = \sum L_k$ where $N_k \doteq H (\rho_k) \chi ([\rho_k, \infty))$ and $L_k \doteq N - N_k^p$ are the the compensated jumps of L. Recall that the convergence holds in the topology of uniform convergence in probability[93]. $L \in \mathcal{H}^q$ so by Burkholder's inequality $[L]^{q/2} \in \mathcal{A}$ and as the compensator N_k^p is continuous

$$[L_k]^{q/2} (\infty) = (\Delta L (\rho_k))^q \leq \left(\sum (\Delta L)^2 (\infty) \right)^{q/2} \leq$$

$$\leq [L]^{q/2} (\infty) \in L^1 (\Omega).$$

This implies that $L_k \in \mathcal{H}^q$. $\mathcal{H}^q$ is a vector space hence $Y_n \doteq \sum_{k=1}^n L_k \in \mathcal{H}^q$. If $n > m$ then

$$\|Y_n - Y_m\|_{\mathcal{H}^q} \doteq \left\| \sup_t |Y_n (t) - Y_m (t)| \right\|_q \leq$$

$$\leq C_p \left\| \sqrt{[Y_n - Y_m] (\infty)} \right\|_q =$$

$$= C_p \left\| \sqrt{\sum_s (\Delta L)^2 (s) \chi (B_n \backslash B_m)} \right\|_q ,$$

where $B_n \doteq \cup_{k=1}^n [\rho_k]$. $\sqrt{\sum (\Delta L)^2}$ is in $L^q (\Omega)$. Therefore if $n, m \to \infty$ then

$$\|Y_n - Y_m\|_{\mathcal{H}^q} \to 0.$$

So (Y_n) is convergent in $\mathcal{H}^q$. Convergence in $\mathcal{H}^q$ implies uniform convergence in probability so obviously $Y_n \xrightarrow{\mathcal{H}^q} L$.

$\square$

---

[93]See: Proposition 4.30, page 243.

5

SOME OTHER THEOREMS

In this chapter we shall discuss some further theorems from the general theory of stochastic processes. First we shall prove the so-called Doob–Meyer decomposition. By the Doob–Meyer decomposition every integrable submartingale is a semimartingale. We shall also prove the theorem of Bichteler and Dellacherie, which states that the semimartingales are the only 'good integrators'.

5.1   The Doob–Meyer Decomposition

If $A \in \mathcal{A}^+$ and $M \in \mathcal{M}$ then $X \stackrel{\circ}{=} A + M$ is a class $\mathcal{D}$ submartingale. Since if τ is a finite valued stopping time then

$$|A(\tau)| = |A(\tau) - A(0)| \leq \mathrm{Var}(A)(\infty) \in L^1(\Omega), \tag{5.1}$$

hence the set

$$\{X(\tau) : \tau < \infty \text{ is a stopping time}\}$$

is uniformly integrable. The central observation of the stochastic analysis is that the reverse implication is also true:

Theorem 5.1 (Doob–Meyer decomposition) *If a submartingale X is in class $\mathcal{D}$ then X has a decomposition*

$$X = X(0) + M + A,$$

where $A \in \mathcal{A}^+$, $M \in \mathcal{M}$ and A is predictable. Up to indistinguishability this decomposition is unique.

5.1.1   The proof of the theorem

We divide the proof into several steps. The proof of the uniqueness is simple. If

$$X(0) + M_1 + A_1 = X(0) + M_2 + A_2$$

292

are two decompositions of X then

$$M_1 - M_2 = A_2 - A_1.$$

$A_2 - A_1$ is a predictable martingale, hence it is continuous[1]. As $A_2 - A_1$ has finite variation by Fisk's theorem[2] $A_1 = A_2$, hence $M_1 = M_2$.

The proof of the existence is a bit more complicated.

Definition 5.2 *We say that a supermartingale P is a potential[3], if*

1. *P is non-negative and*
2. *$\lim_{t \to \infty} \mathbf{E}(P(t)) = 0$.*

Proposition 5.3 (Riesz's decomposition) *If X is a class $\mathcal{D}$ submartingale then X has a decomposition*

$$X = X(0) + M - P \tag{5.2}$$

where P is a class $\mathcal{D}$ potential and M is a uniformly integrable martingale. Up to indistinguishability this decomposition is unique.

Proof. As X is in class $\mathcal{D}$ the set $\{X(t) : t \geq 0\}$ is uniformly integrable, hence it is bounded in $L^1(\Omega)$. Hence

$$\sup_t \mathbf{E}\left(X^+(t)\right) \leq \sup_t \mathbf{E}\left(|X(t)|\right) < K.$$

By the submartingale convergence theorem[4] the limit

$$\lim_{t \to \infty} X(t) = X(\infty) \in L^1(\Omega)$$

exists. Let us define the variables $M(t) \stackrel{\circ}{=} \mathbf{E}(X(\infty) \mid \mathcal{F}_t)$. As the filtration satisfies the usual conditions M has a version which is a uniformly integrable martingale. The process $P \stackrel{\circ}{=} M - X$ is in class $\mathcal{D}$ since it is the difference of two processes of class $\mathcal{D}$. By the submartingale property

$$P(s) \stackrel{\circ}{=} M(s) - X(s) \geq \mathbf{E}(M(t) \mid \mathcal{F}_s) - \mathbf{E}(X(t) \mid \mathcal{F}_s) =$$
$$= \mathbf{E}(M(t) - X(t) \mid \mathcal{F}_s).$$

If $t \to \infty$, then $M(t) - X(t) \stackrel{a.s.}{\to} 0$ and as $(M(t) - X(t))_t$ is uniformly integrable the convergence holds in $L^1(\Omega)$ as well. By the $L^1(\Omega)$-continuity of the

[1]See: Corollary 3.40, page 205.
[2]See: Theorem 2.11. page 117.
[3]Recall that the expected value of the supermartingales is decreasing.
[4]See: Corollary 1.72, page 44.

conditional expectation the right-hand side of the inequality almost surely goes to zero, that is $P(s) \overset{a.s.}{\geq} 0$.

$$\mathbf{E}(P(s)) = \mathbf{E}(M(s)) - \mathbf{E}(X(s)) \to \mathbf{E}(M(\infty)) - \mathbf{E}(X(\infty)) = 0,$$

hence P is a potential. Assume that the decomposition is not unique. Let P_i, M_i, $i = 1, 2$ be two decompositions of X. In this case

$$(P_1 - P_2)(t) = M_1(t) - M_2(t) = \mathbf{E}(M_1(\infty) - M_2(\infty) \mid \mathcal{F}_t).$$

By the definition of the potential $P_i(t) \overset{L_1}{\to} 0$. Hence if $t \to \infty$, then

$$0 = \mathbf{E}(M_1(\infty) - M_2(\infty) \mid \mathcal{F}_\infty) = M_1(\infty) - M_2(\infty),$$

hence $M_1 = M_2$, so $P_1 = P_2$. $\qquad\qquad\qquad\qquad\qquad\qquad\qquad\square$

It is sufficient to proof the Doob–Meyer decomposition for the potential part of the submartingale. One should prove that if P is a class $\mathcal{D}$ potential, then there is one and only one $N \in \mathcal{M}$ and a predictable process $A \in \mathcal{A}^+$ for which

$$P = N - A.$$

If it holds then substituting $-P = -N + A$ into line (5.2) we get the needed decomposition of X. From the definition of the potential

$$\mathbf{E}(A(t)) = \mathbf{E}(N(t)) - \mathbf{E}(P(t)) \leq \mathbf{E}(N(\infty)).$$

$A \in \mathcal{A}^+$, so A is increasing. $0 = A(0) \leq A(t) \nearrow A(\infty)$ where $\mathbf{E}(A(\infty)) < \infty$. Hence by the Monotone Convergence Theorem $A(t) \overset{L^1}{\to} A(\infty)$. By the definition of the potential $P(t) \overset{L^1}{\to} P(\infty) = 0$, hence $A(\infty) = N(\infty)$. So to prove the theorem it is sufficient to prove that there is a predictable process $A \in \mathcal{A}^+$ and $N \in \mathcal{M}$ such that

$$P(t) + A(t) = N(t) = \mathbf{E}(N(\infty) \mid \mathcal{F}_t) = \mathbf{E}(A(\infty) \mid \mathcal{F}_t),$$

which holds if there is an $A \in \mathcal{A}^+$ such that

$$P(t) = \mathbf{E}(A(\infty) - A(t) \mid \mathcal{F}_t).$$

By the definition of the conditional expectation it is equivalent to

$$\mathbf{E}(\chi_F(A(\infty) - A(t))) = \mathbf{E}(\chi_F P(t)) = \mathbf{E}(\chi_F(P(t) - P(\infty))), \quad F \in \mathcal{F}_t.$$

Observe that $S \stackrel{\circ}{=} -P$ is a submartingale and $S(\infty) = 0$, hence the previous line is equivalent to

$$\mathbf{E}(\chi_F (A(\infty) - A(t))) = \mathbf{E}(\chi_F (S(\infty) - S(t))), \quad F \in \mathcal{F}_t. \tag{5.3}$$

For an arbitrary process X on the set of predictable rectangles

$$(s, t] \times F, \quad F \in \mathcal{F}_s$$

let us define the set function

$$\mu_X ((s, t] \times F) \stackrel{\circ}{=} \mathbf{E}(\chi_F (X(t) - X(s))).$$

Recall[5] that the predictable rectangles and the sets $\{0\} \times F$, $F \in \mathcal{F}_0$ generate the σ-algebra of the predictable sets $\mathcal{P}$. Let

$$\mu_X (\{0\} \times F) \stackrel{\circ}{=} 0, \quad F \in \mathcal{F}_0.$$

Definition 5.4 *If a set function μ_X has a unique extension to the σ-algebra $\mathcal{P}$ which is a measure on $\mathcal{P}$ then μ_X is called[6] the* Doléans type measure *of X.*

Observe that the sets in (5.3) are in the σ-algebra generated by the predictable rectangles. Hence to prove the Doob–Meyer decomposition one should prove the following:

Proposition 5.5 *If $S \in \mathcal{D}$ is a submartingale then there is a predictable process $A \in \mathcal{A}^+$ such that the measure μ_S of S on the predictable sets is generated by A, that is there is a predictable process $A \in \mathcal{A}^+$ such that*

$$\mu_A (Y) = \mu_S (Y), \quad Y \in \mathcal{P}. \tag{5.4}$$

As a first step we prove that μ_S is really a measure on $\mathcal{P}$.

Proposition 5.6 *If S is a class $\mathcal{D}$ submartingale then the Doléans type measure μ_S of S can be extended from the semi-algebra of the predictable rectangles to the σ-algebra of the predictable sets.*

Proof. Denote by $\mathcal{C}$ the semi-algebra of the predictable rectangles. We want to use Carathéodory's extension theorem. To do this we should prove that μ_S is a measure on $\mathcal{C}$. As S is a submartingale μ_S is non-negative. μ_S is trivially additive, hence μ_S is monotone on $\mathcal{C}$. For all $C \in \mathcal{C}$, using that μ_S is monotone

---

[5]See: Corollary 1.44, page 26.
[6]See: Definition 2.56, page 151.

and $(0, \infty] \in \mathcal{C}$,

$$
\mu_S\left(C\right) \le \mu_S\left([0, \infty]\right) = \mu_S\left(\{0\} \times \Omega\right) + \mu_S\left((0, \infty]\right) =
$$
$$
= \mu_S\left((0, \infty]\right) \doteq \mathbf{E}\left(S\left(\infty\right) - S\left(0\right)\right) \le
$$
$$
\le \mathbf{E}\left(|S\left(\infty\right)|\right) + \mathbf{E}\left(|S\left(0\right)|\right) < \infty.
$$

Observe that in the last line we used that S is uniformly integrable and therefore $S\left(\infty\right)$ and $S\left(0\right)$ are integrable. As μ_S is finite it is sufficient to prove that whenever $C_n \in \mathcal{C}$, and $C_n \searrow \emptyset$, then $\mu_S\left(C_n\right) \searrow 0$. Let $\varepsilon > 0$ be arbitrary. If $(s, t] \times F \in \mathcal{C}$ then

$$
\left(s + \frac{1}{n}, t\right] \times F \subseteq \left[s + \frac{1}{n}, t\right] \times F \subseteq (s, t] \times F.
$$

S is a submartingale so for every $F \in \mathcal{F}_s$

$$
\mathbf{E}\left(\chi_F\left[S\left(s + \frac{1}{n}\right) - S\left(s\right)\right]\right) \ge 0,
$$
$$
\mathbf{E}\left(\chi_{F^c}\left[S\left(s + \frac{1}{n}\right) - S\left(s\right)\right]\right) \ge 0.
$$

S is uniform integrable, hence for the sum of the two sequences above

$$
\lim_{n \to \infty} \mathbf{E}\left(S\left(s + \frac{1}{n}\right) - S\left(s\right)\right) = \mathbf{E}\left(\lim_{n \to \infty}\left[S\left(s + \frac{1}{n}\right) - S\left(s\right)\right]\right) =
$$
$$
= \mathbf{E}\left(S\left(s+\right) - S\left(s\right)\right) = 0,
$$

hence

$$
\lim_{n \to \infty} \mathbf{E}\left(\chi_F\left[S\left(s + \frac{1}{n}\right) - S\left(s\right)\right]\right) = 0
$$

so

$$
\lim_{n \to \infty} \mu_S\left(\left(s + \frac{1}{n}, t\right] \times F\right) \doteq \lim_{n \to \infty} \mathbf{E}\left(\chi_F\left[S\left(t\right) - S\left(s + \frac{1}{n}\right)\right]\right) =
$$
$$
= \mathbf{E}\left(\chi_F\left[S\left(t\right) - S\left(s\right)\right]\right) \doteq \mu_S\left((s, t] \times F\right).
$$

Hence for every $C_n \in \mathcal{C}$ there are sets K_n and $B_n \in \mathcal{C}$ such that

$$
B_n \subseteq K_n \subseteq C_n,
$$

and for all ω the sections $K_n\left(\omega\right)$ of K_n are compact and

$$
\mu_S\left(C_n\right) < \mu_S\left(B_n\right) + \varepsilon 2^{-n}. \tag{5.5}
$$

Let us introduce the decreasing sequence

$$L_n \overset{\circ}{=} \cap_{k \leq n} B_k.$$

$\mathcal{C}$ is a semi-algebra, hence $L_n \in \mathcal{C}$ for every n. Let $\overline{L}_n$ and $\overline{B}_n$ be the sets in which we close the time intervals of L_n and B_n.

$$\overline{L}_n \subseteq \overline{B}_n \subseteq K_n \subseteq C_n \setminus \emptyset,$$

We prove that if

$$\gamma_n(\omega) \overset{\circ}{=} \inf \{t : (t, \omega) \in L_n\} = \min \{t : (t, \omega) \in \overline{L}_n\} < \infty$$

then $\gamma_n(\omega) \nearrow \infty$ for all ω. Otherwise $\gamma_n(\omega) \leq K$ for some ω and $K < \infty$ and $(\gamma_n(\omega), \omega) \in \overline{L}_n$. The sets $[0, K] \cap \overline{L}_n(\omega)$ are compact and $\gamma_n(\omega) \in [0, K] \cap \overline{L}_n(\omega)$ for all n. Hence their intersection is non-empty. Let γ_∞ be in the intersection. Then $(\gamma_\infty, \omega) \in \overline{L}_n$ for all n so $(\gamma_\infty, \omega) \in \cap_n \overline{L}_n$, which is impossible. Let

$$S = S(0) + M - P$$

be the decomposition of S, where P is the potential part of S. As M is uniformly integrable $\mathbf{E}(M(\infty)) = \mathbf{E}(M(\gamma_n))$. Therefore

$$\mu_S(L_n) \leq \mathbf{E}(S(\infty) - S(\gamma_n)) = \mathbf{E}(P(\gamma_n)).$$

As P is in class $\mathcal{D}$ $(P(\gamma_n \wedge t))$ is uniformly integrable for every t, so as $\gamma_n \nearrow \infty$

$$\lim_{n \to \infty} \mathbf{E}(P(\gamma_n \wedge t)) = \mathbf{E}(P(t)).$$

Using that P is a supermartingale

$$\limsup_{n \to \infty} \mathbf{E}(P(\gamma_n)) \leq \limsup_{n \to \infty} \mathbf{E}(P(\gamma_n \wedge t)) = \mathbf{E}(P(t)).$$

As

$$\lim_{t \to \infty} \mathbf{E}(P(t)) = 0$$

obviously $\mu_S(L_n) \to 0$. By (5.5)

$$\mu_S(L_n) \leq \mathbf{E}(S(\gamma_n) - S(\infty)) \to 0.$$

By (5.5)

$$\mu_S(C_n \setminus L_n) \overset{\circ}{=} \mu_S(C_n \cap (\cap_{k \leq n} B_k)^c) = \mu_S(C_n \cap (\cup_{k \leq n} B_k^c)) \leq$$

$$\leq \sum_{k=1}^{n} \mu_S(C_n \setminus B_k) \leq \sum_{k=1}^{n} \mu_S(C_k \setminus B_k) \leq \varepsilon,$$

hence

$$\limsup_{n\to\infty} \mu_S(C_n) \le \limsup_{n\to\infty} \mu_S(C_n \setminus L_n) + \limsup_{n\to\infty} \mu_S(L_n) \le \varepsilon. \qquad \square$$

Now we can finish the proof of the Doob–Meyer decomposition. Let us recall that by (5.4) one should prove that there is a predictable process A such that

$$\mu_A(Y) = \mu_S(Y), \quad Y \in \mathcal{P}. \tag{5.6}$$

To construct A let us extend μ_S from $\mathcal{P}$ to the product measurable subsets of $\mathbb{R}_+ \times \Omega$ with the definition

$$\mu(Y) \overset{\circ}{=} \mu_S({}^P Y) \overset{\circ}{=} \int_{\mathbb{R}_+ \times \Omega} {}^P \chi_Y d\mu_S. \tag{5.7}$$

Observe that as ${}^P \chi_Y$ is well-defined the set function $\mu(Y)$ is also well-defined. If Y_1 and Y_2 are disjoint then by the additivity of the predictable projection

$$\mu(Y_1 \cup Y_2) \overset{\circ}{=} \mu_S({}^P(Y_1 \cup Y_2)) \overset{\circ}{=} \int_{\mathbb{R}_+ \times \Omega} {}^P \chi_{Y_1 \cup Y_2} d\mu_S =$$

$$= \int_{\mathbb{R}_+ \times \Omega} {}^P(\chi_{Y_1} + \chi_{Y_2}) d\mu_S =$$

$$= \int_{\mathbb{R}_+ \times \Omega} ({}^P \chi_{Y_1} + {}^P \chi_{Y_2}) d\mu_S =$$

$$= \mu_S({}^P Y_1) + \mu_S({}^P Y_2) \overset{\circ}{=} \mu(Y_1) + \mu(Y_2),$$

so μ is additive. It is clear from the Monotone Convergence Theorem for the predictable projection that μ is σ-additive. Hence μ is a measure. μ is absolutely continuous, since if $Y \subseteq \mathbb{R}_+ \times \Omega$ is a negligible set, then there is a set $N \subseteq \Omega$ with probability zero that Y can be covered by the random intervals $[0, \tau_n]$ where

$$\tau_n(\omega) \overset{\circ}{=} \begin{cases} n & \text{if } \omega \in N \\ 0 & \text{if } \omega \notin N \end{cases}.$$

As $\mathbf{P}(N) = 0$ and as the usual conditions hold τ_n is a stopping time for every n. Hence the intervals $[0, \tau_n]$ are predictable, and their Doléans-measure is obviously zero. So

$$\mu(Y) \le \sum_n \mu([0, \tau_n]) = \sum_n \mu_S([0, \tau_n]) = 0.$$

By the generalized Radon–Nikodym theorem[7] we can represent μ with a predictable[8] process $A \in \mathcal{A}^+$. Hence for all predictable Y

$$\mu_A(Y) = \mu(Y) \stackrel{\circ}{=} \mu_S({}^pY) = \mu_S(Y)$$

therefore for this A (5.6) holds. $\qquad\qquad\qquad\qquad\qquad\qquad\square$

5.1.2 Dellacherie's formulas and the natural processes

In some applications of the Doob–Meyer decomposition it is more convenient to assume that in the decomposition the increasing process A is natural.

Definition 5.7 *We say that a process* $V \in \mathcal{V}$ *is* natural *if for every non-negative, bounded martingale* N

$$\mathbf{E}\left(\int_0^t N\,dV\right) = \mathbf{E}\left(\int_0^t N_-\,dV\right). \tag{5.8}$$

Recall that for local martingales ${}^pN = N_-$, hence (5.8) can be written as

$$\mathbf{E}\left(\int_0^t N\,dV\right) = \mathbf{E}\left(\int_0^t {}^pN\,dV\right).$$

Proposition 5.8 (Dellacherie's formula) *If* $V \in \mathcal{A}^+$ *is natural then for every non-negative, product measurable process* X

$$\mathbf{E}\left(\int_0^\infty X\,dV\right) = \mathbf{E}\left(\int_0^\infty {}^pX\,dV\right), \tag{5.9}$$

where the two sides exist or do not exist in the same time.

Proof. If η is non-negative, bounded random variable and $X \stackrel{\circ}{=} \eta \cdot \chi((s,t])$ then

$$\mathbf{E}\left(\int_0^\infty X\,dV\right) = \mathbf{E}\left(\eta\left(V(t) - V(s)\right)\right) =$$

$$= \mathbf{E}\left(\eta \sum_k \left(V\left(t_k^{(n)}\right) - V\left(t_{k-1}^{(n)}\right)\right)\right) =$$

$$= \sum_k \mathbf{E}\left(\mathbf{E}\left(\eta\left(V\left(t_k^{(n)}\right) - V\left(t_{k-1}^{(n)}\right)\right) \mid \mathcal{F}_{t_k^{(n)}}\right)\right) =$$

---

[7]See: Proposition 3.49, page 208.
[8]See: Proposition 3.51, page 211.

$$= \mathbf{E} \left(\sum_k \mathbf{E} \left(\eta \mid \mathcal{F}_{t_k^{(n)}} \right) \left(V \left(t_k^{(n)} \right) - V \left(t_{k-1}^{(n)} \right) \right) \right) \overset{\circ}{=}$$

$$\overset{\circ}{=} \mathbf{E} \left(\sum_k M \left(t_k^{(n)} \right) \left(V \left(t_k^{(n)} \right) - V \left(t_{k-1}^{(n)} \right) \right) \right).$$

By our general assumption the filtration satisfies the usual conditions so $M(t) \overset{\circ}{=} \mathbf{E}(\eta \mid \mathcal{F}_t)$ has a version which is a bounded, non-negative martingale. If

$$\max_k \left| t_k^{(n)} - t_{k-1}^{(n)} \right| \to 0$$

then using that M, as every martingale, is right-continuous,

$$M_n \overset{\circ}{=} \sum_k M \left(t_k^{(n)} \right) \chi \left(t_{k-1}^{(n)}, t_k^{(n)} \right] \to M.$$

η is bounded and $V \in \mathcal{A}^+$, hence the sum behind the expected value is dominated by an integrable variable, so by the Dominated Convergence Theorem

$$\mathbf{E} \left(\int_0^\infty X dV \right) = \lim_{n \to \infty} \mathbf{E} \left(\sum_k M \left(t_k^{(n)} \right) \left(V \left(t_k^{(n)} \right) - V \left(t_{k-1}^{(n)} \right) \right) \right) =$$

$$= \mathbf{E} \left(\lim_{n \to \infty} \sum_k M \left(t_k^{(n)} \right) \left(V \left(t_k^{(n)} \right) - V \left(t_{k-1}^{(n)} \right) \right) \right) =$$

$$= \mathbf{E} \left(\lim_{n \to \infty} \int_s^t M_n dV \right) = \mathbf{E} \left(\int_s^t \lim_{n \to \infty} M_n dV \right) = \mathbf{E} \left(\int_s^t M dV \right).$$

Remember that if $X \overset{\circ}{=} \eta \cdot \chi_I$ then[9]

$$^p X \overset{\circ}{=} {}^p (\eta \cdot \chi_I) = M_- \cdot \chi_I.$$

Using that V is natural

$$\mathbf{E} \left(\int_0^\infty X dV \right) = \mathbf{E} \left(\int_s^t M dV \right) = \mathbf{E} \left(\int_s^t M_- dV \right) =$$

$$= \mathbf{E} \left(\int_0^\infty M_- \chi \left((s, t] \right) dV \right) = \mathbf{E} \left(\int_0^\infty {}^p X dV \right).$$

Hence for this special X (5.9) holds. These processes form a π-system. The bounded processes for which (5.9) is true is a λ-system, hence by the Monotone

---

[9]See: Corollary 3.43, page 206.

Class Theorem one can extend (5.9) to the bounded processes which are measurable with respect to the σ-algebra generated by the processes $X \stackrel{\circ}{=} \eta \cdot \chi\left((s,t]\right)$, hence (5.9) is true if X is a bounded product measurable process. To prove the proposition it is sufficient to apply the Monotone Convergence Theorem. $\qquad\square$

Proposition 5.9 (Dellacherie's formula) *If $A \in \mathcal{V}$ and A is predictable then for any non-negative, product measurable process X*

$$\mathbf{E}\left(\int_0^\infty X\,dA\right) = \mathbf{E}\left(\int_0^\infty {}^p X\,dA\right),$$

where the two sides exist or do not exist in the same time.

Proof. If A is predictable then $\mathrm{Var}\,(A)$ is also predictable. Therefore we can assume that A is increasing. In this case the expressions in the expectations exist and they are non-negative. Define the process

$$\sigma\left(t,\omega\right) \stackrel{\circ}{=} \inf\left\{s : A\left(s,\omega\right) \geq t\right\}.$$

As A is increasing $\sigma\left(t,\omega\right)$ is increasing and right-continuous in t for any fixed ω. As the usual conditions hold σ_t, as a function of ω is a stopping time for any fixed t. Observe that as A is right-continuous $[\sigma_t] \subseteq \{A \geq t\}$, so as A is predictable

$$\mathrm{Graph}\left(\sigma_t\right) = [\sigma_t] = [0, \sigma_t] \cap \{A \geq t\} \in \mathcal{P},$$

hence σ_t is a predictable stopping time[10]. By the definition of the predictable projection

$$\mathbf{E}\left(X\left(\sigma_t\right)\chi\left(\sigma_t < \infty\right)\right) = \mathbf{E}\left({}^p X\left(\sigma_t\right)\chi\left(\sigma_t < \infty\right)\right).$$

Let us remark, that for every non-negative Borel measurable function f

$$\int_0^\infty f\left(u\right)dA\left(u\right) = \int_0^\infty f\left(\sigma_t\right)\chi\left(\sigma_t < \infty\right)dt.$$

To see this let us remark that A is right-continuous and increasing hence

$$\{t \leq A\left(v\right)\} = \{\sigma_t \leq v\}. \tag{5.10}$$

So if $f \stackrel{\circ}{=} \chi\left([0, v]\right)$ then as $A\left(0\right) = 0$

$$\int_0^\infty f\,dA = A\left(v\right) = \int_0^\infty \chi\left(t \leq A\left(v\right)\right)dt =$$

$$= \int_0^\infty \chi\left(\sigma_t \leq v\right)dt = \int_0^\infty f\left(\sigma_t\right)\chi\left(\sigma_t < \infty\right)dt.$$

[10]See: Corollary 3.34, page 199.

One can prove the general case in the usual way. As σ_t is predictable and as $\sigma(t, \omega)$ is product measurable by Fubini's theorem

$$\mathbf{E}\left(\int_0^\infty X\,dA\right) = \mathbf{E}\left(\int_0^\infty X\left(\sigma_t\right)\chi\left(\sigma_t < \infty\right)dt\right) =$$

$$= \int_0^\infty \mathbf{E}\left(X\left(\sigma_t\right)\chi\left(\sigma_t < \infty\right)\right)dt =$$

$$= \int_0^\infty \mathbf{E}\left({}^p X\left(\sigma_t\right)\chi\left(\sigma_t < \infty\right)\right)dt =$$

$$= \mathbf{E}\left(\int_0^\infty {}^p X\,dA\right). \qquad \square$$

Theorem 5.10 (Doléans) *A process $V \in \mathcal{A}^+$ is natural if and only if V is predictable.*

Proof. If V is natural, then by the first formula of Dellacherie if ${}^p X = {}^p Y$, then $\mu_V(X) = \mu_V(Y)$, hence by the uniqueness of the representation of μ_V V is predictable[11]. To see the other implication assume that V is predictable. By the second formula of Dellacherie for every product measurable process X

$$\mathbf{E}\left(\int_0^\infty X\,dV\right) = \mathbf{E}\left(\int_0^\infty {}^p X\,dV\right).$$

If N is a local martingale then[12] ${}^p N = N_-$, hence V is natural. $\qquad \square$

Dellacherie's formulas have an interesting consequence. When the integrator is a continuous local martingale then the stochastic integral is meaningful whenever the integrand is progressively measurable. By Dellacheries's formulas even in this case the set of all possible integral processes is the same as the set of integral processes when the integrands are just predictable. Assume first that $X \in \mathcal{L}^2(M)$. By Jensen's inequality $\left({}^p X\right)^2 \leq {}^p\left(X^2\right)$, hence by the second Dellacherie's formula ${}^p X \in \mathcal{L}^2(M)$. $[M, N]$ is continuous, hence it is predictable also by Dellacherie's formula

$$\mathbf{E}\left(\int_0^\infty X\,d[M, N]\right) = \mathbf{E}\left(\int_0^\infty {}^p X\,d[M, N]\right).$$

Hence during the definition of the stochastic integral the linear functionals

$$N \mapsto \mathbf{E}\left(\int_0^\infty X\,d[M, N]\right), \quad N \mapsto \mathbf{E}\left(\int_0^\infty {}^p X\,d[M, N]\right)$$

---

[11] See: Proposition 3.51, page 211.
[12] See: Proposition 3.38, page 204.

coincide. Hence $X \bullet M = {}^{p}X \bullet M$, and with localization if $X \in \mathcal{L}^{2}_{\mathrm{loc}}(M)$ then ${}^{p}X \in \mathcal{L}^{2}_{\mathrm{loc}}(M)$ and $X \bullet M = {}^{p}X \bullet M$.

5.1.3   The sub- super- and the quasi-martingales are semimartingales

The main problem with the definition of the semimartingales is that it is very formal. An important consequence of the Doob–Meyer decomposition is that we can show some nontrivial examples for semimartingales. The most important direct application of the Doob–Meyer decomposition is the following:

Proposition 5.11 *Every integrable*[13] *sub- and supermartingale* X *is semi-martingale.*

Proof. Let X be integrable submartingale. To make the notation simple we shall assume that $X(0) = 0$.

1. Let us first assume that if X is an integrable submartingale. Let τ be an arbitrary stopping time. We prove that as in the case of martingales, X^{τ} is also a submartingale. Let $s < t$ and $A \in \mathcal{F}_{s}$. Let us define the bounded stopping time

$$\sigma \stackrel{\circ}{=} (\tau \wedge t) \chi_{A^{c}} + (\tau \wedge s) \chi_{A}.$$

As X is integrable one can use the Optional Sampling Theorem, hence as $\sigma \leq \tau \wedge t$

$$\mathbf{E}(X(\sigma)) \stackrel{\circ}{=} \mathbf{E}(X(\tau \wedge t) \chi_{A^{c}} + X(\tau \wedge s) \chi_{A}) \leq$$
$$\leq \mathbf{E}(X(\tau \wedge t)) = \mathbf{E}(X^{\tau}(t) \chi_{A^{c}} + X^{\tau}(t) \chi_{A}),$$

therefore

$$\mathbf{E}(X^{\tau}(s) \chi_{A}) \leq \mathbf{E}(X^{\tau}(t) \chi_{A}),$$

which means that

$$X^{\tau}(s) \leq \mathbf{E}(X^{\tau}(t) \mid \mathcal{F}_{s}),$$

that is X^{τ} is a submartingale.

2. If submartingale X is in class $\mathcal{D}$ then by the Doob–Meyer decomposition X is semimartingale. One should prove that there is a localizing sequence (τ_{n}), for which $X^{\tau_{n}}$ is in class $\mathcal{D}$ for all n , hence as the Doob–Meyer decomposition

---

[13]That is $X(t)$ is integrable for every t.

is unique the decomposition $L_{n+1} + V_{n+1}$ of $X^{\tau_{n+1}}$ on the interval $[0, \tau_n]$ is indistinguishable from the decomposition $L_n + V_n$ of X^{τ_n}. From this it is clear that X has the decomposition

$$L + V \overset{\circ}{=} \lim_n L_n + \lim_n V_n,$$

where L is a local martingale and V has finite variation.

3. Let us define the bounded stopping times

$$\tau_n \overset{\circ}{=} \inf \{t : |X(t)| > n\} \wedge n.$$

As X is integrable by the Optional Sampling Theorem $X(\tau_n) \in L^1(\Omega)$. For all t

$$|X^{\tau_n}(t)| \leq n + |X(\tau_n)| \in L^1(\Omega),$$

hence X^{τ_n} is a class $\mathcal{D}$ submartingale. Obviously $\tau_n \leq \tau_{n+1}$. Assume that for some ω the sequence $(\tau_n(\omega))$ is bounded. In this case $\tau_n(\omega) \nearrow \tau_\infty(\omega) < \infty$. So there is an N such that if $n \geq N$ then $\tau_n(\omega) < n$. Hence $|X(\tau_n(\omega))| \geq n$ by the definition of τ_n, therefore the sequence $(X(\tau_n(\omega)))$ is not convergent, which is a contradiction as by the right-regularity of the submartingales X has finite left limit at $\tau_\infty(\omega)$. $\qquad\square$

The semimartingales form a linear space, therefore if $X \overset{\circ}{=} Y - Z$, where Y and Z are integrable, non-negative supermartingales then X is also a semimartingale. Let us extend X to $t = \infty$. By definition let

$$X(\infty) \overset{\circ}{=} Y(\infty) \overset{\circ}{=} Z(\infty) \overset{\circ}{=} 0.$$

As Y and Z are non-negative, after this extension they remain supermartingales[14]. Hence one can assume that Y, Z and X are defined on $[0, \infty]$. Let

$$\Delta : 0 = t_0 < t_1 < \ldots < t_n < t_{n+1} = \infty \qquad (5.11)$$

be an arbitrary decomposition of $[0, \infty]$. Let us define the expression

$$\sup_\Delta \mathbf{E} \left(\sum_{i=0}^n |\mathbf{E}(X(t_i) - X(t_{i+1}) \mid \mathcal{F}_{t_i})| \right), \qquad (5.12)$$

---

[14]Observed that we used the non-negativity assumption.

where one should calculate the supremum over all possible subdivisions (5.11).

$$\mathbf{E}\left(\sum_i |\mathbf{E}\left(X\left(t_i\right) - X\left(t_{i+1}\right) \mid \mathcal{F}_{t_i}\right)|\right) \leq$$

$$\leq \mathbf{E}\left(\sum_i |\mathbf{E}\left(Y\left(t_i\right) - Y\left(t_{i+1}\right) \mid \mathcal{F}_{t_i}\right)|\right) + \mathbf{E}\left(\sum_i |\mathbf{E}\left(Z\left(t_i\right) - Z\left(t_{i+1}\right) \mid \mathcal{F}_{t_i}\right)|\right).$$

Y is a supermartingale, hence

$$\mathbf{E}\left(Y\left(t_i\right) - Y\left(t_{i+1}\right) \mid \mathcal{F}_{t_i}\right) = Y\left(t_i\right) - \mathbf{E}\left(Y\left(t_{i+1}\right) \mid \mathcal{F}_{t_i}\right) \geq 0.$$

Therefore one can drop the absolute value. By the simple properties of the conditional expectation, using the assumption that Y is integrable

$$\mathbf{E}\left(\sum_{i=0}^{n} |\mathbf{E}\left(Y\left(t_i\right) - Y\left(t_{i+1}\right) \mid \mathcal{F}_{t_i}\right)|\right) = \mathbf{E}\left(Y\left(0\right)\right) - \mathbf{E}\left(Y\left(\infty\right)\right) = \mathbf{E}\left(Y\left(0\right)\right) < \infty.$$

Applying the same to Z one can easily see that if X has the just mentioned decomposition then the supremum (5.12) is finite.

Definition 5.12 *We say that the integrable[15], adapted, right-regular process X is a* quasi-martingale *if the supremum in (5.12) is finite.*

Proposition 5.13 (Rao) *An integrable, right-regular process X defined on $\mathbb{R}_+$ is a quasi-martingale if and only if it has a decomposition*

$$X = Y - Z$$

where Y and Z are non-negative supermartingales.

Proof. We have already proved one implication. We should only show that every quasi-martingale has the mentioned decomposition. X is defined on $\mathbb{R}_+$, hence as above we shall assume that $X\left(\infty\right) \stackrel{\circ}{=} 0$. Let us fix an s. For any decomposition

$$\Delta : t_0 = s < t_1 < t_2 \ldots$$

of $[s, \infty]$ let us define the two variables

$$C_{\Delta}^{\pm}\left(s\right) \stackrel{\circ}{=} \mathbf{E}\left(\sum_i \left(\mathbf{E}\left(X\left(t_i\right) - X\left(t_{i+1}\right) \mid \mathcal{F}_{t_i}\right)\right)^{\pm} \mid \mathcal{F}_s\right).$$

---

[15]That is $X\left(t\right)$ is integrable for every t.

The variables $C_\Delta^\pm(s)$ are $\mathcal{F}_s$-measurable. Let (Δ_n) be an infinitesimal[16] sequence of partitions of $[s, \infty]$, and let us assume that $\Delta_n \subseteq \Delta_{n+1}$, that is let us assume that we get Δ_{n+1} by adding further points to Δ_n. We shall prove that the sequences $(C_{\Delta_n}^\pm(s))$ are almost surely convergent and the limits are almost surely finite. First we prove that if the partition Δ'' is finer than Δ', then

$$C_{\Delta'}^\pm(s) \le C_{\Delta''}^\pm(s), \tag{5.13}$$

which will imply the convergence. By the quasi-martingale property the set of variables $C_\Delta^\pm(s)$ is bounded in $L^1(\Omega)$. From the Monotone Convergence Theorem it is obvious, that $C_{\Delta_n}^\pm(s) \nearrow \infty$ cannot hold on a set which has positive measure. To prove (5.13) let us assume that the new point t is between t_i and t_{i+1}. Let us introduce the variables

$$\xi \overset{\circ}{=} \mathbf{E}\left(X(t_i) - X(t) \mid \mathcal{F}_{t_i}\right), \quad \eta \overset{\circ}{=} \mathbf{E}\left(X(t) - X(t_{i+1}) \mid \mathcal{F}_t\right),$$
$$\zeta \overset{\circ}{=} \mathbf{E}\left(X(t_i) - X(t_{i+1}) \mid \mathcal{F}_{t_i}\right).$$

As $\zeta = \xi + \mathbf{E}(\eta \mid \mathcal{F}_{t_i})$, by Jensen's inequality

$$\zeta^+ \le \xi^+ + \mathbf{E}(\eta \mid \mathcal{F}_{t_i})^+ \le \xi^+ + \mathbf{E}\left(\eta^+ \mid \mathcal{F}_{t_i}\right),$$

hence

$$\mathbf{E}\left(\zeta^+ \mid \mathcal{F}_s\right) \le \mathbf{E}\left(\xi^+ \mid \mathcal{F}_s\right) + \mathbf{E}\left(\eta^+ \mid \mathcal{F}_s\right),$$

from which the inequality (5.13) is trivial. Let us introduce the variables

$$C^\pm(s) \overset{\circ}{=} \lim_{n \to \infty} C_{\Delta_n}^\pm(s).$$

Obviously $C^\pm(s)$ is integrable and $\mathcal{F}_s$-measurable. Let us observe that the variables $C_{\Delta_n}^\pm(s)$ are defined up to a measure-zero set, hence the variables $C^\pm(s)$ are also defined up to a measure-zero set. For arbitrary partitions $\Delta_n \overset{\circ}{=} \left(t_i^{(n)}\right)$ as $X(\infty) \overset{\circ}{=} 0$ and as X is adapted

$$C_{\Delta_n}^+(s) - C_{\Delta_n}^-(s) = \mathbf{E}\left(\sum_i \mathbf{E}\left(X\left(t_i^{(n)}\right) - X\left(t_{i+1}^{(n)}\right) \mid \mathcal{F}_{t_i^{(n)}}\right) \mid \mathcal{F}_s\right) =$$
$$= \sum_i \mathbf{E}\left(X\left(t_i^{(n)}\right) - X\left(t_{i+1}^{(n)}\right) \mid \mathcal{F}_s\right) =$$
$$= \mathbf{E}\left(X(s) \mid \mathcal{F}_s\right) - \mathbf{E}\left(X(\infty) \mid \mathcal{F}_s\right) \overset{a.s}{=} X(s).$$

---

[16] As the length of the $[s, \infty]$ is infinite this property, it means that we map order preservingly $[0, \infty]$ onto $[0, 1]$ and then the $(\Delta_n)_n$ is infinitesimal on $[0, 1]$.

This remains valid after we take the limit, hence for all s

$$C^+ (s) - C^- (s) \overset{a.s}{=} X (s). \tag{5.14}$$

Let us assume that t is in Δ_n for all n. As $s < t$

$$\mathbf{E}\left(C^\pm_{\Delta_n} (t) \mid \mathcal{F}_s\right) = \mathbf{E}\left(\sum_{t^{(n)}_{ii} \geq t} \left(\mathbf{E}\left(X\left(t^{(n)}_i\right) - X\left(t^{(n)}_{i+1}\right) \mid \mathcal{F}_{t^{(n)}_i}\right)\right)^\pm \;\middle|\; \mathcal{F}_s\right) \leq$$

$$\leq \mathbf{E}\left(\sum_{i} \left(\mathbf{E}\left(X\left(t^{(n)}_i\right) - X\left(t^{(n)}_{i+1}\right) \mid \mathcal{F}_{t^{(n)}_i}\right)\right)^\pm \;\middle|\; \mathcal{F}_s\right)$$

$$= C^\pm_{\Delta_n} (s),$$

from which taking the limit and using the Monotone Convergence Theorem for the conditional expectation

$$\mathbf{E}\left(C^\pm (t) \mid \mathcal{F}_s\right) \leq C^\pm (s). \tag{5.15}$$

Let (Δ_n) be an infinitesimal sequence of partitions of $[0, \infty]$. Let S be the union of the points in (Δ_n). Obviously S is dense in $\mathbb{R}_+$. By the above $C^\pm$ are supermartingales on S. As S is countable so on S one can define the trajectories of $C^\pm$ up to a measure zero set. By the supermartingale property except on a measure zero set N for every t the limit

$$D^\pm (t, \omega) \overset{\circ}{=} C^\pm (t+, \omega) \overset{\circ}{=} \lim_{s \searrow t, s \in S} C^\pm (s, \omega)$$

exist and $D^\pm (t)$ is right-regular. X is also right-regular, hence from (5.14) on the N^c for every $t \geq 0$

$$D^+ (t) - D^- (t) = X (t).$$

$D^\pm (t)$ is $\mathcal{F}_{t+1/n}$-measurable for all n, hence $D^\pm (t)$ is $\mathcal{F}_{t+}$-measurable. As $\mathcal{F}$ satisfies the usual conditions $D^\pm (t)$ is $\mathcal{F}_t$ measurable, that is the processes $D^\pm$ are adapted. If $s_n \searrow t$ and $s_n \in S$, then the sequence $(C^\pm (s_n))$ is a reversed supermartingale. Hence for the $L^1 (\Omega)$ convergence of $(C^\pm (s_n))$ it is necessary and sufficient that the sequence is bounded in $L^1 (\Omega)$. By the supermartingale property as (s_n) is decreasing the expected value of $(C^\pm (s_n))_n$ is increasing. By the quasi-martingale property the variables $C^\pm (0)$ are integrable, hence by the non-negativity the sequences $(C^\pm (s_n))$ are bounded in $L^1 (\Omega)$. Hence they are convergent in $L^1 (\Omega)$. From this $D^\pm (t)$ is integrable for all t. The conditional expectation is continuous in $L^1 (\Omega)$ therefore one can take the limit in (5.15) into the conditional expectation. Hence the processes $D^\pm$ are integrable supermartingales on $\mathbb{R}_+$. $\qquad\square$

Corollary 5.14 *Every quasi-martingale is a semimartingale.*

5.2 Semimartingales as Good Integrators

The definition of the semimartingales is quite artificial. In this section we present an important characterization of the semimartingales. We shall prove that the only class of integrators for which one can define a stochastic integral with reasonable properties is the class of the semimartingales. Recall the following definition:

Definition 5.15 *Process E is a* predictable step *process if*

$$E = \sum_{i=0}^{n} \xi_i \chi \left((t_i, t_{i+1}] \right)$$

where

$$0 = t_0 < t_1 < \ldots < t_{n+1}$$

and ξ_i are $\mathcal{F}_{t_i}$-measurable random variables.

If X an arbitrary process then the only reasonable definition of the stochastic integral $E \bullet X$ is[17]

$$(E \bullet X)(t) = \sum_{i} \xi_i \left(X \left(t_{i+1} \wedge t \right) - X \left(t_i \wedge t \right) \right).$$

For an arbitrary stochastic process X the definition obviously makes the integral linear over the linear space of the predictable step processes. On the other hand it is reasonable to say that a linear mapping is an integral if the correspondence has some continuity property. Let us define the topology of uniform convergence in (t, ω) among the predictable step processes and let us define the topology for the random variables with the stochastic convergence.

Definition 5.16 *We say that process X is a* good integrator, *if for every t the correspondence*

$$E \mapsto (E \bullet X)(t)$$

is a continuous, linear mapping from the space of predictable step processes to the set of random variables.

Observe that the required continuity property is very weak, as on the domain of definition we have a very strong, and on the image space we have a very weak, topology. As the integral is linear it is continuous if and only if it is continuous at $E = 0$. This means that if a sequence of step processes is uniformly convergent to zero then for any t the integral on the interval $(0, t]$ is stochastically convergent to zero.

---

[17]See: Theorem 2.88, page 174, line (4.11), page 252. Recall that by definition $(E \bullet X)(t)$ is the integral on $(0, t]$.

Theorem 5.17 (Bichteler–Dellacherie) *An adapted, right-regular process X is a semimartingale if and only if it is a good integrator.*

Proof. If X is a semimartingale, then by the Dominated Convergence Theorem it is obviously a good integrator[18]. Hence we have to prove only the other direction. We split the proof into several steps.

1. As a first step let us separate the 'big jumps' of X, that is let us separate from X the jumps of X which are larger than one. By the assumptions of the theorem the trajectories of X are regular so the 'big jumps' do not have an accumulation point. Hence the decomposition is meaningful. From this trivially follows that the process

$$\sum \Delta X \chi \left(|\Delta X| \geq 1 \right)$$

has finite variations. As the continuity property of the good integrators holds for processes with finite variation $Y \overset{\circ}{=} X - \sum \Delta X \chi \left(|\Delta X| \geq 1 \right)$ is also a good integrator. If we prove that Y is a semimartingale, then we obviously prove that X is a semimartingale as well. Y does not contain 'big jumps hence if it is a semimartingale, then it is a special semimartingale[19]. Therefore the decomposition of Y is unique[20]. As the decomposition is unique it is sufficient to prove that Y is a semimartingale on every interval $[0, t]$.

2. As we have already seen[21] if probability measures $\mathbf{P}$ and $\mathbf{Q}$ are equivalent, that is the measure-zero sets under $\mathbf{P}$ and $\mathbf{Q}$ are the same, then X is a semimartingale under $\mathbf{P}$ if and only if it is a semimartingale under $\mathbf{Q}$. Therefore it is sufficient to prove that if X is a good integrator under $\mathbf{P}$ then one can find a probability measure $\mathbf{Q}$ which is equivalent to $\mathbf{P}$ and X is a semimartingale under $\mathbf{Q}$. Observe that a sequence of random variables is stochastically convergent to some random variable if and only if any subsequence of the original sequence has another subsequence which is almost surely convergent to the same function. Therefore the stochastic convergence depends only on the collection of measure-zero sets, which is not changing during the equivalent change of measure. From this it is obvious that the class of good integrators is not changing under the equivalent change of measure.

3. Let us fix an interval $[0, t]$. As the trajectories of X are regular the trajectories are bounded on any finite interval. Hence $\eta \overset{\circ}{=} \sup_{s \leq t} |X(s)| < \infty$. Again by the regularity of the trajectories it is sufficient to calculate the supremum over the rational points $s \leq t$. Therefore η is a random variable. Let $A_m \overset{\circ}{=} \{ m \leq \eta < m + 1 \}$ and $\zeta \overset{\circ}{=} \sum_m 2^{-m} \chi_{A_m}$. ζ is evidently bounded, and as

---

[18]See: Lemma 2.12, page 118.
[19]See: Example 4.47, page 258.
[20]See: Corollary 3.41, page 205.
[21]See: Corollary 4.58, page 271.

η is finite ζ is trivially positive. As

$$\mathbf{E}\left(\eta\zeta\right) = \sum_m \mathbf{E}\left(\eta 2^{-m}\chi\left(m \le \eta < m+1\right)\right) \le \sum_m \left(m+1\right)2^{-m}$$

it is obvious that $\eta\zeta$ is integrable under $\mathbf{P}$.

$$\mathbf{R}\left(A\right) \overset{\circ}{=} \frac{1}{\mathbf{E}\left(\zeta\right)}\int_A \zeta d\mathbf{P}$$

is a probability measure and as ζ is positive it is equivalent to $\mathbf{P}$. For every $s \le t$

$$\int_\Omega \left|X\left(s\right)\right| d\mathbf{R} \le \int_\Omega \eta d\mathbf{R} = \frac{1}{\mathbf{E}\left(\zeta\right)}\int_\Omega \eta\zeta d\mathbf{P} < \infty,$$

therefore $X\left(s\right)$ is integrable under $\mathbf{R}$ for all s. To make the notation simple we assume that $X\left(s\right)$ are already integrable under $\mathbf{P}$ for all $s \in [0, t]$.

4. Let us define the set

$$B \overset{\circ}{=} \left\{\left(E \bullet X\right)\left(t\right) : |E| \le 1, E \in \mathcal{E}\right\}, \tag{5.16}$$

where $\mathcal{E}$ is the set of predictable step processes over $[0, t]$. Using the continuity property of the good integrators we prove that B is stochastically bounded, that is for every $\varepsilon > 0$ there is a number k, such that $\mathbf{P}\left(|\eta| \ge k\right) < \varepsilon$ for all $\eta \in B$. If it was not true then there were an $\varepsilon > 0$, a sequence of step processes $|E_n| \le 1$ and $k_n \nearrow \infty$, such that

$$\mathbf{P}\left(\frac{\left(E_n \bullet X\right)\left(t\right)}{k_n} \ge 1\right) \ge \varepsilon.$$

The sequence $\left(E_n/k_n\right)$ is uniformly converging to zero, hence by the continuity property of the good integrators

$$\frac{\left(E_n \bullet X\right)\left(t\right)}{k_n} = \left(\frac{E_n}{k_n} \bullet X\right)\left(t\right) \overset{P}{\to} 0,$$

which is, by the indirect assumption, is not true.

5. As a last step of the proof in the next point we shall prove that for every non-empty, stochastically bounded, convex subset B of L^1 there is a probability measure $\mathbf{Q}$ which is equivalent to $\mathbf{P}$ and for which

$$\sup\left\{\int_\Omega \beta d\mathbf{Q} : \beta \in B\right\} \overset{\circ}{=} c < \infty. \tag{5.17}$$

From this the theorem follows as for every partition of $[0, t]$

$$0 = t_0 < t_1 < \ldots < t_{n+1} = t$$

if[22]

$$\xi_i \overset{\circ}{=} \operatorname{sgn} \left(\mathbf{E}^{\mathbf{Q}} \left(X \left(t_{i+1} \right) - X \left(t_i \right) \mid \mathcal{F}_{t_i} \right) \right),$$

and

$$E \overset{\circ}{=} \sum_i \xi_i \chi \left(\left(t_i, t_{i+1} \right] \right)$$

then as $|E| \le 1$

$$\left(E \bullet X \right) \left(t \right) \in B,$$

therefore

$$c \ge \mathbf{E}^{\mathbf{Q}} \left(\left(E \bullet X \right) \left(t \right) \right) = \sum_{i=0}^{n} \mathbf{E}^{\mathbf{Q}} \left(\xi_i \left[X \left(t_{i+1} \right) - X \left(t_i \right) \right] \right) =$$

$$= \sum_{i=0}^{n} \mathbf{E}^{\mathbf{Q}} \left(\mathbf{E}^{\mathbf{Q}} \left(\xi_i \left[X \left(t_{i+1} \right) - X \left(t_i \right) \right] \mid \mathcal{F}_{t_i} \right) \right) =$$

$$= \sum_{i=0}^{n} \mathbf{E}^{\mathbf{Q}} \left(\xi_i \mathbf{E}^{\mathbf{Q}} \left(X \left(t_{i+1} \right) - X \left(t_i \right) \mid \mathcal{F}_{t_i} \right) \right) =$$

$$= \mathbf{E}^{\mathbf{Q}} \left(\sum_{i=0}^{n} \left| \mathbf{E}^{\mathbf{Q}} \left(X \left(t_i \right) - X \left(t_{i+1} \right) \mid \mathcal{F}_{t_i} \right) \right| \right).$$

Hence X is a quasi-martingale under $\mathbf{Q}$. Therefore[23] it is a semimartingale under $\mathbf{Q}$.

6. Let $B \subseteq L^1 \left(\Omega \right)$ be a non-empty stochastically bounded convex convex set[24]. We prove the existence of the equivalent measure $\mathbf{Q}$ in (5.17) with the Hahn–Banach theorem. Let L_+^∞ denote the set of non-negative functions in L^∞.

$$H \overset{\circ}{=} \left\{ \zeta \in L_+^\infty : \sup \left\{ \int_\Omega \beta \zeta d\mathbf{P} : \beta \in B \right\} < \infty \right\}.$$

---

[22] Of course $\mathbf{E}^{\mathbf{Q}}$ denotes the expected value under $\mathbf{Q}$.

[23] See: Corollary 5.14, page 307.

[24] One can assume that B is stochastically bounded above that is for every $\varepsilon > 0$ there is a $k \left(\varepsilon \right)$ such that $\mathbf{P} \left(B \ge k \left(\varepsilon \right) \right) \le \varepsilon$.

It is sufficient to prove that H contains a strictly positive function ζ_0, since in this case

$$\mathbf{Q}(A) \overset{\circ}{=} \frac{1}{\mathbf{E}(\zeta_0)} \int_A \zeta_0 d\mathbf{P}$$

is an equivalent probability measure for which (5.17) holds. Let $\mathcal{G}$ be the set of points of positivity of the functions in H. The set $\mathcal{G}$ is closed under the countable union: if $\zeta_n \in H$, and

$$\sup \left\{ \int_\Omega \beta \zeta_n d\mathbf{P} : \beta \in B \right\} \leq c_n$$

$c_n \geq 1$ then

$$\sum_n \frac{2^{-n}}{c_n \|\zeta_n\|_\infty} \zeta_n \in H.$$

Using the lattice property of $\mathcal{G}$ in the usual way one can prove that $\mathcal{G}$ contains a set D which has maximal measure, that is $\mathbf{P}(G) \leq \mathbf{P}(D)$ for all $G \in \mathcal{G}$. Of course to D there is a $\zeta_D \in H$. We should prove that $\mathbf{P}(D) = 1$, hence in this case $\zeta_D \in H$, as an equivalence class, it is strictly positive. Let us denote by C the complement of D. We shall prove that $\mathbf{P}(C) = 0$. As an indirect assumption let us assume that

$$\mathbf{P}(C) \overset{\circ}{=} \varepsilon > 0. \tag{5.18}$$

As B is stochastically bounded to our $\varepsilon > 0$ in (5.18) there is a k, such that $\mathbf{P}(\beta \geq k) \leq \varepsilon/2$ for all random variable $\beta \in B$. From this $\theta \overset{\circ}{=} 2k\chi_C \notin B$. Of course, if $\vartheta \geq 0$, then $\mathbf{P}(\theta + \vartheta \geq k) \geq \varepsilon$ hence $\theta + \vartheta \notin B$, that is $\theta \notin B - L_+^1$. We can prove a bit more: θ is not even in the closure in $L^1(\Omega)$ of the convex[25] set $B - L_+^1$. That is

$$\theta \notin \mathrm{cl}\left(B - L_+^1\right).$$

If $\gamma_n \overset{\circ}{=} \beta_n - \vartheta_n \to \theta$ in $L^1(\Omega)$, then $\gamma_n \overset{P}{\to} \theta$, but if δ is small enough, then as $\vartheta_n \geq 0$

$$\mathbf{P}(|\gamma_n - \theta| > \delta) \overset{\circ}{=} \mathbf{P}(|\beta_n - \vartheta_n - \theta| > \delta) \geq$$
$$\geq \mathbf{P}(\{\beta_n < k\} \cap \{\theta \geq 2k\}) =$$
$$= \mathbf{P}(\{\beta_n < k\} \cap C) = \mathbf{P}(C \setminus \{\beta_n \geq k\}) \geq \frac{\varepsilon}{2},$$

---

[25]The B is conves hence $B - L_+^1$ is also convex.

which is impossible. By the Hahn–Banach theorem[26] there is a $\zeta \neq 0 \in L^\infty(\Omega)$, such that

$$\int_\Omega (\beta - \vartheta) \, \zeta d\mathbf{P} < \int_\Omega \theta \zeta d\mathbf{P}, \quad \beta \in B, \vartheta \in L^1_+. \tag{5.19}$$

Observe that $\zeta \geq 0$ as if ζ was negative with positive probability then the left-hand side of (5.19) for some $\vartheta \in L^1_+$ would be greater than the fix value on the right-hand side. As $\zeta \neq 0$ obviously ζ is positive on some subset $U \subseteq C$ with positive measure. Taking $\vartheta = 0$

$$\int_\Omega \beta \zeta d\mathbf{P} < \int_\Omega \theta \zeta d\mathbf{P} \stackrel{\circ}{=} c < \infty, \quad \beta \in B$$

that is $\zeta \in H$. Extending the set D with the support of ζ as $U \subseteq C = D^c$ one can get a set in $\mathcal{G}$ which has larger measure than D. This contradicts to the definition of D. $\qquad \square$

Theorem 5.18 (Stricker) *Let X be a semimartingale under a filtration $\mathcal{F}$ and let $\mathcal{G}_t \subseteq \mathcal{F}_t$ for all t for some filtration $\mathcal{G}$. If X is adapted to $\mathcal{G}$ then X is a semimartingale under $\mathcal{G}$ as well.*

Proof. The set of step processes under $\mathcal{G}$ are also step processes under $\mathcal{F}$. $\qquad \square$

Example 5.19 The theorem of Stricker is not valid for local martingales.

Let us remark that the above property holds for martingales as well. The problem with the local martingales comes from the fact that when one shrinks the filtration the set of stopping times can also shrink. Let η be a symmetric random variable which does not have an expected value. Let us assume that the density function of η is continuous and strictly positive. Let

$$X(t) \stackrel{\circ}{=} \begin{cases} 0 & \text{if } t < 1 \\ \eta & \text{if } t \geq 1 \end{cases}.$$

Let the filtration

$$\mathcal{F}_t \stackrel{\circ}{=} \begin{cases} \sigma(|\eta|) & \text{if } t < 1 \\ \sigma(\eta) & \text{if } t \geq 1 \end{cases}.$$

[26]Using that L^∞ is the dual of L^1 and that every convex closed set can be strictly separated from any point of its complement.

The

$$\tau_n(\omega) \overset{\circ}{=} \begin{cases} 0 & \text{if } |\eta| \geq n \\ \infty & \text{if } |\eta| < n \end{cases}$$

is stopping time under $\mathcal{F}$, and as η is symmetric X^{τ_n} is a martingale. The filtration generated by X is

$$\mathcal{G}_t \overset{\circ}{=} \begin{cases} \{0, \Omega\} & \text{if } t < 1 \\ \sigma(\eta) & \text{if } t \geq 1 \end{cases}.$$

The τ_n is not a stopping time under $\mathcal{G}$, as by the assumptions about the density function of η

$$\{\tau_n \leq 0\} = \{\tau_n = 0\} = \{|\eta| \geq n\} \notin \{0, \Omega\} = \mathcal{G}_0.$$

Let τ be a stopping time for the $\mathcal{G}$. If on a set of positive measure $\tau \geq 1$, then almost surely $\tau \geq 1$, therefore $X^\tau(1) = X(1) = \eta$ is not integrable[27], so X is not a local martingale under $\mathcal{G}$. $\qquad\square$

5.3 Integration of Adapted Product Measurable Processes

Let M be a continuous local martingale. If the processes X and Y are almost everywhere equal under the Doléans measure[28] α_M then by the definition of $\mathcal{L}^2_{\text{loc}}(M)$ they belong to the same equivalence class. Hence if the integrals $X \bullet M$ and $Y \bullet M$ exist they are indistinguishable. Using this in certain cases we can extend the integration from progressively measurable processes to adapted product measurable functions.

Proposition 5.20 *If M is continuous local martingale and $\alpha_M \ll \lambda \times \mathbf{P}$ then one can define the stochastic integral $X \bullet M$ for every adapted product measurable process X.*

Proof. The proposition directly follows from the next proposition. $\qquad\square$

Proposition 5.21 *Every adapted product measurable process X is $\lambda \times \mathbf{P}$ equivalent to a progressively measurable process $\hat{X}$.*

Proof. We divide the proof into several steps. One can assume that X is defined on an interval $[a, b)$ and X is bounded.

---

[27] The $\mathcal{G}$ does not satisfy the usual conditions but adding the measure-zero sets is not solving the problem.

[28] See: Definition 2.56, page 151.

1. Introduce the functions $\varphi_n : \mathbb{R} \to \mathbb{R}$

$$\varphi_n(t) \overset{\circ}{=} \frac{k-1}{2^n}, \quad \text{if} \quad t \in \left(\frac{k-1}{2^n}, \frac{k}{2^n} \right].$$

As a first step we prove that for any $s \geq 0$

$$t - \frac{1}{2^n} \leq \varphi_n(t-s) + s < t. \tag{5.20}$$

As $s \geq 0$, there is an integer number $m \geq 0$ such that

$$\frac{m}{2^n} \leq s < \frac{m+1}{2^n}.$$

Hence, if $t \in ((k-1)/2^n, k/2^n]$, then

$$\frac{k-m-2}{2^n} < t - s \leq \frac{k-m}{2^n}.$$

There are two possibilities:

$$\frac{k-m-2}{2^n} < t - s \leq \frac{k-m-1}{2^n}$$

or

$$\frac{k-m-1}{2^n} < t - s \leq \frac{k-m}{2^n}.$$

By the definition of φ_n in the first case

$$\varphi_n(t-s) \overset{\circ}{=} \frac{k-m-2}{2^n},$$

in the second case

$$\varphi_n(t-s) \overset{\circ}{=} \frac{k-m-1}{2^n}.$$

By simple calculation in the first case

$$t - \frac{1}{2^n} \leq \frac{k-m-2}{2^n} + s \overset{\circ}{=} \varphi_n(t-s) + s < t,$$

and in the second case

$$t - \frac{1}{2^n} \leq \frac{k-m-1}{2^n} + s = \varphi_n(t-s) + s < t.$$

2. Fix an $s \geq 0$, and let us define the stochastic processes

$$X_n \left(s, t, \omega \right) \overset{\circ}{=} X \left(\varphi_n \left(t - s \right) + s, \omega \right).$$

As X is perhaps defined only on the interval $[a, b)$, it can happen that $X_n \left(s, t, \omega \right)$ is not defined so to make the notation as simple as possible we assume that if

$$\varphi_n \left(t - s \right) + s < a,$$

then $X_n \left(s, t, \omega \right) \overset{\circ}{=} 0$.

$$\varphi_n \left(t - s \right) + s < t,$$

and $X \left(t, \omega \right)$ is $\mathcal{F}_t$ adapted, hence if $t \in [a, b]$, then $X_n \left(s, t, \omega \right)$ is also $\mathcal{F}_t$-adapted. $X_n \left(s, t, \omega \right)$ is a left-continuous step function on $[a, b]$ for a fixed s and ω. So $X_n \left(s, t, \omega \right)$ is progressively measurable for every s. If $s_1 < s_2$ are two points in the interval $[i/2^n, (i+1)/2^n)$ for some integer $i \geq 0$, then

$$\varphi_n \left(t - s_1 \right) + s_1 \neq \varphi_n \left(t - s_2 \right) + s_2,$$

since otherwise

$$\frac{1}{2^n} > s_2 - s_1 = \varphi_n \left(t - s_2 \right) - \varphi_n \left(t - s_1 \right) > 0$$

which is impossible as $\varphi_n \left(u \right) \overset{\circ}{=} k/2^n$. Using (5.20) one can easily prove that

$$s \mapsto \varphi_n \left(t - s \right) + s$$

is an injective map of $[i/2^n, (i+1)/2^n)$ into $[t - 1/2^n, t)$. Using the definition of this map[29] for all t

$$\int\limits_{i/2^n}^{(i+1)/2^n} |X_n \left(s, t \right) - X \left(t \right)| \, ds \overset{\circ}{=} \int\limits_{i/2^n}^{(i+1)/2^n} |X \left(\varphi_n \left(t - s \right) + s \right) - X \left(t \right)| \, ds \leq$$

$$\leq \int\limits_{t-1/2^n}^{t} |X \left(u \right) - X \left(t \right)| \, du \leq$$

$$\leq \int\limits_{0}^{1/2^n} |X \left(t - h \right) - X \left(t \right)| \, dh.$$

[29]The function $y = x + \sum jumps$ where the number of jumps is finite in the interval.

Using this if

$$I \overset{\circ}{=} \mathbf{E} \left(\int_a^b \int_0^1 |X_n(s,t,\omega) - X(t,\omega)|\, ds dt \right),$$

then by Fubini's theorem, using that X is product measurable

$$I \overset{\circ}{=} \mathbf{E} \left(\int_a^b \sum_{i=0}^{2^n-1} \int_{i/2^n}^{(i+1)/2^n} |X_n(s,t,\omega) - X(t,\omega)|\, ds dt \right) \leq \qquad (5.21)$$

$$\leq 2^n \mathbf{E} \left(\int_a^b \int_0^{1/2^n} |X(t-h,\omega) - X(t,\omega)|\, dh dt \right) =$$

$$= 2^n \int_0^{1/2^n} \mathbf{E} \left(\int_a^b |X(t-h,\omega) - X(t,\omega)|\, dt \right) dh \leq$$

$$\leq \max_{0 \leq h \leq 2^{-n}} \frac{2^n}{2^n} \cdot \mathbf{E} \left(\int_a^b |X(t-h,\omega) - X(t,\omega)|\, dt \right).$$

3. Let f be an integrable function over $\mathbb{R}$ with respect to λ. We show that

$$\lim_{h \to 0} \int_{\mathbb{R}} |f(t-h) - f(t)|\, dt = 0. \qquad (5.22)$$

The relation is trivial if f is continuous. Let $T_h f(t) \overset{\circ}{=} f(t-h)$. Obviously

$$\|T_h f\|_1 = \|f\|_1.$$

As λ is regular the continuous functions are dense in $L^1(\mathbb{R}, \lambda)$. If g is continuous and $\|f - g\|_1 < \varepsilon$ then

$$\|T_h f - f\|_1 \leq \|T_h f - T_h g\|_1 + \|T_h g - g\|_1 + \|g - f\|_1 \leq$$
$$\leq 2\varepsilon + \|T_{\mathbf{h}} g - g\|_1,$$

from which (5.22) is obvious.

4. X is bounded, hence if $n \to \infty$ then by the just proved result and by the Dominated Convergence Theorem, using that $[a, b]$ is finite, the last expression in (5.21) goes to zero. Hence in the space

$$L^1([0,1] \times [a,b] \times \Omega)$$

$X_n(s,t,\omega) \to X(t,\omega)$. For a subsequence $X_{n_k}(s,t,\omega) \overset{a.s.}{\to} X(t,\omega)$. Let H be the set of such points (s,t,ω) where the convergence holds. By Fubini's theorem on a subset of the product space which has complete measures, almost all the sections parallel with the coordinate axes have complete measure. Hence for almost all $s \in [0,1]$ the measure of

$$H_s \stackrel{\circ}{=} \{(t,\omega) : (s,t,\omega) \in H\}$$

is equal to the measure of $[a,b] \times \Omega$. Hence there is a point s for which this holds, that is there is an s such that for almost all (t,ω)

$$\lim_{k \to \infty} X_{n_k}(s,t,\omega) = X(t,\omega).$$

If

$$\widehat{X} \stackrel{\circ}{=} \liminf_{k \to \infty} X_{n_k},$$

then $\widehat{X}$ is progressively measurable. On the other hand $\widehat{X}(t) \stackrel{a.s.}{=} X(t)$ for almost all[30] t. $\qquad \square$

Corollary 5.22 *Let w be a Wiener process. In the definition of the stochastic integral $X \bullet w$ the integrand X can be an arbitrary adapted product measurable process for which*

$$\mathbf{P}\left(\int_0^t X^2(s)ds < \infty\right) = 1.$$

Proof. Let X be an adapted product measurable process and let $\widehat{X}$ be the progressively measurable process in the previous proposition. $X \stackrel{a.s.}{=} \widehat{X}$ with respect to $\lambda \times \mathbf{P}$, so by Fubini's theorem

$$\int_0^t (X - \hat{X})^2 ds \stackrel{a.s.}{=} 0$$

for almost all t. As the integral is continuous in t the relation is valid for every t. Let N be a continuous local martingale. To construct the integral $X \bullet w$ one should show that $X \bullet [w,N]$ is adapted. By the Kunita-Watanabe inequality

$$\mathbf{E}\left(\left(\int_0^t (X - \widehat{X})d\,[w,N]\right)^2\right) \leq \mathbf{E}\left([N](t)\int_0^t (X - \widehat{X})^2 ds\right) = 0.$$

So

$$\int_0^t X d\,[w,N] \stackrel{a.s.}{=} \int_0^t \widehat{X} d\,[w,N]$$

---

[30] And not for all t.

for every t. But as $\widehat{X}$ is progressively measurable, the integral on the right-hand side is adapted. As the filtration contains the measure-zero sets the integral on the left-hand side is also adapted. □

5.4   Theorem of Fubini for Stochastic Integrals

In this section we discuss a generalization of Fubini's theorem to stochastic integrals. Let X be a semimartingale on a stochastic base $(\Omega, \mathcal{A}, \mathbf{P}, \mathcal{F})$ and let $(C, \mathcal{C}, \mu)$ be an arbitrary finite measure space. In this generalization we want to interchange a stochastic and a classical abstract integral:

$$\int_C (H \bullet X)(c) d\mu(c) = \left(\int_C H(c)\, d\mu(c) \right) \bullet X.$$

As in the classical case the discussion is built on the Monotone Class Theorem and on the Dominated Convergence Theorems. First with the Monotone Class Theorem one proves the theorem in the bounded case then one generalizes it to unbounded integrands. Let us shortly discuss the general properties of the expressions in the equality. Assume that H is bounded. As H is bounded and as μ is finite by the classical theorem of Fubini, the classical parametric integral $\int_C H(c, t, \omega)\, d\mu(c)$ is bounded and measurable. Hence the double integral on the right-hand side obviously exists. On the other hand, on the left-hand side the parametric stochastic integral $(H \bullet X)(c)$ is not necessarily bounded. The classical integration is monotone, order preserving operation, but the stochastic integration is not! As $(H \bullet X)(c)$ is not necessarily bounded, it is not obvious that for a fixed t and ω why does the classical integral $\int_C (H \bullet X)(c) d\mu(c)$ exist? The same problem arises when one wants to use the Dominated Convergence Theorem. On the right-hand side, as the classical integral is order preserving, one can easily apply the Dominated Convergence Theorem, but on the left-hand side we cannot directly use it.

Recall that in the general case we have defined the stochastic integral only for predictable processes. On the other hand, if the integrator is continuous then the set of the possible integrable processes is a subset of the progressively measurable processes. If the integrator is a Wiener process then one can assume that the integrand is adapted and product measurable. In this section we shall denote by $\mathcal{G}$ the σ-algebra of the subsets of $\mathbb{R}_+ \times \Omega$ for which the integrands in the stochastic integrals with respect to X are measurable.

As a generalization of the classical approach first we discuss the measurability of the parametric stochastic integrals.

Proposition 5.23 (Measurability of the parametric integrals) *Let X be a semimartingale and let $(C, \mathcal{C})$ be an arbitrary measurable space. If $H(c, t, \omega)$ is measurable with respect to the product σ-algebra $\mathcal{C} \times \mathcal{G}$ and $H(c)$ is integrable with* *respect to X for every c, then the parametric stochastic integral $c \mapsto H(c) \bullet X$ has a version, denoted by $(H \bullet X)(c)$ which is measurable with respect to the product σ-algebra $\mathcal{C} \times \mathcal{B}(\mathbb{R}_+) \times \mathcal{A}$. This means that there is a function $Z(c, t, \omega) \overset{\circ}{=} (H \bullet X)(c, t, \omega)$, which is:*

1. *measurable with respect to the product σ-algebra $\mathcal{C} \times \mathcal{B}(\mathbb{R}_+) \times \mathcal{A}$,*
2. *$Z(c)$ is almost surely right-regular and adapted in t for any fixed c,*
3. *the function $(t, \omega) \mapsto Z(c, t, \omega)$ is indistinguishable from the stochastic integral[31] $H(c) \bullet X$ for any fixed c.*

Proof. The proof is nearly the same as the proof of the classical theorem of Fubini. Let $\mathcal{S}$ be the set of bounded processes on the product $C \times \mathbb{R}_+ \times \Omega$ for which the proposition holds. For an arbitrary set $B \subseteq C$ as the stochastic integration is homogeneous $\chi_B(H \bullet X) = (\chi_B H \bullet X)$, where for every $c \in C$ the two sides are indistinguishable. Of course if $B \in \mathcal{C}$ then for any fixed version of $H \bullet X$

$$Z(c, t, \omega) \overset{\circ}{=} \chi_B(c)(H \bullet X)$$

satisfies the proposition. If $H(c, t, \omega) = H_1(c) H_2(t, \omega)$, where H_1 is a $\mathcal{C}$-measurable step function and H_2 is $\mathcal{G}$-measurable, then by the linearity[32] of stochastic integral the product $H_1(H_2 \bullet X)$ satisfies the proposition. $\mathcal{S}$ is obviously a vector space. One should prove that $\mathcal{S}$ is a λ-system, that is $\mathcal{S}$ is closed under the monotone convergence, hence in this case using the Monotone Class Theorem we can use the property in the proposition for every bounded $(\mathcal{C} \times \mathcal{G})$-measurable process. To prove this monotone class property one needs the following lemma:

Lemma 5.24 *Let $(Z_n(c, t, \omega))_n$ be a sequence of right-regular, $(\mathcal{C} \times \mathcal{B}(\mathbb{R}_+) \times \mathcal{A})$-measurable functions. If for all c the sequence $(Z_n(c))_n$ is uniformly convergent on compacts in probability, then there is a function Z which is*

1. *$(\mathcal{C} \times \mathcal{B}(\mathbb{R}_+) \times \mathcal{A})$-measurable,*
2. *$Z(c)$ is almost surely right-regular in t for every c,*
3. *$Z_n(c) \overset{ucp}{\to} Z(c)$ for every c.*

Proof. Let us denote by $D[0, \infty)$ the space of right-regular functions. $D[0, \infty)$ is a complete metric space with the topology of uniform convergence on compacts. Let us denote by d the metric of $D[0, \infty)$. As the functions in $D[0, \infty)$ are right-regular for every interval $[0, s]$ one can calculate the supremum in the semi-norm $\sup_{t \leq s} |f(t)|$ only for the rational numbers in $[0, s]$, hence the product measurability of the functions Z_n implies the measurability of the value of the

---

[31] Observe that the stochastic integral is defined up to indistinguishability.

[32] Recall that the equality $(aH) \bullet X = a(H \bullet X)$ means that the two side are indistinguishable, hence we can apply the linearity only for finite valued processes H_1.

semi-norms $\sup_{t \leq s} |Z_n(t)|$, therefore for $d(Z_i(c, \omega), Z_j(c, \omega))$ is measurable in (c, ω) all i and j. For all c let $n_0(c) \stackrel{\circ}{=} 1$. With induction let us define the sequence

$$n_k(c) \stackrel{\circ}{=} \inf \left\{ m > n_{k-1}(c) : \sup_{i, j \geq m} \mathbf{P}\left(d\left(Z_i(c), Z_j(c)\right) > 2^{-k}\right) < 2^{-k} \right\}.$$

As we observed the real valued functions $d(Z_i(c, \omega), Z_j(c, \omega))$ are measurable in (c, ω), therefore by Fubini's theorem the probability in the formula depends on c in a measurable way. Hence n_k is a measurable function of c. Let us define the 'stopped variables'

$$Y_k(c, t, \omega) \stackrel{\circ}{=} Z_{n_k(c)}(c, t, \omega).$$

For all open set G

$$\{Y_k \in G\} = \cup_p \{n_k = p, Z_p \in G\},$$

therefore Y_k is also product measurable. For all c

$$\sum_k \sup_{i, j \geq k} \mathbf{P}\left(d\left(Y_i(c), Y_j(c)\right) > 2^{-k}\right) \leq \sum_k 2^{-k} < \infty,$$

hence for every c by the Borel and Cantelli lemma if the indexes i, j are big enough then except on a measure-zero set $\omega \in N(c)$

$$d(Y_i(c, \omega), Y_j(c, \omega)) \leq 2^{-k}.$$

$D[0, \infty)$ is complete, hence $(Y_i(c, \omega))$ is almost surely convergent in $D[0, \infty)$ for all c. The function

$$Z(c, t, \omega) \stackrel{\circ}{=} \begin{cases} \lim_i Y_i(c, t, \omega), & \text{if the limit exists,} \\ 0 & \text{otherwise} \end{cases}$$

is product measurable and Z is right-regular almost surely for all c. For an arbitrary c $(Y_i(c) - Z(c))$ is a subsequence of $(Z_n(c) - Z(c))$, therefore it is stochastically convergent in $D[0, \infty)$. The measure is finite therefore for the metric space valued random variables the almost sure convergence implies the stochastic convergence. Hence $Z(c, \omega)$ is the limit of the sequence $(Z_n(c, \omega))$ for almost all ω. □

Returning to the proof of the proposition let us assume that $H_n \in \mathcal{S}$ and $0 \leq H_n \nearrow H$, where H is bounded. By the Dominated Convergence Theorem $H_n(c) \bullet X \stackrel{ucp}{\to} H \bullet X$ for every c. Hence by the lemma $H \bullet X$ has a $(\mathcal{C} \times \mathcal{B}(\mathbb{R}_+) \times \mathcal{A})$-measurable version. That is $H \in \mathcal{S}$. Hence the proposition is valid for bounded processes. If H is not bounded, then let $H_n \stackrel{\circ}{=} H\chi(|H| \leq n)$. The processes H_n are also $(\mathcal{C} \times \mathcal{G})$-measurable, and of course they are bounded. Therefore the processes $H_n \bullet X$ have the stated version. By the Dominated Convergence

Theorem $H_n(c) \bullet X \overset{ucp}{\rightrightarrows} H(c) \bullet X$ for every c. By the lemma this means that $H(c) \bullet X$ also has a measurable version.

Theorem 5.25 (Fubini's theorem for bounded integrands) *Let X be a semimartingale, and let $(C, \mathcal{C}, \mu)$ be an arbitrary finite measure space. Let $H(c, t, \omega)$ be a function measurable with respect to the product σ-algebra $\mathcal{C} \times \mathcal{G}$. Let us denote by $(H \bullet X)(c)$ the product measurable version of the parametric integral $c \mapsto H(c) \bullet X$. If $H(c, t, \omega)$ is bounded, then*

$$\int_C (H \bullet X)(c) d\mu(c) = \left(\int_C H(c) \, d\mu(c) \right) \bullet X, \qquad (5.23)$$

that is the integral of the parametric stochastic integral on the left side is indistinguishable from the stochastic integral on the right side.

Proof. It is not a big surprise that the proof is built on the Monotone Class Theorem again.

1. By the Fundamental Theorem of Local Martingales semimartingale X has a decomposition $X(0) + V + L$, where $V \in \mathcal{V}$ and $L \in \mathcal{H}_{\text{loc}}^2$. For $V \in \mathcal{V}$ one can prove the equality by the classical theorem of Fubini, hence one can assume that $X \in \mathcal{H}_{\text{loc}}^2$. One can easily localize the right side of (5.23). On the left side one can interchange the localization and the integration with respect to c therefore one can assume that $X(0) = 0$ and $X \in \mathcal{H}^2$. Therefore[33] we can assume that $\mathbf{E}([X](\infty)) < \infty$.

2. Let us denote by $\mathcal{S}$ the set of bounded, $(\mathcal{C} \times \mathcal{G})$-measurable processes for which the theorem holds. If $H \overset{\circ}{=} H_1(c) H_2(t, \omega)$, where H_1 is $\mathcal{C}$-measurable step function and H_2 is $\mathcal{G}$-measurable and H_1 and H_2 are bounded functions, then arguing as in the previous proposition

$$\int_C H \bullet X d\mu \overset{\circ}{=} \int_C (H_1(c) H_2) \bullet X d\mu(c) =$$

$$= \int_C \left(\sum_i \alpha_i \chi_{B_i} H_2 \right) \bullet X d\mu(c) =$$

$$= \sum_i \alpha_i \int_C \chi_{B_i} (H_2 \bullet X) \, d\mu(c) =$$

$$= \int_C H_1(c) \, d\mu(c) (H_2 \bullet X) =$$

$$= \left(\int_C H_1(c) \, d\mu(c) H_2 \right) \bullet X = \left(\int_C H d\mu \right) \bullet X,$$

so $H \in \mathcal{S}$.

---

[33]See: Proposition 3.64, page 223.

3. By the Monotone Class Theorem, one should prove that $\mathcal{S}$ is a λ-system. Let $H_n \in \mathcal{S}$ and let $0 \leq H_n \nearrow H$, where H is bounded. We prove that one can take the limit in the equation

$$\int_C (H_n \bullet X)\, d\mu = \left(\int_C H_n d\mu \right) \bullet X.$$

As H is bounded and μ is finite, therefore on the right-hand side the integrands are uniformly bounded so one can apply the classical and the stochastic Dominated Convergence Theorem, so on the right-hand side

$$\left(\int_C H_n d\mu \right) \bullet X \overset{\text{ucp}}{\rightarrow} \left(\int_C H d\mu \right) \bullet X. \qquad (5.24)$$

4. Introduce the notations $Z_n \overset{\circ}{=} H_n \bullet X$ and $Z \overset{\circ}{=} H \bullet X$. One should prove that the left-hand side is also convergent that is

$$\delta \overset{\circ}{=} \sup_t \left| \int_C Z_n(c)\, d\mu(c) - \int_C Z(c)\, d\mu(c) \right| \overset{P}{\rightarrow} 0.$$

By the inequalities of Cauchy–Schwarz and Doob

$$\mathbf{E}(\delta) \leq \mathbf{E}\left(\int_C \sup_t |Z_n(c) - Z(c)|\, d\mu(c) \right) \leq$$

$$\leq \sqrt{\mu(C)} \sqrt{\mathbf{E}\left(\int_C \left(\sup_t |Z_n(c) - Z(c)| \right)^2 d\mu(c) \right)} =$$

$$= \sqrt{\mu(C)} \sqrt{\int_C \mathbf{E}\left(\left(\sup_t |Z_n(c) - Z(c)| \right)^2 \right) d\mu(c)} \leq$$

$$\leq \sqrt{\mu(C)} \sqrt{\int_C 4 \cdot \mathbf{E}\left((Z_n(c, \infty) - Z(c, \infty))^2 \right) d\mu(c)}.$$

By Itô's isometry[34] the last integral is

$$\int_C \mathbf{E}\left(\int_0^t (H_n - H)^2\, d[X] \right) d\mu. \qquad (5.25)$$

As μ and $\mathbf{E}([X](\infty))$ are finite and as the integrand is bounded and $H_n \to H$ by the classical Dominated Convergence Theorem the (5.25) goes to zero.

[34]See: Proposition 2.64, page 156.

So $\mathbf{E}(\delta) \to 0$, that is

$$\int_C (H_n \bullet X)\, d\mu \stackrel{\circ}{=} \int_C Z_n d\mu \stackrel{\text{ucp}}{\to} \int_C Z d\mu \stackrel{\circ}{=} \int_C (H \bullet X)\, d\mu. \qquad (5.26)$$

Particularly

$$\int_C \sup_t |Z_n(c) - Z(c)|\, d\mu(c) < \infty, \quad \text{a.s.}$$

The expression

$$\left(\int_C H_n(c)\, d\mu(c) \right) \bullet X = \int_C (H_n(c) \bullet X)\, d\mu(c) \stackrel{\circ}{=} \int_C Z_n d\mu$$

is meaningful, therefore for all t and for almost all outcome ω

$$\int_C |(H(c) \bullet X)(t, \omega)|\, d\mu(c) \stackrel{\circ}{=} \int_C |Z(c, t, \omega)|\, d\mu(c) < \infty.$$

Hence the left-hand side of (5.23) is meaningful for H as well. By (5.24) the right-hand side is also convergent, hence from (5.26)

$$\left(\int_C H d\mu \right) \bullet X = \lim_{n \to \infty} \left(\int_C H_n d\mu \right) \bullet X =$$

$$= \lim_{n \to \infty} \int_C (H_n \bullet X)\, d\mu = \int_C (H \bullet X)\, d\mu.$$

The just proved stochastic generalization of Fubini's theorem is sufficient for most of the applications. On the other hand one can still be interested in the unbounded case:

Theorem 5.26 (Fubini's theorem for unbounded integrands) *Let X be a semimartingale and let $(C, \mathcal{C}, \mu)$ be a finite measure space. Let $H(c, t, \omega)$ be a $(\mathcal{C} \times \mathcal{G})$-measurable process, and assume that the expression*

$$\|H(t, \omega)\|_2 \stackrel{\circ}{=} \sqrt{\int_C H^2(c, t, \omega)\, d\mu(c)} < \infty \qquad (5.27)$$

is integrable with respect to X. Under these conditions μ almost surely the stochastic integral $H(c) \bullet X$ exists and if $(H \bullet X)(c)$ denote the measurable version of this parametric integral then

$$\int_C (H \bullet X)(c) d\mu(c) = \left(\int_C H(c)\, d\mu(c) \right) \bullet X. \qquad (5.28)$$

Proof. If on the place of H one puts $H_n \stackrel{\circ}{=} H\chi\left(|H| \leq n\right)$ then the equality holds by the previous theorem. As in the proof of the classical Fubini's theorem one should take a limit on both sides of the truncated equality.

1. Let us first investigate the right-hand side of the equality. By the Cauchy–Schwarz inequality

$$\int_C |H\left(c,t,\omega\right)| \, d\mu\left(c\right) \leq \sqrt{\mu\left(C\right)}\sqrt{\int_C H^2\left(c,t,\omega\right) d\mu\left(c\right)}. \qquad (5.29)$$

By the assumptions μ is finite and $H\left(c,t,\omega\right)$ as a function of c is in the space $L^2\left(\mu\right) \subseteq L^1\left(\mu\right)$, hence by the Dominated Convergence Theorem for all (t,ω)

$$\int_C H_n\left(c,t,\omega\right) d\mu\left(c\right) \to \int_C H\left(c,t,\omega\right) d\mu\left(c\right).$$

By the just proved inequality (5.29) the processes $\int_C H d\mu$ and $\int_C |H| \, d\mu$ are integrable with respect to X, hence by the Dominated Convergence Theorem for stochastic integrals $\left(\int_C H_n d\mu\right) \bullet X \stackrel{\mathrm{ucp}}{\to} \left(\int_C H d\mu\right) \bullet X$. This means that one can take the limit on the right side of the equation.

2. Now let us investigate the left-hand side. We first prove that for almost all c the integral $H\left(c\right) \bullet X$ exists. Let $X \stackrel{\circ}{=} X\left(0\right) + V + L$, where $V \in \mathcal{V}$, $L \in \mathcal{L}$ is the decomposition of X for which the integral $\|H\left(t,\omega\right)\|_2 \bullet X$ exists. One can assume that $V \in \mathcal{V}^+$. Using (5.29) and for every trajectory the theorem of Fubini

$$\int_C \int_0^t |H| \, dV \, d\mu = \int_0^t \int_C |H| \, d\mu \, dV = \int_0^t \|H\|_1 \, dV$$

$$\leq \sqrt{\mu\left(C\right)} \int_0^t \|H\|_2 \, dV < \infty.$$

Therefore for any t for almost every[35] c the integral $\int_0^t H(c)dV$ is finite. Of course if the integral exists for every rational t then it exists for every t, therefore unifying the measure-zero sets it is easy to show that for almost all c the integral $H(c) \bullet V$ is meaningful. Recall that a process G is integrable with respect to the local martingale L if and only if $\sqrt{G^2 \bullet [L]} \in \mathcal{A}_{\mathrm{loc}}^+$. This means that $\|H\|_2 \stackrel{\circ}{=} \sqrt{\int_C H^2\left(c\right) d\mu\left(c\right)}$ is integrable if and only if there is a localizing sequence $\left(\tau_n\right)$

---

[35]Of course with respect to μ.

for which the expected value of

$$
\sqrt{\int_0^{T_n} \int_C H^2\left(c\right) d\mu\left(c\right) d\left[L\right]} = \sqrt{\int_C \int_0^{T_n} H^2\left(c\right) d\left[L\right] d\mu\left(c\right)}
$$

is finite. By Jensen's inequality

$$
\sqrt{\int_C \int_0^{T_n} H^2\left(c\right) d\left[L\right] d\frac{\mu\left(c\right)}{\mu(C)}} \geq \int_C \sqrt{\int_0^{T_n} H^2\left(c\right) d\left[L\right]} d\frac{\mu\left(c\right)}{\mu(C)}.
$$

Therefore by Fubini's theorem

$$
\int_C \mathbf{E}\left(\sqrt{\int_0^{T_n} H^2\left(c\right) d\left[L\right]}\right) d\mu\left(c\right) = \mathbf{E}\left(\int_C \sqrt{\int_0^{T_n} H^2\left(c\right) d\left[L\right]} d\mu\left(c\right)\right) < \infty.
$$

Hence except on a set C_n with $\mu\left(C_n\right) = 0$ the expected value of

$$
\sqrt{\int_0^{T_n} H^2\left(c\right) d\left[L\right]}
$$

is finite. Unifying the measure-zero sets C_n one can easily see that $\sqrt{H^2\left(c\right) \bullet \left[L\right]} \in \mathcal{A}_{\mathrm{loc}}^+$ for almost all[36] c, that is for almost all c the integral $H(c) \bullet L$ exists.

3. If integral $H\left(c\right) \bullet X$ exists, then $H_n\left(c\right) \bullet X \overset{ucp}{\to} H\left(c\right) \bullet X$. Unfortunately, as we mentioned above from the inequality $\left|H_n\left(c\right)\right| \leq \left|H\left(c\right)\right|$ does not follow the inequality $\left|H_n\left(c\right) \bullet X\right| \leq \left|H\left(c\right) \bullet X\right|$, and we do not know that $H\left(c\right) \bullet X$ is μ integrable hence one cannot use the classical Dominated Convergence Theorem for the outer integral with respect to μ. Therefore, as in the proof of the previous theorem, we prove the convergence of the right side with direct estimation. As by the classical Fubini's theorem the theorem is obviously valid if the integrator has finite variation one can assume that $X \in \mathcal{L}$.

4. Let $s \geq 0$. Like in the previous proof introduce the variable

$$
\delta_n \overset{\circ}{=} \sup_{t \leq s} \left|\int_C \left(\left(H_n\left(c\right) - H\left(c\right)\right) \bullet X\right) d\mu\left(c\right)\right|.
$$

---

[36] Of course with respect to μ.

By Davis' inequality

$$\mathbf{E}\left(\delta_{n}\right) \leq \mathbf{E}\left(\int_{C} \sup_{t \leq s}\left|\left(H_{n}\left(c\right) - H\left(c\right)\right) \bullet X\right| d\mu\left(c\right)\right) = \tag{5.30}$$

$$= \int_{C} \mathbf{E}\left(\sup_{t \leq s}\left|\left(H_{n}\left(c\right) - H\left(c\right)\right) \bullet X\right|\right) d\mu\left(c\right) \leq$$

$$\leq K \int_{C} \mathbf{E}\left(\sqrt{\left[\left(H_{n}(c) - H(c)\right) \bullet X\right]}\left(s\right)\right) d\mu =$$

$$= K \int_{C} \mathbf{E}\left(\sqrt{\left(H_{n}(c) - H(c)\right)^{2} \bullet \left[X\right]}\left(s\right)\right) d\mu =$$

$$= \mu(C) K \mathbf{E}\left(\int_{C} \sqrt{\left(\left(H_{n}\left(c\right) - H\left(c\right)\right)^{2} \bullet \left[X\right]\right)}\left(s\right) d\frac{\mu}{\mu(C)}\right) \leq$$

$$\leq \mu(C) K \mathbf{E}\left(\sqrt{\left(\int_{C}\left(H_{n}\left(c\right) - H\left(c\right)\right)^{2} \bullet \left[X\right]\right)\left(s\right) d\frac{\mu}{\mu(C)}}\right) =$$

$$= \sqrt{\mu(C)} K \mathbf{E}\left(\sqrt{\left(\left(\int_{C}\left(H_{n}\left(c\right) - H\left(c\right)\right)^{2} d\mu \bullet \left[X\right]\right)\left(s\right)}\right).$$

$\sqrt{\int_{C} H^{2} d\mu}$ is integrable with respect to X, therefore

$$\sqrt{\int_{C}\left(H_{n}\left(c\right) - H\left(c\right)\right)^{2} d\mu \bullet \left[X\right]} \leq \sqrt{\int_{C} H^{2} d\mu \bullet \left[X\right]} \in \mathcal{A}_{\text{loc}}^{+}.$$

Let $\left(\tau_{m}\right)$ be a localizing sequence. With localization one can assume that the last expected value is finite, that is

$$\mathbf{E}\left(\sqrt{\int_{C} H^{2} d\mu \bullet \left[X^{\tau_{m}}\right]}\right) < \infty.$$

Applying the estimation (5.30) for $X^{\tau_{m}}$ and writing $\delta_{n}^{(m)}$ instead of δ_{n} by the classical Dominated Convergence Theorem $\mathbf{E}\left(\delta_{n}^{(m)}\right) \to 0$. Hence if m is sufficiently large then

$$\mathbf{P}\left(\delta_{n} > \varepsilon\right) \leq \mathbf{P}\left(\delta_{n} > \varepsilon, \tau_{m} > s\right) + \mathbf{P}\left(\tau_{m} \leq s\right) \leq \mathbf{P}\left(\delta_{n}^{(m)} > \varepsilon\right) + \mathbf{P}\left(\tau_{m} \leq s\right)$$

Therefore $\delta_{n} \to 0$ in probability. From this point the proof of the theorem is the same as the proof of the previous one. $\qquad\square$

Corollary 5.27 (Fubini's theorem for local martingales) *Let $(C, \mathcal{C}, \mu)$ be a finite measure space. If L is a local martingale, $H(c, t, \omega)$ is a $(\mathcal{C} \times \mathcal{P})$-measurable function and*

$$\sqrt{\overline{\int_0^t \int_C H^2(c, s) d\mu(c) d[L](s)}} \in \mathcal{A}_{loc}^+,$$

then

$$\int_C \int_0^t H(c, s) \, dL(s) \, d\mu(c) = \int_0^t \int_C H(c, s) \, d\mu(s) \, dL(s). \tag{5.31}$$

If L is a continuous local martingale and H is a $(\mathcal{C} \times \mathcal{R})$-measurable process and

$$\mathbf{P}\left(\int_0^t \int_C H^2(c, s) \, d\mu(c) \, d[L](s) < \infty \right) = 1,$$

then (5.31) holds.

Corollary 5.28 (Fubini's theorem for Wiener processes) *Let $(C, \mathcal{C}, \mu)$ be a finite measure space. If w is a Wiener process, $H(c, t, \omega)$ is an adapted, product measurable process and*

$$\mathbf{P}\left(\int_0^t \int_C H^2(c, s) \, d\mu(c) \, ds < \infty \right) = 1,$$

then

$$\int_C \int_0^t H(c, s) \, dw(s) \, d\mu(c) = \int_0^t \int_C H(c, s) \, d\mu(s) \, dw(s).$$

5.5 Martingale Representation

Let $\mathcal{H}_0^p$ denote the space of $\mathcal{H}^p$ martingales which are zero at time zero. Recall that by definition martingales M and N are orthogonal if their product MN is a local martingale. This is equivalent to the condition that the quadratic variation $[M, N]$ is a local martingale. This implies that if M and N are orthogonal then M^τ and N are also orthogonal for every stopping time τ. The topology in the spaces $\mathcal{H}_0^p$ is given by the norm $\|\sup_t |M(t)|\|_p$. The basic message of the Burkholder–Davis inequality is that this norm is equivalent to the norm

$$\|M\|_{\mathcal{H}_0^p} \doteq \left\| \sqrt{[M](\infty)} \right\|_p. \tag{5.32}$$

In this section we shall use this norm. Observe that if $p \geq 1$ then $\mathcal{H}_0^p$ is a Banach space.

Definition 5.29 *Let $1 \leq p < \infty$. We say that the closed, linear subspace $\mathcal{X}$ of $\mathcal{H}_0^p$ is stable if it is stable under truncation, that is if $X \in \mathcal{X}$ then $X^\tau \in \mathcal{X}$ for every stopping time τ. If $\mathcal{X}$ is a subset of $\mathcal{H}_0^p$ then we shall denote by $\mathrm{stable}_p(\mathcal{X})$ the smallest closed linear subspace of $\mathcal{H}_0^p$ which is closed under truncation and contains $\mathcal{X}$.*

Obviously $\mathcal{H}_0^p$ is a stable subspace. The intersection of stable subspaces is also stable, hence $\mathrm{stable}_p(\mathcal{X})$ is meaningful for every $\mathcal{X} \subseteq \mathcal{H}_0^p$. To make the notation as simple as possible if the subscript p is not important we shall drop it and instead of $\mathrm{stable}_p(\mathcal{X})$ we shall simply write $\mathrm{stable}(\mathcal{X})$.

Lemma 5.30 *Let $1 \leq p < \infty$ and let $\mathcal{X} \subseteq \mathcal{H}_0^p$. Let N be a bounded martingale. If N is orthogonal to $\mathcal{X}$ then N is orthogonal to $\mathrm{stable}(\mathcal{X})$.*

Proof. Let us denote by $\mathcal{Y}$ the set of $\mathcal{H}_0^p$-martingales which are orthogonal to N. Of course $\mathcal{X} \subseteq \mathcal{Y}$ so it is sufficient to prove that $\mathcal{Y}$ is a stable subspace of $\mathcal{H}_0^p$. As we remarked $\mathcal{Y}$ is closed under stopping. Let $M_n \in \mathcal{Y}$ and let $M_n \to M_\infty$ in $\mathcal{H}_0^p$. As N is bounded $M_n N$ is a local martingale which is in class $\mathcal{D}$. Hence it is a uniformly integrable martingale. So $\mathbf{E}\left((M_n N)(\tau)\right) = 0$ for every stopping time τ. Let $k < \infty$ be an upper bound of N.

$$
\begin{aligned}
\left| \mathbf{E}\left((M_\infty N)(\tau)\right) \right| &= \left| \mathbf{E}\left((M_\infty N)(\tau)\right) - \mathbf{E}\left((M_n N)(\tau)\right) \right| \leq \\
&\leq \mathbf{E}\left(\left| \left((M_\infty - M_n)N\right)(\tau) \right| \right) \leq \\
&\leq k \cdot \mathbf{E}\left(\left| (M_\infty - M_n)(\tau) \right| \right) \leq \\
&\leq k \cdot \mathbf{E}\left(\sqrt{[M_\infty - M_n](\infty)} \right) \leq \\
&\leq k \cdot \|M_\infty - M_n\|_{\mathcal{H}_0^p} \to 0.
\end{aligned}
$$

So $M_\infty N$ is also a martingale. Hence $\mathcal{Y} \doteq \{X \in \mathcal{H}_0^p : X \perp N\}$ is closed in $\mathcal{H}_0^p$. $\qquad\square$

Definition 5.31 *Let $1 \leq p < \infty$. We say that the subset $\mathcal{X} \subseteq \mathcal{H}_0^p$ has the Martingale Representation Property if*

$$
\mathcal{H}_0^p = \mathrm{stable}(\mathcal{X}).
$$

Recall that we have fixed a stochastic base $(\Omega, \mathcal{A}, \mathbf{P}, \mathcal{F})$.

Definition 5.32 *Let $1 \leq p < \infty$. Let us say that the probability measure $\mathbf{Q}$ on $(\Omega, \mathcal{A})$ is a $\mathcal{H}_0^p$-measure of the subset $\mathcal{X} \subseteq \mathcal{H}_0^p$ if*

1. $\mathbf{Q} \ll \mathbf{P}$,
2. $\mathbf{Q} = \mathbf{P}$ *on $\mathcal{F}_0$,*
3. *if $M \in \mathcal{X}$ then M is in $\mathcal{H}_0^p$ under $\mathbf{Q}$ as well.*

$\mathfrak{M}^p(\mathcal{X})$ *will denote the set of $\mathcal{H}_0^p$-measures of $\mathcal{X}$.*

Lemma 5.33 $\mathfrak{M}^p(\mathcal{X})$ *is always convex.*

Proof. If $\mathbf{Q}_1, \mathbf{Q}_2 \in \mathfrak{M}^p(\mathcal{X})$ and $0 \leq \lambda \leq 1$ and $\mathbf{Q}_\lambda \stackrel{\circ}{=} \lambda \mathbf{Q}_1 + (1 - \lambda)\mathbf{Q}_2$ then for every $M \in \mathcal{X}$

$$\mathbf{E}^{\mathbf{Q}_\lambda} \left(\left(\sup_t |M(t)| \right)^p \right) =$$

$$= \lambda \mathbf{E}^{\mathbf{Q}_1} \left(\left(\sup_t |M(t)| \right)^p \right) + (1 - \lambda) \mathbf{E}^{\mathbf{Q}_2} \left(\left(\sup_t |M(t)| \right)^p \right) < \infty.$$

If $F \in \mathcal{F}_s$ and $t > s$ then by the martingale property under $\mathbf{Q}_1$ and $\mathbf{Q}_2$

$$\int_F M(t)d\mathbf{Q}_\lambda = \lambda \int_F M(t)d\mathbf{Q}_1 + (1 - \lambda) \int_F M(t)d\mathbf{Q}_2 =$$

$$= \lambda \int_F M(s)d\mathbf{Q}_1 + (1 - \lambda) \int_F M(s)d\mathbf{Q}_2 =$$

$$= \int_F M(s)d\mathbf{Q}_\lambda.$$

Hence M is in $\mathcal{H}_0^p$ under $\mathbf{Q}_\lambda$. $\qquad \square$

Definition 5.34 *If C is a convex set and $x \in C$ then we say that x is an extremal point of C if whenever $u, v \in C$ and $x = \lambda u + (1 - \lambda)v$ for some $0 \leq \lambda \leq 1$ then $x = u$ or $x = v$.*

Proposition 5.35 *Let $1 \leq p < \infty$ and let $\mathcal{X} \subseteq \mathcal{H}_0^p$. If $\mathcal{X}$ has the Martingale Representation Property then $\mathbf{P}$ is an extremal point of $\mathfrak{M}^p(\mathcal{X})$.*

Proof. Assume that

$$\mathbf{P} = \lambda \mathbf{Q} + (1 - \lambda)\mathbf{R},$$

where $0 \leq \lambda \leq 1$ and $\mathbf{Q}, \mathbf{R} \in \mathfrak{M}^p(\mathcal{X})$. As $\mathbf{R} \geq 0$ obviously $\mathbf{Q} \ll \mathbf{P}$ so one can define the Radon–Nikodym derivative $L(\infty) \stackrel{\circ}{=} d\mathbf{Q}/d\mathbf{P} \in L^1(\Omega, \mathbf{P}, \mathcal{F}_\infty)$. Define the martingale

$$L(t) \stackrel{\circ}{=} \mathbf{E}(L(\infty) \mid \mathcal{F}_t).$$

From the definition of the conditional expectation

$$\int_F L(t)\,d\mathbf{P} = \int_F L(\infty)\,d\mathbf{P} = \mathbf{Q}(F), \quad F \in \mathcal{F}_t,$$

so $L(t)$ is the Radon–Nikodym derivative of $\mathbf{Q}$ with respect to $\mathbf{P}$ on the measure space $(\Omega, \mathcal{F}_t)$. Let $X \in \mathcal{X}$. If $s < t$ and $F \in \mathcal{F}_s$ then as X is a martingale under $\mathbf{Q}$

$$\int_F X\left(t\right) L\left(t\right) d\mathbf{P} = \int_F X\left(t\right) \frac{d\mathbf{Q}}{d\mathbf{P}} d\mathbf{P} = \int_F X\left(t\right) d\mathbf{Q} =$$
$$= \int_F X\left(s\right) d\mathbf{Q} = \int_F X\left(s\right) L\left(s\right) d\mathbf{P}$$

so XL is a martingale under $\mathbf{P}$. Obviously $\mathbf{Q} \leq \mathbf{P}/\lambda$ so $0 \leq L \leq 1/\lambda$. Hence L is uniformly bounded. $L\left(0\right)$ is bounded and $\mathcal{F}_0$-measurable so $X \cdot L\left(0\right)$ is a martingale. This implies that $X \cdot \left(L - L\left(0\right)\right)$ is also a martingale under $\mathbf{P}$, that is X and $L - L\left(0\right)$ are orthogonal as local martingales. That is $L - L\left(0\right)$ is orthogonal to $\mathcal{X}$. Hence by the previous lemma $L - L\left(0\right)$ is orthogonal to stable$(\mathcal{X})$. As $\mathcal{X}$ has the Martingale Representation Property $L - L\left(0\right)$ is orthogonal to $\mathcal{H}_0^p$. As $L - L\left(0\right)$ is bounded $L - L\left(0\right) \in \mathcal{H}_0^p$. But this means[37] that $L - L\left(0\right) = 0$. By definition $\mathbf{Q}$ and $\mathbf{P}$ are equal on $\mathcal{F}_0$, hence $L\left(\infty\right) = L\left(0\right) = 1$. Hence $\mathbf{P} = \mathbf{Q}$. $\qquad \square$

Now we want to prove the converse statement for $p = 1$. Let $\mathbf{P}$ be an extremal point of $\mathfrak{M}^p\left(\mathcal{X}\right)$ and assume that $\mathcal{X}$ does not have the Martingale Representation Property, that is stable$(\mathcal{X}) \neq \mathcal{H}_0^p$. As stable$(\mathcal{X})$ is a closed linear space by the Hahn–Banach theorem there is a non-zero linear functional L for which

$$L\left(\text{stable}(\mathcal{X})\right) = 0. \tag{5.33}$$

Assume temporarily that L has the following representation: there is a locally bounded local martingale N such that

$$L\left(M\right) = \mathbf{E}\left(\left[M, N\right]\left(\infty\right)\right), \quad M \in \mathcal{H}_0^p. \tag{5.34}$$

stable$(\mathcal{X})$ is closed under truncation, hence for every stopping time τ

$$\mathbf{E}\left(\left[M, N^\tau\right]\left(\infty\right)\right) = \mathbf{E}\left(\left[M, N\right]^\tau\left(\infty\right)\right) =$$
$$= \mathbf{E}\left(\left[M^\tau, N\right]\left(\infty\right)\right) = L\left(M^\tau\right) = 0$$

whenever $M \in$ stable$(\mathcal{X})$. Hence instead of N we can use N^τ. As N is locally bounded we can assume that N is a uniformly bounded martingale. Instead of N we can also write $N - N\left(0\right)$ so one can assume that $N\left(0\right) = 0$. Let $|N| \leq c$. If

$$d\mathbf{Q} \stackrel{\circ}{=} \left(1 - \frac{N\left(\infty\right)}{2c}\right) d\mathbf{P}, \quad d\mathbf{R} \stackrel{\circ}{=} \left(1 + \frac{N\left(\infty\right)}{2c}\right) d\mathbf{P}$$

then $\mathbf{Q}$ and $\mathbf{R}$ are non-negative measures. As N is a bounded martingale

$$\mathbf{E}\left(N\left(\infty\right)\right) = \mathbf{E}\left(N\left(0\right)\right) = \mathbf{E}\left(0\right) = 0,$$

[37]See: Proposition 4.4, page 228.

so $\mathbf{Q}$ and $\mathbf{R}$ are probability measures and obviously $\mathbf{P} = (\mathbf{Q} + \mathbf{R})/2$. If $X \in \mathcal{X}$ then

$$\int_\Omega \sup_s |X(s)|^p \, d\mathbf{Q} = \int_\Omega \sup_s |X(s)|^p \left(1 - \frac{N(\infty)}{2c} \right) d\mathbf{P} \leq$$

$$\leq 2 \int_\Omega \sup_s |X(s)|^p \, d\mathbf{P} < \infty.$$

If $s < t$ and $F \in \mathcal{F}_s$ then

$$\int_F X(t) \, d\mathbf{Q} \overset{\circ}{=} \int_F X(t) \left(1 - \frac{N(\infty)}{2c} \right) d\mathbf{P} =$$

$$= \int_F X(t) d\mathbf{P} - \frac{1}{2c} \int_F X(t) N(\infty) \, d\mathbf{P} =$$

$$= \int_F X(s) \, d\mathbf{P} - \frac{1}{2c} \int_F X(t) N(\infty) \, d\mathbf{P}.$$

As $F \in \mathcal{F}_s$

$$\sigma(\omega) \overset{\circ}{=} \begin{cases} s & \text{if } \omega \in F \\ \infty & \text{if } \omega \notin F \end{cases}$$

is a stopping time. As $s \leq t$

$$\tau(\omega) \overset{\circ}{=} \begin{cases} t & \text{if } \omega \in F \\ \infty & \text{if } \omega \notin F \end{cases}$$

is also a stopping time. Hence $X^\tau, X^\sigma \in \text{stable}(\mathcal{X})$, so

$$X^\tau - X^s = \left(X^t - X^s \right) \chi_F \in \text{stable}(\mathcal{X}). \tag{5.35}$$

Obviously $\mathcal{H}_0^p \subseteq \mathcal{H}_0^1$ if $p \geq 1$ so

$$|MN| \leq \sup_t |M|(t) \sup_t |N|(t) \in L^1(\Omega).$$

As N is bounded obviouly[38] $N \in \mathcal{H}_0^q$. Hence by the Kunita–Watanabe inequality using also Hölder's inequality

$$|[M, N]| \leq \sqrt{[M](\infty)} \sqrt{[N](\infty)} \in L^1(\Omega).$$

---

[38] Recall the definition of the $\mathcal{H}^p$ spaces! See: (5.32) on page 328. Implicitly we have used the Burkholder–Davis inequality.

By this $MN - [M, N]$ is a class $\mathcal{D}$ local martingale hence it is a uniformly integrable martingale[39]. Hence

$$\mathbf{E}\left(M\left(\infty\right)N\left(\infty\right)\right) = \mathbf{E}\left(M\left(\infty\right)N\left(\infty\right)\right) - L\left(M\right) =$$
$$= \mathbf{E}\left(M\left(\infty\right)N\left(\infty\right) - [M, N]\left(\infty\right)\right) =$$
$$= \mathbf{E}\left(M\left(0\right)N\left(0\right) - [M, N]\left(0\right)\right) = 0$$

so by (5.35)

$$\mathbf{E}\left(N\left(\infty\right)\chi_F X^t\left(\infty\right)\right) = \mathbf{E}\left(N\left(\infty\right)\chi_F X^s\left(\infty\right)\right).$$

Therefore

$$\int_F X\left(t\right)N\left(\infty\right)d\mathbf{P} = \int_F X\left(s\right)N\left(\infty\right)d\mathbf{P}.$$

Hence X is a martingale under $\mathbf{Q}$. This implies that $\mathbf{Q} \in \mathfrak{M}^p\left(\mathcal{X}\right)$. In a similar way $\mathbf{R} \in \mathfrak{M}^p\left(\mathcal{X}\right)$ which is a contradiction.

So one should only prove that if stable$(\mathcal{X}) \neq \mathcal{H}_0^p$ *then there is a locally bounded local martingale N for which (5.33) and (5.34) hold.*

It is easy to see that if $p > 1$ then the dual of $\mathcal{H}_0^p$ is $\mathcal{H}_0^q$, where of course $1/p + 1/q = 1$. The $\mathcal{H}_0^q$ martingales are not locally bounded[40] so the argument above is not valid if $p > 1$. Assume that $p = 1$.

Proposition 5.36 *If L is a continuous linear functional over $\mathcal{H}_0^1$ then (5.34) holds, that is for some locally bounded local martingale N*

$$L\left(M\right) = \mathbf{E}\left([M, N]\left(\infty\right)\right), \quad M \in \mathcal{H}_0^1.$$

Proof. Obviously $\mathcal{H}_0^2 \subseteq \mathcal{H}_0^1$ and $\|M\|_{\mathcal{H}^1} \leq \|M\|_{\mathcal{H}^2}$ so if $c \doteq \|L\|$ then

$$|L\left(M\right)| \leq c \|M\|_{\mathcal{H}_0^1} \leq c \|M\|_{\mathcal{H}_0^2}$$

so L is a continuous linear functional over $\mathcal{H}_0^2$.

1. $\mathcal{H}_0^2$ is a Hilbert space so for some $N \in \mathcal{H}_0^2$

$$L\left(M\right) = \mathbf{E}\left(M\left(\infty\right)N\left(\infty\right)\right), \quad M \in \mathcal{H}_0^2.$$

Let $M \in \mathcal{H}_0^2$. From the Kunita–Watanabe inequality[41]

$$|[M, N]| \leq \sqrt{[M]}\sqrt{[N]} \leq \sqrt{[M]}\left(\infty\right)\sqrt{[N]}\left(\infty\right) \in L^1\left(\Omega\right).$$

---

[39] See: Example 1.144, page 102.

[40] One can easily modify Example 1.138, on page 96 to construct a counter-example.

[41] Observe that we used again that the two definition of $\mathcal{H}_0^2$ spaces are equivalent.

Also as $M, N \in \mathcal{H}_0^2$

$$|(MN)(t)| \leq \sup_t |M(t)| \sup_t |N(t)| \in L^1(\Omega).$$

Therefore $MN - [M, N]$ has an integrable majorant so it is a local martingale from class $\mathcal{D}$. Therefore it is a uniformly integrable martingale. This implies that for some $N \in \mathcal{H}_0^2$

$$L(M) = \mathbf{E}(M(\infty)N(\infty)) = \mathbf{E}([M, N](\infty)), \quad M \in \mathcal{H}_0^2. \tag{5.36}$$

2. Now we prove that for almost all trajectory $|\Delta N| \leq 2c$. Let

$$\tau \stackrel{\circ}{=} \inf\{t : |\Delta N| > 2c\}.$$

As $N(0) = 0$ and N is right-continuous $\tau > 0$. If $\tau(\omega) < \infty$ then $|\Delta N(\tau)|(\omega) > 2c$. Hence we should prove that $\mathbf{P}(|\Delta N(\tau)| > 2c) = 0$. Every stopping time can be covered by countable number totally inaccessible or predictable stopping times, hence one can assume that τ is either predictable or totally inaccessible. If $\mathbf{P}(|\Delta N(\tau)| > 2c) > 0$ then let

$$\xi \stackrel{\circ}{=} \frac{\text{sgn}(\Delta N(\tau))}{\mathbf{P}(|\Delta N(\tau)| > 2c)} \chi(|\Delta N(\tau)| > 2c).$$

$S \stackrel{\circ}{=} \xi\chi([\tau, \infty))$ is adapted, right-continuous and it has an integrable variation. Let $M \stackrel{\circ}{=} S - S^p$. If τ is predictable then the graph $[\tau]$ is a predictable set, hence

$$\Delta(S^p) = {}^p(\Delta S) \stackrel{\circ}{=} {}^p(\xi\chi([\tau])) = ({}^p\xi)\chi([\tau]).$$

where ${}^p(\xi)$ is the predictable projection of the constant process $U(t) \equiv \xi$. By the definition of the predictable projection

$$^p(\xi)(\tau) = \mathbf{E}(\xi \mid \mathcal{F}_{\tau-}).$$

If τ is totally inaccessible then $\mathbf{P}(\tau = \sigma) = 0$ for every predictable stopping time σ. Hence

$$^p(\Delta M)(\sigma) = {}^p(\xi\chi([\tau]))(\sigma) \stackrel{\circ}{=} \widehat{\mathbf{E}}(\xi\chi([\tau])(\sigma) \mid \mathcal{F}_{\sigma-}) =$$
$$= \widehat{\mathbf{E}}(0 \mid \mathcal{F}_{\sigma-}) = 0.$$

Hence $(\Delta M^p) = {}^p(\Delta M) = 0$. Therefore in both cases S^p has just one jump which occurs at τ. This implies that M has finite variation and it has just one jump which occurs at τ. As we have seen

$$\Delta M(\tau) = \begin{cases} \xi - \mathbf{E}(\xi \mid \mathcal{F}_{\tau-}) & \text{if } \tau \text{ is predictable} \\ \xi & \text{if } \tau \text{ is totally inaccessible} \end{cases}.$$

Obviously

$$\|M\|_{\mathcal{H}_0^1} \overset{\circ}{=} \mathbf{E}\left(\sqrt{[M](\infty)}\right) = \mathbf{E}\left(\sqrt{(\Delta M)^2(\tau)}\right) =$$

$$= \mathbf{E}\left(|\Delta M(\tau)|\right) \leq \mathbf{E}\left(|\xi|\right) + \mathbf{E}\left(|\mathbf{E}(\xi \mid \mathcal{F}_{\tau-})|\right) \leq$$

$$\leq 2\mathbf{E}\left(|\xi|\right) = 2.$$

$\int_0^t M_- \, dM$ is a local martingale with localizing sequence (ρ_n). By the integration by parts formula and by Fatou's lemma

$$\mathbf{E}\left(M^2(t)\right) = \mathbf{E}\left(\lim_{n\to\infty} M^2(t \wedge \rho_n)\right) \leq \limsup_{n\to\infty} \mathbf{E}\left(M^2(t \wedge \rho_n)\right) =$$

$$= \limsup_{n\to\infty} \mathbf{E}\left([M](t \wedge \rho_n)\right) \leq \mathbf{E}\left([M](t)\right) \leq \mathbf{E}\left([M](\infty)\right) =$$

$$= \mathbf{E}\left((\Delta M(\tau))^2\right) < \infty.$$

Hence $M \in \mathcal{H}_0^2$. If τ is totally inaccessible then

$$L(M) = \mathbf{E}\left([M, N](\infty)\right) = \mathbf{E}\left((\Delta M(\tau) \Delta N(\tau))\right) =$$

$$= \mathbf{E}\left((\xi \Delta N(\tau))\right) =$$

$$= \frac{\mathbf{E}\left(|\Delta N(\tau)| \chi\left(|\Delta N(\tau)| > 2c\right)\right)}{\mathbf{P}\left(|\Delta N(\tau)| > 2c\right)} >$$

$$> 2c \frac{\mathbf{E}\left(\chi\left(|\Delta N(\tau)| > 2c\right)\right)}{\mathbf{P}\left(|\Delta N(\tau)| > 2c\right)} = 2c \geq c \|M\|_{\mathcal{H}^1}$$

which is impossible. If τ is predictable then

$$\mathbf{E}\left((\Delta M(\tau) \Delta N(\tau))\right) = \mathbf{E}\left((\xi \Delta N(\tau))\right) - \mathbf{E}\left((\mathbf{E}(\xi \mid \mathcal{F}_{\tau-}) \Delta N(\tau))\right).$$

N is a martingale therefore $^p(\Delta N) = 0$ so

$$\mathbf{E}\left(\mathbf{E}(\xi \mid \mathcal{F}_{\tau-}) \Delta N(\tau)\right) = \mathbf{E}\left(\mathbf{E}(\xi \mid \mathcal{F}_{\tau-}) \mathbf{E}(\Delta N(\tau) \mid \mathcal{F}_{\tau-})\right) = 0,$$

and we can get the same contradiction as above. This implies that $|\Delta N| \leq 2c$. Therefore N is locally bounded.

3. To finish the proof we should show that the identity in the theorem holds not only in $\mathcal{H}_0^2$ but in $\mathcal{H}_0^1$ as well. To do this we should prove that $\mathcal{H}_0^2$ is dense in $\mathcal{H}_0^1$ and $\mathbf{E}\left([M, N](\infty)\right)$ is a continuous linear functional in $\mathcal{H}_0^1$. Because these statements have some general importance we shall present them as separate lemmas. □

Lemma 5.37 $\mathcal{H}^2$ *is dense in* $\mathcal{H}^1$.

Proof. If $M \in \mathcal{H}^1$ then $M = M^c + M^d$, where M^c is the continuous part and M^d is the purely discontinuous part of M.

$$[M] = [M^c] + [M^d]$$

so from (5.32) it is obvious that $M^c, M^d \in \mathcal{H}^1$.

1. M^c is locally bounded so there is a localizing sequence (τ_n) that $(M^c)^{\tau_n} \in \mathcal{H}^2$ for all n. Observe that if (τ_n) is a localizing sequence then by the Dominated Convergence Theorem $\|M^{\tau_n} - M\|_{\mathcal{H}^1} \to 0$ for every $M \in \mathcal{H}^1$.

2. For the purely discontinuous part $M^d = \sum_{k=1}^{\infty} L_k$ where L_k are continuously compensated single jumps of M. Recall[42] that the series $\sum L_k$ converges in $\mathcal{H}^1$. Therefore it is sufficient to prove the lemma when $M \stackrel{\circ}{=} S - S^p$ is a continuously compensated single jump. Let τ be the jump-time of M, that is let $S \stackrel{\circ}{=} \Delta M(\tau) \chi([\tau, \infty))$. Let

$$\xi_k \stackrel{\circ}{=} \Delta M(\tau) \chi(|\Delta M(\tau)| \le k).$$

Let $S_k = \xi_k \chi([\tau, \infty))$ and $M_k \stackrel{\circ}{=} S_k - S_k^p$. By the construction of L_k the stopping time τ is either predictable or totally inaccessible. In a same way as in the proof of the proposition just above one can easily prove that M_k has just one jump which occurs at τ. Also as during the previous proof one can easily prove that $M_k \in \mathcal{H}^2$.

$$\|M - M_k\|_{\mathcal{H}^1} = \|\Delta M(\tau) - \Delta M_k(\tau)\|_1.$$

If τ is totally inaccessible then as $\Delta M(\tau)$ is integrable

$$\|\Delta M(\tau) - \Delta M_k(\tau)\|_1 = \|\Delta M(\tau) \chi(|\Delta M(\tau)| > k)\|_1 \to 0.$$

If τ is predictable then we also have the component

$$\|\mathbf{E}(\Delta M(\tau) \chi(|\Delta M(\tau)| > k) \mid \mathcal{F}_{\tau-})\|_1.$$

But if $k \to \infty$ then in $L^1(\Omega)$

$$\lim_{k \mapsto \infty} \mathbf{E}(\Delta M(\tau) \chi(|\Delta M(\tau)| > k) \mid \mathcal{F}_{\tau-}) = \mathbf{E}(\Delta M(\tau) \mid \mathcal{F}_{\tau-}) = 0,$$

from which the lemma is obvious. $\qquad\square$

---

[42]See: Theorem 4.26, page 236 and Proposition 4.30, page 243.

Our next goal is to prove that $\mathbf{E}\left([M, N]\left(\infty\right)\right)$ in (5.36) is a continuous linear functional over $\mathcal{H}_0^1$. To do this we need two lemmas. As a first step we prove the following observation:

Lemma 5.38 *If for some* $N \in \mathcal{H}_0^2$

$$\mathbf{E}\left(|[M, N]\left(\infty\right)|\right) \leq c \cdot \|M\|_{\mathcal{H}_0^1}, \quad M \in \mathcal{H}_0^2$$

then

$$\sup_\tau \left\| \sqrt{\mathbf{E}\left(\left(N\left(\infty\right) - N\left(\tau-\right)\right)^2 \mid \mathcal{F}_\tau\right)} \right\|_\infty < \infty$$

where the supremum is taken over all possible stopping times.

Proof. Let τ be a stopping time and let $M \overset{\circ}{=} N - N^\tau$. Obviously $M \in \mathcal{H}^2$.

$$\mathbf{E}\left([M]\left(\infty\right)\right) \overset{\circ}{=} \mathbf{E}\left([M, M]\left(\infty\right)\right) = \mathbf{E}\left([M, N]\left(\infty\right)\right) \leq c \cdot \|M\|_{\mathcal{H}^1} =$$
$$= c \cdot \mathbf{E}\left(\sqrt{[M]\left(\infty\right)}\right) = c \cdot \mathbf{E}\left(\sqrt{[M]\left(\infty\right)}\chi\left(\tau < \infty\right)\right) \leq$$
$$\leq c \cdot \sqrt{\mathbf{E}\left([M]\left(\infty\right)\right)}\sqrt{\mathbf{P}\left(\tau < \infty\right)}.$$

Hence

$$\mathbf{E}\left([N]\left(\infty\right) - [N]\left(\tau\right)\right) = \mathbf{E}\left([M]\left(\infty\right)\right) \leq c^2 \cdot \mathbf{P}\left(\tau < \infty\right).$$

If $F \in \mathcal{F}_\tau$ and one applies the inequality for the stopping time

$$\tau_F\left(\omega\right) \overset{\circ}{=} \begin{cases} \tau\left(\omega\right) & \text{if} \quad \omega \in F \\ \infty & \text{if} \quad \omega \notin F \end{cases}$$

then we get the inequality

$$\int_F [N]\left(\infty\right) - [N]\left(\tau\right) d\mathbf{P} = \mathbf{E}\left([N]\left(\infty\right) - [N]\left(\tau\right)\right) \leq c^2 \cdot \mathbf{P}\left(F\right).$$

From this

$$\mathbf{E}\left([N]\left(\infty\right) - [N]\left(\tau\right) \mid \mathcal{F}_\tau\right) \leq c^2.$$

$M^2 - [M]$ is a uniformly integrable martingale, hence almost surely

$$\mathbf{E}\left(\left(N\left(\infty\right) - N\left(\tau\right)\right)^2 \mid \mathcal{F}_\tau\right) \overset{\circ}{=} \mathbf{E}\left(M^2\left(\infty\right) \mid \mathcal{F}_\tau\right) =$$
$$= \mathbf{E}\left([M]\left(\infty\right) \mid \mathcal{F}_\tau\right) =$$
$$= \mathbf{E}\left([N]\left(\infty\right) - [N]\left(\tau\right) \mid \mathcal{F}_\tau\right) \leq c^2.$$

During the proof of the proposition we proved that the jumps of N are bounded so

$$\mathbf{E}\left((N(\infty)-N(\tau-))^2 \mid \mathcal{F}_\tau\right) = \mathbf{E}\left((N(\infty)-N(\tau)+\Delta N(\tau))^2 \mid \mathcal{F}_\tau\right) =$$

$$= \mathbf{E}\left((N(\infty)-N(\tau))^2 + (\Delta N(\tau))^2 \mid \mathcal{F}_\tau\right) \leq a,$$

where a is independent of τ. From this the lemma follows. $\square$

Finally we prove the next inequality:

Lemma 5.39 (Fefferman's inequality) *If $M \in \mathcal{H}_0^1$ and $N \in \mathcal{H}_0^2$ then*

$$\mathbf{E}\left(|[M,N](\infty)|\right) \leq \sqrt{2} \cdot \|M\|_{\mathcal{H}_0^1} \cdot \sup_\tau \left\| \sqrt{\mathbf{E}\left((N(\infty)-N(\tau-))^2 \mid \mathcal{F}_\tau\right)} \right\|_\infty$$

where the supremum is taken over all stopping times.

Proof. Let

$$a^\ominus \overset{\circ}{=} \begin{cases} 0 & \text{if} \quad a = 0 \\ a^{-1} & \text{if} \quad a > 0 \end{cases}.$$

From the Kunita–Watanabe inequality

$$\int_0^\infty 1 d\mathrm{Var}\left([M,N]\right) \leq$$

$$\leq \sqrt{\int_0^\infty \left(\sqrt{[M]} + \sqrt{[M]_-}\right)^\ominus d[M]} \sqrt{\int_0^\infty \sqrt{[M]} + \sqrt{[M]_-} d[N]}.$$

Therefore by the Cauchy–Schwarz inequality

$$\left(\mathbf{E}\left(|[M,N](\infty)|\right)\right)^2 \leq$$

$$\leq \mathbf{E}\left(\int_0^\infty \left(\sqrt{[M]} + \sqrt{[M]_-}\right)^\ominus d[M]\right) \mathbf{E}\left(\int_0^\infty 2\sqrt{[M]} d[N]\right).$$

Let

$$a = t_0^{(n)} < t_1^{(n)} < \ldots < t_n^{(n)} = b$$

be an infinitesimal sequence of partitions of $[a, b]$. Let $f > 0$ be a right-regular function with bounded variation on $[a, b]$.

$$\sqrt{f(b)} - \sqrt{f(a)} = \sum_{i=1}^{n} \left(\sqrt{f\left(t_i^{(n)}\right)} - \sqrt{f\left(t_{i-1}^{(n)}\right)} \right) =$$

$$= \sum_{i=1}^{n} \frac{f\left(t_i^{(n)}\right) - f\left(t_{i-1}^{(n)}\right)}{\sqrt{f\left(t_i^{(n)}\right)} + \sqrt{f\left(t_{i-1}^{(n)}\right)}}.$$

f generates a finite measure on $[a, b]$. As f is right-regular and it is positive

$$\frac{1}{\sqrt{f\left(t_i^{(n)}\right)} + \sqrt{f\left(t_{i-1}^{(n)}\right)}}$$

is bounded and for every $t \in \left(t_{i-1}^{(n)}, t_i^{(n)}\right]$

$$\frac{1}{\sqrt{f\left(t_i^{(n)}\right)} + \sqrt{f\left(t_{i-1}^{(n)}\right)}} \rightarrow \frac{1}{\sqrt{f(t)} + \sqrt{f(t-)}}.$$

So by the Dominated Convergence Theorem it is easy to see that if $n \to \infty$ then

$$\sqrt{f(b)} - \sqrt{f(a)} = \int_a^b \frac{1}{\sqrt{f(t)} + \sqrt{f(t-)}} df(t).$$

With the Monotone Convergence Theorem one can easily prove that if f is a right-regular, non-negative, increasing function then[43]

$$\sqrt{f(\infty)} - \sqrt{f(0)} = \int_0^\infty \left(\sqrt{f(t)} + \sqrt{f(t-)} \right)^{\ominus} df(t).$$

Using this

$$\mathbf{E} \left(\int_0^\infty \left(\sqrt{[M]} + \sqrt{[M]_-} \right)^{\ominus} d[M] \right) = \mathbf{E} \left(\sqrt{[M](\infty)} \right) \overset{\circ}{=} \|M\|_{\mathcal{H}_0^1}.$$

[43]See: Example 6.50, page 400.

Let us estimate the second integral. Integrating by parts

$$\mathbf{E}\left(\int_0^\infty \sqrt{[M]}d\,[N]\right) =$$

$$= \mathbf{E}\left(\sqrt{[M]}\,(\infty)\,[N]\,(\infty) - \int_0^\infty [N]_-\,d\sqrt{[M]}\right) =$$

$$= \mathbf{E}\left(\int_0^\infty ([N]\,(\infty) - [N]_-)\,d\sqrt{[M]}\right).$$

It is easy to see that[44]

$$\mathbf{E}\left(\int_0^\infty [N]\,(\infty)\,d\sqrt{[M]}\right) = \mathbf{E}\left([N]\,(\infty)\,\sqrt{[M]}\,(\infty)\right) =$$

$$= \mathbf{E}\left(\sum_k [N]\,(\infty)\left(\sqrt{[M]}\,(s_k) - \sqrt{[M]}\,(s_{k-1})\right)\right) =$$

$$= \mathbf{E}\left(\sum_k \mathbf{E}\,([N]\,(\infty)\mid \mathcal{F}_{s_k})\left(\sqrt{[M]}\,(s_k) - \sqrt{[M]}\,(s_{k-1})\right)\right) =$$

$$= \mathbf{E}\left(\int_0^\infty \mathbf{E}\,([N]\,(\infty)\mid \mathcal{F}_s)\,d\sqrt{[M]}\,(s)\right).$$

So if

$$k \doteq \sup_\tau \left\|\sqrt{\mathbf{E}\left((N\,(\infty) - N\,(\tau-))^2 \mid \mathcal{F}_\tau\right)}\right\|_\infty$$

then

$$\mathbf{E}\left(\int_0^\infty \sqrt{[M]}d\,[N]\right) =$$

$$= \mathbf{E}\left(\int_0^\infty \mathbf{E}\,([N]\,(\infty)\mid \mathcal{F}_s) - [N]\,(s-)\,d\sqrt{[M]}\,(s)\right) =$$

$$= \mathbf{E}\left(\int_0^\infty \mathbf{E}\,([N]\,(\infty)\mid \mathcal{F}_s) - [N]\,(s) + \Delta\,[N]\,(s)\,d\sqrt{[M]}\,(s)\right)$$

$$= \mathbf{E}\left(\int_0^\infty \mathbf{E}\left(N^2\,(\infty) - N^2\,(s) + (\Delta N\,(s))^2 \mid \mathcal{F}_s\right)\,d\sqrt{[M]}\,(s)\right) =$$

---

[44]First one should assume that $[N]\,(\infty)$ is bounded and we should use that $\sqrt{[M]\,(\infty)}$ is integrable. Then with Monotone Convergence Theorem one can drop the assumption that $[N]\,(\infty)$ is bounded.

$$= \mathbf{E} \left(\int_0^\infty \mathbf{E} \left((N(\infty) - N(s-))^2 \mid \mathcal{F}_s \right) d\sqrt{[M]}(s) \right) \leq$$

$$\leq k^2 \cdot \mathbf{E} \left(\sqrt{[M]}(\infty) \right) = k^2 \cdot \|M\|_{\mathcal{H}_0^1}.$$

So

$$\left(\mathbf{E} \left(|[M, N](\infty)| \right) \right)^2 \leq 2 \cdot k^2 \cdot \|M\|_{\mathcal{H}_0^1}^2$$

which proves the inequality. $\qquad\square$

Definition 5.40 *N is a BMO martingale if $N \in \mathcal{H}^2$ and*

$$\sup_\tau \left\| \sqrt{\mathbf{E} \left((N(\infty) - N(\tau-))^2 \mid \mathcal{F}_\tau \right)} \right\|_\infty < \infty.$$

Corollary 5.41 *The BMO martingales are locally bounded.*

Corollary 5.42 (Dual of $\mathcal{H}_0^1$) *L is a continuous linear functional over $\mathcal{H}_0^1$ if and only if for some BMO martingale N*

$$L(M) = \mathbf{E} \left([M, N](\infty) \right).$$

The dual of the Banach space $\mathcal{H}_0^1$ is the space of BMO martingales.

Let us return to the Martingale Representation Problem. We proved the following statement:

Theorem 5.43 (Jacod–Yor) *The set $\mathcal{X} \subseteq \mathcal{H}_0^1$ has the Martingale Representation Property if and only if the underlying probability measure $\mathbf{P}$ is an extremal point of $\mathfrak{M}^1(\mathcal{X})$.*

Proposition 5.44 *Let $1 \leq p < \infty$ and let $\mathcal{X}$ be a closed linear subspace of $\mathcal{H}_0^p$. The following properties are equivalent:*

1. *If $M \in \mathcal{X}$ and $H \bullet M \in \mathcal{H}_0^p$ for some predictable process H then $H \bullet M \in \mathcal{X}$.*
2. *If $M \in \mathcal{X}$ and H is a bounded and predictable process then $H \bullet M \in \mathcal{X}$.*
3. *$\mathcal{X}$ is stable under truncation, that is if $M \in \mathcal{X}$ and τ is an arbitrary stopping time then $M^\tau \in \mathcal{X}$.*
4. *If $M \in \mathcal{X}$, $s \leq t \leq \infty$ and $F \in \mathcal{F}_s$ then $(M^t - M^s) \chi_F \in \mathcal{X}$.*

Proof. Let H be a bounded predictable process and let $|H| \leq c$.

$$[H \bullet M](\infty) = \left(H^2 \bullet [M] \right)(\infty) \leq c^2 [M](\infty)$$

so if $M \in \mathcal{H}_0^p$ then $H \bullet M \in \mathcal{H}_0^p$ and the implication 1. $\Rightarrow$ 2. is obvious. If τ is an arbitrary stopping time then

$$\chi([0,\tau]) \bullet M = 1 \bullet M^\tau = M^\tau - M(0) = M^\tau$$

hence 2. implies 3. If $F \in \mathcal{F}_s$ then

$$\tau(\omega) \stackrel{\circ}{=} \begin{cases} s & \text{if} \quad \omega \in F \\ \infty & \text{if} \quad \omega \notin F \end{cases}$$

is a stopping time. If 3. holds then $M^\tau \in \mathcal{X}$. As $s \le t$

$$\sigma(\omega) \stackrel{\circ}{=} \begin{cases} t & \text{if} \quad \omega \in F \\ \infty & \text{if} \quad \omega \notin F \end{cases}$$

is also a stopping time hence $M^\sigma \in \mathcal{X}$. As $\mathcal{X}$ is a linear space $M^\sigma - M^\tau \in \mathcal{X}$. But obviously

$$M^\sigma - M^\tau = (M^t - M^s)\chi_F,$$

hence 3. implies 4. Now let

$$H = \sum_i \chi_{F_i} \chi\left((t_i, t_{i+1}]\right) \tag{5.37}$$

where $F_i \in \mathcal{F}_{t_i}$. Obviously

$$(H \bullet X)(t) = \sum_i \chi_{F_i}\left(M(t \wedge t_{i+1}) - M(t \wedge t_i)\right)$$

and by 4. $H \bullet M \in \mathcal{X}$.

$$\|H_n \bullet M - H \bullet M\|_{\mathcal{H}_0^p} = \|(H_n - H) \bullet M\|_{\mathcal{H}_0^p} =$$
$$= \left\|\sqrt{[(H_n - H) \bullet M](\infty)}\right\|_p =$$
$$= \left\|\sqrt{(H_n - H)^2 \bullet [M](\infty)}\right\|_p.$$

$M \in \mathcal{H}_0^p$ so $\left\|\sqrt{[M](\infty)}\right\|_p < \infty$. Therefore if $H_n \to H$ is a uniformly bounded sequence of predictable processes then from the Dominated Convergence Theorem it is obvious that

$$\|H_n \bullet M - H \bullet M\|_{\mathcal{H}_0^p} = \left\|\sqrt{(H_n - H)^2 \bullet [M](\infty)}\right\|_p \to 0.$$

$\mathcal{X}$ is closed so if $H_n \bullet M \in \mathcal{X}$ for all n then $H \bullet M \in \mathcal{X}$ as well. Using this property and 4. with the Monotone Class Theorem one can easily show that if H is a bounded predictable process then $H \bullet M \in \mathcal{X}$. If $H \bullet M \in \mathcal{H}_0^p$ for some predictable process H then

$$\left\| \sqrt{(H^2 \bullet [M])(\infty)} \right\|_p < \infty.$$

From this as above it is easy to show that in $\mathcal{H}_0^p$

$$H\left(\chi\left(|H| \le n\right)\right) \bullet M \to H \bullet M,$$

so $H \bullet M \in \mathcal{X}$. $\qquad\square$

Proposition 5.45 *If* $1 \le p < \infty$ *and* $M \in \mathcal{H}_0^p$ *then the set*

$$\mathcal{C} \doteq \{X \in \mathcal{H}_0^p : X = H \bullet M\}$$

is closed in $\mathcal{H}_0^p$.

Proof. It is easy to see that the set of predictable processes H for which[45]

$$\|H\|_{\mathcal{L}^p(M)} \doteq \left\| \sqrt{H^2 \bullet [M](\infty)} \right\|_p < \infty \tag{5.38}$$

is a linear space. In the usual way, as in the classical theory of L^p-spaces[46], one can prove that if $H_1 \sim H_2$ whenever $\|H_1 - H_2\|_{\mathcal{L}^p(M)} = 0$ then the set of equivalence classes, denoted by $\mathcal{L}^p(M)$, is a Banach space. Let $X_n \in \mathcal{C}$ and assume that $X_n \to X$ in $\mathcal{H}_0^p$. Let $X_n = H_n \bullet M$.

$$\|X_n\|_{\mathcal{H}_0^p} \doteq \left\| \sqrt{[X_n](\infty)} \right\|_p = \left\| \sqrt{H^2 \bullet [M](\infty)} \right\|_p \doteq \|H_n\|_{\mathcal{L}^p(M)}.$$

This implies that (H_n) is a Cauchy sequence in $\mathcal{L}^p(M)$, so it is convergent, hence $H_n \to H$ in $\mathcal{L}^p(M)$ for some H and $H_n \bullet M \to H \bullet M$. Therefore $X = H \bullet M$, so $\mathcal{C}$ is closed. $\qquad\square$

Proposition 5.46 *Let* $(M_i)_{i=1}^n$ *be a finite subset of* $\mathcal{H}_0^p$. *Assume that if* $i \ne j$ *then the martingales* M_i *and* M_j *are strongly orthogonal[47] as local martingales, that is* $[M_i, M_j] = 0$ *whenever* $i \ne j$. *In this case*

$$\text{stable}(M_1, M_2, \dots, M_n) = \left\{ \sum_{i=1}^n H_i \bullet M_i : H_i \in \mathcal{L}^p(M_i) \right\}.$$

---

[45]See: Definition 2.57, page 151
[46]See: [80], Theorem 3.11, page 69.
[47]See: Definition 4.1, page 227.

That is the stable subspace generated by a finite set of strongly orthogonal $\mathcal{H}_0^p$-martingales is the linear subspace generated by the stochastic integrals $H_i \bullet M_i$, $H_i \in \mathcal{L}^p(M_i)$.

Proof. Recall that as in the previous proposition $\mathcal{L}^p(M)$ is the set of equivalence classes of progressively measurable processes for which (5.38) hold. Let $\mathcal{I}$ denote the linear space on the right side of the equality. By Proposition 5.44 for all i

$$H_i \bullet M_i \in \text{stable}(M_i) \subseteq \text{stable}(\mathcal{X})$$

hence

$$\mathcal{I} \subseteq \text{stable}(\mathcal{X}).$$

From the stopping rule of the stochastic integrals $\mathcal{I}$ is closed under stopping. $M_i(0) = 0$ and $M_i = 1 \bullet M_i$ so $M_i \in \mathcal{I}$ for all i. By strong orthogonality

$$\left\| \mathbf{E} \sqrt{\left[\sum_{i=1}^n H_i \bullet M_i \right]} \right\|_p = \left\| \sqrt{\sum_{i=1}^n H_i^2 \bullet [M_i]} \right\|_p \leq$$

$$\leq \left\| \sum_{i=1}^n \sqrt{H_i^2 \bullet [M_i]} \right\|_p.$$

From Jensen's inequality it is also easy to show that

$$\frac{1}{\sqrt{n}} \left\| \sum_{i=1}^n \sqrt{H_i^2 \bullet [M_i]} \right\|_p \leq \left\| \mathbf{E} \sqrt{\left[\sum_{i=1}^n H_i \bullet M_i \right]} \right\|_p.$$

This means that the norms $\left\| \mathbf{E} \sqrt{[\sum_{i=1}^n H_i \bullet M_i]} \right\|_p$ and $\left\| \sum_{i=1}^n \sqrt{H_i^2 \bullet [M_i]} \right\|_p$ are equivalent. In a similar way, as in the previous proposition, one can show that $\mathcal{I}$ is a closed linear subspace of $\mathcal{H}_0^2$. Therefore

$$\text{stable}(M_1, \dots, M_n) \subseteq \mathcal{I}. \qquad \square$$

Example 5.47 The assumption about orthogonality is important.

Let w_1 and w_2 be independent Wiener processes. Let $J(t) \stackrel{\circ}{=} t$. If

$$M_1 \stackrel{\circ}{=} w_1, \quad M_2 \stackrel{\circ}{=} (1 - J) \bullet w_1 + J \bullet w_2$$

then

$$[M_1, M_2] = [w_1, (1 - J) \bullet w_1 + J \bullet w_2] = (1 - J)[w_1] = (1 - J)J$$

which is not a local martingale. So the conditions of the above proposition do not hold. We show that

$$\mathcal{I} \stackrel{\circ}{=} \left\{ \sum_{i=1}^{2} H_i \bullet M_i : H_i \in \mathcal{L}^p(M_i) \right\}$$

is not a closed set in $\mathcal{H}_0^p$. Let $\varepsilon > 0$. Obviously

$$H_1^{(\varepsilon)} \stackrel{\circ}{=} \frac{J - 1 + \varepsilon}{J + \varepsilon}, \quad H_2^{(\varepsilon)} \stackrel{\circ}{=} \frac{1}{J + \varepsilon}$$

are bounded predictable processes.

$$X_\varepsilon \stackrel{\circ}{=} H_1^{(\varepsilon)} \bullet M_1 + H_2^{(\varepsilon)} \bullet M_2 =$$
$$= \frac{J - 1 + \varepsilon}{J + \varepsilon} \bullet w_1 + \frac{1 - J}{J + \varepsilon} \bullet w_1 + \frac{J}{J + \varepsilon} \bullet w_2 =$$
$$= \frac{\varepsilon}{J + \varepsilon} \bullet w_1 + w_2 - \frac{\varepsilon}{J + \varepsilon} \bullet w_2.$$

As w_1 and w_2 are independent

$$\left[\frac{\varepsilon}{J + \varepsilon} \bullet w_1 - \frac{\varepsilon}{J + \varepsilon} \bullet w_2 \right](t) = 2 \int_0^t \left(\frac{\varepsilon}{s + \varepsilon} \right)^2 ds \to 0,$$

so $X_\varepsilon \to w_2$ in $\mathcal{H}_0^p$. Assume that for some H_1 and H_2

$$w_2 = H_1 \bullet M_1 + H_2 \bullet M_2 =$$
$$= H_1 \bullet w_1 + H_2 (1 - J) \bullet w_1 + H_2 J \bullet w_2.$$

Reordering

$$(H_1 + H_2 (1 - J)) \bullet w_1 = (1 - H_2 J) \bullet w_2.$$

From this

$$[(1 - H_2 J) \bullet w_2] = [(H_1 + H_2 (1 - J)) \bullet w_1, (1 - H_2 J) \bullet w_2] =$$
$$= (H_1 + H_2 (1 - J))(1 - H_2 J) \bullet [w_1, w_2] = 0,$$

so

$$(H_1 + H_2 (1 - J)) \bullet w_1 = (1 - H_2 J) \bullet w_2 = 0.$$

This implies that

$$1 - H_2 J = (H_1 + H_2 (1 - J)) = 0$$

that is $H_2 = 1/J$ and $H_1 = 1 - 1/J$. But as

$$\int_0^t \left(1 - \frac{1}{s}\right)^2 ds = +\infty$$

$H_1 = 1 - 1/J \notin \mathcal{L}^p (w_1)$. □

Definition 5.48 *Let* $(M_i)_{i=1}^n$ *be a finite subset of* $\mathcal{H}_0^p$. *We say that* $(M_i)_{i=1}^n$ *has the* Integral Representation Property *if for every* $M \in \mathcal{H}_0^p$

$$M = \sum_{i=1}^n H_i \bullet M_i, \quad H_i \in \mathcal{L}^p (M_i).$$

The main result about integral representation is an easy consequence of the Jacod–Yor theorem and the previous proposition:

Theorem 5.49 (Jacod–Yor) *Let* $1 \le p < \infty$ *and let* $\mathcal{X} \doteq (M_i)_{i=1}^n$ *be a finite subset of* $\mathcal{H}_0^p$. *Assume that if* $i \ne j$ *then the martingales* M_i *and* M_j *are strongly orthogonal*[48] *as local martingales, that is* $[M_i, M_j] = 0$ *whenever* $i \ne j$. *If these assumptions hold then* $\mathcal{X}$ *has the Integral Representation Property in* $\mathcal{H}_0^p$ *if and only if* $\mathbf{P} \in \mathfrak{M}^p (\mathcal{X})$.

Proof. If $\mathcal{X}$ has the Integral Representation Property then[49] stable$(\mathcal{X}) = \mathcal{H}_0^p$ so $\mathbf{P}$ is an extremal point of $\mathfrak{M}^p (\mathcal{X})$. Assume that $\mathcal{X}$ does not have the Integral Representation Property. This means that stable$_p(\mathcal{X}) \ne \mathcal{H}_0^p$. We show that in this case stable$_1(\mathcal{X}) \ne \mathcal{H}_0^1$ as well: If stable$_1(\mathcal{X}) = \mathcal{H}_0^1$ then for every $M \in \mathcal{H}_0^p \subseteq \mathcal{H}_0^1$

$$M = \sum_{i=1}^n H_i \bullet M_i, \quad H_i \in \mathcal{L}^1 (M_i).$$

---

[48] See: Definition 4.1, page 227.
[49] See: Proposition 5.35, page 330.

But by the strong orthogonality assumption for every k

$$[M](\infty) = \left[\sum_{i=1}^{n} H_i \bullet M_i\right] = \sum_{i=1}^{n} H_i^2 \bullet [M_i] \geq H_i^2 \bullet [M_i]$$

$\sqrt{[M](\infty)} \in L^p(\Omega)$ so $\sqrt{H_i^2 \bullet [M_i](\infty)} \in L^p(\Omega)$. Hence $H_i \in \mathcal{L}^p(M_i)$ for every i, which is impossible as $\mathcal{X}$ does not have the Integral Representation Property in $\mathcal{H}_0^p$. Hence

$$\text{stable}_p(\mathcal{X}) \subseteq \text{stable}_1(\mathcal{X}) \neq \mathcal{H}_0^1.$$

By the Hahn–Banach theorem there is a continuous linear functional $L \in \left(\mathcal{H}_0^1\right)^*$ that $L\left(\text{stable}_1(\mathcal{X})\right) = 0$. This implies that

$$L\left(\text{stable}_p(\mathcal{X})\right) = 0.$$

L is of course a BMO martingale so it is locally bounded. As we have remarked one can assume that L is bounded. As we already discussed in this case $\mathbf{P}$ is not an extremal point of $\mathfrak{M}^p(\mathcal{X})$. □

The most important example is the following:

Example 5.50 If $\mathcal{X} \stackrel{\circ}{=} (w_k)_{k=1}^n$ are independent Wiener processes and the filtration $\mathcal{F}$ is the filtration generated by $\mathcal{X}$ then $\mathcal{X}$ has the Integral Representation Property on any finite interval.

On any finite interval[50] $w_k \in \mathcal{H}_0^1$. We show that $\mathfrak{M}^1(\mathcal{X}) = \{\mathbf{P}\}$. If $\mathbf{Q} \in \mathfrak{M}^1(\mathcal{X})$ then w_k is a continuous local martingale under $\mathbf{Q}$ for every k. Obviously $[w_k, w_j](t) = \delta_{ij}t$. By Lévy's characterization theorem[51] $\mathbf{X} \stackrel{\circ}{=} (w_1, w_2, \ldots, w_n)$ is an n-dimensional Wiener process under $\mathbf{Q}$ as well. This implies that

$$\int_\Omega f(\mathbf{X}) \, d\mathbf{P} = \int_\Omega f(\mathbf{X}) \, d\mathbf{Q}.$$

for every $\mathcal{F}_\infty$-measurable bounded function f. As $\mathcal{F}$ is the filtration generated by $\mathbf{X}$ this implies that $\mathbf{P}(F) = \mathbf{Q}(F)$ for every $F \in \mathcal{F}_\infty$ so $\mathbf{P} = \mathbf{Q}$. □

Example 5.51 If $\mathcal{X} \stackrel{\circ}{=} (\pi_k)_{k=1}^n$ are independent compensated Poisson processes and the filtration $\mathcal{F}$ is the filtration generated by $\mathcal{X}$ then $\mathcal{X}$ has the Integral Representation Property on any finite interval.

---

[50]On finite interval $[0, s]$ $\|w\|_{\mathcal{H}^1} = \mathbf{E}\left(\sqrt{[w](s)}\right) = \sqrt{s}$. See: Example 1.124, page 87.

[51]See: Theorem 6.13, page 368.

On any finite interval $\pi_k \in \mathcal{H}_0^1$. If two Poisson processes are independent then they do not have common jumps[52] so $[\pi_k, \pi_j] = 0$. So we can apply the Jacod–Yor theorem. We shall prove that again $\mathfrak{M}^1(\mathcal{X}) = \{\mathbf{P}\}$. If X is a compensated Poisson process, then

$$[X](t) - \lambda t \overset{a.s.}{=} X(t). \tag{5.39}$$

Of course this identity holds under any probability measure $\mathbf{Q} \ll \mathbf{P}$. As in the previous example one should show that if X is a local martingale then (5.39) implies that X is a compensated Poisson process with parameter λ. Let us assume that for some process X under some measure (5.39) holds. In this case obviously $(\Delta X)^2 = \Delta X$, that is if $\Delta X \neq 0$ then $\Delta X = 1$. $[X]$ has finite variation, hence X also has finite variation, so $X \in \mathcal{V} \cap \mathcal{L}$. Hence X is purely discontinuous, that is X is a quadratic jump process: $[X] = \sum (\Delta X)^2$. The size of the jumps is constant, so as $[X]$ is finite for every trajectory there is just finite number of jumps on every finite interval. Let $N(t)$ denote the number of jumps in the interval $[0, t]$.

$$N(t) - \lambda t = [X](t) - \lambda t = X(t). \tag{5.40}$$

As X is a local martingale this means that the compensator of N is λt. N is a counting process so

$$\exp(itN(u)) =$$
$$= \sum_{s \leq u} (\exp(itN(s)) - \exp(itN(s-))) =$$
$$= \sum_{s \leq u} (\exp(it(N(s-) + 1)) - \exp(itN(s-))) =$$
$$= \sum_{s \leq u} (\exp(it) - 1) \cdot \exp(itN(s-)) \cdot 1 =$$
$$= (\exp(it) - 1) \sum_{s \leq u} \exp(itN(s-)) [N(s) - N(s-)] =$$
$$= (\exp(it) - 1) \int_0^u \exp(itN(s-)) dN(s).$$

Taking expected value and using elementary properties of the compensator, and that on every finite interval N has only finite number of jumps

$$\varphi_u(t) \overset{\circ}{=} \mathbf{E}(\exp(itN(u))) = (\exp(it) - 1) \mathbf{E} \left(\int_0^u \exp(itN(s-)) dN(s) \right) =$$
$$= (\exp(it) - 1) \mathbf{E} \left(\int_0^u \exp(itN(s-)) dN^p(s) \right) =$$

---

[52]See: Proposition 7.13, page 471.

$$= \lambda \left(\exp \left(it \right) - 1 \right) \mathbf{E} \left(\int_0^u \exp \left(itN \left(s- \right) \right) ds \right) =$$

$$= \lambda \left(\exp \left(it \right) - 1 \right) \mathbf{E} \left(\int_0^u \exp \left(itN \left(s \right) \right) ds \right) =$$

$$= \lambda \left(\exp \left(it \right) - 1 \right) \int_0^u \varphi_s \left(t \right) ds,$$

where $\varphi_u \left(t \right)$ is the Fourier transform of $N \left(u \right)$. Differentiating both sides by u

$$\frac{d}{du} \varphi_u \left(t \right) = \lambda \left(\exp \left(it \right) - 1 \right) \varphi_u \left(t \right).$$

The solution of this equation is

$$\varphi_u \left(t \right) = \exp \left(\lambda u \left(\exp \left(it \right) - 1 \right) \right).$$

Hence $N \left(u \right)$ has a Poisson distribution with parameter λu. By (5.40) X is a compensated Poisson process with parameter λ. Finally recall that Poisson processes are independent if and only if[53] they do not have common jumps. This means that under $\mathbf{Q}$ the processes π_k remain independent Poisson processes. $\qquad\square$

Example 5.52 Continuous martingale which does not have the Integral Representation Property.

Let $\left(\left(w_1, w_2 \right), \mathcal{G} \right)$ be a two-dimensional Wiener process. Let $X \overset{\circ}{=} w_1 \bullet w_2$, and let $\mathcal{F}$ be the filtration generated by X. Evidently $\mathcal{F}_t \subseteq \mathcal{G}_t$. X is obviously a local martingale under $\mathcal{G}$.

$$\mathbf{E} \left([X] \left(T \right) \right) = \mathbf{E} \left(\int_0^T w_1^2 d \left[w_2 \right] \right) = \int_0^T \mathbf{E} \left(w_1^2 \left(t \right) \right) dt < \infty$$

so on every finite interval X is in $\mathcal{H}_0^2$. Hence X is a $\mathcal{G}$-martingale. As X is $\mathcal{F}$-adapted one can easily show that X is an $\mathcal{F}$-martingale. The quadratic variation $[X]$ is $\mathcal{F}$-adapted.

$$[X] \left(t \right) = \int_0^t w_1^2 d \left[w_2 \right] = \int_0^t w_1^2 \left(s \right) ds,$$

[53]See: Proposition 7.11, page 469 and 7.13, page 471

therefore the derivative of $[X]$ is w_1^2. This implies that w_1^2 is also $\mathcal{F}$-adapted. As $[w_1]$ is deterministic

$$Z \overset{\circ}{=} \frac{1}{2} \left(w_1^2 - [w_1] \right) = w_1 \bullet w_1$$

is also $\mathcal{F}$-adapted. Z is an $\mathcal{F}$-martingale: If $s < t$, then using that $Z = w_1^2 - [w_1]$ is a $\mathcal{G}$-martingale[54]

$$\mathbf{M}\left(Z\left(t\right) \mid \mathcal{F}_s\right) = \mathbf{M}\left(\mathbf{M}\left(Z\left(t\right) \mid \mathcal{G}_s\right) \mid \mathcal{F}_s\right) =$$
$$= \mathbf{M}\left(Z\left(s\right) \mid \mathcal{F}_s\right) = Z\left(s\right).$$

If X had the Integral Representation Property then for some Y

$$Z = Y \bullet X \overset{\circ}{=} Y \bullet \left(w_1 \bullet w_2\right) = Y w_1 \bullet w_2.$$

As w_1 and w_2 are independent $[w_1, w_2] = 0$.

$$0 < [Z \bullet Z] = [w_1 \bullet w_1, Y \bullet X] = [w_1 \bullet w_1, Y w_1 \bullet w_2] = Y w_1^2 \bullet [w_1, w_2] = 0,$$

which is impossible. $\qquad\qquad\qquad\qquad\qquad\qquad\qquad\qquad\qquad\qquad\qquad\square$

---

[54] w_1 is in $\mathcal{H}_0^2$.

6

ITÔ's FORMULA

Itô's formula is the most important relation of stochastic analysis. The formula is a stochastic generalization of the Fundamental Theorem of Calculus. Recall that for an arbitrary process X, for an arbitrary differentiable function f and for an arbitrary partition $(t_k^{(n)})$ of an interval $[0, t]$

$$f(X(t)) - f(X(0)) = \sum_k \left(f(X(t_k^{(n)})) - f(X(t_{k-1}^{(n)})) \right) = \qquad (6.1)$$

$$= \sum_k f'(\xi_k^{(n)}) \left(X(t_k^{(n)}) - X(t_{k-1}^{(n)}) \right).$$

where $\xi_k^{(n)} \in (X(t_{k-1}^{(n)}), X(t_k^{(n)}))$. If X is continuous then by the intermediate value theorem $\xi_k^{(n)} = X(\tau_k^{(n)})$, where $\tau_k^{(n)} \in (t_{k-1}^{(n)}, t_k^{(n)})$. If X has finite variation then if $n \nearrow \infty$ the sum on the right-hand side will be convergent and one can easily get the Fundamental Theorem of Calculus:

$$f(X(t)) - f(X(0)) = \int_0^t f'(X(s))dX(s).$$

On the other hand, if X is a local martingale then the telescopic sum on the right-hand side of (6.1) does not necessarily converge to the stochastic integral $\int_0^t f'(X(s))dX(s)$, as one cannot guarantee the convergence unless $\tau_k^{(n)} = t_{k-1}^{(n)}$. If we make a second-order approximation

$$f(X(t_k^{(n)})) - f(X(t_{k-1}^{(n)})) = f'(X(t_{k-1}^{(n)})) \left(X(t_k^{(n)}) - X(t_{k-1}^{(n)}) \right) +$$

$$+ \tfrac{1}{2} f''(\xi_k^{(n)}) \left(X(t_k^{(n)}) - X(t_{k-1}^{(n)}) \right)^2$$

then the sum of the first order terms

$$I_n \stackrel{\circ}{=} \sum_k f' \left(X(t_{k-1}^{(n)}) \right) \left(X(t_k^{(n)}) - X(t_{k-1}^{(n)}) \right)$$

is an approximating sum of the Itô–Stieltjes integral $\int_0^t f'(X(s))dX(s)$. Of course the sum of the second order terms is also convergent, the only question is what is the limit? As

$$(X(t_k^{(n)}) - X(t_{k-1}^{(n)}))^2 \approx \left[X(t_k^{(n)})\right] - \left[X(t_{k-1}^{(n)})\right]$$

one can guess that the limit is

$$\frac{1}{2}\int_0^t f''(X(s))d\,[X]\,(s).$$

This is true if X is continuous as in this case again $\xi_k^{(n)} = X(\tau_k^{(n)})$ and the second order term is 'close' to the Stieltjes-type approximating sum

$$\frac{1}{2}f''\left(X\left(\tau_k^{(n)}\right)\right)\left(\left[X(t_k^{(n)})\right] - \left[X(t_{k-1}^{(n)})\right]\right).$$

The argument just introduced is 'nearly valid' even if X is discontinuous. In this case the first order term is again an Itô–Stieltjes type approximating sum and it is convergent again in Itô–Stieltjes sense and the limit is[1]

$$\int_0^t f'\,(X(s))\,dX(s) = \int_0^t f'\,(X_-(s))\,dX(s).$$

The main difference is that in this case one cannot apply for the second order term the intermediate value theorem. Therefore the second order term is not a simple Stieltjes type approximating sum. If we take only the 'continuous' subintervals, then one gets a Stieljes-type approximating sum and the limit is

$$\frac{1}{2}\int_0^t f''(X_-(s))d\,[X^c]\,.$$

For the remaining terms one can only apply the approximation

$$\frac{1}{2}f''(\xi_k^{(n)})\left(\Delta X(t_k^{(n)})\right)^2 =$$
$$= f(X(t_k^{(n)})) - f(X(t_{k-1}^{(n)})) - f'(X(t_{k-1}^{(n)}))\left(X(t_k^{(n)}) - X(t_{k-1}^{(n)})\right)$$

which converges to

$$f(X(s)) - f\,(X(s-)) - f'(X(s-))\Delta X(s),$$

---

[1]See: Theorem 2.21, page 125. The second integral is convergent in the general sense as well.

so in the limit the second-order term is

$$\frac{1}{2}\int_0^t f''(X_-(s))d\,[X^c] + \sum_{0<s\leq t} (f(X(s)) - f\,(X(s-)) - f'(X(s-))\Delta X(s))\,.$$

6.1   Itô's Formula for Continuous Semimartingales

Recall that for continuous semimartingales one has the following integration by parts formula[2]:

Proposition 6.1 *If X and Y are continuous semimartingales then for every t*

$$X\,(t)\,Y\,(t) - X\,(0)\,Y\,(0) = \int_0^t X\,dY + \int_0^t Y\,dX + [X,Y]\,(t)\,. \qquad (6.2)$$

Theorem 6.2 (Itô's formula) *Let U be an open subset of $\mathbb{R}^n$. If the elements of the vector $\mathbf{X} \stackrel{\circ}{=} (X_1, X_2, \ldots, X_n)$ are continuous semimartingales, $\mathbf{X}\,(t) \in U$ for every t and $f \in C^2\,(U)$, then*

$$f\,(\mathbf{X}\,(t)) - f\,(\mathbf{X}\,(0)) = \sum_{k=1}^n \int_0^t \frac{\partial f}{\partial x_k}\,(\mathbf{X})\,dX_k+ \qquad (6.3)$$

$$+ \frac{1}{2}\sum_{i,j}\int_0^t \frac{\partial^2 f}{\partial x_i \partial x_j}\,(\mathbf{X})\,d\,[X_i, X_j]\,.$$

Proof. We divide the proof into several steps.

1. As a first step we prove the theorem for polynomials. If $f \equiv c$, where c is a constant, then the theorem is trivial. It is sufficient to prove that if the identity is valid for a polynomial f then it is true for the polynomial $g \stackrel{\circ}{=} x_l f$ as well. Assume, that

$$f\,(\mathbf{X}) = f\,(\mathbf{X}\,(0)) + \sum_k \frac{\partial f}{\partial x_k}\,(\mathbf{X}) \bullet X_k + \frac{1}{2}\sum_{i,j} \frac{\partial^2 f}{\partial x_i \partial x_j}\,(\mathbf{X}) \bullet [X_i, X_j]\,.$$

By (6.2)

$$g\,(\mathbf{X}) \stackrel{\circ}{=} X_l f\,(\mathbf{X}) =$$

$$= g\,(\mathbf{X}\,(0)) + X_l \bullet f\,(\mathbf{X}) + f\,(\mathbf{X}) \bullet X_l + [X_l, f\,(\mathbf{X})] =$$

[2]See: Proposition 2.28, page 129.

$$= g\left(\mathbf{X}\left(0\right)\right) + X_l \bullet f\left(\mathbf{X}\left(0\right)\right) + X_l \bullet \left(\sum_k \left(\frac{\partial f}{\partial x_k}\left(\mathbf{X}\right) \bullet X_k\right)\right) +$$

$$+ X_l \bullet \left(\frac{1}{2}\sum_{i,j} \frac{\partial^2 f}{\partial x_i \partial x_j}\left(\mathbf{X}\right) \bullet [X_i, X_j]\right) +$$

$$+ f\left(\mathbf{X}\right) \bullet X_l + [X_l, f\left(\mathbf{X}\right)].$$

Now $X_l \bullet f\left(\mathbf{X}\left(0\right)\right) = 0$, and by the associativity rule for stochastic integrals[3]

$$g\left(\mathbf{X}\right) = g\left(\mathbf{X}\left(0\right)\right) + \sum_k X_l \frac{\partial f}{\partial x_k}\left(\mathbf{X}\right) \bullet X_k +$$

$$+ \frac{1}{2}\sum_{i,j} X_l \frac{\partial^2 f}{\partial x_i \partial x_j}\left(\mathbf{X}\right) \bullet [X_i, X_j] +$$

$$+ f\left(\mathbf{X}\right) \bullet X_l + [X_l, f\left(\mathbf{X}\right)].$$

By the product rule of differentiation

$$\frac{\partial g}{\partial x_k} = \begin{cases} x_l \dfrac{\partial f}{\partial x_k} & \text{if} \quad k \neq l \\[2mm] x_l \dfrac{\partial f}{\partial x_l} + f & \text{if} \quad k = l \end{cases}. \tag{6.4}$$

Substituting it in the formula above,

$$g\left(\mathbf{X}\right) = g\left(\mathbf{X}\left(0\right)\right) + \sum_k \frac{\partial g}{\partial x_k}\left(\mathbf{X}\right) \bullet X_k +$$

$$+ \frac{1}{2}\sum_{i,j} X_l \frac{\partial^2 f}{\partial x_i \partial x_j}\left(\mathbf{X}\right) \bullet [X_i, X_j] + [X_l, f\left(\mathbf{X}\right)].$$

The second partial derivatives of g are

$$\frac{\partial^2 g}{\partial x_i \partial x_j} = \begin{cases} x_l \dfrac{\partial^2 f}{\partial x_i \partial x_j} & \text{if} \quad i,j \neq l \\[2mm] x_l \dfrac{\partial^2 f}{\partial x_l \partial x_j} + \dfrac{\partial f}{\partial x_j} & \text{if} \quad i = l, j \neq l \\[2mm] x_l \dfrac{\partial^2 f}{\partial x_i \partial x_l}/ + \dfrac{\partial f}{\partial x_i} & \text{if} \quad i \neq l, j = l \\[2mm] x_l \dfrac{\partial^2 f}{\partial^2 x_l} + 2\dfrac{\partial f}{\partial x_l} & \text{if} \quad i = j = l \end{cases}, \tag{6.5}$$

---

[3]See: Proposition 2.71, page 160.

that is matrices f'' and g'' are different only in column l and in row l. It is sufficient to prove that

$$[X_l, f(\mathbf{X})] = \sum_{j=1}^{n} \frac{\partial f}{\partial x_j}(\mathbf{X}) \bullet [X_l, X_j].$$

By the induction hypothesis $f(X)$ is a semimartingale. As X_l is continuous the quadratic co-variation of the bounded variation part of $f(X)$ is zero. The quadratic variation of the stochastic integral part is

$$\left[X_l, \sum_{j=1}^{n} \frac{\partial f}{\partial x_k}(\mathbf{X}) \bullet X_k \right] = \sum_{j=1}^{n} \frac{\partial f}{\partial x_j}(\mathbf{X}) \bullet [X_l, X_j].$$

This means that the theorem is valid for polynomials.

2. Let us prove that one can localize the expression. That is, it is sufficient to prove the theorem for $\mathbf{X}^{\tau_n}$ where (τ_n) is some localizing sequence of $\mathbf{X}$. Let τ be an arbitrary stopping time. The integrals in the second line are integrals taken by trajectory, hence obviously

$$\frac{\partial^2 f}{\partial x_i \partial x_j}(\mathbf{X}^\tau) \bullet [X_i^\tau, X_j^\tau] = \frac{\partial^2 f}{\partial x_i \partial x_j}(\mathbf{X}^\tau) \bullet [X_i, X_j]^\tau =$$

$$= \frac{\partial^2 f}{\partial x_i \partial x_j}(\mathbf{X}) \chi([0,\tau]) \bullet [X_i, X_j].$$

In a similar way, using the stopping rule for stochastic integrals

$$\frac{\partial f}{\partial x_k}(\mathbf{X}^\tau) \bullet X_k^\tau = \frac{\partial f}{\partial x_k}(\mathbf{X}^\tau) \chi([0,\tau]) \bullet X_k =$$

$$= \frac{\partial f}{\partial x_k}(\mathbf{X}) \chi([0,\tau]) \bullet X_k.$$

Assume that the theorem is valid for the truncated processes $\mathbf{X}^{\tau_n}$. $f \in C^2(U)$, hence the trajectories of the $\partial f/\partial x_k(\mathbf{X})$ and $\partial^2 f/\partial x_i \partial x_j(\mathbf{X})$ are continuous and therefore they are integrable. Evidently the integrands above are dominated by these common integrable processes. If $\tau_n \to \infty$, then $\chi([0,\tau_n]) \to 1$. Applying the Dominated Convergence Theorem on both sides and using that $f(\mathbf{X}^{\tau_n}) \to f(\mathbf{X})$ one can easily prove the equality.

3. As $\mathbf{X}$ is continuous it is locally bounded. Let (τ_n) be a localizing sequence for which the images of the stopped processes $\mathbf{X}^{\tau_n}$ are bounded. Let $K \subseteq U$ be a compact set which contains the image of $\mathbf{X}^{\tau_n}$. One can prove, that there is a sequence of polynomials (p_n) that in the topology of $C^2(K)$ one has $p_n|_K \to f|_K$. By the definition of the topology of C^2 all the derivatives

are uniformly convergent. As the formula is valid for every polynomial by the Dominated Convergence Theorem it is valid for the function $f \in C^2(U)$ as well. □

Proposition 6.3 *If the semimartingale X_l has finite variation, then it is sufficient to assume that the partial derivative $\partial f / \partial x_l$ exists and it is continuous. In this case in the formula (6.3) one can drop the second-order terms with index l.*

Proof. If X_l has finite variation then as X_i is continuous $[X_l, X_i] = 0$. If f is a polynomial, then the second-order terms with index l are zero, and in the approximation we do not need the second-order terms with index l. □

Corollary 6.4 (Time dependent Itô formula) *If the elements of the vector*

$$\mathbf{X} \overset{\circ}{=} (X_1, X_2, \ldots, X_n)$$

are continuous semimartingales and the image space of $\mathbf{X}$ is part of an open subset $U \subseteq \mathbb{R}^n$ and $f \in C^2(\mathbb{R}_+ \times U)$ then[4]

$$f(t, \mathbf{X}(t)) = f(0, \mathbf{X}(0)) + \int_0^t \frac{\partial f}{\partial s}(s, \mathbf{X}(s))\, ds +$$

$$+ \sum_{i=1}^n \int_0^t \frac{\partial f}{\partial x_i}(s, \mathbf{X}(s))\, dX_i(s) +$$

$$+ \frac{1}{2} \sum_{i=1}^n \sum_{j=1}^n \int_0^t \frac{\partial^2 f}{\partial x_i \partial x_j}(s, \mathbf{X}(s))\, d[X_i, X_j](s).$$

If X and Y are real-valued semimartingales then we can define the object $Z \overset{\circ}{=} X + iY$, which one can call a *complex semimartingale*. Let $f : \mathbb{C} \to \mathbb{C}$ be a holomorphic function. $f(z)$ has the representation $u(x, y) + iv(x, y)$, where u and v are differentiable functions. Recall that

$$\frac{\partial u}{\partial x} = \frac{\partial v}{\partial y} \quad \text{and} \quad \frac{\partial u}{\partial y} = -\frac{\partial v}{\partial x}.$$

If Z is a complex semimartingale then

$$f(Z) = u(X, Y) + iv(X, Y).$$

---

[4]It is sufficient to assume that f is continuously differentiable by the time parameter.

One can apply Itô's formula for u and for v.

$$u(X(t), Y(t)) = u(X(0), Y(0)) +$$

$$+ \int_0^t \frac{\partial u}{\partial x}(X, Y) dX + \int_0^t \frac{\partial u}{\partial y}(X, Y) dY +$$

$$+ \frac{1}{2} \int_0^t \frac{\partial^2 u}{\partial x^2}(X, Y) d[X, X] +$$

$$+ \frac{1}{2} \int_0^t \frac{\partial^2 u}{\partial y^2}(X, Y) d[Y, Y] +$$

$$+ \int_0^t \frac{\partial^2 u}{\partial x \partial y}(X, Y) d[X, Y]$$

and

$$v(X(t), Y(t)) = v(X(0), Y(0)) +$$

$$+ \int_0^t \frac{\partial v}{\partial x}(X, Y) dX + \int_0^t \frac{\partial v}{\partial y}(X, Y) dY +$$

$$+ \frac{1}{2} \int_0^t \frac{\partial^2 v}{\partial x^2}(X, Y) d[X, X] +$$

$$+ \frac{1}{2} \int_0^t \frac{\partial^2 v}{\partial y^2}(X, Y) d[Y, Y] +$$

$$+ \int_0^t \frac{\partial^2 v}{\partial x \partial y}(X, Y) d[X, Y].$$

The sum of the first-order terms is

$$\int_0^t \frac{\partial u}{\partial x}(X, Y) dX + \int_0^t \frac{\partial u}{\partial y}(X, Y) dY +$$

$$+ i \int_0^t \frac{\partial v}{\partial x}(X, Y) dX + i \int_0^t \frac{\partial v}{\partial y}(X, Y) dY.$$

As $u'_x + iv'_x = v'_y - iu'_y = f'$ this sum is

$$\int_0^t f'(Z) dX + \int_0^t f'(Z) d(iY) \stackrel{\circ}{=} \int_0^t f'(Z) dZ.$$

Let us calculate the second-order terms:

$$\int_0^t \frac{\partial^2 u}{\partial x^2} + i\frac{\partial^2 v}{\partial x^2} d[X,X] = \int_0^t \frac{\partial^2 u}{\partial x^2} + i\frac{\partial^2 v}{\partial x^2} d[X,X],$$

$$\int_0^t \frac{\partial^2 u}{\partial y^2} + i\frac{\partial^2 v}{\partial y^2} d[Y,Y] = \int_0^t -\frac{\partial^2 u}{\partial x^2} - i\frac{\partial^2 v}{\partial x^2} d[Y,Y] =$$

$$= \int_0^t \frac{\partial^2 u}{\partial x^2} + i\frac{\partial^2 v}{\partial x^2} d[iY,iY],$$

$$\int_0^t \frac{\partial^2 u}{\partial x \partial y} + i\frac{\partial^2 v}{\partial x \partial y} d[X,Y] = \int_0^t -\frac{\partial^2 v}{\partial^2 x} + i\frac{\partial^2 u}{\partial x^2} d[X,Y] =$$

$$= \int_0^t \frac{\partial^2 u}{\partial x^2} + i\frac{\partial^2 v}{\partial^2 x} d[X,iY].$$

Also by definition

$$[Z] \overset{\circ}{=} [X] + 2i[X,Y] - [Y].$$

Therefore the second order term is

$$\frac{1}{2}\int_0^t \frac{\partial^2 u}{\partial^2 x} + i\frac{\partial^2 v}{\partial^2 x} d[Z] = \frac{1}{2}\int_0^t f''(Z)d[Z].$$

Corollary 6.5 (Itô's formula for holomorphic functions) *If $f(t,z)$ is continuously differentiable in t and it is holomorphic in z and Z is a continuous complex semimartingale then*

$$f(t,Z(t)) = f(0,Z(0)) + \int_0^t \frac{\partial f}{\partial s}(s,Z(s))\,ds+$$

$$+ \int_0^t \frac{\partial f(s,Z(s))}{\partial z}dZ(s) + \frac{1}{2}\int_0^t \frac{\partial^2 f(s,Z(s))}{\partial z^2}d[Z](s).$$

Example 6.6 If $Z \overset{\circ}{=} w_1 + iw_2$ is a planar Brownian motion and f is an entire function then $f(Z)$ is a complex local martingale and

$$f(Z(t)) = f(Z(0)) + \int_0^t f'(Z)dZ.$$

As $[w_1, w_2] = 0$ and $[w_1](t) = [w_2](t) = t$ obviously $[Z] = 0$. $\qquad\qquad\square$

6.2   Some Applications of the Formula

In this section we present some famous and important applications of the formula.

6.2.1   Zeros of Wiener processes

As a first application let us investigate some important properties of the multi-dimensional Wiener processes. By definition assume that the coordinates of a d-dimensional Wiener process $\mathbf{w}$ are independent one-dimensional Wiener processes. To simplify the notation we say that a stochastic process $\mathbf{w}$ is a d-dimensional Wiener process starting from some point $\mathbf{x} \in \mathbb{R}^d$ if it has the representation $\mathbf{w} = \mathbf{x} + \widetilde{\mathbf{w}}$, where $\widetilde{\mathbf{w}}$ is an ordinary d-dimensional Wiener process, obviously starting from the origin. In the same way if $\mathbf{x}$ is an $\mathcal{F}_0$-measurable random vector then one can talk about a Wiener process starting from $\mathbf{x}$. Assume that $\mathbf{w}$ starts from some vector $\mathbf{x}$. Let[5]

$$\vartheta \overset{\circ}{=} \inf \left\{ \|\mathbf{w}(t)\| : t \geq 0 \right\}.$$

What is the distribution of ϑ?

Theorem 6.7 (Return of a Wiener process to the origin) *Every d-dimensional Wiener process $\mathbf{w}$ starting from some vector $\mathbf{x} \neq \mathbf{0}$ satisfies the following*[6]*:*

1. *If $d \geq 2$ then for almost every outcome ω the trajectory $\mathbf{w}(\omega)$ is never zero, that is*

$$\mathbf{P}\left(\|\mathbf{w}(t)\| \neq 0, \forall t > 0\right) = 1.$$

2. *If $d = 2$ then*

$$\mathbf{P}\left(\vartheta = 0\right) = 1,$$

   that is, $\mathbf{w}$ is almost surely never zero, but it hits every neighborhood of the origin almost surely.
3. *If $d = 2$ then the trajectories of $\mathbf{w}$ are almost surely dense in $\mathbb{R}^2$.*
4. *If $d \geq 3$ and $\mathbf{w}(0) = \mathbf{x} \neq \mathbf{0}$ then*

$$\mathbf{P}\left(\vartheta \leq r\right) = \left(\frac{r}{\|\mathbf{x}\|}\right)^{d-2}, \quad if \ \ 0 \leq r \leq \|\mathbf{x}\|.$$

---

[5] In this section $\|\mathbf{x}\|$ denotes the norm $\sqrt{\sum_k x_k^2}$.

[6] See: Corollary B.8. page 565.

Proof. Assume that the twice continuously differentiable function f defined on an open set $U \subseteq \mathbb{R}^d$ satisfies the *Laplace equation*

$$\sum_{k=1}^{d} \frac{\partial^2 f}{\partial x_i^2} = 0, \quad f \in C^2(U). \tag{6.6}$$

Let τ be a stopping time. If a d-dimensional Wiener process $\mathbf{w}$ starting from an $\mathbf{x}$ remains in U then by Itô's formula

$$f(\mathbf{w}^\tau) - f(\mathbf{w}(0)) = \sum_{k=1}^{d} \frac{\partial f}{\partial x_k}(\mathbf{w}^\tau) \bullet w_k^\tau +$$

$$+ \frac{1}{2} \sum_{i,j} \frac{\partial^2 f}{\partial x_i \partial x_j}(\mathbf{w}^\tau) \bullet \left[w_i^\tau, w_j^\tau \right].$$

If $i \neq j$ then[7] $\left[w_i^\tau, w_j^\tau \right] = 0^\tau = 0$. Hence as $\left[w_i^\tau \right](s) = s \wedge \tau$

$$f(\mathbf{w}^\tau(t)) - f(\mathbf{x}) = f(\mathbf{w}^\tau(t)) - f(\mathbf{w}(0)) = \tag{6.7}$$

$$= \sum_{k=1}^{d} \int_0^t \frac{\partial f}{\partial x_k}(\mathbf{w}^\tau) \, dw_k^\tau + \frac{1}{2} \int_0^\tau (\Delta f)(\mathbf{w}^\tau(s)) ds =$$

$$= \sum_{k=1}^{d} \int_0^t \frac{\partial f}{\partial x_k}(\mathbf{w}^\tau) \, dw_k^\tau.$$

Assume that $\tau < \infty$ and $\mathbf{w}$ is bounded on the random interval $[0, \tau]$. In this case the integrands in (6.7) are bounded. As on any finite interval $\mathbf{w}^\tau$ is square-integrable the stochastic integrals are martingales[8]. Hence for every point of time $t < \infty$

$$\mathbf{E}(f(\mathbf{w}^\tau(t))) = \mathbf{E}(f(\mathbf{w}(t \wedge \tau))) = \mathbf{E}(f(\mathbf{w}(0))) = \mathbf{E}(f(\mathbf{x})).$$

By the assumption $\mathbf{w}$ is bounded on $[0, \tau]$, therefore by the Dominated Convergence Theorem one can take the limit $t \to \infty$. Hence we get the so-called *Dynkin's formula*

$$\mathbf{E}(f(\mathbf{w}(\tau))) = \mathbf{E}(f(\mathbf{x})). \tag{6.8}$$

[7]See: Example 2.46, page 144.
[8]See: Proposition 2.59, page 152.

Using Dynkin's formula one can deduce the theorem with some direct calculation:

1. With a simple calculation one can show that the function[9]

$$f(\mathbf{u}) \overset{\circ}{=} \begin{cases} \log \|\mathbf{u}\| & \text{if } d = 2 \\ \|\mathbf{u}\|^{2-d} & \text{if } d \geq 3 \end{cases} \tag{6.9}$$

satisfies the Laplace equation (6.6) on the open set $U \overset{\circ}{=} \mathbb{R}^d \setminus \{0\}$.

2. Assume[10] that $\|\mathbf{x}\|$ lies in between the radii $0 < r < R < \infty$. The trajectories of Wiener processes are continuous and almost surely unbounded on the half-line $t \geq 0$, hence $\mathbf{w}$ almost surely leaves the d-dimensional ring

$$B \overset{\circ}{=} \{\mathbf{u} : r \leq \|\mathbf{u}\| \leq R\}.$$

The only question is whether it leaves the ring first at the outer or at the inner boundary of B.

3. Assume that $d \geq 3$. Let[11]

$$\tau_{\partial B} \overset{\circ}{=} \inf \{t : w(t) \in \partial B\}.$$

By Dynkin's formula

$$\mathbf{E}(f(\mathbf{w}(\tau_{\partial B}))) = \mathbf{E}(f(\mathbf{x})) = f(\mathbf{x}). \tag{6.10}$$

Substituting expression (6.9) for f in (6.10)

$$r^{2-d}\mathbf{P}(\|\mathbf{w}(\tau_{\partial B})\| = r) + R^{2-d}(1 - \mathbf{P}(\|\mathbf{w}(\tau_{\partial B})\| = r)) = \|\mathbf{x}\|^{2-d},$$

that is

$$\mathbf{P}(\|\mathbf{w}(\tau_{\partial B})\| = r) = \frac{\|\mathbf{x}\|^{2-d} - R^{2-d}}{r^{2-d} - R^{2-d}}.$$

If $R \to \infty$ then using that R^{2-d} converges to zero if $R \to \infty$, the limit of the right-hand side is $(r/\|\mathbf{x}\|)^{d-2}$. This is the probability that $\mathbf{w}$ intersects the ball with radius $r \leq \|\mathbf{x}\|$.

4. If $d = 2$ then $\|\mathbf{u}\|^{2-d} \equiv 1$, and one cannot use the previous calculation. In this case $f(\mathbf{u}) \overset{\circ}{=} \log \|\mathbf{u}\|$. Using the fact that in this case

---

[9]Observe that if $\|\mathbf{x}\| \to \infty$ then f behaves differently if $n = 2$ and if $n \geq 3$.

[10]Observe that $\mathbf{x}$ can be random. It is sufficient to assume, that $\|\mathbf{x}\|$ is deterministic.

[11]See: Example 1.32, page 17.

$\lim_{u \to \infty} f(u) = \infty$

$$\lim_{R \to \infty} \mathbf{P}\left(\|\mathbf{w}(\tau_{\partial B})\| = r\right) = \lim_{R \to \infty} \frac{\log\|\mathbf{x}\| - \log R}{\log r - \log R} = 1.$$

Hence with probability one $\mathbf{w}$ intersects the ball with radius r. For some fixed R if $r \searrow 0$ then

$$\lim_{r \searrow 0} \frac{\log\|\mathbf{x}\| - \log R}{\log r - \log R} = 0.$$

This means that $\mathbf{w}$ starting from some $\mathbf{x} \neq \mathbf{0}$ reaches the point $\mathbf{0}$ with probability zero before it leaves the ball with radius R. It is valid for every R so if $\mathbf{x} \neq \mathbf{0}$ then $\mathbf{w}$ can be exactly zero only with probability zero.

5. Assume that $\mathbf{x} = \mathbf{0}$. Let $\varepsilon > 0$.

$$\mathbf{P}\left(\inf_{t \geq \varepsilon} \|\mathbf{w}(t)\| > 0\right) = \int_{\mathbb{R}^d} \mathbf{P}\left(\inf_{t \geq \varepsilon} \|\mathbf{w}(t)\| > 0 \mid \mathbf{w}(\varepsilon) = \mathbf{y}\right) d\rho(\mathbf{y}),$$

where ρ is the distribution of $\mathbf{w}(\varepsilon)$. Let us calculate the conditional probability. As $\mathbf{w}$ has stationary and independent increments

$$\mathbf{P}\left(\inf_{t \geq \varepsilon} \|\mathbf{w}(t)\| > 0 \mid \mathbf{w}(\varepsilon) = \mathbf{y}\right) =$$

$$= \mathbf{P}\left(\inf_{t \geq \varepsilon} \|\mathbf{w}(t) - \mathbf{w}(\varepsilon) + \mathbf{w}(\varepsilon)\| > 0 \mid \mathbf{w}(\varepsilon) = \mathbf{y}\right) =$$

$$= \mathbf{P}\left(\inf_{t \geq \varepsilon} \|\mathbf{w}(t) - \mathbf{w}(\varepsilon) + \mathbf{y}\| > 0\right) =$$

$$= \mathbf{P}\left(\inf_{u \geq 0} \|\mathbf{w}(u) + \mathbf{y}\| > 0\right) =$$

$$= \mathbf{P}\left(\inf_{u \geq 0} \|\mathbf{w}^{\mathbf{y}}(u)\| > 0\right),$$

where $\mathbf{w}^{\mathbf{y}}$ is the Wiener process starting from the point $\mathbf{y}$. By the formula already proved for $\mathbf{x} \neq \mathbf{0}$ in 3. and 4. above

$$\mathbf{P}\left(\inf_{t \geq \varepsilon} \|\mathbf{w}(t)\| > 0\right) = \int_{\mathbb{R}^d} \mathbf{P}\left(\inf_{t \geq 0} \|\mathbf{w}^{\mathbf{y}}(t)\| > 0\right) d\rho(\mathbf{y}) =$$

$$= \int_{\mathbb{R}^d \setminus \{\mathbf{0}\}} \mathbf{P}\left(\inf_{t \geq 0} \|\mathbf{w}^{\mathbf{y}}(t)\| > 0\right) d\rho(\mathbf{y}) =$$

$$= \int_{\mathbb{R}^d \setminus \{\mathbf{0}\}} 1 d\rho(\mathbf{y}) = 1.$$

If $\varepsilon \to 0$ then

$$\mathbf{P}\left(\|\mathbf{w}(t)\| > 0, \forall t > 0\right) = \lim_{\varepsilon \searrow 0} \mathbf{P}\left(\inf_{t \geq \varepsilon} \|\mathbf{w}(t)\| > 0\right) = 1.$$

This means that with probability one $\mathbf{w}$ does not return back to the origin. Hence we have proved the theorem for all initial vectors $\mathbf{x} \in \mathbb{R}^d$.

6. Instead of balls around the origin one can take any ball. If we take the balls with rational centers and rational radii then the two-dimensional Wiener process with probability one intersects all of them. Therefore the trajectories of the Wiener processes are dense in $\mathbb{R}^2$. □

In the same way one can prove the following:

Corollary 6.8 *Let $d \geq 3$ and let $\mathbf{w}$ be a d-dimensional Wiener process starting from some random vector $\mathbf{x}$. If $\|\mathbf{x}\|$ is deterministic then*

$$\mathbf{P}\left(\vartheta \leq r\right) = \left(\frac{r}{\|\mathbf{x}\|}\right)^{d-2}, \quad \text{if} \ \ 0 \leq r \leq \|\mathbf{x}\|.$$

Corollary 6.9 *If $d \geq 3$ and $\mathbf{w}$ is a d-dimensional Wiener process then* $\lim_{t \to \infty} \|\mathbf{w}(t)\| = \infty$.

Proof. Let $r > 0$ be arbitrary and for any $a \geq r$ let

$$\tau_a \doteq \inf\{t : \|\mathbf{w}(t)\| \geq a\}.$$

As almost surely[12]

$$\limsup_{t \to \infty} \|\mathbf{w}(t)\| = \infty$$

obviously $\tau_a < \infty$ almost surely. By the strong Markov property of $\mathbf{w}$

$$\mathbf{w}^*(t) \doteq (\mathbf{w}(t + \tau_a) - \mathbf{w}(\tau_a)) + \mathbf{w}(\tau_a), \quad t \geq 0$$

is a Wiener process starting from the random point

$$\mathbf{w}(\tau_a) \in \{\|\mathbf{u}\| = a\}.$$

Since $d \geq 3$

$$\mathbf{P}\left(\exists t \geq \tau_a, \|\mathbf{w}(t)\| \leq r\right) = \mathbf{P}\left(\exists t \geq 0, \|\mathbf{w}^*(t)\| \leq r\right) = \left(\frac{r}{a}\right)^{d-2}.$$

[12]See: Proposition B.7, page 564.

If $a \nearrow \infty$ then this probability goes to zero. Let $a_n \nearrow \infty$. The probability that $\mathbf{w}(t)$ returns to the ball $\{\|\mathbf{u}\| \leq r\}$ after infinitely many τ_{a_n} is zero. Hence with probability one for any ω there is an $n \stackrel{\circ}{=} n(\omega)$ that

$$\mathbf{w}(t, \omega) \notin \{\|\mathbf{u}\| \leq r\} \quad t \geq \tau_n(\omega).$$

That is with probability one[13] if $t \nearrow \infty$ then $\|w(t, \omega)\| \to \infty$. $\qquad \square$

Example 6.10 Hitting times of open and closed sets in higher dimensions[14].

---

1. Let $B(\mathbf{x}_0, r) \stackrel{\circ}{=} \{\mathbf{x} \in \mathbb{R}^d : \|\mathbf{x} - \mathbf{x}_0\| < r\}$. Let $\mathbf{x}_0 \neq \mathbf{0} \in B(\mathbf{0}, 1)$ and let

$$f(\mathbf{x}) \stackrel{\circ}{=} g(\|\mathbf{x} - \mathbf{x}_0\|) \stackrel{\circ}{=} \begin{cases} \log\|\mathbf{x} - \mathbf{x}_0\| & \text{if} \quad d = 2 \\ \|\mathbf{x} - \mathbf{x}_0\|^{2-d} & \text{if} \quad d \geq 3 \end{cases}.$$

Obviously f satisfies the Laplace equation (6.6) on $\mathbb{R}^d \setminus \{\mathbf{x}_0\}$. If $B(\mathbf{x}_0, r) \subseteq B(\mathbf{0}, 1)$ and

$$B \stackrel{\circ}{=} B(\mathbf{0}, 1) \setminus \text{cl}(B(\mathbf{x}_0, r))$$

then f is bounded on B. Let $\mathbf{w}$ be a d-dimensional Wiener process and let

$$\tau \stackrel{\circ}{=} \inf\{t : \mathbf{w}(t) \in \partial B(\mathbf{0}, 1)\}.$$

As $\limsup_t \|\mathbf{w}(t)\| = \infty$, obviously[15] almost surely $\tau < \infty$. By Itô's formula $X \stackrel{\circ}{=} f(\mathbf{w}^\tau)$ is a bounded local martingale on B, therefore X is a uniformly integrable martingale[16]. Hence if

$$\rho \stackrel{\circ}{=} \inf\{t : \mathbf{w}(t) \in \partial B\},$$

then

$$\mathbf{E}(X(\rho)) = \mathbf{E}(X(0)) = f(\mathbf{0}).$$

If

$$\rho_1 \stackrel{\circ}{=} \inf\{t : \mathbf{w}(t) \in \partial B(\mathbf{0}, 1)\}$$
$$\rho_2 \stackrel{\circ}{=} \inf\{t : \mathbf{w}(t) \in \partial B(\mathbf{x}_0, r)\}$$

---

[13] Take $r \stackrel{\circ}{=} 1, 2, \ldots$.
[14] See: Corollary B.12, page 566.
[15] See: Proposition B.7, page 564.
[16] See: Corollary 1.145, page 103.

then as $\rho = \rho_1 \wedge \rho_2$

$$f\left(0\right) = \mathbf{E}\left(X\left(\rho\right)\right) =$$
$$= \mathbf{E}\left(X\left(\rho_1\right)\chi\left(\rho_1 \leq \rho_2\right)\right) + \mathbf{E}\left(X\left(\rho_2\right)\chi\left(\rho_2 < \rho_1\right)\right).$$

Obviously

$$\mathbf{E}\left(X\left(\rho_2\right)\chi\left(\rho_2 < \rho_1\right)\right) = g\left(r\right) \cdot \mathbf{P}\left(\rho_2 < \rho_1\right)$$

and for some k

$$\mathbf{E}\left(X\left(\rho_1\right)\chi\left(\rho_1 \leq \rho_2\right)\right) \leq k$$

for all $r > 0$. This implies that for any $0 < r < 1$

$$\mathbf{P}\left(\rho_1 > \rho_2\right) = \frac{\log\|\mathbf{x}_0\| - \mathbf{E}\left(X\left(\rho_2\right)\chi\left(\rho_1 > \rho_2\right)\right)}{g\left(r\right)} \leq \frac{g\left(\|\mathbf{x}_0\|\right) - k}{g\left(r\right)}.$$

If $r \searrow 0$, then the right-hand side goes to zero, so for any $\varepsilon > 0$ there is an $r > 0$ such that $\mathbf{P}\left(\rho_2 < \rho_1\right) < \varepsilon$.

2. Let $(\mathbf{q}_i)$ be the non-zero rational points of $B\left(\mathbf{0}, 1\right)$ and for any i let $r_i > 0$ be such that

$$\mathbf{P}\left(\rho_2^{(i)} < \rho_1\right) < 2^{-(i+1)}$$

where of course

$$\rho_2^{(i)} \overset{\circ}{=} \inf\left\{t : \mathbf{w}\left(t\right) \in \partial B\left(\mathbf{q}_i, r_i\right)\right\}.$$

Let $G \overset{\circ}{=} \cup_i B\left(\mathbf{q}_i, r_i\right)$. Obviously G is open and

$$\tau_{\mathrm{cl}(G)} \overset{\circ}{=} \inf\left\{t : \mathbf{w}\left(t\right) \in \mathrm{cl}\left(G\right)\right\} = \inf\left\{t : \mathbf{w}\left(t\right) \in \mathrm{cl}\left(B\left(\mathbf{0}, 1\right)\right)\right\} = 0.$$

On the other hand obviously $\rho_1 > 0$ and if

$$\tau_G \overset{\circ}{=} \inf\left\{t : \mathbf{w}\left(t\right) \in G\right\}$$

then

$$\mathbf{P}\left(\tau_G \geq \tau_1\right) = 1 - \mathbf{P}\left(\tau_G < \tau_1\right) \geq 1 - \sum_i \mathbf{P}\left(\rho_2^{(i)} < \rho_1\right) \geq$$
$$\geq 1 - \sum_i 2^{-(i+1)} \geq \frac{1}{2}.$$

Therefore $\tau_{\mathrm{cl}(G)}$ and τ_G are not almost surely equal. $\square$

6.2.2   Continuous Lévy processes

Let X be a continuous Lévy process. Since X is continuous all the moments of X are finite[17]. Hence $X(t)$ has an expected value for every t. Observe that as on any finite interval the second moments are bounded X is uniformly integrable on these intervals. Therefore $\mathbf{E}(X(t))$ is continuous in t, hence $\mathbf{E}(X(t)) = t\mathbf{E}(X(1))$. Therefore if m denotes the expected value of $X(1)$ then $X(t) - t\cdot m$ is a martingale. This means that X is a continuous semimartingale. To simplify the notation assume that $m = 0$. By the definition of the quadratic variation $[X]$ is also a continuous Lévy process. This again implies that $Y(t) \overset{\circ}{=} [X](t) - \mathbf{E}([X](t))$ is a martingale. As Y obviously has finite variation by Fisk's theorem[18] it is a constant. So $[X](t) = \mathbf{E}([X](t)) = a \cdot t$. By Itô's formula

$$\exp\left(iuX\left(t\right)\right) - 1 = iu \int_0^t \exp\left(iuX\left(s\right)\right) dX\left(s\right) -$$

$$- \frac{u^2}{2} \int_0^t \exp\left(iuX\left(s\right)\right) d\left[X\left(s\right)\right].$$

$\exp\left(iuX\left(t\right)\right)$ is bounded and the quadratic variation of X is deterministic, therefore by the characterization of $\mathcal{H}^2$-martingales[19] the stochastic integral is a martingale. Taking expected value on both sides

$$\mathbf{E}\left(\exp\left(iuX\left(t\right)\right)\right) - 1 = -\frac{u^2}{2} \mathbf{E}\left(\int_0^t \exp\left(iuX\left(s\right)\right) d\left[X\left(s\right)\right]\right) =$$

$$= -\frac{u^2}{2} \mathbf{E}\left(\int_0^t \exp\left(iuX\left(s\right)\right) d\left(as\right)\right) =$$

$$= -a\frac{u^2}{2} \int_0^t \mathbf{E}\left(\exp\left(iuX\left(s\right)\right)\right) ds.$$

If $\varphi\left(u,t\right) \overset{\circ}{=} \mathbf{E}\left(\exp\left(iuX\left(t\right)\right)\right)$ then

$$\varphi\left(u,t\right) - 1 = -a\frac{u^2}{2} \int_0^t \varphi\left(u,s\right) ds.$$

Differentiating w.r.t. t

$$\frac{d\varphi\left(u,t\right)}{dt} = -a\frac{u^2}{2} \cdot \varphi\left(u,t\right).$$

Solving the differential equation,

$$\varphi\left(u,t\right) = \exp\left(-a\frac{u^2}{2} \cdot t\right)$$

[17]See: Proposition 1.111, page 74.
[18]See: Theorem 2.11, page 117.
[19]See: Proposition 2.53, page 148.

for every u. By the formula of the Fourier transform for the normal distribution $X(t) \cong N(0, \sqrt{at})$. Hence $X/\sqrt{a}$ is a Wiener process. In general m is not zero, hence we have proved the next proposition:

Theorem 6.11 *Every continuous Lévy process is a linear combination of a Wiener process and a linear trend.*

One can extend the theorem to processes with independent increments:

Theorem 6.12 *Every continuous process with independent increments is a Gaussian process, that is for every $t_1, t_2, \ldots, t_n$*

$$(X(t_1), X(t_2), \ldots, X(t_n))$$

has Gaussian distribution.

Proof. If X has independent increments then $Z(t) \overset{\circ}{=} X(t+s) - X(s)$ also has independent increments for every s. Therefore it is easy to prove that it is sufficient to show that $X(t)$ has a Gaussian distribution for every t. By the continuity of X all the moments of X are bounded on every finite interval[20]. Therefore the expected value $\mathbf{E}(X(t))$ is finite for every t. As X is bounded in $L^2(\Omega)$ on every finite interval it is uniformly integrable on any finite interval, so $\mathbf{E}(X(t))$ is continuous. Hence it is easy to see that $Y(t) \overset{\circ}{=} X(t) - \mathbf{E}(X(t))$ is a continuous martingale. Therefore one may assume that X is a continuous martingale. As X has independent increments $[X]$ also has independent increments, so $U(t) \overset{\circ}{=} [X](t) - \mathbf{E}([X](t))$ is again a continuous martingale. As $[X]$ is increasing U has finite variation. So by Fisk's theorem almost surely $U \equiv 0$. Therefore one can assume that $[X]$ is deterministic. By Itô's formula

$$\exp(iuX(t)) - 1 = iu \int_0^t \exp(iuX(s)) \, dX(s) -$$

$$- \frac{u^2}{2} \int_0^t \exp(iuX(s)) \, d[X(s)].$$

$\exp(iuX)$ is bounded and on any finite interval $X \in \mathcal{H}^2$, therefore the stochastic integral is a martingale[21]. Taking expected value

$$\mathbf{E}(\exp(iuX(t))) - 1 = -\frac{u^2}{2} \cdot \mathbf{E}\left(\int_0^t \exp(iuX(s)) \, d[X](s)\right). \tag{6.11}$$

[20]See: Proposition 1.114, page 78.
[21]See: Proposition 2.24, page 128.

The quadratic variation is deterministic so one can change the order of the integration:

$$\mathbf{E}\left(\exp\left(iuX\left(t\right)\right)\right) - 1 = -\frac{u^2}{2} \cdot \int_0^t \mathbf{E}\left(\exp\left(iuX\left(s\right)\right)\right) d\left[X\right]\left(s\right).$$

If $\varphi\left(u,t\right) \overset{\circ}{=} \mathbf{E}\left(\exp\left(iuX\left(t\right)\right)\right)$, then φ satisfies the integral equation

$$\varphi\left(u,t\right) - 1 = -\frac{u^2}{2} \cdot \int_0^t \varphi\left(u,s\right) d\left[X\right]\left(s\right). \tag{6.12}$$

If

$$\varphi\left(u,t\right) \overset{\circ}{=} \exp\left(-\frac{u^2}{2}\left[X\left(t\right)\right]\right) \tag{6.13}$$

then, as $\left[X\right]$ is deterministic with finite variation, φ satisfies[22] (6.12). One can easily prove[23] that (6.13) is the only solution of (6.12). Therefore $X\left(t\right)$ has a Gaussian distribution for every t. □

6.2.3 Lévy's characterization of Wiener processes

The characterization theorem of Lévy is similar to the proposition just proved: it characterizes Wiener processes among the continuous local martingales. If $X \in \mathcal{L}$ and if $\left[X\right]\left(t\right) = t$ then by the same argument[24] as above one can prove that $X\left(t\right) \cong N\left(0, \sqrt{t}\right)$. As $X\left(t+s\right) - X\left(s\right) \in \mathcal{L}$ for every s the increments of X are also Gaussian. As $X\left(u\right) - X\left(v\right) \cong N\left(0, \sqrt{u-v}\right)$ it is easy to prove that the increments of X are not correlated. As X has Gaussian increments the increments are independent. Therefore by the same argument as above one can prove that X is a Wiener process with respect to its own filtration[25]. Our goal is to prove that X is a Wiener process with respect to the original filtration[26].

Theorem 6.13 (Lévy's characterization of Wiener processes) *Let us fix a filtration $\mathcal{F}$. If the n-dimensional continuous process*

$$\mathbf{X} \overset{\circ}{=} \left(X_1, X_2, \ldots, X_n\right)$$

is zero at $t = 0$ then the next three statements are equivalent:

1. $\mathbf{X}$ *is an n-dimensional Wiener process with respect to $\mathcal{F}$.*
2. $\mathbf{X}$ *is a local martingale with respect to $\mathcal{F}$ and $\left[X_i, X_j\right]\left(t\right) = \delta_{ij}t$.*

---

[22]See: (6.32), page 398.

[23]See: (6.48), page 416.

[24]Of course $X \in \mathcal{H}_{\text{loc}}^2$ and not $X \in \mathcal{H}^2$ so one can first localize X and then take limit in (6.11) otherwise the argument is nearly the same.

[25]See: Definition B.1, page 559.

[26]See: Definition B.4, page 561.

3. *Whenever $f_k \in L^2(\mathbb{R}_+, \lambda)$, where λ is Lebesgue's measure, then*

$$\mathcal{E}\left(i\left(\mathbf{f} \bullet \mathbf{X}\right)\right)(t) \overset{\circ}{=} \exp\left(i \sum_{k=1}^{n} \int_0^t f_k dX_k + \frac{1}{2} \sum_{k=1}^{n} \int_0^t f_k^2 d\lambda\right)$$

will be a complex martingale with respect to $\mathcal{F}$.

In particular, if X is a continuous local martingale and $Y(t) \overset{\circ}{=} X^2(t) - t$ is a continuous local martingale then X is a Wiener process.

Proof. Let us show that each statement implies the next one.

1. The implication 1. $\Rightarrow$ 2. follows[27] from the relation $[w](t) = t$.

2. The proof of the implication 2. $\Rightarrow$ 3. is the following: Using Itô's formula with a simple calculation one can show that $\mathcal{E}(i\mathbf{f} \bullet \mathbf{X})$ is a local martingale. As $f_k \in L^2(\mathbb{R}_+, \lambda)$

$$\mathcal{E}\left(i\left(\mathbf{f} \bullet \mathbf{X}\right)\right)(t) = \exp\left(i \sum_{k=1}^{n} \int_0^t f_k dX_k\right) \exp\left(\frac{1}{2} \sum_{k=1}^{n} \int_0^t f_k^2 d\lambda\right)$$

is uniformly bounded, hence it is a local martingale in class $\mathcal{D}$. Hence $\mathcal{E}(i\mathbf{f} \bullet \mathbf{X})$ is a martingale[28].

3. Finally we prove the implication 3. $\Rightarrow$ 1. If $\mathbf{u} \in \mathbb{R}^n$, $0 \le r < \infty$ and $\mathbf{f} \overset{\circ}{=} \mathbf{u}\chi([0,r])$ then as $\mathbf{X}(0) = 0$

$$\mathcal{E}\left(i\mathbf{f} \bullet \mathbf{X}\right)(t) = \exp\left(i \sum_{k=1}^{n} \int_0^t u_k \chi([0,r]) dX_k + \frac{1}{2}\|\mathbf{u}\|_2^2 (t \wedge r)\right) =$$

$$= \exp\left(i\left(\mathbf{u}, \mathbf{X}(r \wedge t)\right) + \frac{1}{2}\|\mathbf{u}\|_2^2 (t \wedge r)\right).$$

$\mathcal{E}(i\mathbf{f} \bullet \mathbf{X}) \neq 0$ is a martingale, hence if $s < t < r$ then

$$1 = \mathbf{E}\left(\mathcal{E}(i\mathbf{f} \bullet \mathbf{X})(t)\left(\mathcal{E}(i\mathbf{f} \bullet \mathbf{X})(s)\right)^{-1} \mid \mathcal{F}_s\right) =$$

$$= \mathbf{E}\left(\exp\left(i\left(\mathbf{u}, \mathbf{X}(t) - \mathbf{X}(s)\right) + \frac{1}{2}\|\mathbf{u}\|_2^2 (t - s)\right) \mid \mathcal{F}_s\right),$$

therefore

$$\mathbf{E}\left(\exp\left(i\left(\mathbf{u}, \mathbf{X}(t) - \mathbf{X}(s)\right)\right) \mid \mathcal{F}_s\right) = \exp\left(-\frac{1}{2}\|\mathbf{u}\|_2^2 (t - s)\right),$$

---

[27]See: Example 2.27. page 129, Example 2.46, page 144.
[28]See: Proposition 1.144, page 102.

which means that for any set $F \in \mathcal{F}_s$

$$\int_F \exp\left(i\left(\mathbf{u}, \mathbf{X}\left(t\right) - \mathbf{X}\left(s\right)\right)\right) d\mathbf{P} = \mathbf{P}\left(F\right) \cdot \exp\left(-\frac{1}{2} \|\mathbf{u}\|_2^2 \left(t - s\right)\right).$$

If $F = \Omega$ then this implies that the distribution of $X_i\left(t\right) - X_i\left(s\right)$ is $N\left(0, \sqrt{t - s}\right)$. Therefore

$$\int_F \exp\left(i\left(\mathbf{u}, \mathbf{X}\left(t\right) - \mathbf{X}\left(s\right)\right)\right) d\mathbf{P} = \mathbf{P}\left(F\right) \cdot \int_\Omega \exp\left(i\left(\mathbf{u}, \mathbf{X}\left(t\right) - \mathbf{X}\left(s\right)\right)\right) d\mathbf{P}$$

Since this equality holds for every trigonometric polynomial, by the Monotone Class Theorem for every $B \in \mathbb{R}^n$

$$\mathbf{P}\left(\{\mathbf{X}\left(t\right) - \mathbf{X}\left(s\right) \in B\} \cap F\right) =$$

$$= \int_F \chi_B\left(\mathbf{X}\left(t\right) - \mathbf{X}\left(s\right)\right) d\mathbf{P} = \mathbf{P}\left(F\right) \int_\Omega \chi_B\left(\mathbf{X}\left(t\right) - \mathbf{X}\left(s\right)\right) d\mathbf{P} =$$

$$= \mathbf{P}\left(\{\mathbf{X}\left(t\right) - \mathbf{X}\left(s\right) \in B\}\right) \cdot \mathbf{P}\left(F\right).$$

Hence the increment $\mathbf{X}\left(t\right) - \mathbf{X}\left(s\right)$ is independent of the σ-algebra $\mathcal{F}_s$. So $\mathbf{X}$ is a Wiener process. $\square$

Example 6.14 For every Wiener process w the integral $\text{sgn}\left(w\right) \bullet w$ is a Wiener process.

The process is a continuous local martingale. The quadratic variation of $\text{sgn}\left(w\right) \bullet w$ is

$$\int_0^t \left(\text{sgn}\left(w\right)\right)^2 d\left[w\right] = \int_0^t \left(\text{sgn}\left(w(s)\right)\right)^2 ds = t. \qquad \square$$

Example 6.15 The reflected Wiener process is also a Wiener process.

Let w be a Wiener process and let τ be a stopping time. Define the reflected process

$$\widehat{w}\left(t, \omega\right) \doteq \begin{cases} w\left(t, \omega\right) & \text{if } t \leq \tau\left(\omega\right) \\ 2w\left(\tau\left(\omega\right), \omega\right) - w\left(t, \omega\right) & \text{if } t > \tau\left(\omega\right) \end{cases} = (2w^\tau - w)(t, \omega).$$

Obviously $\widehat{w}\left(0\right) = 0$, and the trajectories of $\widehat{w}$ are continuous. It is also obvious that

$$[\widehat{w}] = [2w^\tau - w] = [2w^\tau] - 2[2w^\tau, w] + [w]$$
$$= 4[w]^\tau - 4[w]^\tau + [w] = [w].$$

As w^τ is a martingale and the sum of martingales is again a margingale $\widehat{w}$ is a continuous local martingale, so by Lévy's theorem it is a Wiener process. □

Let us discuss an interesting relation between exponential martingales and the quadratic variation:

Proposition 6.16 *Let X and A be continuous adapted processes on the half-line $t \geq 0$. If $X\left(0\right) = 0$ then the next statements are equivalent:*

1. *A has finite variation and for every $\alpha \in \mathbb{C}$ the process $Y_\alpha \overset{\circ}{=} \exp\left(\alpha X - \alpha^2 A/2\right)$ is a local martingale,*
2. *$[X] = A$, and X is a local martingale.*

Proof. We prove that each statement implies the other one.

1. Assume that Y_α is a local martingale and let $\left(\sigma_n\right)$ be a localizing sequence of Y_α. Let

$$\tau_n \overset{\circ}{=} \inf\left\{t : |X\left(t\right)| \geq n\right\} \wedge \inf\left\{t : |A\left(t\right)| \geq n\right\} \wedge \sigma_n.$$

$Y_\alpha^{\tau_n}$ is a martingale and obviously

$$|Y_\alpha^{\tau_n}| \leq \exp\left(|\alpha| n + \frac{1}{2}\alpha^2 n\right),$$

$$\left|\frac{d}{d\alpha}Y_\alpha^{\tau_n}\right| \leq |Y_\alpha^{\tau_n}| |X^{\tau_n} - \alpha A^{\tau_n}|,$$

$$\left|\frac{d^2}{d\alpha^2}Y_\alpha^{\tau_n}\right| \leq |Y_\alpha^{\tau_n}| \left|(X^{\tau_n} - \alpha A^{\tau_n})^2 - A^{\tau_n}\right|.$$

It is easy to see that if α is in a bounded neighbourhood of the origin then the expressions on the right-hand side are bounded. Hence in the next calculation one can differentiate under the integral sign at $\alpha = 0$.

If $\alpha = 0$ then

$$\frac{d}{d\alpha} Y_\alpha^{\tau_n} = X^{\tau_n},$$

hence for any $F \in \mathcal{F}_s$

$$\int_F \mathbf{E}\left(X^{\tau_n}(t) \mid \mathcal{F}_s\right) d\mathbf{P} = \int_F \mathbf{E}\left(\frac{d}{d\alpha} Y_\alpha^{\tau_n}(t) \mid \mathcal{F}_s\right) d\mathbf{P} =$$

$$= \int_F \frac{d}{d\alpha} Y_\alpha^{\tau_n}(t) \, d\mathbf{P} =$$

$$= \frac{d}{d\alpha} \int_F Y_\alpha^{\tau_n}(t) \, d\mathbf{P} =$$

$$= \frac{d}{d\alpha} \int_F Y_\alpha^{\tau_n}(s) \, d\mathbf{P} =$$

$$= \int_F \frac{d}{d\alpha} Y_\alpha^{\tau_n}(s) \, d\mathbf{P} = \int_F X^{\tau_n}(s) \, d\mathbf{P},$$

therefore

$$\mathbf{E}\left(X^{\tau_n}(t) \mid \mathcal{F}_s\right) \stackrel{a.s.}{=} X^{\tau_n}(s).$$

Therefore X^{τ_n} is a martingale. Hence X is a local martingale. In a similar way, using that at $\alpha = 0$

$$\frac{d^2 Y_\alpha^{\tau_n}}{d\alpha^2} = (X^{\tau_n})^2 - A^{\tau_n},$$

one can prove that $(X^{\tau_n})^2 - A^{\tau_n}$ is a martingale. This implies[29] that A is increasing and $[X] = A$.

2. The implication 2. $\Rightarrow$ 1. is an easy consequence of Itô's formula. As the quadratic variation of a continuous semimartingale is equal to the quadratic variation of its local martingale part if $Z \stackrel{\circ}{=} \alpha X - \alpha^2 A/2$, then $Y_\alpha = \exp(Z)$ and

$$Y_\alpha - Y_\alpha(0) = Y_\alpha \bullet Z + \frac{1}{2} Y_\alpha \bullet [Z] \stackrel{\circ}{=}$$

$$\stackrel{\circ}{=} Y_\alpha \bullet \left(\alpha X - \alpha^2 \frac{A}{2}\right) + \frac{1}{2} Y_\alpha \bullet \left[\alpha X - \alpha^2 \frac{A}{2}\right] =$$

---

[29]See: Proposition 2.40, page 141.

$$= \alpha Y_\alpha \bullet X - \frac{\alpha^2}{2} Y_\alpha \bullet [X] + \frac{1}{2} Y_\alpha \bullet [\alpha X] =$$

$$= \alpha Y_\alpha \bullet X - \frac{\alpha^2}{2} Y_\alpha \bullet [X] + \frac{\alpha^2}{2} Y_\alpha \bullet [X] =$$

$$= \alpha Y_\alpha \bullet X,$$

which is, as a stochastic integral with respect to a continuous local martingale, a local martingale. □

6.2.4 Integral representation theorems for Wiener processes

In this subsection we return to the Integral Representation Problem. Let w be a Wiener process and let $\mathcal{F}$ be the filtration generated by w. Let L be a local martingale with respect to $\mathcal{F}$. Let assume that $L(0) = 0$. Every local martingale has an $\mathcal{H}^1$-localization[30]. By the integral representation property of Wiener processes[31] $L^{\tau_n} = H \bullet w$ on any finite interval. Hence

$$[L]^{\tau_n} = [L^{\tau_n}] = [H \bullet w] = H^2 \bullet [w].$$

As $[w](t) = t$ it is obvious that $[L]$ is continuous. Therefore L is continuous. So $L \in \mathcal{H}^2_{\mathrm{loc}}$ and one can assume that $L^{\tau_n} \in \mathcal{H}^2$. This implies that $H \in \mathcal{L}^2(w)$. By Itô's isometry[32] H is unique in $\mathcal{L}^2(w)$. Hence $L = H \bullet w$ for some $H \in \mathcal{L}^2_{\mathrm{loc}}(w)$.

Proposition 6.17 *If w is a Wiener process and L is a local martingale with respect to the filtration generated by w then L is continuous and*

$$L = L(0) + H \bullet w$$

with some $H \in \mathcal{L}^2_{loc}(w)$.

Our next statement is an easy consequence of Lévy's characterization theorem.

Proposition 6.18 (Doob) *Let M be a continuous local martingale on a stochastic base $(\Omega, \mathcal{A}, \mathbf{P}, \mathcal{F})$. If the quadratic variation of M has the representation*

$$[M](t, \omega) = \int_0^t \alpha^2(s, \omega)\, ds, \tag{6.14}$$

---

[30]See: Corollary 3.59, page 221.
[31]See: Example 5.50, page 347.
[32]See: Proposition 2.64, page 156.

where $\alpha(t,\omega) > 0$ and α is an adapted and product measurable process, then there is a Wiener process w on $(\Omega, \mathcal{A}, \mathbf{P}, \mathcal{F})$ for which

$$M(t) = M(0) + \int_0^t \alpha(s)\, dw(s).$$

Proof. One can explicitly construct the Wiener process w:

$$w \overset{\circ}{=} \frac{1}{\alpha} \bullet M. \tag{6.15}$$

First we prove that the integral exists. $[M] \ll \lambda$, so if α_M is the Doléans measure of M then $\alpha_M \ll \lambda \times \mathbf{P}$. Therefore the stochastic integrals are defined among adapted product measurable processes[33].

$$\int_0^t \frac{1}{\alpha^2} d[M] = \int_0^t \frac{1}{\alpha^2} \alpha^2 ds = t < \infty.$$

Hence $1/\alpha \in \mathcal{L}^2_{\mathrm{loc}}(M)$. So $1/\alpha$ is integrable with respect to M. That is integral (6.15) exists. As M is continuous w is a continuous local martingale. By (6.14)

$$[w](t) \overset{\circ}{=} \left[\frac{1}{\alpha} \bullet M\right](t) = \left(\frac{1}{\alpha^2} \bullet [M]\right)(t) = t.$$

Therefore by Lévy's theorem w is a Wiener process. By (6.14) α is integrable with respect to w, therefore

$$\alpha \bullet w \overset{\circ}{=} \alpha \bullet \left(\frac{1}{\alpha} \bullet M\right) = \alpha \frac{1}{\alpha} \bullet M = 1 \bullet M = M - M(0).$$

Hence the proposition holds. $\qquad\square$

Corollary 6.19 *Let M be a continuous local martingale on a stochastic base $(\Omega, \mathcal{A}, \mathbf{P}, \mathcal{F})$. If $[M] \ll \lambda$ then there is an extension $\left(\widetilde{\Omega}, \widetilde{\mathcal{A}}, \widetilde{\mathbf{P}}, \widetilde{\mathcal{F}}\right)$ of $(\Omega, \mathcal{A}, \mathbf{P}, \mathcal{F})$ and a Wiener process $\widetilde{w}$ on the extended base space that*

$$M(t) = M(0) + \int_0^t \sqrt{\frac{d[M]}{d\lambda}}\, d\widetilde{w}(s).$$

Proof. Let $\widehat{w}$ be an arbitrary Wiener process on some stochastic base $\left(\widehat{\Omega}, \widehat{\mathcal{A}}, \widehat{\mathbf{P}}, \widehat{\mathcal{F}}\right)$. Let the new stochastic base be the product of $(\Omega, \mathcal{A}, \mathbf{P}, \mathcal{F})$ and

---

[33]See: Proposition 5.20, page 314.

$\left(\widehat{\Omega}, \widehat{\mathcal{A}}, \widehat{\mathbf{P}}, \widehat{\mathcal{F}}\right)$. Obviously $\widehat{w}$ is independent of $\mathcal{A}$. Let us define α by $[M]\,(t) \overset{\circ}{=} \int_0^t \alpha^2\,(s)\,ds$. That is, let

$$\alpha \overset{\circ}{=} \sqrt{\frac{d\,[M]}{d\lambda}}.$$

The process

$$\widetilde{w}\,(t) \overset{\circ}{=} \int_0^t \frac{1}{\alpha}\chi\,(\alpha > 0)\,dM + \int_0^t \chi\,(\alpha = 0)\,d\widehat{w}$$

is a continuous local martingale. The quadratic co-variation of independent local martingales is zero[34], so $[M, \widehat{w}] = 0$. Therefore

$$[\widetilde{w}]\,(t) = \int_0^t \chi\,(\alpha > 0)\,ds + \int_0^t \chi\,(\alpha = 0)\,ds = t.$$

Hence by Lévy's theorem $\widetilde{w}$ is a Wiener process.

$$\alpha \bullet \widetilde{w} \overset{\circ}{=} \alpha \bullet \left(\frac{1}{\alpha}\chi\,(\alpha > 0) \bullet M + \chi\,(\alpha = 0) \bullet \widehat{w}\right) =$$

$$= \chi\,(\alpha > 0) \bullet M.$$

On the other hand

$$[\chi\,(\alpha = 0) \bullet M] = \chi\,(\alpha = 0) \bullet [M] = 0,$$

hence $\chi\,(\alpha = 0) \bullet M = 0$. So

$$\alpha \bullet \widetilde{w} = \chi\,(\alpha > 0) \bullet M + \chi\,(\alpha = 0) \bullet M = 1 \bullet M = M - M\,(0). \qquad \square$$

6.2.5 Bessel processes

As an application of Lévy's theorem let us investigate the Bessel processes. Let $\mathbf{w} \overset{\circ}{=} (w_1, w_2, \ldots, w_d)$ be a d-dimensional Wiener process. Define the Bessel process

$$R \overset{\circ}{=} \|\mathbf{w}\| \overset{\circ}{=} \|\mathbf{w}\|_2 \overset{\circ}{=} \sqrt{\sum_{k=1}^d w_k^2}.$$

We assume that $\mathbf{w}$ starts at $\mathbf{x} \in \mathbb{R}^d$, that is $R\,(0) = \|\mathbf{x}\|$. If it is necessary we shall explicitly indicate the initial value $\mathbf{x}$. Evidently the distribution of R

---

[34]See: Example 2.46, page 144.

depends on $\mathbf{x}$ only through the size of $r \overset{\circ}{=} \|\mathbf{x}\|$: If $\|\mathbf{x}\| = \|\mathbf{y}\|$ then $\mathbf{Q}\mathbf{x} = \mathbf{y}$ for some orthonormal transformation $\mathbf{Q}$. It is easy to show that $\mathbf{Q}\mathbf{w}$ is also a Wiener process and $\mathbf{Q}\mathbf{w}$ starts at $\mathbf{y}$. Obviously $R^{\mathbf{x}} \overset{\circ}{=} \|\mathbf{w}\| = \|\mathbf{Q}\mathbf{w}\| \overset{\circ}{=} R^{\mathbf{y}}$.

Proposition 6.20 *If $d \geq 2$ and $r \geq 0$ then if we start $\mathbf{w}$ from some point $\mathbf{x} \in \mathbb{R}^d$ with $r = \|\mathbf{x}\|$ then $R \overset{\circ}{=} \|\mathbf{w}\|$ satisfies the integral equation*

$$R(t) = r + \int_0^t \frac{d-1}{2R(s)} ds + B(t), \quad 0 \leq t < \infty, \tag{6.16}$$

where B is a Wiener process and

$$B \overset{\circ}{=} \sum_k B^{(k)}, \quad B^{(k)}(s) \overset{\circ}{=} \int_0^s \frac{w_k}{R} dw_k. \tag{6.17}$$

Put another way, $R \overset{\circ}{=} \|w\|$ satisfies the stochastic differential equation

$$dR = \frac{d-1}{2R} dt + dB.$$

Proof. First observe that the expression in (6.16) is meaningful: as $d \geq 2$ the $R(s)$ in the denominator is almost surely not zero for every $t \geq 0$. As the integral in (6.16) is taken by trajectories it is also meaningful. On the other hand

$$\int_0^t \left(\frac{w_k}{R}\right)^2 d[w_k] = \int_0^t \left(\frac{w_k}{R}\right)^2 d\lambda \leq \int_0^t 1 d\lambda = t,$$

hence the stochastic integrals in (6.17) are in $\mathcal{L}^2(w_k)$ on every finite interval. Therefore the stochastic integrals $B^{(k)}$ are also meaningful.

1. By the formula for the quadratic co-variation of the stochastic integrals

$$\left[B^{(k)}, B^{(l)}\right](t) = \int_0^t \frac{w_k w_j}{R^2} d[w_k, w_j] = \delta_{kj} \int_0^t \frac{w_k w_j}{R^2} d\lambda,$$

therefore

$$[B](t) = \sum_k \left[B^{(k)}\right](t) = \int_0^t \sum_k \frac{w_k^2}{R^2} d\lambda = \int_0^t 1 d\lambda = t.$$

The sum of local martingales is again a local martingale. Therefore by the characterization theorem of Lévy B is a Wiener process.

2. The proof of (6.16) uses the integration by parts formula:

$$R^2(t) - R^2(0) \overset{\circ}{=} \left(2 \sum_k w_k \bullet w_k + \sum_k [w_k] \right)(t) =$$

(6.18)

$$= 2 \sum_k \int_0^t w_k dw_k + t \cdot d.$$

The multi-dimensional Wiener processes are almost surely not zero[35], therefore almost surely $R^2 > 0$. Hence one can use Itô's formula with $\sqrt{x}$:

$$R - r = \frac{1}{2\sqrt{R^2}} \bullet R^2 - \frac{1}{2}\frac{1}{2}\frac{1}{2}\frac{1}{(R^2)^{3/2}} \bullet [R^2] =$$

(6.19)

$$= \sum_k \frac{w_k}{R} \bullet w_k + \frac{d}{2R} \bullet \lambda - \frac{1}{8}\frac{1}{R^3} 4 \sum_k w_k^2 \bullet \lambda =$$

$$= \sum_k \frac{w_k}{R} \bullet w_k + \frac{d-1}{2R} \bullet \lambda.$$

$\square$

6.3 Change of measure for continuous semimartingales

The class of semimartingales is remarkably stable under a lot of operations. For example, by Itô's formula a C^2 transform of a semimartingale is a semimartingale again. Later we shall show that convex transforms of semimartingales are also semimartingales. In this section we return to the discussion of the operation of equivalent changes of measure.

6.3.1 Locally absolutely continuous change of measure

If a measure $\mathbf{Q}$ is absolutely continuous with respect to $\mathbf{P}$ then one can define the Radon–Nikodym derivative $d\mathbf{Q}/d\mathbf{P}$. If a filtration $\mathcal{F}$ satisfies the usual conditions then the process

$$\Lambda(t) \overset{\circ}{=} \mathbf{E}\left(\frac{d\mathbf{Q}}{d\mathbf{P}} \mid \mathcal{F}_t \right)$$

is a martingale and as

$$\int_F \Lambda(t) \, d\mathbf{P} = \int_F \frac{d\mathbf{Q}}{d\mathbf{P}} d\mathbf{P} = \mathbf{Q}(F), \quad F \in \mathcal{F}_t$$

[35]Let us remark that this is a critical observation as here we used the assumption that $n \geq 2$. If $n = 1$, then one cannot use Itô's formula as in this case one can only assume that $R^2 \geq 0$ and the function $\sqrt{x}$ for $x \geq 0$ is not a C^2 function. If we formally still apply the formula, then we get the relation $R = \text{sign}(w) \bullet w$. By Example 6.14. this expression is a Wiener process. The left-hand side is non-negative, hence the two sides cannot be equal.

$\Lambda(t)$ is the Radon–Nikodym derivative of $\mathbf{Q}$ on $(\Omega, \mathcal{F}_t, \mathbf{P})$. On the other hand let $\mathbf{Q}(t)$ be the restriction of $\mathbf{Q}$ and let $\mathbf{P}(t)$ be that of $\mathbf{P}$ to $\mathcal{F}_t$. If $\mathbf{Q}(t)$ is absolutely continuous with respect to $\mathbf{P}(t)$ then one can define the derivative

$$\Lambda(t) \stackrel{\circ}{=} \frac{d\mathbf{Q}(t)}{d\mathbf{P}(t)}.$$

If $F \in \mathcal{F}_s \subseteq \mathcal{F}_t$ then

$$\int_F \Lambda(t)\, d\mathbf{P} \stackrel{\circ}{=} \int_F \frac{d\mathbf{Q}(t)}{d\mathbf{P}(t)}\, d\mathbf{P} = \mathbf{Q}(F) = \int_F \frac{d\mathbf{Q}(s)}{d\mathbf{P}(s)}\, d\mathbf{P} \stackrel{\circ}{=} \int_F \Lambda(s)\, d\mathbf{P},$$

hence Λ is a martingale. Of course Λ is not necessarily uniformly integrable, so it can happen that there is no ξ for which $\Lambda(t) = \mathbf{E}(\xi \mid \mathcal{F}_t)$. To put it another way, it can happen that $\mathbf{Q} \ll \mathbf{P}$ on $\mathcal{F}_t$ for every t, but $\mathbf{Q}$ is not absolutely continuous on the σ-algebra $\mathcal{F}_\infty = \sigma(\cup_t \mathcal{F}_t)$. So the derivative $d\mathbf{Q}/d\mathbf{P}$ need not necessarily exist. Recall the following definition:

Definition 6.21 *We say that a measure $\mathbf{Q}$ is* locally absolutely continuous *with respect to a measure $\mathbf{P}$ if $\mathbf{Q}(t) \ll \mathbf{P}(t)$ for every t where $\mathbf{Q}(t)$ is the restriction of $\mathbf{Q}$ and $\mathbf{P}(t)$ is the restriction of $\mathbf{P}$ to $\mathcal{F}_t$. We shall denote this relation by $\mathbf{Q} \stackrel{loc}{\ll} \mathbf{P}$. If $\mathbf{Q} \stackrel{loc}{\ll} \mathbf{P}$ and $\mathbf{P} \stackrel{loc}{\ll} \mathbf{Q}$ then we shall say that $\mathbf{P}$ and $\mathbf{Q}$ are* locally equivalent. *We shall denote this by $\mathbf{P} \stackrel{loc}{\sim} \mathbf{Q}$.*

Definition 6.22 *If $\mathbf{Q} \stackrel{loc}{\ll} \mathbf{P}$ then the* right-regular version *of*

$$\Lambda(t) \stackrel{\circ}{=} \frac{d\mathbf{Q}(t)}{d\mathbf{P}(t)}$$

is called the Radon–Nikodym process *of $\mathbf{P}$ and $\mathbf{Q}$.*

6.3.2 Semimartingales and change of measure

We have already proved the following important observations[36]:

Proposition 6.23 (Invariance of semimartingales) *If $\mathbf{Q} \stackrel{loc}{\ll} \mathbf{P}$ then every semimartingale under $\mathbf{P}$ is a semimartingale under $\mathbf{Q}$.*

Proposition 6.24 (Integration and change of measure) *Let X be an arbitrary semimartingale and assume that the integral $H \bullet X$ exists under the measure $\mathbf{P}$. If $\mathbf{Q} \stackrel{loc}{\ll} \mathbf{P}$ then $H \bullet X$ exists under $\mathbf{Q}$ as well. Under the measure $\mathbf{Q}$ the two processes, the integral under $\mathbf{P}$ and the integral under $\mathbf{Q}$, are indistinguishable.*

---

[36]See: Proposition 4.55, page 266, Corollary 4.58, page 271, Proposition 4.59, page 271.

Proposition 6.25 (Transformation of local martingales) *Let* $\mathbf{Q} \overset{\text{loc}}{\ll} \mathbf{P}$ *and let* Λ *be the Radon–Nikodym process of* $\mathbf{P}$ *and* $\mathbf{Q}$. *If* L *is a continuous local martingale under the measure* $\mathbf{P}$ *then under the measure* $\mathbf{Q}$:

1. Λ^{-1} *is well defined,*
2. *the integral* $\Lambda^{-1} \bullet [L, \Lambda]$ *exists and has finite variation on compact intervals,*
3. *the expression*

$$\widehat{L} \overset{\circ}{=} L - \Lambda^{-1} \bullet [L, \Lambda] \tag{6.20}$$

is a local martingale.

Corollary 6.26 *If* $\mathbf{Q} \overset{\text{loc}}{\sim} \mathbf{P}$ *then* $\Lambda > 0$ *and* Λ^{-1} *is a martingale under* $\mathbf{Q}$.

Proof. One only needs to prove that Λ^{-1} is a martingale under $\mathbf{Q}$. If $F \in \mathcal{F}_s$ and $t > s$ then

$$\int_F \frac{1}{\Lambda}(t) \, d\mathbf{Q} = \int_F \frac{1}{\Lambda}(t) \Lambda(t) \, d\mathbf{P} =$$

$$= \mathbf{P}(F) = \int_F \frac{1}{\Lambda(s)} \Lambda(s) \, d\mathbf{P} =$$

$$= \int_F \frac{1}{\Lambda(s)} d\mathbf{Q}.$$

$\square$

Corollary 6.27 *If* $\mathbf{Q} \overset{\text{loc}}{\ll} \mathbf{P}$ *and* X *and* Y *are semimartingales then* $[X, Y]$ *calculated under* $\mathbf{Q}$ *is indistinguishable under* $\mathbf{Q}$ *from* $[X, Y]$ *calculated under* $\mathbf{P}$. *If* L *is a local martingale and* N *is a continuous semimartingale then*

$$[L, N] = \left[\widehat{L}, N \right]$$

where $\widehat{L}$ *is as in (6.20).*

Proof. As

$$[X, Y] \overset{\circ}{=} XY - X(0)Y(0) - Y_- \bullet X - X_- \bullet Y$$

the first statement is obvious from Proposition 6.24. $\Lambda^{-1} \bullet [L, \Lambda] \in \mathcal{V}$ and N is continuous so

$$\left[\widehat{L}, N \right] \overset{\circ}{=} \left[L - \Lambda^{-1} \bullet [L, \Lambda], N \right] =$$

$$= [L, N] - \left[\Lambda^{-1} \bullet [L, \Lambda], N \right] = [L, N].$$

$\square$

Definition 6.28 $\widehat{L}$ *in (6.20) is called the* Girsanov transform *of* L.

6.3.3 Change of measure for continuous semimartingales

If L is a continuous local martingale then from Itô's formula it is trivial that the *exponential martingale*

$$\mathcal{E}(L) \stackrel{\circ}{=} \exp\left(L - \frac{1}{2}[L]\right)$$

is a positive local martingale.

Proposition 6.29 (Logarithm of local martingales) *If Λ is a positive and continuous local martingale then there is a continuous local martingale*

$$L \stackrel{\circ}{=} \mathrm{Log}\,(\Lambda) \stackrel{\circ}{=} \log \Lambda(0) + \Lambda^{-1} \bullet \Lambda$$

which is the only continuous local martingale for which

$$\Lambda = \mathcal{E}(L) \stackrel{\circ}{=} \exp\left(L - \frac{1}{2}[L]\right).$$

$$\log \Lambda = L - \frac{1}{2}[L] \stackrel{\circ}{=} \mathrm{Log}\,(\Lambda) - \frac{1}{2}[\mathrm{Log}\,(\Lambda)].$$

Proof. If $\Lambda = \mathcal{E}(L_1) = \mathcal{E}(L_2)$, then as $\Lambda > 0$

$$1 = \frac{\Lambda}{\Lambda} = \exp\left(L_1 - L_2 - \frac{1}{2}[L_1] + \frac{1}{2}[L_2]\right),$$

that is $L_1 - L_2 = \frac{1}{2}([L_1] - [L_2])$. Hence the continuous local martingale $L_1 - L_2$ has bounded variation and it is constant. Evidently $L_1(0) = L_2(0)$, therefore $L_1 = L_2$. As $\Lambda > 0$ the expression $\log \Lambda$ is meaningful. By Itô's formula

$$\log \Lambda = \log \Lambda(0) + \Lambda^{-1} \bullet \Lambda - \frac{1}{2}\frac{1}{\Lambda^2} \bullet [\Lambda] \stackrel{\circ}{=}$$

$$\stackrel{\circ}{=} L - \frac{1}{2}\frac{1}{\Lambda^2} \bullet [\Lambda] = L - \frac{1}{2}[L].$$

Therefore

$$\Lambda = \exp(\log \Lambda) = \exp\left(L - \frac{1}{2}[L]\right) \stackrel{\circ}{=} \mathcal{E}(L). \qquad \square$$

Proposition 6.30 (Logarithmic transformation of local martingales) *Assume that $\mathbf{P} \stackrel{\mathrm{loc}}{\sim} \mathbf{Q}$ and let*

$$\Lambda(t) \stackrel{\circ}{=} \frac{d\mathbf{Q}}{d\mathbf{P}}(t)$$

be continuous. If $\Lambda = \mathcal{E}(L)$, that is $L = \mathrm{Log}(\Lambda)$ then

$$\frac{d\mathbf{P}}{d\mathbf{Q}}(t) = \left(\frac{d\mathbf{Q}}{d\mathbf{P}}(t)\right)^{-1} = (\mathcal{E}(L)(t))^{-1} = \mathcal{E}\left(-\widehat{L}\right)(t).$$

If M is a local martingale under measure $\mathbf{P}$ then

$$\widehat{M} = M - [M, L] = M - [M, \mathrm{Log}(\Lambda)] \tag{6.21}$$

is a local martingale under measure $\mathbf{Q}$.

Proof. $\Lambda > 0$ as $\mathbf{P} \overset{\mathrm{loc}}{\sim} \mathbf{Q}$.

$$[M, L] \overset{\circ}{=} [M, \mathrm{Log}(\Lambda)] \overset{\circ}{=} \left[M, \log \Lambda(0) + \Lambda^{-1} \bullet \Lambda\right] =$$

$$= \left[M, \Lambda^{-1} \bullet \Lambda\right] = \Lambda^{-1} \bullet [M, \Lambda].$$

$$\widehat{M} \overset{\circ}{=} M - \Lambda^{-1} \bullet [M, \Lambda] = M - [M, L].$$

$$\mathcal{E}\left(-\widehat{L}\right) \overset{\circ}{=} \exp\left(-\widehat{L} - \frac{1}{2}\left[-\widehat{L}, -\widehat{L}\right]\right) =$$

$$= \exp\left(-L + [L, L] - \frac{1}{2}[L, L]\right) =$$

$$= \exp\left(-\left(L - \frac{1}{2}[L, L]\right)\right) = (\mathcal{E}(L))^{-1}.$$

$\square$

Proposition 6.31 (Girsanov's formula) *If M and $L \in \mathcal{L}$ are continuous local martingales and the process*

$$\Lambda \overset{\circ}{=} \mathcal{E}(L) \overset{\circ}{=} \exp\left(L - \frac{1}{2}[L]\right)$$

is a martingale on the finite or infinite interval $[0, s]$ then under the measure

$$\mathbf{Q}(A) \overset{\circ}{=} \int_A \Lambda(s) \, d\mathbf{P}.$$

the process

$$\widehat{M} \overset{\circ}{=} M - [L, M] = M - \frac{1}{\Lambda} \bullet [\Lambda, M] \tag{6.22}$$

is a continuous local martingale on $[0, s]$.

Proof. $L(0) = 0$, therefore $\Lambda(0) = 1$. Λ is a martingale on $[0, s]$ so

$$\mathbf{Q}(\Omega) = \int_\Omega \Lambda(s) \, d\mathbf{P} = 1.$$

Hence $\mathbf{Q}$ is also a probability measure.

$$\Lambda(t) = \mathbf{E}\left(\Lambda(s) \mid \mathcal{F}_t\right) \overset{\circ}{=} \mathbf{E}\left(\frac{d\mathbf{Q}}{d\mathbf{P}} \mid \mathcal{F}_t\right),$$

that is if $F \in \mathcal{F}_t$ then

$$\int_F \Lambda(t) \, d\mathbf{P} = \int_F \frac{d\mathbf{Q}}{d\mathbf{P}} d\mathbf{P} = \mathbf{Q}(F),$$

so $\Lambda(t) = d\mathbf{Q}(t)/d\mathbf{P}(t)$ on $\mathcal{F}_t$. The other parts of the proposition are obvious from Proposition 6.30. $\qquad\square$

6.3.4 Girsanov's formula for Wiener processes

Let w be a Wiener process under measure $\mathbf{P}$. If $\mathbf{Q} \overset{\text{loc}}{\ll} \mathbf{P}$ then w is a continuous semimartingale[37] under $\mathbf{Q}$. Let $M + V$ be its decomposition under $\mathbf{Q}$. M is a continuous local martingale and $M(0) = 0$. The quadratic variation of M under $\mathbf{Q}$ is[38]

$$[M](t) = [M + V](t) = [w](t) = t.$$

By Lévy's theorem[39] M is therefore a Wiener process under the measure $\mathbf{Q}$. By (6.20)

$$\widehat{w} \overset{\circ}{=} w - \Lambda^{-1} \bullet [w, \Lambda]$$

is a continuous local martingale. As $\Lambda^{-1} \bullet [w, \Lambda]$ has finite variation by Fisk's theorem $M = \widehat{w}$. If $\mathcal{F}$ is the augmented filtration of w then by the integral representation property of the Wiener processes Λ is continuous[40]. If $\mathbf{Q} \overset{\text{loc}}{\sim} \mathbf{P}$ then $\Lambda > 0$ hence for some L

$$\Lambda \overset{\circ}{=} \mathcal{E}(L) \overset{\circ}{=} \exp\left(L - \frac{1}{2}[L]\right).$$

---

[37] See: Proposition 6.23, page 378.
[38] See: Example 2.26, page 129.
[39] See: Theorem 6.13, page 368.
[40] See: Proposition 6.17, page 373.

Therefore by Proposition 6.30

$$M = \widehat{w} = w - [w, L].$$

If $\mathcal{F}$ is the augmented filtration of w then $\mathcal{F}_0$ is the trivial σ-algebra, so $\Lambda(0) = 1$, hence $L(0) = 0$. Again by the integral representation theorem there exists an $X \in \mathcal{L}^2_{\text{loc}}(w)$

$$L = L(0) + X \bullet w = X \bullet w, \quad X \in \mathcal{L}^2_{\text{loc}}(w).$$

Hence

$$M = \widehat{w} = w - [w, L] = w - [w, X \bullet w] =$$
$$= w - X \bullet [w].$$

Hence if $\mathbf{P} \overset{\text{loc}}{\sim} \mathbf{Q}$ then there is an $X \in \mathcal{L}^2_{\text{loc}}(w)$ such that

$$\Lambda(t) \overset{\circ}{=} \exp\left(\int_0^t X(s) \, dw(s) - \frac{1}{2} \int_0^t X^2(s) \, ds\right) \overset{\circ}{=} \tag{6.23}$$

$$\overset{\circ}{=} \exp\left(X \bullet w - \frac{1}{2} X^2 \bullet [w]\right)(t) \overset{\circ}{=} \mathcal{E}(X \bullet w)$$

and

$$\widehat{w}(t) \overset{\circ}{=} w(t) - \int_0^t X(s) \, ds, \quad X \in \mathcal{L}^2_{\text{loc}}(w) \tag{6.24}$$

is a Wiener process under $\mathbf{Q}$. On the other hand, let $X \in \mathcal{L}^2_{\text{loc}}(w, [0, s])$. Assume that Λ in (6.23) is a martingale on $[0, s]$. Define the measure $\mathbf{Q}$ by $d\mathbf{Q}/d\mathbf{P} \overset{\circ}{=} \Lambda(s)$. Obviously the process in (6.24) is a Wiener process under $\mathbf{Q}$.

Theorem 6.32 (Girsanov formula for Wiener processes) *Let w be a Wiener process under measure $\mathbf{P}$ and let $\mathcal{F}$ be the augmented filtration of w. Girsanov's transform $\widehat{w}$ of w has the following properties:*

1. *If $\mathbf{Q} \overset{\text{loc}}{\ll} \mathbf{P}$ then the Girsanov transform of w is a Wiener process under measure $\mathbf{Q}$.*
2. *If $\mathbf{Q} \overset{\text{loc}}{\sim} \mathbf{P}$ then the Girsanov transform of w has the representation (6.24).*
3. *If $X \in \mathcal{L}^2_{\text{loc}}(w)$ and the process Λ in line (6.23) is a martingale over the segment $[0, s]$ then the process $\widehat{w}$ in (6.24) is a Wiener process over $[0, s]$ under the measure $\mathbf{Q}$ where $d\mathbf{Q}/d\mathbf{P} \overset{\circ}{=} \Lambda(s)$.*

Example 6.33 Even on finite intervals $\Lambda \overset{\circ}{=} \mathcal{E}(X \bullet w)$ is not always a martingale.

Let $u = 1$ and let $\tau \overset{\circ}{=} \inf \{t : w^2(t) = 1 - t\}$. If $t = 0$ then almost surely $w^2(t, \omega) < 1 - t$, and if $t = 1$ then almost surely $w^2(t, \omega) > 1 - t$. So by the intermediate value theorem $\mathbf{P}(0 < \tau < 1) = 1$. If

$$X(t) \overset{\circ}{=} \frac{-2w(t)\chi(\tau \geq t)}{(1-t)^2},$$

then as $\tau < 1$

$$\int_0^1 X^2 d[w] = 4 \int_0^\tau \frac{w^2(t)}{(1-t)^4} dt \leq 4 \int_0^\tau \frac{(1-t)^2}{(1-t)^4} dt < \infty.$$

Hence $X \in \mathcal{L}^2_{\mathrm{loc}}(w, [0,1])$. By Itô's formula, if $t < 1$ then

$$\frac{w^2(t)}{(1-t)^2} = \int_0^t \frac{2w^2(s)}{(1-s)^3} ds + \int_0^t \frac{2w(s)}{(1-s)^2} dw(s) + \int_0^t \frac{1}{(1-s)^2} ds.$$

From this

$$I \overset{\circ}{=} \int_0^1 X dw - \frac{1}{2} \int_0^1 X^2 ds = (X \bullet w)^\tau - \frac{1}{2}(X^2 \bullet [w])^\tau =$$

$$= -\frac{w^2(\tau)}{(1-\tau)^2} + \int_0^\tau \frac{2w^2(s)}{(1-s)^3} ds + \int_0^\tau \frac{1}{(1-s)^2} ds - \int_0^\tau \frac{2w^2(s)}{(1-s)^4} ds =$$

$$= -\frac{1}{1-\tau} + \int_0^\tau 2w^2(s) \left(\frac{1}{(1-s)^3} - \frac{1}{(1-s)^4} \right) + \frac{1}{(1-s)^2} ds \leq$$

$$\leq -\frac{1}{1-\tau} + \int_0^\tau \frac{1}{(1-s)^2} ds = -1,$$

Therefore $\Lambda(1) = \exp(I) \leq 1/e$. Hence

$$\mathbf{E}(\Lambda(1)) = \mathbf{E}(\exp(I)) \leq \frac{1}{e} < 1 = \mathbf{E}(\Lambda(0)),$$

so Λ is not a martingale. $\qquad\square$

Example 6.34 If $\widehat{w}(t) \overset{\circ}{=} w(t) - \mu \cdot t$ then there is no probability measure $\mathbf{Q} \ll \mathbf{P}$ on $\mathcal{F}_\infty$ for which $\widehat{w}$ is a Wiener process under $\mathbf{Q}$.

Let $\mu \neq 0$ and let

$$A \overset{\circ}{=} \left\{ \lim_{t \to \infty} \frac{\widehat{w}(t)}{t} = 0 \right\} = \left\{ \lim_{t \to \infty} \frac{w(t)}{t} = \mu \right\}.$$

If $\widehat{w}$ is a Wiener process under $\mathbf{Q}$ then by the law of large numbers,

$$1 = \mathbf{Q}\left(A\right) \neq \mathbf{P}\left(A\right) = 0.$$

Therefore $\mathbf{Q}$ is not absolutely continuous with respect to $\mathbf{P}$ on $\mathcal{F}_\infty$. Observe that the martingale

$$\Lambda\left(t\right) = \exp\left(\mu w\left(t\right) - \frac{1}{2}\mu^2 t\right)$$

is not uniformly integrable. Therefore if $s = \infty$ then Λ is not a martingale on $[0, s]$. $\qquad\square$

Let us discuss the underlying measure-theoretic problem.

Definition 6.35 *Let $(\Omega, \mathcal{F})$ be a filtered space. We say that the probability spaces $(\Omega, \mathcal{F}_t, \mathbf{P}_t)$ are* consistent, *if for any $s < t$ the restriction of $\mathbf{P}_t$ to $\mathcal{F}_s$ is $\mathbf{P}_s$. The filtered space $(\Omega, \mathcal{F})$ is a* Kolmogorov type *filtered space if whenever $(\Omega, \mathcal{F}_t, \mathbf{P}_t)$ are consistent probability spaces for $0 \leq t < \infty$, then there is a probability measure $\mathbf{P}$ on $\mathcal{F}_\infty \stackrel{\circ}{=} \sigma\left(\mathcal{F}_t : t \geq 0\right)$ such that every $\mathbf{P}_t$ is a restriction of $\mathbf{P}$ to $\mathcal{F}_t$.*

Example 6.36 The space $C\left(\left[0, \infty\right)\right)$ with its natural filtration is a Kolmogorov-type filtered space.

One can identify the σ-algebra $\mathcal{F}_t$ with the Borel sets of $C\left(\left[0, t\right]\right)$. Let $\mathcal{C} \stackrel{\circ}{=} \cup_{t \geq 0}\mathcal{F}_t$. If we have a consistent stream of probability spaces over $\mathcal{F}$, then one can define a set function $\mathbf{P}\left(C\right) \stackrel{\circ}{=} \mathbf{P}_t\left(C\right)$ on $\mathcal{C}$. $C\left(\left[0, t\right]\right)$ is a complete, separable metric space so $\mathbf{P}$ is compact regular on $\mathcal{C}$, hence $\mathbf{P}$ is σ-additive on $\mathcal{C}$. By Carathéodory's theorem one can extend $\mathbf{P}$ to $\sigma\left(\mathcal{C}\right) = \mathcal{B}\left(C\left[0, \infty\right)\right) = \mathcal{F}_\infty$. $\qquad\square$

Observe that in Example 6.34 Λ is a martingale so the measure spaces $(\Omega, \mathcal{F}_t, \mathbf{Q}_t)$ are consistent. If we use the canonical representation, that is $\Omega = C\left(\left[0, \infty\right)\right)$, then there is a probability measure $\mathbf{Q}$ on Ω such that $\mathbf{Q}\left(t\right)$ is a restriction of $\mathbf{Q}$ for every t. Obviously $\widehat{w}$ is a Wiener process under $\mathbf{Q}$ with respect to the natural filtration $\mathcal{F}^\Omega$. Recall that by the previous example $\mathbf{Q}$ cannot be absolutely continuous with respect to $\mathbf{P}$. The $\mathbf{P}$-measure of set A is zero so A and all of its subsets are in the augmented filtration $\mathcal{F}^\mathbf{P}$. As $\mathbf{Q}\left(A\right) = 1$ obviously $\widehat{w}$ cannot be a Wiener process under $\mathcal{F}^\mathbf{P}$. If the measures $\mathbf{P}$ and $\mathbf{Q}$ are not equivalent then the augmented filtrations can be different! Hence with the change of the measure one should also change the filtration. Of course one should augment the natural filtration $\mathcal{F}^\Omega$ because $\mathcal{F}^\Omega$ does not satisfy the usual conditions.

There is a simple method to solve this problem. Observe that on every $\mathcal{F}_t^\Omega$ the two measures $\mathbf{P}$ and $\mathbf{Q}$ are equivalent. It is very natural to assume that we augment

$\mathcal{F}_t^{\Omega}$ not with every measure-zero set of $\mathcal{F}_{\infty}^{\Omega}$ but only with the measure-zero sets of the σ-algebras $\mathcal{F}_t^{\Omega}$ for $t \geq 0$. It is not difficult to see that this filtration is right-continuous and most of the results of the stochastic analysis remain valid with this augmented filtration.

There is nothing special in the problem above. Let us show a similar elementary example.

Example 6.37 The filtration generated by the dyadic rational numbers.

Let $(\Omega, \mathcal{A}, \mathbf{P})$ be the interval $[0, 1]$ with Lebesgue's measure as probability $\mathbf{P} \stackrel{\circ}{=} \lambda$. We change the filtration only at points $t = 0, 1, 2, \ldots$. If $n < t < n+1$ then $\mathcal{F}_t \stackrel{\circ}{=} \mathcal{F}_n$. Obviously $\mathcal{F}$ is right-continuous. Let $\mathcal{F}_n$ be the σ-algebra generated by the finite number of intervals $[k2^{-n}, (k+1)2^{-n}]$ where $k = 0, 1, \ldots, 2^n - 1$. Observe that as the intervals are closed $\mathcal{F}_n$ contains all the dyadic rational numbers $0 < k2^{-n} < 1$. It is also worth noting that $\{0\}, \{1\} \notin \mathcal{F}_t$. It is also clear that the dyadic rational numbers $0 < k2^{-n} < 1$ form the only measure-zero subsets of $\mathcal{F}_n$. This implies that if $\mathbf{P}_t$ is the restriction of $\mathbf{P}$ to $\mathcal{F}_t$, then $(\Omega, \mathcal{F}_t, \mathbf{P}_t)$ is complete. $\mathcal{F}_{\infty} \stackrel{\circ}{=} \sigma(\mathcal{F}_t, t \geq 0)$ is the σ-algebra generated by the intervals with dyadic rational endpoints, so $\mathcal{F}_{\infty}$ is the Borel σ-algebra of $[0, 1]$. $\mathcal{B}([0, 1])$ is not complete under Lebesgue's measure. If we complete it, the new measure space is the set of Lebesgue measurable subsets of $[0, 1]$. In the completed space the number of the measure-zero sets is $2^{\mathfrak{c}}$, where $\mathfrak{c}$ denotes the cardinality of the continuum. If we augment $\mathcal{F}_{\infty}$ only with the measure-zero sets of the σ-algebras $\mathcal{F}_t$ then $\mathcal{F}_{\infty}$ does not change. The cardinality of $\mathcal{B}([0, 1])$ is just $\mathfrak{c}$! Let $\mathbf{Q}$ be Dirac's measure δ_0. If $t < \infty$, then the set $\{0\}$ is not in $\mathcal{F}_t$, so if $A \in \mathcal{F}_t$ and $\mathbf{P}_t(A) = 0$, then $\mathbf{Q}(A) = 0$, that is $\mathbf{Q}$ is absolutely continuous with respect to $\mathbf{P}_t$ for every $t < \infty$, that is $\mathbf{Q} \stackrel{\text{loc}}{\ll} \mathbf{P}$. Obviously $\mathbf{Q} \ll \mathbf{P}$ does not hold. $\quad\square$

6.3.5 Kazamaki–Novikov criteria

From Itô's formula it is clear that if L is a continuous local martingale then $\mathcal{E}(L)$ is also a local martingale. It is very natural to ask when $\mathcal{E}(L)$ will be a true martingale on some $[0, T]$. As $\mathcal{E}(L) \geq 0$, from Fatou's lemma it is clear that it is a supermartingale, that is if $t > s$ then

$$\mathbf{E}(\mathcal{E}(L)(t) \mid \mathcal{F}_s) = \mathbf{E}\left(\mathcal{E}\left(\lim_{n \to \infty} L^{\tau_n}\right)(t) \mid \mathcal{F}_s\right) \leq$$
$$\leq \liminf_{n \to \infty} \mathcal{E}(L^{\tau_n})(s) = \mathcal{E}(L)(s).$$

Hence taking expected value on both sides

$$\mathbf{E}(\mathcal{E}(L)(t)) \leq \mathbf{E}(\mathcal{E}(L)(s)) \quad t \geq s.$$

If $L(0) = 0$ then $\mathcal{E}(L)(0) = 1$ and in this case $\mathcal{E}(L)$ is a martingale on some $[0, t]$ if and only if $\mathbf{E}(\mathcal{E}(L)(t)) = 1$. Let us first mention a simple, but very frequently used condition:

Proposition 6.38 *If X is constant and w is a Wiener process then $\Lambda \overset{\circ}{=} \mathcal{E}(X \bullet w)$ is a martingale on any finite interval $[0, t]$. A bit more generally: if X and w are independent then $\Lambda \overset{\circ}{=} \mathcal{E}(X \bullet w)$ is a martingale on any finite interval $[0, t]$.*

Proof. The first part of the proposition trivially follows from the formula of the expected value of the lognormal distribution. Using the second condition one can assume that

$$(\Omega, \mathcal{A}, \mathbf{P}) = (\Omega_1, \mathcal{A}_1, \mathbf{P}_1) \times (\Omega_2, \mathcal{A}_2, \mathbf{P}_2).$$

X depends only on ω_2, hence for every ω_1 the integrand below is a martingale on Ω_1 so

$$\mathbf{E}(\Lambda(t)) =$$

$$= \int_{\Omega_1 \times \Omega_2} \Lambda(t) \, d(\mathbf{P}_1 \times \mathbf{P}_2) \overset{\circ}{=}$$

$$\overset{\circ}{=} \int_{\Omega_2} \left(\int_{\Omega_1} \exp\left(\int_0^t X(\omega_2) \, dw(\omega_1) - \frac{1}{2} \int_0^t X^2(\omega_2) \, d\lambda \right) d\mathbf{P}_1 \right) d\mathbf{P}_2 =$$

$$= \int_{\Omega_2} 1 \, d\mathbf{P}_2 = 1.$$

$\square$

The next condition is more general:

Proposition 6.39 (Kazamaki's criteria) *If for a continuous local martingale $L \in \mathcal{L}$*

$$\sup_{\tau \leq T} \mathbf{E}\left(\exp\left(\frac{1}{2} L(\tau) \right) \right) < \infty, \tag{6.25}$$

where the supremum is taken over all stopping times τ for which $\tau \leq T$ then $\mathcal{E}(L)$ is a uniformly integrable martingale on $[0, T]$. In the case if $T = \infty$ it is also sufficient to assume that the supremum in (6.25) is finite over just the bounded stopping times.

Proof. Observe that if τ is an arbitrary stopping time and (6.25) holds for bounded stopping times then by Fatou's lemma

$$\mathbf{E}\left(\exp\left(\frac{1}{2}L\left(\tau\right)\right)\right) = \mathbf{E}\left(\lim_{n\to\infty}\exp\left(\frac{1}{2}L\left(\tau\wedge n\right)\right)\chi\left(\tau<\infty\right)\right)\leq$$

$$\leq\liminf_{n\to\infty}\mathbf{E}\left(\exp\left(\frac{1}{2}L\left(\tau\wedge n\right)\right)\right)\leq k.$$

1. Let $p>1$ and assume that

$$\sup_{\tau\leq T}\mathbf{E}\left(\exp\left(\frac{\sqrt{p}}{2\left(\sqrt{p}-1\right)}L\left(\tau\right)\right)\right)\overset{\circ}{=}k<\infty, \tag{6.26}$$

where the supremum is taken over all bounded stopping times $\tau\leq T$. We show that $\mathcal{E}\left(L\right)\left(\tau\right)$ is bounded in $L^q\left(\Omega\right)$, where $1/p+1/q=1$. The $L^q\left(\Omega\right)$-bounded sets are uniformly integrable hence if (6.26) holds then $\mathcal{E}\left(L\right)$ is a uniformly integrable martingale. Let

$$r\overset{\circ}{=}\frac{\sqrt{p}+1}{\sqrt{p}-1}.$$

Let s be the conjugate exponent of r. By simple calculation

$$s=\frac{1}{2}\sqrt{p}+1.$$

Obviously

$$\mathcal{E}\left(L\right)^q=\exp\left(\sqrt{\frac{q}{r}}L-\frac{q}{2}\left[L\right]\right)\exp\left(\left(q-\sqrt{\frac{q}{r}}\right)L\right).$$

By Hölder's inequality

$$\mathbf{E}\left(\mathcal{E}\left(L\right)^q\left(\tau\right)\right)\leq\mathbf{E}\left(\mathcal{E}\left(\sqrt{rq}L\left(\tau\right)\right)\right)^{1/r}\mathbf{E}\left(\exp\left(s\left(q-\sqrt{\frac{q}{r}}\right)L\left(\tau\right)\right)\right)^{1/s}.$$

$\mathcal{E}\left(\sqrt{rq}L\right)$ is a non-negative local martingale, so it is a supermartingale. Hence by the Optional Sampling Theorem[41] the first part of the product cannot be larger than 1.

$$s\left(q-\sqrt{\frac{q}{r}}\right)=\frac{\sqrt{p}}{2\left(\sqrt{p}-1\right)},$$

[41]See: Proposition 1.88, page 54.

hence

$$\mathbf{E}\left(\mathcal{E}\left(L\right)^{q}\left(\tau\right)\right) \leq \mathbf{E}\left(\exp\left(\frac{\sqrt{p}}{2\left(\sqrt{p}-1\right)}L\left(\tau\right)\right)\right)^{1/s} \leq k^{1/s}.$$

2. As

$$\exp\left(x\right) \leq \exp\left(x^{+}\right) \leq \exp\left(x\right) + 1$$

one has

$$\mathbf{E}\left(\exp\left(\frac{1}{2}L\left(\tau\right)\right)\right) \leq \mathbf{E}\left(\exp\left(\frac{1}{2}L^{+}\left(\tau\right)\right)\right) \leq$$
$$\leq \mathbf{E}\left(\exp\left(\frac{1}{2}L\left(\tau\right)\right) + 1\right)$$

from which it is obvious that

$$\sup_{\tau \leq T}\mathbf{E}\left(\exp\left(\frac{1}{2}L^{+}\left(\tau\right)\right)\right) < \infty$$

is equivalent to (6.25).

3. Let $0 < a < 1$ and assume that

$$1 < \frac{\sqrt{p}}{\sqrt{p}-1} < \frac{1}{a}.$$

Applying the part of the theorem already proved for aL

$$\mathbf{E}\left(\exp\left(\frac{a\sqrt{p}}{2\left(\sqrt{p}-1\right)}L^{+}\left(\tau\right)\right)\right) \leq \mathbf{E}\left(\exp\left(\frac{1}{2}L^{+}\left(\tau\right)\right)\right) \leq$$
$$\leq k + 1 < \infty,$$

so $\mathcal{E}\left(aL\right)$ is a uniformly integrable martingale.

$$\mathcal{E}\left(aL\right) \overset{\circ}{=} \exp\left(aL - \frac{1}{2}\left[aL\right]\right) = \exp\left(aL - \frac{a^{2}}{2}\left[L\right]\right) =$$
$$= \exp\left(a\left(1-a\right)L\right)\exp\left(a^{2}L - \frac{a^{2}}{2}\left[L\right]\right) =$$
$$= \exp\left(a\left(1-a\right)L\right)\left(\mathcal{E}\left(L\right)\right)^{a^{2}}.$$

Observe that as $\mathcal{E}\left(aL\right)$ and $\mathcal{E}\left(L\right)$ are non-negative supermartingales one can extend the inequality continuously to $T = \infty$. As $L\left(0\right) = 0$ and as $\mathcal{E}\left(aL\right)$ is a uniformly integrable martingale

$$\mathbf{E} \left(\mathcal{E} \left(aL \right) \left(T \right) \right) = 1.$$

By Hölder's inequality

$$1 = \mathbf{E} \left(\mathcal{E} \left(aL \right) \left(T \right) \right) =$$

$$= \mathbf{E} \left(\mathcal{E} \left(L \right)^{a^2} \left(T \right) \exp \left(a \left(1 - a \right) L \left(T \right) \right) \right) \leq$$

$$\leq \left(\mathbf{E} \left(\mathcal{E} \left(L \right) \left(T \right) \right) \right)^{a^2} \left(\mathbf{E} \left(\exp \left(\frac{a \left(1 - a \right)}{1 - a^2} L \left(T \right) \right) \right) \right)^{1 - a^2} =$$

$$= \left(\mathbf{E} \left(\mathcal{E} \left(L \right) \left(T \right) \right) \right)^{a^2} \left(\mathbf{E} \left(\exp \left(\frac{a}{1 + a} L \left(T \right) \right) \right) \right)^{1 - a^2}.$$

From this $L \left(T \right)$ is not everywhere $-\infty$. The function x^y is continuous on the set $x > 0$, hence by the Dominated Convergence Theorem

$$\lim_{a \nearrow 1} \mathbf{E} \left(\exp \left(\frac{a}{1 + a} L \left(T \right) \right) \right)^{1 - a^2} = \left(\mathbf{E} \left(\exp \left(\frac{1}{2} L \left(T \right) \right) \right) \right)^0 = 1.$$

Therefore

$$1 \leq \mathbf{E} \left(\mathcal{E} \left(L \right) \left(T \right) \right)$$

from which, by the supermartingale property of $\mathcal{E} \left(L \right)$, the proposition is obvious.

Corollary 6.40 *If L is a continuous local martingale and $\exp \left(\frac{1}{2} L \right)$ is a uniformly integrable submartingale then $\mathcal{E} \left(L \right)$ is a uniformly integrable martingale.*

Proof. By the uniform integrability one can take $\exp \left(\frac{1}{2} L \right)$ on the closed interval $[0, T]$. By the Optional Sampling Theorem for integrable submartingales[42] if $\tau \leq T$ then

$$\exp \left(\frac{1}{2} L \left(\tau \right) \right) \leq \mathbf{E} \left(\exp \left(\frac{1}{2} L \left(T \right) \right) \mid \mathcal{F}_\tau \right),$$

from which (6.25) holds. □

Corollary 6.41 *If L is a uniformly integrable continuous martingale and $\mathbf{E} \left(\exp \left(\frac{1}{2} L \left(T \right) \right) \right) < \infty$ then $\mathcal{E} \left(L \right)$ is a uniformly integrable martingale.*

---

[42]See: Proposition 1.88, page 54.

Proof. As L is uniformly integrable $L(T)$ is meaningful. A convex function of a martingale is a submartingale.

$$\exp\left(\tfrac{1}{2}L(t)\right) \leq \mathbf{E}\left(\exp\left(\tfrac{1}{2}L(T)\right) \mid \mathcal{F}_t\right).$$

Taking the expected value on both sides, it is clear that $\exp\left(\tfrac{1}{2}L\right)$ is an integrable submartingale. By the Optional Sampling Theorem for submartingales $\exp\left(\tfrac{1}{2}L(\tau)\right)$ is integrable for every τ and (6.25) holds. □

Corollary 6.42 (Novikov's criteria) *If $L \in \mathcal{L}$ is a continuous local martingale on some finite or infinite interval $[0, T]$ and*

$$\mathbf{E}\left(\exp\left(\tfrac{1}{2}[L](T)\right)\right) < \infty, \tag{6.27}$$

and $\Lambda \stackrel{\circ}{=} \mathcal{E}(L)$ then

$$\mathbf{E}(\Lambda(T)) = \mathbf{E}(\Lambda(0)) = 1$$

and Λ is a uniformly integrable martingale on $[0, T]$.

Proof. $\mathcal{E}(L)$ is a non-negative local martingale, hence it is a supermartingale. By the Optional Sampling Theorem[43] for any bounded stopping time τ

$$\mathcal{E}(L(\tau)) \leq \mathcal{E}(L(0)) = 1.$$

By the Cauchy–Schwarz inequality

$$\mathbf{E}\left(\exp\left(\frac{1}{2}L(\tau)\right)\right) \leq$$

$$\leq \sqrt{\mathbf{E}\left(\exp\left(\left(L(\tau) - \frac{[L](\tau)}{2}\right)\right)\right)}\sqrt{\mathbf{E}\left(\exp\left(\frac{[L](\tau)}{2}\right)\right)} \stackrel{\circ}{=}$$

$$\stackrel{\circ}{=} \sqrt{\mathcal{E}(L(\tau))}\sqrt{\mathbf{E}\left(\exp\left(\frac{[L](\tau)}{2}\right)\right)} \leq \sqrt{\mathbf{E}\left(\exp\left(\frac{[L](\tau)}{2}\right)\right)} \leq$$

$$\leq \sqrt{\mathbf{E}\left(\exp\left(\frac{1}{2}[L](T)\right)\right)} < \infty.$$

Hence Kazamaki's criteria holds. □

---

[43]See: Proposition 1.88, page 54.

Corollary 6.43 *If* $L \stackrel{\circ}{=} X \bullet w$, T *is finite and for some* $\delta > 0$

$$\sup_{t \leq T} \mathbf{E} \left(\exp \left(\delta X^2 \left(t \right) \right) \right) < \infty \tag{6.28}$$

then

$$\Lambda \left(t \right) \stackrel{\circ}{=} \exp \left(\int_0^t X dw - \frac{1}{2} \int_0^t X^2 d\lambda \right)$$

is a martingale on $[0, T]$.

Proof. Let $L \stackrel{\circ}{=} X \bullet w$. By Jensen's inequality

$$\exp \left(\frac{1}{2} \left[L \right] \left(T \right) \right) = \exp \left(\frac{1}{T} \int_0^T \frac{T X^2 \left(t \right)}{2} dt \right) \leq$$

$$\leq \frac{1}{T} \int_0^T \exp \left(\frac{T X^2 \left(t \right)}{2} \right) dt.$$

If $T/2 \leq \delta$ then we can continue the estimation

$$\mathbf{E} \left(\exp \left(\frac{1}{2} \left[L \right] \left(T \right) \right) \right) \leq \frac{1}{T} \int_0^T \mathbf{E} \left(\exp \left(\frac{T X^2 \left(t \right)}{2} \right) \right) dt \leq$$

$$\leq \sup_{t \leq T} \mathbf{E} \left(\exp \left(\delta X^2 \left(t \right) \right) \right) < \infty$$

by condition (6.28), so Novikov's criteria holds. Hence $\mathbf{E} \left(\Lambda \left(T \right) \right) = 1$. Let $\left(t_k \right)_{k=0}^n$ be a partition of $[0, T]$. Assume that the size of the intervals $[t_{k-1}, t_k]$ is smaller than 2δ. If

$$\Lambda_k \stackrel{\circ}{=} \exp \left(\int_{t_k}^{t_{k+1}} X \left(s \right) dw \left(s \right) - \frac{1}{2} \int_{t_k}^{t_{k+1}} X^2 \left(s \right) ds \right)$$

then $\Lambda = \prod_k \Lambda_k$, $\mathbf{E} \left(\Lambda_k \right) = 1$ and $\mathbf{E} \left(\Lambda_k \mid \mathcal{F}_{t_k} \right) \stackrel{a.s.}{=} 1$. Hence

$$\mathbf{E} \left(\Lambda \left(T \right) \right) = \mathbf{E} \left(\mathbf{E} \left(\Lambda \left(T \right) \mid \mathcal{F}_{t_{n-1}} \right) \right) =$$

$$= \mathbf{E} \left(\mathbf{E} \left(\Lambda_{n-1} \Lambda \left(t_{n-1} \right) \mid \mathcal{F}_{t_{n-1}} \right) \right) =$$

$$= \mathbf{E} \left(\Lambda \left(t_{n-1} \right) \mathbf{E} \left(\Lambda_{n-1} \mid \mathcal{F}_{t_{n-1}} \right) \right) =$$

$$= \mathbf{E} \left(\Lambda \left(t_{n-1} \right) \right) = \cdots = \mathbf{E} \left(\Lambda \left(t_1 \right) \right) = 1. \qquad \square$$

Corollary 6.44 *If X is a Gaussian process, T is finite and*

$$\sup_{t \leq T} \mathbf{D}\left(X\left(t\right)\right) < \infty,$$

then $\Lambda = \mathcal{E}\left(X \bullet w\right)$ is a martingale on $[0, T]$.

If μ_t and σ_t denote the expected value and the standard deviation of $X\left(t\right)$ then

$$\mathbf{E}\left(\exp\left(\delta X^2\left(t\right)\right)\right) = \frac{1}{\sigma_t \sqrt{2\pi}} \int_{\mathbb{R}} \exp\left(\delta x^2\right) \exp\left(-\frac{1}{2}\left(\frac{x - \mu_t}{\sigma_t}\right)^2\right) dx =$$

$$= \frac{\exp\left(\delta \mu_t^2 / \left(1 - 2\delta\sigma_t\right)\right)}{\sqrt{1 - 2\delta\sigma_t}}.$$

If $\delta < 1 / \left(2 \sup_{t \leq T} \mathbf{D}\left(X\left(t\right)\right)\right)$ then $\mathbf{E}\left(\exp\left(\delta X^2\left(t\right)\right)\right)$ is bounded. $\square$

Example 6.45 Novikov's criteria is an elegant but not a too strong condition.

Let τ be a stopping time. If L is a continuous local martingale, then L^τ is also a continuous local martingale.

$$\mathcal{E}\left(L^\tau\right) = \exp\left(L^\tau - \frac{1}{2}\left[L^\tau\right]\right) = \mathcal{E}\left(L\right)^\tau,$$

so one could write any stopping time $\tau \leq T$ in (6.27) instead of T. If for a stopping time τ

$$\mathbf{E}\left(\exp\left(\frac{1}{2}\tau\right)\right) < \infty \tag{6.29}$$

and w is a Wiener process then Novikov's condition holds as for Wiener processes $[w]\left(\tau\right) = \tau$. Hence if (6.29) holds then

$$\mathbf{E}\left(\exp\left(w\left(\tau\right) - \frac{1}{2}\tau\right)\right) = 1.$$

Perhaps the simplest stopping times for Wiener processes are first passage times. Let $\tau \triangleq \inf\left\{t : w\left(t\right) = 1\right\}$. Observe that condition (6.29) does not hold[44] for τ. It is well-known[45] that the Laplace transform of τ is

$$l\left(s\right) = \exp\left(-\sqrt{2s}\right).$$

---

[44]See: (1.58), page 83.
[45]See: Example 1.118, page 82.

Using this

$$\mathbf{E}\left(\exp\left(w\left(\tau\right)-\frac{1}{2}\tau\right)\right) = \mathbf{E}\left(\exp\left(1-\frac{1}{2}\tau\right)\right) =$$

$$= e \cdot \mathbf{E}\left(\exp\left(-\frac{1}{2}\tau\right)\right) = e \cdot l\left(\frac{1}{2}\right) = 1.$$

Hence if $L = w^{\tau}$ then $\mathcal{E}\left(L\right)$ is a martingale on $[0, \infty]$. Perhaps the main weakness of Novikov's condition is that it depends only on $[L]$ so it holds for L if and only if it holds for $-L$ as well. In our case $L \stackrel{\circ}{=} w^{\tau}$ and as

$$\mathcal{E}\left(L\right)(t) = \exp\left(w^{\tau}\left(t\right)-\frac{1}{2}\tau \wedge t\right) \le \exp\left(w^{\tau}\left(t\right)\right) \le e$$

$\mathcal{E}\left(L\right)$ is bounded. Hence it is a uniformly integrable martingale. However $\mathcal{E}\left(-L\right)$ is not uniformly integrable as

$$\mathbf{E}\left(\mathcal{E}\left(-L\right)(\infty)\right) = \mathbf{E}\left(\exp\left(\left(-L\right)(\infty)-\frac{1}{2}\tau\right)\right) =$$

$$= \mathbf{E}\left(\exp\left(-1-\frac{1}{2}\tau\right)\right) =$$

$$= e^{-1}\mathbf{E}\left(\exp\left(-\frac{1}{2}\tau\right)\right) = e^{-2} < 1. \qquad \square$$

6.4   Itô's Formula for Non-Continuous Semimartingales

In this section we discuss the generalization of Itô's formula to non-continuous semimartingales.

Theorem 6.46 (Itô's formula) *If the coordinates of the vector* $\mathbf{X} \stackrel{\circ}{=} (X_1, X_2, \ldots, X_d)$ *are semimartingales and* $f \in C^2\left(\mathbb{R}^d\right)$ *then*

$$f\left(\mathbf{X}\left(t\right)\right) - f\left(\mathbf{X}\left(0\right)\right) = \qquad\qquad (6.30)$$

$$= \sum_{k=1}^{d} \int_0^t \frac{\partial f}{\partial x_k}\left(\mathbf{X}_-\right) dX_k + \frac{1}{2}\sum_{i,j} \int_0^t \frac{\partial^2 f}{\partial x_i \partial x_j}\left(\mathbf{X}_-\right) d\left[X_i, X_j\right]^c +$$

$$+ \sum_{0 < s \le t}\left(f\left(\mathbf{X}\left(s\right)\right) - f\left(\mathbf{X}\left(s-\right)\right) - \sum_{k=1}^{d}\frac{\partial f}{\partial x_k}\left(\mathbf{X}\left(s-\right)\right)\Delta X_k\left(s\right)\right).$$

If $\mathbf{X}(t) \subseteq U$ *for all* t *where* U *is an open subset of* $\mathbb{R}^d$ *then one can assume that* $f \in C^2(U)$.

Recall[46] that if X is a semimartingale then by definition X^c is the continuous part of the local martingale part of X. If X and Y are arbitrary semimartingales then $[X^c, Y^c] = [X, Y]^c$. So (6.30) can be written as

$$f\left(\mathbf{X}\left(t\right)\right) - f\left(\mathbf{X}\left(0\right)\right) =$$

$$= \sum_{k=1}^{d} \int_0^t \frac{\partial f}{\partial x_k}\left(\mathbf{X}_-\right) dX_k + \frac{1}{2} \sum_{i,j} \int_0^t \frac{\partial^2 f}{\partial x_i \partial x_j}\left(\mathbf{X}_-\right) d\left[X_i^c, X_j^c\right] +$$

$$+ \sum_{0 < s \le t} \left(f\left(\mathbf{X}\left(s\right)\right) - f\left(\mathbf{X}\left(s-\right)\right) - \sum_{k=1}^{d} \frac{\partial f}{\partial x_k}\left(\mathbf{X}\left(s-\right)\right) \Delta X_k\left(s\right) \right).$$

The jumps of the stochastic integral part of the formula are

$$\sum_{k=1}^{d} \frac{\partial f}{\partial x_k}\left(\mathbf{X}\left(s-\right)\right) \Delta X_k\left(s\right).$$

As X does not necessarily have finite variation the sums

$$\sum_{0 < s \le t} \left(f\left(\mathbf{X}\left(s\right)\right) - f\left(\mathbf{X}\left(s-\right)\right)\right)$$

and

$$\sum_{0 < s \le t} \sum_{k=1}^{n} \frac{\partial f}{\partial x_k}\left(\mathbf{X}\left(s-\right)\right) \Delta X_k\left(s\right)$$

are generally not finite, so one *cannot* write Itô's formula as

$$f\left(\mathbf{X}\left(t\right)\right) - f\left(\mathbf{X}\left(0\right)\right) =$$

$$= \sum_{k=1}^{d} \int_0^t \frac{\partial f}{\partial x_k}\left(\mathbf{X}_-\right) dX_k^c + \frac{1}{2} \sum_{i,j} \int_0^t \frac{\partial^2 f}{\partial x_i \partial x_j}\left(\mathbf{X}_-\right) d\left[X_i, X_j\right]^c +$$

$$+ \sum_{0 < s \le t} \left(f\left(\mathbf{X}\left(s\right)\right) - f\left(\mathbf{X}\left(s-\right)\right)\right).$$

On the other hand as $\mathbf{X}_-$ is locally bounded and f'' is continuous, after localization

$$\left| \sum_{i=1}^{d} \sum_{j=1}^{d} \int_0^t \frac{\partial^2 f}{\partial x_i \partial x_j}\left(\mathbf{X}_-\right) d\left[X_i, X_j\right]^d \right| =$$

$$= \left| \sum_{0 < s \le t} \sum_{i=1}^{d} \sum_{j=1}^{d} \frac{\partial^2 f}{\partial x_i \partial x_j}\left(\mathbf{X}(s-)\right) \Delta X_i(s) \Delta X_j(s) \right| \le$$

[46] See: Definition 4.23, page 235.

$$\leq \sum_{0<s\leq t} \sum_{i=1}^{d} \sum_{j=1}^{d} \left| \frac{\partial^2 f}{\partial x_i \partial x_j} (\mathbf{X}(s-)) \Delta X_i(s) \Delta X_j(s) \right| \leq$$

$$\leq K \sum_{i=1}^{d} \sum_{j=1}^{d} [X_i, X_j](t) < \infty.$$

The series

$$\sum_{0<s\leq t} \left(f(\mathbf{X}(s)) - f(\mathbf{X}(s-)) - \sum_{k=1}^{d} \frac{\partial f}{\partial x_k} (\mathbf{X}(s-)) \Delta X_k(s) \right)$$

and

$$\sum_{0<s\leq t} \sum_{i=1}^{d} \sum_{j=1}^{d} \frac{\partial^2 f}{\partial x_i \partial x_j} (\mathbf{X}(s-)) \Delta X_i(s) \Delta X_j(s)$$

are absolutely convergent. Hence one can write the formula in the following way:

Theorem 6.47 *If the coordinates of the vector* $\mathbf{X} \overset{\circ}{=} (X_1, X_2, \ldots, X_d)$ *are semimartingales and* $f \in C^2(\mathbb{R}^d)$ *then*

$$f(\mathbf{X}(t)) - f(\mathbf{X}(0)) = \qquad\qquad (6.31)$$

$$= \sum_{k=1}^{d} \int_0^t \frac{\partial f}{\partial x_k} (\mathbf{X}_-) \, dX_k + \frac{1}{2} \sum_{i,j} \int_0^t \frac{\partial^2 f}{\partial x_i \partial x_j} (\mathbf{X}_-) \, d[X_i, X_j] +$$

$$+ \sum_{0<s\leq t} R(s),$$

where $R(s)$ *is the 'third-order remainder of the approximation of the jumps':*

$$R(s) \overset{\circ}{=} f(\mathbf{X}(s)) - f(\mathbf{X}(s-)) -$$

$$- \sum_{k=1}^{d} \frac{\partial f}{\partial x_k} (\mathbf{X}(s-)) \Delta X_k(s) -$$

$$- \frac{1}{2} \sum_{i,j} \frac{\partial^2 f}{\partial x_i \partial x_j} (\mathbf{X}(s-)) \Delta X_i(s) \Delta X_j(s).$$

If $\mathbf{X}(t) \subseteq U$ *for all* t *where* U *is an open subset of* $\mathbb{R}^d$ *then one can assume that* $f \in C^2(U)$.

As in the continuous case one can reformulate the theorem for holomorphic functions.

Corollary 6.48 (Itô's formula for holomorphic functions) *If f is a holomorphic function and Z is a complex semimartingale then*

$$f(Z(t)) = f(Z(0)) +$$

$$+ \int_0^t f'(Z_-)dZ + \frac{1}{2}\int_0^t f''(Z_-)d[Z]^c +$$

$$+ \sum_{0 < s \le t} (f(Z(s)) - f(Z(s-)) - f'(Z(s-))\Delta Z(s)).$$

Proof. One has to calculate only the jump part. The calculation of the other terms is the same as in the continuous case. If $Z \stackrel{\circ}{=} X + iY$ and $f \stackrel{\circ}{=} u + iv$ then the jump part of the real part is the sum over the jumps of the expressions

$$u(X(s), Y(s)) - u(X(s-), Y(s-)) -$$

$$- \frac{\partial}{\partial x}u(X(s-), Y(s-))\Delta X(s) - \frac{\partial u}{\partial y}(X(s-), Y(s-))\Delta Y(s)$$

and similarly the jump part of the imaginary part is the sum of the expressions

$$v(X(s), Y(s)) - v(X(s-), Y(s-)) -$$

$$- \frac{\partial v}{\partial x}(X(s-), Y(s-))\Delta X(s) - \frac{\partial v}{\partial y}(X(s-), Y(s-))\Delta Y(s).$$

Adding them up and using that

$$f' = \frac{\partial u}{\partial x} + i\frac{\partial v}{\partial x} = \frac{\partial v}{\partial y} - i\frac{\partial v}{\partial x}$$

one can easily get that the jump part is the sum of the expressions

$$f(Z(s)) - f(Z(s-) - f'(Z(s-))\Delta Z(s).$$

□

Example 6.49 Itô's formula and the integration by parts formula.

Itô's formula is a generalization, and also a consequence, of the integration by parts formula. If $f(x, y) \stackrel{\circ}{=} xy$ and X and Y are semimartingales then by Itô's formula

$$XY - X(0)Y(0) = X_- \bullet Y + Y_- \bullet X + \frac{1}{2}[X,Y]^c + \frac{1}{2}[Y,X]^c +$$
$$+ \sum (XY - X_-Y_- - Y_-\Delta X - X_-\Delta Y).$$

The expression after the sum is $\Delta X \Delta Y$ so

$$XY - X(0)Y(0) = X_- \bullet Y + Y_- \bullet X + [X,Y]^c +$$
$$+ \sum \Delta X \Delta Y,$$

that is Itô's formula reduces to the integration by parts formula

$$XY - X(0)Y(0) = X_- \bullet Y + Y_- \bullet X + [X,Y]. \qquad \square$$

6.4.1 Itô's formula for processes with finite variation

1. Let f be a continuously differentiable function. First assume that X is a continuous process with finite variation. In this case Itô's formula has the following simple form:

$$f(X(t)) - f(X(0)) = \int_0^t f'(X)\,dX. \qquad (6.32)$$

In this special case one can prove the formula in the following way: let $\left(t_k^{(n)}\right)$ be an arbitrary infinitesimal sequence of partitions of the interval $[0, t]$. By the mean value theorem and by the intermediate value theorem

$$f(X(t)) - f(X(0)) = \sum_k \left(f\left(X\left(t_k^{(n)}\right)\right) - f\left(X\left(t_{k-1}^{(n)}\right)\right)\right) =$$
$$= \sum_k f'\left(\xi_k^{(n)}\right)\left(X\left(t_k^{(n)}\right) - X\left(t_{k-1}^{(n)}\right)\right) =$$
$$= \sum_k f'\left(X\left(\tau_k^{(n)}\right)\right)\left(X\left(t_k^{(n)}\right) - X\left(t_{k-1}^{(n)}\right)\right),$$

where $\tau_k^{(n)} \in \left[t_{k-1}^{(n)}, t_k^{(n)}\right]$. As f' and X are continuous, $f'(X(s))$ is continuous. Hence $f'(X(s))$ is Stieltjes integrable with respect to X. Therefore if $n \to \infty$ then the right-hand side converges to $\int_0^t f'(X(s))\,dX(s)$.

2. Now assume that X is a right-regular simple jump process with finite variation. Recall that in this case $X(t) - X(0) = \int_0^t \Delta X d\mu$ where μ is the counting

measure. Obviously $[X^c] = 0$.

$$\int_0^t f'(X_-)\,dX = \sum_{0<s\leq t} f'(X_-)\,\Delta X,$$

and the sum is obviously finite. In this case Itô's formula is

$$f(X(t)) - f(X(0)) = \sum_{0<s\leq t} (f(X(s)) - f(X(s-))). \qquad (6.33)$$

One can prove this identity directly: if all the jumps of X are bigger than a given $\varepsilon > 0$ then X has just finitely many of jumps on the interval $[0, t]$ and between the jumps X is constant. In this case (6.33) is a simple telescopic sum, therefore (6.33) holds. If $X^{(\varepsilon)} \triangleq |\Delta X|\chi(|\Delta X| > \varepsilon)$ then (6.33) is valid for $X^{(\varepsilon)}$.

$$\lim_{\varepsilon \to 0} |\Delta X|\chi(|\Delta X| > \varepsilon) = 0$$

therefore by the Dominated Convergence Theorem $X^{(\varepsilon)}(t \pm 0) \to X(t \pm 0)$ for every t. f is continuous, hence

$$f\left(X^{(\varepsilon)}(t)\right) - f\left(X^{(\varepsilon)}(0)\right) \to f(X(t)) - f(X(0)).$$

If f is continuously differentiable then on the interval $[-\varepsilon, \varepsilon]$

$$|f(x) - f(y)| \leq K|x - y|.$$

So

$$\sum_{0<s\leq t, |\Delta X|\leq \varepsilon} |\Delta f(X(s))| \leq \sum_{0<s\leq t} K|\Delta X(s)| < \infty,$$

hence $\Delta f(X) = f(X) - f(X-)$ is also integrable with respect to μ. By the Dominated Convergence Theorem

$$\sum_{0<s\leq t} \left(f\left(X^{(\varepsilon)}(s)\right) - f\left(X^{(\varepsilon)}(s-)\right)\right) \to \sum_{0<s\leq t} (f(X(s)) - f(X(s-))).$$

3. Finally, assume, that X is an arbitrary right-regular process with finite variation. The continuous part of the semimartingale X is the continuous part of the local martingale part so $X^c = 0$, hence in this case $[X^c, X^c] = 0$. If $X \in \mathcal{V}$

then Itô's formula is

$$f(X(t)) - f(X(0)) = \int_0^t f'(X_-)\,dX$$
$$+ \sum_{0 < s \leq t} (f(X(s)) - f(X(s-)) - f'(X(s-))\,\Delta X(s)).$$

Denote by

$$X = \left(X - \sum \Delta X \right) + \sum \Delta X \overset{\circ}{=} X^c + X^d$$

the decomposition of X. Here[47] X^c denotes the continuous part of X. As X has finite variation the decomposition is well-defined. X^c is a continuous and X^d is a jump process. Both of them have finite variation. Of course

$$\int_0^t f'(X_-)\,dX^d = \sum_{0 < s \leq t} f'(X(s-))\,\Delta X(s),$$

which is finite again. Hence Itô's formula simplifies to

$$f(X(t)) - f(X(0)) = \int_0^t f'(X_-)\,dX^c + \tag{6.34}$$
$$+ \sum_{0 < s \leq t} (f(X(s)) - f(X(s-))). \qquad \square$$

Example 6.50 If $X \in \mathcal{V}^+$ and $a < b$ then[48]

$$\sqrt{X(b)} - \sqrt{X(a)} = \int_a^b \frac{1}{\sqrt{X(s)} + \sqrt{X(s-)}}\,dX(s). \tag{6.35}$$

Assume that X is positive. By (6.34)

$$\sqrt{X(b)} - \sqrt{X(a)} = \int_a^b \frac{1}{2\sqrt{X_-}}\,dX^c + \sum_{a < s \leq b} \sqrt{X(s)} - \sqrt{X(s-)}.$$

[47] Of course now the symbol X^c has a double meaning. Hopefully the reader is not confused by the notation.
[48] See: Lemma 5.39, page 338.

As X^c is continuous and the number of the jumps of X is at most countable

$$\int_a^b \frac{1}{2\sqrt{X_-}} dX^c = \int_a^b \frac{1}{\sqrt{X} + \sqrt{X_-}} dX^c.$$

On the other hand

$$\sum_{a < s \le b} \sqrt{X(s)} - \sqrt{X(s-)} = \sum_{a < s \le b} \frac{X(s) - X(s-)}{\sqrt{X(s)} + \sqrt{X(s-)}} =$$

$$= \int_a^b \frac{1}{\sqrt{X(s)} + \sqrt{X(s-)}} dX^d.$$

Adding up we get (6.35). If $X = 0$ on some interval then both sides of (6.35) are, by definition, zero on that interval. $\square$

6.4.2 The proof of Itô's formula

Itô's formula has many proofs. One can prove the general formula by the same method we used in the continuous case: using the integration by parts formula[49] one can first show the formula for polynomials by induction, then using approximation one can show the general case. We show (6.31). For $f \equiv c$ or $f(x) = x$ the formula is trivial. If $f(x) = x^2$ then from Taylor's formula it is clear that $R(s) = 0$, and Itô's formula is just the integration by parts formula. Now let $f(x) = x^3$. By the integration by parts formula again

$$f(X) - f(X(0)) \overset{\circ}{=} X^3 - X^3(0) =$$

$$= X_-^2 \bullet X + X_- \bullet X^2 + [X, X^2] =$$

$$= X_-^2 \bullet X + X_- \bullet \left(2X_- \bullet X + [X] - X^2(0)\right) + [X, X^2] =$$

$$= 3X_-^2 \bullet X + X_- \bullet [X] + \left[X, 2X_- \bullet X + [X] - X^2(0)\right] =$$

$$= 3X_-^2 \bullet X + 3X_- \bullet [X] + [X, [X]] \overset{\circ}{=}$$

$$\overset{\circ}{=} f'(X_-) \bullet X + \frac{1}{2} f''(X_-) \bullet [X] + [X, [X]].$$

Now as $[X] \in \mathcal{V}$ it is purely discontinuous so

$$[X, [X]] = \sum \Delta X \Delta [X] = \sum (\Delta X)^3.$$

From Taylor's formula it is easy to see that if $f(x) = x^3$ then $R(s) = (\Delta X(s))^3$. So Itô's formula is valid if $f(x) = x^3$. Now let $f(\mathbf{x})$ be a polynomial of d variables

---

[49]See: Example 6.49, page 397.

and assume that the formula is valid for $f(\mathbf{X})$:

$$f(\mathbf{X}) - f(\mathbf{X}(0)) = \sum_{i=1}^{d} \frac{\partial f}{\partial x_i}(\mathbf{X}_-) \bullet X_i + \frac{1}{2} \sum_{i,j} \frac{\partial^2 f}{\partial x_i \partial x_j}(\mathbf{X}_-) \bullet [X_i, X_j] +$$

$$+ \sum R(s).$$

Let $g(\mathbf{x}) \stackrel{\circ}{=} x_l f(\mathbf{x})$. Integrating by parts

$$g(\mathbf{X}) - g(\mathbf{X}(0)) = f(\mathbf{X}_-) \bullet X_l + X_{l-} \bullet f(\mathbf{X}) + [X_l, f(\mathbf{X})].$$

Using the formula of the quadratic variation of stochastic integrals

$$[X_l, f(\mathbf{X})] = \sum_{i=1}^{d} \frac{\partial f}{\partial x_i}(\mathbf{X}_-) \bullet [X_i, X_l] +$$

$$+ \frac{1}{2} \sum \Delta X_l \sum_{i,j} \frac{\partial^2 f}{\partial x_i \partial x_j}(\mathbf{X}_-) \Delta X_i \Delta X_j +$$

$$+ \sum \Delta X_l(s) R(s).$$

Using the definition of $R(s)$

$$[X_l, f(\mathbf{X})] = \sum_{i=1}^{d} \frac{\partial f}{\partial x_i}(\mathbf{X}_-) \bullet [X_i, X_l] +$$

$$+ \sum \Delta X_l(s) \left(f(\mathbf{X}(s)) - f(\mathbf{X}(s-)) - \sum_{i=1}^{d} f'(\mathbf{X}(s-) \Delta X_i(s)) \right).$$

Using Itô's formula for $f(\mathbf{X})$

$$(X_l)_- \bullet f(\mathbf{X}) = \sum_{i=1}^{d} X_{l-} \frac{\partial f}{\partial x_i}(\mathbf{X}_-) \bullet X_i +$$

$$+ \frac{1}{2} \sum_{i,j} X_{l-} \frac{\partial^2 f}{\partial x_i \partial x_j}(\mathbf{X}_-) \bullet [X_i, X_j] +$$

$$+ \sum X_l(s-) R(s).$$

Using the same calculation as in the continuous case with (6.4) and (6.5) one can easily get the first part of the formula for g:

$$g\left(\mathbf{X}\right) - g\left(\mathbf{X}\left(0\right)\right) = \sum_{i=1}^{d} \frac{\partial g}{\partial x_i}\left(\mathbf{X}_-\right) \bullet X_i + \frac{1}{2} \sum_{i,j} \frac{\partial^2 g}{\partial x_i \partial x_j}\left(\mathbf{X}_-\right) \bullet \left[X_i, X_j\right] +$$

$$+ \text{ the jump part.}$$

We should finally calculate the value of the jump part:

$$\sum \Delta X_l\left(s\right) \left(f\left(\mathbf{X}\left(s\right)\right) - f\left(\mathbf{X}\left(s-\right)\right) - \sum_{i=1}^{d} \frac{\partial f}{\partial x_i}\left(\mathbf{X}\left(s-\right)\Delta X_i\left(s\right)\right) \right) +$$

$$+ \sum X_l\left(s-\right) R\left(s\right).$$

Obviously it is

$$\sum X_l\left(s\right) \left(f\left(\mathbf{X}\left(s\right)\right) - f\left(\mathbf{X}\left(s-\right)\right) - \sum_{i=1}^{d} \frac{\partial f}{\partial x_i}\left(\mathbf{X}\left(s-\right)\Delta X_i\left(s\right)\right) \right) - \qquad (6.36)$$

$$- \frac{1}{2} \sum X_l\left(s-\right) \sum_{i,j} \frac{\partial^2 f}{\partial x_i \partial x_j}\left(\mathbf{X}_-\right) \Delta X_i\left(s\right) \Delta X_j\left(s\right).$$

Observe that the expression after the sum in the first line is

$$g\left(\mathbf{X}\left(s\right)\right) - g\left(\mathbf{X}\left(s-\right)\right) - f\left(\mathbf{X}\left(s-\right)\right) \Delta X_l\left(s\right) -$$

$$- X_l\left(s-\right) \sum_{i=1}^{d} \frac{\partial f}{\partial x_i}\left(\mathbf{X}\left(s-\right)\Delta X_i\left(s\right)\right)$$

$$- \Delta X_l\left(s\right) \sum_{i=1}^{d} \frac{\partial f}{\partial x_i}\left(\mathbf{X}\left(s-\right)\Delta X_i\left(s\right)\right).$$

Using (6.4) it is just

$$g\left(\mathbf{X}\left(s\right)\right) - g\left(\mathbf{X}\left(s-\right)\right) - \sum_{k=1}^{d} \frac{\partial g}{\partial x_k}\left(\mathbf{X}\left(s-\right)\right) \Delta X_k\left(s\right) - \qquad (6.37)$$

$$- \Delta X_l\left(s\right) \sum_{i=1}^{d} \frac{\partial f}{\partial x_i}\left(\mathbf{X}\left(s-\right)\Delta X_i\left(s\right)\right).$$

Adding the sum over s of the second line of (6.37) to the second line of (6.36) and using (6.5) we get the term

$$-\frac{1}{2}\sum_{i,j}\sum \frac{\partial^2 g}{\partial x_i \partial x_j}\left(\mathbf{X}_-\right)\Delta X_i\left(s\right)\Delta X_j\left(s\right).$$

Hence Itô's formula is valid for polynomials. From this point the proof of the general case is the same as that of the continuous case[50].

A very natural approach to prove Itô's formula is to use Taylor approximation. To make the proof more interesting, let us first introduce the following concepts:

Recall that a measure is *locally finite* if the measure of every compact set is finite and that $\mu_n \to \mu$ in the *vague topology* if

$$\mu_n\left(\left(0,t\right]\right) \to \mu\left(\left(0,t\right]\right)$$

for every point t which is a point of continuity[51] of the limit μ. Let $(0,t]$ be an arbitrary interval and let $r > t$ be a point of continuity of μ.

$$\limsup_{n\to\infty} \mu_n\left(\left(0,t\right]\right) \le \limsup_{n\to\infty} \mu_n\left(\left(0,r\right]\right) = \mu\left(\left(0,r\right]\right).$$

Since the points of continuity of μ are dense in $\mathbb{R}_+$ and as μ is right-continuous

$$\limsup_{n\to\infty} \mu_n\left(\left(0,t\right]\right) \le \mu\left(\left(0,t\right]\right) \qquad (6.38)$$

for every $t \ge 0$. Also recall that μ^c denotes the continuous part of the increasing function $t \mapsto \mu\left(\left(0,t\right]\right)$.

Definition 6.51 *Let (Δ_n) be an infinitesimal[52] sequence of partitions:*

$$\Delta_n : 0 = t_0^{(n)} < t_1^{(n)} < \ldots < t_{k_n}^{(n)} = \infty.$$

1. *We say that a right-regular function f on $[0,\infty)$ has finite quadratic variation with respect to (Δ_n) if the sequence of point measures[53]*

$$\mu_n \overset{\circ}{=} \sum_{t_i^{(n)}\in\Delta_n} \left(f\left(t_{i+1}^{(n)}\right) - f\left(t_i^{(n)}\right)\right)^2 \delta\left(t_i^{(n)}\right) \qquad (6.39)$$

---

[50] One should use the fact that $\mathbf{X}_-$ is locally bounded.

[51] As the points of continuity are dense the limit is unique.

[52] That is, on any finite interval $\max_k \left|t_{k+1}^{(n)} - t_k^{(n)}\right| \to 0$.

[53] Recall that $\delta\left(a\right)$ is Dirac's measure concentrated at point a.

converges in the vague topology to a locally finite measure μ where μ has the decomposition

$$\mu\left((0,t]\right) = \mu^c\left((0,t]\right) + \sum_{s \leq t} \left(\Delta f\left(s\right)\right)^2.$$

We shall denote $\mu\left((0,t]\right)$ by $[f]\left(t\right) \overset{\circ}{=} [f, f]\left(t\right)$.

2. *We say that right-regular functions f and g on $[0, \infty)$ have finite quadratic co-variation with respect to (Δ_n) if $[f]$, $[g]$ and $[f + g]$ exist. In this case*

$$[f, g] \overset{\circ}{=} \frac{1}{2}\left([f + g] - [f] - [g]\right).$$

3. *A function g is (Δ_n)-integrable with respect to some function G if the limit*

$$\lim_{n \to \infty} \sum_{t_i^{(n)} \leq t} g\left(t_i^{(n)}\right)\left(G\left(t_{i+1}^{(n)}\right) - G\left(t_i^{(n)}\right)\right)$$

is finite for every $t \geq 0$. We shall denote this (Δ_n)-integral by

$$\int_0^t g\left(s-\right) dG\left(s\right).$$

Theorem 6.52 (Föllmer) *Let $F \in C^2\left(\mathbb{R}^d\right)$ and let (Δ_n) be an infinitesimal sequence of partitions of $[0, \infty)$. If $\mathbf{f} \overset{\circ}{=} (f_k)_{k=1}^d$ are right-regular functions on $\mathbb{R}_+$ with finite quadratic variation and co-variation with respect to (Δ_n) then for every $t > 0$*

$$F\left(\mathbf{f}\left(t\right)\right) - F\left(\mathbf{f}\left(0\right)\right) =$$

$$= \int_0^t \left(\frac{\partial F}{\partial \mathbf{x}}\left(\mathbf{f}\left(s-\right)\right), d\mathbf{f}\left(s\right)\right) +$$

$$+ \frac{1}{2} \sum_{i,j} \int_0^t \frac{\partial^2 F}{\partial x_i \partial x_j}\left(\mathbf{f}\left(s-\right)\right) d\left[f_i, f_j\right]\left(s\right) -$$

$$- \frac{1}{2} \sum_{s \leq t} \sum_{i,j} \frac{\partial^2 F}{\partial x_i \partial x_j}\left(\mathbf{f}\left(s-\right)\right) \Delta f_i\left(s\right) \Delta f_j\left(s\right) +$$

$$+ \sum_{s \leq t} \left(F\left(\mathbf{f}\left(s\right)\right) - F\left(\mathbf{f}\left(s-\right)\right) - \sum_{i=1}^d \frac{\partial F}{\partial x_i}\left(\mathbf{f}\left(s-\right)\right) \Delta f_i\left(s\right)\right)$$

where

$$\int_0^t \left(\frac{\partial F}{\partial \mathbf{x}} \left(\mathbf{f}\left(s- \right) \right), df\left(s \right) \right) \overset{\circ}{=} \lim_{n \to \infty} \sum_{t_i^{(n)} \le t} \left(\frac{\partial F}{\partial \mathbf{x}} \left(\mathbf{f}\left(t_i^{(n)} \right) \right), \mathbf{f}\left(t_{i+1}^{(n)} \right) - \mathbf{f}\left(t_i^{(n)} \right) \right)$$

where

$$\frac{\partial F}{\partial \mathbf{x}} \overset{\circ}{=} \left(\frac{\partial F}{\partial x_1}, \frac{\partial F}{\partial x_2}, \ldots, \frac{\partial F}{\partial x_d} \right)$$

denotes the gradient vector of F and all the other integrals are (Δ_n)*-integrals.*

If the coordinates of the vector $\mathbf{X} \overset{\circ}{=} (X_1, X_2, \ldots, X_n)$ are semimartingales, then the quadratic variations and co-variations exist and they converge uniformly on compact sets in probability. This implies that for some subsequence they converge uniformly, almost surely. Also, for semimartingales the stochastic integrals

$$\int_0^t \frac{\partial F}{\partial x_k} \left(\mathbf{X}\left(s- \right) \right) dX_k\left(s \right)$$

exist and by the Dominated Convergence Theorem, uniformly on compact intervals in probability,

$$\int_0^t \frac{\partial F}{\partial x_k} \left(\mathbf{X}\left(s- \right) \right) dX_k\left(s \right) = \sum_{t_i^{(n)} \le t} \frac{\partial F}{\partial x_k} \left(\mathbf{X}\left(t_i \right) \right) \left(X_k\left(t_{i+1}^{(n)} \right) - X\left(t_i^{(n)} \right) \right)$$

therefore Föllmer's theorem implies Itô's formula.

Proof. Fix $t > 0$. To simplify the notation we drop the superscript n.

1. If the first point in Δ_n which is larger than t is t_{k_n} then $t_{k_n} \searrow t$. As $\mathbf{f}$ is right-continuous

$$F\left(\mathbf{f}\left(t \right) \right) - F\left(\mathbf{f}\left(0 \right) \right) = \lim_{n \to \infty} F\left(\mathbf{f}\left(t_{k_n} \right) \right) - F\left(\mathbf{f}\left(0 \right) \right) =$$

$$= \lim_{n \to \infty} \sum_i \left(F\left(\mathbf{f}\left(t_{i+1} \right) \right) - F\left(\mathbf{f}\left(t_i \right) \right) \right).$$

To simplify the notation further we drop all the point from Δ_n which are larger than t_{k_n}. By Taylor's formula

$$F\left(\mathbf{f}\left(t_{i+1}\right)\right) - F\left(\mathbf{f}\left(t_{i}\right)\right) = \sum_{k=1}^{d} \frac{\partial F}{\partial x_k}\left(\mathbf{f}\left(t_{i}\right)\right)\left(f_k\left(t_{i+1}\right) - f_k\left(t_{i}\right)\right) +$$

$$+\frac{1}{2}\sum_{k,l} \frac{\partial^2 F}{\partial x_k \partial x_l}\left(\mathbf{f}\left(t_{i}\right)\right)\left(f_k\left(t_{i+1}\right) - f_k\left(t_{i}\right)\right)\left(f_l\left(t_{i+1}\right) - f_l\left(t_{i}\right)\right) +$$

$$+r\left(\mathbf{f}\left(t_{i}\right), \mathbf{f}\left(t_{i+1}\right)\right)$$

where

$$\left|r\left(\mathbf{a}, \mathbf{b}\right)\right| \le \varphi\left(\|\mathbf{b} - \mathbf{a}\|\right)\|\mathbf{b} - \mathbf{a}\|^2.$$

As F is twice continuously differentiable one may assume that φ is increasing and $\lim_{c \searrow 0} \varphi(c) = 0$.

2. Given $\varepsilon > 0$ we split the set of jumps of f into two classes. C_1 is a finite set and C_2 is the set of jumps for which

$$\sum_{s \in C_2, s \le t} \left(\sum_{k=1}^{d} |\Delta f_k(s)|\right)^2 \le \varepsilon.$$

As $\mathbf{f}$ has quadratic variation and co-variation this separation is possible. Since C_1 is finite and as f is right-regular if $\sum_{(1)}$ denotes the sum over the sub-intervals which contain a point from C_1 then

$$\lim_{n \to \infty} \sum_{(1)} \left(F\left(\mathbf{f}\left(t_{i+1}\right)\right) - F\left(\mathbf{f}\left(t_{i}\right)\right)\right) = \sum_{s \in C_1} \left(F\left(\mathbf{f}(s)\right) - F\left(\mathbf{f}(s-)\right)\right). \tag{6.40}$$

Let F' denote the first derivative and F'' the second derivative of F. Adding up the increments of other intervals

$$\sum_{(2)} \left(F\left(\mathbf{f}\left(t_{i+1}\right)\right) - F\left(\mathbf{f}\left(t_{i}\right)\right)\right) = \sum F'\left(\mathbf{f}\left(t_{i}\right)\right)\left(\mathbf{f}\left(t_{i+1}\right) - \mathbf{f}_k\left(t_{i}\right)\right) +$$

$$+\sum \frac{1}{2}F''\left(\mathbf{f}\left(t_{i}\right)\right)\left(\mathbf{f}\left(t_{i+1}\right) - \mathbf{f}\left(t_{i}\right)\right) -$$

$$-\sum_{(1)} \left(F'\left(\mathbf{f}\left(t_{i}\right)\right)\left(\mathbf{f}\left(t_{i+1}\right) - \mathbf{f}\left(t_{i}\right)\right) + \frac{1}{2}F''\left(\mathbf{f}\left(t_{i}\right)\right)\left(\mathbf{f}\left(t_{i+1}\right) - \mathbf{f}\left(t_{i}\right)\right)\right) +$$

$$+\sum_{(2)} r\left(\mathbf{f}\left(t_{i}\right), \mathbf{f}\left(t_{i+1}\right)\right).$$

As C_1 is finite the expression in the third line goes to

$$\sum_{(1)} \left(F' \left(\mathbf{f} \left(s- \right) \right) \Delta \mathbf{f} \left(s \right) + \frac{1}{2} F'' \left(\mathbf{f} \left(s- \right) \right) \left(\Delta \mathbf{f} \left(s \right) \right) \right). \qquad (6.41)$$

One can estimate the last expression as

$$\left| \sum_{(2)} r \left(\mathbf{f} \left(t_i \right), \mathbf{f} \left(t_{i+1} \right) \right) \right| \leq \varphi \left(\max_{(2)} \| \mathbf{f} \left(t_{i+1} \right) - \mathbf{f} \left(t_i \right) \| \right) \sum_{(2)} \| \mathbf{f} \left(t_{i+1} \right) - \mathbf{f} \left(t_i \right) \|^2$$

therefore, using (6.38),

$$\limsup_{k \to \infty} \left| \sum_{(2)} r \left(\mathbf{f} \left(t_i \right), \mathbf{f} \left(t_{i+1} \right) \right) \right| \leq$$

$$\leq \varphi \left(\varepsilon + \right) \limsup_{n \to \infty} \sum_{t_i \leq t} \| \mathbf{f} \left(t_i \right) - \mathbf{f} \left(t_{i+1} \right) \|^2 \leq$$

$$\leq \varphi \left(\varepsilon + \right) \limsup_{n \to \infty} \sum_{k=1}^{d} \mu_n^{(k)} \left((0, t] \right) \leq \varphi \left(\varepsilon + \right) \sum_{k=1}^{d} [f_k] \left(t \right).$$

If $\varepsilon \searrow 0$ then this expression goes to zero and the difference of (6.40) and (6.41) goes to

$$\sum_{s \leq t} \left(F \left(\mathbf{f} \left(s \right) \right) - F \left(\mathbf{f} \left(s- \right) \right) - F' \left(\mathbf{f} \left(s- \right) \right) \left(\Delta \mathbf{f} \left(s \right) \right) - \frac{1}{2} F'' \left(\mathbf{f} \left(s- \right) \right) \left(\Delta \mathbf{f} \left(s \right) \right) \right)$$

3. Let G now be a continuous function. We show that if f is one of the functions f_k or $f_k + f_l$ then

$$\lim_{n \to \infty} \sum G \left(\mathbf{f} \left(t_i \right) \right) \left(f \left(t_{i+1} \right) - f \left(t_i \right) \right)^2 =$$

$$= \int_0^t G \left(\mathbf{f} \left(s- \right) \right) d \left[f \right] \left(s \right).$$

Using the definition of measures related to the quadratic variation this means that

$$\lim_{n \to \infty} \int_0^t G \left(\mathbf{f} \right) d\mu_n = \int_0^t G \left(\mathbf{f} \left(s- \right) \right) d\mu \left(s \right), \qquad (6.42)$$

where the integrals are usual Lebesgue–Stieltjes integrals. Let $\varepsilon > 0$ and let

$$h(u) \stackrel{\circ}{=} \sum_{s \in C_1, s \leq u} \Delta f(s).$$

Let $\mu_n^{(C_1)}$ be the point measure like (6.39) based on h. As C_1 is a finite set it is easy to see that the sequence of point measures $\left(\mu_n^{(C_1)} \right)$ converges to the point measure

$$\mu^{(C_1)} \stackrel{\circ}{=} \sum_{s \in C_1} (\Delta f(s))^2 \delta(s).$$

As C_1 is finite it is also easy to see, that

$$\lim_{n \to \infty} \int_0^t G(\mathbf{f}(s)) \, d\mu_n^{(C_1)}(s) = \int_0^t G(\mathbf{f}(s-)) \, d\mu^{(C_1)}(s). \qquad (6.43)$$

Let $g \stackrel{\circ}{=} f - h$. As $f = h + g$ obviously

$$\sum_{t_i \leq u} (f(t_{i+1}) - f(t_i))^2 = \sum_{t_i \leq u} (h(t_{i+1}) - h(t_i))^2 +$$

$$+ \sum_{t_i \leq u} (g(t_{i+1}) - g(t_i))^2 +$$

$$+ 2 \sum_{t_i \leq u} (g(t_{i+1}) - g(t_i))(h(t_{i+1}) - h(t_i)).$$

C_1 has only a finite number of points and if h is not continuous at some point s then g is continuous at s. Hence the third term goes to zero. Therefore $\mu_n - \mu_n^{(C_1)}$ converges to $\mu - \mu^{(C_1)}$.

$$\left| \int_0^t G(\mathbf{f}(s)) \, d\left(\mu_n - \mu_n^{(C_1)} \right)(s) - \int_0^t G(\mathbf{f}(s-)) \, d\left(\mu - \mu^{(C_1)} \right)(s) \right| \leq$$

$$\leq \left| \int_0^t G(\mathbf{f}(s)) \, d\left(\mu_n - \mu_n^{(C_1)} \right)(s) - \int_0^t G(\mathbf{f}(s)) \, d\left(\mu - \mu^{(C_1)} \right)(s) \right| +$$

$$+ \left| \int_0^t G(\mathbf{f}(s)) - G(\mathbf{f}(s-)) \, d\left(\mu - \mu^{(C_1)} \right)(s) \right|.$$

The total size of the atoms of the measure $\mu - \mu^{(C_1)}$ is smaller than ε^2. The function $G(\mathbf{f})$ is continuous at the point of continuity of $\mu - \mu^{(C_1)}$ so one can estimate the second term by

$$\left| \int_0^t G\left(\mathbf{f}\left(s\right)\right) - G\left(\mathbf{f}\left(s-\right)\right) d\left(\mu - \mu^{(C_1)}\right)(s) \right| \leq 2\varepsilon^2 \sup_{s \leq t} \left| G\left(\mathbf{f}\left(s\right)\right) \right|.$$

Recall that $\mathbf{f}$ is bounded[54], and therefore

$$\sup_{s \leq t} \left| G\left(\mathbf{f}\left(s\right)\right) \right| < \infty.$$

Obviously $\left(\mu - \mu^{(C_1)}\right)(C_1) = 0$. Hence there are finitely many open intervals which cover the points of C_1 with total measure smaller than ε. Let O be the union of these intervals. As the points of continuity are dense one may assume that the points of the boundary of O are points of continuity of $\mu - \mu^{(C_1)}$. By the vague convergence one can assume that for some n sufficiently large $\left(\mu_n - \mu_n^{(C_1)}\right)(O) < \varepsilon$. If one deletes O from $[0, t]$ the jumps of f are smaller than ε then on the compact set $[0, t] \setminus C_1$. G is uniformly continuous on the bounded range[55] of $\mathbf{f}$ so there is a δ such that if $s_1, s_2 \in [0, t] \setminus O$ and $|s_1 - s_2| < \delta$ then

$$\left| G\left(\mathbf{f}\left(s_1\right)\right) - G\left(\mathbf{f}\left(s_2\right)\right) \right| < 2\varepsilon.$$

This means that there is a step function H such that $|H(s) - G(\mathbf{f}(s))| < 2\varepsilon$ on $[0, t] \setminus O$. On may also assume that the points of discontinuities of the step function H are points of continuity of measure $\mu - \mu^{(C_1)}$.

$$\limsup_{n \to \infty} \left| \int_0^t G\left(\mathbf{f}\left(s\right)\right) d\left(\mu_n - \mu_n^{(C_1)}\right)(s) - \int_0^t G\left(\mathbf{f}\left(s\right)\right) d\left(\mu - \mu^{(C_1)}\right)(s) \right| \leq$$

$$\leq 2\varepsilon \sup_{s \leq t} \left| G\left(\mathbf{f}\left(s\right)\right) \right| +$$

$$+ 2\varepsilon \left(\left(\mu_n - \mu_n^{(C_1)}\right)([0, t]) + \left(\mu - \mu^{(C_1)}\right)([0, t]) \right) +$$

$$+ \limsup_{n \to \infty} \left| \int_0^t H(s) d\left(\mu_n - \mu^{(C_1)}\right)(s) - \int_0^t H(s) d\left(\mu - \mu^{(C_1)}\right)(s) \right|.$$

Since the last expression, by the vague convergence goes to zero, for some k independent of ε

$$\limsup_{n \to \infty} \left| \int_0^t G\left(\mathbf{f}\left(s\right)\right) d\left(\mu_n - \mu_n^{(C_1)}\right)(s) - \int_0^t G\left(\mathbf{f}\left(s\right)\right) d\left(\mu - \mu^{(C)}\right)(s) \right| \leq \varepsilon k.$$

---

[54] See: Proposition 1.6, page 5.
[55] See: Proposition 1.7, page 6.

As ε is arbitrary

$$\lim_{n \to \infty} \int_0^t G\left(\mathbf{f}\left(s\right)\right) d\left(\mu_n - \mu_n^{(C_1)}\left(s\right)\right) = \int_0^t G\left(\mathbf{f}\left(s-\right)\right) d\left(\mu - \mu^{(C_1)}\left(s\right)\right).$$

Using (6.43) one can easily show (6.42).

4. Applying this observation and the definition of the co-variation one gets the convergence of

$$\sum F''\left(\mathbf{f}\left(t_i\right)\right)\left(\mathbf{f}\left(t_{i+1}\right) - \mathbf{f}\left(t_i\right)\right) =$$

$$= \sum \sum_{k,l} \frac{\partial^2 F\left(\mathbf{f}\left(t_i\right)\right)}{\partial x_k \partial x_k}\left(f_k\left(t_{i+1}\right) - f_k\left(t_i\right)\right)\left(f_l\left(t_{i+1}\right) - f_l\left(t_i\right)\right)$$

to the sum of integrals

$$\sum_{k,l} \int_0^t \frac{\partial^2 F}{\partial x_k \partial x_l}\left(\mathbf{f}\left(s-\right)\right) d\left[f_k, f_l\right]\left(s\right).$$

5. As all the other terms converge,

$$\sum_i \left(\frac{\partial F\left(\mathbf{f}\left(t_i\right)\right)}{\partial \mathbf{x}}\left(\mathbf{f}\left(t_i\right)\right), \mathbf{f}\left(t_{i+1}\right) - \mathbf{f}\left(t_i\right)\right)$$

also converges and its limit, by definition, is

$$\int_0^t \left(\frac{\partial F\left(\mathbf{f}\left(s-\right)\right)}{\partial \mathbf{x}}, d\mathbf{f}\left(s\right)\right)$$

which proves the formula. $\qquad \square$

6.4.3 Exponential semimartingales

As an application of the general Itô formula let us discuss the exponential semimartingales. Let Z be an arbitrary complex semimartingale, that is let $Z \overset{\circ}{=} X + iY$, where X and Y are real-valued semimartingales. Let us investigate the stochastic integral equation

$$E = 1 + E_- \bullet Z. \qquad (6.44)$$

Definition 6.53 *The equation (6.44) is called the* Doléans *equation.*

The simplest version of the equation is when $Z(s) \equiv s$

$$E\left(t\right) = 1 + \int_0^t E\left(s-\right) ds = 1 + \int_0^t E\left(s\right) ds,$$

which characterizes the exponential function $E(t) = \exp(t)$. This explains the next definition:

Definition 6.54 *The solution of (6.44), denoted by $\mathcal{E}(Z)$, is called the exponential semimartingale of Z.*

Proposition 6.55 (Yor's formula) *If X and Y are arbitrary semimartingales then*

$$\mathcal{E}(X)\mathcal{E}(Y) = \mathcal{E}(X + Y + [X,Y]).$$

Proof. By the formula for the quadratic variation of stochastic integrals

$$[\mathcal{E}(X),\mathcal{E}(Y)] \stackrel{\circ}{=} \left[1 + \mathcal{E}(X)_- \bullet X, 1 + \mathcal{E}(Y)_- \bullet Y\right] =$$
$$= \left(\mathcal{E}(X)_- \mathcal{E}(Y)_-\right) \bullet [X,Y].$$

Integrating by parts

$$\mathcal{E}(X)\mathcal{E}(Y) - 1 = \mathcal{E}(X)_- \bullet \mathcal{E}(Y) + \mathcal{E}(Y)_- \bullet \mathcal{E}(X) + [\mathcal{E}(X),\mathcal{E}(Y)] =$$
$$= \left(\mathcal{E}(X)_- \mathcal{E}(Y)_-\right) \bullet (Y + X + [X,Y]),$$

from which, by the definition of the operator $\mathcal{E}$, Yor's formula is evident. □

In the definition of $\mathcal{E}(Z)$ and during the proof of Yor's formula we have implicitly used the following theorem:

Theorem 6.56 (Solution of Doléans' equation) *Let Z be an arbitrary complex semimartingale.*

1. *There is a process E which satisfies the integral equation (6.44).*
2. *If E_1 and E_2 are two solutions of (6.44) then E_1 and E_2 are indistinguishable.*
3. *If $\tau \stackrel{\circ}{=} \inf\{t : \Delta Z = -1\}$ then $\mathcal{E}(Z) \neq 0$ on $[0,\tau)$, $\mathcal{E}(Z)_- \neq 0$ on $[0,\tau]$ and $\mathcal{E}(Z) = 0$ on $[\tau,\infty)$.*
4. *$\mathcal{E}(Z)$ is a semimartingale.*
5. *If Z has finite variation then $\mathcal{E}(Z)$ has finite variation.*
6. *If Z is a local martingale then $\mathcal{E}(Z)$ is a local martingale.*
7. *E has the following representation:*

$$E \stackrel{\circ}{=} \mathcal{E}(Z) = \exp\left(Z - Z(0) - \frac{1}{2}[Z]^c\right) \times \tag{6.45}$$
$$\times \prod(1 + \Delta Z)\exp(-\Delta Z),$$

where the product in the formula is absolutely convergent.

Proof. The proof of the theorem is a direct and simple, but lengthy calculation. We divide the proof into several steps.

1. The quadratic variation of semimartingales is finite. Hence the sum $\sum_{s \leq t} |\Delta Z(s)|^2$ is convergent. Therefore on the interval $[0, t]$ there are just finitely many moments when $|\Delta Z| > 1/2$. If $|u| \leq 1/2$, then

$$|\ln(1 + u) - u| \leq C |u|^2,$$

hence

$$\ln\left(\prod |1 + \Delta Z| \left|\exp(-\Delta Z)\right|\right) = \sum (\ln(|1 + \Delta Z|) - |\Delta Z|) \leq$$
$$\leq \sum |\ln(1 + |\Delta Z|) - |\Delta Z|| \leq$$
$$\leq C \sum |\Delta Z|^2 < \infty.$$

Therefore the product

$$V(t) \stackrel{\circ}{=} \prod_{s \leq t} (1 + \Delta Z(s)) \exp(-\Delta Z(s))$$

is absolutely convergent. Separating the real and the imaginary parts and taking logarithm, one can immediately see that V is a right-regular process with finite variation. By the definition of the product operation obviously[56]

$$V(0) \stackrel{\circ}{=} \prod_{s \leq 0} (1 + \Delta Z(s)) = 1 + \Delta Z(0) = 1.$$

2. Let us denote by U the expression in the exponent of $\mathcal{E}(Z)$:

$$U(t) \stackrel{\circ}{=} Z - Z(0) - \frac{1}{2} [Z^c].$$

With this notation

$$E \stackrel{\circ}{=} \mathcal{E}(Z) \stackrel{\circ}{=} V \exp(U).$$

By Itô's formula for complex semimartingales, using that $E(0) = 1$, and that V has finite variation, the co-variation $[U, V]^c = [U^c, V^c]$ and

---

[56]See: (1.1) on page 4.

$[V]^c = [V^c]$ are zero and hence

$$E = 1 + E_- \bullet U + \exp(U_-) \bullet V +$$
$$+ \frac{1}{2} E_- \bullet [U]^c +$$
$$+ \sum (\Delta E - V_- \exp(U_-) \Delta U - \exp(U_-) \Delta V).$$

V is a pure jump process and therefore

$$A \overset{\circ}{=} \exp(U_-) \bullet V = \sum \exp(U_-) \Delta V.$$

As $\Delta U = \Delta Z$

$$\Delta E \overset{\circ}{=} E - E_- \overset{\circ}{=} \exp(U) V - \exp(U_-) V_- =$$
$$= \exp(U_- + \Delta U) V_- (1 + \Delta Z) \exp(-\Delta Z) - \exp(U_-) V_- =$$
$$= \exp(U_- + \Delta U) \exp(-\Delta U) V_- (1 + \Delta U) - \exp(U_-) V_- =$$
$$= \exp(U_-) V_- \Delta U \overset{\circ}{=} E_- \Delta U.$$

Substituting the expressions A and ΔE

$$A + \sum (\Delta E - E_- \Delta U - \exp(U_-) \Delta V) = 0.$$

Obviously

$$[U]^c \overset{\circ}{=} \left[Z - Z(0) - \frac{1}{2} [Z]^c \right]^c = [Z^c] = [Z]^c,$$

and therefore

$$E = 1 + E_- \bullet U + \frac{1}{2} E_- \bullet [U]^c =$$
$$= 1 + E_- \bullet \left(U + \frac{1}{2} [Z]^c \right) \overset{\circ}{=}$$
$$\overset{\circ}{=} 1 + E_- \bullet (Z - Z(0)) = 1 + E_- \bullet Z,$$

hence E satisfies (6.44).

3. One has to prove that the solution is unique. Let Y be an arbitrary solution of (6.44). The stochastic integrals are semimartingales so Y is a semimartingale. By Itô's formula $H \overset{\circ}{=} Y \cdot \exp(-U)$ is also a semimartingale. Applying the

multidimensional complex Itô's formula for the complex function $z_1 \cdot \exp(-z_2)$

$$H = 1 - H_- \bullet U + \exp(-U_-) \bullet Y +$$
$$+ \frac{1}{2} H_- \bullet [U]^c - \exp(-U_-) \bullet [U,Y]^c +$$
$$+ \sum (\Delta H + H_- \Delta U - \exp(-U_-) \Delta Y).$$

Y is a solution of the Doléans equation so

$$\exp(-U_-) \bullet Y = \exp(-U_-) Y_- \bullet Z \overset{\circ}{=} H_- \bullet Z.$$
$$[U,Y]^c = [U, (Y_- \bullet Z)]^c = Y_- \bullet [U,Z]^c \overset{\circ}{=}$$
$$\overset{\circ}{=} Y_- \bullet \left[Z - \frac{1}{2}[Z]^c, Z\right]^c = Y_- \bullet [Z]^c.$$
$$\exp(-U_-) \bullet [U,Y]^c = H_- \bullet [Z]^c.$$

Adding up these terms and using that $[U]^c = [Z]^c$

$$H_- \bullet \left(Z + \frac{1}{2}[U]^c - [Z]^c\right) = H_- \bullet U,$$

hence

$$H = 1 + \sum (\Delta H + H_- \Delta U - \exp(-U_-) \Delta Y).$$

Y is a solution of (6.44), so

$$\Delta Y = Y_- \Delta Z = Y_- \Delta U. \tag{6.46}$$

Hence

$$H = 1 + \sum (\Delta H + H_- \Delta U - \exp(-U_-) Y_- \Delta U) \overset{\circ}{=}$$
$$\overset{\circ}{=} 1 + \sum (\Delta H + H_- \Delta U - H_- \Delta U) =$$
$$= 1 + \sum \Delta H.$$

On the other hand, using (6.46) again

$$\Delta H \overset{\circ}{=} H - H_- \overset{\circ}{=} Y \exp(-U) - H_- =$$
$$= \exp(-U_- - \Delta U)(Y_- + \Delta Y) - H_- =$$
$$= \exp(-U_- - \Delta U) Y_- (1 + \Delta Z) - H_- =$$
$$= \exp(-U_-) Y_- \exp(-\Delta U)(1 + \Delta Z) - H_- =$$
$$= H_- (\exp(-\Delta Z)(1 + \Delta Z) - 1)$$

so

$$H = 1 + H_- \bullet R, \tag{6.47}$$

where

$$R \overset{\circ}{=} \sum (\exp(-\Delta Z)(1 + \Delta Z) - 1).$$

For some constant C if $|x| \leq 1/2$

$$|\exp(-x)(1 + x) - 1| \leq Cx^2.$$

Z is a semimartingale so $\sum (\Delta Z)^2 < \infty$ and therefore R is a complex process with finite variation.

4. Let us prove the following simple general observation: if v is a right-regular function with finite variation then the only right-regular function f for which

$$f(h) = \int_0^h f(s-) \, dv(s), \quad h \geq 0 \tag{6.48}$$

is $f \equiv 0$. Let $s \overset{\circ}{=} \inf\{t : f(t) \neq 0\}$. Obviously $f = 0$ on the interval $[0, s)$. Hence by the integral equation (6.48)

$$f(s) = \int_0^s f(t-) \, dv(t) = \int_0^s 0 \, dv = 0.$$

If $s < \infty$ then, as v is right-regular, there is a $t > s$ such that $\text{Var}(v(t)) - \text{Var}(v(s)) \leq 1/2$. If $t \geq u > s$ then

$$f(u) = f(s) + \int_s^u f_- \, dv = \int_s^u f_- \, dv \leq$$
$$\leq \text{Var}(v, s, u) \sup_{s \leq u \leq t} |f(u)| \leq \frac{1}{2} \sup_{s \leq u \leq t} |f(u)|$$

and therefore

$$\sup_{s<u\leq t} |f(u)| \leq \frac{1}{2} \sup_{s<u\leq t} |f(u)|.$$

As f is regular f is bounded on every finite interval so $\sup_{s<u\leq t} |f(u)| < \infty$. Hence f is zero on the interval $(s,t]$ which is impossible by the definition of s.

5. $\mathcal{E}(Z) \stackrel{\circ}{=} V\exp(U)$ is a solution of the Doléans equation so the process

$$V = \mathcal{E}(Z)\exp(-U)$$

satisfies the equation

$$V = 1 + V_- \bullet R. \qquad (6.49)$$

If $G \stackrel{\circ}{=} H - V$ then subtracting equations (6.47) and (6.49) $G = G_- \bullet R$. R is right-regular and has finite variations. Therefore $G \equiv 0$. So $H \equiv V$. Hence

$$Y \stackrel{\circ}{=} H\exp(U) = V\exp(U) = \mathcal{E}(Z),$$

so $\mathcal{E}(Z)$ is the only solution of Doléans' equation.

6. As we have already mentioned, by Itô's formula $\mathcal{E}(Z)$ is a semimartingale. By (6.44) and by the basic properties of stochastic integrals if Z is a local martingale then $\mathcal{E}(Z)$ is also a local martingale, if Z has finite variation then $\mathcal{E}(Z)$ also has finite variation. From (6.45) the other parts of the theorem are obvious. $\qquad\square$

6.5   Itô's Formula For Convex Functions

In this section we present some generalizations of Itô's formula. One particular deficiency of the formula is that one can use it only with smooth functions. But some very important functions like $|x|$ or $x^{\pm}$ are non-differentiable so, e.g., with the differentiable Itô's formula we cannot prove that the absolute value of a semimartingale is a semimartingale again[57]. As we shall see, the key property of the function $|x|$ is that it is a convex function. The main result of this section is that the class of semimartingales is closed under transformation by convex functions.

---

[57]See: Theorem 6.65, page 422.

6.5.1 Derivative of convex functions

We shall use the next elementary, but important observation very often:

Theorem 6.57 (Fundamental Theorem of Calculus for Convex Functions) *Let f be a continuous convex function defined on some finite or infinite interval $[a, b]$. If f'_+ and f'_- denote the right and the left derivatives of f, then*

$$f(b) - f(a) = \int_a^b f'_+(x)\,dx = \int_a^b f'_-(x)\,dx. \qquad (6.50)$$

Proof. f is convex so for an arbitrary $x \in (a, b)$

$$h \mapsto \frac{f(x+h) - f(x)}{h} \qquad (6.51)$$

is meaningful and increasing in some neighborhood of $h = 0$. So the derivatives $f'_\pm(x)$ exist and it is not difficult to show that they are increasing. Every monotone function is Riemann integrable on any finite interval $[c, d]$: if g is monotone then the difference of the upper and the lower approximating sums is bounded by

$$|g(d) - g(c)| \cdot \max_n (x_n - x_{n-1}) \to 0.$$

Let (x_n) be a partition of some $[c, d] \subseteq (a, b)$. As (6.51) is increasing

$$f'_-(x_{n-1}) \le f'_+(x_{n-1}) \le \frac{f(x_n) - f(x_{n-1})}{x_n - x_{n-1}} =$$

$$= \frac{f(x_{n-1}) - f(x_n)}{x_{n-1} - x_n} \le f'_-(x_n) \le f'_+(x_n).$$

Multiplying by $x_n - x_{n-1}$ and adding up for all n

$$\sum_n f'_\pm(x_{n-1})(x_n - x_{n-1}) \le f(b) - f(a) \le$$

$$\le \sum_n f'_\pm(x_n)(x_n - x_{n-1}).$$

The expression on the left is the lower approximating sum, the expression on the right is the upper approximating sum. As the Riemann integral exits on arbitrary compact interval $[c, d] \subseteq (a, b)$ (6.50) holds on $[c, d]$. One can get the general formula for $[a, b]$ by taking limits on both sides[58]. $\qquad \square$

---

[58] Of course on infinite intervals it is possible that the integral is not finite, but in this case $f(b) - f(a)$ is also infinite.

Lemma 6.58 *If f is a convex function on an open interval[59] $I \subseteq \mathbb{R}$ then the left derivative f'_- is left-continuous, the right derivative f'_+ is right-continuous.*

Proof. Let us take for example the function f'_-. One should show that

$$\lim_{x \nearrow w} f'_- (x) = f'_- (w). \tag{6.52}$$

As f is convex

$$f'_- (x) = \sup_{z < x} \frac{f(z) - f(x)}{z - x}.$$

Let $y < x < w$. Since I is open f is continuous on I, so

$$\lim_{x \nearrow w} f'_- (x) = \lim_{x \nearrow w} \sup_{z < x} \frac{f(z) - f(x)}{z - x} \geq \lim_{x \nearrow w} \frac{f(y) - f(x)}{y - x} = \frac{f(y) - f(w)}{y - w}.$$

Taking the supremum on the right-hand side

$$\lim_{x \nearrow w} f'_- (x) \geq \sup_{y < w} \frac{f(y) - f(w)}{y - w} = f'_- (w).$$

On the other hand, since f'_- is increasing $\lim_{x \nearrow w} f'_- (x) \leq f'_- (w)$. Hence (6.52) holds. $\qquad \square$

Definition 6.59 *If f is a convex function then f' will denote the left-hand side derivative of f, that is $f' \overset{\circ}{=} f'_-$.*

Definition 6.60 $\operatorname{sign}(x)$ *is the left-hand side derivative of the convex function* $|x|$, *that is*[60]

$$\operatorname{sign}(x) \overset{\circ}{=} \begin{cases} 1 & if \quad x > 0 \\ -1 & if \quad x \leq 0 \end{cases}.$$

Definition 6.61 *Let I be an open interval. We say that function $g : I \to \mathbb{R}$ is a generalized derivative[61] of function $h : I \to \mathbb{R}$, if for arbitrary[62] $\phi \in C_c^\infty (I)$ the integration by parts formula holds, that is, if $\phi \in C_c^\infty (I)$ then*

$$\int_I h \phi' d\lambda = [h\phi]_I - \int_I g \phi d\lambda = - \int_I g \phi d\lambda.$$

It is not too surprising that if f is a convex function on an open interval I then f'_- and f'_+ are generalized derivatives of f: The support of ϕ is in I and f

---

[59]As I is open f is continuous in I. See: (6.51) above.

[60]Observe that it is a bit of an unusual definition of the sign function.

[61]Recall that we are dealing with three different definitions of the derivative.

[62]$C_c^\infty (U)$ is the set of continuously differentiable functions with compact support.

is continuous, hence it is bounded on the support of ϕ. For example, by the Dominated Convergence Theorem

$$\int_I f\phi'\,d\lambda = \int_I f(x)\lim_{h\to0}\frac{\phi(x+h)-\phi(x)}{h}\,dx =$$

$$= \lim_{h\to0}\int_I f(x)\frac{\phi(x+h)-\phi(x)}{h}\,dx =$$

$$= \lim_{h\to0}\int_I -\frac{f(x-h)-f(x)}{-h}\phi(x)\,dx =$$

$$= \lim_{w\to0}\int_I -\frac{f(x+w)-f(x)}{w}\phi(x)\,dx = -\int_I f'_{\pm}\phi\,d\lambda.$$

One can think about the generalized derivative of $f' \stackrel{\circ}{=} f'_{-}$ or f'_{+} as the generalized second derivative of f. For convex functions the generalized second derivative f'' is generally not a function: it is the measure μ generated by the left-continuous increasing function f'_{-} or by the right-continuous function f'_{+}. By the integration by parts formula, using that $\phi \in C_c^{\infty}(I)$, hence it is zero around the endpoints of I

$$\int_I f'_{\pm}\phi'\,d\lambda = \int_I f'_{\pm}\,d\phi = -\int_I \phi\,df'_{\pm} \stackrel{\circ}{=} -\int_I \phi\,d\mu.$$

Example 6.62 The generalized second derivative of function $|x|$ is $2\delta_0$. The generalized second derivative of functions $x^{\pm}$ are δ_0.

$f(x) \stackrel{\circ}{=} |x|$ is a convex function and

$$f'_{-}(x) = \text{sign}(x) \stackrel{\circ}{=} \begin{cases} 1 & \text{if } x > 0 \\ -1 & \text{if } x \le 0 \end{cases},$$

and the measure generated by f'_{-} is $2\delta_0$. The proof of the other relation is similar. □

Lemma 6.63 Let the measure $\mu \stackrel{\circ}{=} f''$ be the generalized second derivative of a convex function f. If the support of μ is a subset of a compact interval $[a,b]$, then with some constants α and β

$$f(x) = \alpha x + \beta + \frac{1}{2}\int_a^b |x-t|\,d\mu(t) = \tag{6.53}$$

$$= \alpha x + \beta + \frac{1}{2}\int_{\mathbb{R}} |x-t|\,d\mu(t).$$

Proof. For convex functions one can use the Fundamental Theorem of Calculus. If $x \in [a, b]$, then integrating by parts[63]

$$f(x) - f(a) = \int_{[a,x)} f'_-(t)\, dt = \int_a^x f'_+(t)\, dt =$$

$$= xf'_+(x) - af'_+(a) - \int_a^x t\, df'_+(t) =$$

$$= xf'_+(a) - af'_+(a) + \int_a^x x - t\, df'_+(t),$$

that is with some constants α_1, β_1

$$f(x) = \alpha_1 x + \beta_1 + \int_a^x |x - t|\, d\mu(t).$$

With the same calculation from the other side

$$f(x) = \alpha_2 x + \beta_2 + \int_x^b |x - t|\, d\mu(t).$$

Adding up these identities and dividing by two one gets the representation (6.53).

Example 6.64 Representation of function $f(x) = x^+$.

The generalized second derivative of $f(x) = x^+$ is δ_0, so

$$\frac{1}{2} \int_{\mathbb{R}} |x - t|\, d\mu(t) = \frac{1}{2} |x|$$

and

$$x^+ = \frac{1}{2}x + \frac{1}{2} |x|,$$

that is in the representation $\alpha = 1/2$ and $\beta = 0$. $\qquad\square$

---

[63] f'_- is left continuous so $\mu([a, b)) = f'_-(b) - f'_-(a)$. f'_+ is right continuous so $\mu((a, b]) = f'_+(b) - f'_+(a)$. Recall that in this book $\int_a^b h\, d\mu \overset{\circ}{=} \int_{(a,b]} h\, d\mu$.

6.5.2   Definition of local times

The most important result of the present section is the following:

Theorem 6.65 *If $f : \mathbb{R} \to \mathbb{R}$ is a convex function, X is a semimartingale then $f(X)$ is also a semimartingale. $f(X)$ has the following decomposition:*

$$f(X(t)) - f(X(0)) = \int_0^t f'(X(s-)) \, dX(s) + A(t), \qquad (6.54)$$

where $A \in \mathcal{V}^+$. For the jumps of A

$$\Delta A = f(X) - f(X_-) - f'(X_-) \Delta X.$$

Proof. $f' \overset{\circ}{=} f'_-$ is increasing, hence it is Borel measurable. X_- is locally bounded therefore $f'(X_-)$ is predictable and locally bounded. This means that the stochastic integral in (6.54) exists. The main idea of the proof is that one can approximate the convex function f by C^2 functions and for the approximating C^2 functions one can use Itô's formula.

1. Let $g \in C_c^\infty((-\infty, 0])$ be a non-negative function for which $\int_\mathbb{R} g \, d\lambda = 1$. For every n let

$$f_n(x) \overset{\circ}{=} \int_\mathbb{R} f\left(x + \frac{y}{n}\right) g(y) \, dy =$$

$$= n \int_\mathbb{R} f(z) g(n(z - x)) \, dz.$$

f is convex on $\mathbb{R}$, hence it is continuous, so on every finite interval it is bounded. Therefore by the last formula one can differentiate under the integral sign:

$$f'_n(x) = -n^2 \int_\mathbb{R} f(z) g'(n(z - x)) \, dz$$

and obviously $f_n \in C^\infty$. By the just proved version of the Fundamental Theorem of Calculus[64]

$$f(t) = f(s) + \int_s^t f'(v) \, dv.$$

---

[64]See: Theorem 6.57, page 418.

If we integrate by parts and $h(z) \stackrel{\circ}{=} ng(n(z-x))$ then

$$0 = [hf]_{-\infty}^{\infty} = \int_{\mathbb{R}} h \, df + \int_{\mathbb{R}} f \, dh = \int_{\mathbb{R}} hf' d\lambda + \int_{\mathbb{R}} fh' d\lambda.$$

Therefore

$$f_n'(x) = \int_{\mathbb{R}} fh' d\lambda = n \int_{\mathbb{R}} f'(z) g(n(z-x)) dz = \qquad (6.55)$$

$$= \int_{\mathbb{R}} f'\left(x + \frac{y}{n}\right) g(y) \, dy.$$

f' is increasing, hence f' is locally bounded. So, if $n \to \infty$ one may take the limit under the integral sign. The support of g is in the set of non-positive numbers, so if $n \to \infty$ then the limit of the integrand is $f'(x-0) g(y)$.

$$\lim_{n \to \infty} f_n'(x) = \int_{\mathbb{R}} \lim_{n \to \infty} f'\left(x + \frac{y}{n}\right) g(y) \, dy = f'(x-0) \int_{\mathbb{R}} g \, d\lambda =$$

$$= f'(x-0) = f_-'(x-0) = f_-'(x) \stackrel{\circ}{=} f'(x).$$

2. Let us apply Itô's formula to the functions $f_n \in C^2$:

$$f_n(X(t)) - f_n(X(0)) = \int_0^t f_n'(X_-) \, dX + A_n(t), \qquad (6.56)$$

where

$$A_n(t) \stackrel{\circ}{=} \sum_{0 < s \le t} \left(f_n(X(s)) - f_n(X(s-)) - f_n'(X(s-)) \Delta X(s) \right) +$$

$$+ \frac{1}{2} \int_0^t f_n''(X(s-)) \, d[X^c(s)].$$

f_n is also convex so A_n is increasing. f_n is continuous so the left-hand side of (6.56) is right-continuous. The integral on the right-hand side is, by the definition of the stochastic integrals, right-continuous. Hence the function A_n is right-continuous and the formula for the jumps of A in the statement of the theorem is trivially valid.

3. If $n \to \infty$, then the limit on the left side is

$$f(X(t)) - f(X(0)).$$

f' is increasing, hence if $y \le 0$, then $f'(x + y/n) \le f'(x)$, and therefore by (6.55) one has that $f_1' \le f_n' \le f'$. Assume that $X(0) = 0$. As X_- is locally bounded and as the derivative of a convex function is increasing $f'(X_-)$ and $f_1'(X_-)$ are locally bounded, hence the sequence $(f_n'(X_-))$ is dominated by a locally bounded process. Therefore if $n \to \infty$ then uniformly on compact intervals in probability

$$\lim_{n \to \infty} f_n'(X_-) \bullet X = f'(X_-) \bullet X.$$

From this it follows that A_n is stochastically convergent, and the limit is increasing.

4. We show that A is right-continuous, and the condition for the jumps holds. It is sufficient to show that the convergence $A_n \to A$ on every compact interval is uniform in probability. For this one should prove that $f_n(X) \overset{\text{ucp}}{\to} f(X)$. If $n \to \infty$, then the convergence $f_n \to f$ is uniform on every compact interval[65]. Let (τ_m) be a localizing sequence of X_-. Obviously one can localize the line (6.54) so it is sufficient to prove the relation for the truncated processes X^{τ_m}.

$$f_n(X^{\tau_m}) - f_n(X(0)) = f_n(X_-^{\tau_m}) - f_n(X(0)) + \\ + \Delta f_n(X(\tau_m)).$$

If $n \to \infty$, then as $X_-^{\tau_m}$ is bounded, by the uniform convergence on compacts of (f_n), the convergence of the $f_n(X_-^{\tau_m})$ is uniform for every trajectory. The convergence of $\Delta f_n(X(\tau_m))$ is a convergence of random variables, hence the convergence

$$f_n(X^{\tau_m}) \to f(X^{\tau_m})$$

is uniform on any compact interval.

5. If $X(0) \ne 0$ then

$$f(X) = f(X - X(0) + X(0)) \overset{\circ}{=} g(X - X(0), X(0)).$$

One can approximate g by

$$g_n(X - X(0), X(0)) = f_n(X - X(0) + X(0)).$$

Using the multi-dimensional Itô formula one can prove the theorem as in the case $X(0) = 0$. □

Our next goal is to investigate the properties of A.

---

[65]It is generally true, see [35], page 105, that for convex functions, pointwise convergence implies uniform convergence. Now, using the definition of f_n one can directly prove the uniform convergence.

Corollary 6.66 *For any a*

$$|X(t) - a| = |X(0) - a| + \int_0^t \text{sign}(X(s-) - a) \, dX(s) + A^a(t). \qquad (6.57)$$

Definition 6.67 *For arbitrary a the continuous part of A^a in (6.57) that is the expression*

$$L(a, t) \stackrel{\circ}{=} A^a(t) - \sum_{0 < s \le t} (\Delta |X(s) - a| - \text{sign}(X(s-) - a) \Delta X(s))$$

is called the local time of X at point a.

By the construction of local times $L(a, t)$ is continuous in the time parameter t. On the other hand, at the moment we cannot say anything about the spatial parameter a. The reason for this is that during the construction of A we used Itô's formula, therefore for each value of a the two sides of (6.57) are just indistinguishable and they are not the same. The number of possible values of parameter a is not countable so it is not clear how one can unify the exceptional zero sets. To say it in another way, the main problem is that for any fixed a the stochastic integral part is defined up to indistinguishability. By Fubini's theorem for stochastic integrals[66] one can assume that the parametric stochastic integral in (6.57) is product measurable. This implies the following:

Proposition 6.68 *The local time $L(a, t, \omega)$ has a version which is product measurable[67] in (a, t, ω).*

From now on we shall assume that L is product measurable.

Example 6.69 If X is a semimartingale, then

$$(X(t) - a)^+ - (X(0) - a)^+ = \int_0^t \chi(X_- > a) \, dX +$$
$$+ \sum_{0 < s \le t} \chi(X(s-) > a)(X(s) - a)^- +$$
$$+ \sum_{0 < s \le t} \chi(X(s-) \le a)(X(s) - a)^+ +$$
$$+ \frac{1}{2} L(a, t),$$

---

[66]See: Proposition 5.23 page 319. Observe that the integrand is uniformly bounded.

[67]Later we show that for continuous local martingales the local time $L(a, t, \omega)$ has a version which is continuous in (a, t).

or

$$(X(t) - a)^- - (X(0) - a)^- = - \int_0^t \chi(X_- \le a) \, dX +$$

$$+ \sum_{0 < s \le t} \chi(X(s-) > a)(X(s) - a)^- +$$

$$+ \sum_{0 < s \le t} \chi(X(s-) \le a)(X(s) - a)^+ +$$

$$+ \frac{1}{2} L(a, t).$$

These formulas are called *Tanaka's formulas.*

Let us apply the generalization of Itô's formula (6.54) for convex functions $f(x) \overset{\circ}{=} (x - a)^+$ and $g(x) \overset{\circ}{=} (x - a)^-$:

$$f(X(t)) = f(X(0)) + \int_0^t f'(X_-) \, dX + A^{(+)}(t),$$

$$g(X(t)) = g(X(0)) + \int_0^t g'(X_-) \, dX + A^{(-)}(t).$$

Subtracting the two lines above and using that

$$f'(x) = \chi(x > a), \quad g'(x) = -\chi(x \le a)$$

one gets

$$X(t) - X(0) = \int_0^t 1 dX + A^{(+)}(t) - A^{(-)}(t).$$

This implies that $A^{(+)}(t) = A^{(-)}(t)$. If

$$B^{(+)}(t) \overset{\circ}{=} A^{(+)}(t) - \sum_{0 < s \le t} (f(X(s)) - f(X(s-)) - f'(X(s-)) \Delta X(s))$$

$$B^{(-)}(t) \overset{\circ}{=} A^{(-)}(t) - \sum_{0 < s \le t} (g(X(s)) - g(X(s-)) - g'(X(s-)) \Delta X(s))$$

then by the definition of the local time $B^{(+)}(t) + B^{(-)}(t) = L(a, t)$. As the difference of the sums above is zero $B^{(+)}(t) = B^{(-)}(t)$, hence $B^{(+)} = B^{(-)} = L(a)$, so the formula is valid. $\qquad \square$

For any process X one can introduce the occupation times measure

$$\mu_t(B) \triangleq \lambda \left(s \leq t : X(s) \in B \right).$$

Later we shall see[68] that for Wiener processes the local time $L(a,t)$ is the density function of μ_t. With the usual interpretation of the density functions for Wiener processes one can think about $L(a,t)\,da$ as the time during the time interval $[0,t]$ a Wiener process is infinitely closely around a.

Example 6.70 The Green function and the local time of Wiener processes.

Let $w^{(x)}$ be a Wiener process starting from point x. Let $x \in I \triangleq (a,b)$ be a bounded interval, and let $\tau_I^{(x)}$ be the exit-time of $w^{(x)}$ from I. Let us calculate the expected value $\mathbf{E}\left(L^{(x)}(y,\tau_I)\right)$. By the definition of local time $L^{(x)}$

$$\left|w^{(x)}(t)-y\right| = \left|w^{(x)}(0)-y\right| + \int_0^t \mathrm{sign}\left(w^{(x)}-y\right)dw^{(x)} + L^{(x)}(y,t).$$

If we truncate $w^{(x)}$ by $\tau_I^{(x)}$ then the truncated process is bounded. If we truncate both sides with $\tau_I^{(x)}$ then the truncated integrator is in $\mathcal{H}^2$. By Itô's isometry the integral is also in $\mathcal{H}^2$. Therefore the stochastic integral is a uniformly integrable martingale. By the Optional Sampling Theorem the expected value of the stochastic integral is zero, so

$$\mathbf{E}\left(\left|w^{(x)}\left(\tau_I^{(x)}\right)-y\right|\right) = |x-y| + \mathbf{E}\left(L^{(x)}\left(y,\tau_I^{(x)}\right)\right).$$

$w^{(x)}$ leaves the bounded set $[a,b]$ almost surely so

$$\mathbf{E}\left(\left|w^{(x)}\left(\tau_I^{(x)}\right)-y\right|\right) = |a-y|\,\mathbf{P}\left(w\left(\tau_I^{(x)}\right)=a\right)$$
$$+ |b-y|\,\mathbf{P}\left(w\left(\tau_I^{(x)}\right)=b\right).$$

With the Optional Sampling Theorem one can easily calculate the probabilities[69]. Obviously

$$\mathbf{P}\left(w\left(\tau_I^{(x)}\right)=a\right) + \mathbf{P}\left(w\left(\tau_I^{(x)}\right)=b\right) = 1,$$

---

[68]See: Corollary 6.75, page 435.
[69]See: Example 1.116, page 81.

and

$$x = \mathbf{E}\left(w^{(x)}(0)\right) = \mathbf{E}\left(w^{(x)}\left(\tau_I^{(x)}\right)\right)$$
$$= a\mathbf{P}\left(w\left(\tau_I^{(x)}\right) = a\right) + b\mathbf{P}\left(w\left(\tau_I^{(x)}\right) = b\right).$$

Solving the equations

$$\mathbf{P}\left(w\left(\tau_I^{(x)}\right) = a\right) = \frac{b-x}{b-a}, \quad \mathbf{P}\left(w\left(\tau_I^{(x)}\right) = b\right) = \frac{x-a}{b-a}.$$

Substituting back

$$\mathbf{E}\left(L^{(x)}\left(y, \tau_I^{(x)}\right)\right) = |a-y|\frac{b-x}{b-a} + |b-y|\frac{x-a}{b-a} - |x-y|.$$

With elementary calculation

$$\mathbf{E}\left(L^{(x)}\left(y, \tau_I^{(x)}\right)\right) = \frac{2}{b-a}\begin{cases}(x-a)(b-y) & \text{if} \quad a \le x \le y \le b \\ (y-a)(b-x) & \text{if} \quad a \le y \le x \le b\end{cases}.$$

If we introduce the so-called *Green function*

$$G_I(x,y) \overset{\circ}{=} \frac{1}{b-a}\begin{cases}(x-a)(b-y) & \text{if} \quad a \le x \le y \le b \\ (y-a)(b-x) & \text{if} \quad a \le y \le x \le b\end{cases}$$

then

$$\mathbf{E}\left(L^{(x)}\left(y, \tau_I^{(x)}\right)\right) = 2G_I(x,y). \qquad \square$$

Example 6.71 If $0 < a < b$ then before reaching point b a Wiener process starting from $x = 0$ on average spends $2(b-a)\,da$ time units in the da neighbourhood point a.

Let w be a Wiener process and let $0 < a < b$. Let us denote by τ_b the first passage time of point b. Using the interpretation of the local times one should calculate the expected value $\mathbf{E}(L(a, \tau_b))$. Using the same method as in the previous example

$$|b-a| = \mathbf{E}(|w(\tau_b) - a|) =$$
$$= |a| + \mathbf{E}\left(\int_0^{\tau_b} \text{sign}(w(s) - a)\,dw(s)\right) + \mathbf{E}(L(a, \tau_b)).$$

Observe that now w^{τ_b} is not bounded, so it is not in $\mathcal{H}^2$ so the stochastic integral is not a uniformly integrable martingale. If $c < 0 < a < b$, then as in the previous example

$$\mathbf{E}\left(L\left(a, \tau_b \wedge \tau_c\right)\right) = 2G_{(c,b)}\left(0, a\right).$$

If $c \searrow -\infty$, then the limit on the right-hand side is $2\left(b - a\right)$. On the left-hand side $\tau_b \wedge \tau_c \nearrow \tau_b$ and as $t \mapsto L\left(a, t\right)$ is increasing and continuous by the Monotone Convergence Theorem

$$\mathbf{E}\left(L\left(a, \tau_b\right)\right) = 2\left(b - a\right).$$

$\square$

6.5.3 Meyer–Itô formula

Theorem 6.72 *Let X be a semimartingale. If $L\left(a\right)$ is the local time of X at point a then for almost all outcome ω the support of the measure generated by the increasing function $t \mapsto L\left(a, t, \omega\right)$ is in the set*

$$\left\{s : X\left(s-, \omega\right) = X\left(s, \omega\right) = a\right\}.$$

Proof. By the definition of local times $L\left(a, t, \omega\right)$ is continuous in time parameter t. This implies that the measure of every single point, with respect to the measure generated by $L\left(a, t, \omega\right)$, is zero. For every trajectory the number of the jumps of X is maximum countable, so it is sufficient to prove that the support of the measure generated by $L\left(a, t, \omega\right)$ is a subset of

$$\left\{s : X\left(s-, \omega\right) = a\right\}$$

for almost all outcome ω. As convex functions of semimartingales are semimartingales $Y \overset{\circ}{=} |X - a|$ is a semimartingale.

$$Y^2 = Y^2\left(0\right) + 2Y_- \bullet Y + [Y].$$

$Z \overset{\circ}{=} X - a$ is also a semimartingale.

$$Y^2 = Z^2 = Y^2\left(0\right) + 2Z_- \bullet Z + [Z].$$

Obviously $[Z] = [Y]$, therefore $Y_- \bullet Y = Z_- \bullet Z$. As $Y = |Z|$

$$Y\left(t\right) = Y\left(0\right) + \int_0^t \text{sign}\left(Z_-\right) dZ + A^a\left(t\right).$$

By the associativity rule

$$\int_0^t Y_- dY = \int_0^t Y_- \text{sign}\left(Z_-\right) dZ + \int_0^t Y_- dA^a.$$

By the definition of sign

$$Y_- \operatorname{sign}(Z_-) \stackrel{\circ}{=} |Z_-| \operatorname{sign}(Z_-) = Z_-. \tag{6.58}$$

Therefore

$$\int_0^t Z_- dZ = \int_0^t Y_- dY = \int_0^t Z_- dZ + \int_0^t Y_- dA^a.$$

Hence, by the definition of $L(a, t, \omega)$

$$0 = \int_0^t Y_- dA^a = \tag{6.59}$$

$$[4pt] = \int_0^t Y_- dL^a + \sum_{0 < s \le t} Y(s-) (\Delta |Z(s)| - \operatorname{sign}(Z(s-)) \Delta Z(t)).$$

Observe that by (6.58) the expression after the sum is finite and has the form

$$|a|(|b| - |a|) - a(b - a) = |a||b| - a^2 - ab + a^2 =$$
$$= |ab| - ab \ge 0.$$

L^a is increasing, therefore the integral $\int_0^t Y_- dL^a$ is non-negative. This implies that the sum and the integral in (6.59) are zero. But as the integral is zero the support of the measure generated by L^a is part of the set

$$\{Y(s-) = 0\} \stackrel{\circ}{=} \{|X(s-) - a| = 0\} = \{X(s-) = a\}. \qquad \square$$

Example 6.73 If L is the local time of a Wiener process and τ_b is the first passage time of a point b and $0 \le a < b$ then $L(a, \tau_b)$ has an exponential distribution with parameter[70] $\lambda \stackrel{\circ}{=} (2(b-a))^{-1}$.

We show that the Laplace transform of the random variable $L(a, \tau_b)$ is

$$l(s) \stackrel{\circ}{=} \mathbf{E}(\exp(-s \cdot L(a, \tau_b))) = \frac{1}{1 + 2s \cdot (b-a)}. \tag{6.60}$$

[70]See: Example 6.71, page 428.

As the Laplace transform of an exponentially distributed random variable is

$$\frac{1}{1 + s/\lambda}$$

this implies the statement.

1. The main idea of the proof is to show that

$$X(t) \stackrel{\circ}{=} \left(\frac{1}{2} + s \cdot (w(t) - a)^+\right) \exp\left(-s \cdot L(a, t)\right)$$

is a local martingale. As

$$X^{\tau_b} = \left(\frac{1}{2} + s \cdot (w^{\tau_b} - a)^+\right) \exp\left(-s \cdot L^{\tau_b}(a)\right) \tag{6.61}$$

is bounded, X^{τ_b} is a bounded local martingale. Hence (6.61) is a uniformly integrable martingale. Therefore by the Optional Sampling Theorem as $0 \le a < b$

$$\frac{1}{2} = \mathbf{E}\left(\left(\frac{1}{2} + s \cdot (w(0) - a)^+\right) \exp\left(-s \cdot L(a, 0)\right)\right) =$$

$$= \mathbf{E}\left(\left(\frac{1}{2} + s \cdot (w(\tau_b) - a)^+\right) \exp\left(-s \cdot L(a, \tau_b)\right)\right) =$$

$$= \left(\frac{1}{2} + s \cdot (b - a)\right) l(s),$$

from which (6.60) is trivial.

2. Let us return to process X. Let

$$U(t) \stackrel{\circ}{=} \frac{1}{2} + s \cdot (w(t) - a)^+, \quad V(t) \stackrel{\circ}{=} \exp\left(-sL(a, t)\right).$$

Integrating by parts

$$X(t) = U(t) V(t) = X(0) + \int_0^t U \, dV + \int_0^t V \, dU + [U, V].$$

U is continuous, V has finite variation so $[U, V] = 0$. By the previous theorem the support of the measure generated by V is in $\{w = a\}$, so

$$\int_0^t U \, dV = \left(\frac{1}{2} + s \cdot (a - a)^+\right)(V(t) - V(0)) = \frac{1}{2}(V(t) - 1).$$

By Tanaka's formula

$$U(t) \stackrel{\circ}{=} H(t) + \frac{1}{2} s \cdot L(a, t),$$

where H is a continuous local martingale. V is continuous so it is locally bounded so $Z(t) \stackrel{\circ}{=} \int_0^t V dH$ is a local martingale. On the other hand, by the Fundamental Theorem of Calculus[71]

$$\int_0^t V d\left(\frac{1}{2} s \cdot L\right) = \frac{s}{2} \int_0^t \exp(-s \cdot L(a, u)) L(a, du) =$$

$$= \frac{s}{2} \left[\frac{\exp(-s \cdot L(a, u))}{-s}\right]_0^t =$$

$$= -\frac{1}{2} (\exp(-s \cdot L(a, u)) - 1) = -\frac{1}{2} (V(t) - 1).$$

Hence

$$X(t) = X(0) + Z(t) + \frac{1}{2}(V(t) - 1) - \frac{1}{2}(V(t) - 1) =$$

$$= X(0) + Z(t),$$

that is, X is a local martingale. $\qquad\square$

Theorem 6.74 (Meyer–Itô formula) *Let X be a semimartingale and let f be a convex function. If $f' \stackrel{\circ}{=} f'_-$ denotes the left derivative of f and μ is the second generalized derivative of f and L is the local time of X, then*

$$f(X(t)) - f(X(0)) = \qquad\qquad (6.62)$$

$$= \int_0^t f'(X_-) dX +$$

$$+ \sum_{0 < s \leq t} (f(X(s)) - f(X(s-)) - f'(X(s-)) \Delta X(s)) +$$

$$+ \frac{1}{2} \int_{\mathbb{R}} L(a, t) d\mu(a).$$

Proof. Recall that the second generalized derivative of $|x|$ is $2\delta_0$. So if $f(x) = |x|$, then by the theorem one gets just the definition of local times.

1. Let us first assume that the support of μ is compact. In this case the representation (6.53) holds. If $f(x) = \alpha x + \beta$ then the theorem is trivially true,

---

[71] See: (6.32), page 398. Or, if one likes, by Itô's formula.

therefore one can assume that

$$f(x) = \frac{1}{2} \int_{\mathbb{R}} |x - a| \, d\mu(a).$$

With the Dominated Convergence Theorem one can differentiate under the integral sign

$$f'(x) \overset{\circ}{=} f'_-(x) = \frac{1}{2} \int_{\mathbb{R}} \text{sign}(x - a) \, d\mu(a).$$

If

$$J(a, t) \overset{\circ}{=} \sum_{0 < s \leq t} (|X(s) - a| - |X(s-) - a| - \text{sign}(X(s-) - a) \Delta X(s)),$$

then by the Monotone Convergence Theorem

$$\frac{1}{2} \int_{\mathbb{R}} J(a, t) \, d\mu(a) =$$

$$= \frac{1}{2} \sum_{0 < s \leq t} \int_{\mathbb{R}} (|X(s) - a| - |X(s-) - a| - \text{sign}(X(s-) - a) \Delta X(s)) \, d\mu(a)$$

$$= \sum_{0 < s \leq t} (f(X(s)) - f(X(s-)) - f'(X(s-)) \Delta X(s)).$$

Similarly if

$$H(a, t) \overset{\circ}{=} |X(t) - a| - |X(0) - a|,$$

then

$$f(X(t)) - f(X(0)) = \frac{1}{2} \int_{\mathbf{R}} H(a, t) \, d\mu(a).$$

Let

$$Z(a, t) \overset{\circ}{=} \int_0^t \text{sign}(X(s-) - a) \, dX(s)$$

and let us take a $\mathcal{B}(\mathbb{R}) \times \mathcal{B}(\mathbb{R}_+) \times \mathcal{A}$ measurable version of this parametric integral[72]. By Fubini's theorem for stochastic integrals[73]

$$\frac{1}{2} \int_{\mathbb{R}} Z(a,t) \, d\mu(a) = \int_0^t \frac{1}{2} \int_{\mathbb{R}} \operatorname{sign}(X(s-) - a) \, d\mu(a) \, dX(s) =$$

$$= \int_0^t f'(X(s-)) \, dX(s).$$

By the definition of local times $L = H - J - Z$, that is

$$\frac{1}{2} H = \frac{1}{2} J + \frac{1}{2} Z + \frac{1}{2} L.$$

Integrating by μ and using the already proved formulas one can easily prove the theorem.

2. Let us take the general case and let

$$f_n(x) \doteq \begin{cases} f(-n) + f'(-n)(x+n) & \text{if} & x \le -n \\ f(x) & \text{if} & -n < x < n \\ f(n) + f'(n)(x-n) & \text{if} & x \ge n \end{cases}.$$

f_n is also convex. Let μ_n be the generalized second derivative of f_n. Obviously the support of μ_n is in $[-n, n]$ and the measure μ_n is finite. Hence we can use the already proved part of the theorem. Let

$$\tau_n \doteq \inf \{t : |X(t)| \ge n\},$$

and let us consider the stopped processes X^{τ_m}. By the already proved part of the theorem

$$f_n(X^{\tau_n}(t)) - f_n(X^{\tau_n}(0)) =$$

$$= \int_0^t f'_n(X^{\tau_n}(s-)) \, dX^{\tau_n}(s) +$$

$$+ \sum_{0 < s \le t} (\Delta f_n(X^{\tau_n}(s)) - f'_n(X^{\tau_n}(s-)) \Delta X^{\tau_n}(s)) +$$

$$+ \frac{1}{2} \int_{\mathbb{R}} L_n(a,t) \, d\mu_n(a),$$

where obviously $L_n(a)$ denotes the local time of X^{τ_n}. Observe that $|X^{\tau_n}| \le n$ on $[0, \tau_n)$. Therefore on $[0, \tau_n)$ one can write f instead f_n. The support of the measure generated by $L_n(a)$ is in the set $\{X^{\tau_n}(s-) = a\}$, that is if $|a| \ge n$, then

---

[72]See: Proposition 5.23, page 319.
[73]See: Theorem 5.25, page 322.

$L_n(a, t) = 0$ for all t. The measure μ and μ_n are equal on the interval $[-n, n]$ so in the integral containing the local time one can write μ instead of μ_n. That is

$$\int_{\mathbb{R}} L_n(a, t) \, d\mu_n(a) = \int_{\mathbb{R}} L_n(a, t) \, d\mu(a).$$

From the definition of the local time it is evident that the local time of X^{τ_n} is L^{τ_n}. Hence if $t \leq \tau_n$, then

$$\int_{\mathbb{R}} L_n(a, t) \, d\mu_n(a) = \int_{\mathbb{R}} L_n(a, t) \, d\mu(a) = \int_{\mathbb{R}} L^{\tau_n}(a, t) \, d\mu(a) =$$

$$= \int_{\mathbb{R}} L(a, t) \, d\mu(a).$$

If $n \to \infty$, then $\tau_n \nearrow \infty$, and the theorem holds in the general case as well. $\qquad \square$

Corollary 6.75 (Occupation Times Formula) *If X is a semimartingale and L is the local time of X then for every bounded Borel measurable function $g : \mathbb{R} \to \mathbb{R}$ and for all t for almost all outcomes*

$$\int_{\mathbb{R}} L(a, t) g(a) \, da = \int_0^t g(X(s-)) \, d[X^c](s). \qquad (6.63)$$

The identity is meaningful and it is also valid if g is a non-negative Borel measurable function.

Proof. Let f be convex and let $f \in C^2$. In this case one can use Itô's formula. Comparing Itô's formula with (6.62)

$$\int_{\mathbb{R}} L(a, t, \omega) f''(a) \, da = \int_0^t f''(X(s-)) \, d[X^c].$$

Of course instead of f'' one can write any g non-negative, continuous function. By the Monotone Class Theorem the identity is valid for every bounded Borel measurable function. With the Monotone Convergence Theorem one can extend the identity to non-negative Borel measurable functions. $\qquad \square$

Let $X = w$ be a Wiener process. In this case $[X](s) = s$ and by (6.63) for every Borel measurable set B

$$\int_B L(a, t) \, da = \int_0^t \chi(w(s) \in B) \, ds = \lambda(s \leq t : w(s) \in B).$$

The last variable gives the time w is in the set B. For fixed t this occupation time is a measure on the time-line and $L(a, t)$ is the Radon–Nikodym derivative of this occupation time measure. By the interpretation of the density functions $L(a,t)\,da$ is the time w is around a during the time interval $[0,t]$.

Corollary 6.76 *If X is a semimartingale and L is the local time of X then*

$$[X^c](t) = \int_{\mathbb{R}} L(a,t)\,da.$$

Corollary 6.77 (Meyer–Tanaka formula) *If X is a continuous semimartingale and L denotes the local time of X then*[74]

$$|X| = |X(0)| + \operatorname{sign}(X) \bullet X + L(0).$$

By Itô's formula and by the Itô–Meyer formula the class of semimartingales is closed for a quite broad class of transformations. That is why the next example is interesting.

Example 6.78 If $X \neq 0$ is a continuous local martingale, $X(0) = 0$ and $0 < \alpha < 1$ then $|X|^\alpha$ is not a semimartingale.

---

1. The example is a bit surprising because $|X|$ is a semimartingale and by the Itô–Meyer formula a concave function of a semimartingale is again a semimartingale. But recall that in Theorem 6.65 the domain of definition of F is the whole real line, or at least an open convex set containing the range of X. Now this is not true. Let us also observe that the function $|x|^\alpha$ is not concave on the whole line.

2. Let L be the local time of X. Assume that $L(0) \equiv 0$. By the Meyer–Tanaka formula

$$|X| = \operatorname{sign}(X) \bullet X + L(0) = \operatorname{sign}(X) \bullet X.$$

On the right-hand side the integral is a local martingale, hence $|X|$ is a non-negative local martingale so by Fatou's lemma it is a supermartingale[75]. As $|X|(0) = 0$

$$0 = \mathbf{E}(|X(0)|) \geq \mathbf{E}(|X(t)|),$$

which implies that if $L(0) \equiv 0$ then $|X| = 0$.

3. Now we prove that if $Y \stackrel{\circ}{=} |X|^\alpha$ is a semimartingale then $L(0) \equiv 0$. With localization one can assume that $X \in \mathcal{H}_0^2$. The support of $L(0)$ is in $\{X(s) = 0\}$

---

[74] Obviously $L(0)$ denotes the process $t \mapsto L(0,t)$.
[75] See: page 386.

so by the Meyer–Tanaka formula

$$L(0,t) = \int_0^t 1 dL = \int_0^t \chi(X(s) = 0) dL(0,s) =$$

$$= \int_0^t \chi(X(s) = 0) d|X(s)| - \int_0^t \chi(X(s) = 0) \operatorname{sign}(X(s)) dX(s).$$

Let us first investigate the second integral

$$Z(t) \stackrel{\circ}{=} \int_0^t \chi(X(s) = 0) \operatorname{sign}(X(s)) dX(s).$$

By Itô's isometry and by (6.63)

$$\mathbf{E}(Z^2(t)) = \mathbf{E}\left(\int_0^t \chi(X(s) = 0) d[X](s)\right) =$$

$$= \mathbf{E}\left(\int_{\mathbb{R}} \chi(\{0\})(a) L(a,t) da\right) = \mathbf{E}\left(\int_{\{0\}} L(a,t) da\right) = 0$$

hence $Z = 0$. Now let us calculate the first integral

$$\int_0^t \chi(X(s) = 0) d|X(s)|.$$

$0 < \alpha < 1$ so $\beta \stackrel{\circ}{=} 1/\alpha > 1$. If $\beta \geq 2$ then by Itô' formula for C^2 functions

$$|X| = Y^\beta = \beta Y^{\beta-1} \bullet Y + \frac{\beta(\beta-1)}{2} Y^{\beta-2} \bullet [Y].$$

Using that $\{X(s) = 0\} = \{Y(s) = 0\}$

$$I(t) \stackrel{\circ}{=} \int_0^t \chi(Y(s) = 0) d|X(s)| =$$

$$= \beta \int_0^t \chi(Y(s) = 0) Y^{\beta-1} dY +$$

$$+ \frac{\beta(\beta-1)}{2} \int_0^t \chi(Y(s) = 0) Y^{\beta-2} d[Y].$$

The integrand in the first integral is zero, so the integral is zero. If $\beta > 2$ then the integrand in the second integral is also zero, so the second integral is zero again. If $\beta = 2$, then using (6.63)

$$\int_0^t \chi(Y(s) = 0) d[Y] = \int_{\mathbb{R}} L(a,t) \chi(\{0\}) da = \int_{\{0\}} L(a,t) da = 0.$$

Let $2 > \beta > 1$. The function

$$g(x) \overset{\circ}{=} \begin{cases} x^\beta & \text{if} \quad x > 0 \\ 0 & \text{if} \quad x \leq 0 \end{cases}$$

is a convex function on $\mathbb{R}$. Hence by Itô's formula for convex functions

$$|X| = g(Y) = Y^\beta = g'(Y) \bullet Y + \frac{1}{2} \int_{\mathbb{R}} H(a) \, d\mu(a),$$

where H is the local time of Y. In this case again

$$\int_0^t \chi(X = 0) \, g'(Y) \, dY = \int_0^t \chi(X = 0) \, \beta Y^{\beta-1} dY = 0.$$

Let us calculate the integral

$$\int_0^t \chi(Y(s) = 0) \, d \left(\int_{\mathbb{R}} H(a, s) \, d\mu(a) \right). \tag{6.64}$$

μ is defined by the increasing function

$$g'_-(x) \overset{\circ}{=} \begin{cases} \beta x^{\beta-1} & \text{if} \quad x > 0 \\ 0 & \text{if} \quad x \leq 0 \end{cases} = \int_{-\infty}^x h(t) \, dt,$$

where

$$h(x) \overset{\circ}{=} \begin{cases} \beta(\beta-1) x^{\beta-2} & \text{if} \quad x > 0 \\ 0 & \text{if} \quad x \leq 0 \end{cases}.$$

H is the local time of Y so

$$\int_{\mathbb{R}} H(a, s) \, d\mu(a) = \int_0^\infty H(a, s) \, h(a) \, da = \int_0^s h(Y) \, d[Y],$$

therefore (6.64) is

$$\int_0^t \chi(Y(s) = 0) \, h(Y) \, d[Y] = 0.$$

This means that if Y is a semimartingale then $L(0) = 0$, hence $X = 0$.   □

6.5.4   Local times of continuous semimartingales

Observe that for every a the local time $L(a, t, \omega)$ is defined only up to indistinguishability. This means that for every a one can modify $L(a, t, \omega)$ on a set with probability zero. The local time is always continuous in parameter t so one can think about L as an $C([0, \infty))$ valued stochastic process: $(a, \omega) \mapsto L(a, \omega)$, where $L(a,\omega)$ denotes the trajectory of L in t. As this function valued process is defined only almost surely one can use any of its modification as local time. In this subsection we prove that under some restrictions on semimartingale X, the process $L(a,t,\omega)$ has a version which is right-regular in a. To do this we shall use the next result:

Proposition 6.79 (Kolmogorov's criteria) *Let I be an interval in $\mathbb{R}$ and let X be a Banach space valued stochastic process on I. If for some positive constants a, b and c*

$$\mathbf{E}\left(\|X(u) - X(v)\|^a\right) \le c\|u - v\|^{1+b},$$

then X has a continuous modification.

Proposition 6.80 *If X is a continuous local martingale then the local time $L(a,t,\omega)$ of X has a modification in a which is continuous in (a,t).*

Proof. One can localize the proposition as if L is the local time of X and τ is a stopping time then the local time of X^τ is L^τ. Therefore one can assume that $X - X(0) \in \mathcal{H}_0^2$. By definition

$$L(a,t) = |X(t) - a| - |X(0) - a| - \int_0^t \text{sign}(X(s) - a)\,dX(s).$$

Let us introduce the notation[76]

$$\widehat{M}(a,u) \stackrel{\circ}{=} \int_0^u \text{sign}(X(s) - a)\,dX^c(s).$$

It is sufficient to show that $\widehat{M}$ has a continuous version. We want to apply Kolmogorov's criterion. $C([0,t])$ is a Banach space for arbitrary fix t. Obviously if a function $g : I \to C([0,t])$ is continuous then it defines a continuous function over $I \times [0,t]$. We show that for all t

$$\mathbf{E}\left(\left\|\widehat{M}(a) - \widehat{M}(b)\right\|^4_{C([0,t])}\right) \stackrel{\circ}{=} \mathbf{E}\left(\sup_{s \le t}\left|\widehat{M}(a,s) - \widehat{M}(b,s)\right|^4\right) \le \qquad (6.65)$$

$$\le k \cdot |a - b|^2.$$

[76]Of course now instead of X^c one can write X. But later we shall re-use this part of the proof in a bit different situation.

By Burkholder's and by Jensen's inequality, using the Occupation Times Formula

$$\mathbf{E}\left(\sup_{s \leq t} \left|\widehat{M}(a, s) - \widehat{M}(b, s)\right|^4\right) \leq c \cdot \mathbf{E}\left(\left[\widehat{M}(a, t) - \widehat{M}(b, t)\right]^2\right) = \quad (6.66)$$

$$= c \cdot \mathbf{E}\left(\left(\int_0^t 4\chi\left(a < X(s) \leq b\right) d\left[X^c\right](s)\right)^2\right) =$$

$$= 4c \cdot \mathbf{E}\left(\left(\int_a^b L(x, t)\, dx\right)^2\right) =$$

$$= 4c \cdot (b - a)^2\, \mathbf{E}\left(\left(\int_a^b L(x, t)\frac{dx}{b - a}\right)^2\right) \leq$$

$$\leq 4c \cdot (b - a)^2\, \mathbf{E}\left(\left(\frac{1}{b - a}\int_a^b L^2(x, t)\, dx\right)\right).$$

Changing the integrals by Fubini's theorem one can estimate the last line with the following expression:

$$4c \cdot (b - a)^2 \sup_x \mathbf{E}\left(L^2(x, t)\right). \quad (6.67)$$

Using the definition of the local times and the elementary inequalities

$$||X(t) - a| - |X(0) - a|| \leq |X(t) - X(0)|.$$

$$(z_1 - z_2)^2 \leq 2\left(z_1^2 + z_2^2\right)$$

$$L^2(x, t) \leq 2\left(X(t) - X(0)\right)^2 + 2\left(\int_0^t \text{sign}\left(X(s) - x\right) dX(s)\right)^2.$$

One can estimate the expected value in (6.67) by

$$2\|X - X(0)\|_{\mathcal{H}^2}^2 + 2\|\text{sign}\left(X - x\right) \bullet X\|_{\mathcal{H}^2}^2.$$

By Itô's isometry

$$\|\text{sign}\left(X - x\right) \bullet X\|_{\mathcal{H}^2}^2 = \mathbf{E}\left(\int_0^\infty 1\, d\left[X\right]\right) =$$

$$= \|1 \bullet X\|_{\mathcal{H}^2}^2 = \|X - X(0)\|_{\mathcal{H}^2}^2,$$

so the estimation of $\mathbf{E}\left(L^2\left(x,t\right)\right)$ is independent of x. So by (6.67) inequality (6.65) follows. $\qquad\square$

Definition 6.81 *If X is a continuous local martingale then $L\left(a,t,w\right)$ denotes the version which is continuous in $\left(a,t\right)$.*

Corollary 6.82 *If X is a continuous local martingale then almost surely for every value of parameters a and t*

$$L\left(a,t\right)=\lim_{\varepsilon\searrow 0}\frac{1}{2\varepsilon}\int_0^t\chi\left(-\varepsilon+a<X\left(s\right)<a+\varepsilon\right)d\left[X\right]\left(s\right).\qquad(6.68)$$

Proof. By the occupation times formula for any interval I

$$\frac{1}{\lambda\left(I\right)}\int_0^t\chi_I\left(X\left(s\right)\right)d\left[X\right]\left(s\right)=\frac{1}{\lambda\left(I\right)}\int_{\mathbb{R}}L\left(a,t\right)\chi_I\left(a\right)da=$$

$$=\frac{1}{\lambda\left(I\right)}\int_I L\left(a,t\right)da.$$

L is continuous in a hence if $a_0\in I$ and $\lambda\left(I\right)\to 0$ then

$$\frac{1}{\lambda\left(I\right)}\int_I L\left(a,t\right)da\to L\left(a_0,t\right),$$

from which (6.68) is evident. $\qquad\square$

Corollary 6.83 *If w is a Wiener process then the occupation time measure*

$$\mu_t\left(B\right)\overset{\circ}{=}\lambda\left(s\le t:w\left(s\right)\in B\right)$$

almost surely has a differentiable distribution function and the derivative of this function is $L\left(a,t\right)$.

Definition 6.84 *A semimartingale X satisfies the so-called hypothesis A if for every t almost surely*

$$\sum_{0<s\le t}\left|\Delta X\left(s\right)\right|<\infty.$$

Proposition 6.85 *If semimartingale X satisfies hypothesis A then the local time $L\left(a,t,w\right)$ has a $\mathcal{B}\left(\mathbb{R}\right)\times\mathcal{P}$-measurable equivalent modification which is almost surely continuous in t and right-regular in a.*

Proof. If X satisfies hypothesis A then process ΔX has finite variation. In this case $X - \sum \Delta X$ is meaningful and it is a continuous semimartingale. Let $J \doteq \sum \Delta X$. As $Y \doteq X - J$ is a continuous semimartingale it has a unique decomposition $M + V$, where M is a continuous local martingale, V is a continuous process with finite variation. By the definition of local times

$$|X(t) - a| = |X(0) - a| +$$
$$+ \int_0^t \text{sign}(X(s-) - a)\, dX(s) +$$
$$+ \sum_{0 < s \leq t} (\Delta |X(s) - a| - \text{sign}(X(s-) - a)\,\Delta X(s)) +$$
$$+ L(a, t).$$

For every s by the triangle inequality

$$|\Delta |X(s) - a|| \leq |\Delta X(s)|. \tag{6.69}$$

Therefore by hypothesis A the sums

$$\sum_{0 < s \leq t} \text{sign}(X(s-) - a)\,\Delta X(s)$$

and

$$\sum_{0 < s \leq t} \Delta |X(s) - a|$$

are finite. Hence one can separate the terms in

$$\sum_{0 < s \leq t} (\Delta |X(s) - a| - \text{sign}(X(s-) - a)\,\Delta X(s)). \tag{6.70}$$

For every semimartingale Z let

$$\widehat{Z}(a, t) \doteq \int_0^t \text{sign}(X(s-) - a)\, dZ(s).$$

Observe that the second term of the sum (6.70) is $-\widehat{J}(a, t)$. Using the decomposition $X \doteq M + V + J$

$$|X(t) - a| = |X(0) - a| + \widehat{M}(a, t) + \widehat{V}(a, t) + \widehat{J}(a, t) - \widehat{J}(a, t) +$$
$$+ \sum_{0 < s \leq t} \Delta |X(s) - a| + L(a, t),$$

that is

$$L(a,t) = |X(t) - a| - |X(0) - a| - \widehat{M}(a,t) - \widehat{V}(a,t) - \qquad (6.71)$$

$$- \sum_{0 < s \le t} \Delta |X(s) - a|.$$

By (6.69) and by hypothesis A $\Delta |X(s) - a|$ is continuous by a and it is dominated by an integrable variable with respect to the counting measure. By the Dominated Convergence Theorem

$$\lim_{u \to a} \sum_{0 < s \le t} \Delta |X(s) - u| = \sum_{0 < s \le t} \Delta |X(s) - a|,$$

so the sum is continuous with respect to a. One should show that the proposition is valid for $\widehat{M}(a,t)$ and $\widehat{V}(a,t)$. V has finite variation on any finite interval, the bounded function $\text{sign}(X(s-) - u)$ is right-regular with respect to u. By the Dominated Convergence Theorem $\widehat{V}$ is right-regular with respect to a. Finally let us consider $\widehat{M}$. The continuous part of semimartingale X is $X^c = M$, so repeating the proof of the previous proposition one can easily prove that $M(a,t)$ has a continuous version $\widehat{M}(a,t)$. $\qquad\square$

Corollary 6.86 *If a semimartingale X satisfies hypothesis A and if $M + V$ is the decomposition of $X - \sum \Delta X$ then*

$$\Delta L(a,t) \stackrel{\circ}{=} L(a,t) - L(a-,t-) = L(a,t) - L(a-,t) =$$

$$= 2 \int_0^t \chi(X(s-) = a) \, dV(s) =$$

$$= 2 \int_0^t \chi(X(s) = a) \, dV(s).$$

Proof. By the proof of the previous proposition only $\widehat{V}(a,t)$ is not continuous so

$$\Delta L(a) = -\Delta \widehat{V}(a) = - \int_0^t \text{sign}(X(s-) - a) - \text{sign}(X(s-) - a-) \, dV(s) =$$

$$= 2 \int_0^t \chi(X(s-) = a) \, dV(s).$$

V continuous and $X(s-) = X(s)$ outside countable number of points s, so

$$2 \int_0^t \chi(X(s-) = a) \, dV(s) = 2 \int_0^t \chi(X(s) = a) \, dV(s). \qquad\square$$

Example 6.87 Even for continuous semimartingales the local time can be discontinuous.

1. Let w be a Wiener process and let $X \stackrel{\circ}{=} |w|$. As the support of the measure generated by $L(a)$ is in the set $\{X = a\}$ if $a < 0$, then $L(a, t) = 0$. Let $a = 0$. L is right-continuous in parameter a therefore using the occupation times formula

$$L(0, t) = \lim_{\varepsilon \searrow 0} \frac{1}{\varepsilon} \int_0^\varepsilon L(a, t) \, da = \lim_{\varepsilon \searrow 0} \frac{1}{\varepsilon} \int_{\mathbb{R}} \chi(0 \le a < \varepsilon) L(a, t) \, da =$$

$$= \lim_{\varepsilon \searrow 0} \frac{1}{\varepsilon} \int_0^t \chi(|w| < \varepsilon) \, d[[w]].$$

By Tanaka's formula

$$|w| = \operatorname{sign}(w) \bullet w + L_w(0).$$

$L_w(0)$ is continuous and increasing so $[|w|] = [\operatorname{sign}(w) \bullet w] = [w]$. Hence using again that L_w is continuous

$$L(0, t) = \lim_{\varepsilon \searrow 0} \frac{1}{\varepsilon} \int_0^t \chi(-\varepsilon < w < \varepsilon) \, d[w] =$$

$$= \lim_{\varepsilon \searrow 0} \frac{1}{\varepsilon} \int_\varepsilon^\varepsilon L_w(a) \, da = 2L_w(0) \ne 0.$$

This implies that the local time $L(a, t)$ is not left-continuous in parameter a.

2. On the other hand it is interesting to discuss the case $a > 0$. Again by the right-continuity

$$L(a, t) = \lim_{\varepsilon \searrow 0} \frac{1}{\varepsilon} \int_0^t \chi(|w| \in [a, a + \varepsilon))(s) \, ds =$$

$$= \lim_{\varepsilon \searrow 0} \frac{1}{\varepsilon} \int_0^t \chi(w \in [a, a + \varepsilon))(s) \, ds +$$

$$+ \lim_{\varepsilon \searrow 0} \frac{1}{\varepsilon} \int_0^t \chi(-w \in [a, a + \varepsilon))(s) \, ds.$$

The first limit is $L_w(a, t)$ and the second is $L_w(-a, t)$. Hence

$$L(a, t) = L_w(-a, t) + L_w(a, t).$$

This expression is continuous on the set $a \ge 0$. $\qquad\square$

6.5.5 Local time of Wiener processes

In this subsection we shall investigate the local times of Wiener processes.

Definition 6.88 *If w is a Wiener process then L denotes the local time of w at point $a = 0$. That is $L \overset{\circ}{=} L_w(0)$. We shall very often refer to L as the local time of w.*

Example 6.89 Tanaka's formula for Wiener processes.

If w is a Wiener process and $L \overset{\circ}{=} L_w(0)$ is the local time of w then by Tanaka's formula

$$|w| = \text{sign}(w) \bullet w + L \overset{\circ}{=} \beta + L. \tag{6.72}$$

$\text{sign}(w) \bullet w$ is a continuous local martingale with quadratic variation $(\text{sign}(w))^2 \bullet [w] = [w]$. By Lévy's characterization theorem[77] $\beta \overset{\circ}{=} \text{sign}(w) \bullet w$ is also a Wiener process. □

Our goal is to describe the distribution of L. To do this we shall need the next simple lemma:

Lemma 6.90 (Skorohod) *If y is a continuous function defined on $\mathbb{R}_+$ and $y(0) \geq 0$ then there are functions on $\mathbb{R}_+$ denoted by z and a for which:*

1. *$z = y + a$,*
2. *z is non-negative,*
3. *a is increasing, continuous and $a(0) = 0$ and the support of the measure generated by a is in the set $\{z = 0\}$.*

Functions a and z are unique and

$$a(t) = \sup_{s \leq t} \{y^-(s)\} \overset{\circ}{=} \sup_{s \leq t} \max(-y(s), 0). \tag{6.73}$$

Proof. First we show that the decomposition is unique. Let (a_1, z_1) and (a_2, z_2) be two decompositions satisfying the conditions of the lemma.

$$y = z_1 - a_1 = z_2 - a_2,$$

---
[77]See: Theorem 6.13, page 368.

so $z_1 - z_2 = a_1 - a_2$. As a_1 and a_2 are increasing $z_1 - z_2$ and $a_1 - a_2$ have finite variation. Integrating by parts

$$0 \leq (z_1 - z_2)^2 (t) = 2 \int_0^t z_1 (s) - z_2 (s) \, d (z_1 - z_2) (s) =$$

$$= 2 \int_0^t z_1 (s) - z_2 (s) \, d (a_1 - a_2) (s) .$$

By the assumption about the support of measures generated by functions a_1 and a_2 and as $z_1 \geq 0$ and $z_2 \geq 0$ the last integral is

$$-2 \int_0^t z_1 (s) \, da_2 - 2 \int_0^t z_2 da_1 \leq 0.$$

Hence $z_1 = z_2$. As a second step we show that a in (6.73) and $z \stackrel{\circ}{=} y + a$ satisfy the conditions of the lemma. a is trivially increasing. By the assumptions y is continuous, hence y^- is also continuous. It is easy to show that a is continuous. For every t

$$z (t) \stackrel{\circ}{=} y (t) + a (t) \geq y (t) + y^- (t) = y^+ (t) \geq 0.$$

One should prove that the support of the measure generated by a is in the set $\{z = 0\}$, that is

$$\int_{\mathbb{R}_+} \chi (z > 0) \, da = \lim_{n \to \infty} \int_{\mathbb{R}_+} \chi \left(z > \frac{1}{n} \right) \, da = 0.$$

This means that one should prove that for every $\varepsilon > 0$

$$\int_{\mathbb{R}_+} \chi (z > \varepsilon) \, da = 0.$$

z is continuous, hence for every $\varepsilon > 0$ the set $\{z > \varepsilon\}$ is open, hence $\{z > \varepsilon\}$ is a union of countable number of open intervals. Let (u, v) be one of these intervals. It is sufficient to prove that $a (v) = a (u)$. If $s \in (u, v)$ then

$$-y (s) \stackrel{\circ}{=} a (s) - z (s) \leq a (v) - \varepsilon.$$

From this

$$a (v) = \max \left(a (u), \ \sup_{u \leq s \leq v} y^- (s) \right) \leq \max (a (u), a (v) - \varepsilon) .$$

This can happen only if $a (v) \leq a (u)$, that is $a (v) = a (u)$. $\qquad \square$

Proposition 6.91 *The distribution of* $L(t,\omega) \overset{\circ}{=} L(0,t,\omega)$ *is the same as the distribution of the maximum of a Wiener process on the interval* $[0,t]$. *Hence the density function of* $L(t)$ *is*

$$f_t(x) \overset{\circ}{=} \frac{2}{\sqrt{2\pi t}} \exp\left(-\frac{x^2}{2t}\right), \quad x > 0.$$

Proof. By Tanaka's formula

$$|w| = \beta + L,$$

where β is a Wiener process and the two sides are equal up to indistinguishability. The support of the measure generated by L is in the set $\{|w| = 0\}$. Hence by Skorohod's lemma

$$L(t) \overset{a.s.}{=} \sup_{s \leq t} \beta^-(s) = \sup_{s \leq t}(-\beta(s)) \overset{\circ}{=} S_{-\beta}(t), \tag{6.74}$$

from which by the symmetry of Wiener process the proposition is evident[78]. $\square$

Proposition 6.92 *The augmented filtration generated by* $\beta \overset{\circ}{=} \mathrm{sign}(w) \bullet w$ *is the same as the augmented filtration generated by* $|w|$.

Proof. Let $\mathcal{F}^\beta$ and $\mathcal{F}^{|w|}$ be the augmented filtration generated by β and by $|w|$. By (6.74) L is adapted with respect $\mathcal{F}^\beta$. By Tanaka's formula $|w|$ is $\mathcal{F}^\beta$-adapted. Hence $\mathcal{F}^{|w|} \subseteq \mathcal{F}^\beta$. On the other hand for Wiener processes $L(a,t)$ is almost surely continuous in a so by (6.68) and by the occupation times formula

$$L(t) = \lim_{\varepsilon \searrow 0} \frac{1}{2\varepsilon} \int_{-\varepsilon}^{\varepsilon} L(a,t,\omega)\, da =$$

$$= \lim_{\varepsilon \searrow 0} \frac{1}{2\varepsilon} \int_{\mathbb{R}} L(a,t,\omega)\, \chi((-\varepsilon,\varepsilon))(a)\, da =$$

$$= \lim_{\varepsilon \searrow 0} \frac{1}{2\varepsilon} \int_0^t \chi(|w(s)| < \varepsilon)\, ds.$$

Hence L is $\mathcal{F}^{|w|}$-adapted. Therefore β is $\mathcal{F}^{|w|}$-adapted, so $\mathcal{F}^\beta \subseteq \mathcal{F}^{|w|}$. $\square$

Proposition 6.93 *If* $L(a,\infty,\omega)$ *denote the limit* $\lim_{t\to\infty} L(a,t,\omega)$ *then for every* a

$$\mathbf{P}(L(a,\infty) = \infty) = 1.$$

---

[78]See: Example 1.123, page 87 and Proposition B.7, page 564.

Proof. By definition

$$|w(t) - a| \overset{\circ}{=} |a| + \beta(t) + L(a,t).$$

where

$$\beta \overset{\circ}{=} \operatorname{sign}(w - a) \bullet w.$$

By Lévy's theorem β is a Wiener process. Again by Skorohod's lemma

$$L(a,t) = \sup_{s \leq t} (\beta(t) + |a|)^-.$$

Hence

$$\mathbf{P}(L(a,\infty) = \infty) = 1. \qquad \qquad \square$$

Finally we show that for Wiener processes the support of the measure generated by $t \mapsto L(t,\omega)$ is not only almost surely in the set

$$\mathcal{Z}(\omega) \overset{\circ}{=} \{t : w(t,\omega) = 0\}$$

but the two sets are almost surely equal.

Proposition 6.94 *For almost all outcome ω the set $\mathcal{Z}(\omega)$ is closed and has empty interior.*

Proof. The trajectories of Wiener processes are continuous which immediately implies that $\mathcal{Z}(\omega)$ is closed. We show that almost surely the Lebesgue measure of $\mathcal{Z}(\omega)$ is zero. This will imply that $\mathcal{Z}(\omega)$ does not contain a segment with positive length. By Fubini's theorem, using that for every $t > 0$ the value of a Wiener process has non-degenerated Gaussian distribution so $\mathbf{P}(w(t) = 0) = 0$ for every $t > 0$

$$\mathbf{E}(\lambda(\mathcal{Z}(\omega))) = \mathbf{E}\left(\int_0^\infty \chi(\mathcal{Z}(\omega))(t)\,dt\right) =$$
$$= \int_0^\infty \mathbf{E}(\chi(\mathcal{Z}(\omega))(t))\,dt = 0$$

hence $\lambda(\mathcal{Z}(\omega)) = 0$ almost surely. $\qquad \square$

Definition 6.95 *If w is a Wiener process then the intervals in the open set $\mathcal{Z}(\omega)^c = \{|w(\omega)| > 0\}$ are called the excursion intervals of w.*

For every t let

$$\sigma_t (\omega) \doteq \inf \{s > 0 : L(s) \geq t\},$$
$$\rho_t (\omega) \doteq \inf \{s > 0 : L(s) > t\}.$$

σ_t and ρ_t are obviously stopping times. $[\sigma_t, \rho_t]$ is the largest closed interval where L is constantly t. Let

$$\mathcal{O}(\omega) \doteq \cup_t (\sigma_t (\omega), \rho_t (\omega)).$$

$\mathcal{O}(\omega)$ is an open set in $\mathbb{R}$ so by the structure of the open sets of the real line $\mathcal{O}(\omega)$ is the union of maximum countable many disjoint intervals. As L is increasing it is easy to see that if $t_1 \neq t_2$ then

$$\left(\sigma_{t_1} (\omega), \rho_{t_1} (\omega) \right) \cap \left(\sigma_{t_2} (\omega), \rho_{t_2} (\omega) \right) = \emptyset.$$

Hence $\mathcal{O}(\omega)$ is maximum countable union of some intervals $(\sigma_t (\omega), \rho_t (\omega))$. Obviously $\mathcal{O}(\omega)$ is the maximum countable number of intervals where L is constant.

Proposition 6.96 *If w is a Wiener process and L is the local time of w at zero then almost surely $\mathcal{O}(w)$ is the union of the excursion intervals of w, that is*

$$\mathcal{O}(\omega) \overset{a.s}{=} \{|w(\omega)| > 0\} = \mathcal{Z}(\omega)^c.$$

Proof. The proof uses several interesting properties of the Wiener processes.

1. Observe that with probability one the maximum of a Wiener process β on any two disjoint, compact interval is different: If $a < b < c < d < \infty$ then by the definition of the conditional expectation using the independence of the increments

$$\mathbf{P}\left(\sup_{a \leq t \leq b} \beta(t) \neq \sup_{c \leq t \leq d} \beta(t) \right) =$$

$$= \mathbf{P}\left(\sup_{a \leq t \leq b} (\beta(t) - \beta(b)) + \beta(b) \neq \sup_{c \leq t \leq d} (\beta(t) - \beta(c)) + \beta(c) \right) =$$

$$\mathbf{P}\left(\beta(c) - \beta(b) \neq \sup_{a \leq t \leq b} (\beta(t) - \beta(b)) - \sup_{c \leq t \leq d} (\beta(t) - \beta(c)) \right) =$$

$$= \int_{\mathbb{R}} \int_{\mathbb{R}} \mathbf{P}(\beta(c) - \beta(b) \neq x - y) \, dF(x) \, dG(y) =$$

$$= \int_{\mathbb{R}} \int_{\mathbb{R}} 1 \, dF(x) \, dG(y) = 1.$$

Unifying the measure-zero sets one can prove the same result for every interval with rational endpoints.

2. This implies that with probability one every local maximum of a Wiener process has different value.

3. By Tanaka's formula

$$|w| = L - \beta \tag{6.75}$$

for some Wiener process β. Recall that by Skorohod's lemma[79] L is the running maximum of β. This and (6.75) implies that L is constant on any interval[80] where $|w| > 0$. As with probability one, the local maximums of β are different on the flat segments of L with probability one w is not zero. Hence the excursion intervals of w and the flat parts of L are almost surely equal. □

Proposition 6.97 *Let w be a Wiener process. For almost all ω the following three sets are equal[81]:*

1. *the sets of zeros of w;*
2. *the complement of the $\mathcal{O}(\omega)$;*
3. *support of the measure generated by local time $L(\omega)$.*

Proof. Let $\mathcal{S}(\omega)$ denote the support of the measure generated by $L(\omega)$. By definition $\mathcal{S}(\omega)$ is the complement of the largest open set $\mathcal{G}(\omega)$ with $L(\mathcal{G}(\omega)) = 0$. L is constant on the components of $\mathcal{O}$, so $L(\mathcal{O}) = 0$ that is $\mathcal{O}(\omega) \subseteq \mathcal{G}(\omega)$. Hence

$$\mathcal{S}(\omega) \overset{\circ}{=} \mathcal{G}^c(\omega) \subseteq \mathcal{O}^c(\omega).$$

Let I be an open interval with $I \cap \mathcal{O}(\omega) = \emptyset$. If $s_1 < s_2$ are in I then $L(s_1, \omega) = L(s_2, \omega)$ is impossible, so the measure of I with respect to $L(\omega)$ is positive, hence $\mathcal{O}(\omega)$ is the maximal open set with zero measure, that is $\mathcal{O}(\omega) = \mathcal{G}(\omega)$. Hence the equivalence of the last two sets is evident. By the previous proposition $(\mathcal{Z}(\omega))^c = \mathcal{O}(\omega) = \mathcal{S}^c(\omega)$ so $\mathcal{Z}(\omega) = \mathcal{S}(\omega)$. □

6.5.6 Ray–Knight theorem

Let b be an arbitrary number and let τ_b be the hitting time of b. On $[0, b]$ one can define the process

$$Z(a, \omega) \overset{\circ}{=} L(b - a, \tau_b(\omega), \omega), \quad a \in [0, b]. \tag{6.76}$$

---

[79]See: Proposition 6.91, page 447.
[80]See: Proposition 6.97, page 450.
[81]See: Example 7.43, page 494.

If $a > 0$ then $Z(a)$ has an exponential distribution[82] with parameter $\lambda \overset{\circ}{=} 1/(2a)$. In this subsection we try to find some deep reason for this surprising result. Let us first prove some lemmas.

Lemma 6.98 *Let $\mathcal{Z} \overset{\circ}{=} (\mathcal{Z}_a)$ be the filtration generated by (6.76). If $\xi \in L^2(\Omega, \mathcal{Z}_a, \mathbf{P})$, then ξ has the following representation:*

$$\xi = \mathbf{E}(\xi) + \int_0^\infty H \cdot \chi(b \geq w > b - a)\, dw. \tag{6.77}$$

In the representation H is a predictable process and

$$\mathbf{E}\left(\int_0^\infty H^2 \chi(b \geq w > b - a)\, d[w]\right) < \infty.$$

Proof. Let us emphasize that predictability of H means that H is predictable with respect to the filtration $\mathcal{F}$ generated by the underlying Wiener process.

1. Let $\mathcal{U}$ be the set of random variables ξ with representation (6.77).

$$\chi(b \geq w > b - a)$$

is a left-regular process, so the processes

$$U \overset{\circ}{=} H \cdot \chi(b \geq w > b - a), \quad H \in \mathcal{L}^2(w)$$

form a closed subset of $\mathcal{L}^2(w)$. From Itô's isometry it is clear that the random variables satisfying (6.77) form a closed subset of $L^2(\Omega, \mathcal{F}_\infty, \mathbf{P})$. Obviously $\mathcal{Z}_a \subseteq \mathcal{F}_\infty$ and so the set of variables with the given property is a closed subspace of $L^2(\Omega, \mathcal{Z}_a, \mathbf{P})$.

2. Let

$$\eta_g \overset{\circ}{=} \exp\left(-\int_0^a g(s) Z(s)\, ds\right), \quad g \in C_c^1([0, a])$$

where $C_c^1([0, a])$ denotes the set of continuously differentiable functions which are zero outside $[0, a]$. Z is continuous so the σ-algebra generated by the variables η_g is equal $\mathcal{Z}_a$. Let

$$U(t) \overset{\circ}{=} \exp\left(-\int_0^t g(b - w(s))\, ds\right) \overset{\circ}{=} \exp(-K(t)).$$

[82]See: Example 6.73, page 430.

g is bounded so U is bounded. By the Occupation Times Formula

$$\eta_g \overset{\circ}{=} \exp\left(-\int_0^a g\,(s)\,Z\,(s)\,ds\right) \overset{\circ}{=} \exp\left(-\int_0^a g\,(s)\,L\,(b-s,\tau_b)\,ds\right) =$$

$$= \exp\left(-\int_{b-a}^b g\,(b-v)\,L\,(v,\tau_b)\,dv\right) =$$

$$= \exp\left(-\int_{\mathbb{R}} g\,(b-v)\,L\,(v,\tau_b)\,dv\right) =$$

$$= \exp\left(-\int_0^{\tau_b} g\,(b-w\,(v))\,dv\right) = U\,(\tau_b)\,.$$

Let $f \in C^2$,

$$M \overset{\circ}{=} f\,(w)\exp\,(-K) \overset{\circ}{=} f\,(w)\,U.$$

K is continuously differentiable so it has finite variation so by Itô's formula

$$M - M\,(0) = f'\,(w)\,U \bullet w - f\,(w)\,U \bullet K +$$

$$+ \frac{1}{2} U f''\,(w) \bullet [w]\,.$$

Let f' be zero on $(-\infty, b-a]$, $f\,(b) = 1$ and $f''\,(x) = 2g\,(b-x)\,f\,(x)$. The third integral is

$$\frac{1}{2} U f''\,(w) \bullet [w] = U g\,(b-w)\,f\,(w) \bullet [w] = U f\,(w) \bullet K$$

hence the second and the third integrals are the same. Hence

$$M - M\,(0) = f'\,(w)\,U \bullet w.$$

As $f'\,(x) = f'\,(x)\,\chi\,(x > b-a)$

$$\eta_g = U\,(\tau_b) = \frac{M\,(\tau_b)}{f\,(w\,(\tau_b))} = \frac{M\,(\tau_b)}{f\,(b)} = M\,(\tau_b) =$$

$$= M\,(0) + \int_0^{\tau_b} U\,(s)\,f'\,(w\,(s))\,dw\,(s) =$$

$$= M\,(0) + \int_0^{\tau_b} U\,(s)\,f'\,(w\,(s))\,\chi\,(w\,(s) > b-a)\,dw\,(s) \overset{\circ}{=}$$

$$\overset{\circ}{=} \mathbf{E}\,(\eta_g) + \int_0^{\tau_b} H\chi\,(w > b-a)\,dw.$$

So for η_g the representation (6.77) is valid. As (η_g) generates $\mathcal{Z}_a$ and the set of variables for which (6.77) is valid is a closed set the lemma holds. $\qquad\square$

Lemma 6.99 *If the filtration is given by $\mathcal{Z}$ then $Z(a) - 2a$ is a continuous martingale on $[0, b]$.*

Proof. Obviously $Z(a) - 2a$ is continuous in a. By Tanaka's formula

$$(w(t) - (b - a))^{+} = \int_0^t \chi(w(s) > b - a)\,dw(s) + \frac{1}{2}L(b - a, t).$$

If $t = \tau_b$, then

$$Z(a) - 2a \stackrel{\circ}{=} L(b - a, \tau_b) - 2a =$$

$$= -2\int_0^{\tau_b} \chi(w(s) > b - a)\,dw(s) =$$

$$= -2\int_0^{\infty} \chi(b \geq w(s) > b - a)\,dw(s).$$

From this $Z(a)$ is integrable and its expected value is $2a$. If $u < v$, then for every $\mathcal{Z}_u$-measurable bounded variable ξ, by the previous lemma and by Itô's isometry

$$\mathbf{E}\left((Z(v) - 2v)\,\xi\right) =$$

$$= -2\mathbf{E}\left(\int_0^{\infty} \chi(b \geq w > b - v)\,dw \int_0^{\infty} H\chi(b \geq w > b - u)\,dw\right) =$$

$$= -2\mathbf{E}\left(\int_0^{\infty} \chi(b \geq w(s) > b - v)\,H\chi(b \geq w(s) > b - u)\,ds\right) =$$

$$= -2\mathbf{E}\left(\int_0^{\infty} H\chi(b \geq w(s) > b - u)\,ds\right) = \mathbf{E}\left((Z(u) - 2u)\,\xi\right).$$

Hence $Z(a) - 2a$ is a martingale. $\qquad\square$

Lemma 6.100 *If X is a continuous local martingale and $\sigma \geq 0$ is a random variable, then the quadratic variation of the stochastic process $L_\sigma(a, \omega) \stackrel{\circ}{=} L(a, \sigma(\omega), \omega)$ is finite. If $u < v$ then the quadratic variation of L_σ on the interval $[u, v]$ is*

$$[L_\sigma]_u^v \stackrel{a.s.}{=} 4\int_u^v L(a, \sigma)\,da.$$

Proof. Of course, by definition, the random variable ξ is the quadratic variation of L_σ on the interval $[u, v]$ if for arbitrary infinitesimal partition $\left(a_k^{(n)}\right)_{k,n}$ of $[u, v]$

if $n \to \infty$ then

$$\sum_k \left(L_\sigma \left(a_k^{(n)} \right) - L_\sigma \left(a_{k-1}^{(n)} \right) \right)^2 \xrightarrow{P} \xi.$$

1. Let us fix t. Let

$$\widehat{X}(a) \doteq \int_0^t \operatorname{sign}\left(X(s) - a \right) dX(s).$$

By the definition of local times

$$L(a,t) = |X(t) - a| - |X(0) - a| - \widehat{X}(a,t).$$

Let us remark that if f is a continuous and g is a Lipschitz continuous function then

$$|[f,g]| \le \limsup_{n\to\infty} \max_k \left| f\left(a_k^{(n)} \right) - f\left(a_{k-1}^{(n)} \right) \right| \sum_k \left| g\left(a_k^{(n)} \right) - g\left(a_{k-1}^{(n)} \right) \right| \le$$

$$\le \limsup_{n\to\infty} \max_k \left| f\left(a_k^{(n)} \right) - f\left(a_{k-1}^{(n)} \right) \right| K \sum_k \left| a_k^{(n)} - a_{k-1}^{(n)} \right| = 0.$$

The process

$$F_\sigma(a) \doteq |X(\sigma) - a| - |X(0) - a|$$

is obviously Lipschitz continuous in parameter a. X is a continuous local martingale so $\widehat{X}$ is continuous[83] in a so for every outcome

$$\left[F_\sigma + \widehat{X}_\sigma, F_\sigma \right] = 0 \quad \text{and} \quad [F_\sigma] = 0.$$

Therefore

$$[L_\sigma] = \left[F_\sigma + \widehat{X}_\sigma \right] = \left[\widehat{X}_\sigma \right].$$

2. By Itô's formula

$$\left(\widehat{X}\left(a_k^{(n)} \right) - \widehat{X}\left(a_{k-1}^{(n)} \right) \right)^2 =$$

$$= 2\left(\widehat{X}\left(a_k^{(n)} \right) - \widehat{X}\left(a_{k-1}^{(n)} \right) \right) \bullet \left(\widehat{X}\left(a_k^{(n)} \right) - \widehat{X}\left(a_{k-1}^{(n)} \right) \right) +$$

$$+ \left[\widehat{X}\left(a_k^{(n)} \right) - \widehat{X}\left(a_{k-1}^{(n)} \right) \right].$$

[83]See: Proposition 6.80, page 439.

By the Occupation Times Formula for every t almost surely

$$\left[\widehat{X}\left(a_k^{(n)}\right) - \widehat{X}\left(a_{k-1}^{(n)}\right)\right] \stackrel{\circ}{=}$$

$$\stackrel{\circ}{=} \left[\left(\operatorname{sign}\left(X - a_k^{(n)}\right) - \operatorname{sign}\left(X - a_{k-1}^{(n)}\right)\right) \bullet X\right] =$$

$$= \left[-2\chi\left(a_{k-1}^{(n)} < X \le a_k^{(n)}\right) \bullet X\right] =$$

$$= 4\chi\left(a_{k-1}^{(n)} < X \le a_k^{(n)}\right) \bullet [X] = 4\int_{a_{k-1}^{(n)}}^{a_k^{(n)}} L(a)\,da.$$

Hence almost surely

$$\sum_k \left[\widehat{X}\left(a_k^{(n)}\right) - \widehat{X}\left(a_{k-1}^{(n)}\right)\right](\sigma) = 4\int_u^v L(a,\sigma)\,da = 4\int_u^v L_\sigma(a)\,da.$$

3. Finally we should calculate the limit of the sum of first terms. The sum of the stochastic integrals is

$$-\left(\sum_k 2\left(\widehat{X}\left(a_k^{(n)}\right) - \widehat{X}\left(a_{k-1}^{(n)}\right)\right)\chi\left(a_{k-1}^{(n)} < X \le a_k^{(n)}\right)\right) \bullet X.$$

As $\widehat{X}$ is continuous if $n \to \infty$ the integrand goes to zero. The integrand is locally bounded so the stochastic integral goes to zero uniformly on compact intervals in probability. □

Theorem 6.101 (Ray–Knight) *There is a Wiener process β with respect to the filtration $\mathcal{Z}$, that $Z(a) \stackrel{\circ}{=} L(b-a, \tau_b)$ satisfies the equation*

$$Z(a) - 2a = 2\int_0^a \sqrt{Z}d\beta, \quad a \in [0, b]. \tag{6.78}$$

Proof. $L(u, t)$ is positive for every $t > 0$, so $Z(a) > 0$. The quadratic variation of $Z(a) - 2a$ is $4\int_0^a Z(s)\,ds$. By Doob's representation theorem[84] there is a Wiener process β with respect to filtration generated by Z for which (6.78) valid. □

$Z(a)$ is a continuous semimartingale. By Itô's formula

$$\exp(-sZ(a)) - 1 = \int_0^a \exp(-sZ)\,d(-sZ) +$$

$$+ \frac{1}{2}\int_0^a \exp(-sZ)\,d[-sZ].$$

---

[84]See: Proposition 6.18, page 373.

$Y(u) \overset{\circ}{=} Z(u) - 2u$ is a martingale $Z \geq 0$ so, $\exp(-sZ) \leq 1$

$$\mathbf{E}\left(\int_0^a (\exp(-sZ))^2 \, d[-sZ]\right) \leq \mathbf{E}\left(\int_0^a d[-sZ]\right) =$$

$$= 4s^2 \mathbf{E}\left(\int_0^a Z(s) \, ds\right) =$$

$$= 4s^2 \int_0^a \mathbf{E}(Z(s)) \, ds =$$

$$= 8s^2 \int_0^a s \, ds < \infty.$$

Hence the integral

$$\int_0^a \exp(-sZ(u)) \, d(-s(Z(u) - 2u))$$

is a martingale. Let

$$\mathcal{L}(a, s) \overset{\circ}{=} \mathbf{E}(\exp(-sZ(a))).$$

Taking expected value on both sides of Itô's formula and using the martingale property of the above integral

$$\mathcal{L}(s, a) - 1 = \mathbf{E}\left(\int_0^a \exp(-sZ(u)) \, d(-2su)\right) +$$

$$+ \frac{1}{2}\mathbf{E}\left(\int_0^a \exp(-sZ) \, d[-sZ]\right).$$

Let us calculate the second integral. Using (6.78)

$$2s^2 \mathbf{E}\left(\int_0^a \exp(-sZ(u)) Z(u) \, du\right) = 2s^2 \int_0^a \mathbf{E}(\exp(-sZ(u)) Z(u)) \, du =$$

$$= -2s^2 \int_0^a \mathbf{E}\left(\frac{d}{ds}\exp(-sZ(u))\right) du.$$

Changing the expected value and differentiating by a

$$\frac{\partial \mathcal{L}}{\partial a} = -2s\mathcal{L}(a, s) - 2s^2 \mathbf{E}\left(\frac{d}{ds}\exp(-sZ(a))\right).$$

For Laplace transforms one can change the differentiation and the integration so

$$\frac{\partial \mathcal{L}}{\partial a} = -2s\mathcal{L}(a, s) + 2s^2 \frac{\partial \mathcal{L}}{\partial s}, \quad \mathcal{L}(a, 0) = 1.$$

With direct calculation one can easily verify that

$$\mathcal{L}(a, s) = \frac{1}{1 + 2sa}$$

satisfies the equation. The Laplace transform $\mathcal{L}(a, s)$ is necessarily analytic so by the theorem of Cauchy and Kovalevskaja $1/(1 + 2sa)$ is the unique solution of the equation. This implies that $Z(a)$ has an exponential distribution with parameter $\lambda = 1/(2a)$.

6.5.7 Theorem of Dvoretzky Erdős and Kakutani

First let us introduce some definitions:

Definition 6.102 *Let f be a real valued function on an interval $I \subseteq \mathbb{R}$.*

1. *We say that t is a point of increase of f if there is a $\delta > 0$ such that*

$$f(s) \leq f(t) \leq f(u)$$

whenever $s, u \in I \cap (-\delta + t, t + \delta)$ and $s < t < u$.
2. *We say that t is a point of strict increase of f if there is a $\delta > 0$ such that*

$$f(s) < f(t) < f(u)$$

whenever $s, u \in I \cap (-\delta + t, t + \delta)$ and $s < t < u$.

A striking feature of Wiener processes is the following observation:

Theorem 6.103 (Dvoretzky–Erdős–Kakutani) *Almost surely the trajectories of Wiener processes do not have a point of increase.*

Proof. Let w be a Wiener process.

1. One should show that

$$\mathbf{P}(\{\omega : w(\omega) \text{ has a point of increase}\}) = 0.$$

Obviously sufficient to prove that for an arbitrary $v > 0$

$$\mathbf{P}(\{\omega : w(\omega) \text{ has a point of increase in } [0, v]\}) = 0.$$

By Girsanov's theorem there is a probability measure $\mathbf{P} \sim \mathbf{Q}$ on $(\Omega, \mathcal{F}_v)$ such that $\widetilde{w}(t) \overset{\circ}{=} w(t) + t$ is a Wiener process on $[0, v]$ under $\mathbf{Q}$. Every point of increase of w is a strict point of increase of $\widetilde{w}$. Therefore it is sufficient to prove that

$$\mathbf{P}(\{\omega : w(\omega) \text{ has a point of strict increase in } [0, v]\}) = 0.$$

Of course this is the same as

$$\mathbf{P}\left(\{\omega : w\left(\omega\right) \text{ has a point of strict increase}\}\right) = 0.$$

To prove this it is sufficient to show that $\mathbf{P}\left(\Omega_{p,q}\right) = 0$ for every rational numbers p and q where

$$\Omega_{p,q} \stackrel{\circ}{=} \left\{ \begin{array}{c} \omega : \exists t \text{ such that } w\left(s, \omega\right) < w\left(t, \omega\right) < w\left(u, \omega\right), \\ \text{for every } s, u \in \left(p, q\right), s < t < u \end{array} \right\}.$$

Using the strong Markov property of w one can assume that $p = 0$.

2. Let L be the local time of w. We show that for every b almost surely

$$Z\left(a\right) \stackrel{\circ}{=} L\left(b - a, \tau_b\left(\omega\right), \omega\right) > 0, \quad \forall a \in \left(0, b\right].$$

As we know[85] if $a > 0$ then $Z\left(a\right)$ has an exponential distribution so it is almost surely positive for every fixed $a \in \left(0, b\right]$. $Z\left(a\right)$ is continuous so if Ω_n is the set of outcomes ω for which $Z\left(a, \omega\right) \geq 1/n$ for every rational a then $Z\left(a, \omega\right) \geq 1/n$ for every $a \in \left(0, b\right]$. If $\Omega' \stackrel{\circ}{=} \cup_n \Omega_n$ then $\mathbf{P}\left(\Omega'\right) = 1$ and if $\omega \in \Omega'$ then $Z\left(a, \omega\right) > 0$ for every $a \in \left(0, b\right]$.

3. Now it is obvious that there is an Ω^* with $\mathbf{P}\left(\Omega^*\right) = 1$ that whenever $\omega \in \Omega^*$ then

a. $L\left(a, t, \omega\right)$ is continuous in $\left(a, t\right)$;

b. the support of $L\left(a, \omega\right)$ is $\{w\left(\omega\right) = a\}$ for every rational number a;

c. $Z\left(a\right) \stackrel{\circ}{=} L\left(b - a, \tau_b\left(\omega\right), \omega\right) > 0$ whenever $0 < a \leq b$ for every rational number b.

4. Let $\omega \in \Omega^*$ and let $\omega \in \Omega_{p,q} = \Omega_{0,q}$. This means that for some t

$$w\left(s, \omega\right) < w\left(t, \omega\right) < w\left(u, \omega\right), \quad 0 \leq s < t < u \leq q. \tag{6.79}$$

Let us fix a rational number $w\left(t, \omega\right) < b < w\left(q, \omega\right)$. Let $\left(b_n\right)$ be a sequence of rational numbers for which $b_n \nearrow w\left(t, \omega\right)$. As $w\left(t, \omega\right) < b$ and b is rational by c.

$$L\left(w\left(t, \omega\right), \tau_b\left(\omega\right), \omega\right) = L\left(b - \left(b - w\left(t, \omega\right)\right), \tau_b\left(\omega\right), \omega\right) > 0.$$

L is continuous so the measure of every single point is zero so by b. Obviously $L\left(b_n, \tau_{b_n}, \omega\right) = 0$. So

$$\begin{aligned} L\left(w\left(t, \omega\right), \tau_b\left(\omega\right), \omega\right) = {} & L\left(w\left(t, \omega\right), \tau_b\left(\omega\right), \omega\right) - L\left(b_n, \tau_b\left(\omega\right), \omega\right) + \\ & + L\left(b_n, \tau_b\left(\omega\right), \omega\right) - L\left(b_n, t, \omega\right) + \\ & + L\left(b_n, t, \omega\right) - L\left(b_n, \tau_{b_n}, \omega\right). \end{aligned}$$

---

[85]See: Example 6.73, page 430.

By the construction as t is a point of increase

$$b_n < w(t, \omega) < w(a, \omega) < b, \quad a \in (t, \tau_b).$$

By *b.* the support of the measure generated by $L(b_n, \omega)$ is $\{w(\omega) = b_n\}$. Hence the second line in the above estimation is zero. t is a point of increase so by (6.79) if $n \to \infty$ then $\tau_{b_n} \to t$. Therefore using *a.*

$$0 < L(w(t, \omega), \tau_b(\omega), \omega) = L(w(t, \omega), \tau_b(\omega), \omega) - L(w(t, \omega), \tau_b(\omega), \omega) +$$
$$+ L(w(t, \omega), t, \omega) - L(w(t, \omega), t, \omega) = 0.$$

This is a contradiction so if $\omega \in \Omega^*$ then $\omega \notin \Omega_{p,q}$. Hence $\mathbf{P}(\Omega_{p,q}) = 0$. $\qquad\square$

7

PROCESSES WITH INDEPENDENT INCREMENTS

In this chapter we discuss the classical theory of processes with independent increments. In the first section we return to the theory of Lévy processes. The increments of Lévy processes are not only independent but they are also stationary. Lévy processes are semimartingales, but the same is not true for processes with independent increments. In the second part of the chapter we show the generalization of the Lévy–Khintchine formula to processes with just independent increments. The main difference between the theory of Lévy processes and the more general theory of processes with independent increments is that every Lévy process is continuous in probability. This property does not hold for the more general class. This implies that processes with independent increments can have jumps with positive probability.

7.1   Lévy processes

In this section we briefly return to the theory of Lévy processes. The theory of Lévy processes is much simpler than the more general theory of processes with independent increments. Recall that Lévy processes have *stationary* and *independent* increments. The main consequence of these assumptions is that if $\varphi_t(u)$ denotes the Fourier transform of $X(t)$ then for every u

$$\varphi_{t+s}(u) = \varphi_t(u)\varphi_s(u), \tag{7.1}$$

so $\varphi_t(u)$ for every u satisfies Cauchy's functional equation[1]. As the Fourier transforms of distributions are always bounded the solutions of equation (7.1) have the form

$$\varphi_t(u) = \exp\left(t\phi(u)\right), \tag{7.2}$$

[1]See: line (1.40), page 62.

460

for some ϕ. One of our main goals is to find the proper form[2] of $\phi(u)$. Representation (7.2) has two very important consequences:

1. $\varphi_t(u) \neq 0$ for every u and t,
2. $\varphi_t(u)$ is continuous in t.

As φ_t is continuous in t, if $t_n \nearrow t$, then $\varphi_{t_n}(u) \to \varphi_t(u)$ for every u. Hence $X(t_n) - X(t) \overset{w}{\to} 0$, that is $X(t_n) - X(t) \overset{P}{\to} 0$. Hence for some subsequence $X(t_{n_k}) \overset{a.s.}{\to} X(t)$. Therefore $X(t-) \overset{a.s.}{=} X(t)$. Hence if X is a Lévy process then it is continuous in probability and, as a consequence of this continuity, for every moment of time t the probability of a jump at t is zero, that is

$$\mathbf{P}(\Delta X(t) \neq 0) = 0$$

for every t. As $\varphi_t(u) \neq 0$ for every u one can define the exponential martingale

$$Z_t(u, \omega) \overset{\circ}{=} \frac{\exp(iuX(t, \omega))}{\varphi_t(u)}. \tag{7.3}$$

Recall that, applying the Optional Sampling Theorem to (7.3), one can prove that every Lévy process is a strong Markov process[3].

7.1.1   Poisson processes

Let us recall that a Lévy process X is a Poisson process if its trajectories are increasing and the image of trajectories is almost surely the set of integers $\{0, 1, 2, \ldots\}$. One should emphasize that all the non-negative integers have to be in the image of the trajectories, so Poisson processes do not have jumps which are larger than one. To put it another way: Poisson processes are the Lévy type counting processes.

Definition 7.1 *A process is a* counting process *if its image space is the set of integers* $\{0, 1, \ldots\}$. *X is a* Poisson process *with respect to a filtration $\mathcal{F}$ if it is a counting Lévy process with respect to the filtration $\mathcal{F}$.*

Since the values of the process are integers and as the trajectories are right-regular there is always a positive amount of time between the jumps. That is if $X(t, \omega) = k$ then $X(t + u, \omega) = k$, whenever $0 \leq u \leq \delta$ for some $\delta(t, \omega) > 0$. As the trajectories are defined for every $t \geq 0$ and the values of the trajectories are finite at every t the jumps of the process cannot accumulate. Let

$$\tau_1(\omega) \overset{\circ}{=} \inf\{t \colon X(t, \omega) = 1\} = \inf\{t \colon X(t, \omega) > 0\} < \infty.$$

---

[2]This is the famous Lévy–Khintchine formula.
[3]See: Proposition 1.109, page 70.

τ_1 is obviously a stopping time. We show that τ_1 is exponentially distributed: if $u, v \geq 0$ then

$$\mathbf{P}\left(\tau_1 > u + v\right) = \mathbf{P}\left(X\left(u+v\right) = 0\right) =$$
$$= \mathbf{P}\left(X\left(u\right) = 0, X\left(u+v\right) - X\left(u\right) = 0\right) =$$
$$= \mathbf{P}\left(X\left(u\right) = 0\right) \cdot \mathbf{P}\left(X\left(u+v\right) - X\left(u\right) = 0\right) =$$
$$= \mathbf{P}\left(X\left(u\right) = 0\right) \cdot \mathbf{P}\left(X\left(v\right) = 0\right),$$

hence if $f\left(t\right) \stackrel{\circ}{=} \mathbf{P}\left(\tau_1 > t\right)$ then

$$f\left(u+v\right) = f\left(u\right) \cdot f\left(v\right), \quad u, v \geq 0.$$

$f \equiv 0$ and $f \equiv 1$ cannot be solutions as X cannot be a non-trivial Lévy process[4], so for some $0 < \lambda < \infty$

$$\mathbf{P}\left(\tau_1 > t\right) = \mathbf{P}\left(X\left(t\right) = 0\right) = \exp\left(-\lambda t\right).$$

By the strong Markov property of Lévy processes[5] the distribution of

$$X_1^*\left(t\right) \stackrel{\circ}{=} X\left(\tau_1 + t\right) - X\left(\tau_1\right)$$

is the same as the distribution of $X\left(t\right)$ so if

$$\tau_2\left(\omega\right) \stackrel{\circ}{=} \inf\left\{t\colon X\left(t + \tau_1\left(\omega\right), \omega\right) = 2\right\} = \inf\left\{t\colon X_1^*\left(t, \omega\right) > 0\right\} < \infty$$

then τ_1 and τ_2 are independent and they have the same distribution[6].

Proposition 7.2 *If λ denotes the common parameter, then for every $t \geq 0$*

$$\mathbf{P}\left(\sum_{k=1}^{n+1} \tau_k > t \geq \sum_{k=1}^{n} \tau_k\right) = \mathbf{P}\left(X\left(t\right) = n\right) = \frac{\left(\lambda t\right)^n}{n!} \exp\left(-\lambda t\right).$$

Proof. Recall that a non-negative variable has gamma distribution $\Gamma\left(a, \lambda\right)$ if the density function of the distribution is

$$f_{a,\lambda}\left(x\right) \stackrel{\circ}{=} \frac{\lambda^a}{\Gamma\left(a\right)} x^{a-1} \exp\left(-\lambda x\right), \quad x > 0.$$

First we show that if ξ_i are independent random variables with distribution $\Gamma\left(a_i, \lambda\right)$, then the distribution of $\sum_{i=1}^{n} \xi_i$ is $\Gamma\left(\sum_{i=1}^{n} a_i, \lambda\right)$. It is sufficient to

---

[4]If $f \equiv 1$ then $\tau_1 = \infty$, hence $X \equiv 0$ and the image of trajectories is $\{0\}$ only and not the set of integers.

[5]See: Proposition 1.109, page 70.

[6]Let us recall that τ_1 is $\mathcal{F}_{\tau_1}$-measurable and by the strong Markov property X_1^* is independent of $\mathcal{F}_{\tau_1}$. See Proposition 1.109, page 70.

show the calculation for two variables. If the distribution of ξ_1 is $\Gamma(a, \lambda)$, and the distribution of ξ_2 is $\Gamma(b, \lambda)$, and if they are independent, then the density function of $\xi_1 + \xi_2$ is the convolution of the density functions of ξ_1 and ξ_2

$$h(x) \overset{\circ}{=} \int_{-\infty}^{\infty} f_{a,\lambda}(x-t) f_{b,\lambda}(t) dt =$$

$$= \int_0^x \frac{\lambda^a (x-t)^{a-1}}{\Gamma(a)} \exp(-\lambda(x-t)) \frac{\lambda^b t^{b-1}}{\Gamma(b)} \exp(-\lambda t) dt =$$

$$= \frac{\lambda^{a+b}}{\Gamma(a)\Gamma(b)} \exp(-\lambda x) \int_0^x (x-t)^{a-1} t^{b-1} dt =$$

$$= \frac{\lambda^{a+b}}{\Gamma(a)\Gamma(b)} \exp(-\lambda x) \int_0^1 (x-xz)^{a-1} (xz)^{b-1} x dz =$$

$$= \frac{\lambda^{a+b}}{\Gamma(a)\Gamma(b)} \exp(-\lambda x) x^{a+b-1} \int_0^1 (1-z)^{a-1} z^{b-1} dz =$$

$$= \frac{\lambda^{a+b}}{\Gamma(a+b)} \exp(-\lambda x) x^{a+b-1}.$$

Hence the distribution of $\xi_1 + \xi_2$ is $\Gamma(a+b, \lambda)$. The density function of $\Gamma(1, \lambda)$ is

$$\frac{\lambda^1}{\Gamma(1)} x^{1-1} \exp(-\lambda x) = \lambda \exp(-\lambda x), \quad x > 0,$$

so $\Gamma(1, \lambda)$ is the exponential distribution with parameter λ. If $\sigma_m \overset{\circ}{=} \sum_{k=1}^m \tau_k$ then σ_m has gamma distribution $\Gamma(m, \lambda)$.

$$\mathbf{P}(X(t) < n+1) =$$

$$= \mathbf{P}(\sigma_{n+1} > t) = \int_t^{\infty} \frac{\lambda^{n+1}}{\Gamma(n+1)} x^n \exp(-\lambda x) dx =$$

$$= \left[-\frac{(\lambda x)^n}{\Gamma(n+1)} \exp(-\lambda x) \right]_t^{\infty} + \int_t^{\infty} n \frac{\lambda^n x^{n-1}}{\Gamma(n+1)} \exp(-\lambda x) dx =$$

$$= \frac{(\lambda t)^n}{n!} \exp(-\lambda t) + \mathbf{P}(X(t) < n).$$

Hence

$$\mathbf{P}(X(t) = n) = \mathbf{P}(X(t) < n+1) - \mathbf{P}(X(t) < n) = \frac{(\lambda t)^n}{n!} \exp(-\lambda t).$$

$\square$

7.1.2   Compound Poisson processes generated by the jumps

Let X now be a Lévy process and let Λ be a Borel measurable set.

$$\tau_1(\omega) \overset{\circ}{=} \inf\{t\colon \Delta X(t,\omega) \in \Lambda\}.$$

Since $(\Omega, \mathcal{A}, \mathbf{P}, \mathcal{F})$ satisfies the usual conditions τ_1 is a stopping time[7]. As τ_1 is measurable

$$\mathbf{P}(\tau_1 > t) = \mathbf{P}(\Delta X(u) \notin \Lambda, \ \forall u \in [0,t])$$

is meaningful. Assume that the closure of Λ denoted by $\mathrm{cl}(\Lambda)$ does not contain the point 0, that is Λ is in the complement of a ball with some positive radius $r > 0$. As X is right-continuous and as $X(0) = 0$ obviously $0 < \tau_1 \le \infty$. In a similar way as in the previous subsection, using that the jumps in Λ cannot accumulate[8]

$$\mathbf{P}(\tau_1 > t_1 + t_2) =$$

$$= \mathbf{P}(\Delta X(u) \notin \Lambda, \ u \in (0, t_1 + t_2]) =$$

$$= \mathbf{P}(\Delta X(u) \notin \Lambda, \ u \in (0, t_1]) \cdot \mathbf{P}(\Delta X(u) \notin \Lambda, \ u \in (t_1, t_1 + t_2]) =$$

$$= \mathbf{P}(\Delta X(u) \notin \Lambda, \ u \in (0, t_1]) \cdot \mathbf{P}(\Delta X(u) \notin \Lambda, \ u \in (0, t_2]) =$$

$$= \mathbf{P}(\tau_1 > t_1) \cdot \mathbf{P}(\tau_1 > t_2).$$

So τ_1 has an exponential distribution. Let us observe that now we cannot guarantee that $\lambda > 0$ as $\tau_1 \equiv \infty$ is possible. Let us assume that $\tau_1 < \infty$. Let

$$X^*(t) \overset{\circ}{=} X(\tau_1 + t) - X(\tau_1)$$

and let

$$\tau_2 \overset{\circ}{=} \inf\{t : \Delta X^*(t) \in \Lambda\},$$

etc. If $\tau_1 < \infty$ then $\tau_k < \infty$ for all k. Let $\sigma_n \overset{\circ}{=} \sum_{k=1}^{n} \tau_k$. As $0 \notin \mathrm{cl}(\Lambda)$ and as X has limits from left the almost surely[9] strictly increasing sequence (σ_n) almost surely cannot have a finite accumulation point. So almost surely $\sigma_n \nearrow \infty$. As on every trajectory the number of jumps is at most countable one can define the

---

[7]See: Corollary 1.29, page 16, Example 1.32, page 17.

[8]As $0 \notin \mathrm{cl}(\Lambda)$ all the jumps are larger than some $r > 0$. τ_1 is a stopping time so the sets below are measurable.

[9]The trajectories of a Poisson process are just almost surely nice. For example, with probability zero $N(\omega) \equiv 0$ is possible.

process N^Λ which counts the jumps of X with $\Delta X \in \Lambda$.

$$N^\Lambda (t) \overset{\circ}{=} \sum_{0 < s \leq t} \chi_\Lambda (\Delta X (s)) = \sum_{n=1}^\infty \chi \{\sigma_n \leq t\}. \qquad (7.4)$$

$N^\Lambda (t) - N^\Lambda (s)$ is the number of jumps in Λ during the time interval $(s, t]$ so it is evidently measurable with respect to the σ-algebra generated by the increments of X. Hence[10] $N^\Lambda (t) - N^\Lambda (s)$ is independent of the σ-algebra $\mathcal{F}_s$. So N^Λ has independent increments. It is also easy to prove that the distribution of $N^\Lambda (t) - N^\Lambda (s)$ is the same as the distribution of $N^\Lambda (t - s)$. It is trivial from the definition that N^Λ is a right-regular counting process. Hence N^Λ is a counting Lévy process. Therefore we have proved the following:

Lemma 7.3 *If $0 \notin \mathrm{cl}(\Lambda)$ then N^Λ is a Poisson process.*

Definition 7.4 *A stopping time σ is a* jump time *of a process X if $\Delta X (\sigma) \neq 0$ almost surely.*

Example 7.5 The jump times of Lévy processes are totally inaccessible.

Let τ be a predictable stopping time and let $\mathbf{P} (\Delta X (\tau) \neq 0) > 0$. We can assume that $\mathbf{P} (|\Delta X (\tau)| \geq \varepsilon) > 0$ for some $\varepsilon > 0$. If $\Lambda \overset{\circ}{=} \{|x| \geq \varepsilon\}$ and if (σ_n) are the stopping times of the Poisson process N^Λ then $\mathbf{P} (\sigma_n = \tau) > 0$ for some n. But this is impossible as σ_n is totally inaccessible[11] for every n. Therefore if τ is predictable then $\mathbf{P} (\Delta X (\tau) \neq 0) = 0$. $\qquad \square$

With N^Λ one can define the process

$$J^\Lambda (t, \omega) \overset{\circ}{=} \sum_{0 < s \leq t} \Delta X (s, \omega) \chi_\Lambda (\Delta X (s, \omega)) = \qquad (7.5)$$

$$= \sum_{n=1}^{N^\Lambda(t)} \Delta X (\sigma_n) = \sum_{n=1}^\infty \Delta X (\sigma_n) \chi \{\sigma_n \leq t\}.$$

Lemma 7.6 *If $0 \notin \mathrm{cl}(\Lambda)$ then J^Λ is a compound Poisson process that is:*

1. $J^\Lambda (0) = 0$.
2. J^Λ *has countable many jumps.*
3. *After every jump J^Λ has an exponentially distributed waiting time. After this waiting time J^Λ jumps again. The time between the jumps are independent and they have the same distribution.*

[10]See: Proposition 1.97, page 61.
[11]See: Example 3.7, page 183.

4. *The sizes of the jumps are independent of the waiting times up to the jumps.*
5. *The sizes of the jumps have the same distribution and they are independent random variables.*

Proof. If $\eta_n \triangleq \Delta X(\sigma_n)$ then by the strong Markov property the variables (η_n) are independent and they have the same distribution. One need only prove that (σ_n) and (η_n) are independent. Let $\tau_n \triangleq \sigma_n - \sigma_{n-1}$.

1. If $s > t$, then

$$\{\eta_1 < a, \ \sigma_1 > s\} = \{\sigma_1 > t\} \cap \left\{\eta_1^{(t)} < a, \sigma_1^{(t)} > s - t\right\},$$

where $\eta_1^{(t)}$ and $\sigma_1^{(t)}$ are the size and the time of the first jump of $X^*(u) = X(u+t) - X(t)$. As σ_1 is a stopping time $\{\sigma_1 > t\} \in \mathcal{F}_t$. Hence by the strong Markov property $\{\sigma_1 > t\}$ is independent of

$$\left\{\eta_1^{(t)} < a, \sigma_1^{(t)} > s - t\right\}.$$

Hence again by the strong Markov property

$$\mathbf{P}(\eta_1 < a, \sigma_1 > s) = \mathbf{P}\left(\{\sigma_1 > t\} \cap \left\{\eta_1^{(t)} < a, \sigma_1^{(t)} > s - t\right\}\right) =$$
$$= \mathbf{P}(\sigma_1 > t)\, \mathbf{P}\left(\eta_1^{(t)} < a, \sigma_1^{(t)} > s - t\right) =$$
$$= \mathbf{P}(\sigma_1 > t)\, \mathbf{P}(\eta_1 < a, \sigma_1 > s - t).$$

If $s \searrow t$ then using that $0 \notin \mathrm{cl}(\Lambda)$ and therefore $\mathbf{P}(\sigma_1 > 0) = 1$,

$$\mathbf{P}(\eta_1 < a, \sigma_1 > t) = \mathbf{P}(\sigma_1 > t)\, \mathbf{P}(\eta_1 < a, \sigma_1 > 0) =$$
$$= \mathbf{P}(\sigma_1 > t) \cdot \mathbf{P}(\eta_1 < a).$$

Hence $\sigma_1 \triangleq \tau_1$ and η_1 are independent. In a similar way, using the strong Markov property again one can prove that τ_n is independent of η_n for every n.

3. By the strong Markov property (η_n, τ_n) is independent of $\mathcal{F}_{\sigma_{n-1}}$. Hence

$$\mathbf{E}\left(\exp\left(i\sum_{m=1}^{N} u_m \eta_m + i\sum_{n=1}^{N} v_n \tau_n\right)\right) =$$
$$= \mathbf{E}\left(\mathbf{E}\left(\exp\left(i\sum_{m=1}^{N} u_m \eta_m + i\sum_{n=1}^{N} v_n \tau_n\right) \mid \mathcal{F}_{\sigma_{N-1}}\right)\right) =$$

$$= \mathbf{E}\left(\exp\left(i\sum_{m=1}^{N-1} u_m\eta_m + i\sum_{n=1}^{N-1} v_n\tau_n\right)\mathbf{E}\left(\exp\left(iu_N\eta_N + iv_N\tau_N\right)\mid \mathcal{F}_{\sigma_{N-1}}\right)\right) =$$

$$= \mathbf{E}\left(\exp\left(i\sum_{m=1}^{N-1} u_m\eta_m + i\sum_{n=1}^{N-1} v_n\tau_n\right)\right)\cdot\mathbf{E}\left(\exp\left(iu_N\eta_N + iv_N\tau_N\right)\right) =$$

$$= \mathbf{E}\left(\exp\left(i\sum_{m=1}^{N-1} u_m\eta_m + i\sum_{n=1}^{N-1} v_n\sigma_n\right)\right)\cdot\mathbf{E}\left(\exp\left(iu_N\eta_N\right)\right)\cdot\mathbf{E}\left(\exp\left(iv_N\tau_N\right)\right) =$$

$$= \cdots = \prod_{m=1}^{N} \mathbf{E}\left(\exp\left(iu_m\eta_m\right)\right)\prod_{m=1}^{N}\mathbf{E}\left(\exp\left(iv_m\tau_m\right)\right).$$

This implies[12] that the σ-algebras generated by (η_m) and (τ_n) are independent. Hence (η_m) and (σ_n) are also independent. □

Lemma 7.7 *The Fourier transform of $J^\Lambda(s)$ is*

$$\mathbf{E}\left(\exp\left(iu\cdot J^\Lambda(s)\right)\right) = \exp\left(\lambda s\left(\int_{\mathbb{R}}\left(\exp\left(iux\right)-1\right)dF(x)\right)\right)$$

where λ is the parameter of the Poisson part and F is the common distribution function of the jumps.

Proof. Let G be the distribution function of $N^\Lambda(s)$.

$$\varphi(u) \overset{\circ}{=} \mathbf{E}\left(\exp\left(iu\cdot\sum_{k=1}^{N^\Lambda(s)}\Delta X(\sigma_k)\right)\right) =$$

$$= \int_{\mathbb{R}}\mathbf{E}\left(\exp\left(iu\cdot\sum_{k=1}^{N^\Lambda(s)}\Delta X(\sigma_k)\right)\mid N^\Lambda(s)=n\right)dG(n).$$

$N^\Lambda(s)$ has a Poisson distribution. As $N^\Lambda(s)$ and the variables $(\Delta X(\sigma_k))$ are independent one can substitute and drop the condition $N^\Lambda(s)=k$:

$$\varphi(u) = \sum_{n=0}^{\infty}\mathbf{E}\left(\exp\left(iu\cdot\sum_{k=1}^{n}\Delta X(\sigma_k)\right)\right)\frac{(\lambda s)^n}{n!}\exp\left(-\lambda s\right) =$$

$$= \sum_{n=0}^{\infty}\left(\int_{\mathbb{R}}\exp\left(iux\right)dF(x)\right)^n\frac{(\lambda s)^n}{n!}\exp\left(-\lambda s\right) =$$

$$= \exp\left(\lambda s\left(\int_{\mathbb{R}}\left(\exp\left(iux\right)-1\right)dF(x)\right)\right). \qquad\square$$

[12]See: Lemma 1.96, page 60.

Lemma 7.8 *If X is a Lévy process with respect to some filtration $\mathcal{F}$ and $0 \notin$ $\mathrm{cl}\,(\Lambda)$ then J^Λ and $X - J^\Lambda$ are also Lévy processes with respect to $\mathcal{F}$.*

Proof. First recall[13] that if X is a Lévy process then the σ-algebra $\mathcal{G}_t$ generated by the increments

$$X\,(u) - X\,(v)\,, \quad u \geq v \geq t$$

is independent of $\mathcal{F}_t$ for all t. Observe that for all t increments of J^Λ and $X - J^\Lambda$ of this type are $\mathcal{G}_t$-measurable. So these processes have independent increment with respect to $\mathcal{F}$. From the strong Markov property it is clear that the increments of these processes are stationary. As J^Λ obviously has right-regular trajectories the processes in the lemma are Lévy processes as well. $\square$

Lemma 7.9 *If X is a Lévy process, Λ is a Borel measurable set and $0 \notin \mathrm{cl}\,(\Lambda)$ then the variables $J^\Lambda\,(t)$ and $\left(X - J^\Lambda\right)(t)$ are independent for every $t \geq 0$.*

Proof. Let us fix a t. To prove the independence of the variables $J^\Lambda\,(t)$ and $X\,(t) - J^\Lambda\,(t)$ it is sufficient to prove[14] that

$$\varphi\,(u, v) \overset{\circ}{=} \mathbf{E}\left(\exp\left(i\left[u \cdot J^\Lambda\,(t) + v \cdot \left(X\,(t) - J^\Lambda\,(t)\right)\right]\right)\right) = \qquad (7.6)$$
$$= \mathbf{E}\left(\exp\left(iu \cdot J^\Lambda\,(t)\right)\right) \cdot \mathbf{E}\left(\exp\left(iv \cdot \left(X\,(t) - J^\Lambda\,(t)\right)\right)\right).$$

Let us emphasize that as $0 \notin \mathrm{cl}\,(\Lambda)$ on every finite interval the number of jumps in Λ is finite so J^Λ has trajectories with finite variation. That is $J^\Lambda \in \mathcal{V}$. Let

$$M\,(s, \omega, u) \overset{\circ}{=} \frac{\exp\left(iu \cdot J^\Lambda\,(s, \omega)\right)}{\mathbf{E}\left(\exp\left(iu \cdot J^\Lambda\,(s, \omega)\right)\right)},$$

$$N\,(s, \omega, v) \overset{\circ}{=} \frac{\exp\left(iv \cdot \left[X\,(s, \omega) - J^\Lambda\,(s, \omega)\right]\right)}{\mathbf{E}\left(\exp\left(iv \cdot \left[X\,(s, \omega) - J^\Lambda\,(s, \omega)\right]\right)\right)}$$

be the exponential martingale of J^Λ and $X - J^\Lambda$. The Fourier transforms in the denominators are never zero and they are continuous, hence the expressions are meaningful and the jumps of these processes are the jumps of the numerators. Integrating by parts

$$M\,(t)\,N\,(t) - M\,(0)\,N\,(0) = \int_0^t M_-\,dN + \int_0^t N_-\,dM +$$
$$+ [M, N]\,(t)\,.$$

---

[13]See: Proposition 1.97, page 61.
[14]See: Lemma 1.96, page 60.

The Fourier transforms in the denominators are never zero and they are continuous so their absolute value have a positive minimum on the compact interval $[0, t]$. The numerators are bounded, so the integrators are bounded on any finite interval. Hence the stochastic integrals above are real martingales[15]. So their expected value is zero. We show that $[M, N] = 0$. As $J^\Lambda(t)$ has a compound Poisson distribution one can explicitly write down its Fourier transform:

$$\mathbf{E}\left(\exp\left(iu \cdot J^\Lambda(s)\right)\right) = \exp\left(\lambda s\left(\int_{\mathbb{R}}\left(\exp\left(iux\right) - 1\right) dF(x)\right)\right) \overset{\circ}{=}$$
$$\overset{\circ}{=} \exp\left(s \cdot \phi(u)\right)$$

As $J^\Lambda \in \mathcal{V}$ obviously $M \in \mathcal{V}$. So M is purely discontinuous. Hence[16]

$$[M, N] = \sum \Delta M \Delta N.$$

J^Λ and $X - J^\Lambda$ do not have common jumps, therefore

$$[M, N](t) = \sum_{0 < s \leq t} \Delta M(s) \Delta N(s) = 0.$$

Hence

$$\mathbf{E}\left(M(t) N(t)\right) = \mathbf{E}\left(M(0) N(0)\right) = 1.$$

From which (7.6) trivially holds. □

If N_1 and N_2 are Poisson processes and N_1 and N_2 do not have common jumps then

$$[N_1, N_2] = \sum \Delta N_1 \Delta N_2 = 0.$$

Using this one can prove in a similar way as above the following observation:

Lemma 7.10 *If N_1 and N_2 are Poisson processes with respect to some filtration $\mathcal{F}$ and N_1 and N_2 do not have common jumps almost surely then $N_1(t)$ and $N_2(t)$ are independent for every t.*

Proposition 7.11 *If (N_i) are finitely many Poisson processes with respect to some filtration then they do not have common jumps almost surely if and only if the variables $(N_i(t))$ are independent[17] for every t.*

---

[15]See: Proposition 2.24, page 128.
[16]See: Corollary 4.34, page 245.
[17]See: Example 2.29, page 130.

Proof. If the values of Poisson processes are independent then the same is true for the compensated Poison processes. By the independence on every finite time interval the compensated Poisson processes are orthogonal in the Hilbert space $\mathcal{H}_0^2$. Hence they are orthogonal as local martingales[18]. Therefore their quadratic variation is a uniformly integrable martingale[19]. This implies that the expected value of the quadratic co-variation

$$[N_1, N_2] = \sum \Delta N_1 \Delta N_2$$

is almost surely zero. As $\Delta N_1 \Delta N_2 \geq 0$ the quadratic co-variation is almost surely zero. Hence the two processes do not have common jumps almost surely. The proof of the other part of the proposition is clear from the previous lemma. $\qquad\square$

Theorem 7.12 (Decomposition of Lévy processes) *If X is a Lévy process, Λ is a Borel measurable set and $0 \notin \text{cl}(\Lambda)$ then J^Λ and $X - J^\Lambda$ are independent Lévy processes.*

Proof. Recall that by definition two processes are independent if they are independent as sets of random variables. As we proved[20] $J^\Lambda(t)$ and $\left(X - J^\Lambda\right)(t)$ are independent for every t. From the Markov property it is clear that if $h > 0$ then the increments

$$J^\Lambda(t + h) - J^\Lambda(t)$$

and

$$\left(X - J^\Lambda\right)(t + h) - \left(X - J^\Lambda\right)(t)$$

are also independent. Let (t_k) be a time sequence. Let (α_k) denote the corresponding increments of J^Λ and let (β_k) denote the corresponding increments of $X - J^\Lambda$. Let $\mathcal{G}_t$ be the σ-algebra generated by the increments of X after t. Observe that α_k and β_k are $\mathcal{G}_{t_k}$-measurable. Hence the linear combination $u_k\alpha_k + v_k\beta_k$ is also $\mathcal{G}_{t_k}$-measurable. So $u_k\alpha_k + v_k\beta_k$ is independent[21] of $\mathcal{F}_{t_k}$. Using these one can easily decompose the joint Fourier transform:

$$\varphi(\mathbf{u}, \mathbf{v}) \doteq \mathbf{E}\left(\exp\left(\sum_{k=1}^n iu_k\alpha_k + \sum_{k=1}^n iv_k\beta_k\right)\right) =$$

$$= \mathbf{E}\left(\exp\left(\sum_{k=1}^n i\left(u_k\alpha_k + v_k\beta_k\right)\right)\right) =$$

---

[18] See: Proposition 4.15, page 230.
[19] See: Proposition 2.84, page 170.
[20] See: Lemma 7.9, page 468.
[21] See: Proposition 1.97, page 61.

$$= \mathbf{E}\left(\mathbf{E}\left(\exp\left(\sum_{k=1}^{n} i\left(u_k\alpha_k + v_k\beta_k\right)\right) \mid \mathcal{F}_{t_{n-1}}\right)\right) =$$

$$= \mathbf{E}\left(\exp\left(\sum_{k=1}^{n-1} i\left(u_k\alpha_k + v_k\beta_k\right)\right) \mathbf{E}\left(\exp\left(i\left(u_n\alpha_n + v_n\beta_n\right)\right)\right)\right) =$$

$$= \cdots = \prod_{k=1}^{n} \mathbf{E}\left(\exp\left(i\left(u_k\alpha_k + v_k\beta_k\right)\right)\right) =$$

$$= \prod_{k=1}^{n} \left(\mathbf{E}\left(\exp\left(iu_k\alpha_k\right)\right) \cdot \mathbf{E}\left(\exp\left(iv_k\beta_k\right)\right)\right) = \varphi_1\left(\mathbf{u}\right) \cdot \varphi_2\left(\mathbf{v}\right).$$

This means that the sets of variables (α_k) and (β_k) are independent. Hence the σ-algebras generated by the increments, that is by the processes, are independent. Therefore the processes $X - J^\Lambda$ and J^Λ are independent. $\qquad\square$

With nearly the same method one can prove the following proposition.

Proposition 7.13 *If (N_i) are finitely many Poisson processes with respect to some common filtration then they do not have common jumps almost surely if and only if the processes are independent.*

Proof. Let $\mathcal{F}$ be the common filtration of N_1 and N_2 and let U and V be the exponential martingales of N_1 and N_2. As N_1 and N_2 do not have a common jumps the quadratic co-variation of U and V is zero. Hence they are orthogonal. That is UV is a local martingale with respect to $\mathcal{F}$. On every finite interval $U, V \in \mathcal{H}^2$, therefore

$$|UV(t)| \leq \sup_s |U(s)| \sup_s |V(s)| \in L^1(\Omega).$$

Hence UV is a martingale. Therefore

$$\mathbf{E}\left(UV\left(t_k\right) \mid \mathcal{F}_{t_{k-1}}\right) = UV\left(t_{k-1}\right).$$

If we use the notation of the proof of the previous proposition then with simple calculation one can write this as

$$\mathbf{E}\left(\exp\left(i\left(u_k\alpha_k + v_k\beta_k\right)\right) \mid \mathcal{F}_{t_{n-1}}\right) = \mathbf{E}\left(\exp\left(iu_k\alpha_k\right)\right) \cdot \mathbf{E}\left(\exp\left(iv_k\beta_k\right)\right).$$

From this the proof of the proposition is obvious. $\qquad\square$

Corollary 7.14 *If (N_i) are countably many independent Poisson processes then they do not have common jumps almost surely.*

Proof. Let N_1 and N_2 be independent Poisson processes and let $\mathcal{F}^{(1)}$ and $\mathcal{F}^{(2)}$ be the filtration generated by the processes. Let U and V be the exponential martingales of N_1 and N_2. U and V are martingales with respect to filtrations $\mathcal{F}^{(1)}$ and $\mathcal{F}^{(2)}$. Let $\mathcal{F}$ be the filtration generated by the two processes N_1 and N_2. Using the independence of N_1 and N_2 we show that U and V are martingales with respect to $\mathcal{F}$ as well. If $F_1 \in \mathcal{F}_s^{(1)}$ and $F_2 \in \mathcal{F}_s^{(2)}$ where $s < t$ then

$$\int_{F_1 \cap F_2} U(t) \, d\mathbf{P} = \mathbf{E}\left(\chi_{F_1}\chi_{F_2}U(t)\right) = \mathbf{E}\left(\chi_{F_2}\right)\mathbf{E}\left(\chi_{F_1}U(t)\right) =$$

$$= \mathbf{E}\left(\chi_{F_2}\right)\mathbf{E}\left(\chi_{F_1}U(s)\right) = \mathbf{E}\left(\chi_{F_2}\chi_{F_1}U(s)\right) =$$

$$= \int_{F_1 \cap F_2} U(s) \, d\mathbf{P}.$$

With the Monotone Class Theorem one can prove that the equality holds for every

$$F \in \sigma\left\{F_1 \cap F_2 \colon F_1 \in \mathcal{F}_s^{(1)}, \; F_2 \in \mathcal{F}_s^{(2)}\right\} = \mathcal{F}_s,$$

that is $\mathbf{E}\left(U(t) \mid \mathcal{F}_s\right) = U(s)$. Hence U is a martingale with respect to $\mathcal{F}$. $\square$

Example 7.15 Poisson processes without common jumps which are not independent.

Let (σ_k) be the jump times generating some Poisson process. Obviously variables $(2\sigma_k)$ also generate a Poisson process. As the probability that two independent continuous random variable is equal is zero the jump times of the two processes are almost surely never equal. But as they generate the same non-trivial σ-algebra they are obviously not independent. $\square$

Proposition 7.16 *If X is a Lévy process and (Λ_k) are finitely many disjoint Borel measurable sets with $0 \notin \mathrm{cl}(\Lambda_k)$ for all k, then processes $\left(N^{\Lambda_k}\right)$ are independent. The same is true for $\left(J^{\Lambda_k}\right)$.*

Proof. It is sufficient to show the second part of the proposition. If $X \stackrel{\circ}{=} J^{\cup_{i=1}^n \Lambda_k}$ then $J^{\cup_{i=2}^n \Lambda_k} = X - J^{\Lambda_1}$ and J^{Λ_1} are independent. From this the proposition is obvious. $\square$

7.1.3 Spectral measure of Lévy processes

First let us prove a very simple identity.

Definition 7.17 *Let* $(X, \mathcal{A})$ *and* $(Y, \mathcal{B})$ *be measurable spaces. A function* μ : $X \times \mathcal{B} \to [0, \infty]$ *is a random measure if:*

1. *for every* $B \in \mathcal{B}$ *the function* $x \mapsto \mu(x, B)$ *is* $\mathcal{A}$-*measurable,*
2. *for every* $x \in X$ *the set function* $B \mapsto \mu(x, B)$ *is a measure on* $(Y, \mathcal{B})$.

Proposition 7.18 *Let* $(X, \mathcal{A})$ *and* $(Y, \mathcal{B})$ *be measurable spaces and let*

$$\mu : X \times \mathcal{B} \to [0, \infty]$$

be a random measure. If ρ *is a measure on* $(X, \mathcal{A})$ *and*

$$\nu(B) \stackrel{\circ}{=} \int_X \mu(x, B) \, d\rho(x),$$

then ν *is a measure on* $(Y, \mathcal{B})$. *If* f *is a measurable function on* $(Y, \mathcal{B})$ *then*

$$\int_Y f(y) \, d\nu(y) = \int_X \int_Y f(y) \mu(x, dy) \, d\rho(x),$$

whenever the integral on the left-hand side $\int_Y f d\nu$ *is meaningful.*

Proof. ν is non-negative and if (B_n) are disjoint sets then by the Monotone Convergence Theorem

$$\nu(\cup_n B_n) \stackrel{\circ}{=} \int_X \mu(x, \cup_n B_n) \, d\rho(x) = \int_X \sum_n \mu(x, B_n) \, d\rho(x) =$$

$$= \sum_n \int_X \mu(x, B_n) \, d\rho(x) \stackrel{\circ}{=} \sum_n \nu(B_n),$$

so ν is really a measure. If $f = \chi_B$, $B \in \mathcal{B}$, then

$$\int_Y f(y) \, d\nu(y) = \nu(B) \stackrel{\circ}{=} \int_X \mu(x, B) \, d\rho(x) =$$

$$= \int_X \int_Y \chi_B(y) \mu(x, dy) \, d\rho(x) =$$

$$= \int_X \int_Y f(y) \mu(x, dy) \, d\rho(x).$$

In the usual way, using the linearity of the integration and the Monotone Convergence Theorem the formula can be extended to non-negative measurable functions. If f is non-negative and $\int_Y f d\nu$ is finite then almost surely w.r.t. ρ

the inner integral is also finite. Let $f = f^+ - f^-$ and assume that the integral of f^- w.r.t. ν is finite. In this case, as we remarked, the integral $\int_Y f^- (y) \, \mu (x, dy)$ is finite for almost all x and the integral

$$\int_Y f (y) \, \mu (x, dy) = \int_Y f^+ (y) \, \mu (x, dy) - \int_Y f^- (y) \, \mu (x, dy)$$

is almost surely meaningful. The integral of the second part with respect to ρ is finite, hence

$$\int_Y f \, d\nu \overset{\circ}{=} \int_Y f^+ \, d\nu - \int_Y f^- \, d\nu =$$

$$= \int_X \int_Y f^+ (y) \, \mu (x, dy) \, d\rho (x) - \int_X \int_Y f^- (y) \, \mu (x, dy) \, d\rho (x) =$$

$$= \int_X \left(\int_Y f^+ (y) \, \mu (x, dy) - \int_Y f^- (y) \, \mu (x, dy) \right) d\rho (x) \overset{\circ}{=}$$

$$\overset{\circ}{=} \int_X \int_Y f (y) \, \mu (x, dy) \, d\rho (x) . \qquad \square$$

Let us fix a moment t. For an arbitrary ω define the counting measure supported by the jumps of $s \mapsto X (s, \omega)$ in $[0, t]$. Denote this random measure by $\mu^X (t, \omega, \Lambda) = \mu_t^X (\omega, \Lambda)$. That is

$$\mu_t^X (\omega, \Lambda) \overset{\circ}{=} \sum_{0 < s \leq t} \chi_\Lambda (\Delta X (s, \omega)) = N^\Lambda (t, \omega) . \qquad (7.7)$$

In general the process X is fixed so in order to simplify the notation as much as possible we shall drop the superscript X and instead of μ^X we shall simply write μ. If $0 \notin \mathrm{cl} (\Lambda)$ then by (7.7) $\mu_t (\omega, \Lambda)$ is measurable in ω. Obviously if $\Lambda \subseteq \mathbb{R} \setminus \{0\}$ then

$$\mu (t, \omega, \Lambda) = \lim_{n \to \infty} \mu (t, \omega, \Lambda \cap [-1/n, 1/n]^c) ,$$

so $\mu_t (\omega, \Lambda)$ is also measurable in ω for any Borel measurable subset Λ of $\mathbb{R} \setminus \{0\}$. This implies that $\mu_t (\omega, \Lambda)$ is a random measure over $\mathbb{R} \setminus \{0\}$. Hence

$$\Lambda \mapsto \nu_t (\Lambda) \overset{\circ}{=} \mathbf{E} (\mu_t (\Lambda)) \overset{\circ}{=} \int_\Omega \mu_t (\omega, \Lambda) \, d\mathbf{P} (\omega) , \quad \Lambda \in \mathcal{B} (\mathbb{R} \setminus \{0\})$$

is a measure on $(\mathbb{R} \setminus \{0\}, \mathcal{B} (\mathbb{R} \setminus \{0\}))$. If $0 \notin \mathrm{cl} (\Lambda)$ then $\nu_t (\Lambda)$ is the expected value of a Poisson process at a fixed time, therefore $\nu_t (\Lambda) < \infty$. Therefore ν_t is σ-finite for every t.

Definition 7.19 *The measures*

$$\nu_t(\Lambda) \stackrel{\circ}{=} \mathbf{E}(\mu_t(\Lambda)), \quad \Lambda \in \mathcal{B}(\mathbb{R} \setminus \{0\})$$

are called the spectral measures *of X. To simplify the notation let* $\nu \stackrel{\circ}{=} \nu_1$.

Lemma 7.20 $\nu_t(\Lambda) = t \cdot \nu_1(\Lambda) \stackrel{\circ}{=} t \cdot \nu(\Lambda)$.

Proof. If $0 \notin \mathrm{cl}(\Lambda)$ then N^Λ is a Poisson process. In this case

$$\nu_t(\Lambda) \stackrel{\circ}{=} \mathbf{E}\left(N^\Lambda(t)\right) = t \cdot \mathbf{E}\left(N^\Lambda(1)\right) \stackrel{\circ}{=} t\nu(\Lambda).$$

In the general case by the Monotone Convergence Theorem

$$\nu_t(\Lambda) = \mathbf{E}\left(\lim_{n \to \infty} \mu_t\left(\Lambda \cap [-1/n, 1/n]^c\right)\right) =$$
$$= \lim_{n \to \infty} \mathbf{E}\left(\mu_t\left(\Lambda \cap [-1/n, 1/n]^c\right)\right) =$$
$$= \lim_{n \to \infty} t \cdot \nu\left(\Lambda \cap [-1/n, 1/n]^c\right) = t \cdot \nu(\Lambda).$$

$\square$

Proposition 7.21 (L^1-identity) *If X is a Lévy process then for every Borel measurable function* $f : \mathbb{R} \setminus \{0\} \to \mathbb{R}$

$$\mathbf{E}\left(\int_{\mathbb{R} \setminus \{0\}} f d\mu_t\right) = \mathbf{E}\left(\sum_{0 < s \leq t} f(\Delta X(s)) \chi(\Delta X(s) \neq 0)\right)$$

$$= \int_{\mathbb{R} \setminus \{0\}} f d\nu_t = t \int_{\mathbb{R} \setminus \{0\}} f d\nu, \tag{7.8}$$

whenever the integral $\int_{\mathbb{R} \setminus \{0\}} f d\nu$ *is meaningful.*

Proof. As $\mu_t(\omega, \Lambda)$ is a counting measure for ever Borel measurable function f

$$\int_{\mathbb{R} \setminus \{0\}} f(x) \mu_t(\omega, dx) = \sum_{0 < s \leq t} f(\Delta X(s, \omega)) \chi(\Delta X(s) \neq 0).$$

The other parts of (7.8) are direct consequences of the previous proposition. $\square$

Corollary 7.22 *Let X be a Lévy process. If* $0 \notin \mathrm{cl}(\Lambda)$ *and* $\int_\Lambda x d\nu(x)$ *is finite then*

$$J^\Lambda(t) - \mathbf{E}\left(J^\Lambda(t)\right) = J^\Lambda(t) - t \int_\Lambda x d\nu(x) \tag{7.9}$$

is a martingale. In particular if Λ *is bounded and* $0 \notin \mathrm{cl}(\Lambda)$ *then (7.9) is a martingale.*

Proof. As $\int_\Lambda x d\nu(x) \doteq \int_{\mathbb{R}\setminus\{0\}} x \chi_\Lambda(x) d\nu(x)$ is finite by the L^1-identity with $f(x) \doteq x\chi_\Lambda(x)$

$$\mathbf{E}\left(J^\Lambda(t)\right) \doteq \mathbf{E}\left(\sum_{0\le s\le t} \Delta X(s)\chi_\Lambda(\Delta X(s))\right) =$$

$$= t\int_{\mathbb{R}\setminus\{0\}} x\chi_\Lambda(x) d\nu(x) = t\int_\Lambda x d\nu(x).$$

X is a Lévy process so J^Λ has independent increments. As $0 \notin \mathrm{cl}(\Lambda)$ the jumps in Λ cannot accumulate. So J^Λ has right-regular trajectories. This implies that $J^\Lambda(t) - \mathbf{E}\left(J^\Lambda(t)\right)$ is a martingale. $\square$

Let $\mathcal{P}$ denote the σ-algebra of the predictable sets. By the martingale property of the compensated jumps it is clear that if $0 \notin \mathrm{cl}(\Lambda)$, $F \in \mathcal{F}_s$ and $s < t$ then

$$\int_F \mu(t,\omega,\Lambda) - t\cdot\nu(\Lambda) d\mathbf{P}(\omega) = \int_F \mu(s,\omega,\Lambda) - s\cdot\nu(\Lambda) d\mathbf{P}(\omega).$$

This means that as $\nu(\Lambda) < \infty$

$$\int_F \mu(t,\omega,\Lambda) - \mu(s,\omega,\Lambda) d\mathbf{P}(\omega) = \int_F (t-s)\cdot\nu(\Lambda) d\mathbf{P}(\omega),$$

that is if

$$H(u,\omega,e) \doteq \chi_\Lambda(e)\chi_F(\omega)\chi_{(s,t]}(u)$$

then

$$\mathbf{E}\left(\int_0^\infty \int_{\mathbb{R}\setminus\{0\}} H\mu(du,\omega,de)\right) = \mathbf{E}\left(\int_0^\infty \int_{\mathbb{R}\setminus\{0\}} H d\nu(e) du\right).$$

The meaning of the left-hand side is the following. For every ω let $\mu(\omega, D)$ denote[22] the counting measure of the jumps of X, that is if $D \in \mathcal{B}(\mathbb{R}_+) \times \mathcal{B}(\mathbb{R}\setminus\{0\})$ then let $\mu(\omega, D)$ be the number of jumps in D. First we integrate by this measure and then, if it is meaningful, we take the expected value. If the time interval is finite and we restrict μ to a set with $\nu(\Lambda) < \infty$ then the set of bounded processes for which the formula is valid is a linear space. From this in the usual way, using the Monotone Class Theorem and the Monotone

---

[22]See: Definition 7.44, page 496.

Convergence Theorem, one can prove the following:

Proposition 7.23 (General L^1-identity) *If $H \geq 0$ is measurable with respect to $\mathcal{P} \times \mathcal{B}(\mathbb{R} \setminus \{0\})$ then*

$$\mathbf{E}\left(\int_0^\infty \int_{\mathbb{R}\setminus\{0\}} H(u,\omega,e)\,\mu(du,\omega,de)\right) = \mathbf{E}\left(\int_0^\infty \int_{\mathbb{R}\setminus\{0\}} H(u,\omega,e)\,d\nu(e)\,du\right).$$

Example 7.24 The Lévy–Khintchine formula for compound Poisson processes.

Let X be a Lévy process and let $0 \notin \mathrm{cl}(\Lambda)$. Let J^Λ be the compound Poisson process of the jumps of X. The Fourier transform of $J^\Lambda(s)$ is[23]

$$\exp\left(\lambda s\left(\int_{\mathbb{R}} (\exp(iux) - 1)\,dF(x)\right)\right),$$

where F is the common distribution function of the jumps, and λ is the parameter of the underlying Poisson process. What is the relation between F and ν? If $B \in \mathcal{B}(\mathbb{R}\setminus\{0\})$ and τ is the time of the first jump in Λ then by the general L^1-identity using that $\chi([0,\tau])$ is predictable

$$F(B) = \mathbf{P}(\Delta X(\tau) \in B \cap \Lambda) = \mathbf{E}(\chi_{B\cap\Lambda}(\Delta X(\tau))) =$$

$$= \mathbf{E}\left(\int_0^\infty \chi([0,\tau])\,\mu(du, B \cap \Lambda)\right) =$$

$$= \mathbf{E}\left(\int_0^\infty \int_{\mathbb{R}\setminus\{0\}} \chi_{B\cap\Lambda}(e)\,\chi([0,\tau])\,\mu(du,de)\right) =$$

$$= \mathbf{E}\left(\int_0^\infty \int_{\mathbb{R}\setminus\{0\}} \chi_{B\cap\Lambda}(e)\,\chi([0,\tau])\,d\nu(e)\,du\right) =$$

$$= \nu(B \cap \Lambda)\,\mathbf{E}\left(\int_0^\infty \chi([0,\tau])\,du\right) =$$

$$= \nu(B \cap \Lambda)\,\mathbf{E}(\tau) = \frac{\nu(B \cap \Lambda)}{\lambda}.$$

That is the Fourier transform of $J^\Lambda(s)$ is

$$\exp\left(s\left(\int_\Lambda (\exp(iux) - 1)\,d\nu(x)\right)\right). \qquad \square$$

---

[23]See: Lemma 7.7, page 467.

Definition 7.25 *Let* $(E, \mathcal{E}, \nu)$ *be a measure space and let* $(\Omega, \mathcal{A}, \mathbf{P})$ *be a proba-bility space. We say that the random measure* $\mu : \Omega \times \mathcal{E} \to [0, \infty]$ *is a random* Poisson measure *with* control measure ν *if:*

1. *whenever the sets* (Λ_k) *are disjoint the variables* $\mu(\omega, \Lambda_k)$ *are independent and*
2. *whenever* $\nu(\Lambda) < \infty$ *the variable* $\mu(\omega, \Lambda)$ *has a Poisson distribution with parameter* $\nu(\Lambda)$.

Proposition 7.26 *Let* X *be a Lévy process. For every* t *the counting measure* $\mu_t(\omega, \Lambda)$ *is a random Poisson measure. The control measure of* μ_t *is the spectral measure* ν_t.

Proof. For every $\Lambda \subseteq \mathbb{R} \setminus \{0\}$ let $\Lambda_n \triangleq \Lambda \cap \left[-\frac{1}{n}, \frac{1}{n}\right]^c$. Obviously $0 \notin \mathrm{cl}(\Lambda_n)$ and

$$\mu(t, \omega, \Lambda) = \lim_{n \to \infty} \mu(t, \omega, \Lambda_n).$$

As $0 \notin \mathrm{cl}(\Lambda_n)$ the variable

$$\omega \mapsto \mu(t, \omega, \Lambda_n) = N^{\Lambda_n}(t, \omega)$$

has a Poisson distribution. The Fourier transform of this variable is

$$\exp(t\nu(\Lambda_n)(\exp(iu) - 1)).$$

The convergence for every ω implies the weak convergence, so if $\nu(\Lambda) < \infty$, then as $\nu(\Lambda) = \lim_{n \to \infty} \nu(\Lambda_n)$ the Fourier transform of $\omega \mapsto \mu(t, \omega, \Lambda)$ is

$$\exp(t\nu(\Lambda)(\exp(iu) - 1)).$$

Hence it has a Poisson distribution. If the sets

$$\Lambda_k = \cup_n \Lambda_n^{(k)} \triangleq \cup_n \left(\Lambda_k \cap \left[-\frac{1}{n}, \frac{1}{n}\right]^c\right)$$

are disjoint then the sets $\Lambda_n^{(k)}$ are also disjoint for every n. Hence the variables

$$\mu\left(t, \omega, \Lambda_n^{(k)}\right)$$

are independent. The limit of independent variables is independent, so if the sets (Λ_k) are disjoint, then the variables $\mu(t, \omega, \Lambda_k)$ are independent.

$\square$

Definition 7.27 *Let H be a Hilbert space and let $(C, \mathcal{C}, \nu)$ be a measure space and let $\mathcal{S} \subseteq \mathcal{C}$ denote the subsets of $\mathcal{C}$ with finite measure. $\pi : \mathcal{S} \to H$ is a vector measure with control measure ν if for every $S \in \mathcal{S}$:*

1. *$\pi(S) \in H$ is defined,*
2. *$\|\pi(S)\|_H^2 = \nu(S)$,*
3. *if S_1 and S_2 are disjoint sets in $\mathcal{S}$ then the vectors $\pi(S_1)$ and $\pi(S_2)$ are orthogonal.*

We say that a function $f : C \to \mathbb{R}$ is integrable with respect to π if there is a sequence of finite valued step functions $(s_n) = \left(\sum_k c_{nk} \chi_{C_{nk}} \right)$ with:

1. *$s_n \to f$ in $L^2(\nu)$ and*
2. *$I_n \doteq \sum_k c_{nk} \pi(C_{nk})$ is a Cauchy sequence in H. If $I \doteq \lim_{n \to \infty} I_n$, then we shall call this limit I the integral of f with respect to π. We shall denote this integral by $\int_C f(x) \, d\pi(x)$ or simply $\int_C f \, d\pi$.*

Proposition 7.28 *If $f \in L^2(C, \mathcal{C}, \nu)$ and π is a vector measure with control measure $(C, \mathcal{C}, \nu)$ then f is integrable with respect to π and*

$$\left\| \int_C f \, d\pi \right\|_H = \|f\|_2 \doteq \sqrt{\int_C f^2 \, d\nu}. \tag{7.10}$$

Proof. Let $s \doteq \sum_k c_k \cdot \chi_{C_k}$ where C_k are disjoint and in $\mathcal{S}$. By conditions 3. and 2.

$$\left\| \int_C s \, d\pi \right\|_H^2 = \left\| \sum_k c_k \cdot \pi(C_k) \right\|_H^2 = \tag{7.11}$$

$$= \sum_k c_k^2 \cdot \|\pi(C_k)\|_H^2 =$$

$$= \sum_k c_k^2 \cdot \nu(C_k) = \int_C s^2 \, d\nu = \|s\|_2^2.$$

As the step functions are dense in L^2 there is a sequence $s_n \doteq \sum_k c_{nk} \chi_{C_{nk}}$ with $s_n \to f$ in $L^2(\nu)$. From (7.11) $I_n \doteq \sum_k c_{nk} \pi(C_{nk})$ is a Cauchy sequence in H. From this the proposition is obvious. $\qquad \square$

Corollary 7.29 *If $f \in L^2(C, \mathcal{C}, \nu)$ and π is a vector measure with control measure ν then the value of the vector integral $\int_C f \, d\pi$ is independent of the approximating sequence (s_n).*

Proposition 7.30 *If X is a Lévy process and $H \doteq L^2(\Omega)$ then for every $t \geq 0$*

$$\pi_t(\Lambda) \doteq N^\Lambda(t) - \nu_t(\Lambda) = N^\Lambda(t) - t \cdot \nu(\Lambda) \tag{7.12}$$

is a a Hilbert space valued vector measure over $(\mathbb{R} \setminus \{0\}, \mathcal{B}(\mathbb{R} \setminus \{0\}), \nu_t)$. *The same is true if* $H \doteq \mathcal{H}_0^2$ *on the time interval* $[0, t]$ *and*

$$(\pi(\Lambda))(s) \doteq N^\Lambda(s) - s \cdot \nu(\Lambda), \quad s \leq t < \infty.$$

Proof. As we have already proved, if $\Lambda \subseteq \mathbb{R} \setminus \{0\}$ and $\nu_t(\Lambda) < \infty$ then the Fourier transform of $N^\Lambda(t)$ is

$$\exp(\nu_t(\Lambda)(\exp(iu) - 1)).$$

Hence if $\nu_t(\Lambda) < \infty$, then $N^\Lambda(t)$ has a Poisson distribution with parameter $\nu_t(\Lambda)$. This implies that the expected value of (7.12) is zero and $\|\pi_t\|_H^2 = \nu_t(\Lambda)$. As we have also proved that if $\Lambda_1 \cap \Lambda_2 = \emptyset$ then $N^{\Lambda_1}(t)$ and $N^{\Lambda_2}(t)$ are independent[24]. So

$$(\pi_t(\Lambda_1), \pi_t(\Lambda_2)) \doteq \int_\Omega \pi_t(\Lambda_1)\pi_t(\Lambda_2)\,d\mathbf{P} = 0. \qquad \square$$

7.1.4 Decomposition of Lévy processes

Now we are ready to prove that Lévy processes are semimartingales.

Proposition 7.31 *If X is a Lévy process then:*

1. *X is a semimartingale,*
2. *X has a decomposition*

$$X = V + M$$

where:

3. *V and M are independent Lévy processes,*
4. *M is a martingale with bounded jumps and on every finite interval $M \in \mathcal{H}_0^2$,*
5. *$V \in \mathcal{V}$, that is on every finite interval the trajectories of V have finite variation.*

Proof. If $\Lambda \doteq \{|x| \geq 1\}$ then the jumps of $Y \doteq X - J^\Lambda$ are bounded. Y is a Lévy process with bounded jumps[25]. This implies that $Y(t)$ has an expected value[26] for every t. Therefore

$$M(t) \doteq Y(t) - \mathbf{E}(Y(t)) =$$
$$= X(t) - J^\Lambda(t) - t \cdot \mathbf{E}(X(1) - J^\Lambda(1)) \doteq$$
$$\doteq X(t) - J^\Lambda(t) - t \cdot \gamma$$

---

[24]See: Proposition 7.16. page 472.

[25]See: Lemma 7.8, page 468.

[26]See: Proposition 1.111, page 74.

is a Lévy process with zero expected value. Hence M is a martingale. The martingale M has finite moments, so on any finite interval M is in $\mathcal{H}_0^2$. Therefore M satisfies 4. Obviously

$$V\left(t\right) \overset{\circ}{=} J^{\Lambda}\left(t\right) + \mathbf{E}\left(Y\left(t\right)\right) \overset{\circ}{=} J^{\Lambda}\left(t\right) + \gamma \cdot t$$

satisfies 5. As $X - J^{\Lambda}$ and J^{Λ} are independent[27] the proposition holds.
$\square$

Corollary 7.32 *The spectral measure ν has the following properties*

$$\nu\left(|x| \geq 1\right) < \infty, \quad \int_{0 < |x| < 1} x^2 \, d\nu\left(x\right) < \infty.$$

That is

$$\int_{\mathbb{R} \setminus \{0\}} \left(1 \wedge x^2\right) \, d\nu\left(x\right) < \infty. \tag{7.13}$$

Proof. $M \in \mathcal{H}_0^2$ therefore $M^2 - [M]$ is a martingale[28]. Let $\Lambda \overset{\circ}{=} \{0 < |x| < 1\}$. By the L^1-identity (7.8)

$$t \int_{\Lambda} x^2 \, d\nu\left(x\right) = \mathbf{E}\left(\int_{\Lambda} x^2 \, d\mu_t\left(x\right)\right) = \mathbf{E}\left(\sum_{s \leq t} \left(\Delta X\left(s\right)\right)^2 \chi\left(\Delta X\left(s\right) \in \Lambda\right)\right) =$$

$$= \mathbf{E}\left(\sum_{s \leq t} \left(\Delta M\left(s\right)\right)^2\right) \leq \mathbf{E}\left([M]\left(t\right)\right) = \mathbf{E}\left(M^2\left(t\right)\right) < \infty.$$

The other relations are obvious.
$\square$

Every element of $\mathcal{H}_0^2$ has a unique decomposition into a sum of a continuous and of a purely discontinuous martingale[29]. For Lévy processes one can prove a bit more:

Proposition 7.33 *The martingale M in the decomposition $X = V + M$ has a decomposition*

$$M = M^c + M^d \tag{7.14}$$

[27]See: Theorem 7.12, page 470.
[28]See: Proposition 2.84, page 170.
[29]See: Corollary 4.18, page 232.

where

1. M^c and M^d are independent Lévy processes,
2. the trajectories of M^c are almost surely continuous,
3. M^d is purely discontinuous and for every t, in $L^2(\Omega)$-convergence

$$M^d(t) = \lim_{\varepsilon \searrow 0} \int_{\varepsilon \leq |x| < 1} x \, d\pi_t(x) = \lim_{\varepsilon \searrow 0} \int_{\varepsilon \leq |x| < 1} x \, d\left(\mu_t(x) - \nu_t(x)\right). \quad (7.15)$$

Proof. Let

$$\Lambda_k \stackrel{\circ}{=} \left\{ \frac{1}{k+1} \leq |x| < \frac{1}{k} \right\}, \quad \Lambda_{n,m} \stackrel{\circ}{=} \left\{ \frac{1}{m+1} \leq |x| < \frac{1}{n} \right\}.$$

1. On any finite interval the processes

$$M^{\Lambda_k}(t) \stackrel{\circ}{=} J^{\Lambda_k}(t) - \mathbf{E}\left(J^{\Lambda_k}(t)\right)$$

are independent $\mathcal{H}_0^2$ martingales. If $U \in \mathcal{H}_0^2$ then $U^2 - [U]$ is a martingale[30] so

$$\mathbf{E}\left(U^2(t)\right) = \mathbf{E}\left([U](t)\right).$$

Hence by Doob's inequality and by the L^1-identity (7.8) if $n < m$

$$\|M_n - M_m\|_{\mathcal{H}^2}^2 \leq 4 \|M_n(t) - M_m(t)\|_2^2 = 4\mathbf{E}\left([M_n - M_m]^2(t)\right) =$$

$$= 4\mathbf{E}\left(\sum_{s \leq t} (\Delta M(s))^2 \chi(\Delta M(s) \in \Lambda_{n,m})\right) =$$

$$= 4\mathbf{E}\left(\int_{\Lambda_{n,m}} x^2 \, d\mu_t(x)\right) = 4t \int_{\Lambda_{n,m}} x^2 \, d\nu(x).$$

By (7.13) and by the Dominated Convergence Theorem (M_n) is a Cauchy sequence in $\mathcal{H}_0^2$. As $\mathcal{H}_0^2$ is a Hilbert space there is a martingale $M_\infty \in \mathcal{H}_0^2$ with $M_n \to M_\infty$. As the processes (M^{Λ_k}) are independent their sum M_n is a Lévy process. By the convergence of the trajectories this implies that M_∞ is also a Lévy process. M^{Λ_k} is a purely discontinuous martingale. The set of purely discontinuous martingales is closed linear subspace in $\mathcal{H}_0^2$, so M^d is purely discontinuous. Let $M \stackrel{\circ}{=} M^c + M^d$ be the decomposition of M. By the uniform convergence of the trajectories $\Delta M_\infty = \Delta M$. As $\Delta M^d = \Delta M$ obviously

---

[30]See: Proposition 2.84, page 170.

$\Delta M^d = \Delta M_\infty$. As M^d and M_∞ are purely discontinuous they are equal[31]. As $M - M_n$ and M_n are independent by the uniform convergence of the trajectories $M^c = M - M^d$ and M^d are also independent. This also implies that M^c is a Lévy-process.

2. $0 \notin \mathrm{cl}\,(\Lambda_k)$ so $\nu\,(\Lambda_k) < \infty$, and Λ_k is bounded. Hence $\int_{\Lambda_k} x d\nu\,(x)$ is meaningful. By the L^1-identity (7.8)

$$M^{\Lambda_k}\,(t) \overset{\circ}{=} J^{\Lambda_k(t)} - \mathbf{E}\left(J^{\Lambda_k}\,(t)\right) = \tag{7.16}$$

$$= \sum_{s \le t} \chi\,(\Delta X\,(s) \in \Lambda_k)\,\Delta X\,(s) - \mathbf{E}\left(J^{\Lambda_k}\,(t)\right) =$$

$$= \int_{\mathbb{R}\backslash\{0\}} \chi_{\Lambda_k}\,(x)\,x\,d\mu_t\,(x) - \mathbf{E}\left(\int_{\mathbb{R}\backslash\{0\}} \chi_{\Lambda_k}\,(x)\,x\,d\mu_t\,(x)\right) =$$

$$= \int_{\Lambda_k} x\,d\mu_t\,(x) - \int_{\Lambda_k} x d\nu_t\,(x) = \int_{\Lambda_k} x\,d\,(\mu_t - \nu_t)\,(x).$$

For every t, in $L^2\,(\Omega)$-convergence

$$M^d\,(t) = \sum_k M^{\Lambda_k}\,(t) = \sum_k \int_{\Lambda_k} x\,d\,(\mu_t\,(x) - \nu_t\,(x)) =$$

$$= \lim_{k \to \infty} \int_{1/k \le |x| < 1} x\,d\,(\mu_t\,(x) - \nu_t\,(x)),$$

and from which (7.15) is evident. $\qquad\qquad\qquad\qquad\qquad\qquad\qquad\square$

Corollary 7.34 *One can also think about* (7.15) *as a $\mathcal{H}_0^2$-valued vector integral*

$$M^d\,(t) = \int_{0 < |x| < 1} x\,d\pi_t\,(x).$$

Proof. Let $\Lambda \overset{\circ}{=} \{0 < |x| < 1\}$ and let $f\,(x) \overset{\circ}{=} x$. As $\int_\Lambda x^2\,d\nu\,(x) < \infty$ obviously $f \in L^2\,(\Lambda, \nu)$. This implies that the vector integral $\int_\Lambda f\,d\pi_t \overset{\circ}{=} \int_\Lambda x\,d\pi_t\,(x)$ is meaningful. Since $f \in L^2\,(\Lambda, \nu_t)$, by the Dominated Convergence Theorem

$$\lim_{k \to \infty} \sqrt{\int_{0 < |x| < 1/k} f^2\,(x)\,d\nu_t\,(x)} = 0.$$

[31] See: Corollary 4.7, page 228.

From (7.10) using that the vector integral is obviously additive

$$\left\| \int_{1/k \leq |x| < 1} x \, d\pi_t (x) - \int_{0 < |x| < 1} x \, d\pi_t (x) \right\|_2 =$$

$$= \left\| \int_{0 < |x| < 1/k} x \, d\pi_t (x) \right\|_2 = \sqrt{\int_{0 < |x| < 1/k} x^2 \, d\nu_t (x)} \to 0.$$

Hence in $L^2 (\Omega)$

$$\lim_{k \nearrow \infty} \int_{1/k \leq |x| < 1} x \, d\pi_t (x) = \int_{0 < |x| < 1} x \, d\pi_t (x).$$

That is

$$\left\| M^d (t) - \int_\Lambda x \, d\pi_t \right\|_2 = \left\| \lim_{k \nearrow \infty} \int_{1/k \leq |x| < 1} x \, d\pi_t (x) - \int_\Lambda x \, d\pi_t \right\|_2 = 0$$

$\square$

Example 7.35 For some Lévy process the pathwise integral

$$\int_{0 < |x| < 1} x \, d (\mu_t - \nu_t) (x)$$

is meaningless[32].

It is natural to ask whether one can define the random signed measure

$$\rho (t, \omega, \Lambda) \doteq \mu_t (\omega, \Lambda) - \nu_t (\Lambda), \quad \Lambda \in \mathcal{B} (\mathbb{R} \backslash \{0\})$$

and whether one can express the limit in (7.15) as an ordinary pathwise integral over $\{0 < |x| < 1\}$ with respect to ρ. Recall that if ρ is a signed measure then by definition

$$\int_A f \, d\rho \doteq \int_A f \, d\rho^+ - \int_A f \, d\rho^-$$

and of course we assume that the integral on the left is meaningful if both integrals on the right are meaningful and one does not get an expressions[33] of

---

[32]See: Example 4.20, page 233.

[33]Of course one does not have this problem if ν is a finite measure. But in general ν is just σ-finite, so $\nu (\Lambda) = \infty$ is possible.

the type $\infty - \infty$. Therefore it is sufficient to show that for some Lévy process M

$$\lim_{\varepsilon \searrow 0} \int_{\varepsilon < |x| < 1} x \, d \left(\mu_t (x) - \nu_t (x) \right)$$

is finite, but

$$\int_{0 < |x| < 1} x \, d\nu_t (x) = \int_{0 < |x| < 1} x \, d\mu_t (x) =$$

$$= \sum_{s \leq t} \chi \left(|\Delta M| < 1 \right) \Delta M (s) \equiv \infty.$$

Let (N_i) be a sequence of independent Poisson processes with $\lambda = 1$. For any t the compensated Poisson processes on the finite time horizon $[0, t]$

$$M_i (t) \stackrel{\circ}{=} N_i (t) - \lambda t = N_i (t) - t$$

are in $\mathcal{H}_0^2$. As they are independent, they are also orthogonal in $\mathcal{H}_0^2$. As $\sum_i 1/i^2 < \infty$ the sequence

$$M \stackrel{\circ}{=} \sum_{i=1}^{\infty} \frac{1}{i} M_i$$

is convergent in the Hilbert space $\mathcal{H}_0^2$. As the processes N_i are independent they almost surely do not have common jumps[34], so all the jumps are not larger than one and obviously they are non-negative. By Fubini's theorem and by the Monotone Convergence Theorem

$$\int_0^1 x \, d\mu_t = \sum_{s \leq t} \Delta M (s) = \sum_{s \leq t} \sum_{i=1}^{\infty} \frac{\Delta M_i (s)}{i} = \sum_{i=1}^{\infty} \sum_{s \leq t} \frac{\Delta M_i (s)}{i} =$$

$$= \sum_{i=1}^{\infty} \sum_{s \leq t} \frac{\Delta N_i (s)}{i} = \sum_{i=1}^{\infty} \frac{N_i (t)}{i}.$$

The variables $N_i (t) - t$ are independent, they have zero expected value. So for any t the sequence

$$R_n \stackrel{\circ}{=} \sum_{i=1}^{n} \frac{N_i (t) - t}{i}$$

is a discrete time martingale. Obviously (R_n) is bounded in $L^2 (\Omega)$ so by the Martingale Convergence Theorem it is convergent almost surely. As $\sum_i 1/i = \infty$

---

[34]See: Corollary 7.14, page 471.

obviously

$$\sum_{s \leq t} \Delta M(s) = \sum_{i=1}^{\infty} \frac{N_i(t)}{i} = \infty.$$

The spectral measure ν of M is $\nu \{1/i\} \overset{\circ}{=} 1$. Therefore $\nu((0,1]) = \infty$ and

$$\int_0^1 x \, d\nu(x) = \sum_{i=1}^{\infty} \frac{1}{i} = \infty$$

but

$$\int_0^1 x^2 \, d\nu(x) = \sum_{i=1}^{\infty} \frac{1}{i^2} < \infty.$$

From Kolmogorov's zero–one law it is also clear, that if $t > 0$ then $\mu_t(\omega, (0,1)) = \infty$ almost surely, which implies that the signed measure ρ_t is almost surely meaningless. $\square$

7.1.5   Lévy–Khintchine formula for Lévy processes

1. If $0 \notin \mathrm{cl}(\Lambda)$ then $N^\Lambda(t)$ has a Poisson distribution with parameter $\lambda \overset{\circ}{=} t\nu(\Lambda)$. So its Fourier transform is

$$\varphi_t(u) \overset{\circ}{=} \mathbf{E}\left(\exp\left(iuN^\Lambda(t)\right)\right) = \exp\left(t\nu(\Lambda)(\exp(iu) - 1)\right).$$

2. If $0 \notin \mathrm{cl}(\Lambda)$, e.g. if $\Lambda \overset{\circ}{=} \{|x| \geq 1\}$, then by the general L^1-identity the Fourier transform of J^Λ is[35]

$$\exp\left(t \int_\Lambda (\exp(iux) - 1) \, d\nu(x)\right).$$

3. If $\nu(\Lambda)$ is finite and if Λ is bounded then $\int_\Lambda x \, d\nu(x)$ is finite. So in this case the Fourier transform of

$$J^\Lambda(t) - t \int_\Lambda x \, d\nu(x) \tag{7.17}$$

is

$$\exp\left(t \int_\Lambda (\exp(iux) - 1 - iux) \, d\nu(x)\right).$$

[35]See: Example 7.24, page 477.

4. By (7.15) M^d is a limit in L^2 of processes in (7.17). L^2-convergence implies weak convergence, hence the Fourier transform of M^d is

$$\lim_{\varepsilon \searrow 0} \exp\left(t \cdot \phi_\varepsilon\left(u\right)\right) = \exp\left(\lim_{\varepsilon \searrow 0} t \cdot \phi_\varepsilon\left(u\right)\right),$$

where

$$\phi_\varepsilon\left(u\right) \overset{\circ}{=} \int_{\varepsilon < |x| < 1} \exp\left(iux\right) - 1 - iux \, d\nu\left(x\right).$$

On any bounded interval

$$|\exp\left(iux\right) - 1 - iux| \leq k x^2.$$

By (7.13) $f\left(x\right) \overset{\circ}{=} x^2$ is ν-integrable on $0 < |x| < 1$, hence by the Dominated Convergence Theorem

$$\lim_{\varepsilon \searrow 0} \int_{\varepsilon < |x| < 1} \exp\left(iux\right) - 1 - iux \, d\nu\left(x\right) = \int_{0 < |x| < 1} \exp\left(iux\right) - 1 - iux \, d\nu\left(x\right).$$

5. This implies that the Fourier transform of the jump part of X is

$$\varphi_t\left(u\right) = \exp\left(t \cdot \phi\left(u\right)\right), \tag{7.18}$$

where

$$\phi\left(u\right) \overset{\circ}{=} \int_{|x| \geq 1} \exp\left(iux\right) - 1 \, d\nu\left(x\right) + \int_{0 < |x| < 1} \exp\left(iux\right) - 1 - iux \, d\nu\left(x\right) =$$

$$= \int_{\mathbb{R} \setminus \{0\}} \exp\left(iux\right) - 1 - iux\chi\left(|x| < 1\right) \, d\nu\left(x\right).$$

6. As we know every continuous Lévy process is a linear combination of a Wiener process and a linear trend[36]. As M^c is a martingale $M^c = \sigma w$ where w is a Wiener process.

Recall that if X is a Lévy process then

$$X\left(t\right) = J^{\{|x| \geq 1\}}\left(t\right) + \gamma t + M^d\left(t\right) + M^c\left(t\right)$$

[36]See: Theorem 6.11, page 367.

where γ is the expected value of the small jumps. So we have proved the next famous theorem.

Theorem 7.36 (Lévy–Khintchine formula) *Let X be a Lévy process. If* $\varphi_t(u) = \exp(t\phi(u))$ *is the Fourier transform of $X(t)$ then*

$$\phi(u) \stackrel{\circ}{=} iu\gamma - \frac{\sigma^2 u^2}{2} + \tag{7.19}$$

$$+ \int_{\mathbb{R}\backslash\{0\}} \exp(iux) - 1 - iux \cdot \chi(|x| < 1)\, d\nu(x).$$

where ν is the spectral measure of X and

$$\int_{\mathbb{R}\backslash\{0\}} x^2 \wedge 1 d\nu(x) < \infty. \tag{7.20}$$

Proof. It is sufficient to remark that if w is a Wiener process then its characteristic function at time t is

$$\exp\left(-t\frac{u^2}{2}\right)$$

and if X is a linear trend that is $X(t) = \gamma t$ then the characteristic function of $X(t)$ is

$$\exp(i\gamma ut). \qquad \qquad \square$$

Definition 7.37 *The triplet (γ, σ, ν) is called the* characteristics *of the Lévy process X.*

Corollary 7.38 *If X is a non-negative Lévy process then the Laplace transform of $X(t)$ has the representation*

$$L_t(s) = \exp\left(t\left(-\beta s + \int_0^\infty \exp(-sx) - 1 d\nu(x)\right)\right), \quad s \geq 0$$

where $\beta \geq 0$ and ν is the spectral measure of X. In this case

$$\nu((-\infty, 0)) = 0, \quad \int_0^\infty x \wedge 1 d\nu_t(x) < \infty. \tag{7.21}$$

Proof. As $X(t) \geq 0$ its Laplace transform

$$L_t(s) \stackrel{\circ}{=} \mathbf{E}(\exp(-sX(t))), \quad s \geq 0$$

is finite. As $X \geq 0$ and as X has stationary increments almost surely the trajectories of X are increasing. Hence

$$\int_0^\infty x \wedge 1 d\nu_t(x) = \mathbf{E}\left(\sum_{s \leq t} \Delta X(s) \wedge 1\right) \leq \mathbf{E}\left(\left(X - J^{\{x \geq 1\}}\right)(t)\right) < \infty$$

as the jumps of $X - J^{\{x \geq 1\}}$ are bounded. From this (7.21) is clear. By (7.21) one can separate the integral in (7.19) as

$$\int_0^\infty \exp(iux) - 1 d\nu(x) - iu \int_0^1 x d\nu(x)$$

and we can join the second term into the constant. It is easy to see that

$$\int_0^\infty \exp(-sx) - 1 d\nu(x)$$

is finite for every $s \geq 0$ so one can extend the Fourier transform analytically to the complex half-plane $\{-s + iu, s \geq 0\}$. That is one can put $-s$ on the place of iu and the Laplace transform of $X(t)$ has the representation

$$\exp\left(t\left(-\beta s + \frac{\sigma^2 s^2}{2} + \int_0^\infty \exp(-sx) - 1 d\nu(x)\right)\right).$$

As X is increasing in t the Laplace transform is decreasing for every s. But it can happen only if $\sigma = 0$. Hence $L_t(s)$ has the stated representation. $\square$

7.1.6 Construction of Lévy processes

Let us assume that (γ, σ, ν) satisfies (7.20). We want to construct[37] a Lévy process X with characteristics (γ, σ, ν). To do this one needs to construct a right-continuous process with Fourier transform satisfying the Lévy–Khintchine formula with the given triplet.

1. As a first step one should construct a Wiener process[38] w. If $X^c(t) \stackrel{\circ}{=} \gamma t + \sigma w(t)$ then the Fourier transform of $X^c(t)$ is

$$\exp\left(t \cdot \left(iu\gamma - \frac{\sigma^2 u^2}{2}\right)\right).$$

---

[37]One can easily see that the probability space $(\Omega, \mathcal{A}, \mathbf{P})$ should be rich enough to carry a countable number of independent random variables. One can assume that $(\Omega, \mathcal{A}, \mathbf{P}) = ([0, 1], \mathcal{B}, \lambda)$.

[38]See: Theorem B.13, page 567.

This means that it will be sufficient to construct a Lévy process X^d with Fourier transform

$$\exp\left(t\int_{\mathbb{R}\setminus\{0\}}\exp\left(iux\right)-1-iux\cdot\chi\left(|x|<1\right)d\nu\left(x\right)\right).$$

2. As a second step let us construct a random Poisson measure with control measure ν. First assume that ν is finite. Let (ξ_k) be a sequence of independent random variables with distribution $F\left(B\right)\overset{\circ}{=}\nu\left(B\right)/\nu\left(\mathbb{R}\setminus\{0\}\right)$ and let N be a Poisson process independent from the sequence (ξ_k) with parameter $\lambda=\nu\left(\mathbb{R}\setminus\{0\}\right)$. If $U\overset{\circ}{=}\sum_{k=1}^{N}\xi_k$ then its Fourier transform is

$$\varphi_t\left(u\right)=\exp\left(t\left(\lambda\int_{\mathbb{R}}\exp\left(iux\right)-1dF\left(x\right)\right)\right)=\exp\left(t\int_{\mathbb{R}}\exp\left(iux\right)-1d\nu\left(x\right)\right).$$

As we have seen[39] the random measure generated by the jumps of U is a random Poisson measure with control measure ν.

3. If $0\notin\mathrm{cl}\left(\Lambda\right)$ then for some $\varepsilon>0$

$$\nu\left(\Lambda\right)=\frac{\varepsilon^2}{\varepsilon^2}\nu\left(\Lambda\right)\le\frac{1}{\varepsilon^2}\int_\Lambda x^2 d\nu\left(x\right)<\infty,$$

and by the previous step one can easily construct a process J^Λ with Fourier transform

$$\exp\left(t\int_\Lambda\exp\left(iux\right)-1d\nu\left(x\right)\right).$$

As a special case one can construct a Lévy process with Fourier transform

$$\exp\left(t\int_{|x|\ge1}\exp\left(iux\right)-1d\nu\left(x\right)\right).$$

4. The only problem is the convergence of the compensated sum

$$\sum_k\left(J^{\Lambda_k}\left(t\right)-t\int_{\Lambda_k}xd\nu\left(x\right)\right)\overset{\circ}{=}\sum_k M^{\Lambda_k}\left(t\right)$$

where

$$\Lambda_k\overset{\circ}{=}\left\{\frac{1}{k+1}<|x|\le\frac{1}{k}\right\}.$$

[39] See: Example 7.24, page 477.

By Doob's inequality, if $\beta < \alpha$ then[40]

$$\left\| \sup_{0 \le s \le t} \left| M^{(\alpha,1)}(s) - M^{(\beta,1)}(s) \right| \right\|_2^2 \le 4 \left\| M^{(\alpha,1)}(t) - M^{(\beta,1)}(t) \right\|_2^2 =$$

$$= 4 \left\| M^{(\beta,\alpha]}(t) \right\|_2^2 = 4t \int_\beta^\alpha x^2 d\nu(x).$$

As $\int_{0 < |x| < 1} x^2 d\nu(x) < \infty$ on every finite interval $[0, t]$ the processes $\left(M^{(1/n,1)} \right)$ form a Cauchy sequence in $\mathcal{H}_0^2$. So on every finite interval $\left(M^{(1/n,1)} \right)$ has a limit in $\mathcal{H}_0^2$. The limit is a Lévy process with Fourier transform $\varphi_t(u) = \exp(t\phi(u))$ where ϕ is the function in (7.19).

Example 7.39 Symmetric stable processes.

Perhaps the simplest construction of a non-trivial Lévy process is the following: assume that ν is symmetric, that is if $A \subseteq \mathbb{R}_+ \setminus \{0\}$ then $\nu(-A) = \nu(A)$. Let us also assume that $\nu((x, \infty)) \stackrel{\circ}{=} x^{-\alpha}$ with some $\alpha > 0$. In this case

$$\nu((a, b]) = a^{-\alpha} - b^{-\alpha}.$$

By (7.20)

$$\int_0^1 x^2 d\nu(x) = \int_0^1 \alpha x^{-\alpha-1} x^2 dx = \alpha \int_0^1 x^{-\alpha+1} dx < \infty.$$

This happens only if $-\alpha + 1 > -1$, that is if $\alpha < 2$. One can prove[41] that if $(\gamma, \sigma, \nu) = (0, 0, \nu)$ then the Fourier transform is $\varphi_t(u) = \exp(-t \cdot (c |u|^\alpha))$, that is the distribution of the increments is α-stable. □

7.1.7 Uniqueness of the representation

Sometimes the Lévy–Khintchine formula is written in a different way. Instead of the representation of $\phi(u)$ above one can write

$$\phi(u) \stackrel{\circ}{=} iu\gamma' - \frac{\sigma^2 u^2}{2} + \int_{\mathbb{R} \setminus \{0\}} \left(\exp(iux) - 1 - \frac{iux}{1 + x^2} \right) d\nu(x)$$

[40]See: Corollary 7.32, page 481.
[41]See: [23].

since the difference of the two integrals is $iu\Delta\gamma$, where

$$\Delta\gamma \triangleq \left| \int_{\mathbb{R}\setminus\{0\}} x \left(\chi\left(|x| < 1\right) - \frac{1}{1+x^2} \right) d\nu\left(x\right) \right| \leq$$

$$\leq \int_{0<|x|<1} x^2 d\nu\left(x\right) + \int_{|x|\geq 1} 1 d\nu\left(x\right) < \infty.$$

Since

$$\frac{x^2}{1+x^2} \leq \min\left(1, x^2\right) \leq 2 \cdot \frac{x^2}{1+x^2}$$

$\min\left(1, x^2\right)$ is ν-integrable if and only if

$$\rho\left(A\right) \triangleq \int_A \frac{x^2}{1+x^2} d\nu, \quad A \in \mathcal{B}\left(\mathbb{R}\setminus\{0\}\right)$$

is a finite measure.

Definition 7.40 *The kernel function*

$$H\left(u, x\right) \triangleq \begin{cases} \left(\exp\left(iux\right) - 1 - \dfrac{iux}{1+x^2}\right) \dfrac{1+x^2}{x^2} & \text{if } x \neq 0 \\ -u^2/2 & \text{if } x = 0 \end{cases}$$

is called the Lévy–Khintchine kernel.

The Lévy–Khintchine kernel has some useful properties. For a fixed u

$$\left(\exp\left(iux\right) - 1 - \frac{iux}{1+x^2}\right) \frac{1+x^2}{x^2} \tag{7.22}$$

is obviously bounded in x outside any neighborhood of $x = 0$.

$$\lim_{x\to 0} H\left(u, x\right) = \lim_{x\to 0} \left(\frac{\exp\left(iux\right) - 1}{x^2} - \frac{iux}{x^2\left(1+x^2\right)} \right) \left(1 + x^2\right) =$$

$$= \lim_{x\to 0} \left(\frac{\exp\left(iux\right) - 1 - iux}{x^2} - iux \left(\frac{1}{x^2\left(1+x^2\right)} - \frac{1}{x^2} \right) \right) =$$

$$= \lim_{x\to 0} \frac{\exp\left(iux\right) - 1 - iux}{x^2} = -\frac{u^2}{2}$$

so $x \mapsto H\left(u, x\right)$ is continuous in $x = 0$ for every u. Therefore H is bounded and it is continuous in x on $\mathbb{R}$. This implies that if ρ is a finite measure then for every

u one can define the integral

$$\int_{\mathbb{R}} H\left(u,x\right) d\rho\left(x\right) = \int_{\mathbb{R}} \left(\exp\left(iux\right) - 1 - \frac{iux}{1+x^2}\right) \frac{1+x^2}{x^2} \, d\rho\left(x\right). \qquad (7.23)$$

We can also assume that $\rho\left(\{0\}\right) \stackrel{\circ}{=} 0$. If

$$\delta\left(A\right) \stackrel{\circ}{=} \begin{cases} \sigma^2 & \text{if } 0 \in A \\ 0 & \text{if } 0 \notin A \end{cases}$$

then

$$-\frac{\sigma^2 u^2}{2} = \int_{\mathbb{R}} H\left(u,x\right) d\delta\left(x\right),$$

which is a very useful relation as this implies that if $\pi \stackrel{\circ}{=} \rho + \delta$ and

$$\phi\left(u\right) \stackrel{\circ}{=} iu\gamma' - \frac{\sigma^2 u^2}{2} + \int_{\mathbb{R}\setminus\{0\}} \left(\exp\left(iux\right) - 1 - \frac{iux}{1+x^2}\right) d\nu\left(x\right)$$

then

$$\phi\left(u\right) = iu\gamma' + \int_{\mathbb{R}} H\left(u,x\right) d\pi\left(x\right).$$

Obviously π is a finite measure on $\mathbb{R}$. If π is any finite measure and

$$\sigma^2 \stackrel{\circ}{=} \pi\left(\{0\}\right)$$

and

$$\nu\left(A\right) \stackrel{\circ}{=} \int_A \frac{1+x^2}{x^2} d\pi\left(x\right), \quad A \in \mathcal{B}\left(\mathbb{R}\setminus\{0\}\right),$$

then (γ, σ, ν) with an arbitrary γ forms a triplet which satisfies (7.20).

Theorem 7.41 *The representation (7.19) is unique. The same is true for the representation*

$$\phi\left(u\right) = iu\gamma' + \int_{\mathbb{R}} H\left(u,x\right) d\pi\left(x\right).$$

That is, if π_1 and π_2 are finite measures and

$$iu\gamma_1' + \int_{\mathbb{R}} H\left(u,x\right) d\pi_1\left(x\right) = iu\gamma_2' + \int_{\mathbb{R}} H\left(u,x\right) d\pi_2\left(x\right)$$

for all u then $\gamma_1' = \gamma_2'$ and $\pi_1 = \pi_2$.

The theorem is an easy consequence of a more general statement which we shall prove later[42].

Corollary 7.42 (Uniqueness) *The Fourier transform of a Lévy process uniquely determines the triplet (γ, σ, ν).*

Proof: Let $\varphi(t, u) = \exp(t\phi(u))$ be the Fourier transform of some Lévy process. It is sufficient to prove that if $\exp(\phi_1(u)) = \exp(\phi_2(u))$ and $\phi_1(u)$ and $\phi_2(u)$ are continuous with $\phi_1(0) = \phi_2(0) = 0$ then $\phi_1(u) = \phi_2(u)$. Obviously $\phi_1(u) = \phi_2(u) + i2\pi k(u)$ where $k(u)$ is always an integer number. As $k(u)$ is always an integer it can be continuous if and only if $0 = k(0) = k(u)$. $\qquad\square$

Example 7.43 Lévy's representation of local time $L(0)$.

---

1. Let S be the process of the first passage times[43] of some Wiener process. If $a \geq 0$ then the Laplace transform of $S(a)$ is[44] $\exp(-a\sqrt{s})$. With simple calculation

$$\int_0^\infty \frac{\exp(-sx) - 1}{x^{3/2}}\,dx = \int_0^\infty \frac{\exp(-u) - 1}{u^{3/2}} s^{3/2}\frac{du}{s} =$$

$$= \sqrt{s}\int_0^\infty \frac{\exp(-u) - 1}{u^{3/2}}\,du.$$

The value of the integral is

$$\int_0^\infty \frac{\exp(-u) - 1}{u^{3/2}}\,du = -\int_0^\infty \frac{1}{u^{1/2}}\int_0^1 \exp(-ux)\,dx\,du =$$

$$= -\int_0^1 \int_0^\infty \frac{1}{u^{1/2}} \exp(-ux)\,du\,dx =$$

$$= -\int_0^1 \int_0^\infty \frac{\sqrt{x}}{y^{1/2}} \exp(-y)\frac{dy}{\sqrt{x}}\,dx =$$

$$= -\int_0^1 \frac{1}{\sqrt{x}}\Gamma\left(\frac{1}{2}\right)\,dx = -2\sqrt{\pi}$$

Hence the Laplace transform of $S(a)$ has the representation

$$\exp\left(a\frac{1}{\sqrt{2\pi}}\int_0^\infty \frac{\exp(-sx) - 1}{x^{3/2}}\,dx\right).$$

---

[42] See: Theorem 7.85, page 530.
[43] See: Example 1.126, page 90.
[44] See: Example 1.118, page 82.

As the representation of the Fourier transform with the characteristics is unique, the same is true for the Laplace transform. Hence the spectral measure of S is

$$\nu\left(\Lambda\right) = \frac{1}{\sqrt{2\pi}} \int_\Lambda \frac{1}{x^{3/2}} dx.$$

Fix a t and let $N\left(s\right)$ be the number of jumps of S in $\left(0, t\right]$ which are bigger than $1/s^2$. Obviously $N\left(s\right)$ has a Poisson distribution with expected value

$$t\nu\left(\left(\frac{1}{s^2}, \infty\right)\right) = t\sqrt{\frac{2}{\pi}} s \tag{7.24}$$

Hence N is a Poisson process. So by the law of large numbers for Poisson processes[45] almost surely

$$t\sqrt{\frac{2}{\pi}} = \lim_{s \to \infty} \frac{N\left(s\right)}{s} = \lim_{\varepsilon \searrow 0} \sqrt{\varepsilon} N^{\Lambda\left(\varepsilon\right)}\left(t\right). \tag{7.25}$$

where $\Lambda\left(\varepsilon\right) \stackrel{\circ}{=} \left\{\varepsilon > 0\right\}$ and $N^{\Lambda\left(\varepsilon\right)}\left(t\right)$ is the number of jumps of S bigger than ε during the time interval $\left(0, t\right]$. Unifying the measure-zero sets one can assume that (7.25) holds with probability one for every t.

2 Recall that the flat parts of $L\left(0\right)$ and the excursion intervals of w are almost surely equal[46].

3. Now let S be the first passage time of β. By (7.24)

$$L_t\left(0\right) = \lim_{\varepsilon \searrow 0} \sqrt{\frac{\varepsilon\pi}{2}} N^{\Lambda\left(\varepsilon\right)}\left(L_t\left(0\right)\right). \tag{7.26}$$

$N^{\Lambda\left(\varepsilon\right)}\left(L_t\left(0\right)\right)$ is the number of jumps of S during the time interval

$$\left(0, L_t\left(0\right)\right] = \left(0, \max_{s \le t} \beta\left(s\right)\right]$$

which are bigger than ε. The jumps of $S\left(a\right)$ are exactly the flat parts of the maximum process of the corresponding Wiener process. Hence $N^{\Lambda\left(\varepsilon\right)}\left(L_t\left(0\right)\right)$ is exactly the number of flat parts of $\max_s \beta\left(s\right)$ which are bigger than ε and started during the time interval $\left(0, t\right]$. Hence $N^{\Lambda\left(\varepsilon\right)}\left(L_t\left(0\right)\right)$ is the number of excursion intervals of w which are bigger than $\varepsilon > 0$ and started *before* t. If we denote this

---

[45]The proof is nearly the same as for the Wiener processes. See: B.9, page 565. If we assume that $\lambda = 1$ then instead w one can write in the proof the compensated Poisson process.

[46]See: Proposition 6.96, page 449.

number by $E(\varepsilon, t)$ then by (7.26) almost surely for every t

$$L_t(0) = \lim_{\varepsilon \searrow 0} \sqrt{\frac{\varepsilon \pi}{2}} \cdot E(\varepsilon, t).$$

This relation is the so called Lévy's representation of $L(0)$. Observe that instead of the above definition we can define $E(\varepsilon, t)$ as the number of excursions bigger than ε *during* the time interval $(0, t]$ as the two numbers differ only by maximum one and the difference goes to zero when $\varepsilon \searrow 0$. $\qquad\square$

7.2 Predictable Compensators of Random Measures

The key of the Lévy–Khintchine formula is the decomposition of Lévy processes into three parts. The most noticeable object in the decomposition is the spectral measure of the process. Let us fix a measurable space $(E, \mathcal{E})$. One can think about E as the space of the possible jumps of some n-dimensional process X. Hence in this chapter $E \doteq \mathbb{R}^n \setminus \{0\}$ and $\mathcal{E}$ is the set of Borel sets of E. To simplify the notation if x and y are n-dimensional vectors then yx will denote the scalar product of x and y.

The main result of this section is that in some sense one can define a spectral measure for every right-regular process. Of course it is not clear how one can define the spectral measure in this general setting. The main idea is that the spectral measure of a right regular process is a measure which satisfies the general L^1-identity[47]. Observe that the spectral measure of a Lévy process is deterministic. So if $H(t, \omega, e)$ is a non-negative function of three variables and it is measurable with respect to $\mathcal{P} \times \mathcal{B}(\mathbb{R} \setminus \{0\})$ then by Fubini's theorem

$$\int_{\mathbb{R} \setminus \{0\}} H(t, \omega, e)\, d\nu_t(e) = t \int_{\mathbb{R} \setminus \{0\}} H(t, \omega, e)\, d\nu(e)$$

is $\mathcal{P}$-measurable.

Definition 7.44 *Let X be a right-regular stochastic process with jumps in $E \doteq \mathbb{R}^n \setminus \{0\}$ and let μ^X be the random counting measure of the jumps of the trajectories of X: If $D \in \mathcal{B}(\mathbb{R}_+) \times \mathcal{E}$ then let*

$$\mu^X(\omega, D) \doteq \text{the number of jumps of } X \text{ in } D \text{ for trajectory } X(\omega). \quad (7.27)$$

The main result of this section is the following[48]:

Theorem 7.45 (Existence of predictable compensator) *If X is a right-regular process on $\mathbb{R}^n$ and μ^X is the random counting measure in (7.27) then μ^X*

---

[47]See: Proposition 7.23, page 477.
[48]See: Corollary 7.63, page 508.

has a predictable compensator

$$\nu(\omega, D), \quad D \in \mathcal{B}(\mathbb{R}_+) \times \mathcal{E}.$$

By definition this means that $\nu(\omega, D)$ *is a random measure over* $\mathcal{B}(\mathbb{R}_+) \times \mathcal{E}$ *and for every non-negative,* $(\mathcal{P} \times \mathcal{E})$-*measurable function* H *on* $\mathbb{R}_+ \times \Omega \times E$

1. *the parametric integral*

$$I(t, \omega) \overset{\circ}{=} \int_{(0,t] \times E} H(s, \omega, e)\, \nu(ds, \omega, de)$$

is $\mathcal{P}$-*measurable and*

2. $\mathbf{E}\left(\int_{(0,\infty) \times E} H d\mu^X\right) \overset{\circ}{=} \mathbf{E}(I(\infty)) = \mathbf{E}\left(\int_{(0,\infty) \times E} H d\nu\right).$

If ν_1 *and* ν_2 *are two predictable compensators of* μ^X *then they are almost surely equal.*

Definition 7.46 *Let* μ^X *be the counting measure generated by the jumps of a right-regular process* X. *The predictable compensator* ν *of* μ^X *is called the* spectral measure *of* X.

7.2.1 Measurable random measures

Let us start with some definitions:

Definition 7.47 *To make the notation simple let* $\widetilde{\Omega} \overset{\circ}{=} \mathbb{R}_+ \times \Omega \times E.$

Definition 7.48 *In the following under random measure we shall always mean some random measure defined on the measurable space*

$$(\mathbb{R}_+ \times E, \mathcal{B}(\mathbb{R}_+) \times \mathcal{E}), \tag{7.28}$$

that is if μ *is a random measure then* $\mu(\omega)$ *is a measure on* (7.28) *for every* ω.

Definition 7.49 *Let* $(\mathbb{R}_+ \times \Omega, \mathcal{Z})$ *be some measurable space. To make the terminology as simple as possible if a function* $f : \widetilde{\Omega} \to \overline{\mathbb{R}}$ *is measurable with respect to the product* σ-*algebra* $\widetilde{\mathcal{Z}} \overset{\circ}{=} \mathcal{Z} \times \mathcal{E}$ *then we say that function* f *is* $\mathcal{Z}$-*measurable. Therefore one can talk about product measurable, adapted, predictable etc. functions on* $\widetilde{\Omega}$.

If f is product measurable on $\widetilde{\Omega}$ that is if f is measurable with respect to the σ-algebra $\mathcal{B}(\mathbb{R}_+) \times \mathcal{A} \times \mathcal{E}$ then by Fubini's theorem $(t, e) \mapsto f(t, \omega, e)$ is measurable with respect to $\mathcal{B}(\mathbb{R}_+) \times \mathcal{E}$ for every fixed ω.

Definition 7.50 *If f is product measurable on $\widetilde{\Omega}$ and μ is a random measure then one can define the pathwise stochastic integral*

$$(f \bullet \mu)(t, \omega) \stackrel{\circ}{=} \int_{(0,t] \times E} f_\omega(s, e) \, d\mu_\omega(s, e) \stackrel{\circ}{=} \int_{(0,t] \times E} f(s, \omega, e) \, \mu(ds, \omega, de).$$

More generally if $f \geq 0$ then sometimes[49] $f \bullet \mu$ will denote the random measure $f d\mu$ that is the random measure

$$B \to \int_B f_\omega(s, e) \, d\mu_\omega(s, e), \quad B \in \mathcal{B}(\mathbb{R}_+) \times \mathcal{E}.$$

Example 7.51 *If X is an $\mathbb{R}^n$-valued right-regular stochastic process and $f \colon \mathbb{R}^n \to \mathbb{R}$ is Borel measurable then*

$$\left(f \bullet \mu^X\right)(t, \omega) = \sum_{0 < s \leq t} f(\Delta X(s, \omega)).$$

If the stochastic process $(f \bullet \mu)$ is finite for every (t, ω) then it has right-regular trajectories with finite variation. In the general case we do not know anything about the measurability properties of μ. Therefore we do not know anything about the measurability of the process $(f \bullet \mu)$.

Definition 7.52 *One can define the measurability and integrability properties of a random measure μ via the properties of the integrals $f \bullet \mu$:*

1. *A random measure μ is $\mathcal{Z}$-measurable if $f \bullet \mu$ is a $\mathcal{Z}$-measurable stochastic process on $\mathbb{R}_+ \times \Omega$ for every predictable, non-negative function $f : \widetilde{\Omega} \to \overline{\mathbb{R}}_+$.*
2. *We say that μ is a finite random measure if the expected value of $\omega \mapsto \mu(\omega, \mathbb{R}_+ \times E)$ is finite.*
3. *We shall denote the set of adapted, finite random measures by $\mathcal{A}^+$.*
4. *A random measure μ is σ-finite if there is a sequence of predictable sets (Z_n) with $Z_n \subseteq \widetilde{\Omega}$ and $\cup_n Z_n = \widetilde{\Omega}$, with $\chi_{Z_n} \bullet \mu$ is finite for all n.*
5. *$\mu \in \mathcal{A}_{\text{loc}}^+$ if μ is σ-finite and adapted.*

Example 7.53 *If X is an arbitrary adapted right-regular process then $\mu^X \in \mathcal{A}_{\text{loc}}^+$.*

---

[49] In most cases $f \bullet \mu$ is a stochastic process and not a random measure. Using standard measure theory one can easily prove that for every $f, g \geq 0$ product measurable functions $(gf) \bullet \mu = g \bullet (f \bullet \mu)$, where of course $(f \bullet \mu)$ is a random measure and not a stochastic process.

By definition this means that μ^X is adapted and it is σ-finite. If $s < t$, $F \in \mathcal{F}_s$, $\Lambda \in \mathcal{E}$ and $0 \notin \mathrm{cl}\,(\Lambda)$ then it is easy so see that

$$\chi\left((s,t] \times F \times \Lambda\right) \bullet \mu^X$$

is an adapted process. In the usual way with the Monotone Class Theorem and with the Monotone Convergence Theorem it is easy to prove that $H \bullet \mu^X$ is adapted for every non-negative predictable H. The second property is a direct and easy consequences of the right-regularity of X: for every m the jumps of X larger than $1/m$ cannot have an accumulation point. So for every m one can define the sequence of stopping times $(\tau_n^{(m)})$ covering the jumps of X which are larger than $1/m$. If

$$P_{nm} \overset{\circ}{=} \left[0, \tau_n^{(m)}\right] \times \{\|x\| \geq 1/m\} \in \mathcal{P} \times \mathcal{E},$$

then

$$\mathbf{E}\left(\left(\chi_{P_{nm}} \bullet \mu^X\right)(\infty)\right) \leq n < \infty,$$

hence μ^X is σ-finite. $\qquad\square$

Lemma 7.54 *Let μ be a random measure. $\mu \in \mathcal{A}_{\mathrm{loc}}^+$ if and only if $V \bullet \mu \in \mathcal{A}^+$ for some positive, predictable function V.*

Lemma 7.55 *If $\mu\,(\omega, B) < \infty$ for every ω and $B \in \mathcal{B}\,(\mathbb{R}_+) \times \mathcal{E}$ then μ is a predictable random measure if and only if*

$$(t, \omega) \mapsto \mu\,(\omega, (0, t] \times \Lambda) \qquad (7.29)$$

is a predictable function for every $\Lambda \in \mathcal{E}$. The same is true for adapted random measures.

Proof. Assume that (7.29) is predictable. If

$$H \overset{\circ}{=} \chi\left((s, t]\right) \cdot \chi_F \cdot \chi_\Lambda,$$

where $\Lambda \in \mathcal{E}$ and $F \in \mathcal{F}_s$ then

$$(H \bullet \mu)\,(u, \omega) = \mu\,(\omega, (0, u] \cap (s, t] \times \Lambda) \cdot \chi_F\,(\omega)\,\chi\left((s, \infty)\right)(u)$$

which is a predictable process. As μ is finite the set of bounded processes H for which $H \bullet \mu$ is predictable is a linear space. The lemma follows from the Monotone Class Theorem and from the Monotone Convergence Theorem. The other direction is obvious. The same argument is valid for adapted random measures. $\qquad\square$

Example 7.56 Predictable random measures generated by predictable kernels.

Let $K(s, \omega, \Lambda)$ be a predictable kernel on $\mathcal{E}$. By definition this means that $K(s, \omega, \Lambda)$ is a $\mathcal{P}$-measurable process for every fixed $\Lambda \in \mathcal{E}$ and K is a measure on $(E, \mathcal{E})$ for every fixed (s, ω). Let assume that K is bounded. Let $A \in \mathcal{A}^+$ and let assume that A is predictable. Let

$$\alpha(\omega, C) \doteq \int_0^\infty \left(\int_E \chi_C(s, e) K(s, \omega, de) \right) A(ds, \omega), \quad C \in \mathcal{B}(\mathbb{R}_+) \times \mathcal{E}.$$

If C is a measurable rectangle then the inner integral is product measurable. Let $\mathcal{L}$ be the set of bounded functions for which this property holds. As K is bounded $\mathcal{L}$ is a λ-system. With the Monotone Class Theorem one can easily prove that the inner integral is product measurable for every product measurable set. Hence by Fubini's theorem α is well-defined. We show that α is a predictable random measure. If $C = (0, t] \times \Lambda$ then

$$\alpha(\omega, C) = K(t, \omega, \Lambda)(A(t, \omega) - A(0, \omega))$$

which is finite and predictable. Hence by the previous lemma α is a predictable random measure. $\qquad \square$

Lemma 7.57 *If $\mu \in \mathcal{A}_{loc}^+$ then $H \bullet \mu$ is adapted for every non-negative progressively measurable process H.*

Proof. As $\mu \in \mathcal{A}_{loc}^+$ there is a predictable process $V > 0$ for which $V \bullet \mu \in \mathcal{A}^+$. This implies that the random measure

$$\rho(A) \doteq \int_A V d\mu, \quad A \in \mathcal{B}(\mathbb{R}_+) \times \mathcal{E}$$

is finite and adapted. As we are integrating by trajectories it is easy too see that

$$H \bullet \mu = HV^{-1}V \bullet \mu = HV^{-1} \bullet \rho.$$

Hence one can assume that μ is finite. Let us fix a t. If $I \doteq (a, b] \subseteq (0, t]$ and $F \in \mathcal{F}_t$ then for every $\Lambda \in \mathcal{E}$

$$(\chi_I \chi_F \chi_\Lambda \bullet \mu)(t, \omega) = \mu(\omega, (a, b] \times \Lambda) \chi_F$$

which is $\mathcal{F}_t$-measurable. Let $\mathcal{L}$ be the set of bounded processes H for which $(H \bullet \mu)(t)$ is $\mathcal{F}_t$-adapted. As $\mu((0, t] \times E) < \infty$ obviously $\mathcal{L}$ is a linear space and it is obviously a λ-system. As the processes $\chi_I \chi_F \chi_\Lambda$ form a π-system by the Monotone Class Theorem $\mathcal{L}$ contains the bounded processes measurable with respect to the product σ-algebra $\mathcal{B}((0, t])$

$\times \mathcal{F}_t \times \mathcal{E}$. Therefore if $H \geq 0$ is progressively measurable then $H \bullet \mu$ is $\mathcal{F}_t$-measurable for every t. □

Definition 7.58 *Let f be a measurable function on $\widetilde{\Omega}$ and let μ be a random measure.*

1. *f is integrable with respect to μ if $|f| \bullet \mu \in \mathcal{A}^+$.*
2. *f is locally integrable with respect to μ if $|f| \bullet \mu \in \mathcal{A}^+_{\mathrm{loc}}$.*

Definition 7.59 $L^1(\widetilde{\Omega}, \mathcal{Z}, \mu) \overset{\circ}{=} L^1(\widetilde{\Omega}, \widetilde{\mathcal{Z}}, \mu)$ *denotes the set of $\mathcal{Z}$-measurable μ-integrable functions.* $L^1_{\mathrm{loc}}(\widetilde{\Omega}, \mathcal{Z}, \mu) \overset{\circ}{=} L^1_{\mathrm{loc}}(\widetilde{\Omega}, \widetilde{\mathcal{Z}}, \mu)$ *denotes the set of $\mathcal{Z}$-measurable functions locally integrable with respect to μ.*

7.2.2 Existence of predictable compensator

Let X be a right-regular process and let f be a non-negative deterministic function. $Y \overset{\circ}{=} f \bullet \mu^X = \sum f(\Delta X)$ is an increasing, right-regular stochastic process. Obviously in general Y is not predictable, but if $Y \in \mathcal{A}_{\mathrm{loc}}$ then Y has a predictable compensator Y^p. It is not a great surprise that one can generalize this theorem to random measures. One can call this generalization the Extended Edition of the Doob–Meyer decomposition.

Definition 7.60 *Let μ be a random measure. A predictable random measure μ^p is the predictable compensator of μ if one of the next conditions holds:*

1. *if $\widetilde{H} \in L^1_{\mathrm{loc}}\left(\widetilde{\Omega}, \widetilde{\mathcal{P}}, \mu\right)$ then $\widetilde{H} \in L^1_{\mathrm{loc}}\left(\widetilde{\Omega}, \widetilde{\mathcal{P}}, \mu^p\right)$ and*

$$\widetilde{H} \bullet \mu - \widetilde{H} \bullet \mu^p \in \mathcal{L}. \tag{7.30}$$

2. *if $\widetilde{H}$ is a non-negative and predictable function on $\widetilde{\Omega}$ then*

$$\mathbf{E}\left(\int_{(0,\infty)\times E} \widetilde{H} d\mu\right) \overset{\circ}{=} \mathbf{E}\left(\left(\widetilde{H} \bullet \mu\right)(\infty)\right) = \tag{7.31}$$

$$= \mathbf{E}\left(\left(\widetilde{H} \bullet \mu^p\right)(\infty)\right) = \mathbf{E}\left(\int_{(0,\infty)\times E} \widetilde{H} d\mu^p\right).$$

Let assume that μ is finite and $\mu(\omega, (0,t] \times \Lambda)$ is adapted. Let $\Lambda \in \mathcal{E}$ and let

$$A(t,\omega) \overset{\circ}{=} \mu(\omega, (0,t] \times \Lambda).$$

As μ is a finite random measure A is increasing and right-regular in t. As $A \in \mathcal{A}^+$ it has a predictable compensator A^p. In the Lévy process case the spectral measure $\nu_t(\Lambda)$ is the expected value of the number of the jumps in Λ during the time period $(0, t]$. By the elementary properties of the predictable compensator of the locally integrable processes

$$\mathbf{E}\left(\mu\left((0, t] \times \Lambda\right)\right) \stackrel{\circ}{=} \mathbf{E}\left(A\left(t\right)\right) = \mathbf{E}\left(A^p\left(t\right)\right).$$

More generally if for some $D \in \widetilde{\mathcal{P}}$

$$A\left(t, \omega\right) \stackrel{\circ}{=} \left(\chi_D \bullet \mu\right)\left(t, \omega\right) \stackrel{\circ}{=} \int_{(0, t] \times E} \chi_D d\mu,$$

then $A \in \mathcal{A}^+$. Let A^p denotes the predictable compensator of A. As μ^p is a predictable measure and χ_D is a predictable process, by the definition of the predictability of the random measures $\chi_D \bullet \mu^p$ is a predictable, right-continuous, increasing process. By (7.30) $\chi_D \bullet \mu^p$ is the predictable compensator of $\chi_D \bullet \mu$ for any $D \in \widetilde{\mathcal{P}}$. Therefore

$$\chi_D \bullet \mu^p = A^p = \left(\chi_D \bullet \mu\right)^p, \quad D \in \widetilde{\mathcal{P}}.$$

Hence its is quite natural to try the definition

$$\mu^p\left(\omega, D\right) = \left(\chi_D \bullet \mu^p\right)\left(\infty, \omega\right) \stackrel{\circ}{=} \left(\chi_D \bullet \mu\right)^p\left(\infty, \omega\right), \quad D \in \mathcal{B}\left(\mathbb{R}_+\right) \times \mathcal{E}.$$

This expression is nearly a random measure. The main problem with this definition is that the predictable compensator $A^p \stackrel{\circ}{=} \left(\chi_D \bullet \mu\right)^p$ is defined up to an event with zero probability. So for every D one can define $\mu^p\left(\omega, D\right)$ only up to an event with zero probability. The situation is very similar to the situation one has during the construction of the conditional probabilities. Do we have a regular version? The answer depends on the topological properties of E. It is not a great surprise that during the proof of the existence of predictable compensators we use the existence of regular conditional distributions on $(E, \mathcal{E})$. The technical details are in the next lemma:

Lemma 7.61 *If μ is a finite and adapted measure then there is a predictable measure μ^p such that*

$$\mathbf{E}\left(\int_{(0, \infty) \times E} \widetilde{H} d\mu\right) = \mathbf{E}\left(\int_{(0, \infty) \times E} \widetilde{H} d\mu^p\right)$$

for every $\widetilde{H} \geq 0$ predictable process on $\widetilde{\Omega}$.

Proof. As the main problem is some micro surgery on the level of measure-zero sets, it is not a great surprise, that the proof is quite technical. Let $\widetilde{\mathcal{P}} \stackrel{\circ}{=} \mathcal{P} \times \mathcal{E}$.

Let us define on the measurable space $(\widetilde{\Omega}, \widetilde{\mathcal{P}})$ the finite measure

$$\widetilde{P}\left(\widetilde{D}\right) \overset{\circ}{=} \mathbf{E}\left(\left(\chi_{\widetilde{D}} \bullet \mu\right)(\infty)\right) \overset{\circ}{=} \mathbf{E}\left(\int_{(0,\infty)\times E} \chi_{\widetilde{D}} d\mu\right), \quad \widetilde{D} \in \widetilde{\mathcal{P}}.$$

As μ is adapted $\left(\chi_{\widetilde{D}} \bullet \mu\right)(\infty)$ is measurable[50]. Hence $\widetilde{P}$ is well-defined. Observe that $\widetilde{P}$ is a natural generalization of the Doléans measure. As μ is finite $\widetilde{P}$ is a finite measure. Therefore one can assume that $\widetilde{P}$ is a probability measure on the measurable space $(\widetilde{\Omega}, \widetilde{\mathcal{P}})$. Let us denote by $\widetilde{E}(\cdot \mid \widetilde{\mathcal{F}})$ the conditional expectation operator generated by $\widetilde{P}$ and by some σ-algebra $\widetilde{\mathcal{F}} \subseteq \widetilde{\mathcal{P}}$. One can embed σ-algebra $\mathcal{P}$ into $\widetilde{\mathcal{P}} \overset{\circ}{=} \mathcal{P} \times \mathcal{E}$ with the mapping

$$D \mapsto D \times E, \quad D \in \mathcal{P}.$$

Denote this σ-algebra by $\widehat{\mathcal{P}}$. Let $\xi(t, \omega, e) \overset{\circ}{=} e$. ξ is an E-valued random variable on $(\widetilde{\Omega}, \widetilde{\mathcal{P}})$. E is a complete separable metric space so the conditional distribution $\widetilde{E}(\chi(\xi \in \Lambda) \mid \widehat{\mathcal{P}})$ has a regular version $p(\widetilde{\omega}, \Lambda) \overset{\circ}{=} p(t, \omega, e, \Lambda)$. By the definition of the conditional expectation p is $\widehat{\mathcal{P}}$-measurable. So p is constant in e. Hence we can drop the third component and we can denote the conditional distribution

$$\widetilde{P}\left(\Lambda \mid \widehat{\mathcal{P}}\right) \overset{\circ}{=} \widetilde{E}\left(\chi(\xi \in \Lambda) \mid \widehat{\mathcal{P}}\right), \quad \Lambda \in \mathcal{E}.$$

as $p(t, \omega, \Lambda)$. Let $H(t, \omega)$ be a $\mathcal{P}$-measurable, therefore $\widehat{\mathcal{P}}$-measurable process and let $\Lambda \in \mathcal{E}$. If $\widetilde{H} \overset{\circ}{=} H\chi(\xi \in \Lambda)$ then almost surely by $\widetilde{P}$

$$\widetilde{E}\left(\widetilde{H} \mid \widehat{\mathcal{P}}\right) \overset{\circ}{=} \widetilde{E}\left(H\chi(\xi \in \Lambda) \mid \widehat{\mathcal{P}}\right) = H \cdot \widetilde{E}\left(\chi(\xi \in \Lambda) \mid \widehat{\mathcal{P}}\right) \overset{\circ}{=} H \cdot \widetilde{P}\left(\Lambda \mid \widehat{\mathcal{P}}\right).$$

p is a regular version of the conditional distribution so

$$\widetilde{P}\left(\Lambda \mid \widehat{\mathcal{P}}\right) = \int_{\Lambda} dp = \int_{E} \chi_{\Lambda}(e) \, dp(e), \quad \Lambda \in \mathcal{E}.$$

Therefore

$$\widetilde{E}\left(\widetilde{H} \mid \widehat{\mathcal{P}}\right) = H \int_{E} \chi_{\Lambda}(e) \, dp(e) = \int_{E} H\chi_{\Lambda}(e) \, dp(e) \overset{\circ}{=} \int_{E} \widetilde{H} dp(e).$$

[50]See: Lemma 7.57, page 500.

In the usual way one can extend this identity to every $\widetilde{\mathcal{P}}$-measurable function $\widetilde{H}$. Hence the disintegration formula

$$\widetilde{E}\left(\widetilde{H} \mid \widehat{\mathcal{P}}\right) = \int_E \widetilde{H} dp\left(e\right)$$

is valid $\widehat{P}$-almost surely for every predictable function $\widetilde{H}$ on $\widetilde{\Omega}$. Let

$$P\left(A\right) \stackrel{\circ}{=} \widetilde{P}\left(A \times E\right) = \mathbf{E}\left(\int_{(0,\infty) \times E} \chi_{A \times E} d\mu\right), \quad A \in \mathcal{P}.$$

By the definition of the conditional expectation using that $\widetilde{E}(\chi_{\widetilde{D}} \mid \widehat{\mathcal{P}})$ is $\widehat{\mathcal{P}}$-measurable, hence it does not depend on e

$$\widetilde{P}\left(\widetilde{D}\right) = \widetilde{E}\left(\chi_{\widetilde{D}}\right) = \int_{\widetilde{\Omega}} \widetilde{E}\left(\chi_{\widetilde{D}} \mid \widehat{\mathcal{P}}\right) d\widehat{P} = \tag{7.32}$$

$$= \int_{\mathbb{R}_+ \times \Omega} \widetilde{E}\left(\chi_{\widetilde{D}} \mid \widehat{\mathcal{P}}\right) dP = \int_{\mathbb{R}_+ \times \Omega} \left(\int_E \chi_{\widetilde{D}} dp\left(e\right)\right) dP =$$

$$= \int_{(0,\infty) \times \Omega} \left(\int_E \chi_{\widetilde{D}} dp\left(e\right)\right) dP.$$

If $D \in \mathcal{P}$ then by the elementary properties of the predictable compensator

$$P\left(D\right) \stackrel{\circ}{=} \widetilde{P}\left(D \times E\right) \stackrel{\circ}{=} \mathbf{E}\left(\left(\chi_{D \times E} \bullet \mu\right)\left(\infty\right)\right) \stackrel{\circ}{=} \mathbf{E}\left(\int_{(0,\infty) \times E} \chi_{D \times E} d\mu\right) =$$

$$= \mathbf{E}\left(\int_{(0,\infty) \times E} \chi_D \chi_E d\mu\right) = \mathbf{E}\left(\left(\chi_D \bullet \left(\chi_E \bullet \mu\right)\right)\left(\infty\right)\right) =$$

$$= \mathbf{E}\left(\left(\chi_D \bullet \left(\chi_E \bullet \mu\right)^p\right)\left(\infty\right)\right) \stackrel{\circ}{=} \mathbf{E}\left(\left(\chi_D \bullet A^p\right)\left(\infty\right)\right),$$

where

$$A\left(t, \omega\right) \stackrel{\circ}{=} \left(\chi_E \bullet \mu\right)\left(t, \omega\right) = \mu\left(\omega, (0, t] \times E\right).$$

As μ is adapted A is adapted. μ is finite so A has integrable variation. Therefore $Y \stackrel{\circ}{=} A^p$ exists. Hence if $D \in \mathcal{P}$ then

$$P\left(D\right) = \mathbf{E}\left(\int_0^\infty \chi_D\left(t, \omega\right) Y\left(dt, \omega\right)\right). \tag{7.33}$$

In the usual way one can extend (7.33) to every predictable process. So if H is a predictable process then

$$\int_{(0,\infty)\times\Omega} H(t,\omega)\,dP(t,\omega) = \int_{\Omega}\int_0^{\infty} H(t,\omega)\,Y(dt,\omega)\,d\mathbf{P}(\omega).$$

$H(t,\omega) \overset{\circ}{=} \tilde{E}\left(\chi_{\tilde{D}} \mid \hat{\mathcal{P}}\right) = \int_E \chi_{\tilde{D}}\,dp$ is predictable, hence by (7.32)

$$\mathbf{E}\left((\chi_{\tilde{D}} \bullet \mu)(\infty)\right) \overset{\circ}{=} \tilde{P}\left(\tilde{D}\right)$$

$$= \int_{(0,\infty)\times\Omega} \left(\int_E \chi_{\tilde{D}}\,dp\,(e)\right) dP(t,\omega) = \tag{7.34}$$

$$= \int_{\Omega}\int_0^{\infty} \left(\int_E \chi_{\tilde{D}}\,p\,(t,\omega,de)\right) Y(dt,\omega)\,d\mathbf{P}(\omega) \overset{\circ}{=} \mathbf{E}\left((\chi_{\tilde{D}} \bullet \mu^p)(\infty)\right),$$

where

$$\mu^p(\omega,C) \overset{\circ}{=} \int_C p(t,\omega,de)\,Y(dt,\omega), \quad C \in \mathcal{B}(\mathbb{R}_+) \times \mathcal{E}.$$

$Y \overset{\circ}{=} A^p$ is predictable $p(t,\omega,\Lambda)$ is a predictable kernel so by the above example[51] μ^p is a predictable random measure. From (7.34) the lemma is obvious. $\quad\square$

Theorem 7.62 (Predictable compensator of random measures) *If $\mu \in \mathcal{A}_{\text{loc}}^+$ then:*

1. *μ has a predictable compensator denoted by μ^p and*
2. *$\mu^p \in \mathcal{A}_{\text{loc}}^+$.*

If μ_1^p and μ_2^p are predictable compensators of some μ then almost surely $\mu_1^p = \mu_2^p$. That is the predictable compensator is unique up to indistinguishability.

Proof. By definition $\mu \in \mathcal{A}_{\text{loc}}^+$ if μ is adapted and σ-finite. The second property means that $V \bullet \mu \in \mathcal{A}^+$ for some positive predictable process V.

1. Let us prove that the first condition in the definition of the predictable compensator implies the second one. Let $\tilde{H}$ be a non-negative predictable function on $\tilde{\Omega}$. Let us first assume that $\tilde{H} \in L_{\text{loc}}^1(\tilde{\Omega}, \tilde{\mathcal{P}}, \mu)$ that is $\tilde{H} \bullet \mu \in \mathcal{A}_{\text{loc}}^+$. This means that the process $\tilde{H} \bullet \mu$ has a predictable compensator $(\tilde{H} \bullet \mu)^p$. μ^p is predictable, which by definition means that $\tilde{H} \bullet \mu^p$ is a predictable, increasing process. By the first condition (7.30) and by the definition of the predictable compensator of locally integrable processes $(\tilde{H} \bullet \mu)^p = \tilde{H} \bullet \mu^p$. By the basic properties of the

---

[51]See: Example 7.56, page 500.

predictable compensator of locally integrable processes[52]

$$\mathbf{E}\left(\left(\tilde{H} \bullet \mu\right)(\infty)\right) = \mathbf{E}\left(\left(\tilde{H} \bullet \mu\right)^{p}(\infty)\right) = \mathbf{E}\left(\left(\tilde{H} \bullet \mu^{p}\right)(\infty)\right),$$

which is just the second condition (7.31). Now let $\tilde{H}$ be an arbitrary non-negative, predictable function on $\tilde{\Omega}$. Let $V > 0$ be a predictable function with $V \bullet \mu \in \mathcal{A}^{+}$. $\tilde{H}_{n} \triangleq \tilde{H}\chi(\tilde{H} \leq nV)$ is also predictable and as

$$\tilde{H}_{n} \bullet \mu \leq nV \bullet \mu$$

$\tilde{H}_{n}$ is integrable. Hence by the just proved part of the observation $\tilde{H}_{n}$ satisfies (7.31). Using Monotone Convergence Theorem one can show that $\tilde{H}$ also satisfies (7.31).

2. Now we show that (7.31) implies (7.30). Let $\tilde{H} \in L^{1}_{\text{loc}}(\tilde{\Omega}, \tilde{\mathcal{P}}, \mu)$. This means that $|\tilde{H}| \bullet \mu \in \mathcal{A}^{+}_{\text{loc}}$. Let (τ_{n}) be a localizing sequence of process $|\tilde{H}| \bullet \mu$. We are integrating by trajectories so trivially

$$\left|\tilde{H}\right|\chi\left([0, \tau_{n}]\right) \bullet \mu = \left(\left|\tilde{H}\right| \bullet \mu\right)^{\tau_{n}} \in \mathcal{A}^{+}.$$

From this, using condition (7.31)

$$\left(\left|\tilde{H}\right| \bullet \mu^{p}\right)^{\tau_{n}} = \left|\tilde{H}\right|\chi\left([0, \tau_{n}]\right) \bullet \mu^{p} \in \mathcal{A}^{+}.$$

Hence $\tilde{H} \in L^{1}_{\text{loc}}(\tilde{\Omega}, \tilde{\mathcal{P}}, \mu^{p})$. Let τ be a stopping time. As by (7.31)

$$\mathbf{E}\left(\left(\tilde{H}^{\pm}\chi\left([0, \tau_{n} \wedge \tau]\right) \bullet \mu\right)(\infty)\right) = \mathbf{E}\left(\left(\tilde{H}^{\pm}\chi\left([0, \tau_{n} \wedge \tau]\right) \bullet \mu^{p}\right)(\infty)\right) < \infty$$

the expected value of the stopped variable

$$\left(\tilde{H}\chi\left([0, \tau_{n}]\right) \bullet \mu - \tilde{H}\chi\left([0, \tau_{n}]\right) \bullet \mu^{p}\right)(\tau)$$

is zero. Hence

$$\tilde{H}\chi\left([0, \tau_{n}]\right) \bullet \mu - \tilde{H}\chi\left([0, \tau_{n}]\right) \bullet \mu^{p}$$

is a martingale. This means that $\tilde{H} \bullet \mu - \tilde{H} \bullet \mu^{p}$ is a local martingale. So condition (7.31) implies (7.30).

[52]See: Theorem 3.52, page 213.

3. As $\mu \in \mathcal{A}_{\text{loc}}^{+}$ it is σ-finite. This means that $V \bullet \mu \in \mathcal{A}^{+}$ with some predictable function $V > 0$ on $\widetilde{\Omega}$. As the second condition of the definition holds obviously $V \bullet \mu^{p} \in \mathcal{A}^{+}$, so μ^{p} is also σ-finite with respect to $\mathcal{P}$.

4. Let us show that μ^{p} is unique. Assume that μ_{1}^{p} and μ_{2}^{p} are two predictable compensators of μ. As $\mathcal{E}$ is the Borel σ-algebra of a separable metric space it has a countable base (Λ_{n}). Obviously $V\chi_{\Lambda_{n}}$ is $(\mathcal{P} \times \mathcal{E})$-measurable so it is a predictable function on $\widetilde{\Omega}$. Obviously

$$V\chi_{\Lambda_{n}} \in L_{\text{loc}}^{1}\left(\widetilde{\Omega}, \widetilde{\mathcal{P}}, \mu\right) = L_{\text{loc}}^{1}\left(\widetilde{\Omega}, \widetilde{\mathcal{P}}, \mu_{1}^{p}\right) = L_{\text{loc}}^{1}\left(\widetilde{\Omega}, \widetilde{\mathcal{P}}, \mu_{2}^{p}\right).$$

So by the first condition

$$V\chi_{\Lambda_{n}} \bullet \mu_{1}^{p} - V\chi_{\Lambda_{n}} \bullet \mu_{2}^{p} \in \mathcal{L} \cap \mathcal{V}$$

is predictable. Hence by Fisk's theorem[53] it is almost surely zero. Let N be the union of the exceptional sets. As (Λ_{n}) is countable $\mathbf{P}(N) = 0$.

$$V\chi\left([0,t] \times \Lambda_{n}\right) \bullet \mu_{1}^{p}(\omega) = V\chi\left([0,t] \times \Lambda_{n}\right) \bullet \mu_{2}^{p}(\omega), \quad \omega \notin N.$$

By the Monotone Class Theorem for every non-negative and $(\mathcal{B}(\mathbb{R}_{+}) \times \mathcal{E})$-measurable function H

$$VH \bullet \mu_{1}^{p}(\omega) = VH \bullet \mu_{2}^{p}(\omega), \quad \omega \notin N.$$

Hence if $\omega \notin N$ then $\mu_{1}^{p}(\omega, A) = \mu_{2}^{p}(\omega, A)$ for every set $A \in \mathcal{B}(\mathbb{R}_{+}) \times \mathcal{E}$. Therefore the predictable compensator is unique.

5. Let

$$\mu_{V}(B) \overset{\circ}{=} \int_{B} V\chi_{B} \, d\mu, \quad B \in \mathcal{B}(\mathbb{R}_{+}) \times \mathcal{E}.$$

μ_{V} is an adapted, finite measure. So by the lemma, if $\widetilde{H} \geq 0$ is predictable, then

$$\mathbf{E}\left(\left(\widetilde{H}V^{-1} \bullet \mu_{V}\right)(\infty)\right) = \mathbf{E}\left(\left(\widetilde{H}V^{-1} \bullet \mu_{V}^{p}\right)(\infty)\right).$$

We are integrating by trajectories, so

$$\mathbf{E}\left(\left(\widetilde{H} \bullet \mu\right)(\infty)\right) = \mathbf{E}\left(\left(\widetilde{H} \bullet \left(VV^{-1}\mu\right)\right)(\infty)\right) =$$
$$= \mathbf{E}\left(\left(\widetilde{H}V^{-1} \bullet (V \bullet \mu)\right)(\infty)\right) \overset{\circ}{=}$$

[53]See: Theorem 2.11, page 117.

$$\stackrel{\circ}{=} \mathbf{E}\left(\left(\tilde{H}V^{-1}\bullet\mu_V\right)(\infty)\right) = \mathbf{E}\left(\left(\tilde{H}V^{-1}\bullet\mu_V^p\right)(\infty)\right) =$$

$$= \mathbf{E}\left(\left(\tilde{H}\bullet\left(V^{-1}\bullet\mu_V^p\right)\right)(\infty)\right).$$

V^{-1} is predictable, therefore $\left(V^{-1}\bullet\mu_V^p\right) = V^{-1}d\mu_V^p$ is also predictable. Hence μ has a predictable compensator. $\qquad\square$

Corollary 7.63 *Theorem 7.45 is true, that is if X is a right-regular process on $\mathbb{R}^n$ then the counting measure μ^X in (7.27) has a predictable compensator ν.*

7.3 Characteristics of Semimartingales

The characteristics of semimartingales are generalizations of the characteristics of Lévy processes. Let X be an n-dimensional semimartingale. Let us fix the so-called *truncating function h*. h can be any bounded function $\mathbb{R}^n \to \mathbb{R}^n$ with compact support and with the property that $h(x) = x$ in some neighborhood of the origin. The simplest example of truncating function is $h(x) \stackrel{\circ}{=} x\chi(\|x\| < 1)$. To make the notation as simple as possible we shall use $h(x) \stackrel{\circ}{=} x\chi(\|x\| < 1)$ as truncating function, but the specific form of h does not really matter. If the jumps of X are small that is if ΔX is in the neighbourhood related to h then $\Delta X - h(\Delta X) = \Delta X - \Delta X = 0$. Let

$$\widehat{X}(h)(t) \stackrel{\circ}{=} \sum_{s\le t}(\Delta X(s) - h(\Delta X(s)))$$

$$X(h) \stackrel{\circ}{=} X - \widehat{X}(h).$$

$\widehat{X}(h)$ is the process of the big jumps of X and $X(h)$ is the process from which we deleted the big jumps of X. The big jumps cannot accumulate therefore $\widehat{X}(h)$ has bounded variation on finite intervals. As $\Delta(X(h)) = h(\Delta X)$ and as h has compact support, $X(h)$ is a semimartingale with bounded jumps, that is $X(h)$ is a special semimartingale[54]. As $X(h)$ is a special semimartingale it has a unique decomposition

$$X(h) = X(0) + B(h) + L(h) \tag{7.35}$$

where $L(h)$ is a local martingale and $B(h)$ is a predictable process with finite variation.

[54]See: Example 4.46, page 258.

Definition 7.64 *If X is a semimartingale and h is a truncating function then we call the triplet (B, C, ν) the characteristics of X under h where:*

1. $B \doteq B(h)$ *is the predictable process with finite variation in line (7.35),*
2. $C = (C_{ij})$, *where* $C_{ij} \doteq \left[X_i^c, X_j^c \right] = [X_i, X_j]^c$, *where X_i^c is the continuous part[55] of semimartingale X_i,*
3. ν *is the spectral measure of X, that is ν is the predictable compensator of μ^X.*

Example 7.65 Characteristics of Lévy processes.

Every Lévy process X has a representation

$$X(t) = M^c(t) + M^d(t) + \widehat{X}(h)(t) + \gamma t$$

where γt is the expected value of the small jumps. Obviously

$$X(h) \doteq X - \widehat{X}(h) = M^c(t) + M^d(t) + \gamma t$$

is a special semimartingale, with $B(h)(t) = \gamma t$. As we have seen $M^c = \sigma w$, where w is a Wiener process. Hence $C(t) = [\sigma w](t) = \sigma^2 t$. Let ρ be the predictable compensator of μ^X. For every $0 \notin \mathrm{cl}(\Lambda)$

$$\chi_\Lambda \bullet \mu^X - \chi_\Lambda \bullet \rho \in \mathcal{L},$$
$$\chi_\Lambda \bullet \mu^X - t\nu(\Lambda) = N^\Lambda(t) - t\nu(\Lambda) \in \mathcal{L}.$$

From this $\chi_\Lambda \bullet \rho - t\nu(\Lambda) \in \mathcal{L} \cap \mathcal{V}$. As $\chi_\Lambda \bullet \rho - t\nu(\Lambda)$ is predictable

$$\int_{(0,t] \times \Lambda} d\rho = t\nu(\Lambda).$$

Hence $\rho((0, t] \times \Lambda) = t\nu(\Lambda)$. As the predictable compensator is unique $(\mu^X)^p = t\nu$. So the semimartingale characteristics of X is the triplet $(\gamma t, \sigma^2 t, t\nu)$. $\qquad \Box$

Proposition 7.66 *If X is a semimartingale and (B, C, ν) is the characteristics of X then:*

1. $(\|x\|^2 \wedge 1) \bullet \nu \in \mathcal{A}^+_{\mathrm{loc}}$ *and*
2. *X is a special semimartingale if and only if*

$$\left(\|x\|^2 \wedge \|x\| \right) \bullet \nu \in \mathcal{A}^+_{\mathrm{loc}}.$$

---

[55]See: Theorem 4.21, page 234.

In this case if

$$X = X(0) + A + L$$

is the canonical decomposition of X and if h denotes the truncating function of the characteristics then

$$A = B + (x - h(x)) \bullet \nu.$$

Proof. One can prove the proposition in several steps:

1. For every semimartingale on any finite interval the quadratic variation is finite so $\sum \|\Delta X\|^2 = \|x\|^2 \bullet \mu^X \in \mathcal{V}^+$. The right-regular process $\sum \|\Delta X\|^2 \wedge 1$ has bounded jumps therefore it is locally bounded[56]. As it is increasing

$$\left(\|x\|^2 \wedge 1 \right) \bullet \mu^X = \sum \left(\|\Delta X\|^2 \wedge 1 \right) \in \mathcal{A}_{\text{loc}}^+.$$

$\|x\|^2 \wedge 1$ is a deterministic, non-negative function so it is predictable, therefore by (7.31), using that $\chi([0, \tau_n])$ is also predictable

$$\mathbf{E}\left(\left(\|x\|^2 \wedge 1 \right) \bullet \nu(\tau_n) \right) = \mathbf{E}\left(\left(\|x\|^2 \wedge 1 \right) \chi[0, \tau_n] \bullet \nu(\infty) \right) =$$
$$= \mathbf{E}\left(\left(\|x\|^2 \wedge 1 \right) \chi[0, \tau_n] \bullet \mu^X(\infty) \right) =$$
$$= \mathbf{E}\left(\left(\|x\|^2 \wedge 1 \right) \bullet \mu^X(\tau_n) \right) < \infty.$$

Hence

$$\left(\|x\|^2 \wedge 1 \right) \bullet \nu \in \mathcal{A}_{\text{loc}}^+,$$

which proves the first statement.

2. $X(h) \stackrel{\circ}{=} X - \widehat{X}(h)$ is always a special semimartingale[57]. Therefore X is a special semimartingale if and only if the process of big jumps $\widehat{X}(h)$ is a special semimartingale. On the other hand

$$\widehat{X}(h) \stackrel{\circ}{=} \sum (\Delta X - h(\Delta X)) = (x - h(x)) \bullet \mu^X \in \mathcal{V}.$$

Therefore[58] $\widehat{X}(h)$ is a special semimartingale if and only if

$$\widehat{X}(h) = (x - h(x)) \bullet \mu^X \in \mathcal{A}_{\text{loc}}.$$

---

[56]See: Proposition 1.152, page 107.
[57]See: Example 4.47, page 258.
[58]See: Theorem 4.44, page 257.

$x - h(x)$ is deterministic so it is predictable. Hence using (7.31) on the positive and on the negative parts of the coordinates one can prove that X is a special semimartingale if and only if

$$\|x - h(x)\| \bullet \nu \in \mathcal{A}_{\mathrm{loc}}^{+}. \tag{7.36}$$

Assume that $h(x) \overset{\circ}{=} x\chi(\|x\| < 1)$.

$$\|x - h(x)\| \leq \|x\| \wedge \|x\|^2 \leq \|x - h(x)\| + \|x\|^2 \wedge 1.$$

By the first part of the proposition for every semimartingale $(\|x\|^2 \wedge 1) \bullet \nu \in \mathcal{A}_{\mathrm{loc}}^{+}$. Which implies that the first part of the second statement under the truncation function $h(x) \overset{\circ}{=} x\chi(\|x\| < 1)$ holds. One can prove the general case in a similar way.

3. If X has the canonical decomposition $X = X(0) + A + L$ and $X(h) = X(0) + B + N$ then

$$\widehat{X}(h) \overset{\circ}{=} X - X(h) = A - B + L - N.$$

$L - N \in \mathcal{L}$ and $A - B$ is predictable. This implies[59] that $\widehat{X}(h) \in \mathcal{A}_{\mathrm{loc}}$ and the predictable compensator of $\widehat{X}(h) = (x - h(x)) \bullet \mu^X \in \mathcal{A}_{\mathrm{loc}}$ is $A - B$. Hence

$$A - B = \left((x - h(x)) \bullet \mu^X \right)^P = (x - h(x)) \bullet \left(\mu^X \right)^P \overset{\circ}{=} (x - h(x)) \bullet \nu,$$

from which the second part of the second statement is evident. □

Example 7.67 A compound Poisson process X is a special semimartingale, if and only if the distribution of the absolute value of the size of the jumps has finite expected value.

As[60] $\nu = \lambda t F$, where F is the distribution function of the jumps

$$\left(\|x\|^2 \wedge \|x\| \bullet \nu \right)(t, \omega) = \lambda t \int_{\mathbb{R}^n} \|x\|^2 \wedge \|x\| \, dF(x).$$

X is a special semimartingale if and only if $(\|x\|^2 \wedge \|x\|) \bullet \nu \in \mathcal{A}_{\mathrm{loc}}^{+}$. ν is deterministic so X is a special semimartingale if and only if

$$\int_{\mathbb{R}^n} \|x\|^2 \wedge \|x\| \, dF(x) < \infty,$$

which happens if and only if $\int_{\mathbb{R}^n} \|x\| \, dF(x)$ is finite. □

---

[59]See: Proposition 3.35, page 200.
[60]See: Example 7.24, page 477.

Corollary 7.68 *Let X be a semimartingale. X is a local martingale if and only if*

$$\|x - x\chi\left(\|x\| < 1\right)\| \bullet \nu \stackrel{\circ}{=} \|x - h\left(x\right)\| \bullet \nu \in \mathcal{A}_{\text{loc}}^{+}$$

and

$$0 = B + \left(x - x\chi\left(\|x\| < 1\right)\right) \bullet \nu \stackrel{\circ}{=} B + \left(x - h\left(x\right)\right) \bullet \nu.$$

Proof. Every local martingale is a special semimartingale with canonical decomposition

$$X = X\left(0\right) + 0 + L$$

where $L \in \mathcal{L}$. During the proof of the previous proposition[61] we have seen that X is a special semimartingale if and only if

$$\|x - h\left(x\right)\| \bullet \nu \in \mathcal{A}_{\text{loc}}^{+}.$$

The second condition is equivalent to the assumption that in the canonical decomposition the finite variation part is zero. □

Example 7.69 Symmetric stable processes.

Recall[62] that a Lévy process X is a symmetric stable process if its characteristics is $\left(0, 0, t|x|^{\alpha}\right)$, where $0 < \alpha < 2$. By the just proved result X is a local martingale if and only if

$$\left(\|x - h\left(x\right)\| \bullet \nu\right)\left(t, \omega\right) = 2t \int_{1}^{\infty} x dx^{-\alpha} < \infty.$$

The integral is finite if and only if $\alpha > 1$. As we shall prove every local martingale with independent increments is a martingale[63]. So X is a martingale if and only if $\alpha > 1$. If $\alpha \leq 1$ then X is a semimartingale[64], but it is not even a special semimartingale. □

[61]See: line (7.36) page 511.
[62]See: Example 7.39, page 491.
[63]See: Theorem 7.97, page 545.
[64]Every Lévy process is a semimartingale.

7.4 Lévy–Khintchine Formula for Semimartingales with Independent Increments

In this section we prove the generalization of the Lévy–Khintchine formula. Recall that if X is a Lévy process then

$$\varphi(u,t) = \exp(t\phi(u))$$

where

$$\phi(u) \overset{\circ}{=} iu\gamma - \frac{\sigma^2 u^2}{2} + \int_{\mathbb{R}\setminus\{0\}} (\exp(iux) - 1 - x\chi(|x| < 1))\, d\nu(x).$$

As $\Psi(u,t) \overset{\circ}{=} t\phi(u)$ is a continuous process with finite variation it is also clear that

$$\varphi(u,t) = \mathcal{E}(\Psi(u,t)). \tag{7.37}$$

Let X be a semimartingale. Using the characteristics of X one can define the exponent $\Psi(u,t)$ in a very straightforward[65] way. Our goal is to show that (7.37) is true for semimartingale with independent increments[66]. There are two major steps in the proof. The first one, and perhaps the more difficult one, is to show that if X is a semimartingale with independent increments then $\Psi(u,t)$ is deterministic. As an other major step with Itô's formula we shall prove that if X is a semimartingale then $Y(u) \overset{\circ}{=} \exp(iuX) - \exp(iuX-) \bullet \Psi(u)$ is a local martingale for every u. If Ψ is deterministic then $Y(u)$ is bounded on any finite interval so it is a martingale. Using Fubini's theorem one can easily show that

$$\mathbf{E}(\exp(iuX(t))) - \mathbf{E}(\exp(iuX(t)-)) \bullet \Psi(u) = 1.$$

That is for every u

$$\varphi(u) - \varphi_-(u) \bullet \Psi(u) = 1.$$

By definition this means that (7.37) holds.

7.4.1 Examples: probability of jumps of processes with independent increments

As we have seen every Lévy process is continuous in probability. This implies that the probability of a jump of a Lévy process at every moment of time is zero. This property does not hold for processes with independent increments. Perhaps this is the most remarkable property of the class of processes with independent

---

[65]See: Definition 7.76, page 518.
[66]See: Definition 1.93, page 58.

increments. To correctly fix the ideas of the reader in this subsequence we show some examples.

Later we shall prove that for processes with independent increments the spectral measure $\nu \stackrel{\circ}{=} \left(\mu^X\right)^p$ has a deterministic version[67]. We shall use this fact several times in the examples of this subsection.

Example 7.70 If X is an arbitrary right-regular process and if ν is the spectral measure of X then for every $\Lambda \in \mathcal{E} \stackrel{\circ}{=} \mathcal{B}\left(\mathbb{R}^n \setminus \{0\}\right)$

$$\nu\left(\{t\} \times \Lambda\right) \stackrel{a.s.}{=} \mathbf{P}\left(\Delta X\left(t\right) \in \Lambda \mid \mathcal{F}_{t-}\right).$$

If X has independent increments then

$$\nu\left(\{t\} \times \Lambda\right) = \mathbf{P}\left(\Delta X\left(t\right) \in \Lambda\right).$$

A process with independent increments has a jump with positive probability at time t if and only if

$$\nu\left(\{t\} \times \left(\mathbb{R}^n \setminus \{0\}\right)\right) > 0.$$

Let $H \stackrel{\circ}{=} \chi\left(\{t\} \times \Lambda\right)$. H is deterministic so it is predictable and obviously $H \bullet \mu^X \in \mathcal{A}_{\text{loc}}$. By (7.30) $H \bullet \nu = \left(H \bullet \mu^X\right)^p$. By the formula for the the jumps of predictable compensators almost surely

$$\nu\left(\{t\} \times \Lambda\right) = \left(\Delta\left(H \bullet \nu\right)\right)(t) = \left(\Delta\left(\left(H \bullet \mu^X\right)^p\right)\right)(t) =$$
$$= \left({}^p\left(\Delta\left(H \bullet \mu^X\right)\right)\right)(t) = \mathbf{E}\left(\Delta\left(H \bullet \mu^X\right)(t) \mid \mathcal{F}_{t-}\right) =$$
$$= \mathbf{P}\left(\Delta X\left(t\right) \in \Lambda \mid \mathcal{F}_{t-}\right).$$

If X has independent increments then $\Delta X\left(t\right)$ is independent of $\mathcal{F}_{t-1/n}$ for any n. Hence it is independent of $\mathcal{F}_{t-}$. So in this case

$$\nu\left(\{t\} \times \Lambda\right) = \nu\left(\omega, \{t\} \times \Lambda\right) \stackrel{a.s.}{=} \mathbf{P}\left(\Delta X\left(t\right) \in \Lambda\right). \qquad \square$$

Definition 7.71 Let $J \stackrel{\circ}{=} \{t \colon \nu\left(\{t\} \times \left(\mathbb{R}^n \setminus \{0\}\right)\right) > 0\}$.

Example 7.72 Processes with independent increments which are not continuous in probability.

---

[67]See: Corollary 7.88, page 532.

1. Perhaps the simplest example is the following. Let ξ be an arbitrary random variable. Let

$$X(t) \overset{\circ}{=} \begin{cases} 0 & \text{if } t < 1 \\ \xi & \text{if } t \geq 1 \end{cases}.$$

It is easy to see, that X is a process with independent increments. If $\xi \neq 0$ then X is not continuous in probability and $J = \{1\}$. Let F be the distribution of ξ. The only non-zero part of the spectral measure of X is

$$\nu(\{1\} \times \Lambda) = F(\Lambda), \quad \Lambda \in \mathcal{B}(\mathbb{R}) \setminus \{0\}.$$

Obviously the Fourier transform of $X(t)$ is

$$\varphi(t, u) = \begin{cases} 1 & \text{if } t < 1 \\ \int_{\mathbb{R}} \exp(iux) \, dF(x) & \text{if } t \geq 1 \end{cases}.$$

Obviously

$$\nu(\{1\} \times (\mathbb{R} \setminus \{0\})) = \mathbf{P}(\xi \neq 0) = 1 - \mathbf{P}(\xi = 0).$$

From this[68]

$$\int_{\mathbb{R}} \exp(iux) \, dF(x) = 1 \cdot \mathbf{P}(\xi = 0) + \int_{\mathbb{R} \setminus \{0\}} \exp(iux) \, \nu(\{1\} \times dx) =$$

$$= 1 + \int_{\mathbb{R} \setminus \{0\}} (\exp(iux) - 1) \, \nu(\{1\} \times dx).$$

2. Assume that ξ has uniform distribution over $[-1, 1]$. In this case

$$\int_{\mathbb{R}} \exp(iux) \, dF(x) = \frac{1}{2} \int_{-1}^{1} \cos ux \, dx = \frac{\sin u}{u}.$$

For certain values of u the Fourier transform $\varphi(t, u)$ is never zero, but for certain u at $t = 1$ it jumps to zero.

3. To make the example a bit more complicated let (ξ_k) be a sequence of independent random variables and let $t_k \nearrow \infty$. Let

$$X(t) \overset{\circ}{=} \sum_{t_k \leq t} \xi_k. \tag{7.38}$$

---

[68]See: Corollary 7.91, page 535.

It is easy to see again that X is a process with independent increments and $J = \{(t_k)\}$. The Fourier transform of X is[69]

$$\varphi_X (t, u) = \prod_{t_k \leq t} \int_{\mathbb{R}} \exp (iux) \, dF_k (x) =$$

$$= \prod_{t_k \leq t} \left(1 + \int_{\mathbb{R} \setminus \{0\}} (\exp (iux) - 1) \, \nu (\{t_k\} \times dx) \right).$$

4. Let $B (t)$ be a deterministic, right-regular function. Obviously it is a process with independent increments. Its Fourier transform is $\varphi_B (t, u) = \exp (iuB (t))$.

5. Let us now investigate the process $V \stackrel{\circ}{=} X + B$, where X is the process in line (7.38). As X and B are independent the Fourier transform of V is $\varphi_B \varphi_X$. But let us observe that the spectral measure of V is different from the spectral measure of X as the jumps of B introduce some new jumps for V. Therefore

$$\varphi_V (t, u) = \exp (iuB (t)) \times$$

$$\times \prod_{t_k \leq t} \left(\exp (-iu \Delta B (t_k)) \left(1 + \int_{\mathbb{R} \setminus \{0\}} (\exp (iux) - 1) \, \nu (\{t_k\} \times dx) \right) \right)$$

where of course ν denotes the spectral measure of V. Which one can write as

$$\exp (iuB (t)) \times$$

$$\times \prod_{0 < r \leq t} \left(\exp (-iu \Delta B (r)) \left(1 + \int_{\mathbb{R} \setminus \{0\}} (\exp (iux) - 1) \, \nu (\{r\} \times dx) \right) \right),$$

where r can be any jump-time of B or one of the points t_k. $\square$

Example 7.73 If X is an adapted, right-regular process and if ν is the spectral measure of X then for every predictable stopping time τ and for every set $\Lambda \in \mathcal{E} \stackrel{\circ}{=} \mathcal{B} (\mathbb{R}^n \setminus \{0\})$ on the set $\{\tau < \infty\}$

$$\nu ([\tau] \times \Lambda) \stackrel{a.s.}{=} \mathbf{P} (\Delta X (\tau) \in \Lambda \mid \mathcal{F}_{\tau-}) \stackrel{\circ}{=} \mathbf{E} (\chi (\Delta X (\tau) \in \Lambda) \mid \mathcal{F}_{\tau-})$$

that is $\nu (\omega, \{t\} \times \Lambda)$ is a version of $^p (\chi (\Delta X \in \Lambda))$ for every $\Lambda \in \mathcal{E} \stackrel{\circ}{=} \mathcal{B} (\mathbb{R}^n \setminus \{0\})$.

If $V (t, \omega) \stackrel{\circ}{=} \chi (\{t\} \times \Lambda)$ then V is predictable. As

$$\nu (\omega, \{t\} \times \Lambda) = (\Delta (V \bullet \nu)) (t, \omega),$$

[69]See: Theorem 7.90, page 534.

and $V \bullet \nu$ is predictable $\nu(\omega, \{t\} \times \Lambda)$ is a predictable process. Let $H \stackrel{\circ}{=} \chi([\tau] \times \Lambda)$. As τ is a predictable stopping time $[\tau]$ is a predictable subset[70]. So $[\tau] \times \Lambda \in \widetilde{\mathcal{P}}$ and H is predictable. Obviously $H \bullet \mu^X \in \mathcal{A}_{\text{loc}}$. By (7.30) $H \bullet \nu = \left(H \bullet \mu^X\right)^p$. By the formula for the the jumps of predictable compensators almost surely on the set $\{\tau < \infty\}$

$$\nu([\tau] \times \Lambda) = (\Delta(H \bullet \nu))(\tau) = \left(\Delta\left(\left(H \bullet \mu^X\right)^p\right)\right)(\tau) =$$

$$= \left({}^p\left(\Delta\left(H \bullet \mu^X\right)\right)\right)(\tau) = \mathbf{E}\left(\Delta\left(H \bullet \mu^X\right)(\tau) \mid \mathcal{F}_{\tau-}\right) =$$

$$= \mathbf{P}\left(\Delta X(\tau) \in \Lambda \mid \mathcal{F}_{\tau-}\right). \qquad \square$$

Example 7.74 If X is an adapted, right-regular process and if ν is the spectral measure of X then every jump time of X is totally inaccessible if and only if $\nu(\omega, \{t\} \times (\mathbb{R}^n \setminus \{0\})) \equiv 0$ up to indistinguishability.

If τ is a predictable jump time then

$$\mathbf{P}(\Delta X(\tau) \neq 0) = \mathbf{E}(\nu([\tau] \times (\mathbb{R}^n \setminus \{0\}))) = 0.$$

On the other hand if X does not have a predictable jump time then ${}^p(\chi(\Delta X \neq 0)) = 0$. As the predictable projection is unique $\nu(\omega, \{t\} \times E) = 0$ up to indistinguishability. $\qquad \square$

Example 7.75 Jump times of a process with independent increments are totally inaccessible if and only if the process is continuous in probability.

---

1. Let τ be a predictable stopping time. From the just proved proposition using that ν has a deterministic version[71]

$$\mathbf{P}(\Delta X(\tau) \neq 0) = \mathbf{E}(\nu([\tau] \times E)) = \int_\Omega \nu(\omega, \{\tau(\omega)\} \times E)\, d\mathbf{P}(\omega) =$$

$$= \int_\Omega \nu(\{\tau(\omega)\} \times E)\, d\mathbf{P}(\omega) =$$

$$= \int_\Omega \mathbf{P}(\Delta X(\tau(\omega)) \neq 0)\, d\mathbf{P}(\omega) = \int_\Omega 0\, d\mathbf{P}(\omega) = 0,$$

---

[70]See: Corollary 3.34, page 199.
[71]See: Corollary 7.88, page 532.

where in the last line we used that X is continuous in probability so

$$\mathbf{P}\left(\Delta X\left(t\right)\neq 0\right)=0$$

for every t. Therefore $\mathbf{P}\left(\Delta X\left(\tau\right)\neq 0\right)=0$, which implies that τ, or any part of τ, is not a jump time of X.

2. On the other hand if X is not continuous in probability then $\mathbf{P}\left(\Delta X\left(t\right)\neq 0\right)>0$ for some t. Obviously $\tau\equiv t$ is a predictable stopping time. $\qquad\square$

7.4.2 Predictable cumulants

Definition 7.76 *If (B,C,ν) is a characteristic of some semimartingale X then as in the Lévy–Khintchine formula let us introduce the exponent*

$$\Psi\left(u,t\right)\overset{\circ}{=}iuB\left(t\right)-\frac{1}{2}uC\left(t\right)u+\left(L\left(u,x\right)\bullet\nu\left(x\right)\right)\left(t\right)\tag{7.39}$$

where

$$L\left(u,x\right)\overset{\circ}{=}\exp\left(iux\right)-1-iuh\left(x\right)$$

is the so called Lévy kernel. *We shall call Ψ the* predictable cumulant[72] *of X.*

Observe that L is deterministic and

$$\left|L\left(u,x\right)\right|\leq k\left(u\right)\cdot\left(\left\|x\right\|^{2}\wedge 1\right).\tag{7.40}$$

Therefore the integral in (7.39) exists[73] and $L\bullet\nu\in\mathcal{A}_{\mathrm{loc}}$. As $uC\left(t\right)u$ is a continuous increasing process and as every right-regular predictable process is locally bounded[74] it is clear from the definition that $\Psi\left(u,t\right)\in\mathcal{A}_{\mathrm{loc}}$ for every u.

First we prove an important technical observation:

Lemma 7.77 *X is an n-dimensional semimartingale if and only if $\exp\left(iuX\right)$ is a semimartingale for every u.*

Proof. If X is a semimartingale, then by Itô's formula $\exp\left(iuX\right)$ is a semimartingale. On the other hand assume that $\exp\left(iuX\right)$ is a semimartingale for every u. This implies that $\sin\left(uX_{j}\right)$ is a semimartingale for every u and for

---

[72] As ν and B are predictable and C is continuous Ψ is predictable.

[73] See: Proposition 7.66, page 509.

[74] See: Proposition 3.35, page 200.

every coordinate X_j of X . Let $f \in C^2 (\mathbb{R})$ be such that $f (\sin x) = x$ on the set $|x| \leq 1 < \pi/2$. Let us introduce the stopping times

$$\tau_n \overset{\circ}{=} \inf \{t : |X_j (t)| > n\} = \inf \left\{ t: \left| \frac{1}{n} X_j (t) \right| > 1 \right\}.$$

$\left| \frac{1}{n} X_j (t) \right| \leq 1$ on $[0, \tau_n)$, so on this random interval

$$X_j = n \frac{X_j}{n} = nf \left(\sin \left(\frac{X_j}{n} \right) \right).$$

By Itô's formula the right-hand side is always a semimartingale. Therefore by the next lemma X_j is a semimartingale. □

Lemma 7.78 *Let* (τ_n) *be a localizing sequence and let* (Y_n) *be a sequence of semimartingales. If* $X = Y_n$ *on* $[0, \tau_n)$ *for every* n *then* X *is a semimartingale.*

Proof. To make the notation simple let $\tau_0 = 0$. As $\tau_n \nearrow \infty$ for every t

$$X (t) = \lim_{n \to \infty} Y_n (t).$$

Hence X is adapted and it is obviously right-regular. If

$$Z_n \overset{\circ}{=} Y_n^{\tau_n} + (X (\tau_n) - Y_n (\tau_n)) \chi ([\tau_n, \infty)),$$

then X and Z_n are equal on $[0, \tau_n]$. As X is adapted and right-regular the second component is adapted hence it is in $\mathcal{V}$. The first expression is a stopped semimartingale, so the sum, Z_n, is a semimartingale. Let $Z_n = X (0) + L_n + V_n$ be a decomposition of Z_n. Then

$$X = X (0) + L + V,$$

where $L \overset{\circ}{=} \sum_n L_n \chi ((\tau_{n-1}, \tau_n])$ and $V \overset{\circ}{=} X - X (0) - L$.

$$L^{\tau_n} = \sum_{p=1}^{n} L_p \chi ((\tau_{p-1}, \tau_p]) = \sum_{p=1}^{n} \left(L_p^{\tau_p} - L_p^{\tau_{p-1}} \right) \in \mathcal{M},$$

so $L \in \mathcal{L}$. The proof of $V \in \mathcal{V}$ is similar. □

Proposition 7.79 (Characterization of predictable cumulants) *Let* X *be an n-dimensional right-regular process. The next statements are equivalent:*

1. X *is a semimartingale and* Ψ *is the predictable cumulant of* X.
2. $\exp (iuX) - \exp (iuX_-) \bullet \Psi (u)$ *is a complex valued local martingale for every* u.

Proof. The main part of the proof is an application of Itô's formula.

1. Assume that the first statement holds. Using the definition of the characteristics X has a decomposition

$$X = B + L + \sum \left(\Delta X - h\left(\Delta X\right)\right),$$

where L is the local martingale part of the special semimartingale $X\left(h\right)$. Let $f \in C^2\left(\mathbb{R}^n\right)$. By Itô's formula

$$f\left(X\right) - f\left(X\left(0\right)\right) = \sum_{j=1}^{n} \frac{\partial f}{\partial x_j}\left(X_-\right) \bullet B_j +$$

$$+ \sum_{j=1}^{n} \frac{\partial f}{\partial x_j}\left(X_-\right) \bullet L_j +$$

$$+ \sum_{j=1}^{n} \frac{\partial f}{\partial x_j}\left(X_-\right) \bullet \sum\left(\Delta X_j - h\left(\Delta X\right)\right) +$$

$$+ \frac{1}{2}\sum_{j=1}^{n}\sum_{k=1}^{n} \frac{\partial^2 f}{\partial x_j \partial x_k}\left(X_-\right) \bullet \left[X_j^c, X_k^c\right] +$$

$$+ \sum \left(f\left(X\right) - f\left(X_-\right) - \sum_{j=1}^{n} \frac{\partial f}{\partial x_j}\left(X_-\right)\Delta X_j\right).$$

One can write the third line as

$$\sum \sum_{j=1}^{n} \frac{\partial f}{\partial x_j}\left(X_-\right)\left(\Delta X_j - h\left(\Delta X\right)\right)$$

Let us introduce the predictable process

$$H\left(t, \omega, e\right) \stackrel{\circ}{=} \sum_{j=1}^{n} \frac{\partial f}{\partial x_j}\left(X_-\right)\left(e_j - h\left(e\right)\right) +$$

$$+ f\left(X\left(t-\right) + e\right) - f\left(X\left(t-\right)\right) - \sum_{j=1}^{n} \frac{\partial f}{\partial x_j}\left(X\left(t-\right)\right)e_j =$$

$$= f\left(X\left(t-\right) + e\right) - f\left(X\left(t-\right)\right) - \sum_{j=1}^{n} \frac{\partial f}{\partial x_j}\left(X_-\right)h\left(e\right).$$

With this notation

$$f(X) - f(X(0)) = \sum_{j=1}^{n} \frac{\partial f}{\partial x_j}(X_-) \bullet B_j +$$

$$+ \frac{1}{2} \sum_{j=1}^{n} \sum_{k=1}^{n} \frac{\partial^2 f}{\partial x_j \partial x_k}(X_-) \bullet [X_j^c, X_k^c] +$$

$$+ H \bullet \mu^X +$$

$$+ \sum_{j=1}^{n} \frac{\partial f}{\partial x_j}(X_-) \bullet L_j.$$

Let us assume that f is bounded. In this case the left-hand side is a bounded semimartingale, hence it is a special semimartingale. The first and the second expressions on the right-hand side are obviously predictable and have finite variation. The fourth expression is a local martingale. This implies that the third expression on the right-hand side is also a special semimartingale. Hence[75] $H \bullet \mu^X \in \mathcal{A}_{\mathrm{loc}}$. H is predictable so by definition $H \in L_{\mathrm{loc}}^1(\widetilde{\Omega}, \widetilde{\mathcal{P}}, \mu^X)$. Therefore by the elementary properties of the predictable compensator[76] of μ^X

$$H \bullet \mu^X - H \bullet \nu \in \mathcal{L}.$$

2. Let $f(x) \stackrel{\circ}{=} \exp(iux)$.

$$\frac{\partial f}{\partial x_j} = iu_j f, \quad \frac{\partial^2 f}{\partial x_j \partial x_k} = -u_j u_k f.$$

In this case

$$H(t, \omega, e) \stackrel{\circ}{=} \exp(iu(X(t-))) \exp(iue) -$$

$$- \exp(iuX(t-)) - \sum_{j=1}^{n} iu_j \exp(iuX(t-)) h_j(e),$$

that is

$$H(t, \omega, e) = \exp(iu(X(t-))) \cdot (\exp(iue) - 1 - iuh(e)).$$

Hence

$$f(X) - f(X(0)) - \sum_{j=1}^{n} iu_j f(X_-) \bullet B_j + \frac{1}{2} \sum_{j=1}^{n} \sum_{k=1}^{n} u_j u_k f(X_-) \bullet C_{jk} - H \bullet \nu$$

[75] See: Theorem 4.44, page 257.
[76] See: line (7.30), page 501.

is a local martingale. One can write the last three expression as

$$f\left(X_{-}\right) \bullet \left(iuB - \frac{1}{2}uCu + \left(\exp\left(iux\right) - 1 - iuh\left(x\right)\right) \bullet \nu\left(x\right)\right)$$
$$\stackrel{\circ}{=} f\left(X_{-}\right) \bullet \Psi\left(u\right)$$

hence

$$\exp\left(iuX\right) - \exp\left(iuX_{-}\right) \bullet \Psi\left(u\right)$$

is a local martingale.

3. Assume that the second statement holds. First we prove that X is a semimartingale. $\exp\left(iuX_{-}\right) \bullet \Psi\left(u\right)$ has finite variation for every u as the integrand is bounded and $\Psi\left(u,t\right)$ has finite variation in t. By the assumption $\exp\left(iuX\right) - \exp\left(iuX_{-}\right) \bullet \Psi\left(u\right)$ is a local martingale. Therefore $\exp\left(iuX\right)$ is a semimartingale for every u. Hence by the lemma above X is a semimartingale.

4. Finally we prove that the predictable cumulant of X is Ψ. By the already proved part of the proposition if $\widetilde{\Psi}$ denotes the predictable cumulant of X then $\exp\left(iuX\right) - \exp\left(iuX_{-}\right) \bullet \widetilde{\Psi}\left(u\right)$ is a local martingale. Hence

$$Y\left(u\right) \stackrel{\circ}{=} \exp\left(iuX_{-}\right) \bullet \left(\Psi\left(u\right) - \widetilde{\Psi}\left(u\right)\right) =$$
$$= \exp\left(iuX_{-}\right) \bullet \Psi\left(u\right) - \exp\left(iuX_{-}\right) \bullet \widetilde{\Psi}\left(u\right)$$

is also a local martingale. Y has finite variation on any finite interval and as $\Psi\left(u\right) - \widetilde{\Psi}\left(u\right)$ is predictable it is also predictable[77]. Therefore by Fisk's theorem $Y\left(u\right)$ with probability one[78] is zero for every u. Therefore

$$0 = \exp\left(-iuX_{-}\right) \bullet Y\left(u\right) =$$
$$= \exp\left(-iuX_{-}\right) \bullet \left(\exp\left(iuX_{-}\right) \bullet \left(\Psi\left(u\right) - \widetilde{\Psi}\left(u\right)\right)\right) =$$
$$= 1 \bullet \left(\Psi\left(u\right) - \widetilde{\Psi}\left(u\right)\right) = \Psi\left(u\right) - \widetilde{\Psi}\left(u\right).$$

So with probability one $\widetilde{\Psi}\left(u,t,\omega\right) = \Psi\left(u,t,\omega\right)$ in t for every u. The expressions are continuous in u, hence one can unify the zero sets. So $\widetilde{\Psi}\left(u\right) = \Psi\left(u\right)$ with probability one for every u. $\square$

---

[77] See: Example 7.56, page 500.
[78] See: Corollary 3.40, page 205.

From the last part of the proof of the proposition the next statement is trivial:

Corollary 7.80 *Let* $\Phi(u, t, \omega)$ *be predictable, continuous in u and right-continuous with finite variation in t. If for every u*

$$\exp(iuX) - \exp(iuX_-) \bullet \Phi(u)$$

is a local martingale then X is a semimartingale and $\Phi - \Phi(0)$ is the predictable cumulant of X.

Corollary 7.81 *If Ψ is the predictable cumulant of a semimartingale X and τ is a stopping time then the predictable cumulant of X^τ is Ψ^τ.*

Proof. For every u

$$(\exp(iuX) - \exp(iuX_-) \bullet \Psi(u))^\tau = \exp(iuX^\tau) - \exp(iuX^\tau_-) \bullet \Psi(u)^\tau$$

is a local martingale. $\qquad\qquad\square$

7.4.3 Semimartingales with independent increments

Every Lévy process is a semimartingale. This is not true for processes with independent increments.

Proposition 7.82 *A deterministic right-regular process S is a semimartingale if and only if it has finite variation on any finite interval $[0, t]$.*

Proof. If S has the stated properties then S is obviously a semimartingale. Now let S be a deterministic semimartingale. As S is a semimartingale one can define the continuous linear functional[79] $(f \bullet S)(t)$ on $C([0, t])$. By the Riesz representation theorem[80] there is a function V with finite variation that

$$(f \bullet S)(t) = \int_0^t f dV, \quad f \in C([0, t]).$$

From the Dominated Convergence Theorem it is clear that $f - f(0) = V - V(0)$ on $[0, t]$ so f has finite variation. $\qquad\qquad\square$

As every right-regular, deterministic function starting from the origin is a process with independent increments there are processes with independent increments which are not semimartingales. When is a process with independent increments a semimartingale?

---

[79]As $f \bullet S$ is also an Itô–Stieltjes integral, it is deterministic.
[80]See: [80].

Theorem 7.83 (Characterization of semimartingales with independent increments) *An n-dimensional process X with independent increments is a semimartingale if and only if the Fourier transform of X*

$$\varphi(u,t) \overset{\circ}{=} \mathbf{E}\left(\exp\left(iuX(t)\right)\right)$$

has finite variation on every finite interval in variable t for every u.

Proof. Observe that by definition X is right-regular. Therefore by the Dominated Convergence Theorem φ is also right-regular in t.

1. Let us fix parameter u. As the increments are not stationary[81] it can happen that $\varphi(u,t) = 0$ for some t. Let

$$t_0(u) \overset{\circ}{=} \inf\{t \colon \varphi(u,t) = 0\}.$$

$\varphi(u,0) = 1$ and as $\varphi(u,t)$ is right-regular in t obviously $t_0(u) > 0$. Obviously $|\varphi(u,t)|$ is positive on $[0, t_0(u))$. We show that $t \mapsto |\varphi(u,t)|$ is decreasing on $\mathbb{R}_+$ and it is zero on $[t_0(u), \infty)$. Let

$$h(u,s,t) \overset{\circ}{=} \mathbf{E}\left(\exp\left(iu\left(X(t) - X(s)\right)\right)\right).$$

X has independent increments, so if $s < t$ then

$$\varphi(u,t) = \varphi(u,s)\,h(u,s,t). \tag{7.41}$$

$|h(u,t,s)| \leq 1$, therefore as we said $|\varphi|$ is decreasing. By the right-regularity

$$\varphi(u, t_0(u)) = 0.$$

So as $|\varphi| \geq 0$ and as it is decreasing φ is zero on the interval $[t_0(u), \infty)$. As $\varphi(u,t)$ is right-regular in t if $t_0(u) < \infty$ then $\varphi(u, t_0(u)-)$ is well-defined. We show that it is not zero. By (7.41) if $s < t_0(u)$ then

$$\varphi(u, t_0(u)-) = \varphi(u,s)\,h(u,s,t_0(u)-).$$

$\varphi(u,s) \neq 0$ by the definition of $t_0(u)$. So if $\varphi(u, t_0(u)-) = 0$ then

$$h(u,s,t_0(u)-) = 0$$

---

[81]See: Proposition 1.99, page 63.

for every $s < t_0(u)$.

$$0 = \lim_{s \nearrow t_0(u)} h(u, s, t_0(u) -) =$$

$$= \lim_{s \nearrow t_0(u)} \mathbf{E}\left(\exp\left(iuX\left(t_0(u) -\right) - iuX(s)\right)\right) =$$

$$= \mathbf{E}\left(\exp(0)\right) = 1,$$

which is impossible. Therefore $\varphi(u, t_0(u) -) \neq 0$.

2. Let

$$Z(u, t) \stackrel{\circ}{=} \begin{cases} \exp(iuX(t))/\varphi(u, t) & \text{if } t < t_0(u) \\ \exp(iuX(t_0(u) -))/\varphi(u, t_0(u) -) & \text{if } t \geq t_0(u) \end{cases}.$$

X has independent increments, so $Z(u)$ is a martingale on $t < t_0(u)$. As

$$|\varphi(u, t_0(u) -)| > 0$$

in the next calculation one can use the Dominated Convergence Theorem

$$\mathbf{E}\left(Z(u, t_0(u)) \mid \mathcal{F}_s\right) = \mathbf{E}\left(\lim_{t \nearrow t_0(u)} Z(u, t) \mid \mathcal{F}_s\right) =$$

$$= \lim_{t \nearrow t_0(u)} \mathbf{E}\left(Z(u, t) \mid \mathcal{F}_s\right) = Z(u, s)$$

for every $s < t_0(u)$. So Z is a martingale on $\mathbb{R}_+$.

3. By Itô's formula

$$\varphi(u, t) = \frac{\exp(iuX)}{Z(u, t)} \chi(t < t_0(u))$$

is also a semimartingale. Hence $\varphi(u, t)$ is a deterministic semimartingale, so by the just proved proposition it has finite variation,

5. Now we prove the other implication: Assume that $\varphi(u, s)$ has finite variation on every finite interval $[0, t]$. One should show that X is a semimartingale. Let us fix a t. If $u \to 0$ then $\varphi(u, t) \to 1$, so there is a $b > 0$ such that if $|u| \leq b$ then $|\varphi(u, t)| > 0$. By the first part of the proof $|\varphi|$ is decreasing, so if $s \leq t$ and $|u| \leq b$ then

$$\exp(iuX(s)) = Z(u, s) \cdot \varphi(u, s).$$

$Z(u, s)$ is a martingale, $\varphi(u, s)$ has finite variation, so by Itô's formula the stopped process $\exp(iuX^t)$ is a semimartingale. If $|u| > b$ then for some m large enough $|u/m| \leq b$.

$$\exp(iuX) = \left(\exp\left(i\frac{u}{m}X\right)\right)^m.$$

Therefore by Itô's formula the stopped process $\exp(iuX^t)$ is again a semimartingale for every u. Hence X^t is a semimartingale[82] for every t. Using the trivial localization property of semimartingales[83] it is easy to show that X is a semimartingale. □

Theorem 7.84 (Predictable cumulants and independent increments)
Let X be an n-dimensional semimartingale and let $\Psi(u, t, \omega)$ be the predictable cumulant of X. If X has independent increments then Ψ is deterministic and

$$\varphi = \mathcal{E}(\Psi). \tag{7.42}$$

Proof. Let us fix an u. Let again

$$t_0(u) \stackrel{\circ}{=} \inf\{t \colon \varphi(u, t) = 0\}.$$

$\varphi(u, t)$ is right-continuous and $\varphi(0, u) = 1$ therefore $t_0(u) > 0$.

1. Let

$$U \stackrel{\circ}{=} \begin{cases} X(t) & \text{if } t < t_0(u) \\ X(t_0(u)-) & \text{if } t \geq t_0(u) \end{cases}$$

Let

$$\gamma \stackrel{\circ}{=} \varphi_U(u, t) \stackrel{\circ}{=} \mathbf{E}(\exp(iuU(t))).$$

Recall that $\varphi(u, t_0(u)-) \neq 0$ therefore $\gamma \neq 0$. Let

$$A \stackrel{\circ}{=} \frac{1}{\gamma-} \bullet \gamma.$$

Observe that as φ has finite variations the integral is well-defined[84].

$$\gamma = 1 + 1 \bullet \gamma = 1 + \frac{\gamma_-}{\gamma_-} \bullet \gamma =$$

$$= 1 + \gamma_- \bullet \left(\frac{1}{\gamma_-} \bullet \gamma\right) \stackrel{\circ}{=} 1 + \gamma_- \bullet A.$$

As the Doléans equations have unique solution[85] $\gamma = \mathcal{E}(A)$. That is

$$\varphi(u, t) = \gamma(t) = \mathcal{E}(A), \quad t \in [0, t_0(u)).$$

---

[82] See: Lemma 7.77, page 518.
[83] See: Lemma 7.78, page 519.
[84] See: Proposition 1.151, page 106.
[85] See: Theorem 6.56, page 412.

We prove that $A = \Psi$ on $[0, t_0(u))$. Let

$$Y(t, u) \stackrel{\circ}{=} \exp(iuX(t)), \quad Z(u, t) \stackrel{\circ}{=} \frac{\exp(iuX(t))}{\varphi(u, t)}, \quad t < t_0(u).$$

X has independent increments, so $Z(u)$ is a martingale. Integrating by parts on the interval $[0, t_0(u))$ and using that φ has finite variation[86]

$$Y - Y_- \bullet A \stackrel{\circ}{=} Y - Y_- \bullet \left(\frac{1}{\gamma_-} \bullet \varphi\right) =$$

$$= Y - Y_- \bullet \left(\frac{1}{\varphi_-} \bullet \varphi\right) =$$

$$= Y - \frac{Y_-}{\varphi_-} \bullet \varphi \stackrel{\circ}{=}$$

$$\stackrel{\circ}{=} Z\varphi - Z_- \bullet \varphi =$$

$$= Z(0)\varphi(0) + \varphi_- \bullet Z + [Z, \varphi] =$$

$$= Z(0)\varphi(0) + \varphi_- \bullet Z + \sum \Delta Z \Delta \varphi =$$

$$= Z(0)\varphi(0) + \varphi_- \bullet Z + \Delta \varphi \bullet Z =$$

$$= Z(0)\varphi(0) + \varphi \bullet Z.$$

φ is locally bounded, hence $\varphi \bullet Z$ is a local martingale on $[0, t_0(u))$. As $Y \stackrel{\circ}{=} \exp(iuX)$

$$\exp(iuX) - \exp(iuX_-) \bullet A$$

is a local martingale on $[0, t_0(u))$. So[87] $A(u) = \Psi(u)$ and $\Psi(u)$ is deterministic on $[0, t_0(u))$.

2. Let

$$V(t) \stackrel{\circ}{=} \begin{cases} 0 & \text{if } t < t_0(u) \\ \Delta X(t_0(u)) & \text{if } t \geq t_0(u) \end{cases}.$$

Obviously processes V has independent increments. The spectral measure of V is[88]

$$\nu(\{t_0(u)\} \times \Lambda) \stackrel{a.s}{=} \mathbf{P}(\Delta X(t_0(u)) \in \Lambda) \stackrel{\circ}{=} F(\Lambda).$$

[86]See: Example 4.39, page 249. Let us recall that $\varphi \in \mathcal{V}$ is deterministic therefore it is predictable.

[87]See: Corollary 7.80, page 523.

[88]See: Example 7.70, page 514.

If $t < t_0(u)$ then obviously $\varphi_V(u,t) = 1$. With simple calculation[89] if $t \geq t_0(u)$

$$\varphi_V(u,t) \stackrel{\circ}{=} \mathbf{E}\left(\exp\left(iu\Delta X(t_0)\right)\right) =$$

$$= \int_{\mathbb{R}^n} \exp\left(iux\right) dF(x) =$$

$$= 1 + \int_{\mathbb{R}^n \setminus \{0\}} \exp\left(iux\right) - 1 dF(x) =$$

$$= 1 + \int_{\mathbb{R}^n \setminus \{0\}} \exp\left(iux\right) - 1\nu\left(\{t_0(u)\} \times dx\right) =$$

$$= 1 + \int_{\mathbb{R}^n \setminus \{0\}} iuh(x)\,\nu\left(\{t_0(u)\} \times dx\right) +$$

$$+ \int_{\mathbb{R}^n \setminus \{0\}} L(u,x)\,\nu\left(\{t_0(u)\} \times dx\right).$$

where $\nu\left(\{t_0(u)\} \times \Lambda\right)$ is deterministic. It is easy to see that in the decomposition of $V(h)$ the local martingale part is

$$\begin{cases} 0 & \text{if } t < t_0(u) \\ h(\Delta X(t_0(u))) - \mathbf{E}(h(\Delta X(t_0(u)))) & \text{if } t \geq t_0(u) \end{cases}$$

so

$$B = \begin{cases} 0 & \text{if } t < t_0(u) \\ \mathbf{E}(h(\Delta X(t_0))) & \text{if } t \geq t_0(u) \end{cases} =$$

$$= \begin{cases} 0 & \text{if } t < t_0(u) \\ \int_{\mathbb{R}^n \setminus \{0\}} h(x)\,\nu\left(\{t_0(u)\} \times dx\right) & \text{if } t \geq t_0(u) \end{cases}.$$

Hence

$$\Psi_V(u,t) = \begin{cases} 0 & \text{if } t < t_0(u) \\ \varphi_V(u,t) - 1 & \text{if } t \geq t_0(u) \end{cases}$$

that is

$$\varphi_V(u,t) = 1 + \Psi_V(u,t)$$

where $\Psi_V(u,t)$ is deterministic.

[89] See: Example 7.72, page 514.

3. Obviously U and V are independent and $X = U + V$ on $[0, t_0(u)]$. From the definition of the predictable cumulant one can easily prove that on $[0, t_0(u)]$

$$\Psi_X(u,t) = \Psi_U(u,t) + \Psi_V(u,t).$$

Therefore if $t \in [0, t_0(u)]$ then $\Psi_X(u,t)$ is deterministic.

4. Let

$$W(u,t) = \begin{cases} 0 & \text{if } t < t_0(u) \\ X(t) - X(t_0(u)) & \text{if } t \geq t_0(u) \end{cases}.$$

W is a semimartingale with independent increments and with same argument as above one can show that there is a $t_0(u) < t_1(u)$ that the predictable cumulant of W is deterministic on $[0, t_1(u)]$. If $t \in [0, t_1(u))$ then

$$\Psi_X(u,t) = \Psi_V(u,t) + \Psi_U(u,t) + \Psi_W(u,t),$$

therefore $\Psi_X(u,t)$ is almost surely deterministic if $t \in [0, t_1(u))$.

5. Let $t_\infty(u)$ be the supremum of the time-parameters $t(u)$ for which $\Psi_X(u,t)$ is almost surely deterministic on $t \in [0, t(u)]$. Let $t_n(u) \nearrow t_\infty(u)$. If $t_\infty(u) = \infty$ then unifying the zero sets one can easily show that almost surely $\Psi_X(u,t)$ has a deterministic version. If $t_\infty(u) < \infty$ then as above one can prove that $\Psi_X(u,t)$ is deterministic $[0, t_\infty(u)]$ and one can find a $\delta > 0$ such that $\Psi_X(u,t)$ is deterministic on $[0, t_\infty(u) + \delta]$ which is impossible. Now for every u with rational coordinates let us construct a deterministic version of $\Psi_X(u,t)$. Unifying the measure-zero sets and using that $\Psi_X(u,t)$ is continuous in u one can construct an almost surely deterministic version of $\Psi_X(u,t)$.

6. As Ψ is deterministic the local martingale

$$\exp(iuX) - \exp(iuX_-) \bullet \Psi_X(u,t)$$

is bounded on any finite interval. So it is a martingale. Using Fubini's theorem one can easily show that

$$\mathbf{E}(\exp(iuX(t))) - \mathbf{E}(\exp(iuX(t)-)) \bullet \Psi(u) = 1.$$

That is for every u

$$\varphi(u) - \varphi_-(u) \bullet \Psi(u) = 1.$$

Hence $\varphi(u) = \mathcal{E}(\Psi(u))$. $\qquad\qquad\qquad\qquad\qquad\qquad\qquad\square$

7.4.4 Characteristics of semimartingales with independent increments

Now we show that the characteristics of semimartingales with independent increments are deterministic. The main step is the next famous classical observation:

Theorem 7.85 *Let b be an n-dimensional vector, C a positive semidefinite matrix and let ν be a measure on $\mathbb{R}^n$ with*

$$\nu\left(\{0\}\right) = 0 \quad and \quad \int_{\mathbb{R}^n \setminus \{0\}} \left(\|x\|^2 \wedge 1\right) d\nu\left(x\right) < \infty.$$

Let h be an arbitrary truncating function. The function

$$\phi\left(u\right) \stackrel{\circ}{=} iub - \frac{1}{2}uCu + \int_{\mathbb{R}^n \setminus \{0\}} L\left(x, u\right) d\nu\left(x\right)$$

determines the triplet (b, C, ν).

Proof: Let $v \in \mathbb{R}^n \setminus \{0\}$

$$\psi_v\left(u\right) \stackrel{\circ}{=} \phi\left(u\right) - \frac{1}{2} \int_{-1}^{1} \phi\left(u + tv\right) dt.$$

With simple calculation

$$\psi_v\left(u\right) = \frac{1}{4}\left(vCv\right) \int_{-1}^{1} t^2 dt +$$

$$+ \frac{1}{2} \int_{-1}^{1} \int_{\mathbb{R}^n \setminus \{0\}} \exp\left(iux\right)\left(1 - \exp\left(itvx\right) + itvh\left(x\right)\right) d\nu\left(x\right) dt.$$

By the integrability assumption one can use Fubini's theorem to change the order of the integration. Hence

$$\psi_v\left(u\right) = \frac{1}{6}\left(vCv\right) +$$

$$+ \frac{1}{2} \int_{\mathbb{R}^n \setminus \{0\}} \exp\left(iux\right) \int_{-1}^{1} \left(1 - \exp\left(itvx\right)\right) dt d\nu\left(x\right)$$

$$= \frac{1}{6}\left(vCv\right) + \int_{\mathbb{R}^n \setminus \{0\}} \exp\left(iux\right) \left(1 - \frac{\sin vx}{vx}\right) d\nu\left(x\right).$$

By the integrability condition

$$\sigma_v\left(\Lambda\right) \stackrel{\circ}{=} \frac{1}{6}\left(vCv\right) \delta_0\left(\Lambda\right) + \int_{\Lambda} \left(1 - \frac{\sin vx}{vx}\right) d\nu\left(x\right), \quad \Lambda \in \mathcal{B}\left(\mathbb{R}^n\right)$$

is a finite measure and $\psi_v(u)$ is the Fourier transform of σ_v. If we know σ_v then

$$vCv = 6\sigma_v(\{0\}).$$

If $vx \neq 0$ then

$$1 - \frac{\sin vx}{vx} \neq 0$$

and v and σ_v are equivalent on the set $\{vx \neq 0\}$. Hence the set of measures $\{\sigma_v\}$ determines[90] v. If $\Rightarrow$ denotes the relation that the left-hand side uniquely determines the right-hand side then obviously

$$\phi \Rightarrow \psi_v \Rightarrow \sigma_v \Rightarrow (C, v).$$

Obviously b is determined by (C, v) and ϕ. Therefore ϕ determines (b, C, v).

$\square$

Corollary 7.86 *If (B_1, C_1, ν_1) and (B_2, C_2, ν_2) are different characteristics, Ψ_1 and Ψ_2 are the corresponding predictable cumulants then $\Psi_1 \neq \Psi_2$.*

Proof. Assume, that $\Psi_1 = \Psi_2$. Let us fix an ω and a t and let $u \neq 0 \in \mathbb{R}^n$. L depends only on x so if

$$\tilde{\nu}_i(\Lambda) \stackrel{\circ}{=} \nu(\omega, (0, t] \times \Lambda), \quad i = 1, 2$$

then one can write the integral in the definition of the predictable characteristics as

$$\int_{\mathbb{R}^n \setminus \{0\}} L(u, x) \, d\tilde{\nu}_i(x), \quad i = 1, 2.$$

From the previous theorem for every t and ω

$$uB_1(t, \omega) = uB_2(t, \omega), \quad uC_1(t, \omega)u = uC_2(t, \omega)u$$

and

$$\nu_1(\omega, (0, t] \times \Lambda) = \nu_2(\omega, (0, t] \times \Lambda).$$

Hence $(B_1, C_1, \nu_1) = (B_2, C_2, \nu_2)$.

$\square$

---

[90]$\nu(\{0\}) = 0$, and by the theorem on separating hyperplanes one can calculate the measure of every closed convex set in $\mathbb{R}^n \setminus \{0\}$.

Corollary 7.87 *If X is an n-dimensional semimartingale with independent increments then the characteristics of X are deterministic.*

Corollary 7.88 *If X is an n-dimensional process with independent increments then its spectral measure ν, that is the predictable compensator of μ^X, is deterministic. In this case $\nu\left((0,t] \times \Lambda\right)$ is the expected value of the jumps belonging to Λ during the time period $(0,t]$.*

Proof. Let ν be the spectral measure of X. We show that ν is deterministic. Let $g \doteq \chi_\Lambda$, where $\Lambda \in \mathcal{B}\left(\mathbb{R}^n \setminus \{0\}\right)$ with $0 \notin \mathrm{cl}\left(\Lambda\right)$. As $0 \notin \mathrm{cl}\left(\Lambda\right)$ the jumps in Λ cannot accumulate so $g \bullet \mu^X$ is a finite valued process. For an arbitrary s if $t > s$ then

$$\left(g \bullet \mu^X\right)(t) - \left(g \bullet \mu^X\right)(s) = \sum_{s < r \le t} g\left(\Delta X_r\right),$$

is independent of the σ-algebra $\mathcal{F}_s$. $g \ge 0$ therefore $g \bullet \mu^X$ is increasing so it is a semimartingale. Hence the spectral measure of $g \bullet \mu^X$ is deterministic. As g is bounded $g \bullet \mu^X$ is increasing with bounded jumps. Hence $g \bullet \mu^X$ is locally bounded. If $H \ge 0$ is a predictable process then

$$\mathbf{E}\left(H \bullet \left(g \bullet \mu^X\right)\right) = \mathbf{E}\left(Hg \bullet \mu^X\right) = \mathbf{E}\left(Hg \bullet \nu\right) = \mathbf{E}\left(H \bullet \left(g \bullet \nu\right)\right).$$

$g \bullet \nu$ is predictable, hence the compensator of $g \bullet \mu^X \in \mathcal{A}_{\mathrm{loc}}^+$ is $g \bullet \nu$. This implies that $\left(g \bullet \nu\right)(t) = \nu\left((0,t] \times \Lambda\right)$ is deterministic. ν is defined on $\mathcal{B}\left(\mathbb{R}_+\right) \times \mathcal{B}\left(\mathbb{R}^n \setminus \{0\}\right)$. Hence ν is deterministic. $\qquad\square$

Theorem 7.89 (Characteristics and independent increments) *Let X be an n-dimensional semimartingale with $X(0) = 0$. The characteristics of X have a deterministic version if and only if X has independent increments. In this case if $s < t$ then*

$$h(u,s,t) \doteq \mathbf{E}\left(\exp\left(iu\left(X(t) - X(s)\right)\right)\right) = \mathcal{E}\left(\Psi(u,t) - \Psi(u,s)\right),$$

where Ψ denotes the predictable cumulant of X.

Proof. One should only prove that if the characteristic are deterministic then X has independent increments. Let X be a semimartingale with $X(0) = 0$ and let Ψ be the predictable cumulant of X. If the characteristics are deterministic then Ψ is deterministic. As we have proved[91]

$$U(u) \doteq \exp\left(iuX\right) - \exp\left(iuX_-\right) \bullet \Psi \tag{7.43}$$

---

[91]See: Proposition 7.79, page 519.

is a local martingale. But

$$|U\left(u\right)| \leq 1 + \text{Var}\left(\Psi\left(u\right)\right) < \infty.$$

As the characteristics are deterministic $\text{Var}\left(\Psi\right)$ is deterministic, so it is obviously integrable by **P**. Hence on any finite interval $U\left(u\right) \in \mathcal{D}$. Therefore $U\left(u\right)$ is a martingale. If $0 \leq s < t$ then using (7.43)

$$\frac{\exp\left(iuX\left(t\right)\right)}{\exp\left(iuX\left(s\right)\right)} \overset{\circ}{=}$$

$$\overset{\circ}{=} \frac{U\left(u,t\right)}{\exp\left(iuX\left(s\right)\right)} + \frac{\left(\exp iu\left(X_-\right) \bullet \Psi\right)\left(t\right)}{\exp\left(iuX\left(s\right)\right)} =$$

$$= \frac{U\left(u,t\right) - U\left(u,s\right)}{\exp\left(iuX\left(s\right)\right)} + \frac{\left(\exp iu\left(X_-\right) \bullet \Psi\right)\left(t\right) + U\left(u,s\right)}{\exp\left(iuX\left(s\right)\right)} =$$

$$= \frac{U\left(u,t\right) - U\left(u,s\right)}{\exp\left(iuX\left(s\right)\right)} + \frac{\int_s^t \exp\left(iu\left(X\left(r-\right)\right)\right) d\Psi\left(r\right) + \exp\left(iuX\left(s\right)\right)}{\exp\left(iuX\left(s\right)\right)}.$$

Multiplying by χ_F where $F \in \mathcal{F}_s$

$$\chi_F \exp\left(iu\left(X\left(t\right) - X\left(s\right)\right)\right) =$$
$$= \chi_F \exp\left(-iuX\left(s\right)\right)\left(U\left(u,t\right) - U\left(u,s\right)\right) +$$
$$+ \int_s^t \chi_F \exp\left(iu\left(X\left(r-\right) - X\left(s\right)\right)\right) d\Psi\left(r\right) + \chi_F.$$

U is a martingale, hence by the elementary properties of the conditional expectation the expected value of the first term on the right-hand side is zero. Let

$$f\left(r\right) \overset{\circ}{=} \mathbf{E}\left(\chi_F \exp\left(iu\left(X\left(r\right) - X\left(s\right)\right)\right)\right).$$

Taking expected value and by Fubini's theorem changing the integrals[92]

$$f\left(t\right) = \mathbf{P}\left(F\right) + \int_s^t f\left(r-\right) d\Psi\left(r\right) =$$

$$= \mathbf{P}\left(F\right) + \int_0^t f\left(r-\right) d\left(\Psi\left(r\right) - \Psi\left(r \wedge s\right)\right).$$

[92]Observe that we have used that Ψ is determinstic.

As every Doléans equation has just one solution[93]

$$f(t) = \mathbf{P}(F) \cdot \mathcal{E}(\Psi(t) - \Psi(s \wedge t)) = \mathbf{P}(F) \cdot \mathcal{E}(\Psi(t) - \Psi(s)).$$

Therefore for every u

$$\mathbf{E}(\chi_F \exp(iu(X(t) - X(s)))) = \mathbf{P}(F) \cdot \mathcal{E}(\Psi(u,t) - \Psi(u,t)).$$

This means that $X(t) - X(s)$ is independent of $\mathcal{F}_s$ and

$$\mathbf{E}(\exp(iu(X(t) - X(s)))) = \mathcal{E}(\Psi(u,t) - \Psi(u,s)). \qquad \square$$

7.4.5 The proof of the formula

Theorem 7.90 (Lévy–Khintchine formula) *If X is an n-dimensional semimartingale with independent increments then*

$$\mathbf{E}(\exp(iu(X(t) - X(s)))) = \exp(U) \cdot V, \qquad (7.44)$$

where

$$U \overset{\circ}{=} iu[B(t) - B(s)] - \frac{1}{2}u[C(t) - C(s)]u +$$

$$+ \int_{(s,t] \times (\mathbb{R}^n \setminus \{0\})} (\exp(iux) - 1 - iuh(x)) \chi_{J^c}(r) \, d\nu(r,x)$$

and

$$V = \prod_{s < r \leq t} \left(\exp(-iu\Delta B(r)) \left(1 + \int_{\mathbb{R}^n \setminus \{0\}} (\exp(iux) - 1) \nu(\{r\} \times dx) \right) \right),$$

where

$$J \overset{\circ}{=} \{r \colon \nu(\{r\} \times (\mathbb{R}^n \setminus \{0\})) > 0\}.$$

Proof. The formula in (7.44) is a direct consequence of the formula of the solution of the Doléans equation [94]. $Z(t) \overset{\circ}{=} \Psi(t) - \Psi(s)$ has finite variation so

$$\mathcal{E}(Z) = \exp(Z) \prod (1 + \Delta Z) \exp(-\Delta Z).$$

---

[93]See: line (6.44), page 411.

[94]See: Theorem 6.56, page 412.

The jumps of ΔZ are the sum of the jumps of B and the jumps of the integral in Ψ. The integral in Ψ has a jump at r if and only if $r \in J$. Hence

$$\Delta Z\left(r\right) = iu\Delta B\left(r\right) +$$

$$+ \int_{\mathbb{R}^n\setminus\{0\}} \left(\exp\left(iux\right) - 1 - iuh\left(x\right)\right) \nu\left(\{r\} \times dx\right).$$

As in the integral in U one has χ_{J^c} we have cancelled the jumps

$$\int_{\mathbb{R}^n\setminus\{0\}} \left(\exp\left(iux\right) - 1 - iuh\left(x\right)\right) \nu\left(\{r\} \times dx\right)$$

from (7.44), which proves the formula for U. ν and B are deterministic so

$$\int_{\mathbb{R}^n\setminus\{0\}} h\left(x\right) \nu\left(\{r\} \times dx\right) = \mathbf{E}\left(h\left(\Delta X\left(r\right)\right)\right) =$$

$$= \mathbf{E}\left(\Delta B\left(r\right) + \Delta L\left(r\right)\right) = \Delta B\left(r\right) + \mathbf{E}\left(\Delta L\left(r\right)\right),$$

where L is the local martingale in the decomposition of $X\left(h\right)$. If (τ_n) is a localizing sequence of L then $\mathbf{E}\left(\Delta L^{\tau_n}\left(r\right)\right) = 0$. As $|\Delta L\left(r\right)| \leq 1 + |B\left(r\right)|$ by the Dominated Convergence Theorem $\mathbf{E}\left(\Delta L\left(r\right)\right) = 0$. Therefore

$$\Delta Z\left(r\right) = \int_{\mathbb{R}^n\setminus\{0\}} \left(\exp\left(iux\right) - 1\right) \nu\left(\{r\} \times dx\right),$$

which proves the formula for V. □

Corollary 7.91 *If X is an n-dimensional semimartingale with independent increment then the Fourier transform of $\Delta X\left(t\right)$ is*

$$\mathbf{E}\left(\exp\left(iu\Delta X\left(t\right)\right)\right) = 1 + \int_{\mathbb{R}^n\setminus\{0\}} \left(\exp\left(iux\right) - 1\right) \nu\left(\{t\} \times dx\right),$$

so $\Delta X\left(t\right)$ is not zero with positive probability if and only if $\nu\left(\{t\} \times \left(\mathbb{R}^n \setminus \{0\}\right)\right) > 0$, that is if $t \in J$.

Proof. If $s \nearrow t$ in (7.44) then all the other terms disappear. □

Corollary 7.92 *If X is an n-dimensional semimartingale with independent increment then X is continuous in probability if and only if the probability of a jump at time t is zero for every t. In this case the Fourier transform of $X\left(t\right)$ is*

$$\varphi\left(u, t\right) = \exp\left(\Psi\left(u, t\right)\right).$$

Proof. As X is right-regular X is continuous in probability if and only if $\Delta X(t) \overset{a.s.}{=} 0$. This means that X is stochastically continuous if and only if

$$\mathbf{E}\left(\exp\left(iu\Delta X(t)\right)\right) \equiv 1$$

for all t. By the previous corollary in this case $J = \varnothing$ and in this case $V = 1$ in (7.44). $\qquad\square$

Example 7.93 If X is an n-dimensional semimartingale with independent increments and X is continuous in probability and $0 \notin \mathrm{cl}(\Lambda)$ then for every t the number of jumps in Λ during the time interval $(0,t]$ has a Poisson distribution with parameter $\nu\left((0,t] \times \Lambda\right)$.

Let N^Λ be the process counting the jumps in Λ. As $0 \notin \mathrm{cl}(\Lambda)$ obviously N^Λ has right-regular trajectories and $N^\Lambda(t)$ is finite for every t and it is a process with independent increments. As X is continuous in probability N^Λ does not have fixed time of discontinuities so it is also continuous in probability. By the Lévy–Khintchine formula the Fourier transform of $N^\Lambda(t)$ has the representation

$$\varphi(u,t) = \exp\left(\Psi^\Lambda(u,t)\right)$$

where

$$\Psi^\Lambda(u,t) \overset{\circ}{=} iuB^\Lambda(t) - \frac{1}{2}uC^\Lambda(t)u+$$

$$+ \int_{(0,t]\times(\mathbb{R}^n\setminus\{0\})} \left(\exp\left(iux\right) - 1 - iuh(x)\right) d\nu^\Lambda(r,x).$$

N^Λ is continuous in probability an it has bounded jumps, hence all the moments of $N^\Lambda(t)$ are finite[95]. Therefore the expected value of $N^\Lambda(t)$ is finite so

$$\nu^\Lambda\left((0,t] \times (\mathbb{R}^n \setminus \{0\})\right) = \mathbf{E}\left(N^\Lambda(t)\right) < \infty.$$

Therefore one can write the integral as

$$\int_{(0,t]\times(\mathbb{R}^n\setminus\{0\})} \left(\exp\left(iux\right) - 1\right) d\nu^\Lambda(r,x) - iu \int_{(0,t]\times(\mathbb{R}^n\setminus\{0\})} h(x) d\nu^\Lambda(r,x)$$

---

[95]See: Proposition 1.114, page 78.

and the predictable cumulant has the representation

$$\Psi^\Lambda\left(u,t\right) \overset{\circ}{=} iuD^\Lambda\left(t\right) - \frac{1}{2}uC^\Lambda\left(t\right)u +$$

$$+ \int_{(0,t]\times(\mathbb{R}^n\backslash\{0\})} \left(\exp\left(iux\right) - 1\right)d\nu^\Lambda\left(r,x\right).$$

The derivative of the Fourier transform at $u = 0$ is the expected value of the distribution multiplied by i. So as ν is deterministic

$$i\mathbf{E}\left(N^\Lambda\left(t\right)\right) = \varphi_u'\left(0,t\right) = iD^\Lambda\left(t\right) + i\int_{(0,t]\times(\mathbb{R}^n\backslash\{0\})} x\exp\left(i0x\right)d\nu^\Lambda\left(r,x\right) =$$

$$= iD^\Lambda\left(t\right) + \mathbf{E}\left(\int_{(0,t]\times(\mathbb{R}^n\backslash\{0\})} xdN^\Lambda\left(r,x\right)\right) =$$

$$= iD^\Lambda\left(t\right) + i\mathbf{E}\left(N^\Lambda\left(t\right)\right).$$

Hence $D^\Lambda\left(t\right) = 0$. Differentiating φ twice

$$\mathbf{E}\left(\left(N^\Lambda\left(t\right)\right)^2\right) = C^\Lambda\left(t\right) + \mathbf{E}\left(\left(N^\Lambda\left(t\right)\right)^2\right).$$

Hence $C^\Lambda\left(t\right) = 0$. So as Ψ^Λ is deterministic

$$\Psi^\Lambda\left(u,t\right) = \mathbf{E}\left(\int_{(0,t]\times(\mathbb{R}^n\backslash\{0\})} \left(\exp\left(iux\right) - 1\right)d\nu^\Lambda\left(r,x\right)\right) =$$

$$= \mathbf{E}\left(\int_{(0,t]\times(\mathbb{R}^n\backslash\{0\})} \left(\exp\left(iux\right) - 1\right)dN^\Lambda\left(r,x\right)\right) =$$

$$= \mathbf{E}\left(\sum_{s\leq t}\left(\exp\left(iu\Delta N^\Lambda\left(s\right)\right) - 1\right)\right) =$$

$$= \mathbf{E}\left(\sum_{s\leq t}\left(\exp\left(iu\right) - 1\right)\right) =$$

$$= \left(\exp\left(iu\right) - 1\right)\cdot\mathbf{E}\left(N^\Lambda\left(t\right)\right) = \left(\exp\left(iu\right) - 1\right)\cdot\nu^\Lambda\left((0,t]\times\Lambda\right).$$

Therefore $N^\Lambda\left(t\right)$ has a compound Poisson distribution with parameter

$$\lambda \overset{\circ}{=} \nu^\Lambda\left((0,t]\times\Lambda\right).$$

As $\nu\left((0,t] \times \Lambda\right)$ is the expected value of the number of jumps of X in Λ during the time interval $(0,t]$ obviously

$$\lambda \overset{\circ}{=} \nu^{\Lambda}\left((0,t] \times \Lambda\right) = \nu\left((0,t] \times \Lambda\right). \qquad \square$$

7.5 Decomposition of Processes with Independent Increments

As we have remarked, not every process with independent increments is a semimartingale. On the other hand we have the next nice observation:

Theorem 7.94 *Every n-dimensional process X with independent increments has a decomposition $X = F + S$ where F is a right-regular deterministic process and S is a semimartingale with independent increments.*

Proof. The main idea of the proof is the following: we shall decompose X into several parts. During the decomposition we successively remove the different types of jumps of X. The decomposition procedure is nearly the same as the decomposition of Lévy processes. The only difference is that now we can have jumps which occur with positive probability. When the increments are not stationary one should classify the jumps of X by two different criteria:

1. one can take the jumps which occur with positive or with zero probability at a fixed moment of time t,

2. one can take the large and the small jumps.

Let W be the process which is left after we removed all the jumps of X. We shall prove that all the removed jump-components are semimartingales. As X is not necessarily a semimartingale W is also not necessarily a semimartingale. Process X can have jumps occurring with positive probability, therefore as we shall see, W is not necessarily continuous: when we remove the jumps of X occurring with positive probability we can introduce some new jumps. But very importantly the new jumps have deterministic size and they can occur only at fixed moments of time. Let W' be independent of W with the same distribution as W. As the jumps of W, and of course the jumps of W', are deterministic and they occur at the same moments of time $\widetilde{W} \overset{\circ}{=} W - W'$ is continuous as the jumps of W and W' cancel each other out. As W and W' are independent $\widetilde{W}$ has independent increments. If the Fourier transform of W is $\varphi_W\left(u,t\right)$ then the Fourier transform of $\widetilde{W}$ is $\left|\varphi_W\left(u,t\right)\right|^2$. As we have already observed[96] in this case $\left|\varphi_W\left(u,t\right)\right|^2$ is decreasing. So it has finite variation. This implies that $\widetilde{W}$ is a semimartingale[97].

---

[96] See line (7.41), page 524.

[97] See: Theorem 7.83, page 524.

$\widetilde{W}$ is continuous, hence its spectral measure is zero. From the Lévy–Khintchine formula[98] it is clear that $\widetilde{W}$ is a Gaussian process. By Cramér's theorem[99] W has a Gaussian distribution. Therefore $W(t)$ has an expected value for every t. If $F(t) \doteq \mathbf{E}(W(t))$ then $W(t) - F(t)$ has independent increment and it has zero expected value so it is a martingale and of course F is just the right-regular deterministic process in the theorem.

1. Obviously

$$\widehat{X}(h)(t) \doteq \widehat{X}(t) \doteq \sum_{s \leq t} (\Delta X(s) - h(\Delta X(s)))$$

is the process of large jumps, where $h(x) \doteq x\chi(\|x\| < 1)$. X is right-regular, so the large jumps do not have an accumulation point. Hence $\widehat{X}$ has finite variation on finite intervals. If $s \leq t$ then $\widehat{X}(t) - \widehat{X}(s)$ is the large jumps of X during the time period $(s, t]$ so it is independent of the σ-algebra $\mathcal{F}_s$. Hence $\widehat{X}$ has independent increments. As a first step let us separate from X the process of large jumps $\widehat{X}$. To make the notation simple let us denote by X the process $X - \widehat{X}$. As $\widehat{X}$ has finite variation X is a semimartingale.

2. As a second step we separate the small jumps of X which are not in

$$J \doteq \{t: \nu(\{t\} \times \mathbb{R}^n \setminus \{0\}) > 0\}.$$

The construction is basically the same as the construction we have seen in the Lévy process case: Let

$$Y_m \doteq x\chi_{J^c}\chi\left(\frac{1}{m} \geq \|x\| > \frac{1}{m+1}\right) \bullet \mu^X \doteq g \bullet \mu^X.$$

Y_m is a process of some jumps of X, so it has independent increments. As the jumps are larger than $1/(m+1)$ they cannot have an accumulation point. Hence Y_m has finite variation on finite intervals. As g is bounded $Y_m \in \mathcal{A}_{\text{loc}}$. Let

$$Y_m^p = g \bullet \nu \doteq x\chi_{J^c}\chi\left(\frac{1}{m} \geq \|x\| > \frac{1}{m+1}\right) \bullet \nu$$

---

[98]See: Corollary 7.90, page 534, Theorem 6.12, page 367.
[99]See: Theorem A.14, page 551.

be the compensator of Y_m. If $t \in J^c$ then[100]

$$\nu\left(\{t\} \times \mathbb{R}^n \setminus \{0\}\right) = \mathbf{P}\left(\Delta X\left(t\right) \neq 0\right) = 0.$$

This implies that Y_m^p is continuous[101] in t. Let $L_m \stackrel{\circ}{=} Y_m - Y_m^p$ be the local martingale of the compensated jumps of Y_m. As Y_m^p is continuous

$$\Delta L_m \stackrel{\circ}{=} \Delta Y_m = \chi_{J^c} \chi \left(\frac{1}{m} \geq \|\Delta X\| > \frac{1}{m+1}\right).$$

L_m has finite variation, so it is a pure quadratic jump process[102]. Obviously for every coordinate i

$$\left[L_p^{(i)}, L_q^{(i)}\right] = 0, \quad p \neq q. \tag{7.45}$$

It is also obvious that

$$\left(\sum_{m=1}^{\infty} \left[L_m^{(i)}\right]\right)(t) = \sum_{s \leq t} \left(\Delta X^{(i)}\left(s\right)\right)^2 \chi_{J^c}\left(s\right) = |x_i|^2 \chi_{J^c} \bullet \mu^X.$$

We want to prove that

$$\sqrt{\sum_{m=1}^{\infty} \left[L_m^{(i)}\right]} = \sqrt{|x_i|^2 \chi_{J^c} \bullet \mu^X} \in \mathcal{A}_{\mathrm{loc}}^+. \tag{7.46}$$

By Jensen's inequality

$$\mathbf{E}\left(\sqrt{\sum_{m=1}^{\infty} \left[L_m^{(i)}\right]}\right) \leq \sqrt{\mathbf{E}\left(\sum_{m=1}^{\infty} \left[L_m^{(i)}\right]\right)}$$

so it is sufficient to show that

$$\sum_{m=1}^{\infty} \left[L_m^{(i)}\right] \in \mathcal{A}_{\mathrm{loc}}^+.$$

Observe that the jumps of $\sum_{m=1}^{\infty}[L_m^{(i)}]$ are smaller than one, so if it is finite then it is a right-regular increasing process with bounded jumps, so it is locally bounded[103]. Therefore the main point is that $\sum_{m=1}^{\infty}[L_m^{(i)}]$ is almost surely finite

---

[100]See: Example 7.70, page 514.
[101]As χ_{J^c} is in the integrand of the integral describing it.
[102]See: Example 4.12, page 229.
[103]See: Proposition 1.152, page 107.

for every moment of time t. As $L_m^{(i)}$ is a pure quadratic jump process

$$\sum_{i=1}^{n} \sum_{m=1}^{\infty} \left[L_m^{(i)} \right] = \|x\|^2 \chi_{J^c} \bullet \mu^X.$$

So it is almost surely finite for every t if[104]

$$\mathbf{E} \left(\|x\|^2 \chi_{J^c} \bullet \mu^X \right)(t) = \mathbf{E} \left(\|x\|^2 \chi_{J^c} \bullet \nu \right)(t) = \qquad (7.47)$$

$$= \left(\|x\|^2 \chi_{J^c} \bullet \nu \right)(t) < \infty.$$

We shall show this in the next, third point of the proof. From[105] (7.45) and (7.46) there is a local martingale $L = \sum_{m=1}^{\infty} L_m$. The convergence holds uniformly in probability on compact intervals, so for some subsequence outside a measure-zero set the trajectories converge uniformly on every compact interval. So almost surely

$$\Delta L = \Delta X \chi \left(\|\Delta X\| \le 1 \right) \chi_{J^c}.$$

Processes $X - Y_m$ have independent increments. As X has independent increments ν is deterministic. Hence Y_m^p is deterministic. Therefore

$$X - \sum_{k=1}^{m} L_k \overset{\circ}{=} X - \sum_{k=1}^{m} (Y_k - Y_k^p)$$

has independent increments for every m. So the limit $X - L$ has independent increments.

3. Now we prove (7.47) that is we show that $\|x\|^2 \chi_{J^c}$ is ν-integrable. Of course we already deleted the large jumps of X! So we want to prove that for any process with independent increments

$$\left(\left(\|x\|^2 \wedge 1 \right) \chi_{J^c} \right) \bullet \nu < \infty.$$

Let now X be a process with independent increments and let ν be the spectral measure of ν. Let X' be a process which has the same distribution[106] as X but

---

[104]ν is deterministic as X has independent increments.

[105]See: Theorem 4.26, page 236. In fact the convergence holds in $\mathcal{H}^2$, so one can also use Doob's inequality and the completeness of $\mathcal{H}^2$.

[106]Of course infinite dimensional distribution.

independent of X. To construct X' let us consider the product

$$\left(\tilde{\Omega}, \tilde{\mathcal{A}}, \tilde{\mathbf{P}}, \tilde{\mathcal{F}}\right) \doteq (\Omega, \mathcal{A}, \mathbf{P}, \mathcal{F}) \times (\Omega, \mathcal{A}, \mathbf{P}, \mathcal{F}).$$

Let $X(\omega, \omega') \doteq X(\omega)$ and let $X'(\omega, \omega') \doteq X(\omega')$. It is easy to see that X and X' have independent increments with respect to the filtration

$$\tilde{\mathcal{F}} \doteq \mathcal{F} \times \mathcal{F} \doteq \mathcal{F} \times \mathcal{F}'.$$

For any s the σ-algebra $\mathcal{G}_s$ generated by

$$X(u) - X(v), \quad u, v \geq s$$

is independent of $\mathcal{F}_s$ and the same is true for X'. The increments of $\tilde{X} \doteq X - X'$ are measurable with respect to $\mathcal{G}_s \times \mathcal{G}'_s$ and this σ-algebra is independent of $\tilde{\mathcal{F}}_s$. So $\tilde{X}$ has independent increments on the extended space. If φ_X denotes the Fourier transform of X then the Fourier transform of $\tilde{X}$ is $|\varphi_X|^2$. Function $|\varphi_X|^2$ is decreasing[107], so it has finite variation. Hence $\tilde{X}$ is a semimartingale. As $\tilde{X}$ has independent increments its spectral measure $\tilde{\nu}$ is deterministic[108]. By the semimartingale property[109] $(\|x\|^2 \wedge 1) \bullet \tilde{\nu} < \infty$. Unfortunately as the jumps of X and X' can interfere, this does not imply[110] that $(\|x\|^2 \wedge 1) \bullet \nu < \infty$. If $t \in J$ then X has a jump in t with positive probability and if $t \in J^c$ then the probability of a jump of X in t is zero[111]. Let (τ_k) be the sequence of stopping times covering the jumps of X. Let G_k be the distribution of τ_k. By the definition of the conditional expectation

$$\mathbf{P}\left(\Delta X'(\tau_k) \neq 0, \tau_k \notin J\right) =$$
$$= \mathbf{E}\left(\chi\left(\Delta X'(\tau_k) \neq 0\right) \chi\left(\tau_k \notin J\right)\right) \doteq$$
$$\doteq \int_{\mathbb{R}_+} \mathbf{E}\left(\chi\left(\Delta X'(\tau_k) \neq 0\right) \chi\left(\tau_k \notin J\right) \mid \tau_k = s\right) dG_k(s).$$

$\Delta X'$ is independent of τ_k, as τ_k is measurable with respect to the σ-algebra generated by X and X' is independent of X. The distribution of X and X' are the same so the moments of time where they jump with positive probability are equal. So if $s \in J^c$ then

$$\mathbf{E}\left(\chi\left(\Delta X'(s) \neq 0\right)\right) = \mathbf{P}\left(\Delta X'(s) \neq 0\right) = 0.$$

---

[107] See: line (7.41), page 524.

[108] See: Corollary 7.88, page 532.

[109] See: Proposition 7.66, page 509.

[110] Consider the case when X is a deterministic process with independent increments. Then $\tilde{X} \doteq X - X' = 0$!

[111] See: Example 7.70, page 514.

Hence by the independence

$$\mathbf{E} \left(\chi \left(\Delta X' \left(\tau_k \right) \neq 0 \right) \chi \left(\tau_k \notin J \right) \mid \tau_k = s \right) = \mathbf{E} \left(\chi \left(\Delta X' \left(s \right) \neq 0 \right) \chi \left(s \notin J \right) \right) = 0.$$

This implies that outside a set with zero probability X and X' do not have common jumps in J^c. Hence

$$\left(\| x \|^2 \wedge 1 \right) \chi_{J^c} \bullet \mu^X \stackrel{a.s}{\leq} \left(\| x \|^2 \wedge 1 \right) \bullet \mu^{\widetilde{X}}.$$

Using again that $\widetilde{\nu}$ and ν are deterministic

$$\left(\| x \|^2 \wedge 1 \right) \chi_{J^c} \bullet \nu = \mathbf{E} \left(\left(\| x \|^2 \wedge 1 \right) \chi_{J^c} \bullet \mu^X \right) \leq \tag{7.48}$$
$$\leq \mathbf{E} \left(\left(\| x \|^2 \wedge 1 \right) \chi_{J^c} \bullet \mu^{\widetilde{X}} \right) \leq$$
$$\leq \mathbf{E} \left(\left(\| x \|^2 \wedge 1 \right) \bullet \mu^{\widetilde{X}} \right) = \left(\| x \|^2 \wedge 1 \right) \bullet \widetilde{\nu} < \infty.$$

And that is what we wanted to prove.

4. Now take process $Z \stackrel{\circ}{=} X - L$.

$$\Delta Z = \Delta X \chi \left(\| \Delta X \| < 1 \right) \chi_J.$$

As ν is σ-finite there are maximum countable number of points (t_m) such that $\Delta Z \left(t_m \right) \neq 0$. Define the martingales

$$U_m \left(t \right) \stackrel{\circ}{=} \begin{cases} 0 & \text{if } t < t_m \\ \Delta Z \left(t_m \right) - \mathbf{E} \left(\Delta Z \left(t_m \right) \right) & \text{if } t \geq t_m \end{cases}.$$

$\mathbf{E} \left(\Delta Z \left(t_m \right) \right)$ is meaningful as $\| \Delta Z \left(t_m \right) \| \leq 1$. Obviously for any $i = 1, 2, \ldots, n$

$$\left[U_p^{(i)}, U_q^{(i)} \right] = 0, \quad p \neq q.$$

We should show again that

$$\sqrt{\sum_{m=1}^{\infty} \left[U_m^{(i)} \right]} \in \mathcal{A}_{\text{loc}}^+. \tag{7.49}$$

As above let Z' be independent of Z and let the distribution of Z' be the same as the distribution of Z. $\widetilde{Z} \stackrel{\circ}{=} Z - Z'$ is again a semimartingale with independent increments. If again $\widetilde{\nu}$ is the spectral measure of $\widetilde{Z}$ then as $\widetilde{Z}$ is a semimartingale

$\left(\|z\|^2 \wedge 1\right) \bullet \tilde{\nu} < \infty.$ Hence

$$\sum_{s \le t, s \in J} \mathbf{E}\left(\left\|\Delta \tilde{Z}(s)\right\|^2\right) = \mathbf{E}\left(\sum_{s \le t, s \in J} \left\|\Delta \tilde{Z}(s)\right\|^2\right) \le$$

$$\le \mathbf{E}\left(\left(\left(\|z\|^2 \wedge 1\right) \bullet \mu^{\tilde{Z}}\right)(t)\right) =$$

$$= \left(\left(\|z\|^2 \wedge 1\right) \bullet \tilde{\nu}\right)(t) < \infty.$$

By the definition of U_m

$$\mathbf{E}\left(\|U(t_m)\|^2\right) = \sum_{i=1}^{n} \mathbf{D}^2\left(\Delta Z^{(i)}(t_m)\right) =$$

$$= \sum_{i=1}^{n} \frac{\mathbf{D}^2\left(\Delta Z^{(i)}(t_m)\right) + \mathbf{D}^2\left(-\Delta Z^{(i)\prime}(t_m)\right)}{2} =$$

$$= \sum_{i=1}^{n} \frac{\mathbf{D}^2\left(\Delta Z^{(i)}(t_m) - \Delta Z^{(i)\prime}(t_m)\right)}{2} = \frac{1}{2}\mathbf{E}\left(\left\|\Delta \tilde{Z}(t_m)\right\|^2\right).$$

Hence

$$\mathbf{E}\left(\sum_{t_m \le t} \|U(t_m)\|^2\right) = \sum_{t_m \le t} \mathbf{E}\left(\|U(t_m)\|^2\right) < \infty \tag{7.50}$$

which as above implies (7.49). Let U be the limit of (U_m). If we subtract U from $X - L$ then $W \doteq X - L - U$ has independent increments[112] and

$$\Delta W = \chi_J \cdot \mathbf{E}(\Delta Z). \tag{7.51}$$

5. By (7.51) the jumps of W are fixed and they are deterministic. As we remarked in the introductory part of the proof this implies that the expected value of $W(t)$ is finite. If $F(t) \doteq \mathbf{E}(W(t))$ then as W has independent increments $W - F$ satisfies the martingale condition. As the filtration satisfies the usual conditions $W - F$ has a right-regular version. As W is already right-regular F is also right-regular.

6. Observe that $X = S + F$, where S is a semimartingale. $\qquad \square$

Let us explicitly state some important observations proved above.

---

[112]The jumps of ΔZ disappear but we bring in the expected values of ΔZ.

Corollary 7.95 *If X is a continuous process with independent increments then X is a Gaussian process*[113].

Corollary 7.96 *If X is a process with independent increments then X has a decomposition*

$$X = \widehat{X} + H + G + F$$

where the processes in the decomposition are independent and:

1. $\widehat{X}$ *is the large jumps of X,*
2. H *is a martingale with the small jumps of X and $H \in \mathcal{H}_0^2$ on any finite interval,*
3. G *is a Gaussian martingale and,*
4. F *is a deterministic process.*

Proof. From the proof of the previous theorem it is clear that for every t

$$\mathbf{E}\left(\sum_{i=1}^{n}\left[L^{(i)}\right](t)\right) = \mathbf{E}\left(\sum_{s \le t}\|\Delta X\|^2 \chi_{J^c}\right) = \mathbf{E}\left(\left(\|x\|^2 \chi_{J^c} \bullet \mu^X\right)(t)\right) =$$

$$= \left(\|x\|^2 \chi_{J^c} \bullet \nu\right)(t) < \infty,$$

and

$$\mathbf{E}\left(\sum_{i=1}^{n}\left[U^{(i)}\right](t)\right) = \left(\|x\|^2 \chi_J \bullet \nu\right)(t) < \infty.$$

Hence[114] L and U are in $\mathcal{H}_0^2$ on any finite interval. Therefore $H \triangleq L + U$ is a martingale. $\qquad\square$

Theorem 7.97 (Characterization of local martingales with independent increments) *If X is a local martingale with independent increments then X is a martingale.*

Proof. By the previous corollary

$$X = \widehat{X} + H + G + F.$$

---

[113]See: Theorem 6.12, page 367.

[114]See: Proposition 2.84, page 170. As in the Lévy process case we could use Doob's inequality to construct L and U and directly prove that L and U are in $\mathcal{H}^2$ on any finite interval.

1. X is a local martingale so[115]

$$\left(\|x\|^2 \wedge \|x\|\right) \bullet \nu \in \mathcal{A}^+_{\text{loc}}.$$

Therefore for every t

$$\left(\|x\| \chi \left(\|x\| \geq 1\right) \bullet \nu\right)(t) < \infty.$$

X has independent increments so ν is deterministic[116] and the expression above is deterministic. By the definition of the spectral measures[117]

$$\mathbf{E}\left(\left\|\widehat{X}(t)\right\|\right) \doteq \mathbf{E}\left(\left\|\sum_{0<s\leq t} \Delta X(s) \chi \left(\|\Delta X(s)\| \geq 1\right)\right\|\right) \leq$$

$$\leq \mathbf{E}\left(\sum_{0<s\leq t} \|\Delta X(s)\| \chi \left(\|\Delta X(s)\| \geq 1\right)\right) =$$

$$= \mathbf{E}\left(\left(\|x\| \chi \left(\|x\| \geq 1\right) \bullet \mu^X\right)(t)\right) =$$

$$= \mathbf{E}\left(\left(\|x\| \chi \left(\|x\| \geq 1\right) \bullet \nu\right)(t)\right) =$$

$$= \left(\|x\| \chi \left(\|x\| \geq 1\right) \bullet \nu\right)(t) < \infty.$$

Hence $\widehat{X}(t)$ is integrable for every t.

2. Let $\widehat{m}(t) \doteq \mathbf{E}\left(\widehat{X}(t)\right)$. $\widehat{X}$ is the process of the large jumps of X so $\widehat{X}$ has independent increments. Hence $\widehat{X} - \widehat{m}$ has independent increments and the expected value of the increments is zero. So $\widehat{X} - \widehat{m}$ is a martingale.

3. As X is a local martingale

$$F + \widehat{m} = X - \left(\widehat{X} - \widehat{m} + H + G\right)$$

is a deterministic local martingale. This implies that $F + \widehat{m}$ is a constant. Hence X is a martingale. $\qquad\square$

---

[115]See: Corollary 7.68, page 512.
[116]See: Corollary 7.88, page 532.
[117]See: (7.31), page 501.

Appendix A

RESULTS FROM MEASURE THEORY

A.1 The Monotone Class Theorem

In describing the structure of measurable sets and functions the main tool is the so-called Monotone Class Theorem. The theorem has many forms. In this section we present a simple proof of this important statement.

First we give the necessary definitions:

Definition A.1 *The set $\mathcal{L}$ of bounded real-valued functions defined on a set X is said to be a λ-system if:*

1. *$1 \in \mathcal{L}$, that is the constant function 1 is in $\mathcal{L}$,*
2. *$\mathcal{L}$ is a vector space,*
3. *if $0 \leq f_n \nearrow f$, $f_n \in \mathcal{L}$ and f is bounded then $f \in \mathcal{L}$.*

Definition A.2 *The set $\mathcal{V}$ of real-valued functions on a set X is said to be a vector lattice if:*

1. *$\mathcal{V}$ is a vector space, and*
2. *whenever $f \in \mathcal{V}$ then $|f| \in \mathcal{V}$.*[1]

Definition A.3 *The lattice $\mathcal{V}$ in the previous definition is called a* Stone-lattice *if for every $f \in \mathcal{V}$ one has $f \wedge 1 \in \mathcal{V}$.*

Definition A.4 *The set $\mathcal{P}$ of real-valued functions on a set X is said to be a π-system if whenever $f, g \in \mathcal{P}$ then $fg \in \mathcal{P}$, where fg denotes the product of f and g.*

Lemma A.5 *If $\mathcal{H}$ is an arbitrary set of bounded functions on a set X then there is a set of bounded functions on X, denoted by $\mathcal{L}(\mathcal{H})$, which contains $\mathcal{H}$ and which is the smallest λ-system on X containing $\mathcal{H}$.*

---

[1]If $\mathcal{V}$ is a vector lattice, then for any $f, g \in \mathcal{V}$ $f \vee g \stackrel{\circ}{=} \frac{f+g+|f-g|}{2} \in \mathcal{V}$ and $f \wedge g \stackrel{\circ}{=} \frac{f+g-|f-g|}{2} \in \mathcal{V}$. For any function f $f^+ \stackrel{\circ}{=} f \vee 0$, $f^- \stackrel{\circ}{=} -f \vee 0$. It is evident from the definition, that $f = f^+ - f^-$. If $\mathcal{V}$ is a vector lattice and $f \in \mathcal{V}$, then $f^{\pm} \in \mathcal{V}$.

Proof. The set of all bounded functions on X is a λ-system, the intersection of λ-systems is again a λ-system, hence the intersection of all λ-systems containing $\mathcal{H}$ is well defined and it is trivially the smallest λ-system containing $\mathcal{H}$. □

Lemma A.6 *If $\mathcal{P}$ is a set of bounded functions which is a π-system, then $\mathcal{L}(\mathcal{P})$ is also a π-system.*

Proof. For any f let us introduce

$$\mathcal{L}_f \doteq \{g \in \mathcal{L}(\mathcal{P}) : gf \in \mathcal{L}(\mathcal{P})\}.$$

1. It is easy to see that $\mathcal{L}_f$ is a λ-system for any f. Only the third condition in the definition is not trivial. Let $0 \leq g_n \nearrow g$ where g is bounded and $g_n f \in \mathcal{L}(\mathcal{P})$. As f is bounded there is an α such that $f + 1\alpha \geq 0$, so $0 \leq g_n (f + \alpha 1) \nearrow g(f + \alpha 1)$.

$$g_n (f + \alpha 1) = g_n f + \alpha g_n \in \mathcal{L}(\mathcal{P}),$$

and $\mathcal{L}(\mathcal{P})$ is a λ-system, so $g, g(f + \alpha 1) \in \mathcal{L}(\mathcal{P})$, hence

$$gf = g(f + \alpha 1) - \alpha g \in \mathcal{L}(\mathcal{P}),$$

that is $g \in \mathcal{L}_f$.

2. Fix an $f \in \mathcal{P}$ and let $g \in \mathcal{P}$. As $\mathcal{P}$ is product closed $fg \in \mathcal{P} \subseteq \mathcal{L}(\mathcal{P})$, that is $g \in \mathcal{L}_f$, hence $\mathcal{P} \subseteq \mathcal{L}_f$. $\mathcal{L}_f$ is a λ-system, so

$$\mathcal{L}(\mathcal{P}) \subseteq \mathcal{L}_f, \quad \text{if} \quad f \in \mathcal{P}. \tag{A.1}$$

3. Fix a $g \in \mathcal{L}(\mathcal{P})$ and let again $f \in \mathcal{P}$. In this case from (A.1) $fg \in \mathcal{L}(\mathcal{P})$, which means, that for any $f \in \mathcal{P}$ and $g \in \mathcal{L}(\mathcal{P})$ one has $f \in \mathcal{L}_g$, that is $\mathcal{P} \subseteq \mathcal{L}_g$. $\mathcal{L}_g$ is a λ-system, hence

$$\mathcal{L}(\mathcal{P}) \subseteq \mathcal{L}_g, \quad \text{if} \quad f \in \mathcal{L}(\mathcal{P}).$$

4. By definition this means that for any $f, g \in \mathcal{L}(\mathcal{P})$ one has $fg \in \mathcal{L}(\mathcal{P})$, that is $\mathcal{L}(\mathcal{P})$ is a π-system. □

Lemma A.7 *For any $N > 0$ there is a sequence of polynomials (p_n) such that $0 \leq p_n(x) \nearrow |x|$ for any $|x| \leq N$.*

Proof. We shall approximate $N - |x|$ by a decreasing sequence. Let $y_0 \equiv N$ and by induction define the iteration

$$y_{n+1}(x) \doteq \frac{1}{2N} \left(y_n^2(x) + N^2 - x^2\right).$$

It is clear that $y_n(x)$ is a polynomial in x for all n and $y_n \geq 0$ on $[-N, N]$. We prove by induction that (y_n) is a decreasing sequence. By definition

$$y_1(x) = N - \frac{x^2}{2N},$$

and trivially $y_0 \geq y_1 \geq 0$. If $y_{n-1} \geq y_n \geq 0$ for some n, then

$$y_n(x) - y_{n+1}(x) = \frac{1}{2N} \left(y_{n-1}^2(x) - y_n^2(x) \right) =$$

$$= \frac{1}{2N} \left(y_{n-1}(x) - y_n(x) \right) \left(y_{n-1}(x) + y_n(x) \right) \geq 0.$$

Every sequence of real numbers bounded from below has a limit. If $N \geq y^*(x) \geq 0$ is the limit of $(y_n(x))$, then

$$y^*(x) = \frac{1}{2N} \left((y^*(x))^2 + N^2 - x^2 \right).$$

Hence $x^2 = (y^*(x) - N)^2$, which means that $|x| = |y^*(x) - N| = N - y^*(x)$. $\square$

Proposition A.8 *If a set of bounded functions $\mathcal{P}$ is a π-system then $\mathcal{L}(\mathcal{P})$ is a Stone-lattice.*

Proof. $\mathcal{L}(\mathcal{P})$ is a π- and a λ-system. If $f \in \mathcal{L}(\mathcal{P})$, then f is bounded so there is an N such that $|f| \leq N$. If (p_n) is a sequence of polynomials on $[-N, N]$ for which $p_n(x) \nearrow |x|$, then $0 \leq p_n(f(x)) \nearrow |f(x)|$. $\mathcal{L}(\mathcal{P})$ is a λ-system, $(p_n(f)) \subseteq \mathcal{L}(\mathcal{P})$, so $|f| \in \mathcal{L}(\mathcal{P})$, that is $\mathcal{L}(\mathcal{P})$ is a lattice. As $1 \in \mathcal{L}(\mathcal{P})$, $\mathcal{L}(\mathcal{P})$ is a Stone-lattice. $\square$

Proposition A.9 *If $\mathcal{L}$ is a λ-system and a Stone-lattice, then $\mathcal{L}$ is exactly the set of $\sigma(\mathcal{L})$-measurable functions[2], where $\sigma(\mathcal{L})$ is the σ-algebra generated by $\mathcal{L}$.*

Proof. Every function in $\mathcal{L}$ is $\sigma(\mathcal{L})$-measurable, so it is sufficient to prove that every $\sigma(\mathcal{L})$-measurable function is in $\mathcal{L}$. Let

$$\mathcal{G} \doteq \{A : \chi_A \in \mathcal{L}\}.$$

$\mathcal{L}$ is a lattice, hence if $A, B \in \mathcal{G}$, then $A \cup B, A \cap B \in \mathcal{G}$. $1 \in \mathcal{L}$, and as $\mathcal{L}$ is a linear space $\chi_{A^c} = 1 - \chi_A$ is in $\mathcal{L}$ so $\mathcal{G}$ is an algebra. $\mathcal{L}$ is a λ system, so if $A_n \nearrow A$ and $A_n \in \mathcal{G}$, then $A \in \mathcal{G}$, hence $\mathcal{G}$ is a σ-algebra. We shall prove that $\sigma(\mathcal{L}) \subseteq \mathcal{G}$. Let $f \in \mathcal{L}$. $\mathcal{L}$ is a Stone–lattice, so if

$$f_n \doteq 1 \wedge (n (f - 1 \wedge f)),$$

---

[2] In the definition of the λ-systems we have assumed that the elements of the λ-system are bounded. In this proposition the boundedness of the functions is not used.

then $f_n \in \mathcal{L}$. $0 \le f_n \nearrow \chi(f > 1)$. $\mathcal{L}$ is a λ-system, hence $\chi(f > 1) \in \mathcal{L}$, that is $\{f > 1\} \in \mathcal{G}$. $\mathcal{L}$ is a linear space so for any $\alpha > 0$, $\{f > \alpha\} \in \mathcal{G}$. This means that f^+ is $\mathcal{G}$-measurable. The same argument implies that f^- is also $\mathcal{G}$-measurable, that is f is $\mathcal{G}$-measurable. Hence $\sigma(\mathcal{L}) \subseteq \mathcal{G}$. $\mathcal{G} \subseteq \sigma(\mathcal{L})$ is trivial so $\mathcal{G} = \sigma\mathcal{L}$. Therefore $\mathcal{L}$ contains the characteristic functions of the elements of $\sigma(\mathcal{L})$. $\mathcal{L}$ is a linear space so it contains the $\sigma(\mathcal{L})$-measurable step functions. As $\mathcal{L}$ is closed for the monotone limit it contains all the measurable functions. □

Theorem A.10 (Monotone Class Theorem) *If $\mathcal{P}$ is a π-system and $\mathcal{L}$ is a λ-system and $\mathcal{P} \subseteq \mathcal{L}$, then $\mathcal{L}$ contains all the $\sigma(\mathcal{P})$-measurable bounded real valued functions.*

Proof. Trivially $\mathcal{L}(\mathcal{P}) \subseteq \mathcal{L}$, $\mathcal{L}(\mathcal{P})$ is a λ-system and Stone–lattice, so $\mathcal{L}$ contains the $\sigma(\mathcal{L}(\mathcal{P}))$, hence the $\sigma(\mathcal{P})$-measurable bounded functions. □

Example A.11 One cannot drop the assumption that the functions in the theorem have to be bounded.

Assume, that one could prove the theorem for unbounded functions. Let F and G be two probability distributions on $\mathbb{R}$ and assume, that the moments of F and G are equal. Let $\mathcal{L}$ be the set of functions f for which $\int_{\mathbb{R}} f dG = \int_{\mathbb{R}} f dF$, where the integrals on both sides can be infinite or undefined at the same time. The set of possible functions f is a λ-system. The set $\mathcal{P}$ of polynomials forms a π-system and as the moments of F and G are equal $\mathcal{P} \subseteq \mathcal{L}$. By the assumption $\mathcal{B}(\mathbb{R}) = \sigma(\mathcal{P}) \subseteq \mathcal{L}$, which is impossible since it can happen, that $F \ne G$, but all the moments of the two distributions are the same. □

A.2   Projection and the Measurable Selection Theorems

During the discussion of stochastic analysis we should assume the completeness of the space $(\Omega, \mathcal{A}, \mathbf{P})$ as we use several times the next two theorems[3].

Theorem A.12 (Projection Theorem) *If the space $(\Omega, \mathcal{A}, \mathbf{P})$ is complete and*

$$U \in \mathcal{B}(\mathbb{R}^n) \times \mathcal{A},$$

then

$$\mathrm{proj}_\Omega U \doteq \{x : \exists t \text{ such that } (t, x) \in U\} \in \mathcal{A}.$$

[3]See: [11], [42].

Theorem A.13 (Measurable Selection Theorem) *If the space* $(\Omega, \mathcal{A}, \mathbf{P})$ *is complete and*

$$U \in \mathcal{B}(\mathbb{R}_+) \times \mathcal{A}$$

then there is an $\mathcal{A}$-measurable function $f : \Omega \to [0, \infty]$ for which

$$\text{Graph}(f) \stackrel{\circ}{=} \{(t, \omega) : t = f(\omega) < \infty\} \subseteq U$$

and $\{f < \infty\} = \text{proj}_\Omega U$.

A.3   Cramér's Theorem

Theorem A.14 (Cramér) *If ξ and η are independent random variables and $\xi + \eta$ has Gaussian distribution then ξ and η also have Gaussian distribution.*

Without loss of generality one can assume that the distribution of $\xi + \eta$ is $N(0, 1)$. Let

$$\mathcal{M}_\xi(z) \stackrel{\circ}{=} \int_\mathbb{R} \exp(zx)\, dF_\xi(x), \quad \mathcal{M}_\eta(z) \stackrel{\circ}{=} \int_\mathbb{R} \exp(zx)\, dF_\eta(x)$$

be the complex moment-generating functions of ξ and η. $\mathcal{M}_{\xi+\eta}(z) = \exp(z^2/2)$. As ξ and η are independent $\mathcal{M}_\xi(z)\mathcal{M}_\eta(z) = \exp(z^2/2)$, whenever $\mathcal{M}_\xi(z)$ and $\mathcal{M}_\eta(z)$ are defined. One should prove that $\mathcal{M}_\xi(z)$ and $\mathcal{M}_\eta(z)$ has the form $\exp(\sigma^2 z^2/2)$.

Lemma A.15 $\mathcal{M}_\xi$ *and* $\mathcal{M}_\eta$ *defined on the whole complex plane.*

Proof. From the definition of the complex moment-generating functions it is clear that $\mathcal{M}_\xi$ and $\mathcal{M}_\eta$ are defined on the strips parallel with the imaginary axis based on the domain of finiteness of

$$M_\xi(s) \stackrel{\circ}{=} \int_\mathbb{R} \exp(sx)\, dF_\xi(x), \quad s \in \mathbb{R},$$

$$M_\eta(s) \stackrel{\circ}{=} \int_\mathbb{R} \exp(sx)\, dF_\eta(x), \quad s \in \mathbb{R}.$$

As ξ and η are independent and as for non-negative independent variables the product rule for expected values holds

$$\infty > \exp\left(\frac{s^2}{2}\right) = M_{\xi+\eta}(s) \stackrel{\circ}{=} \mathbf{E}(\exp(s(\xi+\eta))) =$$

$$= \mathbf{E}(\exp(s\xi))\,\mathbf{E}(\exp(s\eta)) \stackrel{\circ}{=} M_\xi(s)\, M_\eta(s).$$

As every real moment-generating function is positive[4] $M_\xi(s)$ and $M_\eta(s)$ are finite for every $s \in \mathbb{R}$. $\square$

As a consequence all the moments of ξ and η are finite. Hence as the expected value of the sum is zero one can assume that $\mathbf{E}(\xi) = \mathbf{E}(\eta) = 0$. Using this and the convexity of $\exp(|x|)$ by Jensen's inequality

$$
\begin{aligned}
|\mathcal{M}_\xi(z)| &\overset{\circ}{=} |\mathbf{E}(\exp(z\xi))| \le \mathbf{E}(|\exp(z\xi)|) \le \mathbf{E}(\exp(|z\xi|)) = \\
&= \mathbf{E}(\exp(|z||\xi + \mathbf{E}(\eta)|)) = \\
&= \int_\mathbb{R} \exp\left(|z|\left|x + \int_\mathbb{R} y\, dF_\eta(y)\right|\right) dF_\xi(x) \le \\
&\le \int_\mathbb{R}\int_\mathbb{R} \exp(|z||x+y|)\, dF_\eta(y)\, dF_\xi(x) = \mathbf{E}(\exp(|z||\xi + \eta|)) = \\
&= \frac{1}{\sqrt{2\pi}} \int_\mathbb{R} \exp(|z||u|) \exp\left(-\frac{u^2}{2}\right) du = \\
&= \frac{2}{\sqrt{2\pi}} \int_0^\infty \exp(|z|u) \exp\left(-\frac{u^2}{2}\right) du \le \\
&\le \frac{2}{\sqrt{2\pi}} \int_{-\infty}^\infty \exp(|z|u) \exp\left(-\frac{u^2}{2}\right) du = 2M_{N(0,1)}(|z|) = \\
&= 2\exp\left(\frac{|z|^2}{2}\right).
\end{aligned}
$$

In a similar way $|\mathcal{M}_\eta(z)| \le 2\exp\left(|z|^2/2\right)$. As

$$
\mathcal{M}_\xi(z)\,\mathcal{M}_\eta(z) = \mathcal{M}_{\xi+\eta}(z) = \exp\left(\frac{z^2}{2}\right) \ne 0, \quad \mathcal{M}_\xi(0) = \mathcal{M}_\eta(0) = 1,
$$

one can define the complex logarithms[5] $g_\xi(z) \overset{\circ}{=} \log \mathcal{M}_\xi(z)$ and $g_\eta(z) \overset{\circ}{=} \log \mathcal{M}_\eta(z)$.

$$
|\mathcal{M}_\xi(z)| = |\exp(g_\xi(z))| = \exp(\mathrm{Re}(g_\xi(z))) \le 2\exp\left(\frac{|z|^2}{2}\right),
$$

$$
|\mathcal{M}_\eta(z)| = |\exp(g_\eta(z))| = \exp(\mathrm{Re}(g_\eta(z))) \le 2\exp\left(\frac{|z|^2}{2}\right).
$$

---

[4] Possibly $+\infty$.

[5] If $f(z) \ne 0$ and if f is continuously differentiable then $g(x) \overset{\circ}{=} \int_0^z f'(z)/f(z)\, dz$ is well-defined and in the whole complex plane $\exp(g(x)) \equiv f(x)$.

Taking the real logarithm of both sides

$$\operatorname{Re}\left(g_{\xi}\left(z\right)\right) \leq \ln 2 + \frac{|z|^2}{2} \leq 1 + \frac{|z|^2}{2}. \tag{A.2}$$

and of course

$$\operatorname{Re}\left(g_{\eta}\left(z\right)\right) \leq 1 + \frac{|z|^2}{2}.$$

Lemma A.16 *If in the circle* $|z| < r_0$

$$f\left(z\right) \overset{\circ}{=} \sum_{n=0}^{\infty} a_n z^n,$$

and

$$A\left(r\right) \overset{\circ}{=} \max_{|z|=r} \operatorname{Re}\left(f\left(z\right)\right),$$

then for all $n > 0$ *and* $0 < r < r_0$

$$|a_n| r^n \leq 4 A^+\left(r\right) - 2 \operatorname{Re}\left(f\left(0\right)\right).$$

Proof. Let $z \overset{\circ}{=} r \exp\left(i\theta\right)$.

$$f\left(z\right) = \sum_{n=0}^{\infty} a_n r^n \exp\left(in\theta\right), \quad r < r_0.$$

Hence if $r < r_0$ then

$$\sum_{n=0}^{\infty} r^n \left|a_n \exp\left(in\theta\right)\right| = \sum_{n=0}^{\infty} r^n \left|a_n\right| < \infty.$$

By the Weierstrass criteria for any $r < r_0$ the next convergence is uniform in θ

$$\operatorname{Re} f\left(z\right) = \sum_{n=0}^{\infty} r^n \operatorname{Re}\left(a_n \exp\left(in\theta\right)\right) = \tag{A.3}$$

$$= \sum_{n=0}^{\infty} r^n \left[\operatorname{Re}\left(a_n\right) \cos n\theta - \operatorname{Im}\left(a_n\right) \sin n\theta\right].$$

Multiplying (A.3) by some $\cos n\theta$ and by $\sin n\theta$ and integrating by θ over $[0, 2\pi]$, by the uniform convergence and by the orthogonality of the trigonometric functions if $n > 0$

$$r^n \operatorname{Re} a_n = \frac{1}{\pi} \int_0^{2\pi} \operatorname{Re} f(z) \cos n\theta d\theta, \quad r^n \operatorname{Im} a_n = -\frac{1}{\pi} \int_0^{2\pi} \operatorname{Re} f(z) \sin n\theta d\theta,$$

that is

$$\left| a_n r^n \right| = \left| \frac{1}{\pi} \int_0^{2\pi} \operatorname{Re} f(z) \exp(-in\theta) d\theta \right| \leq \frac{1}{\pi} \int_0^{2\pi} \left| \operatorname{Re}(f(z)) \right| d\theta.$$

Integrating (A.3)

$$\operatorname{Re}(f(0)) = \operatorname{Re} a_0 = \frac{1}{2\pi} \int_0^{2\pi} \operatorname{Re}(f(z)) d\theta.$$

Hence

$$\left| a_n r^n \right| + 2 \operatorname{Re}(f(0)) \leq \frac{1}{\pi} \int_0^{2\pi} \left| \operatorname{Re}(f(z)) \right| + \operatorname{Re}(f(z)) d\theta =$$

$$= \frac{1}{\pi} \int_0^{2\pi} 2 \left(\operatorname{Re}(f(z)) \right)^+ d\theta \leq$$

$$\leq \frac{1}{\pi} \int_0^{2\pi} 2A^+(r) d\theta = 4A^+(r). \qquad \square$$

g_ξ is analytic in the whole complex plane that is $g_\xi(z) = \sum_{k=0}^\infty a_n z^n$. By (A.2) and by the lemma if $r > 0$, then

$$\left| a_n \right| r^n \leq 4 \left(1 + \frac{r^2}{2} \right) - 2 \cdot 0.$$

Hence if $n > 2$, then $a_n = 0$, that is

$$g_\xi(z) = a_0 + a_1 z + a_2 z^2.$$

But

$$1 = \mathcal{M}_\xi(0) = \exp(a_0),$$

so $a_0 = 0$ and as

$$a_1 = \mathcal{M}'_\xi(0) = \mathbf{E}(\xi) = 0, \quad a_2 = \mathcal{M}''_\xi(0) = \mathbf{D}^2(\xi) > 0,$$

$\mathcal{M}_\xi(z) = \exp\left(\sigma_\xi^2 z^2 \right)$. In a similar way $\mathcal{M}_\eta(z) = \exp\left(\sigma_\eta^2 z^2 \right)$. $\qquad \square$

A.4   Interpretation of Stopped σ-algebras

If X is an arbitrary stochastic process, then the interpretations of the stopped variables and stopped processes X_τ and X^τ are quite obvious and appealing. On the other hand the definition of the stopped σ-algebras $\mathcal{F}_\tau$ are a bit formal. The usual interpretation of $\mathcal{F}_\tau$ is that it contains the events which happened before τ. But in the abstract model of stochastic analysis, it is not clear from the definition how subsets of $(\Omega, \mathcal{A}, \mathbf{P})$ are related to time, and what does it mean that an abstract event happened before τ? In the canonical model the outcomes in Ω explicitly depend on the time parameter, hence the idea that for some function $\omega(t)$ something happened before time $t = \tau(\omega)$ is perhaps more plausible.

To make the next discussion as simple as possible let us assume that Ω is a subset of the right-continuous functions. The restriction that the functions be right-continuous is a bit too restrictive as the topological or measure theoretic properties of the functions in Ω will play practically[6] no role below, so with this assumption we just fix the space of possible trajectories. Perhaps the most specific operation of stochastic analysis is truncation. We shall assume that if X is a stochastic process then the truncated process X^τ is also a stochastic process[7]. In the canonical setup this means that the trajectories of X^τ are in Ω. This happens if $g(t) \stackrel{\circ}{=} f(t \wedge \gamma) \in \Omega$ for arbitrary number γ and for arbitrary $f \in \Omega$. Of course this is a very mild but slightly unusual assumption. If Ω is the set of all right-regular or continuous or increasing functions or Ω is the set of functions which has fixed size of jumps etc. then the condition is satisfied. Let X be the coordinate process $X(t, \omega) \stackrel{\circ}{=} \omega(t)$, and let assume that the filtration $\mathcal{F}$ is generated by Ω that is $\mathcal{F} = \mathcal{F}^\Omega = \mathcal{F}^X$. Let τ be a stopping time of $\mathcal{F}$. Beside $\mathcal{F}_\tau$ let us define two other σ-algebras. One of them is

$$\mathcal{G}_\tau \stackrel{\circ}{=} \sigma\left(X^\tau\right) = \sigma\left(X\left(\tau \wedge t\right) : t \in \Theta\right). \tag{A.4}$$

To define the other, let us introduce on the space Ω an equivalence relation $\sim_\tau$: The outcomes ω' and ω'' are equivalent with respect to τ, if and only if $X^\tau\left(\omega'\right) = X^\tau\left(\omega''\right)$. That is the outcomes ω' and ω'' are equivalent if the trajectories of X for ω' and for ω'' are the same up to the random time τ. $\sim_\tau$ is trivially an equivalence relation on Ω. For every B let $[B]_\tau$ be the set of outcomes which are equivalent to some outcome from B that is

$$[B]_\tau \stackrel{\circ}{=} \left\{\omega'' : \exists \omega' \in B \text{ such that } \omega'' \sim_\tau \omega'\right\}.$$

The obvious interpretation of the elements of the partition generated by $\sim_\tau$ is that they are the outcomes of the experience of the observation of X up to

---

[6]For instance one can also assume that the trajectories are left-continuous. We need some restriction on the trajectories as we should guarantee that the truncated processes remain adapted.

[7]See: Definition 1.128, page 93.

time τ. As the trajectories of X are right-continuous $\omega' \sim_\tau \omega''$ if and only if

$$X\left(\tau\left(\omega'\right) \wedge r, \omega'\right) = X\left(\tau\left(\omega''\right) \wedge r, \omega''\right)$$

for every rational number $r \geq 0$. If $\omega \in \Omega$, then as X is progressively measurable[8]

$$[\omega]_\tau = \left\{\omega' : X\left(\tau\left(\omega'\right) \wedge r, \omega'\right) = X\left(\tau\left(\omega\right) \wedge r, \omega\right), r \in \mathbb{Q}_+\right\} =$$
$$= \cap_{r \in \mathbb{Q}_+} \left\{\omega' : X\left(\tau \wedge r\right)\left(\omega'\right) = X\left(\tau \wedge r\right)\left(\omega\right)\right\} \in \mathcal{F}_\tau \subseteq \mathcal{F}_\infty.$$

Hence all the equivalence classes of $\sim_\tau$ are $\mathcal{F}_\infty$-measurable subsets of Ω. Let us denote by $\mathcal{H}_\tau$ the set of $\mathcal{F}_\infty$-measurable subsets of Ω which are the union of some collection of subsets from the partition generated by $\sim_\tau$. $\mathcal{H}_\tau$ is obviously a σ-algebra and one can naturally interpret the sets in $\mathcal{H}_\tau$ as the events from $\mathcal{F}_\infty$ which happened before τ. Obviously $H \in \mathcal{H}_\tau$ if and only if $H \in \mathcal{F}_\infty$ and $[H]_\tau = H$.

Proposition A.17 *In the just specified model* $\mathcal{F}_\tau = \mathcal{H}_\tau = \mathcal{G}_\tau$.

Proof. We shall prove that $\mathcal{F}_\tau \subseteq \mathcal{H}_\tau \subseteq \mathcal{G}_\tau \subseteq \mathcal{F}_\tau$.

1. The 'hard' part of the proof is the relation $\mathcal{F}_\tau \subseteq \mathcal{H}_\tau$. Assume that $\tau \equiv s$. Let

$$\mathcal{L} \overset{\circ}{=} \left\{B \in \mathcal{F}_\tau : [B]_\tau = B\right\} = \left\{B \in \mathcal{F}_s : [B]_s = B\right\}.$$

$\mathcal{L}$ is trivially a λ-system. Let us consider the sets

$$B \overset{\circ}{=} \cap_k \left\{X(s_k) \leq \gamma_k\right\}, \quad s_k \leq s,$$

which obviously form a π-system. By the definition of $\sim_\tau$ trivially $B = [B]_\tau$. By the Monotone Class Theorem

$$\mathcal{F}_s \overset{\circ}{=} \mathcal{F}_s^X \overset{\circ}{=} \sigma\left(X(s_k) \leq \gamma_k, s_k \leq s\right) \subseteq \mathcal{L},$$

which is exactly what we wanted to proof. Now let τ be an arbitrary stopping time of $\mathcal{F}$ and let $A \in \mathcal{F}_\tau$. We prove that $A \in \mathcal{H}_\tau$. As $A \in \mathcal{F}_\tau \subseteq \mathcal{F}_\infty$ one should only prove that $A = [A]_\tau$. Let $\omega \in A$ and $\omega \sim_\tau \omega'$. One should prove that $\omega' \in A$. If $s \overset{\circ}{=} \tau\left(\omega\right)$ and $t \leq s$, then

$$X(t, \omega) = X^\tau(t, \omega) = X^\tau(t, \omega') = X(t, \omega'),$$

so $\omega \sim_s \omega'$, where obviously $\sim_s$ denotes the equivalence relation defined by the stopping time s.

$$\omega \in A \cap \{\tau \leq s\} \overset{\circ}{=} B \in \mathcal{F}_s.$$

By the case $\tau \equiv s$, just proved $\mathcal{F}_s = \mathcal{H}_s$. Therefore $\omega' \in [B]_s = B \subseteq A$.

---

[8]See: Example 1.18, page 11.

2. By the structure of Ω every stopped trajectory $X^\tau(\omega)$ is in Ω. As Ω is just the set of all trajectories of X, for every ω there is an $\alpha(\omega)$ such that $X^\tau(\omega) = X(\alpha(\omega))$. Let us denote this mapping by α. We shall prove that the mapping

$$\alpha : (\Omega, \mathcal{G}_\tau) \to (\Omega, \mathcal{F}_\infty) \tag{A.5}$$

is measurable. If

$$B \overset{\circ}{=} \{\omega : X(t, \omega) \leq \gamma\}$$

is one of the sets generating the σ-algebra $\mathcal{F}_\infty$, then

$$\alpha^{-1}(B) \overset{\circ}{=} \{\omega : \alpha(\omega) \in B\} = \{\omega : X(t, \alpha(\omega)) \leq \gamma\} =$$
$$= \{\omega : X^\tau(t, \omega) \leq \gamma\} = \{X(\tau \wedge t) \leq \gamma\} \in \mathcal{G}_\tau,$$

from which the (A.5) measurability of α is evident. Assume that $A \in \mathcal{H}_\tau$, that is $A \in \mathcal{F}_\infty$ and $A = [A]_\tau$. We prove that $\alpha^{-1}(A) = A$. Hence by the just proved measurability of α obviously $A \in \mathcal{G}_\tau$. Which implies that $\mathcal{H}_\tau \subseteq \mathcal{G}_\tau$. If $\omega \in A$, then by definition $\alpha(\omega) \sim_\tau \omega$. As $A = [A]_\tau$, one has $\alpha(\omega) \in A$ and hence $\omega \in \alpha^{-1}(A)$, so $A \subseteq \alpha^{-1}(A)$. On the other hand if $\omega \in \alpha^{-1}(A)$, then $\alpha(\omega) \in A$. But as $\omega \sim_\tau \alpha(\omega)$, one has $\omega \in [A]_\tau = A$. Therefore $\alpha^{-1}(A) \subseteq A$.

3. X is right-continuous so X is progressively measurable. Hence the variables $X(\tau \wedge t)$ are $\mathcal{F}_{\tau \wedge t} \subseteq \mathcal{F}_\tau$-measurable and so $\mathcal{G}_\tau \subseteq \mathcal{F}_\tau$. $\square$

Obviously one can use this proposition only when space Ω is big enough. Let us assume that the trajectories of X are just right-continuous. Let Ω be the set of all possible trajectories of X and let $\mathcal{F} \overset{\circ}{=} \mathcal{F}^\Omega \overset{\circ}{=} \mathcal{F}^X$, that is let us represent X by its canonical model. This space $(\Omega, \mathcal{F}^\Omega)$ is called the *minimal representation* of X. Let denote by Φ the set of all right-continuous functions. On the set Φ let define the filtration $\mathcal{F}^\Phi$. Obviously if $f \in \Phi$ then $f^\gamma \overset{\circ}{=} f \wedge \gamma \in \Phi$ for all γ. Of course $\Omega \subseteq \Phi$ and obviously $\mathcal{F}_t^\Omega \subseteq \mathcal{F}_t^\Phi$ for every t.

Lemma A.18 *Let τ be a stopping time of the minimal representation $(\Omega, \mathcal{F}^\Omega)$. If we extend τ to the space Φ with*

$$\tau(\phi) \overset{\circ}{=} \begin{array}{ll} \tau(\omega) & if \quad \omega \in \Omega \\ +\infty & if \quad \phi \in \Phi \setminus \Omega \end{array},$$

then the extended function τ is a stopping time of $(\Phi, \mathcal{F}^\Phi)$.

Proof. $\{\tau \leq t\} \in \mathcal{F}_t^\Omega \subseteq \mathcal{F}_t^\Phi$ for every t. $\square$

By the proposition just proved $\mathcal{H}_\tau^\Phi = \mathcal{F}_\tau^\Phi$. This means that $A \in \mathcal{F}_\tau^\Phi$ if and only if $A = [A]_\tau$ and $A \in \mathcal{F}_\infty^\Phi$. If $f \in \Omega$ and $g \in \Phi \setminus \Omega$, then f and g cannot be $\sim_\tau$ equivalent by the definition of τ. From this it is clear that there are two types of measurable sets in $(\Phi, \mathcal{F}_\tau^\Phi)$. One type of set is formed by the functions from Ω

and the other type of set contains functions only from $\Phi \setminus \Omega$. In the second case the equivalence classes generated by τ are singletons.

1. If $A \in \mathcal{F}_\tau^\Omega \subseteq \mathcal{F}_\infty^\Omega$, then by definition $A \cap \{\tau \le t\} \in \mathcal{F}_t^\Omega \subseteq \mathcal{F}_t^\Phi$, hence $A \in \mathcal{F}_\tau^\Phi$. By the proposition above $A = [A]_\tau$ in $\dot{\Phi}$. Hence in Ω. Hence $\mathcal{F}_\tau^\Omega \subseteq \mathcal{H}_\tau^\Omega$.

2. On the other hand, let assume that $A \in \mathcal{H}_\tau^\Omega$. That is let us assume that $A \in \mathcal{F}_\infty^\Omega$ and $A = [A]_\tau$ in Ω. In this case $A \in \mathcal{F}_\infty^\Phi$ and as the outcomes from Ω and from $\Phi \setminus \Omega$ are never equivalent $A = [A]_\tau$ in Φ as well. By the proposition $A \in \mathcal{F}_\tau^\Phi$, that is $A \cap \{\tau \le t\} \in \mathcal{F}_t^\Phi$. But $A \cap \{\tau \le t\} \subseteq \Omega$ for every finite t, so $A \cap \{\tau \le t\} \in \mathcal{F}_t^\Phi \cap \Omega$. $\mathcal{F}_t^\Phi$ is generated by the set $\{\phi : \phi(s) \le \gamma\}$, and

$$\{\phi : \phi(s) \le \gamma\} \cap \Omega = \{\phi \in \Omega : \phi(s) \le \gamma\} = \{\omega : X(s, \omega) \le \gamma\}.$$

From this it is not difficult to prove that $\mathcal{F}_t^\Phi \cap \Omega = \mathcal{F}_t^\Omega$. Therefore $A \cap \{\tau \le t\} \in \mathcal{F}_t^\Omega$. So $A \in \mathcal{F}_\tau^\Omega$. This means that $\mathcal{H}_\tau^\Omega \subseteq \mathcal{F}_\tau^\Omega$. Hence the following proposition is true:

Proposition A.19 *Let assume that the trajectories of a stochastic process X are right-continuous. If τ is a stopping time of the minimal representation of X then $A \in \mathcal{F}_\tau^X$ if and only if $A \in \mathcal{F}_\infty^X$ and $A = [A]_\tau$. That is, in this case the sets in $\mathcal{F}_\tau^X$ are the 'events before' τ.*

Appendix B
WIENER PROCESSES

Perhaps the most interesting processes are the Wiener processes. The number of theorems about Wiener processes is huge. In this appendix we summarize the simplest properties of this class of processes.

B.1   Basic Properties

It is worth emphasizing that the name Wiener process refers not to a single process but to a class of processes.

Definition B.1 *Process* $\{w(t, \omega)\}_{t \geq 0}$ *is a Wiener process if it satisfies the next assumptions:*

1. $w(0) \equiv 0$,
2. w *has independent increments,*
3. *if* $0 \leq s < t$ *then the distribution of* $w(t) - w(s)$ *is* $N\left(0, \sqrt{t-s}\right)$, *that is the density function of* $w(t) - w(s)$ *is*

$$g_{t-s}(x) \triangleq \frac{1}{\sqrt{2\pi(t-s)}} \exp\left(\frac{-x^2}{2(t-s)}\right).$$

4. *w is continuous that is for any outcome ω the trajectory $w(\omega)$ is continuous.*

By the formula for moments of the normal distribution the next lemma is obvious.

Lemma B.2 *For arbitrary* $0 \leq s < t$

$$\mathbf{E}\left([w(t) - w(s)]^n\right) = \begin{cases} 1 \cdot 3 \cdot \ldots \cdot (n-1) \cdot (t-s)^{n/2} & \text{if} \quad n = 2k \\ 0 & \text{if} \quad n = 2k+1 \end{cases}.$$

Lemma B.3 *If* $t_1 < t_2 < \ldots < t_k$ *then the distribution of*

$$(w(t_1), w(t_2), \ldots, w(t_k)) \tag{B.1}$$

has a density function f and

$$f(x_1, x_2, \ldots, x_k) = \prod_{i=1}^{k} \frac{1}{\sqrt{2\pi (t_i - t_{i-1})}} \exp\left(\frac{-(x_i - x_{i-1})^2}{2(t_i - t_{i-1})}\right)$$

where $t_0 \stackrel{\circ}{=} x_0 \stackrel{\circ}{=} 0$.

Proof. Let $t_0 \stackrel{\circ}{=} 0$. By definition

$$(\Delta w(t_0), \Delta w(t_1), \ldots, \Delta w(t_{k-1}))$$

is a vector with independent coordinates. So its density function is

$$g(u_1, u_2, \ldots, u_k) = \prod_{i=1}^{k} \frac{1}{\sqrt{2\pi (t_i - t_{i-1})}} \exp\left(\frac{-u_i^2}{2(t_i - t_{i-1})}\right).$$

The determinant of the linear mapping $A : \mathbb{R}^k \to \mathbb{R}^k$

$$u_1 = x_1$$

$$u_2 = x_2 - x_1$$

$$\ldots$$

$$u_k = x_k - x_{k-1}$$

is 1. If f is the density function of (B.1) then

$$P \stackrel{\circ}{=} \int_H f(x)\, dx_1 \ldots dx_k = \mathbf{P}\left((w(t_1), w(t_2), \ldots, w(t_k)) \in H\right).$$

By the integral transformation theorem

$$P \stackrel{\circ}{=} \mathbf{P}\left(A(w(t_1), w(t_2), \ldots, w(t_k)) \in AH\right) =$$

$$= \mathbf{P}\left((\Delta w(t_0), \Delta w(t_1), \ldots, \Delta w(t_{k-1})) \in AH\right) =$$

$$= \int_{AH} g(u_1, u_2, \ldots, u_k)\, du_1 \ldots du_k =$$

$$= \int_H g(Ax)\, |\det(A)|\, dx_1 \ldots dx_k = \int_H g(Ax)\, dx_1 \ldots dx_k.$$

Hence

$$f(x) = g(Ax) = \prod_{i=1}^{k} \frac{1}{\sqrt{2\pi (t_i - t_{i-1})}} \exp\left(\frac{-(x_i - x_{i-1})^2}{2(t_i - t_{i-1})}\right),$$

where $x_0 \stackrel{\circ}{=} 0$. □

As we remarked several times, one should assume that the filtration satisfies the usual assumptions. Every Wiener process is a Lévy process, therefore if we augment the filtration generated with the measure-zero sets then the augmented filtration satisfies the usual conditions. If the filtration is already given then one should use the following definition.

Definition B.4 *We say that a stochastic process w defined on $\Theta \overset{\circ}{=} [0, \infty)$ is a Wiener process on the stochastic base $(\Omega, \mathcal{A}, \mathbf{P}, \mathcal{F})$ if:*

1. $w(0) = 0$,
2. *for every $t \in \Theta$ and $h > 0$ the increments $w(t+h) - w(t)$ are independent of $\mathcal{F}_t$,*
3. *for every $0 \le s < t$ the distribution of $w(t) - w(s)$ is $N\left(0, \sqrt{t-s}\right)$,*
4. *the trajectories $w(\omega)$ are continuous for every outcome ω,*
5. *$\mathcal{F}$ satisfies the usual conditions.*

Example B.5 If w is a Wiener process under a filtration $\mathcal{F}$ then it is not necessarily a Wiener process under a larger filtration $\mathcal{G}$.

If w is a Wiener process under $\mathcal{F}$ and $\mathcal{G}_t \overset{\circ}{=} \sigma(\mathcal{F}_t \cup \mathcal{F}_1)$ then w is not a Wiener process under $\mathcal{G}$ as if $s \overset{\circ}{=} 1 > t$ then as $w(s)$ is $\mathcal{F}_t$-measurable the martingale property

$$\mathbf{E}(w(s) \mid \mathcal{G}_t) = w(t)$$

does not hold. □

Perhaps the most well-known property of Wiener processes is the following.

Theorem B.6 (Paley–Wiener–Zygmund) *For almost all ω the trajectory $w(\omega)$ is nowhere differentiable.*

Proof. It is sufficient to prove that almost surely $w(t, \omega)$ does not have a right-derivative for any t.

If f is a real function then for any t let

$$D^+ f(t) \overset{\circ}{=} \limsup_{h \searrow 0} \frac{f(t+h) - f(t)}{h}, \quad D_+ f(t) \overset{\circ}{=} \liminf_{h \searrow 0} \frac{f(t+h) - f(t)}{h}.$$

Obviously f is differentiable at time t from the right if $D^+ f(t) = D_+ f(t)$ and the common value is finite. To make the notation simple let $[a, b] = [0, 1]$.

Let $j, k \geq 1$ integers and let

$$A_{jk} \doteq \bigcup_{t \in [0,1]} \bigcap_{h \in (0,1/k]} \left\{ \left| \frac{w(t+h) - w(t)}{h} \right| \leq j \right\} =$$

$$= \bigcup_{t \in [0,1]} \bigcap_{h \in [0,1/k]} \{ |w(t+h) - w(t)| \leq hj \}.$$

Obviously $B \doteq \cup_{j=1}^{\infty} \cup_{k=1}^{\infty} A_{jk}$ contains the outcomes ω for which there is a time t, that

$$-\infty < D_+ w(t, \omega) \quad \text{and} \quad D^+ w(t, \omega) < +\infty.$$

To prove the theorem it is sufficient to show that $\mathbf{P}(B) = 0$. To show this it is enough to show that $\mathbf{P}(A_{jk}) = 0$ for any k and j. Let us fix a j and a k. Let $\omega \in A_{jk}$ and let t be a moment of time belonging to ω. By definition if $0 < h \leq 1/k$ then

$$\frac{|w(t+h, \omega) - w(t, \omega)|}{h} \leq j,$$

which is the same as

$$|w(t+h, \omega) - w(t, \omega)| \leq hj$$

whenever $0 < h \leq 1/k$. Let $n \geq 4k$ and let us partition interval $[0,1]$ into n equal parts. Let $t \in [(i-1)/n, i/n]$ for some i.

Firstly

$$\left| w\left(\frac{i+1}{n}\right) - w\left(\frac{i}{n}\right) \right| \leq \left| w\left(\frac{i+1}{n}\right) - w(t) \right| + \left| w\left(\frac{i}{n}\right) - w(t) \right| \leq$$

$$\leq \frac{2j}{n} + \frac{j}{n}$$

as

$$t \in \left[\frac{i-1}{n}, \frac{i}{n}\right] \quad \text{and} \quad \frac{4}{n} \leq \frac{1}{k} \tag{B.2}$$

therefore

$$0 < \frac{i+1}{n} - t = \left(\frac{i}{n} - t\right) + \frac{1}{n} \leq \frac{2}{n} \leq \frac{1}{k},$$

$$0 < \frac{i}{n} - t \leq \frac{1}{n} \leq \frac{1}{k}.$$

Secondly

$$\left| w\left(\frac{i+2}{n}\right) - w\left(\frac{i+1}{n}\right) \right| \leq \left| w\left(\frac{i+2}{n}\right) - w(t) \right| + \left| w\left(\frac{i+1}{n}\right) - w(t) \right| \leq$$
$$\leq \frac{3j}{n} + \frac{2j}{n}$$

since by (B.2)

$$0 < \frac{i+2}{n} - t = \left(\frac{i}{n} - t\right) + \frac{2}{n} \leq \frac{3}{n} \leq \frac{1}{k},$$
$$0 < \frac{i+1}{n} - t \leq \frac{2}{n} \leq \frac{1}{k}$$

Thirdly

$$\left| w\left(\frac{i+3}{n}\right) - w\left(\frac{i+2}{n}\right) \right| \leq \left| w\left(\frac{i+3}{n}\right) - w(t) \right| + \left| w\left(\frac{i+2}{n}\right) - w(t) \right| \leq$$
$$\leq \frac{4j}{n} + \frac{3j}{n}$$

since again by (B.2)

$$0 < \frac{i+3}{n} - t = \frac{i}{n} - t + \frac{3}{n} \leq \frac{4}{n} \leq \frac{1}{k},$$
$$0 < \frac{i+2}{n} - t \leq \frac{3}{n} \leq \frac{1}{k}.$$

Let C_{in} be the set

$$\cap_{m=1}^{3} \left\{ \omega : \left| w\left(\frac{i+m}{n}, \omega\right) - w\left(\frac{i+m-1}{n}, \omega\right) \right| \leq \frac{2m+1}{n} j \right\}.$$

If $\omega \in A_{jk}$ then $t \in [(i-1)/n, i/n]$ for some t and i. Hence by the just proved three inequalities

$$A_{jk} \subseteq \cup_{i=1}^{n} C_{in}.$$

Hence it is sufficient to show that

$$\lim_{n \to \infty} \mathbf{P}\left(\cup_{i=1}^{n} C_{in}\right) = 0.$$

Let us estimate the probability of C_{in}. By the definition of Wiener processes the distribution of

$$\xi_m \stackrel{\circ}{=} \sqrt{n}\left(w\left(\frac{i+m}{n}\right) - w\left(\frac{i+m-1}{n}\right)\right)$$

is $N(0,1)$. Hence

$$\mathbf{P}\left(|\xi_m| \leq \alpha\right) = \frac{1}{\sqrt{2\pi}} \int_{-\alpha}^{\alpha} \exp\left(-\frac{1}{2}x^2\right) dx \leq \frac{1}{\sqrt{2\pi}} 2\alpha \leq \alpha.$$

Using that the Wiener processes have independent increments for every i

$$\mathbf{P}\left(C_{in}\right) = \mathbf{P}\left(\cap_{m=1}^{3}\left\{\left|\sqrt{n}w\left(\frac{i+m}{n}\right) - w\left(\frac{i+m-1}{n}\right)\right| \leq \frac{2m+1}{\sqrt{n}}j\right\}\right) =$$

$$= \prod_{m=1}^{3} \mathbf{P}\left(\left\{\left|\sqrt{n}w\left(\frac{i+m}{n}\right) - w\left(\frac{i+m-1}{n}\right)\right| \leq \frac{2m+1}{\sqrt{n}}j\right\}\right) \leq$$

$$\leq \frac{3 \cdot 5 \cdot 7 \cdot j^3}{n^{3/2}}.$$

Hence

$$\limsup_{n\to\infty} \mathbf{P}\left(\cup_{i=1}^{n}C_{in}\right) \leq \limsup_{n\to\infty} \sum_{i=1}^{n} \mathbf{P}\left(C_{in}\right) \leq \lim_{n\to\infty} n\frac{105j^3}{n^{3/2}} = 0. \qquad \square$$

Proposition B.7 *If w is a Wiener process then almost surely*

$$\limsup_{t\to\infty} w(t) = \infty, \quad \liminf_{t\to\infty} w(t) = -\infty.$$

Proof. We prove only the first relation. As $t \mapsto w(t+s) - w(s)$ is a Wiener process for every s one should only prove that

$$\eta \stackrel{\circ}{=} \sup_{t\geq 0} w(t) \stackrel{a.s.}{=} \infty \tag{B.3}$$

for every Wiener process w. Let w be a Wiener process. It is trivial from the definition that if $c \neq 0$ then $w_c \stackrel{\circ}{=} cw\left(t/c^2\right)$ is also a Wiener process. As w is continuous it is sufficient to take the supremum in (B.3) at rational points of time, so η is a random variable. Obviously

$$\sup_{t} w_c(t) \stackrel{\circ}{=} \sup_{t} c \cdot w\left(\frac{t}{c^2}\right) = c \cdot \eta.$$

The distribution of the supremum of some process depends only on the infinite dimensional distribution of the process. Hence η and $c \cdot \eta$ have the same distribution. Therefore η can be almost surely either 0 or ∞. $w(t+1) - w(1)$

is also a Wiener process, therefore $\sup_{t \geq 1} (w(t) - w(1))$ is almost surely either zero or $+\infty$.

$$\mathbf{P}(\eta = 0) \stackrel{\circ}{=} \mathbf{P}\left(\sup_{t \geq 0} w(t) = 0\right) \leq \mathbf{P}\left(\sup_{t \geq 1} w(t) \leq 0\right) =$$

$$= \mathbf{P}\left(w(1) + \sup_{t \geq 0}(w(t+1) - w(1)) \leq 0\right) =$$

$$= \mathbf{P}\left(w(1) \leq 0, \sup_{t \geq 0}(w(t+1) - w(1)) = 0\right).$$

The two events in the last probability are independent so

$$p \stackrel{\circ}{=} \mathbf{P}(\eta = 0) \leq \mathbf{P}(w(1) \leq 0) \cdot \mathbf{P}\left(\sup_{t \geq 0}\{w(t+1) - w(1)\} = 0\right) = \frac{1}{2}p$$

so $p = 0$. $\qquad\qquad\square$

Corollary B.8 *For every number a the set $\{t : w(t) = a\}$ is not bounded from above. Particularly the one dimensional Wiener process returns to the origin infinitely many times.*

Proposition B.9 (Law of large numbers) *If w is a Wiener process then*

$$\lim_{t \to \infty} \frac{w(t)}{t} = 0.$$

Proof. By Doob's inequality

$$\mathbf{E}\left(\sup_{2^n \leq t \leq 2^{n+1}} \left(\frac{w(t)}{t}\right)^2\right) \leq \frac{1}{2^{2n}} \mathbf{E}\left(\sup_{2^n \leq t \leq 2^{n+1}} w^2(t)\right) \leq$$

$$\leq \frac{1}{2^{2n}} \cdot 4 \cdot \mathbf{E}\left(w^2\left(2^{n+1}\right)\right) = \frac{4}{2^{2n}} \mathbf{D}^2 w^2\left(2^{n+1}\right) = \frac{4}{2^{2n}} 2^{n+1} = \frac{8}{2^n}.$$

By Markov's inequality

$$\mathbf{E}\left(\sup_{2^n \leq t \leq 2^{n+1}} \frac{|w(t)|}{t} > \varepsilon\right) \leq \frac{8}{2^n \varepsilon^2}.$$

By the Borel–Cantelli lemma almost surely except for some finite number of n

$$\sup_{2^n \leq t \leq 2^{n+1}} \frac{|w(t)|}{t} \leq \varepsilon,$$

which proves the proposition. $\qquad\qquad\square$

Corollary B.10 *If w is a Wiener process then*

$$\widetilde{w}(t) \stackrel{\circ}{=} \begin{cases} t \cdot w(1/t) & \text{if } t > 0 \\ 0 & \text{if } t = 0 \end{cases}$$

is indistinguishable from a Wiener process.

Corollary B.11 *If w is a Wiener process then for every $r > 0$*

$$\mathbf{P}(A) \stackrel{\circ}{=} \mathbf{P}(w(t) \leq 0 : \forall t \in [0, r]) = 0$$

and for almost all ω there is an $\varepsilon(\omega, r) > 0$ such that

$$(-\varepsilon(\omega, r), \varepsilon(\omega, r)) \subseteq w([0, r], \omega).$$

Proof. The second part easily follows from the first part. As w is a Wiener process one can assume that

$$\widetilde{w}(t) \stackrel{\circ}{=} \begin{cases} t \cdot w(1/t) & \text{if } t > 0 \\ 0 & \text{if } t = 0 \end{cases}$$

is also a Wiener process. But the trajectories of $\widetilde{w}$ are bounded on A which implies[1] that $\mathbf{P}(A) = 0$. $\qquad\square$

Corollary B.12 *If w is a Wiener process then[2]*

$$\inf\{t : w(t) \in \text{cl}(G)\} \stackrel{a.s.}{=} \inf\{t : w(t) \in G\}. \tag{B.4}$$

Proof. Recall, that the random variable on the left-hand side is a stopping time[3]. One can assume that $G \neq \emptyset$ otherwise the statement is trivial. From Proposition B.7 it is easy to see that

$$\tau \stackrel{\circ}{=} \inf\{t : w(t) \in \text{cl}(G)\} = \min\{t : w(t) \in \text{cl}(G)\} < \infty.$$

By the strong Markov property[4] it is clear that

$$w^*(t) \stackrel{\circ}{=} w(\tau + t) - w(\tau)$$

is also a Wiener process. As $w(\tau) \in \text{cl}(G)$ by the previous statement for any rational number $r > 0$ almost surely there is an $t(\omega, r)$ such that

$$w(\tau(\omega) + t(\omega, r), \omega) \in G.$$

From this (B.4) is obvious. $\qquad\square$

---

[1] See: Proposition B.7, page 564.
[2] See: Example 6.10, page 364.
[3] See: Example 1.32, page 17.
[4] See: Proposition 1.109, page 70.

B.2 Existence of Wiener Processes

We defined the Wiener processes with their properties. We now show that these properties are consistent.

Theorem B.13 *One can construct a stochastic process w which is a Wiener process.*

Proof. First we construct w on the time interval $[0,1]$, later we shall extend the construction to $\mathbb{R}_+$.

1. Let $(\Omega, \mathcal{A}, \mathbf{P})$ be a probability space[5] where there are countable number of independent random variables (ξ_n) with distribution $N(0,1)$. Let $H \subseteq L^2(\Omega)$ be the closed linear space generated by (ξ_n). As the linear combination of independent Gaussian variables is again Gaussian and as convergence in L^2 implies weak convergence, all the vectors in H are Gaussian. H is a Hilbert space with the orthonormal bases (ξ_n). Let (e_n) be an orthonormal bases in the Hilbert space $L^2([0,1])$. Let us define the continuous linear isomorphism T determined by the correspondence $e_n \longleftrightarrow \xi_n$:

$$T : \sum_k a_n e_n \longmapsto \sum_k a_n \xi_n.$$

$\chi([0,t]) \in L^2([0,1])$ for every t. If $\chi([0,t]) = \sum_k a_k(t) e_k$ then let

$$w(t) \overset{\circ}{=} T(\chi([0,t])) = T\left(\sum_k a_k(t) e_k\right) = \sum_k a_k(t) T(e_k) =$$

$$= \sum_k a_k(t) \xi_k = \sum_k (\chi[0,t], e_k) \cdot \xi_k = \sum_k \left(\int_0^t e_k d\lambda\right) \cdot \xi_k.$$

For any u and v

$$\eta \overset{\circ}{=} u \cdot w(t) + v \cdot (w(t+h) - w(t)) = T(u \cdot \chi([0,t]) + v \cdot \chi((t,t+h]))$$

is Gaussian. Hence the Fourier transform of η at $s = 1$ is

$$\mathbf{E}\left(\exp\left(iuw(t) + iv(w(t+h) - w(t))\right)\right) =$$

$$= \exp\left(-\frac{\mathbf{D}^2(\eta)}{2}\right) = \exp\left(-\frac{\|u \cdot \chi([0,t]) + v \cdot \chi((t,t+h])\|_2^2}{2}\right) =$$

$$= \exp\left(-\frac{u^2 t + v^2 h}{2}\right) = \exp\left(-\frac{u^2 t}{2}\right) \exp\left(-\frac{v^2 h}{2}\right).$$

[5] For example $(\Omega, \mathcal{A}, \mathbf{P}) = ([0,1], \mathcal{B}([0,1]), \lambda)$.

Therefore $w(t)$ and $w(t+h) - w(t)$ are independent[6] and

$$w(t+h) - w(t) \cong N\left(0, \sqrt{h}\right).$$

In a same way one can easily show that w has independent and stationary increments. This means that w is nearly a Wiener process. The only problem is that the trajectories $w(\omega)$ are not necessarily continuous[7]!

2. We show that if the orthonormal basis in $L^2([0,1])$ is the set of *Haar's functions* then w has a version which is almost surely continuous. Let $I(n)$ be the set of odd numbers between 0 and 2^n, that is let $I(0) \overset{\circ}{=} \{1\}$, $I(1) \overset{\circ}{=} \{1\}$, $I(2) \overset{\circ}{=} \{1,3\}$, etc. Let $H_1^{(0)}(t) \overset{\circ}{=} 1$ and if $n \geq 1$ then let

$$H_k^{(n)}(t) \overset{\circ}{=} \begin{cases} +2^{(n-1)/2} & \text{if } t \in [(k-1)/2^n, k/2^n) \\ -2^{(n-1)/2} & \text{if } t \in [k/2^n, (k+1)/2^n) \\ 0 & \text{if } t \notin [(k-1)/2^n, (k+1)/2^n) \end{cases}.$$

The Haar's functions $\left(H_k^{(n)}\right)_{k,n}$ form a complete orthonormal system of $L^2([0,1])$: One can show the orthonormality with simple calculation. The proof of the completeness is the following: Let f be orthogonal to every $H_k^{(n)}$ and let $F(x) \overset{\circ}{=} \int_0^x f(t)\,dt$. $F(1) - F(0) = \left(f, H_1^{(0)}\right) = 0$, hence $F(1) = F(0) = 0$. Similarly

$$0 = \left(f, H_k^{(n)}\right) = \int_0^1 f H_1^{(1)} d\lambda = \int_0^{1/2} f d\lambda - \int_{1/2}^1 f d\lambda =$$
$$= F(1/2) - F(0) - F(1) + F(1/2) = 2F(1/2).$$

Hence $F(1/2) = 0$. In a similar way one can prove that $F(k/2^n) = 0$, that is $F \equiv 0$. With Monotone Class Theorem one can easily prove that $\int_A f d\lambda = 0$ for every $A \in \mathcal{B}([0,1])$, which implies that $f \overset{a.s.}{=} 0$.

3. For every $k \in I(n)$ let

$$a_k^{(n)}(t) \overset{\circ}{=} \int_0^t H_k^{(n)} d\lambda = \left(\chi([0,t]), e_k^{(n)}\right).$$

Let

$$w_n(t) \overset{\circ}{=} \sum_{m=0}^n \sum_{k \in I(m)} a_k^{(m)}(t) \xi_k^{(m)}.$$

As functions $a_k^{(n)}$ are obviously continuous, the trajectories of $w_n(\omega)$ are continuous for any ω. We show that for almost all ω the series $(w_n(t, \omega))$ is uniformly

---

[6]See: Lemma 1.96, page 60.

[7]And $w(t)$ is defined just as a vector from H, that is $w(t)$ defined up to measure-zero sets.

convergent in t. Let

$$b_n \overset{\circ}{=} \max_{k \in I(n)} \left| \xi_k^{(n)} \right|.$$

$\xi_k^{(n)} \cong N(0, 1)$, so if $x > 0$ then

$$\mathbf{P}\left(\left| \xi_k^{(n)} \right| > x \right) = 2 \frac{1}{\sqrt{2\pi}} \int_x^\infty \exp\left(-\frac{u^2}{2} \right) du \le$$

$$\le \sqrt{\frac{2}{\pi}} \int_x^\infty \frac{u}{x} \exp\left(-\frac{u^2}{2} \right) du = \sqrt{\frac{2}{\pi}} \frac{\exp\left(-x^2/2 \right)}{x},$$

hence

$$\mathbf{P}\left(b_n > n \right) = \mathbf{P}\left(\bigcup_{k \in I(n)} \left\{ \left| \xi_k^{(n)} \right| > n \right\} \right) \le 2^n \sqrt{\frac{2}{\pi}} \frac{\exp\left(-n^2/2 \right)}{n}.$$

$\sum_{n=1}^\infty 2^n \exp\left(-n^2/2 \right) / n < \infty$, so by the Borel–Cantelli lemma

$$\mathbf{P}\left(\limsup_{n \to \infty} \left\{ b_n > n \right\} \right) = 0,$$

that is for almost all ω there is an $n_0(\omega)$ such that if $n \ge n_0(\omega)$ then $b_n(\omega) \le n$. Observe that the supports of the non-negative functions $a_k^{(n)}$ are disjoint hence for any k

$$\left| \sum_{k \in I(n)} a_k^{(n)}(t) \right| \le \max a_k^{(n)} = 2^{(n-1)/2} \cdot 2^{-n} = 2^{-(n+1)/2}.$$

From this for almost all ω from an index n large enough

$$\left| w_n(t, \omega) - w_{n-1}(t, \omega) \right| = \left| \sum_{k \in I(n)} \xi_k^{(n)}(\omega) a_k^{(n)}(t) \right| \le$$

$$\le n \left| \sum_{k \in I(n)} a_k^{(n)}(t) \right| \le n 2^{-(n+1)/2}.$$

$\sum_{n=1}^\infty n 2^{-(n+1)/2} < \infty$, therefore the series $(w_n(t, \omega))$ is uniformly convergent for almost all ω. Hence its limit $w(t, \omega)$ is almost surely continuous in t. By the construction w is defined and continuous up to a measure-zero set. So one can set to zero the trajectories where w is not continuous or where it is not defined.

4. Finally one should extend w from $[0,1]$ to $[0,\infty)$. Let $w^{(n)}$, $(n = 1, 2, \ldots)$ be countable number of independent Wiener processes on $[0,1]$. One can construct such processes as we assumed that there are countable number of independent $N(0,1)$ variables on Ω and one can form an infinite two-dimensional matrix from these variables. Let $w(0) \overset{\circ}{=} 0$ and let

$$w(t) \overset{\circ}{=} w(n) + w^{(n+1)}(t-n) \quad \text{if} \quad t \in [n, n+1).$$

$w^{(n)}(0) = 0$, for every n so w is continuous on $[0,\infty)$. With direct calculation it is easy to check that w is a Wiener process on $\mathbb{R}_+$. $\qquad\square$

On the space $C(\mathbb{R}_+)$ let us define the topology of uniform convergence on compacts. Using the Stone–Weierstrass theorem it is easy to see that $C(\mathbb{R}_+)$ is a complete separable metric space. Let $\mathcal{B} \overset{\circ}{=} \mathcal{B}(C(\mathbb{R}_+))$ be the Borel σ-algebra of $C(\mathbb{R}_+)$. It is easy to see that the σ-algebra generated by the process

$$X(\omega, t) = w(t), \quad \omega \in C(\mathbb{R}_+)$$

is equal to $\mathcal{B}$.

Definition B.14 *A measure W on the measurable space $(C(\mathbb{R}_+), \mathcal{B})$ is a Wiener measure if process*

$$w(t, \omega) \overset{\circ}{=} w(t), \quad \omega \in C(\mathbb{R}_+)$$

satisfies the following conditions:

1. $w(0) \overset{a.s.}{=} 0$,
2. w *has independent increments,*
3. *if $t > s$ the distribution of $w(t) - w(s)$ is $N\left(0, \sqrt{t-s}\right)$.*

Proposition B.15 *There is a Wiener measure.*

Proof. Let w be a Wiener process on some stochastic base $(\Omega, \mathcal{A}, \mathbf{P}, \mathcal{F})$ and let $F(\omega) = w(\omega)$. Obviously $F : \Omega \to C(\mathbb{R}_+)$. As for every Borel measurable set $B \subseteq \mathbb{R}$

$$\{w(t) \in B\} = F^{-1}(\{w(t) \in B\}) \in \mathcal{F}_t$$

F is $(\Omega, \mathcal{F}_\infty) \to (C(\mathbb{R}_+), \mathcal{B})$ measurable. The distribution of F

$$W(A) \overset{\circ}{=} \mathbf{P}\left(F^{-1}(A)\right), \quad A \in \mathcal{B}$$

defines a Wiener measure on $(C(\mathbb{R}_+), \mathcal{B})$. $\qquad\square$

As the sets $\{w(t) \in B\}$ generate $\mathcal{B}$ one can easily prove the next observation.

Proposition B.16 *The Wiener measure is unique.*

B.3 Quadratic Variation of Wiener Processes

It is a natural question to ask, when does the quadratic variation of a Wiener process converge almost surely?

Theorem B.17 (Quadratic variation of Wiener processes) *Let w be a Wiener process and let $P_n \triangleq \left(t_k^{(n)}\right)$ be an infinitesimal sequence of partitions of an interval $[a,b]$. In the topology of convergence in $L^2\left(\Omega\right)$*

$$\lim_{n\to\infty} \sum_k \left(w\left(t_k^{(n)}\right) - w\left(t_{k-1}^{(n)}\right)\right)^2 = b - a. \tag{B.5}$$

In a similar way if w_1 and w_2 are independent Wiener processes then

$$\lim_{n\to\infty} \sum_k \left(w_1\left(t_k^{(n)}\right) - w_1\left(t_{k-1}^{(n)}\right)\right)\left(w_2\left(t_k^{(n)}\right) - w_2\left(t_{k-1}^{(n)}\right)\right) = 0. \tag{B.6}$$

If P_{n+1} is a refinement of P_n for all n then the convergence holds almost surely.

Proof. Let $\Delta w\left(t_k\right) \triangleq w\left(t_k\right) - w\left(t_{k-1}\right)$.

1. By the definition of Wiener processes

$$\mathbf{E}\left(\sum_k \left(\Delta w\left(t_k^{(n)}\right)\right)^2\right) = \sum_k \left(t_k^{(n)} - t_{k-1}^{(n)}\right) = b - a.$$

Recall that if the distribution of ξ is $N\left(0,1\right)$ then

$$\mathbf{D}^2\left(\xi^2\right) = \mathbf{D}^2\left(\chi_1^2\right) = 2.$$

w has independent increments and the expected value of the increments is zero, hence as the sequence of partitions is infinitesimal

$$\left\|\sum_k \left(\Delta w\left(t_k^{(n)}\right)\right)^2 - (b-a)\right\|_2^2 = \mathbf{D}^2\left(\sum_k \left(\Delta w\left(t_k^{(n)}\right)\right)^2\right) =$$

$$= \sum_k \mathbf{D}^2\left(\left(\Delta w\left(t_k^{(n)}\right)\right)^2\right) = \sum_k 2\cdot\left(t_k^{(n)} - t_{k-1}^{(n)}\right)^2 \leq$$

$$\leq 2\cdot(b-a)\cdot\max_k\left(t_k^{(n)} - t_{k-1}^{(n)}\right) \to 0.$$

If w_1 and w_2 are independent Wiener processes then $(w_1 \pm w_2)/\sqrt{2}$ are also Wiener processes. (B.6) follows from the identity

$$\sum_k \frac{\Delta w_1\left(t_k\right) \Delta w_2\left(t_k\right)}{4} = \sum_k \left[\Delta\left(w_1 + w_2\right)\left(t_k\right)\right]^2 -$$

$$- \sum_k \left[\Delta\left(w_1 - w_2\right)\left(t_k\right)\right]^2.$$

2. The proof of the almost sure convergence is a bit more complicated. The quadratic variation depends only on the trajectories so one can assume that

$$(\Omega, \mathcal{A}, \mathbf{P}) = \left(C\left(\mathbb{R}_+\right), \mathcal{B}, W\right),$$

where $\mathcal{B}$ is the σ-algebra of the Borel measurable sets of the function space $C\left(\mathbb{R}_+\right)$, W is the *Wiener measure*, that is the common distribution of the Wiener processes. Fix an interval $[0, u]$ and a partition P_m with n points. Let us consider the 2^n signs ± 1 corresponding to the n points. To every sequence of signs $(s_k)_{k=1}^n$ and to every $f \in C\left[0, \infty\right)$ let us map the function

$$\tilde{f}(t) \overset{\circ}{=} \tilde{f}\left(t_{k-1}^{(n)}\right) + s_k\left(f(t) - f\left(t_{k-1}^{(n)}\right)\right), \quad t \in \left[t_{k-1}^{(n)}, t_k^{(n)}\right).$$

We shall call the correspondence $f \mapsto \tilde{f}$ as alternation. The Gaussian distributions are symmetric, so if w is a Wiener process then the alternated process $\omega \mapsto \tilde{w}\left(\omega\right)$ is also a Wiener process. The Wiener measure is the common, therefore unique distribution of every Wiener process, so W is invariant under all the alternations $f \mapsto \tilde{f}$. Let $\mathcal{B}_n \subseteq \mathcal{B}$ be the set of events which are invariant under the whole 2^n alternations. It is easy to see that $\mathcal{B}_n$ is a σ-algebra. As the $(n+1)$-th partition is refining the n-th one every alternation corresponding to the partition P_{n+1} is an alternation corresponding to P_n. Hence $\mathcal{B}_{n+1} \subseteq \mathcal{B}_n$. If $i \neq j$ then

$$\mathbf{E}\left(\Delta w\left(t_i^{(n)}\right) \Delta w\left(t_j^{(n)}\right) \mid \mathcal{B}_n\right) = 0,$$

as if B is invariant under an alternation of the i-th interval then by the integral transformation theorem, using that W is invariant under the alternations

$$\int_B \Delta w\left(t_i^{(n)}\right) \Delta w\left(t_j^{(n)}\right) dW = -\int_B \Delta w\left(t_i^{(n)}\right) \Delta w\left(t_j^{(n)}\right) dW.$$

On the other hand $\left(\Delta w\left(t_i^{(n)}\right)\right)^2$ is invariant under any possible alternations so it is $\mathcal{B}_n$-measurable. By the energy equality

$$
\mathbf{E}\left(w^2\left(T\right) \mid \mathcal{B}_n\right) = \mathbf{E}\left(\left(\sum_k \Delta w\left(t_k^{(n)}\right)\right)^2 \mid \mathcal{B}_n\right) =
$$

$$
= \mathbf{E}\left(\sum_k \left(\Delta w\left(t_k^{(n)}\right)\right)^2 \mid \mathcal{B}_n\right) = \sum_k \left(\Delta w\left(t_k^{(n)}\right)\right)^2.
$$

By Lévy's theorem about reversed martingales[8] the expression on the left-hand side is almost surely convergent. Hence the sum on the right-hand side is also convergent for almost all ω. If the partition is infinitesimal, then by the just proved convergence in $L^2\left(\Omega\right)$ one can easily prove that the almost sure limit is the quadratic variation. □

It is very natural to ask what does happen if partition P_{n+1} is not refining partition P_n?

Theorem B.18 (Almost sure convergence of the quadratic variation)
If for a sequence of partitions $P_n \stackrel{\circ}{=} \left(t_k^{(n)}\right)$

$$
l_n \stackrel{\circ}{=} \max_k \left(t_{k+1}^{(n)} - t_k^{(n)}\right) = o\left(\frac{1}{\log n}\right)
$$

then for every Wiener process w *the sequence in (B.5) is almost surely convergent.*

Proof. To make the notation simply let $[a, b] = [0, 1]$. Let $\left(t_k^{(n)}\right)_{k=0}^{N(n)}$ be a partition of $[0, 1]$ and let $c_k \stackrel{\circ}{=} t_k^{(n)} - t_{k-1}^{(n)}$.

1. Recall that the moment-generating function of distribution χ_1^2 is

$$
M_1\left(s\right) = \mathbf{E}\left(\exp\left(sN\left(0, 1\right)^2\right)\right) = \frac{1}{\sqrt{2\pi}} \int_{-\infty}^{\infty} \exp\left(sx^2\right) \exp\left(-\frac{x^2}{2}\right) dx =
$$

$$
= \frac{1}{\sqrt{2\pi}} \int_{-\infty}^{\infty} \exp\left(-\frac{x^2}{2}\left(1 - 2s\right)\right) dx = \frac{1}{\sqrt{1 - 2s}}.
$$

2. Let $\varepsilon > 0$ and let $0 < a \stackrel{\circ}{=} 1 - \varepsilon$ and let $s < 0$. By Markov's inequality, using the formula for the moment-generating function of $\chi_1^2 \stackrel{\circ}{=} N\left(0, 1\right)^2$ and using that

[8]See: Theorem 1.75, page 46

if $x \geq 0$ then $\ln (1 + x) \geq x - x^2/2$

$$p_n^{(1)} (a) \doteq \mathbf{P} \left(\sum_k \left(\Delta w \left(t_k^{(n)} \right) \right)^2 - 1 \leq -\varepsilon \right) =$$

$$= \mathbf{P} \left(\sum_k \left(\Delta w \left(t_k^{(n)} \right) \right)^2 \leq a \right) =$$

$$= \mathbf{P} \left(s \sum_k \left(\Delta w \left(t_k^{(n)} \right) \right)^2 \geq sa \right) \leq$$

$$\leq \frac{\mathbf{E} \left(\exp \left(s \sum_k c_k N (0, 1)^2 \right) \right)}{\exp (sa)} = \frac{\prod_k 1/\sqrt{1 - 2sc_k}}{\exp (sa \sum_k c_k)} =$$

$$= \prod_k \exp \left(-sac_k - \frac{1}{2} \ln (1 - 2sc_k) \right) \leq$$

$$\leq \prod_k \exp \left(-s (ac_k - c_k) + s^2 c_k^2 \right) =$$

$$= \exp \left(-s (a - 1) + s^2 \sum_k c_k^2 \right) \leq$$

$$\leq \exp \left(-s (a - 1) + s^2 \left(\max_k c_k \right) \sum_k c_k \right) =$$

$$= \exp \left(-s (a - 1) + s^2 l_n \right).$$

The minimum over $s < 0$ is obtained at

$$s = \frac{a - 1}{2l_n}.$$

Substituting it back

$$p_n^{(1)} (a) \leq \exp \left(-\frac{(a - 1)^2}{2l_n} + \frac{(a - 1)^2}{4l_n} \right) = \tag{B.7}$$

$$= \exp \left(-\frac{(a - 1)^2}{4l_n} \right) \doteq \exp \left(-\frac{K_1 (a)}{l_n} \right),$$

where $K_1 (a) > 0$.

3. Now let $a \overset{\circ}{=} 1 + \varepsilon$, where $s > 0$.

$$p_n^{(2)}(a) \overset{\circ}{=} \mathbf{P}\left(\sum_k \left(\Delta w\left(t_k^{(n)}\right)\right)^2 - 1 \geq \varepsilon\right) =$$

$$= \mathbf{P}\left(\sum_k \left(\Delta w\left(t_k^{(n)}\right)\right)^2 \geq a\right) =$$

$$= \mathbf{P}\left(s\sum_k \left(\Delta w\left(t_k^{(n)}\right)\right)^2 \geq sa\right) \leq$$

$$\leq \frac{\mathbf{E}\left(\exp\left(s\sum_k \left(\Delta w\left(t_k^{(n)}\right)\right)^2\right)\right)}{\exp(sa)} =$$

$$= \frac{\prod_k \mathbf{E}\left(\exp\left(s\left(\Delta w\left(t_k^{(n)}\right)\right)^2\right)\right)}{\exp(sa)} =$$

$$= \frac{\prod_k 1/\sqrt{1-2sc_k}}{\exp(sa)} =$$

$$= \exp\left(-\frac{1}{2}\sum_k \ln(1-2sc_k) - sa\right) \overset{\circ}{=} \exp(f(s)).$$

Obviously $f(0) = 0$ and as $a > 1$

$$f'(0) = 1 - a < 0.$$

If $s \to 1/(2\max c_k)$ then $f(s) \to \infty$. Therefore it has a minimum at point $s^* > 0$ where

$$f'(s^*) = \sum_k \frac{c_k}{1-2s^*c_k} - a = 0. \tag{B.8}$$

Hence if $x_k \overset{\circ}{=} 2s^*c_k \leq 1$

$$p_n^{(2)}(a) \leq \exp\left(-\frac{1}{2}\sum_k \left(\ln(1-x_k) + \frac{x_k}{1-x_k}\right)\right).$$

Now we want to estimate

$$\ln(1-x) + \frac{x}{1-x}$$

over $x \in (0, 1)$.

$$\ln(1-x) + \frac{x}{1-x} = \ln(1-x) - \ln 1 + \frac{1}{1-x} - 1 =$$

$$= \int_0^x \frac{-1}{1-u} + \frac{1}{(1-u)^2} du =$$

$$= \int_0^x \frac{u}{(1-u)^2} du \geq \int_0^x \frac{u - u^2/2}{(1-u)^2} du =$$

$$= \frac{1}{2} \int_0^x \frac{2u - u^2 - 1 + 1}{(1-u)^2} du =$$

$$= \frac{1}{2} \int_0^x \frac{1}{(1-u)^2} - 1 du = \frac{1}{2} \left(\frac{x}{1-x} - x \right).$$

Hence using (B.8)

$$p_n^{(2)}(a) \leq \exp\left(-\frac{1}{4} \left(\sum_k \frac{2s^* c_k}{1 - 2s^* c_k} - \sum_k 2s^* c_k \right) \right) =$$

$$= \exp\left(-\frac{1}{4} \sum_k 2s^* a - 2s^* \right) = \exp\left(-\frac{s^*}{2}(a-1) \right)$$

But again by (B.8)

$$\frac{1}{1 - 2l_n s^*} = \frac{\sum_k c_k}{1 - 2l_n s^*} \geq \sum_k \frac{c_k}{1 - 2s^* c_k} = a$$

That is $s^* \geq (a-1)/(2al_n)$. Using this

$$p_n^{(2)}(a) \leq \exp\left(-\frac{(a-1)^2}{4al_n} \right) \overset{\circ}{=} \exp\left(-\frac{K_2(a)}{l_n} \right). \tag{B.9}$$

4. By the assumption of the theorem $l_n = o(1/\ln n)$. So for some $\varepsilon_n \to 0$

$$b_n \overset{\circ}{=} \exp\left(-\frac{K}{l_n} \right) = \exp\left(-\frac{K \ln n}{\varepsilon_n} \right) = \left(\frac{1}{n} \right)^{K/\varepsilon_n}.$$

So $\sum_n b_n < \infty$. Using this and the just proved estimations (B.7) and (B.9) for all m

$$\sum_{n=1}^\infty \mathbf{P}\left(\left| \sum_k \left(\Delta w \left(t_k^{(n)} \right) \right)^2 - 1 \right| \geq \frac{1}{m} \right) < \infty,$$

hence by the Borel–Cantelli lemma

$$\sum_k \left(\Delta w\left(t_k^{(n)}\right)\right)^2 \xrightarrow{a.s.} 1.$$

$\square$

One can ask whether we can improve the estimation of the order of l_n. The answer is no.

Example B.19 There is a sequence of partitions with $l_n = O\left(1/\log n\right)$ for which (B.5) is not almost surely convergent.

For every integer $p \geq 1$ let Π_p be the set of partitions of $[0, 1]$ formed from the intervals

$$J_p^k \stackrel{\circ}{=} \left[\frac{2k}{2^p}, \frac{2k+2}{2^p}\right] = \left[\frac{k}{2^{p-1}}, \frac{k+1}{2^{p-1}}\right],$$

and

$$I_p^{2k} \stackrel{\circ}{=} \left[\frac{2k}{2^p}, \frac{2k+1}{2^p}\right], \quad I_p^{2k+1} \stackrel{\circ}{=} \left[\frac{2k+1}{2^p}, \frac{2k+2}{2^p}\right]$$

where in both cases $k = 1, \ldots, 2^{p-1}$. During the construction of a partition 2^{p-1} times one should choose between one J_p^k and a pair of $\left(I_p^{2k}, I_p^{2k+1}\right)$ so for any p the number of partitions in Π_p is $2^{2^{p-1}}$. For a p for one partition, when we are using just I_p^k type intervals, the maximal length is 2^{-p} and for the other $2^{2^{p-1}} - 1$ partitions the length of the maximal interval is $2^{-(p-1)}$. If we take any sequence of partitions from Π_p then the index of a partition from Π_p is maximum

$$\sum_{1 \leq q \leq p} 2^{2^{q-1}} < 2^{1+2^{p-1}}.$$

Observe that if l_n is the size of the largest interval in the n-th partition then

$$l_n \ln n \leq 2^{-(p-1)} \left(1 + 2^{p-1}\right) \leq 3,$$

that is $l_n = O\left(1/\log n\right)$. Let $Q(\pi)$ be the approximating sum of the quadratic variation formed with partition π. Let

$$M_p \stackrel{\circ}{=} \max\{Q(\pi) : \pi \in \Pi_p\}.$$

The $\limsup$ of sequence $(Q(\pi))$ is the same as the $\limsup$ of sequence (M_p). Let

$$M_p^{(k)} \stackrel{\circ}{=} \max\left(\left(\Delta w\left(I_p^{2k}\right)\right)^2 + \left(\Delta w\left(I_p^{2k+1}\right)\right)^2, \left(\Delta w\left(J_p^k\right)\right)^2\right).$$

Obviously

$$M_p = \sum_{0 \leq k \leq 2^{p-1}-1} M_p^{(k)}.$$

The variables $\Delta w \left(I_p^{2k} \right)$ and $\Delta w \left(I_p^{2k+1} \right)$ are independent and their distribution is

$$N \left(0, \sigma^2 \right) = N \left(0, 2^{-p} \right).$$

$\Delta w \left(J_p^k \right)$ is the sum of two independent variables $\Delta w \left(I_p^{2k} \right)$ and $\Delta w \left(I_p^{2k+1} \right)$. If ξ and η are independent variables with distribution $N \left(0, \sigma^2 \right)$ and

$$\zeta \overset{\circ}{=} \max \left(\xi^2 + \eta^2, (\xi + \eta)^2 \right)$$

then one can find constants $a, b > 0$ such that

$$\mathbf{E} \left(\zeta \right) = (1 + a) \sigma^2 \quad \text{and} \quad \mathbf{D}^2 \left(\zeta \right) = b \sigma^4.$$

for all σ. With these constants

$$\mathbf{E} \left(M_p^{(k)} \right) = (1 + a) 2^{-(p-1)} \quad \text{and} \quad \mathbf{D}^2 \left(M_p^{(k)} \right) = b 2^{-2p}.$$

The number of variables in M_p is 2^{p-1}, so for their sum M_p

$$\mathbf{E} \left(M_p \right) = 1 + a \quad \text{and} \quad \mathbf{D}^2 \left(M_p \right) = b 2^{-p-1}.$$

$b 2^{-(p-1)} \to 0$, hence by Chebyshev's inequality

$$\lim_{p \to \infty} M_p = \limsup_{n \to \infty} Q \left(\pi_n \right) = 1 + a > 1,$$

so $\left(Q \left(\pi_n \right) \right)$ cannot converge almost surely to 1. $\qquad \square$

Appendix C
POISSON PROCESSES

Let us first define the point processes

Definition C.1 *Let $\mathcal{F}$ be a filtration and let (τ_n) be a sequence of stopping times. (τ_n) generates a* point process *if it satisfies the next assumptions:*

1. $\tau_0 = 0$,
2. $\tau_n \leq \tau_{n+1}$,
3. *if $\tau_n(\omega) < \infty$ then $\tau_n(\omega) < \tau_{n+1}(\omega)$.*

The investigation of a point process (τ_n) is equivalent to the investigation of the counting process

$$N(t) \overset{\circ}{=} \sum_{n=1}^{\infty} \chi(\tau_n \leq t).$$

N is finite on the interval $[0, \tau_\infty)$, where of course $\tau_\infty \overset{\circ}{=} \lim_n \tau_n$. As we defined the stochastic processes only on deterministic intervals we assume that $\tau_\infty = \infty$. That is we assume that $N(t)$ is finite for every t. Otherwise we should restrict our counting processes to some intervals $[0, u]$ where $u < \tau_\infty$. The trajectories of N are increasing, so N is regular. By the second and the third assumptions N is right-continuous. As the functions τ_n are $\mathcal{F}$-stopping times, N is $\mathcal{F}$-adapted as whenever $a \geq 0$

$$\{N(t) \leq a\} = \left\{\tau_{[a]+1} > t\right\} \in \mathcal{F}_t.$$

Very often the filtration $\mathcal{F}$ is not given explicitly and the point process is defined just by the random variables (τ_n). With (τ_n) one can define the counting process N and the filtration $\mathcal{F}$ is defined by the filtration generated by N.

Definition C.2 *The pair $(N, \mathcal{F}^N)$ is called the* minimal representation *of the point process (τ_n).*

Proposition C.3 *If the trajectories of a process X are right-regular and for every t and ω the trajectory $X(\omega)$ is constant on an interval $[t, t + \delta]$, where $\delta > 0$ can depend on ω and t then the filtration $\mathcal{F}^X$ generated by X is right-continuous.*

Proof. It is well-known that for an arbitrary collection of random variables $\mathcal{X} \doteq (\xi_\gamma)_{\gamma \in \Gamma}$ any set C from the σ-algebra generated by $\mathcal{X}$ has a representation $C = \Psi^{-1}(B)$, where $B \in \mathcal{B}(\mathbb{R}^\infty)$ and $\Psi(\omega) \doteq \left(\xi_{\gamma_k}(\omega)\right)_k$ and the number of indexes $(\gamma_k) \subseteq \Gamma$ is maximum countable. Let us fix a moment t. The trajectories $X(\omega)$ are right-regular and they are constant on an interval after t, hence for every n there is a set $A_n \subseteq \Omega$, such that if $\omega \in A_n$ then the trajectory $X(\omega)$ is constant on the closed interval $[t, t + 1/n]$. If $C \in \mathcal{F}_{t+}$ then $C \in \mathcal{F}_{t+1/n}$ for every n. By the just mentioned property of the generated σ-algebras $C = \Psi^{-1}(B_n)$ where

$$\Psi(\omega) \doteq (X(t_1, \omega), \dots, X(t_k, \omega), \dots), \quad B_n \in \mathcal{B}(\mathbf{R}^\infty), \quad t_k \le t + \frac{1}{n}.$$

Let $t_k^{(n)} \doteq t_k \wedge t$ and let Ψ_n be the analogous correspondence defined by $\left(t_k^{(n)}\right)_k$. As $t_k^{(n)} \le t$ obviously $C_n \doteq \Psi_n^{-1}(B_n) \in \mathcal{F}_t^X$. If $\omega \in A_n$ then by the structure of the trajectories $\Psi_n(\omega) = \Psi(\omega)$ and therefore

$$C_n \Delta C \subseteq A_n^c. \tag{C.1}$$

$A_n \nearrow \Omega$, hence $A_n^c \searrow \emptyset$. If

$$C_\infty \doteq \limsup_{n \to \infty} C_n \doteq \cap_n \cup_{m=n}^\infty C_m \in \mathcal{F}_t^X,$$

then as $A_n^c \searrow \emptyset$ by (C.1) $C = C_\infty \in \mathcal{F}_t^X$. This means that the filtration $\mathcal{F}^X$ is right-continuous. $\square$

Corollary C.4 *If N is a counting process of a point process then the filtration $\mathcal{F}^N$ is right-continuous and the jump times (τ_n) are stopping times with respect to $\mathcal{F}^N$.*

Proof. To prove the last statement it is sufficient to remark that τ_n is the hitting time[1] of the open set $(n - 1/2, \infty)$. $\square$

Filtration $\mathcal{F}^N$ is right-continuous, but the usual assumptions do not hold: one should complete the space $(\Omega, \mathcal{F}_\infty^N)$ with respect to $\mathbf{P}$ and add the measure-zero sets to the σ-algebras $\mathcal{F}_t^N$.

Lemma C.5 *If τ is a stopping time of the augmented filtration then $\tau = \tau' + \tau''$, where τ' is a stopping time of the filtration $\mathcal{F}^N$ and τ'' is almost surely zero.*

---

[1] See: Example 1.32, page 17.

Proof. Let τ be a stopping time of the augmented filtration. For every $t \geq 0$ there is an $A_t \in \mathcal{F}_t^N$ such that $\{\tau < t\} \overset{a.s.}{=} A_t$. If

$$\tau'(\omega) \overset{\circ}{=} \inf\{r \in \mathbb{Q}_+ : \omega \in A_r\},$$

then

$$\{\tau' < t\} = \cup_{s \in \mathbb{Q}_+, s < t} A_s \in \mathcal{F}_t^N,$$

so τ' is a weak stopping time of $\mathcal{F}^N$. $\mathcal{F}^N$ is right-continuous so τ' is a stopping time of $\mathcal{F}^N$. For an arbitrary t

$$\{\tau < t\} = \cup_{s \in \mathbb{Q}, s < t}\{\tau < s\} \overset{a.s.}{=} \{\tau' < t\},$$

so $\tau \overset{a.s.}{=} \tau'$. $\qquad\qquad\qquad\qquad\qquad\qquad\qquad\qquad\qquad\qquad\qquad\square$

Proposition C.6 (Representation of stopping times) *If τ is a stopping time of the minimal representation of a point process (τ_n) then there is a sequence of real-valued Borel measurable functions (φ_n), where $\varphi_n : \mathbb{R}^n \to \mathbb{R}$ and a constant φ_0 such that on the set $\{\tau < \infty\}$*

$$\tau = \varphi_0 \chi(\tau_0 \leq \tau < \tau_1) + \sum_{n=1}^{\infty} \chi(\tau_n \leq \tau < \tau_{n+1}) \varphi_n(\tau_0, \ldots, \tau_n) \overset{\circ}{=}$$

$$\overset{\circ}{=} \sum_{n=0}^{\infty} \chi(\tau_n \leq \tau < \tau_{n+1}) \varphi_n(\tau_0, \ldots, \tau_n).$$

Proof. First assume that Ω is the canonical space of the point processes that is Ω is the space of right-continuous functions which have jumps of size one and which are constant between the jumps. Ω is closed under truncation and therefore $\mathcal{G}_\tau = \mathcal{F}_\tau$, where[2]

$$\mathcal{G}_\tau = \sigma(N(\tau \wedge t) : t \geq 0) \qquad\qquad\qquad\qquad (\text{C.2})$$

is the σ-algebra defined in line (A.4). τ is $\mathcal{F}_\tau$-measurable, hence τ is $\mathcal{G}_\tau$-measurable. By (C.2) for some Borel measurable function[3] $\varphi : \mathbb{R}^\infty \to \overline{\mathbb{R}}$

$$\tau = \varphi(N(\tau \wedge t_1), N(\tau \wedge t_2), \ldots, N(\tau \wedge t_k), \ldots) =$$

$$= \left(\sum_{n=1}^{\infty} \chi(\tau_{n-1} \leq \tau < \tau_n) + \chi(\tau_\infty \leq \tau)\right) \varphi(N(\tau \wedge t_k), k \in \mathbb{N}).$$

---

[2]See: Proposition A.17, page 556.
[3]It is a simple consequence of the Monotone Class Theorem.

On the set $\{\tau_n \leq \tau < \tau_{n+1}\}$

$$N\left(\tau \wedge t_k\right) = N\left(\tau_n \wedge t_k\right) = \sum_{l=1}^{n} \chi\left(\tau_l \leq t_k\right),$$

on the set the set $\{\tau_\infty \leq \tau\}$

$$N\left(\tau \wedge t_k\right) = N\left(t_k\right) = \sum_{l=1}^{\infty} \chi\left(\tau_l \leq t_k\right)$$

from which the representation is evident.

Now let us assume that τ is a stopping time of the minimal representation. One can obviously embed the minimal representation to the canonical representation. Let us extend τ with the definition $\tau(\omega) \overset{\circ}{=} \infty$ for every new outcome ω. As we already discussed[4] τ remains a stopping time, therefore one can construct the needed representation. $\qquad\square$

Example C.7 Point process with a single jump.

Let $0 \leq \sigma$ be a random time. One can define a point process with $\tau_1 \overset{\circ}{=} \sigma$ and $\tau_k \overset{\circ}{=} \infty$, for $k \geq 2$. In this case $N\left(t, \omega\right) \overset{\circ}{=} \chi\left(\sigma\left(\omega\right) \leq t\right)$. By the just proved proposition if $\tau \geq 0$ is a stopping time of the filtration $\mathcal{F} \overset{\circ}{=} \mathcal{F}^N$ then

$$\tau = \varphi_0 \chi\left(\tau < \sigma\right) + \chi\left(\sigma \leq \tau < \infty\right)\varphi_1\left(\sigma\right) + \chi\left(\tau = \infty\right) \cdot \infty.$$

If $\tau < \infty$ then

$$\tau = \varphi_0 \chi\left(\tau < \sigma\right) + \chi\left(\sigma \leq \tau\right)\varphi_1\left(\sigma\right). \tag{C.3}$$

Therefore τ is constant on the set $\{\tau < \sigma\}$. $\qquad\square$

N^{τ_n} is bounded for every n, so trivially $N \in \mathcal{A}_{\text{loc}}^+$. This means that N has a compensator N^p.

Example C.8 Predictable compensator of point processes.

Let N be a counting process and let assume that we study N in its minimal representation. Assume that we added the measure-zero sets to the filtration and let assume that $\mathcal{F}_\infty$ is complete. In this case the usual conditions hold and N has a compensator N^p. Let

$$F_1\left(t\right) \overset{\circ}{=} \mathbf{P}\left(\tau_1 \leq t\right), \quad F_k\left(t\right) \overset{\circ}{=} \mathbf{P}\left(\tau_k \leq t \mid \tau_1, \dots, \tau_{k-1}\right), \tag{C.4}$$

---

[4]See: Lemma A.18, page 557.

be the conditional distributions of the jumps. Assume that the conditional distributions are regular. Define the so called integrated conditional hazard rates

$$A_k\left(t\right) \stackrel{\circ}{=} \int_0^{t \wedge \tau_k} \frac{dF_k\left(u\right)}{1 - F_k\left(u-\right)}. \tag{C.5}$$

We show that

$$N^p = \sum_i A_i \stackrel{\circ}{=} B. \tag{C.6}$$

To prove this it is sufficient to show that B is predictable and $N^{\tau_k} - B^{\tau_k}$ is a uniformly integrable martingale. Before the proof, let us discuss the interpretation of the formula. τ_k is larger than τ_{k-1}, and the conditional probability of the random segment $[0, \tau_{k-1}]$ with respect to F_k is zero. Hence the measures related to F_k are concentrated on the set $(\tau_{k-1}, \infty]$. One should pay the compensation fee A_k for the k-th jump only on the interval $(\tau_{k-1}, \tau_k]$. After every jump the amount of the compensation can change. The expression in the integral, the hazard rate, the amount of compensation one should pay at moment u is

$$\frac{dF_k\left(u\right)}{1 - F_k\left(u-\right)}$$

which is the probability that the k-th jump of the process will happen at time u under the condition that there was no jump before u. If the distribution F_k is continuous then[5]

$$\int_0^{t \wedge \tau_k} \frac{dF_k\left(u\right)}{1 - F_k\left(u-\right)} = -\ln\left(1 - F_k\left(t \wedge \tau_k\right)\right).$$

1. Let us first prove that $N^{\tau_k} - B^{\tau_k}$ is a uniformly integrable martingale. To prove this we show first that

$$\mathbf{E}\left(N\left(\theta \wedge \tau_k\right)\right) = \mathbf{E}\left(B\left(\theta \wedge \tau_k\right)\right) \tag{C.7}$$

for every stopping time θ. Using the representation of the stopping times of the minimal representation

$$\theta \stackrel{a.s.}{=} \sum_{k=1}^\infty \chi\left(\tau_{k-1} \le \theta < \tau_k\right) \varphi_{k-1}\left(\tau_0, \dots, \tau_{k-1}\right), \quad \theta < \infty.$$

By this one can define a Borel measurable functions with s variables

$$\theta_s \stackrel{\circ}{=} \theta_s\left(\tau_1, \dots, \tau_s\right)$$

[5]See: (6.32), page 398.

that

$$\theta \wedge \tau_k = \theta_{k-1} \wedge \tau_k.$$

The measure generated by F_k is concentrated on the set $(\tau_{k-1}, \infty]$. Hence

$$\mathbf{E}\left(B\left(\theta \wedge \tau_n\right)\right) \stackrel{\circ}{=} \sum_{k=1}^{\infty} \mathbf{E}\left(A_k\left(\theta \wedge \tau_n\right)\right) = \sum_{k=1}^{n} \mathbf{E}\left(A_k\left(\theta \wedge \tau_k\right)\right) =$$

$$= \sum_{k=1}^{n} \mathbf{E}\left(A_k\left(\theta_{k-1} \wedge \tau_k\right)\right) =$$

$$= \sum_{k=1}^{n} \mathbf{E}\left(\mathbf{E}\left(A_k\left(\theta_{k-1} \wedge \tau_k\right) \mid \tau_1, \tau_2, \ldots, \tau_{k-1}\right)\right).$$

Let us calculate the condition expectation. Let

$$I \stackrel{\circ}{=} \mathbf{E}\left(A_k\left(\theta_{k-1} \wedge \tau_k\right) \mid \tau_1, \tau_2, \ldots, \tau_{k-1}\right) =$$

$$= \mathbf{E}\left(\int_0^{\theta_{k-1} \wedge \tau_k} \frac{dF_k\left(u\right)}{1 - F_k\left(u-\right)} \mid \tau_1, \ldots, \tau_{k-1}\right).$$

Using the definition of A_k and the regularity of the conditional expectations

$$I = \int_0^{\infty} \int_0^{\theta_{k-1} \wedge s} \frac{dF_k\left(u\right)}{1 - F_k\left(u-\right)} dF_k\left(s\right) =$$

$$= \int_0^{\theta_{k-1}} \int_0^{s} \frac{dF_k\left(u\right)}{1 - F_k\left(u-\right)} dF_k\left(s\right) + \int_{\theta_{k-1}}^{\infty} \int_0^{\theta_{k-1} \wedge s} \frac{dF_k\left(u\right)}{1 - F_k\left(u-\right)} dF_k\left(s\right).$$

If $s > \theta_{k-1}$ then in the second term the inner integral is not changing, that is the expression is

$$\left(1 - F_k\left(\theta_{k-1}\right)\right) \int_0^{\theta_{k-1}} \frac{dF_k\left(u\right)}{1 - F_k\left(u-\right)}.$$

Integrating by parts in the first expression and using that $F_k\left(0\right) = 0$, if $k \geq 1$ the first integral is

$$F_k\left(\theta_{k-1}\right) \int_0^{\theta_{k-1}} \frac{dF_k\left(u\right)}{1 - F_k\left(u-\right)} - \int_0^{\theta_{k-1}} \frac{F_k\left(u-\right)}{1 - F_k\left(u-\right)} dF_k\left(u\right).$$

Reordering

$$\left(F_k\left(\theta_{k-1}\right) - 1\right) \int_0^{\theta_{k-1}} \frac{dF_k\left(u\right)}{1 - F_k\left(u-\right)} + F_k\left(\theta_{k-1}\right).$$

Adding up the two integrals

$$\mathbf{E}\left(\int_0^{\theta_{k-1}\wedge\tau_k} \frac{dF_k(u)}{1-F_k(u-)} \mid \tau_1,\dots,\tau_{k-1}\right) = F_k(\theta_{k-1}),$$

hence

$$\mathbf{E}(B(\theta\wedge\tau_n)) = \sum_{k=1}^n F_k(\theta_{k-1}).$$

On the other hand using that $\tau_0 = 0$ and $N(0) = 0$

$$\mathbf{E}(N(\theta\wedge\tau_n)) = \sum_{i=1}^n (\mathbf{E}(N(\theta\wedge\tau_i) - N(\theta\wedge\tau_{i-1}))) =$$

$$= \sum_{i=1}^n (\mathbf{E}(N(\theta_{i-1}\wedge\tau_i) - N(\theta_{i-1}\wedge\tau_{i-1}))) =$$

$$= \mathbf{E}(\chi(\tau_1 \le \theta_0)) + \sum_{i=2}^n \mathbf{E}(\chi(\tau_i \le \theta_{i-1})) = \sum_{i=1}^n F_i(\theta_{i-1}),$$

therefore (C.7) is valid.

2. Observe that this does not imply that the truncated process is a uniformly integrable martingale, as both sides can be infinite.

$$(N^{\tau_n}(t) - B^{\tau_n}(t)) \le N^{\tau_n}(t) + B^{\tau_n}(t) \le n + B^{\tau_n}(t).$$

By the just proved statements

$$\mathbf{E}(B^{\tau_n}(t)) = \mathbf{E}(B(\tau_n\wedge t)) \le \mathbf{E}(B(\tau_n)) = \mathbf{E}(N(\tau_n)) = n, \qquad \text{(C.8)}$$

therefore the truncated process is really a uniformly integrable martingale.

3. One should prove that B is predictable. Every natural process is predictable hence one should show that B is natural[6]. By definition this means that for any non-negative, bounded martingale[7] M

$$\mathbf{E}\left(\int_0^t M\,dB\right) = \mathbf{E}\left(\int_0^t M_-\,dB\right).$$

[6] See: Theorem 5.10, page 302.
[7] See: Definition 5.7, page 299.

Let M be a non-negative bounded martingale. F_1 is a distribution function[8] so by the non-negativity of M one can apply Fubini's theorem

$$\mathbf{E}\left(\int_0^t M dA_1\right) \stackrel{\circ}{=} \mathbf{E}\left(\int_0^t \frac{\chi(u \le \tau_1)}{1 - F_1(u-)} M(u) dF_1(u)\right) =$$

$$= \int_0^t \mathbf{E}\left(\frac{\chi(u \le \tau_1)}{1 - F_1(u-)} M(u)\right) dF_1(u) =$$

$$= \int_0^t \mathbf{E}\left(\mathbf{E}\left(\frac{1 - \chi(\tau_1 < u)}{1 - F_1(u-)} M(u) \mid \mathcal{F}_{u-}\right)\right) dF_1(u) =$$

$$= \int_0^t \mathbf{E}\left(\frac{1 - \chi(\tau_1 < u)}{1 - F_1(u-)} \mathbf{E}(M(u) \mid \mathcal{F}_{u-})\right) dF_1(u) =$$

$$= \int_0^t \mathbf{E}\left(\frac{\chi(u \le \tau_1)}{1 - F_1(u-)} M_-(u)\right) dF_1(u) \stackrel{\circ}{=} \mathbf{E}\left(\int_0^t M_- dA_1\right).$$

The case $i > 1$ is a bit more complicated as in this case for the conditional distribution functions

$$F_i(u) \stackrel{\circ}{=} \mathbf{P}(\tau_i \le u \mid \tau_1, \ldots, \tau_{i-1})$$

one cannot apply Fubini's theorem.

4. Let $\mathcal{G}$ be a σ-algebra. Assume that V is right-regular with finite variation and $V(t)$ is $\mathcal{G}$-measurable for every t. If $X \stackrel{\circ}{=} \sum_{k=1}^n \xi_k \chi_{I_k}$ is a step-function, where I_k are intervals in $[0, t]$ then

$$\mathbf{E}\left(\int_0^t X dV\right) = \mathbf{E}\left(\mathbf{E}\left(\int_0^t X(u) dV(u) \mid \mathcal{G}\right)\right) =$$

$$= \mathbf{E}\left(\mathbf{E}\left(\sum_k \xi_k(V(t_k) - V(t_{k-1})) \mid \mathcal{G}\right)\right) =$$

$$= \mathbf{E}\left(\sum_i \mathbf{E}(\xi_k \mid \mathcal{G})(V(t_k) - V(t_{k-1}))\right) =$$

$$= \mathbf{E}\left(\int_0^t \mathbf{E}(X(u) \mid \mathcal{G}) dV(u)\right).$$

The set of processes X for which the above identity holds form a λ-system. So by the Monotone Class Theorem the identity holds for any product measurable, bounded process X. With the Monotone Convergence Theorem one can extend the identity to any non-negative product measurable process X.

---

[8]Not a conditional distribution function.

5. Using that F_i is $\mathcal{F}_{\tau_{i-1}}$-measurable

$$\mathbf{E}\left(\int_0^t M\,dA_i\right) \overset{\circ}{=} \mathbf{E}\left(\int_0^t \frac{\chi\left(\tau_{i-1} < u \le \tau_i\right) M\left(u\right)}{1 - F_i\left(u-\right)}dF_i\left(u\right)\right) = \qquad\qquad\text{(C.9)}$$

$$= \mathbf{E}\left(\int_0^t \mathbf{E}\left(\frac{\chi\left(\tau_{i-1} < u \le \tau_i\right) M\left(u\right)}{1 - F_i\left(u-\right)} \;\middle|\; \mathcal{F}_{\tau_{i-1}}\right) dF_i\left(u\right)\right) =$$

$$= \mathbf{E}\left(\int_0^t \frac{\mathbf{E}\left(\chi\left(\tau_{i-1} < u \le \tau_i\right) M\left(u\right) \mid \mathcal{F}_{\tau_{i-1}}\right)}{1 - F_i\left(u-\right)}dF_i\left(u\right)\right).$$

Let us calculate the conditional expectation under the integral.

$$\chi\left(\tau_{i-1} < u \le \tau_i\right) = \chi\left(\tau_{i-1} < u\right)\left(1 - \chi\left(\tau_i < u\right)\right) =$$

$$= \lim_{n\nearrow\infty}\left(\chi\left(\tau_{i-1} < u - \frac{1}{n}\right)\left(1 - \chi\left(\tau_i < u - \frac{1}{n}\right)\right)\right) =$$

$$= \lim_{n\nearrow\infty}\left(\chi\left(\tau_{i-1} < u - \frac{1}{n}\right)\left(1 - \chi\left(\tau_i < u - \frac{1}{n}\right)\right)\right)$$

Let $F \in \mathcal{F}_{\tau_{i-1}}$ and let

$$F_n \overset{\circ}{=} F \cap \left\{\tau_{i-1} < u - \frac{1}{n}\right\} \cap \left\{\tau_i < u - \frac{1}{n}\right\}^c \in \mathcal{F}_{u-}.$$

M is bounded so by the Dominated Convergence Theorem using that $M\left(u-\right) = \mathbf{E}\left(M\left(u\right) \mid \mathcal{F}_{u-}\right)$

$$\int_F \chi\left(\tau_{i-1} < u \le \tau_i\right) M\left(u\right) d\mathbf{P} = \int_{F\cap\{\tau_{i-1}<u\le\tau_i\}} M\left(u\right) d\mathbf{P} =$$

$$= \lim_{n\nearrow\infty}\int_{F_n} M\left(u\right) d\mathbf{P} = \lim_{n\nearrow\infty}\int_{F_n} M\left(u-\right) d\mathbf{P} =$$

$$= \int_F \chi\left(\tau_{i-1} < u \le \tau_i\right) M\left(u-\right) d\mathbf{P}.$$

Hence

$$\mathbf{E}\left(\int_0^t M\,dA_i\right) = \mathbf{E}\left(\int_0^t \frac{\mathbf{E}\left(\chi\left(\tau_{i-1} < u \le \tau_i\right) M\left(u-\right) \mid \mathcal{F}_{\tau_{i-1}}\right)}{1 - F_i\left(u-\right)}dF_i\left(u\right)\right) =$$

$$= \mathbf{E}\left(\int_0^t \frac{\chi\left(\tau_{i-1} < u \le \tau_i\right) M\left(u-\right)}{1 - F_i\left(u-\right)}dF_i\left(u\right)\right) \overset{\circ}{=} \mathbf{E}\left(\int_0^t M_-\,dA_i\right).$$

$\square$

Example C.9 Processes with a single jump.

Let $N(t) \overset{\circ}{=} \chi(\tau \leq t)$ and let F be the distribution of τ. If F is continuous then

$$N^p(t) \overset{\circ}{=} \int_0^{t \wedge \tau} \frac{dF(u)}{1 - F(u)} = -\ln(1 - F(t \wedge \tau)).$$
$$N^p(\infty) = N^p(\tau) = -\log(1 - F(\tau)).$$

If τ has an exponential distribution with parameter $\lambda = 1$, then

$$N^p(t) = -\ln(\exp(-t \wedge \tau)) = t \wedge \tau.$$

If F is strictly increasing and continuous, then

$$\mathbf{P}(F(\tau) < x) = \mathbf{P}(\tau < F^{-1}(x)) = F(F^{-1}(x)) = x,$$

so $F(\tau)$ is uniformly distributed. Hence

$$\mathbf{P}(N^p(\infty) < x) = \mathbf{P}(-\log(1 - U) < x) =$$
$$= \mathbf{P}(U < 1 - \exp(-x)) = 1 - \exp(-x).$$

Hence in this case $N^p(\infty)$ is exponentially distributed with parameter $\lambda = 1$.

□

Example C.10 Predictable compensator of Poisson processes.

Let $\pi(t)$ be a Poisson process with parameter λ. As π is a Lévy process and

$$\mathbf{E}(\pi(t)) = \lambda t$$

$\pi(t) - \lambda t$ is a martingale. So by the definition of the predictable compensator $\pi^p(t) = \lambda t$. On the other hand the distribution of the time between the jumps is exponential with parameter λ so

$$F_k(x \mid \tau_1, \tau_2, \ldots, \tau_{k-1}) = \begin{cases} 0 & \text{if} \quad x \leq \tau_{k-1} \\ 1 - \exp(-\lambda(x - \tau_{k-1})) & \text{if} \quad x > \tau_{k-1} \end{cases} =$$
$$= 1 - \exp(-\lambda(\max(0, x - \tau_{k-1}))).$$

From this if $t > 0$

$$A_i(t) = -\ln(1 - F_i(t \wedge \tau_i)) =$$
$$= -\ln(\exp(-\lambda \max(0, t \wedge \tau_k - \tau_{k-1}))) =$$
$$= \lambda \max(0, (t \wedge \tau_i - \tau_{i-1})),$$

From this the predictable compensator is

$$\pi^p(t) = \lambda(t \wedge \tau_1 + \max(0, t \wedge \tau_2 - \tau_1) + \ldots) = \lambda t. \qquad \square$$

Example C.11 Counting process with Weibull distribution.

Let the length of the jump times be independent with Weibull distribution. In this case

$$F_k(x) = 1 - \exp(-\lambda(\max(0, x - \tau_{k-1}))^\alpha).$$

The integrated hazard rate is

$$\Lambda_k(t) = -\ln(\exp(-\lambda(\max(0, t \wedge \tau_k - \tau_{k-1}))^\alpha)) =$$
$$= \lambda(\max(0, t \wedge \tau_k - \tau_{k-1}))^\alpha.$$

The compensator is

$$\lambda((t \wedge \tau_1)^\alpha + \max(0, t \wedge \tau_2 - \tau_1)^\alpha + \ldots)$$

If $\alpha = 1$ then the compensator is λt, which is the compensator of the Poisson process otherwise the compensator is not deterministic. $\qquad \square$

Definition C.12 *The counting process N is an extended Poisson process if the increments $N(t) - N(s)$ are independent of the σ-algebra $\mathcal{F}_s$ for all $s < t$.*

If N is a counting process then it is a semimartingale. As all the jumps have the same size

$$\nu((0, t] \times \Lambda) = \nu((0, t] \times \{1\}) = N^p(t).$$

If N^p denotes the measure generated by N^p then it is easy to see that the characteristics of semimartingale N are $(0, 0, \nu) = (0, 0, N^p)$.

Proposition C.13 *The compensator N^p of a counting process N is deterministic if and only if N is an extended Poisson process. In this case*

$$N^p(t) = \mathbf{E}(N(t)) < \infty. \qquad (C.10)$$

Proof. Let N be an extended Poisson process. As N has independent increments the spectral measure ν of N is deterministic[9]. If $\Lambda \overset{\circ}{=} \{1\}$ then

$$\mathbf{E}\left(N\left(t\right)\right) = \mathbf{E}\left(\sum_{s \leq t} \Delta N\left(s\right) \chi\left(\Delta N\left(s\right) \in \Lambda\right)\right) = \mathbf{E}\left(\left(\chi_\Lambda x \bullet \mu^N\right)(t)\right) =$$

$$= \left(x\chi_\Lambda \bullet \nu\left(x\right)\right)(t) \leq \left(\left(x^2 \wedge 1\right) \bullet \nu\left(x\right)\right)(t) < \infty.$$

Therefore, using the independence of the increments

$$\mathbf{E}\left(N\left(t\right) - N\left(s\right) \mid \mathcal{F}_s\right) = \mathbf{E}\left(N\left(t\right) - N\left(s\right)\right) = \mathbf{E}\left(N\left(t\right)\right) - \mathbf{E}\left(N\left(s\right)\right),$$

and so $N\left(t\right) - \mathbf{E}\left(N\left(t\right)\right)$ is trivially a martingale and (C.10) holds. On the other hand from the general theory of processes with independent increments we know[10] that if the spectral measure ν, that is the measure generated by N^p, is deterministic then N has independent increments. So the proposition is true. □

Proposition C.14 *The Fourier transform of the increment of an extended Poisson process N is*

$$\mathbf{E}\left(\exp\left(iu\left(N\left(t\right) - N\left(s\right)\right)\right)\right) = \tag{C.11}$$
$$= \exp\left(\left(\exp\left(iu\right) - 1\right)\left(\left(N^p\left(t\right)\right)^c - \left(N^p\left(s\right)\right)^c\right)\right) \times$$
$$\times \left(\prod_{s < r \leq t} \left(1 + \left(\exp\left(iu\right) - 1\right) \Delta N^p\left(r\right)\right)\right),$$

where $(N^p)^c$ is the continuous part[11] of the compensator N^p.

Proof. It is a special case of the Lévy–Khintchine formula[12]. Recall that $B = 0$ and $C = 0$ and ν is the measure generated by N^p. So

$$\mathbf{E}\left(\exp\left(iu\left(N\left(t\right) - N\left(s\right)\right)\right)\right) = \exp\left(U\right)V,$$

where

$$U \overset{\circ}{=} \int_{(s,t] \times (\mathbb{R} \setminus \{0\})} \left(\exp\left(iux\right) - 1 - iuh\left(x\right)\right) \chi_{J^c}\left(r\right) d\nu\left(r, x\right)$$

$$V = \prod_{s < r \leq t} \left(1 + \int_{\mathbb{R} \setminus \{0\}} \left(\exp\left(iux\right) - 1\right) \nu\left(\{r\} \times dx\right)\right)$$

[9]See: Corollary 7.88, page 532.
[10]See: Theorem 7.89, page 532.
[11]N^p is an increasing process and $(N^p)^c$ its continuous part.
[12]See: Theorem 7.90, page 534.

If δ denotes the Dirac delta for $x = 1$ then $\nu\left(\{r\} \times \Lambda\right) = \Delta N^p\left(r\right) \delta\left(\Lambda\right)$ so the integrals in the formula are[13]

$$U \overset{\circ}{=} \left(\exp\left(iu\right) - 1\right) \left(\left(N^p\left(t\right)\right)^c - \left(N^p\left(s\right)\right)^c\right)$$
$$V = \prod_{s < r \leq t} \left(1 + \left(\exp\left(iu\right) - 1\right) \Delta N^p\left(r\right)\right). \qquad \square$$

Proposition C.15 *An extended Poisson process N has a jump with positive probability at time t if and only if[14] its compensator N^p is discontinuous at time t.*

Proof. If $s \nearrow t$ in the Fourier transform (C.11) then

$$\mathbf{E}\left(\exp\left(iu\Delta N\left(t\right)\right)\right) = 1 + \left(\exp\left(iu\right) - 1\right) \Delta N^p\left(t\right).$$

The left-hand side is one if and only if $\Delta N^p\left(t\right) = 0$. $\qquad \square$

Definition C.16 *We say that the counting process N is a* generalized Poisson process *if it has independent increments and $N\left(t\right)$ has Poisson distribution for every t.*

Proposition C.17 *If the predictable compensator N^p of a counting process N is deterministic and continuous then $N\left(t\right)$ has a Poisson distribution with parameter $\lambda\left(t\right) \overset{\circ}{=} N^p\left(t\right)$. Hence under these conditions N is a generalized Poisson process[15].*

Proof. Let us recall that the Fourier transform of the Poisson distribution with parameter λ is

$$\exp\left(\lambda\left(\exp\left(iu\right) - 1\right)\right).$$

N^p is continuous so the proposition follows from line (C.11). $\qquad \square$

The jump times of Poisson processes are totally inaccessible[16]. Our goal is to prove the same result for generalized Poisson processes. First we prove a simple, but interesting general result:

Lemma C.18 *If $A \in \mathcal{A}^+$ then A^p is almost surely continuous if and only if A is regular in the following sense: If for some sequence of stopping times $\sigma_n \nearrow \sigma$ then*

$$\mathbf{E}\left(A\left(\sigma_n\right)\right) \to \mathbf{E}\left(A\left(\sigma\right)\right). \qquad (C.12)$$

---

[13]Observe that $h\left(x\right) \overset{\circ}{=} x\chi\left(\|x\| < 1\right)$ so $h\left(1\right) = 0$.
[14]See: Corollary 7.91, page 535.
[15]See: Example 7.93, page 536.
[16]See: Example 3.7, page 183.

Proof. As $A \in \mathcal{A}^+$ by the elementary properties[17] of the predictable compensator $A - A^p$ is a uniformly integrable martingale. Hence by the Optional Sampling Theorem $\mathbf{E}\left(A\left(\sigma_n\right)\right) = \mathbf{E}\left(A^p\left(\sigma_n\right)\right)$. If A^p is continuous then by the Monotone Convergence Theorem

$$\lim_{n \to \infty} \mathbf{E}\left(A\left(\sigma_n\right)\right) = \lim_{n \to \infty} \mathbf{E}\left(A^p\left(\sigma_n\right)\right) = \mathbf{E}\left(\lim_{n \to \infty} A^p\left(\sigma_n\right)\right) =$$
$$= \mathbf{E}\left(A^p\left(\lim_{n \to \infty} \sigma_n\right)\right) = \mathbf{E}\left(A^p\left(\sigma\right)\right) = \mathbf{E}\left(A\left(\sigma\right)\right).$$

On the other hand let us assume that (C.12) holds. If $t_n \nearrow t$ then

$$\mathbf{E}\left(A^p\left(t_n\right)\right) = \mathbf{E}\left(A\left(t_n\right)\right) \nearrow \mathbf{E}\left(A\left(t\right)\right) = \mathbf{E}\left(A^p\left(t\right)\right).$$

So

$$\mathbf{E}\left(\left|A^p\left(t\right) - A^p\left(t_n\right)\right|\right) = \mathbf{E}\left(A^p\left(t\right) - A^p\left(t_n\right)\right) = \mathbf{E}\left(A^p\left(t\right)\right) - \mathbf{E}\left(A^p\left(t_n\right)\right) \to 0.$$

Hence $A^p\left(t_n\right) \nearrow A^p\left(t\right)$ almost surely. Let $P \subseteq \mathbb{R}_+ \times \Omega$ be the set of discontinuities of A^p. If $\mathbf{P}\left(\mathrm{proj}_\Omega P\right) = 0$ then A^p is almost surely continuous. P is a predictable set. So if $\mathbf{P}\left(\mathrm{proj}_\Omega P\right) > 0$ then there is[18] a predictable stopping time σ such that $\mathrm{Graph}(\sigma) \subseteq P$ and $0 < \mathbf{P}\left(\sigma < \infty\right)$. If $\left(\sigma_n\right)$ is announcing σ then

$$\lim_{n \to \infty} \mathbf{E}\left(A\left(\sigma_n\right)\right) = \lim_{n \to \infty} \mathbf{E}\left(A^p\left(\sigma_n\right)\right) = \mathbf{E}\left(A^p\left(\sigma-\right)\right) < \mathbf{E}\left(A^p\left(\sigma\right)\right) = \mathbf{E}\left(A\left(\sigma\right)\right).$$

Hence as A is regular A^p is almost surely continuous. $\qquad\square$

Lemma C.19 *If $A \in \mathcal{A}_{loc}$ then A^p is continuous if and only if for every sequence of stopping times $\sigma_n \nearrow \sigma$*

$$\lim_{n \to \infty} A\left(\sigma_n\right) \stackrel{a.s.}{=} A\left(\lim_{n \to \infty} \sigma_n\right) = A\left(\sigma\right). \qquad (\mathrm{C}.13)$$

Proof. Let $\left(\tau_k\right)$ be the localizing sequence of A. $A^{\tau_k} \in \mathcal{A}^+$, so

$$\mathbf{E}\left(A^{\tau_k}\left(\sigma_n\right) - A^{\tau_k}\left(\sigma\right)\right) = \mathbf{E}\left(\left(A^p\right)^{\tau_k}\left(\sigma_n\right) - \left(A^p\right)^{\tau_k}\left(\sigma\right)\right).$$

If A^p is continuous then one can prove again that $A^{\tau_k}\left(\sigma_n\right) \stackrel{a.s.}{\to} A^{\tau_k}\left(\sigma\right)$. This obviously implies that $A\left(\sigma_n\right) \stackrel{a.s.}{\to} A\left(\sigma_n\right)$. On the other hand if (C.13) holds then by the Monotone Convergence Theorem $\mathbf{E}\left(A^{\tau_k}\left(\sigma_n\right)\right) \nearrow \mathbf{E}\left(A^{\tau_k}\left(\sigma\right)\right)$. Hence by the previous lemma[19] $\left(A^{\tau_k}\right)^p = \left(A^p\right)^{\tau_k}$ is almost surely continuous. So A^p is almost surely continuous. $\qquad\square$

---

[17]See: Property 4, page 217.
[18]See: Proposition 3.32, page 195.
[19]See: Property 5, page 217.

Proposition C.20 *The compensator N^p of a point process N is almost surely continuous if and only if the jump times of N are totally inaccessible[20].*

Proof. If N^p is continuous and τ is a jump time and with positive probability $\rho_n \nearrow \tau$ then by the previous lemma

$$N\left(\tau-\right) = N\left(\lim_{n\to\infty} \rho_n\right) = N\left(\tau\right).$$

That is on a set with positive probability $N\left(\tau-\right) = N\left(\tau\right)$ which is impossible as τ is a jump time of N. On the other hand if $\rho_n \nearrow \rho$ and with positive probability $\rho_n < \rho$ then ρ cannot be a jump time of N as all the jump times are totally inaccessible. Therefore $N\left(\rho_n\right) \to N\left(\rho-\right) = N\left(\rho\right)$. So by the previous lemma N^p is almost surely continuous. $\square$

---

[20]See: Example 7.74, page 517.

Notes and Comments

There are many good books on stochastic analysis: [4], [6], [21], [22], [48], [45], [53], [57], [59], [58], [63], [73], [74], [77], [78], [79]. A small part of the literature deals with the general theory where the integrators are general semimartingales, some of the books describe the theory when the integrators are continuous semimartingales. There are many books, and one can find a lot of lecture notes on the internet introducing the theory when the integrator is a Wiener process. Perhaps from pedagogical point of view, in an introductory course the simplest and most resolute approach is when the integrators are continuous semimartingales. This approach is sufficiently abstract to cover the most important results and from this perspective one can easily see the most elegant aspects of the stochastic analysis. The main advantage of this approach is that one can avoid the concept of predictability and every continuous local martingale is locally square-integrable. It is also very important that in this case the predictable quadratic variation and the quadratic variation are equal. On the other hand, the Wiener process case is a bit too elementary. It hides some very important aspects of the stochastic integration: mainly the role of the quadratic variation and especially if we introduce the stochastic integrals only for Wiener processes we cannot integrate when the integrator is already an integral process with respect to some Wiener process. Of course the popularity of the stochastic integration theory comes from its application in mathematical finance, and very often the stochastic integration is part of some courses on derivative prices, and already this very simple approach is a bit too demanding for an audience interested mainly in elementary financial applications. Perhaps the main disadvantage of the approach based on continuous semimartingales, is that in some sense it hides the most important aspects of stochastic analysis: its relation to other parts of probability theory. The canonical examples for semimartingales are the Lévy processes and the Lévy processes are mainly discontinuous. One can find a good account of the history of stochastic integration in [49].

Chapter 1

Filtration and stopping times were first investigated systematically in [19]. A good source about discrete-time martingales is [72]. The concept of predictability was introduced in [66]. As a general introduction one can also use [29] and [30] or [82]. Theorem 1.28 comes from [11]. Proposition 1.109 was borrowed from [74], while Proposition 1.112 is from [28]. The results about the first passage time

of the Wiener process were taken from [78], [53] and [27]. The treatment of localization was taken from [45]. The definition of local martingales was introduced
in [37]. The theorem on the quadratic variation of discrete time martingales was
taken from [5] and [61].

Chapter 2

There are several approaches to stochastic integration. The theory started by Itô
[38], [39], [40]. Our introduction is mainly based on [78], which is based on [56].
See also [7] and [8]. The main problem with this approach is that one can first
construct the quadratic variation or the predictable quadratic variation of the
process and one can apply this construction only for locally square-integrable
martingales. As every continuous local martingale is locally bounded, perhaps
this approach is the most economical one in the continuous case. One can construct the predictable quadratic variation with the Doob–Meyer decomposition
then construct the stochastic integral with the Hilbert space method of this
chapter, and with the stochastic integral construct the quadratic variation as
the correction term in the integration by parts formula. [9], [10], [45], [69]. The
concept of semimartingales appears in [15]. Fisk's theorem comes from [25].

Chapter 3

In this chapter I follow [45] and [74]. The Fundamental Theorem of Local
Martingales is due to J. A. Yan. See: [43], [70], [17].

Chapter 4

In this chapter I also followed [45] and [74]. The proof of the discrete-time Davis's
inequality comes from [5]. The proof of Burkholder's inequality was borrowed
from [51]. The invariance of semimartingales under change of measure was studied in [87] and [47]. The discussion of the properties of the stochastic integration
was based on [81]. Theorem 4.26 was taken from [57].

Chapter 5

The Doob–Meyer decomposition appears in [64], [65]. The proof was simplified by
[75]. See also [36], [74] and [53]. The theory of quasimartingales was developed by
[25], [76], [86] and [62]. The Bichteler–Dellacherie theorem was proved in [2], [3]
and [14]. It appeared for the first time in [48]. [74] builds the theory of stochastic
integration on this theorem. The theory of parametric stochastic integrals comes
from [85], [18]. See also [19], [48], [52] and [50]. The present discussion of the
integral representation builds on [74]. The Jacod–Yor theorem appears in [46].
See also: [12] and [13]. [44]. One can find the theorem in [45] and [57]. The final
version of $\mathcal{H}^1$ BMO duality appears in [67] and [68]. One can find a simple proof
for the Brownian case is in [73]. About BMO spaces see also: [54].

Chapter 6

Most of the material is taken from [78], [53] and [74]. See also: [69] and [56]. The proof of Lévy's theorem for the continuous case was borrowed from [56], see: [9]. Föllmer's theorem comes from [26]. The discussion of Doléans' equation comes from [45], see: [16] and [17]. The discussion of local times is mainly based on [74], see [69]. I borrowed the proof of theorem of Dvoretzky–Erdős–Kakutany from [53] which follows [55].

Chapter 7

In this chapter I followed mainly [45] and [74]. One can also consult [28], [83], [84] or [51]. The theory of Lévy processes and the Lévy–Khintchine formula has a long history. [31], [60]. The idea of characteristics goes back to [41]. Later it was studied in [88], [1], [33], [47], [32]. The characterization of processes with independent increments comes from [48], [34].

Appendix

There is a vast literature on the Brownian motion: [78], [53]. The condition for almost sure convergence of the quadratic variation comes from [20]. Example B.19 comes from [24].

References

[1] BENVENISTE, A., AND JACOD, J. Systèmes de Lévy des processus de Markov. *Invent. Math. 21* (1973), 183–198.

[2] BICHTELER, K. Stochastic integrators. *Bull. Amer. Math. Soc. (N.S.) 1*, 5 (1979), 761–765.

[3] BICHTELER, K. Stochastic integration and L^p-theory of semimartingales. *Ann. Probab. 9*, 1 (1981), 49–89.

[4] BICHTELER, K. *Stochastic Integration with Jumps*, vol. 89 of *Encyclopedia of Mathematics and its Applications*. Cambridge University Press, Cambridge, 2002.

[5] CHOW, Y. S., AND TEICHER, H. *Probability Theory: Independence, interchangeability, martingales*, third ed. Springer Texts in Statistics. Springer-Verlag, New York, 1997.

[6] CHUNG, K. L., AND WILLIAMS, R. J. *Introduction to Stochastic Integration*, second ed. Probability and its Applications. Birkhäuser Boston Inc., Boston, MA, 1990.

[7] COURRÈGE, P. Intégrales stochastiques associées à une martingale de carré intégrable. *C. R. Acad. Sci. Paris 256* (1963), 867–870.

[8] COURRÈGE, P. Intégrale stochastiques et martingales de carré intégrable. In *Séminaire de Théorie du Potentiel, dirigé par M. Brelot, G. Choquet et J. Deny, 1962/63, No. 7*. Secrétariat mathématique, Paris, 1964, p. 20.

[9] DELLACHERIE, C., AND MEYER, P.-A. *Probabilités et Potentiel*. Hermann, Paris, 1975. Chapitres I à IV, Édition entièrement refondue, Publications de l'Institut de Mathématique de l'Université de Strasbourg, No. XV, Actualités Scientifiques et Industrielles, No. 1372.

[10] DELLACHERIE, C., AND MEYER, P.-A. *Probabilités et Potentiel. Chapitres V à VIII*, revised ed., vol. 1385 of *Actualités Scientifiques et Industrielles [Current Scientific and Industrial Topics]*. Hermann, Paris, 1980. Théorie des martingales. [Martingale theory].

[11] DELLACHERIE, C. *Capacités et processus stochastiques*. Springer-Verlag, Berlin, 1972. Ergebnisse der Mathematik und ihrer Grenzgebiete, Band 67.

[12] DELLACHERIE, C. Intégrales stochastiques par rapport aux processus de Wiener ou de Poisson. In *Séminaire de Probabilités, VIII (Univ. Strasbourg, année universitaire 1972-1973)*. Springer, Berlin, 1974, pp. 25–26. Lecture Notes in Math., Vol. 381.

[13] DELLACHERIE, C. Correction à: "Intégrales stochastiques par rapport aux processus de Wiener ou de Poisson" (*séminaire de probabilités, viii* (Univ. Strasbourg, année universitaire 1972-1973), pp. 25–26, Lecture Notes in

Math., Vol. 381, Springer, Berlin, 1974). In *Séminaire de Probabilités, IX (Seconde Partie, Univ. Strasbourg, Strasbourg, Années Universitaires 1973/1974 et 1974/1975)*. Springer, Berlin, 1975, pp. p. 494. Lecture Notes in Math., Vol. 465.

[14] DELLACHERIE, C. Un survol de la théorie de l'intégrale stochastique. *Stochastic Process. Appl. 10*, 2 (1980), 115–144.

[15] DOLÉANS-DADE, C., AND MEYER, P.-A. Intégrales stochastiques par rapport aux martingales locales. In *Séminaire de Probabilités, IV (Univ. Strasbourg, 1968/69)*. Lecture Notes in Mathematics, Vol. 124. Springer, Berlin, 1970, pp. 77–107.

[16] DOLÉANS-DADE, C. Quelques applications de la formule de changement de variables pour les semimartingales. *Z. Wahrscheinlichkeitstheorie und Verw. Gebiete 16* (1970), 181–194.

[17] DOLÉANS-DADE, C. On the existence and unicity of solutions of stochastic integral equations. *Z. Wahrscheinlichkeitstheorie und Verw. Gebiete 36*, 2 (1976), 93–101.

[18] DOLÉANS, C. Intégrales stochastiques dépendant d'un paramètre. *Publ. Inst. Statist. Univ. Paris 16* (1967), 23–33.

[19] DOOB, J. L. *Stochastic Processes*. Wiley Classics Library. John Wiley & Sons Inc., New York, 1990. Reprint of the 1953 original, A Wiley-Interscience Publication.

[20] DUDLEY, R. M. Sample functions of the Gaussian process. *Ann. Probability 1*, 1 (1973), 66–103.

[21] DURRETT, R. *Stochastic Calculus*. Probability and Stochastics Series. CRC Press, Boca Raton, FL, 1996. A practical introduction.

[22] ELLIOTT, R. J. *Stochastic Calculus and Applications*, vol. 18 of *Applications of Mathematics (New York)*. Springer-Verlag, New York, 1982.

[23] FELLER, W. *An Introduction to Probability Theory and its Applications. Vol. II.* Second edition. John Wiley & Sons Inc., New York, 1971.

[24] FERNANDEZ DE LA VEGA, W. On almost sure convergence of quadratic Brownian variation. *Ann. Probability 2* (1974), 551–552.

[25] FISK, D. L. Quasi-martingales. *Trans. Amer. Math. Soc. 120* (1965), 369–389.

[26] FÖLLMER, H. Calcul d'Itô sans probabilités. In *Seminar on Probability, XV (Univ. Strasbourg, Strasbourg, 1979/1980) (French)*, vol. 850 of *Lecture Notes in Math.* Springer, Berlin, 1981, pp. 143–150.

[27] FREEDMAN, D. *Brownian Motion and Diffusion*. Springer-Verlang, New York, 1983.

[28] GIHMAN, I. I., AND SKOROHOD, A. V. *The Theory of Stochastic Processes. II.* Springer-Verlag, New York, 1975. Translated from the Russian by Samuel Kotz, Die Grundlehren der Mathematischen Wissenschaften, Band 218.

[29] GIHMAN, I. I., AND SKOROHOD, A. V. *The Theory of Stochastic Processes. III.* Springer-Verlag, Berlin, 1979. Translated from the Russian by Samuel Kotz, With an appendix containing corrections to Volumes I and II, Grundlehren der Mathematischen Wissenschaften, 232.

[30] GIHMAN, I. I., AND SKOROHOD, A. V. *The Theory of Stochastic Processes. I*, English ed., vol. 210 of *Grundlehren der Mathematischen Wissenschaften [Fundamental Principles of Mathematical Sciences]*. Springer-Verlag, Berlin, 1980. Translated from the Russian by Samuel Kotz.

[31] GNEDENKO, B. V., AND KOLMOGOROV, A. N. *Limit Distributions for Sums of Independent Random Variables*. Translated from the Russian, annotated, and revised by K. L. Chung. With appendices by J. L. Doob and P. L. Hsu. Revised edition. Addison-Wesley Publishing Co., Reading, Mass.-London-Don Mills., Ont., 1968.

[32] GRIGELIONIS, B. The Markov property of random processes. *Litovsk. Mat. Sb. 8* (1968), 489–502.

[33] GRIGELIONIS, B. The representation of integer-valued random measures as stochastic integrals over the Poisson measure. *Litovsk. Mat. Sb. 11* (1971), 93–108.

[34] GRIGELIONIS, B. Martingale characterization of random processes with independent increments. *Litovsk. Mat. Sb. 17*, 1 (1977), 75–86, 212.

[35] HIRIART-URRUTY, J.-B., AND LEMARÉCHAL, C. *Fundamentals of Convex Analysis*. Springer-Verlag, Berlin, 2001.

[36] IKEDA, N., AND WATANABE, S. *Stochastic Differential Equations and Diffusion Processes*, vol. 24 of *North-Holland Mathematical Library*. North-Holland Publishing Co., Amsterdam, 1981.

[37] ITÔ, K., AND WATANABE, S. Transformation of Markov processes by multiplicative functionals. *Ann. Inst. Fourier (Grenoble) 15*, fasc. 1 (1965), 13–30.

[38] ITÔ, K. Stochastic integral. *Proc. Imp. Acad. Tokyo 20* (1944), 519–524.

[39] ITÔ, K. On a stochastic integral equation. *Proc. Japan Acad. 22*, nos. 1-4 (1946), 32–35.

[40] ITÔ, K. On the stochastic integral. *Sûgaku 1* (1948), 172–177.

[41] ITÔ, K. On stochastic differential equations. *Mem. Amer. Math. Soc. 1951*, 4 (1951), 51.

[42] JACOBS, K. *Measure and Integral*. Academic Press [Harcourt Brace Jovanovich Publishers], New York, 1978. Probability and Mathematical Statistics, With an appendix by Jaroslav Kurzweil.

[43] JACOD, J., AND MÉMIN, J. Caractéristiques locales et conditions de continuité absolue pour les semi-martingales. *Z. Wahrscheinlichkeitstheorie und Verw. Gebiete 35*, 1 (1976), 1–37.

[44] JACOD, J., AND MÉMIN, J. Un théorème de représentation des martingales pour les ensembles régénératifs. In *Séminaire de Probabilités, X (Première partie, Univ. Strasbourg, Strasbourg, Année Universitaire 1974/1975)*. Springer, Berlin, 1976, pp. 24–39. Lecture Notes in Math., Vol. 511.

[45] JACOD, J., AND SHIRYAEV, A. N. *Limit Theorems for Stochastic Processes*, second ed., vol. 288 of *Grundlehren der Mathematischen Wissenschaften [Fundamental Principles of Mathematical Sciences]*. Springer-Verlag, Berlin, 2003.

[46] JACOD, J., AND YOR, M. Étude des solutions extrémales et représentation intégrale des solutions pour certains problèmes de martingales. *Z. Wahrscheinlichkeitstheorie und Verw. Gebiete 38*, 2 (1977), 83–125.

[47] JACOD, J. Multivariate point processes: predictable projection, Radon-Nikodým derivatives, representation of martingales. *Z. Wahrscheinlichkeitstheorie und Verw. Gebiete 31* (1974/75), 235–253.

[48] JACOD, J. *Calcul Stochastique et Problèmes de Martingales*, vol. 714 of *Lecture Notes in Mathematics*. Springer, Berlin, 1979.

[49] JARROW, R., AND PROTTER, P. A short history of stochastic integration and mathematical finance: the early years, 1880–1970. In *A festschrift for Herman Rubin*, vol. 45 of *IMS Lecture Notes Monogr. Ser.* Inst. Math. Statist., Beachwood, OH, 2004, pp. 75–91.

[50] KAILATH, T., SEGALL, A., AND ZAKAI, M. Fubini-type theorems for stochastic integrals. *Sankhyā Ser. A 40*, 2 (1978), 138–143.

[51] KALLENBERG, O. *Foundations of Modern Probability*. Probability and its Applications (New York). Springer-Verlag, New York, 1997.

[52] KALLIANPUR, G., AND STRIEBEL, C. Stochastic differential equations occurring in the estimation of continuous parameter stochastic processes. *Teor. Verojatnost. i Primenen 14* (1969), 597–622.

[53] KARATZAS, I., AND SHREVE, S. E. *Brownian Motion and Stochastic Calculus*, 2 ed. Graduate Texts in Mathematics 113. Springer-Verlang, New York, 1991.

[54] KAZAMAKI, N. *Continuous Exponential Martingales and BMO*, vol. 1579 of *Lecture Notes in Mathematics*. Springer-Verlag, Berlin, 1994.

[55] KNIGHT, F. B. *Essentials of Brownian Motion and Diffusion*, vol. 18 of *Mathematical Surveys*. American Mathematical Society, Providence, R.I., 1981.

[56] KUNITA, H., AND WATANABE, S. On square integrable martingales. *Nagoya Math. J. 30* (1967), 209–245.

[57] LIPTSER, R. S., AND SHIRYAEV, A. N. *Theory of Martingales*. Mathematics and Its Applications. Kluwer Academic Publishers, Dordrecht, 1989.

[58] LIPTSER, R. S., AND SHIRYAEV, A. N. *Statistics of Random Processes II: Applications*, 2 ed. Applications of Mathematics 6. Springer-Verlang, Berlin, 2001.

[59] LIPTSER, R. S., AND SHIRYAEV, A. N. *Statistics of Random Processes I: General Theory*, 2 ed. Applications of Mathematics 5. Springer-Verlang, Berlin, 2001.

[60] LOÈVE, M. *Probability Theory. I-II*, fourth ed. Springer-Verlag, New York, 1978. Graduate Texts in Mathematics, Vol. 46.

[61] MALLIAVIN, P. *Integration and Probability*. Graduate Text in Mathematics 157. Springer-Verlang, New York, 1995.

[62] MÉTIVIER, M., AND PELLAUMAIL, J. On Doleans-Föllmer's measure for quasi-martingales. *Illinois J. Math. 19*, 4 (1975), 491–504.

[63] MÉTIVIER, M. *Semimartingales*, vol. 2 of *de Gruyter Studies in Mathematics*. Walter de Gruyter & Co., Berlin, 1982. A course on stochastic processes.

[64] MEYER, P.-A. A decomposition theorem for supermartingales. *Illinois J. Math. 6* (1962), 193–205.

[65] MEYER, P.-A. Decomposition of supermartingales: the uniqueness theorem. *Illinois J. Math. 7* (1963), 1–17.

[66] MEYER, P.-A. Intégrales stochastiques. I, II, III, IV. In *Séminaire de Probabilités (Univ. Strasbourg, Strasbourg, 1966/67), Vol. I.* Springer, Berlin, 1967, pp. 72–94, 95–117, 118–141, 142–162.

[67] MEYER, P.-A. Le dual de 'H^1' est 'BMO' (cas continu). In *Séminaire de Probabilités, VII (Univ. Strasbourg, Année Universitaire 1971–1972)*. Springer, Berlin, 1973, pp. 136–145. Lecture Notes in Math., Vol. 321.

[68] MEYER, P.-A. Complément sur la dualité entre H^1 et BMO: 'Le dual de "H^1" est "BMO" (cas continu)' (Séminaire de Probabilités, VII (Univ. Strasbourg, année universitaire 1971–1972), pp. 136–145, Lecture Notes in Math., Vol. 321, Springer, Berlin, 1973). In *Séminaire de Probabilités, IX (Seconde Partie, Univ. Strasbourg, Strasbourg, années universitaires 1973/1974 et 1974/1975)*. Springer, Berlin, 1975, pp. 237–238. Lecture Notes in Math., Vol. 465.

[69] MEYER, P.-A. Un cours sur les intégrales stochastiques. In *Séminaire de Probabilités, X (Seconde Partie: Théorie des Intégrales Stochastiques, Univ. Strasbourg, Strasbourg, Année Universitaire 1974/1975)*. Springer, Berlin, 1976, pp. 245–400. Lecture Notes in Math., Vol. 511.

[70] MEYER, P.-A. Notes sur les intégrales stochastiques. II. le théorème fondamental sur les martingales locales. In *Séminaire de Probabilités, XI (Univ. Strasbourg, Strasbourg, 1975/1976)*. Springer, Berlin, 1977, pp. 463–464. Lecture Notes in Math., Vol. 581.

[71] NEVEU, J. *Mathematical Foundations of the Calculus of Probability.* Translated by Amiel Feinstein. Holden-Day Inc., San Francisco, Calif., 1965.

[72] NEVEU, J. *Discrete-Parameter Martingales*, revised ed. North-Holland Publishing Co., Amsterdam, 1975. Translated from the French by T. P. Speed, North-Holland Mathematical Library, Vol. 10.

[73] ØKSENDAL, B. *Stochastic Differential Equations*, sixth ed. Universitext. Springer-Verlag, Berlin, 2003. An introduction with applications.

[74] PROTTER, P. E. *Stochastic Integration and Differential Equations*, second ed., vol. 21 of *Applications of Mathematics (New York)*. Springer-Verlag, Berlin, 2004. Stochastic Modelling and Applied Probability.

[75] RAO, K. M. On decomposition theorems of Meyer. *Math. Scand. 24* (1969), 66–78.

[76] RAO, K. M. Quasi-martingales. *Math. Scand. 24* (1969), 79–92.

[77] RAO, M. M. *Stochastic Processes and Integration.* Martinus Nijhoff Publishers, The Hague, 1979.

[78] REVUZ, D., AND YOR, M. *Continuous Martingales and Brownian Motion.* Grundlehren der mathematischen Wissenschaften 293. Springer-Verlang, Berlin, 1999.

[79] ROGERS, L. C. G., AND WILLIAMS, D. *Diffusions, Markov Processes, and Martingales. Vol. 1.* Cambridge Mathematical Library. Cambridge University

Press, Cambridge, 2000. Foundations, Reprint of the second (1994) edition.

[80] RUDIN, W. *Real and Complex Analysis*, third ed. McGraw-Hill Book Co., New York, 1987.

[81] SHIRYAEV, A. N., AND S., C. A. A vector stochastic integral and the fundamental theorem of asset pricing. *Tr. Mat. Inst. Steklova 237*, Stokhast. Finans. Mat. (2002), 12–56.

[82] SHIRYAEV, A. N. *Probability*, second ed., vol. 95 of *Graduate Texts in Mathematics*. Springer-Verlag, New York, 1996. Translated from the first (1980) Russian edition by R. P. Boas.

[83] SKOROHOD, A. V. *Processes with independent increments (in Russian)*. Izdat. "Nauka", Moscow, 1964.

[84] SKOROKHOD, A. V. *Studies in the Theory of Random Processes*. Translated from the Russian by Scripta Technica, Inc. Addison-Wesley Publishing Co., Inc., Reading, Mass., 1965.

[85] STRICKER, C., AND YOR, M. Calcul stochastique dépendant d'un paramètre. *Z. Wahrsch. Verw. Gebiete 45*, 2 (1978), 109–133.

[86] STRICKER, C. Quasimartingales, martingales locales, semimartingales et filtration naturelle. *Z. Wahrscheinlichkeitstheorie und Verw. Gebiete 39*, 1 (1977), 55–63.

[87] VAN SCHUPPEN, J. H., AND WONG, E. Transformation of local martingales under a change of law. *Ann. Probability 2* (1974), 879–888.

[88] WATANABE, S. On discontinuous additive functionals and Lévy measures of a Markov process. *Japan. J. Math. 34* (1964), 53–70.

Index

603

9 780199 215256